7

A DESIGNER'S GUIDE
TO
VHDL SYNTHESIS

A DESIGNER'S GUIDE
TO
VHDL SYNTHESIS

Douglas E. Ott

Thomas J. Wilderotter

ITT Avionics

KLUWER ACADEMIC PUBLISHERS
Boston / Dordrecht / London

Distributors for North America:
Kluwer Academic Publishers
101 Philip Drive
Assinippi Park
Norwell, Massachusetts 02061 USA

Distributors for all other countries:
Kluwer Academic Publishers Group
Distribution Centre
Post Office Box 322
3300 AH Dordrecht, THE NETHERLANDS

Library of Congress Cataloging-in-Publication Data

A C.I.P. Catalogue record for this book is available
from the Library of Congress.

To our wives,
Carolyn and Maureen,
whose continued love and support
is the foundation for our work.

Table of Contents

Chapter 3

Chapter 4

Chapter 5

Chapter 6

Chapter 7

Chapter 8

Chapter 9

Chapter 10

Chapter 11

Appendix A

xii

List of Figures

List of Examples

List of Tables

Glossary of Terms

ASIC - Application Specific Integrated Circuit, so named because it is a type of integrated circuit that can be customized for a user's specific application. The term is usually applied to IC devices, such as gate arrays, in which the customization process can be done relatively quickly and at a reasonable cost, by the use of computer based design tools.

FPGA - A Field-Programmable Gate Array. This type of ASIC can be customized by the user with a hardware programming system, an operation that can be done "in the field" rather than by a semiconductor vendor.

RTL - Register Transfer Language, a term used to describe the flow of signals, data, and operations in digital systems.

Test Bench - Used during the VHDL simulation process, the "test bench" is the name given to the VHDL code that generates the input signals for the design being simulated and monitors the desired outputs. The test bench also handles disk file operations using the VHDL TEXTIO commands.

VHDL - The VHSIC Hardware Design Language, a specialized language for describing and simulating electronic systems that was a development of the US Government's VHSIC program in the 1980s.

VHSIC - The Very High Speed Integrated Circuit project, whose goal was to advance the technology for development of large high speed ICs.

Preface

This book is intended for both design engineers who want to use VHDL-based logic synthesis to design ASICs and for managers who need to gain a practical understanding of the issues involved in using this technology. The emphasis has been placed more on practical applications of VHDL and synthesis based on actual experiences, rather than on a more theoretical approach to the language.

The need for new approaches to digital design, such as VHDL and synthesis, has been driven by technology, time to market, and productivity issues. In terms of technology, the semiconductor industry has been advancing the capabilities of ASIC technology at a rapid rate over the past ten years, to the point where large complex system functions can now be fabricated as a single chip. The benefits from use of this technology are significant in terms of improved performance, power, size, weight, and reliability, but have made the management of the design process ever more demanding. The industry in general has recognized the need for tools to help manage the complexity of large ASIC designs while keeping the cost and development schedule within reasonable limits. Driven by that need, two of the more significant technological resources that have become available in the early 1990s are the VHDL design and simulation language, coupled with the process of automated logic synthesis.

VHDL and logic synthesis tools provide very powerful capabilities for ASIC design, but are also very complex and represent a radical departure from traditional design methods. This situation has made it difficult to get started in using this technology for both designers and management, since a major learning effort and "culture" change is required. This book has been written to help design engineers and other professionals successfully make the transition to a design methodology based on VHDL and logic synthesis instead of the more traditional schematic based approach. While there are a number of texts on the VHDL language and its use in simulation, little has been written from a designer's viewpoint on how to use VHDL and logic synthesis to design real ASIC systems. The material in this book is based on experience gained in successfully using these techniques for ASIC design at ITT Avionics, and relies heavily on realistic examples to demonstrate the principles involved.

The extensive design examples used are as vendor independent as possible, and the differences to be aware of when using various synthesis tools are noted. The design style has not simply been reduced to the lowest common denominator of available toolsets however, since that would dilute the information for users of more capable systems. All of the examples have been simulated and synthesized, and are provided on a floppy disk enclosed with the book.

Overview of the Book

The book is divided into 11 chapters.

The first two chapters present the issues of VHDL synthesis from a management viewpoint, emphasizing the process more than the technical details. The first chapter is an introductory chapter that discusses what synthesis is and why it has been developed, the advantages of using VHDL based synthesis, and some of the facts, the hype, and the misconceptions that may be confusing to someone trying to assess the process.

Chapter 2 continues by comparing the ASIC design flow using the traditional schematic-based design to the VHDL synthesis-based approach and shows how to modify your design methodology to make the transition smoothly to the language based design system.

Chapter 3 is primarily for ASIC designers with little or no experience with VHDL and covers the VHDL language features used for synthesis, in preparation for succeeding chapters which demonstrate their usage in various applications. The VHDL constructs that cannot be synthesized will be explained, along with a discussion of behavioral, RTL, and structural styles of VHDL.

Chapter 4 introduces the synthesis of sequential logic functions, using counters to illustrate various RTL, dataflow, and state methods. The material begins very simply and becomes more complex as each new function is considered. Chapter 5 continues with the design of more complex sequential applications based on principles learned in Chapter 4.

The design of control logic, sequencers, and state machines is covered in Chapter 6, and several different design techniques are shown.

Chapter 7 continues with a discussion of various data processing functions, which present a different type of design problem than signal processing designs.

Chapter 8 presents the issues of logic optimization and minimization, with examples showing various combinational logic functions. This section discusses how synthesis can control and optimize a design for minimum gate count or maximum speed, or both, and includes ASIC layout concerns.

Chapter 9 shows how the various pieces covered in the earlier chapters are combined to design a complete ASIC, and the design and use of test benches are shown.

Chapter 10 covers the issues involved in evaluating and selecting a logic synthesis tool, while Chapter 11 concludes the book by looking at some of the future directions of synthesis tools and ASIC design.

Acknowledgements

This book has been based on the successful experiences in using VHDL and logic synthesis for ASIC development at ITT Avionics. Many people have been involved in helping to make the transition to this new methodology, and in reviewing this book. In particular, credit should be given to Ray Harrison for fostering an environment that encouraged the use of advanced design tools and methods, and to Keith Thompson for his outstanding efforts in developing and using those tools. We would also like to thank ITT management for their support in undertaking this book and in reviewing the material, especially Doug Frederiks, George Scherer, and Al Kaufman. A significant amount of credit also goes to the engineers and managers in the Digital Design group who contributed their ideas and corrected errors, and to Pete Ditore, who continually pushes the envelope of new ideas.

Thanks also to the many industry reviewers who contributed their ideas to the book, notably Dr. John Hines and Greg Tumbush of Wright Patterson Labs, Brent Gregory and Stan Mazor of Synopsys, Cary Ussery of Cadence, Jeff Fox of Viewlogic, and Tom Scott of Orcad, and to Carl Harris of Kluwer Academic Press for his encouragement and support.

A DESIGNER'S GUIDE
TO
VHDL SYNTHESIS

Chapter 1

Introduction

Logic synthesis tools and hardware description languages are beginning to create a revolution in the way ASIC systems are designed. This major change is primarily a result of rapidly moving developments in semiconductor technology and design processes as well as the general need for improving both design quality and productivity. This introductory chapter will discuss, from a management viewpoint, what synthesis is and why it has been developed, the advantages of using VHDL based synthesis, and some of the facts, the hype, and the misconceptions that may be confusing to someone trying to assess the process. The rest of the book is then devoted to the issues of how to put VHDL synthesis successfully into your current design practice.

The ASIC Technology Explosion

Any discussion of logic synthesis and VHDL must necessarily also include developments in both semiconductor technology and design methodology, since these subjects are all very closely interrelated. Historically the rapid advances in integrated circuit (IC) technology over the past fifteen years have driven the need for more capable design tools, and as those tools have developed, they in turn have made it possible to design even larger and more complex ICs. In a classic case of synergy, the computers and workstations that are used by engineers to design the current and future generations of ICs are themselves a product of the same semiconductor technology. It is the general class of ICs called ASICs (Application Specific Integrated Circuits) that, combined with these new design tools, have transformed the technology and made it possible for large numbers of designers to develop integrated circuits tailored to their specific applications. All of these factors have had a great effect on how digital design engineers develop electronic systems, not only in terms of system performance but also in the related issues of time-to-market, management of the overall design process and schedule, and the cost of development.

In order to maintain the pace of development, the design methods and tools used for ASIC and IC design must be pushed forward as rapidly as the semiconductor technology itself. Faced with these problems in the 1980s, a number of visionary people within the electronics design community, universities, and the Department of Defense realized that conventional design tools and methods would be inadequate to handle the growing complexity and size of electronics systems. Their focus was and continues to be on making continuous improvements in the design process that can both reduce the likelihood of errors throughout the development cycle while simultaneously increasing the design engineer's productivity. Two of the major advances resulting from that work are the development of Hardware Description Languages (HDLs), and their use with powerful logic synthesis systems.

Background

ASIC Technology

Advances in ASIC technology have ushered in a whole new way of designing electronic systems with significant advantages, but at the expense of placing severe demands on design engineers, tool developers, and ASIC semiconductor vendors. There are several reasons for this situation, mainly stemming from the astounding growth in the capabilities of gate arrays and other ASIC technologies that are being used in nearly all electronic applications.

As an example of this growth, consider that in the early 1980s the semiconductor industry's first CMOS gate arrays had a capacity of a few thousand gates. In the short span of only ten years the usable gate density has increased by over a hundred times! Those first ASICs could be used to replace roughly one small circuit board of standard logic. In contrast, today's devices can hold large parts of entire system functions and allow major performance and testability capabilities to be added to a design with little increase in physical size or parts cost. During the past ten years roughly four generations of ASIC technology have been introduced, each providing both higher speed and gate density along with major improvements in power, reliability, and cost. The positive impact of this technology has been obvious in nearly all aspects of our lives, from making possible the widespread development of computers to the ever increasing use of electronics in such varied applications as automobiles, household products, aircraft, communications, games and entertainment, and scientific and military systems. And all indications are that this growth will continue well into the future!

The benefits provided by the use of ASICs, as compared with using more conventional discrete ICs, include greatly reduced size, weight, power, and cost, increased performance, and improved reliability and testability. Secondary benefits also result from such factors as being able to use smaller and less costly power supplies and from reduced cooling requirements. On the other side of the coin, there are a number of design challenges that must be overcome. These can be grouped into two general categories - the need to produce a design that works correctly the first time, and the need to accomplish that task within acceptable cost and schedule constraints. And to do all of this while the technology continues to move forward! This is where VHDL and logic synthesis, along with other design tools, make their contribution to the success of the overall ASIC design process.

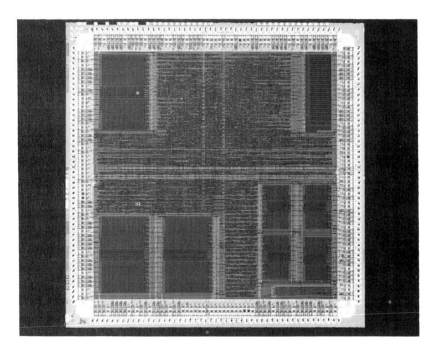

Microphotograph of an ITT Avionics Synthesized ASIC
Photo Courtesy of LSI Logic Corp.

Logic Synthesis

Logic synthesis is a process that is primarily intended to be used in the design of digital integrated circuits, and, in particular, ASIC devices such as gate arrays. The synthesis process covered in this book consists of software tools and a language based design methodology that can be used to design virtually any type of digital logic function. This distinction is made because the term synthesis is sometimes broadened to include the more specialized tools, such as silicon compilers and function generators, that are used by ASIC vendors to generate very specific, regular logic structures such as RAM and ROM memories and FIFOs. It is instead the overall class of general or "random" logic functions that commercial logic synthesis tools can support, and for which the benefits to the design process are the greatest.

In contrast with the traditional schematic-based digital design process, logic synthesis software tools are used with a specialized language such as VHDL to efficiently describe and simulate the desired operation of the circuit. If used properly, the synthesis tools can then automatically generate the gate level logic details and schematics, based on the language description. As we shall see in later chapters, very often a single line of VHDL code can be used to generate large circuit functions consisting of hundreds of gates! Synthesis therefore is a tool that allows logic designers to move up to a higher level design methodology through the use of VHDL, very much the way software engineers adopted higher order languages to replace the need to code the bit level details of the computer's instructions. The fact that much of the detail work is automated leads to an improvement in the overall design process in terms of helping to manage the complexity, improve productivity and accuracy, and shorten schedules. We will see in later sections that these tools are very powerful but still must be used properly and combined with both good engineering and logic design experience in order to be successful.

VHDL

The development of VHDL, the **VHSIC Hardware Description Language**, was sponsored by the US Government and the Air Force during the 1980s, and those organizations have continued to support and nurture its use throughout the electronic design communities. In 1987, the VHDL language was adopted by the IEEE as a standard hardware description language and has since achieved wide spread industry acceptance.

The VHDL language has powerful capabilities that have several possible uses depending on its application. The language can be used to describe and specify a wide variety of electronic systems, at levels of abstraction ranging from pure behavioral down to gate and switch level details. In addition to the description capability, systems modeled in VHDL can also be simulated at any of those same levels in order to verify their functional operation and performance parameters, and a number of very capable commercial VHDL simulators are available in the CAE marketplace. Finally, the VHDL description of a desired logic function can also be used to drive the logic synthesis process, with the constraint that the VHDL code be part of a fairly flexible but non-standardized subset of the language. This combination of the VHDL language used for design description and simulation, coupled with logic synthesis tools, is considered part of an overall design methodology often referred to as top-down design and will be the recommended approach for ASIC design in this book.

Why is VHDL Synthesis Needed?

Well, the good news about ASICs is indeed that systems that required hundreds of discrete ICs in the past can now be designed into an IC that is about 1/4 inch square. On the other side of the coin however, unlike the old song, the best things in life are **not** free! Developing an ASIC requires a fairly large engineering cost and schedule, and if you should make a mistake, unlike a printed circuit board design, you simply cannot add any "cuts and jumpers" to correct the problem. Any design errors discovered during ASIC or system test will require redesign of the circuit and the need to repeat the verification, layout, fab and test cycle again, leading to further cost and schedule impacts. Clearly one of the keys to successful ASIC development therefore is having the proper design tools and a methodology that minimizes the possibility of design errors at any level from getting into the fabricated parts.

The major design tool used today to analyze and verify ASIC and system design is logic simulation. In past years when discrete IC technology was used, designs were often built as a breadboard in order to debug errors and measure system performance. Today, building a conventional prototype or breadboard for a system consisting of multiple 25000 gate ASICs operating with sub-nanosecond delays, just to prove its correctness, is impractical if not impossible. Furthermore a breadboard does little to verify the system's performance over worst case temperature, process and voltage conditions. For this reason, today's standard design methodology uses schematic-based logic design tools that in turn create netlists used as the input to a gate level logic simulator. The operation of the circuit is then simulated and any design errors are corrected by redesigning the logic, redrawing the schematics, and repeating the process until the circuit performs properly.

While there is nothing intrinsically wrong with this approach, it does suffer from one major drawback. The simulation effort cannot take place until the design is completed and schematics are drawn, which means that the checking and elimination of errors is delayed until nearly the end of the design cycle. A significant amount of effort will have been spent by that time on both the detailed logic design as well as the mechanics of entering the schematics, and correcting the errors can be time consuming at a point where the schedule pressures are often the greatest. A much more desirable approach would allow the designer to detect and correct errors as early as possible in the overall design cycle, where minimal cost and schedule impacts will be experienced. The VHDL-based design methodology has been developed to help solve this type of problem, among others, and will be covered in more detail in Chapter 2.

A second aspect of ASIC development involves the schedule and "time-to-market" issues. When designs were smaller and less complex, it was possible to develop a new ASIC using the conventional schematic based logic design process in roughly 6-9 months. In most applications this is a reasonable development cycle for a new product, but consider what happens when the circuit complexity goes up by 5 or 10 times. First, the number of schematic pages usually increases in proportion to the size of the design, with several cases of 25k to 30k gate ASICs requiring over 250 pages. Clearly, the effort to draw this many schematics as well as to manage the complexity of the circuitry and interconnects results in a significant technical, cost, and schedule problem. If this situation were to continue, the cost and schedule would increase to the point where it would no longer be feasible or cost effective to develop larger circuits.

The process of logic synthesis attacks this problem directly by automating much of the detailed logic design mechanics, such as netlist and schematic generation, and by allowing designers to focus on defining and simulating the functional operation of the system logic. The VHDL synthesis process essentially shortens the design cycle by moving the logic definition and simulation phase to a point earlier in the cycle and leaving the mechanics to the automated synthesis tools. As shown in Chapter 2 this process produces quality designs within acceptable cost and schedule limits and makes it feasible to develop larger and more complex ASICs.

The Advantages

There are a number of advantages to be gained by moving to a VHDL synthesis methodology for ASIC design. Many of these issues have been touched on above and will also be discussed in later chapters but are summarized here to provide an overview of the technology.

Shorter Design Cycles :   The overall ASIC design cycle can be significantly reduced by the use of VHDL synthesis partly due to the elimination of the schematic generation and maintenance tasks, and partly due to the reduction in design errors by using higher level VHDL simulation early in the cycle. Other efficiencies are gained from the fact that the size of the VHDL code needed to describe a desired logic function is usually not proportional to the number of gates in the final circuit, unlike the case with traditional schematic design. For example, a large design featuring wide multiplexed data busses and/or arithmetic logic can be described in a few lines of VHDL code and synthesized relatively quickly in most cases.

Part of every design cycle is also the need to accommodate the inevitable changes that are identified after the detailed design has begun. With VHDL synthesis, considerable savings are gained from the relative ease of making rapid design changes throughout the design process.

Improved Design Quality : Overall design quality can be improved due to two factors, the capability to easily explore different design techniques and the ability of the synthesis tool to selectively optimize a design for improved performance or minimum gate count. In the first case, using VHDL to describe and simulate a design at a higher level makes it relatively easy to try different architectures and different circuit configurations in order to arrive at the best approach. This is usually not feasible using a schematic-based design due to the greater time and effort involved to draw the schematics and generate the simulation netlists.

In the second case, synthesis provides a powerful capability to optimize a design for either minimum gate count or for maximum speed, under control of the designer. In order to do this, most synthesis tools have built-in timing analyzers that can automatically calculate worst case time delays and setup and hold conditions, and can use that information to selectively optimize the circuit where needed. The result is a design that has minimum gates where speed is not required, and maximum performance where it is needed.

Vendor and Technology Independence : Most synthesis tool vendors provide libraries for a number of different ASIC foundries and technologies. When a design is described in VHDL, it is first synthesized into generic logic building blocks and subsequently optimized using one of the specific libraries chosen by the designer. The net result is that the choice of the ASIC vendor or vendors can be made later in the design cycle, providing for more schedule and pricing control. Alternatively, the same VHDL design can be synthesized into more than one vendor's library in order to provide multiple sources for the ASIC parts, or can be synthesized into FPGAs for rapid prototyping.

Lower Design Cost : Reduced design costs result from the same factors that contribute to shortened schedules, ie, the elimination of the mechanics of schematic generation and the reduction in design errors. Experienced synthesis designers have cut the detailed logic design effort by 50% compared with that for the conventional methodology. In addition, due to the productivity gains from the use of higher level tools, it is feasible for a single design engineer to handle a design in the 30k-40k gate range within a reasonable period of time, thereby lowering the total development cost.

Design re-usability is another factor in reducing cost that is made more feasible by synthesis. Existing VHDL designs can be easily reused in different configurations or modified to accommodate different requirements, and then synthesized into the technology suitable for the new application.

Growth with the Technology : As of the early 1990s, ASICs with gate counts over 100k gates are being designed, with projections of million gate devices being readily available within 5-10 years. It is obvious that advanced tools are needed to make designs of this complexity feasible. The capabilities of synthesis tools will continue to expand to keep pace with the demands of the ASIC technology and will provide even higher levels of synthesis in the future. The tools themselves will also be able to benefit from the higher performance of future workstations that will be needed to efficiently design larger and more complex ICs.

Design Management :   The process of designing larger and more complex ASICs within schedule constraints may require a team of designers working on a single ASIC. Similarly, on a large project several ASIC designers will often work under the guidance of a project leader whose job is to coordinate and review the individual designs. As an aid to maintaining good design practices and documentation, the VHDL language contains many of the structured design capabilities that can be found in most software programming languages. This includes the ability to easily partition a design into functions, verify each function individually, and then interconnect the functions to verify the entire design. By the proper use of VHDL and synthesis, these tasks can be readily analyzed by the design team and the project leader throughout the design cycle.

Conformance with US Federal Standards (where applicable) : VHDL has become a requirement for any US Federal Government work under the Federal Information Processing Standard (FIPS) 172 agreement. In the past, conformance with these standards was considered more of a documentation issue that was often satisfied by simply providing VHDL formatted netlists, but with today's more powerful VHDL tools, the intent is to truly assure the use of VHDL throughout the design process.

And the Disadvantages ...

Due to the radically different nature of this new and complex design methodology, there has been difficulty in understanding exactly how to apply it in real design situations. The "disadvantages" of this methodology mainly fall into the areas of learning and adapting to a new way of doing things.

A Change of Culture : There is no doubt that making the transition from a schematic-based design system to a language-based synthesis system requires a new way of thinking about the logic design process and significant learning of both the language and the tools. It certainly helps to have PASCAL or other computer language experience, since VHDL is a modern, structured language with many of the same types of features. On the positive side, VHDL has been developed from the beginning with digital design in mind, and most of the basic concepts are rooted in actual design processes. This book has been written from the viewpoint of the logic designer and thereby should help to ease the transition to the new methodology.

Acceptance of the language based approach depends on gaining experience and confidence in the process, just like learning any other new subject.   It has been our experience that once the initial VHDL simulations are run, and the first small trials of synthesis are accomplished successfully, most designers quickly begin to develop an understanding and a faith that the technology can indeed be trusted. In time, the level of VHDL knowledge can be expanded into other useful areas such as the use of abstract behavioral modelling for top down design.

Cost of Getting Started : There are definite startup costs in learning VHDL and synthesis and in evaluating and subsequently purchasing design tools. This book should help in both areas since, in its coverage of both the VHDL language and logic synthesis, you will gain an understanding of the processes involved. In the chapter on evaluating tools there are descriptions of the capabilities that are necessary for successful synthesis.

Additional costs to be accounted for are in the preparation of training materials and in the tailoring of the VHDL synthesis methodology to integrate with your other existing design tools, company standards and procedures.

Learning and Training : As with any new technology, the process of learning and user training is vital to the successful use of this methodology. The use of logic synthesis and VHDL for ASIC design involves more than simply learning the syntax of the language and then beginning to write code. The process is an entire methodology that covers many design disciplines in addition to VHDL language skills.

From experience, it is recommended that initially an expert individual or team of experienced digital design engineers with programming background be established as the pioneers in learning and applying this technology. The tasks of learning and evaluating the necessary tools, purchasing and putting those tools into place, and gaining hands on experience, are essential for a smooth transition into language-based design. The experts then can provide practical training for new users, including a high degree of initial support and confidence building.

From a practical viewpoint, it is not necessary to learn the entire VHDL language in order to be able to use the language for synthesis design, and that fact shortens the learning curve considerably.

Selecting Design Tools : Design tools in general are fairly costly, and considering the state of most budgets, it seems clear that tools should be selected carefully. The idea behind the use of these tools is after all to improve efficiency and productivity while assuring a high quality design, and tool selection is a key factor in the overall success of the process. Some related tool assessment issues have been mentioned above and will also be covered elsewhere in this book.

Debugging Design Problems : One of the "disadvantages" associated with VHDL synthesis is that the output is an automated synthesized product, which means that the resulting gate-level design has been computer generated, not human created. In an ideal situation, the designer should not need to examine this gate-level design, but, until synthesis tools are tightly integrated with ASIC layout tools, there is always the possibility that a timing problem could occur due to layout effects. In addition, issues related to circuit speed, buffering, and testability can introduce gate delays that often require a detailed analysis of the circuit. The problem faced by the ASIC designer at this point arises because the detailed logic was designed by someone else, the computer in this case, and the automated schematic generation of this "foreign" design results in a schematic that is not easily recognizable. Analyzing at the gate level, therefore, becomes a more complicated situation, similar to taking over a design task in the middle of the project. Fortunately this problem can be minimized by following good synthesis design guidelines and by consideration of testability design early in the design cycle, and is made easier by other ASIC timing analysis tools that help identify worst case paths and timing violations.

Facts, Fears, and Misconceptions

As with any new process, there are uncertainties and concerns about making the change to a new technology, especially one that is both different and complex and yet promises great things. The CAE industry has also contributed to this problem through overenthusiastic marketing claims and has created some misconceptions about VHDL and synthesis that add to the confusion. The following is intended to restore some perspective to this situation.

"Just Code It in VHDL and the Synthesis Tool Will Design the Logic" : This type of marketing hype is prevalent in the CAE press as well as advertising and unfortunately leads to a management and technical misunderstanding of the real capabilities of VHDL synthesis. While it is true that these tools can be very powerful when used correctly, the implication that synthesis takes over the design tasks automatically is simply not the case. First, as this book will emphasize throughout, logic design experience is definitely necessary in order to be able to use VHDL synthesis techniques, and the tools are very much under the control of the design engineer. Secondly, it is important how you code the VHDL - poorly done or ambiguous VHDL will result in an inefficient, or incorrect, synthesized design! In practice, most of the normal engineering and logic design decisions are still made, but as part of the VHDL planning and coding process rather than as part of the schematic generation process.

There is No Magic! :   Synthesis tools cannot make certain high level design decisions for you. For example, when you need to multiply two numbers together, even though conceptually you may be thinking A = B times C, as a hardware designer you must decide whether to use a serial or parallel approach, two's complement or unsigned, which of the many algorithms to use, handling of sign and overflow conditions, rounding of the result, etc. These are the decisions that you normally make at the detailed block diagram level, and they must still be made when using logic synthesis. The synthesizer will be able to generate the multiplier circuitry, but essentially you will control which technique to use and the synthesizer will handle the gate details.

On the other hand, when you consider that a synthesizer can create a binary counter circuit from one line of VHDL code, regardless of the number of bits in the counter, maybe there is a little "magic" in the process!

Synthesis just can't be as good as a human designer, can it? :   We all have a tendency to be somewhat skeptical about new ideas, especially when they are both complex and accompanied by a lot of promises but very little practical data. In this regard, logic synthesis suffers from the same problem, and perhaps a historical analogy will provide some useful perspective. Prior to the early 1990s, synthesis work was largely confined to academic and industrial research labs, and most papers presented at technical conferences focussed on problems of combinational gate minimization. While interesting, most logic designers working within industry held the opinion that synthesis was probably one of those blue sky activities that would never be capable of designing real logic circuits and was similar to the efforts to develop chess playing computers. The earliest work on these problems was indeed somewhat primitive and bore out those conclusions, but steady and tenacious development has continued to improve both logic synthesis and chess machines to the point that today they have become a genuine force to be reckoned with. The chess skeptics have certainly been quieted by the creative and strong play even at the grandmaster level by the best chess computers, and by the availability of portable master strength machines for only a few hundred dollars.

In a similar manner, logic designers are finding that the synthesis tools have emerged from the labs as powerful commercial products that can be used to design a wide variety of complex circuits. But are they as good as a person? Well, while synthesis may not be at the "grandmaster" level yet, the answer is that when used properly by a good designer, synthesis can indeed produce very efficient circuits in a relatively short time. The task is therefore to learn how to use and control the synthesis tools so that they can quickly do exactly what you want them to do. And that is essentially the key to successful logic synthesis.

Summary

This chapter has presented some of the background for the development of gate arrays and other ASICs and the reasons why you can benefit from using VHDL synthesis tools for future designs. In Chapter 2 the issues involved in actually making the transition to a VHDL language-based design methodology will be discussed, in preparation for subsequent chapters covering the more technical details of practical VHDL synthesis design.

Chapter 2

Making the Transition to VHDL Synthesis

This chapter will discuss the various aspects of moving to a VHDL synthesis based design methodology as well as some of the problems in getting started that were touched upon in Chapter 1. The focus here is on developing an understanding of how to proceed from the system design phase to the detailed ASIC design, and on how to modify your existing design process to successfully use synthesis tools. Some of the concerns involved in making the culture change to language based logic design will also be discussed, along with other issues such as tool evaluation and training. This chapter focusses on the basic process and how to get started, while Chapter 3 will begin to cover the technical details of using VHDL for design and synthesis.

What Can Be Synthesized?

If you examine a wide range of different systems, you find that almost all digital designs consist of variations of a limited number of basic building blocks for logic functions, regardless of the actual system application. This fact can be seen in practice by the continued long term success of TTL and other standard logic families. These basic functions in turn are implemented using even more primitive logic cells such as gates and flip flops. With today's ASICs capable of holding many thousands of gates, the role of logic synthesis is to allow designers to quickly and efficiently build the specific larger functions from the primitive cells.

Essentially, using today's advanced logic synthesis tools, almost any ASIC circuit can be synthesized that you would design using basic logic cells such as gates and flip flops. Both combinational and sequential logic can be synthesized and, beginning in Chapter 4, the VHDL coding techniques to generate basic as well as more complex functions will be presented.

The specific types of logic functions that can be synthesized can be grouped into those that are strictly combinational and a larger group that are sequential, consisting of both clocked storage devices and combinational gates. Examples of the types of functions that can be synthesized include :

Combinational functions

Multiplexers
Decoders
Encoders
Comparators
Gating logic
        Lookup tables
Adders, subtractors, and ALUs
Multipliers
PLA structures

Parity generators.

Sequential logic can be broken into three groups based on their functions. The groups can be identified as follows:

Counter based functions

> Counters
> > Binary, BCD, Johnson, Gray
> > Up, down, up/down
> Gate generators and pulse generators
> Timing and clock generators
> Event counters
> Memory Address counters
> FIFO memory pointers
> Prescalers

Register and latch functions

> Data registers and latches
> Shift registers
> Register files
> Accumulators
> Parallel/serial converters

Control logic

> Sequencers or Controllers
> Finite state machines
> Edge detectors
> Synchronizers

Digital systems, and particularly those that are implemented using ASIC devices, usually consist of combinations of "random logic" blocks, memory devices, and large regular structures often referred to as megafunctions. Most ASIC vendors use or provide specialized compiler tools to design fast and efficient RAM and ROM memories, plus other custom software used to design more specialized functions such as flow-through FIFOs. Except in relatively small applications it is not efficient to design these types of functions using gates and flip flops, and therefore they would not normally be synthesized either. On the other hand the great bulk of any design lies in the "random logic" blocks that perform the desired function, and this is where logic synthesis is used most effectively. These logic functions are of course not random, but typically consist of the fundamental gates, flip flops, and other primitive logic cells that are used by designers to implement any function. These are the building blocks used by logic synthesis tools.

It should be pointed out that while synthesis can be used in a wide variety of applications, not every ASIC design should be synthesized. In the situation where the planned design is pushing the speed of the technology to its limits, which is a risky business under any circumstances, a synthesis tool may not be able to guarantee that the timing requirements will be met. This is not so much a shortcoming of the synthesis tool itself but is due to the fact that the timing delays resulting from chip layout are not available to the synthesizer when it is trying to optimize the circuit. This could lead to a lot of manual design intervention and as a result might be inefficient for critical applications. The general problem of layout dependency and synthesis is being worked on by industry and will be discussed later.

Some Basic Synthesis Premises

Before beginning the discussion of the ASIC design flow, it is a good idea to briefly cover some of the basic premises of VHDL synthesis. As mentioned in Chapter 1, this new design approach involves describing and simulating a design using VHDL, and then having the synthesis system convert the VHDL description into a detailed logic circuit. Although most of the design effort will be spent on the VHDL generation and simulation, the ability of the synthesizer to generate a quality design greatly depends on how the VHDL is written.

One of the fundamental things to keep in mind about using VHDL for simulation and logic synthesis is that it is a design process, not just a programming or coding process. As we shall see, because VHDL has the rich capabilities and features of a modern programming language, it can be used to solve problems in many different ways. While this richness allows for creative coding approaches from a language viewpoint, it is important to realize that just because a system can be described and simulated in VHDL does not automatically mean that it can be synthesized. There are rules for synthesis as well as design guidelines that need to be used to produce synthesizable logic designs that are both efficient and that match the simulation results. Improper VHDL coding can make it very difficult for a synthesis tool to derive a logic circuit that will perform that function and can result in long synthesis runs and/or inefficient logic.

In general, it is most important to have experience in logic design techniques and in good ASIC design practices in order to be able to use synthesis successfully. The VHDL design guidelines and examples presented in later chapters make use of this experience to ensure that the synthesized logic is both efficient and that it functions correctly.

As we shall see, the second basic rule for successful synthesis is to code the VHDL to do what you want to do, and not force the synthesizer to figure it out for itself. In other words, you should plan the logic you are trying to design in advance, and then use the language to describe that function for the synthesizer in a clear and straightforward manner. As will be shown throughout the book, the emphasis should be on the logic design process, with the VHDL code simply being used to implement that logic via synthesis.

The ASIC Design Process

Background

The design process for gate arrays and other ASICs has been steadily refined since the early 1980s, driven by both the changing requirements of the technology as well as by advances in design tools and computer systems. For instance, disciplines such as testability design and test generation, coupled with the use of tools for fault simulation and timing analysis, have been added to the standard design practice in order to continue to improve the ASIC success rate.

In general, the two major concerns of ASIC designers are to produce a design that works correctly the first time in the system, and to ensure that the resulting physical ICs are fully tested and functioning with no manufacturing defects. Both of these requirements are key to the success of ASIC development, regardless of whether VHDL synthesis is used or not, and involve interrelated design, test, and simulation disciplines that must be planned for throughout the design cycle.

As a starting point to see how VHDL synthesis fits into the process, consider the basic ASIC design flow widely used today, shown in Figure 2-1. This diagram has been simplified to show only major design functions and applies equally well to both the schematic-based as well as the synthesis-based design approach. The areas that will be affected by the synthesis process include the block diagram generation, the logic design, and the logic simulation steps.

System Planning and Definition

In the early phases of every new system design is the effort to analyze the basic system requirements and develop an approach that will satisfy those requirements. A tradeoff of the algorithms to use and partitioning of the design solution between hardware and software is usually done at this time, resulting in a high level block diagram as well as more detailed definition and specifications for the lower level design modules. Although outside of the scope of this book, top down design tools such as VHDL can be effectively used at this point to help make the early tradeoffs and evaluations, and in addition can be used to produce "simulatable" specs for various system blocks in addition to conventional paper specs. These concepts are part of a continuing effort on the part of the design community to reduce errors throughout the design process and are being strongly sponsored by both ARPA and military development organizations at Wright Laboratories and Fort Monmouth.

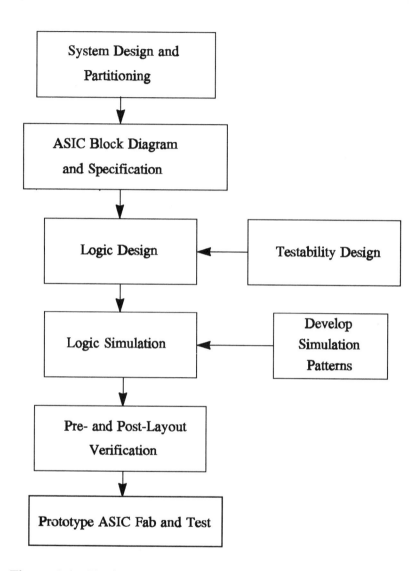

Figure 2-1 : Typical ASIC Development Flow

Block Diagram and Specification

Following the top level design phase, at which point the system has been partitioned into the desired hardware functions, there is a further design effort needed to break down the overall digital logic of the system into realizable pieces such as boards, modules, ASICs, and discrete ICs. These decisions are based on overall system budgets for size, power, weight, cost, and performance as well as schedule constraints andcan have a significant impact on the design technology used in a system. The key output for each of the digital ASICs during this stage of the design process is a specification and block diagram that defines not only what the ASIC is to do in the system, but also how it is to be implemented in terms of logic functions to be used, interfaces with the rest of the system, and details such as timing constraints and clock speeds. These types of specs and block diagrams may be generated formally or informally, but their intent is mainly to provide a firm basis for beginning the detailed logic design process. Keep in mind the "no cuts and jumpers" rule for ASIC design - if the requirements cannot be well defined, maybe that logic should not be included in an ASIC where it will be "cast in silicon". In that case, implementing the design as one or more FPGAs would be preferable until the design requirements are firm.

Depending on the complexity of the ASIC, the "block diagram" may actually consist of several levels of diagrams showing additional detail. For example, for a large design you may generate a very general top level diagram showing major functions (CPU, UART, control logic, timing generator) and then have additional blocks for each of those functions. At this lower level you will normally identify generic registers, counters, or multiplexers that need to be designed during the logic design phase, and will have also determined the specifics of data word lengths and bit assignments, clocking requirements and clock speeds, and the planned operation of any control logic functions. Figure 2-2 shows the type of detail in a typical high level block diagram that identifies major functional blocks, but without any design level details.

In order to begin the logic design, each of the major blocks is normally broken down into simpler recognizable logic blocks. Figure 2-3 shows the typical level of detail, for one of the "signal generator" blocks, that can be used as a basis for developing a synthesizable VHDL description of the function.

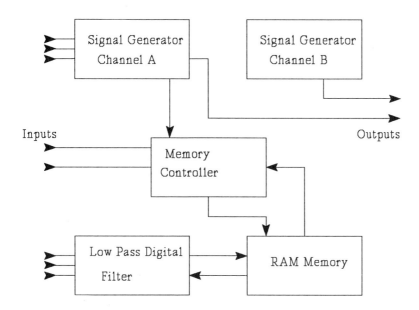

Figure 2-2 : A Typical High Level Block Diagram

It is recommended that you prepare detailed block diagrams at this level for all functional blocks. Although initially these might be informal sketches, this discipline will lead to a clearer understanding of the design throughout the VHDL coding and simulation phase. In addition, this type of design documentation will be used for planning the testability design and simulation, and will directly relate to the generation of VHDL descriptions for synthesis. In that regard it is recommended that the names of various functions and signals be made consistent throughout the design process. This will make it quite straightforward to identify names in the VHDL code with the same functions on the block diagrams and specifications and will simplify design reviews.

An additional output of the block diagram phase should be an estimate of the gate count, based on experience, so that system planning and cost estimates can be refined. This data will also be used during the synthesis process as a "sanity check" on the output produced by the synthesizer and will be discussed in later sections.

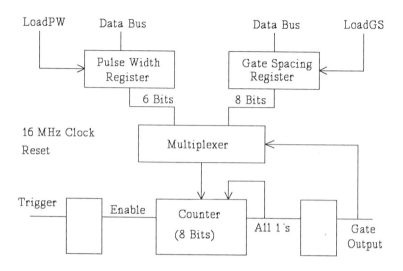

Figure 2-3 : Detailed Block of Signal Generator "A"

Testability and Simulation Planning

In addition to defining the logic blocks needed for the system, the important task of testability planning and design should also be addressed at this time. The ability to fully test an integrated circuit usually requires that additional logic be designed in to improve the IC's controllability and observability. These factors also affect the number of test patterns needed to verify that all of the gates in the IC are functioning properly. Because of the relatively low wafer yield of gate arrays and other ASICs, inadequate testability can result in a large number of chips that pass the physical tests but still contain undetected flaws that will only show up in the final system testing. In most cases, this is a very costly problem to correct, and is best dealt with by means of proper testability planning early in the design cycle.

There are many types of ASIC testability approaches that can be used, ranging from the use of full scan or partial scan logic to less formal or "ad hoc" approaches, such as providing additional output multiplexers for increased observability and input signals for improved controllability. Decisions on whether to use JTAG or other boundary scan approaches for board or system level testing, whether to incorporate built in self test (or BIST) mechanisms, and any other special testability requirements should also be considered at this point. Some design architectures are inherently more testable than others, but regardless of the approach selected, it should be addressed early in the design cycle so that any required testability logic can be efficiently designed from the beginning.

Related to testability planning is the simulation planning effort which simply considers how the design will be simulated at various stages of the development. The general issue is to define how to simulate the ASIC design to verify its correct behavior in its system application, which should tie in with the ASIC specification and block diagram. Typical questions that arise are how to accurately generate the input signals to the ASIC, whether to simulate smaller functional blocks of logic in addition to simulating the entire array, how to handle multiple ASIC/board level simulation, and the means for checking the results for accuracy. These issues need to be addressed whether the detailed logic design is done with conventional schematics or by means of VHDL synthesis, but in the synthesis process they must be determined earlier in the design cycle since the VHDL simulation effort will normally begin at that point.

Other less global test and simulation issues affect how ASIC logic is designed and are normally part of the design-for-test rules. For example, if a 24 bit binary counter is needed, it will take 2^{24} clocks to count through all of its states and verify that the circuit works correctly in the ASIC. Since this would take an enormous amount of time to simulate, and would likely exceed the ASIC tester's capacity, a typical design practice is to break the counter into smaller sections to allow testing with a smaller number of clocks, or to provide a means for presetting the counter to a desired state. These types of problems should also be addressed before the detailed design begins, since they will obviously affect the final logic structure.

Detailed Logic Design

The detailed logic design effort normally begins at this point, using the block diagrams and specification to direct the design. This phase is the part of the overall ASIC development process that is most affected by the use of VHDL synthesis, and is where most of the schedule and cost benefits are realized. Figure 2-4 shows how the detailed design and simulation tasks for VHDL synthesis compare with the conventional approach.

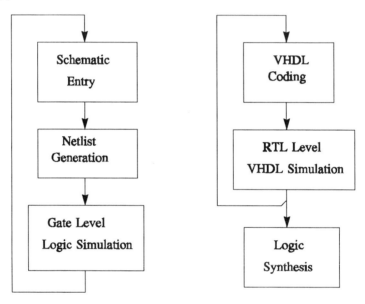

Figure 2-4 : Comparison of Detailed Design Flows

Using the traditional system, you normally begin with the detailed block diagrams and for each logic function roughly sketch out the circuit details prior to drawing the schematics on your CAE system. The amount of detail done at this point depends generally on how complex the circuit is and on your previous design experience. Simple registers for instance may be designed directly with the CAE tools while you would probably do any control logic or complex counters or ALUs in greater detail. In terms of organization or partitioning, logic that belongs together is generally kept together on the same page of the schematic or on closely related pages, and signal names are kept in agreement with the block diagrams for clarity.

Working from the block diagram, another task normally done early in the design phase is to identify any commonly used circuit functions that you may want to design as a "macro" to be used throughout the ASIC. As an example, you may have a special multiplier circuit that will be used throughout a digital filter ASIC. These types of functions can be designed once and used as library components where needed in the design, or even in other ASICs, in order to save time and work.

These steps are basically similar when using VHDL. Starting with the detailed block diagrams, the overall ASIC is roughly partitioned into sections of logic that belong together, which form the basis for generating VHDL blocks called Entities. (As discussed further in Chapter 3, it is desirable to limit the size of these blocks to several thousand gates if possible due to considerations of both ASIC floorplanning and synthesis workstation performance, but this is simply a practical guideline that will most likely grow in the future). Each of these blocks will have inputs and outputs connected to other blocks in the ASIC, plus all of the internal functions within the block. Since the detailed block diagrams show those functions as registers, counters, multiplexers, etc, it is a straightforward task to associate them with VHDL "processes", using the same names for signals and functions on both the diagram and the VHDL code. As in the conventional design approach, more complex functions such as control logic or sequencers and timing generators should have additional definition, not in terms of gates but in terms of what they do in response to their inputs. A good practice is to plan on including this design information as comments in the VHDL code for ease of understanding and review, and which will be shown further in Chapter 3.

The next step is to generate the VHDL code for each function and block, along with the simulation test patterns. Experience has shown that this step can be done very rapidly once the block diagrams are completed, and simulation and debug of the design is usually underway quite early in the design cycle. The reason that this can be accomplished so quickly is that the synthesizable VHDL code is written at a fairly high level compared with the gate by gate details required on a schematic, and this in itself takes much less time. This is a great advantage over the conventional design process in which the detailed schematic drawings for a function must be completed before netlists and simulation files can be generated, since that tends to move the simulation to a later point in the design cycle. The earlier that simulation begins and design problems are found and corrected, the better!

Continuing with the process flow comparison, the major difference is that in the conventional methodology there is a significant effort to get the logic diagrams drawn so that the simulation can begin, whereas with the VHDL approach the simulation begins immediately and the schematics are generated at the end of the cycle, if at all.

The issue of schematics is in fact one of the significant culture changes resulting from the use of synthesis. As a designer you are probably comfortable with using logic diagrams or schematics to represent your design, and use them to analyze and review the circuit before and during simulation. Synthesized schematics however are generally less aesthetic than those drawn by a person, and the gates and flip flops in the circuit may not be arranged in the same way that you would draw them. However, when using VHDL you will learn that with a little experience you will be able to review and debug your simulations from the block diagrams and the VHDL code directly without the need for gate level schematics. From practical experience, you will probably look at the schematics during the learning phase of VHDL synthesis in order to build confidence that the process really works. After you have gained experience however, the schematics are likely to be referred to only if you are curious or want to check a critical circuit, or if a timing problem shows up as a result of chip layout effects.

Simulation

At this point in the design cycle, the block diagrams have been generated and the VHDL descriptions of the logic have been prepared and coded. The next step in the process is to begin to test and debug the design using the VHDL simulator. Note that in contrast to the conventional approach there is no need for gate level details yet, and the design will be tested at the higher level of behavior called Register Transfer Level, or RTL, as explained in Chapter 3. The code for the VHDL descriptions and test patterns will normally be typed as ASCII text files which are the input source for the VHDL simulation tools. Commercial tools also exist that can assist in creating portions of the VHDL code and are useful in reducing the amount of effort in entering I/O signals and the structure of the major design blocks. In either case, the simulation process is similar to traditional logic simulation in that the source files are compiled and checked for errors, then are linked together with any other blocks that are part of the design. The simulation process flow is the same in that you monitor the desired input and output signals, either as graphical waveforms or as tabular listings depending on the application, in order to verify the proper behavior of the circuit. Most commercial VHDL simulators provide the basic capabilities to allow the simulations to be run for short or long time intervals, to be stepped through operations one step at a time, and to run to some desired breakpoint conditions.

When circuit performance problems are detected, the conventional design is debugged by determining why the outputs are incorrect and checking gate by gate through the schematic until the source of the problem is found. In the VHDL case, the procedure is to check the VHDL code description of that function for the reason the behavior is wrong. In both cases additional internal signals are usually monitored during the simulations until the cause of the problem is found, and as a result the circuit is redesigned or the VHDL code is modified. This process continues until the entire design works correctly in the simulator.

There are practical tradeoffs that are usually made during the simulation process that apply to either design approach. One common decision that must be made is whether to simulate portions of a design separately or to simulate the entire array all together. It is a good idea to simulate smaller portions of a design if they are very complex or if detailed checking of its operation is required, or if you are still learning the VHDL synthesis process. For example, if you have an adder or arithmetic unit deep within the ASIC you probably won't want to have to verify all of its combinations while running simulations of the entire array, simply because it will take longer and may be harder to control. It would be better if you knew that each of the smaller modules such as the adder worked properly before you tested them all together as the final design. On the other hand, simulating at the module level means that you may have to generate test patterns for that module that may not be usable at the higher level, resulting in extra effort for that task.

This process is part of the VHDL learning curve when moving to the VHDL synthesis methodology. At first when your experience level is low, you should probably simulate and synthesize each of the modules in order to develop confidence in the process. The situation is a little different later on when you've had some experience, at which point the decision of when to simulate at the module level and when not to is a choice that depends more on the specific design situation. The benefit of doing a simulation plan early in the design cycle is that these issues can be addressed at the beginning, and if, for instance, the overall ASIC simulation requires complete testing of a certain function, then it may be unnecessary to repeat that testing at the module level. In any event, you should be careful not to shortchange this effort since a little extra time spent early in the cycle could save a lot of time towards the end.

Logic Synthesis

The synthesis of the gate level logic normally follows the VHDL simulation effort. The simulation has been used to confirm that the design works correctly as a function, although until this point the delays through the "logic" will have been simply estimates. Although synthesis is a fairly automated process, a few additional details must be provided to the tools. First is the decision of which ASIC vendor will be used, since the vendor specific library of cells and parts will be needed in order to generate the gate level design. This library normally contains the details of the individual gate delays and the rules for computing loading delays due to inputs and due to estimated capacitance and wire length. The second input to the synthesizer is information that will be used to constrain the design to your requirements. This typically includes the clock rate and pulse width, assumptions about operating temperature, voltage, and process variations, output loading, and limits on permissible propagation delays through critical paths.

The synthesizer initially processes the VHDL input code into general lower level logic building blocks such as multiplexers, decoders, registers, and ALUs, from which it can determine whether logic blocks can be shared between functions for efficiency. The second step is the conversion of the generic functions into the vendor-specific library cells, followed by optimization steps to achieve the speed constraints and logic minimization for that logic where speed is not critical.

The outputs of the synthesis tools typically include a vendor specific netlist, reports on timing, gate count, and critical paths, and plotted schematic diagrams. The netlist and schematics are subsequently used by the ASIC vendor's design verification tools for chip bonding and layout, gate level logic simulation, and final timing checks both before and after routing. The reports are used during the synthesis process to control and understand what is being synthesized, as well as to manage any changes to the VHDL source code. This could happen for instance if the critical path delays cannot be met, thus requiring a modification to the VHDL description.

During the learning period of these tools, you will probably want to check the gate counts and logic to see if it is what you expected, as well as to develop confidence in the process. Even after long term usage, it is still important to check the gate count for several reasons. First, you should always expect the synthesis tools to do nearly as good a job as you would, and checking the gate count verifies how well the synthesizer is actually performing. Second, it is possible to describe a circuit with VHDL in several different ways and, depending on the circumstances, some techniques may be better than others. This book will try to show good ways to design most common logic functions, but there are so many possible variations that it always pays to be aware of the synthesized results. Third, there could be a bug in the synthesis software that could result in excess gates in the design, even though it might work correctly, and you certainly will want to know about that so the vendor can correct the problem. The best general course of action is to keep an eye on the synthesized results in terms of gate efficiency. As far as checking the logic, there is little likelihood that a design error will get into the final ASIC because the subsequent gate level logic simulations and other verification checks should catch the problem. This however reinforces the importance of thorough simulation planning and testing, and accurate checking of simulation and verification results.

It is prudent to run the synthesis tools briefly on a design even before the VHDL simulations are complete just to check that the code is synthesizable and no other problems are found that could cause you to have to rework the VHDL description. The reason this could happen is that every synthesis system has some limitations on the VHDL code it can synthesize into logic, even though the code may simulate perfectly well. Fortunately a number of the commercial VHDL simulators can do synthesis guideline checking during the compilation process which reduces the need for the additional check. Chapter 10 expands on the general topic of tool related issues and evaluation guidelines.

ASIC Verification

This step consists of the final gate level checks that precede the actual layout and routing of the ASIC, sometimes known as pre-route signoff. Although these procedures are somewhat vendor dependent, the basic process is similar. The circuit files or netlists, derived either from the synthesis process or from schematic capture, and the simulation pattern files, are the inputs to the verification process. This step normally consists of gate level logic simulation, timing analysis, design rule checking, electrical rule checking, and I/O pin assignments. These checks may be done using a specific ASIC vendor's software system or a vendor approved general purpose CAE system, but in either case the intent is to make sure there are no errors either prior to layout or following layout.

As mentioned at the beginning of the chapter, the layout and routing of the chip has an effect on the gate delays throughout the ASIC. The libraries and rules used during the synthesis process are normally jointly developed by the ASIC vendors and the synthesis vendors and therefore can be reasonably accurate for synthesis and timing optimization. The problem is that the gross ASIC layout planning, or floorplanning, can affect the estimated timing before routing, and most certainly the post-route timing of most gates and flip flops will be affected by the wirelength and capacitance effects of the circuit. This does not necessarily cause design problems in most cases unless the speed of the circuit is very critical, but can result in gate spikes or propagation delay problems in some designs.

The procedure to correct timing or other problems at this point is to use the reports from the ASIC verification system along with the netlist, block diagrams, and schematics to analyze the cause of the problem. Depending on where the problem lies, the fix might be to resynthesize a block to minimize the timing delays or even to modify the function slightly. One thing that should be avoided except in a schedule crisis is to manually edit the netlist, since this eventually causes problems at a later time with the design documentation and with configuration control. If at all possible, make any changes in the source VHDL or synthesis control files, then resynthesize and resimulate the design.

A Few Observations

From the discussion of the overall design flow, it can be seen that the VHDL synthesis methodology involves many of the same tasks as the traditional approach and requires the same high level of discipline in order to produce successful results. Several points should be observed relative to VHDL and logic synthesis :

A. Although VHDL is a language and may be thought of as software, using VHDL and synthesis to design ASICs is very much a logic design process. You should be careful not to think of this work as being just software programming. From the design flow discussion, you can see that most of the decisions made during the development cycle are based on logic design criteria and hardware issues. This is part of the reason that preparation of detailed block diagrams has been stressed - you should know what the basic logic structure should be, before you begin any coding of the VHDL. The most successful approach to synthesis is to code the VHDL to produce what **you** want, rather than having the synthesizer try to figure out what you want. This type of problem can be avoided if you put the emphasis on VHDL as a hardware description language rather than a programming language. To reinforce this point, examples of poorly planned VHDL logic design will be shown in later chapters.

B. Simulation is heavily used throughout the ASIC design cycle and typically uses a combination of VHDL for the actual design and the ASIC vendor's certified or "golden" logic simulator for the final design verification. For all of these activities, the same test patterns should be used for both sets of simulators, and the output results should be compared, to ensure that the final gate level ASIC design matches the original VHDL design. The simplest means of accomplishing this is to generate both sets of test patterns from the VHDL "test bench" used to simulate the VHDL design. This procedure will be covered in Chapter 9.

Getting Started

The additional tasks required to make the transition have been mentioned in the introduction and include the learning and training effort, evaluation of design tools, developing confidence in the process, and integration with your other design tools and procedures. The tasks are normally done in that order, since each succeeding task builds on the previous experience. For instance, it would be difficult to make a good choice of VHDL simulation and synthesis tools without having some background in the VHDL language and its usage. The procedure therefore is to attack the four tasks with a methodical plan.

The learning process can be broken down into separate phases. The VHDL language syntax and rules should be learned first, followed by learning how to apply the language in practical design and logic synthesis work. This book has been written to provide that background in both the language and its application to realistic ASIC design work. The other texts listed in the references provide additional emphasis on the VHDL language by itself and should be used in the long term to develop your knowledge in other areas not related to synthesis, such as higher level simulations and test bench development. For larger organizations or design groups, it is sensible to establish a few key local experts who can lead the development and subsequently assist the other designers in using the technology. In the best scenario, the local experts should have logic design and ASIC design experience, as well as background in the use of computer languages and design tools.

The tool evaluation phase includes assessment of both VHDL simulation and logic synthesis tools. Chapter 10 covers most of the general criteria to use in making a selection, but the factors to consider include the tool's technical performance, integration with other tools, cost and vendor relationships, speed and accuracy, and documentation. An accurate evaluation is likely to require several months of effort when you consider that both simulation and synthesis tools and processes must be learned, while simultaneously making judgements on how well the tools work for actual design work. The relatively high cost of design tools and their importance to the success of the ASIC design process make this a task to be taken seriously.

During the evaluation and tool selection process, you will use many of the examples in this book and your own logic functions to explore the capabilities of the VHDL simulator and logic synthesis systems. The synthesis judgements being made will include how well the synthesizer works, how efficient is the gate level logic design, how well you can optimize the design for speed or gate count, and whether it has limitations in terms of the VHDL used. During this period, you will develop the confidence that VHDL based synthesis can not only work, but that it can work well! It is probably good to be somewhat skeptical at first, until you can see for yourself how well the process works. In practice, following some training with the examples, it has been effective to use small portions of an actual design for the learning and confidence building phase.

The last task to be tackled is the integration of the VHDL synthesis tools with the rest of your ASIC design process and tools. This work is dependent on both the ASIC and tool vendors and can include mundane tasks such as translators or data formatters for netlists and test patterns, or more critical issues such as library accuracy and compatibility between synthesis and ASIC verification systems, and potential post-rout timing problems. The main question to address here is how rapidly and efficiently you can move from synthesis to verification and back, since this will be an important loop in the overall ASIC development process.

Summary

The first two chapters have covered the basic issues involved in VHDL synthesis and how it is used in the ASIC development flow. Chapter 3 will begin with a discussion of the VHDL language and the synthesis concepts used in the examples in subsequent chapters.

Chapter 3

VHDL Background for Synthesis

This chapter will review the basic VHDL concepts used for simulation and synthesis of ASICs, in preparation for succeeding chapters which show their usage in various applications. The material presented is intended for designers with little or no previous VHDL experience, and covers both the language and various logic design and synthesis concepts as they relate to VHDL usage. Experienced VHDL users can skip over some of the VHDL syntax portions but should read the sections on synthesis of clocked and combinational logic, latches and registers, since they deal with general logic synthesis guidelines.

The information will emphasize how VHDL relates to practical design applications, rather than presenting the complete VHDL syntax, and will serve as a convenient reference to supplement the material contained in other VHDL textbooks listed in the references. The material covered here should allow you to understand how digital logic is related to VHDL code, and to understand the VHDL jargon and terminology.

VHDL for Synthesis

The VHDL language has been given a wide range of capabilities to permit the accurate description and modelling of a system, from the highest levels of abstract behavior to the lowest gate level structures. Because of that fact, it has many different syntax features to provide the flexibility to handle general design problems and to enable the implementation of complete top down system design methodologies. The full capabilities of VHDL are much greater than those needed for logic synthesis however, and some of the more abstract modelling features have no counterpart in digital logic. For these reasons logic synthesis uses a subset of the language, although a very powerful and complete subset. At this time, the VHDL features supported by logic synthesis tools are not standardized and therefore vary slightly from vendor to vendor.

Even though synthesis uses a subset of VHDL, it is still important to understand enough of the language so that you can handle non-synthesis related tasks, such as the generation of test benches and test patterns, and so that you can simulate your designs accurately. As an overview, in order to see where synthesis fits into the spectrum of VHDL's capabilities, let's consider the three broad types of system modelling that VHDL can handle. We will find that all three styles of VHDL will be used at some point during the normal ASIC development cycle.

Abstract Behavioral Level

This is the highest level of system modelling, in which VHDL is used to describe **how** the system should behave in response to input signals but without any specific hardware implementation necessarily in mind. There are several reasons for using this type of VHDL simulation, such as to be able to assess how various system design algorithms will work and how well they will work under expected input conditions, and to assist in coming up with a hardware implementation scheme and specifications for the more detailed design stages.

Normally, during this phase many details such as clocks, word lengths and bit assignments are not used, and system operations are handled more abstractly, often using VHDL's ability to describe signals with user defined "enumerated types". With this for example, a computer operation can be called "ADD" without having to define a specific binary code to represent the operation. Also signals such as clocks may not be specifically used, and data busses are often treated as integer or real values without any specific number of bits. As an example, the system's behavior may be to wait for a certain type of message to arrive on a data bus and, when it occurs, to set some system parameters and output signals to desired states and then wait for 80 microseconds. Exactly **how** to wait for 80 usec is part of the detailed hardware implementation and is not needed at this level of abstraction, nor are the details of how the outputs or system states are set.

Generally the use of this style of VHDL is most like using a conventional programming language such as PASCAL or ADA, except for VHDL's ability to handle the concepts of time and "concurrent" operations, which will be covered below. The behavioral coding style is also used within VHDL simulation test benches to provide maximum flexibility, and since the test bench does not get synthesized it can use the full range of the VHDL language.

Register Transfer Level

This level of system description corresponds to the detailed block diagram level in traditional logic design, in which general hardware functions are identified along with specific input and output signals and data busses. Clock and reset signal characteristics are defined and data busses and storage devices (registers, counters, memories, etc) will have specific numbers of bits assigned. The level of abstraction in describing the logic functions is still well above the gate level details of a conventional schematic and can range from the use of Boolean equations to the higher level style referred to as **Register Transfer Level**, or RTL. The use of the term RTL has changed since it was first coined and today the term is loosely applied to logic signals that may or may not have actual registers associated with them. Using the RTL style, any logic signal or bus assumes a certain state or value based on its inputs, and the logic values in a system are said to be transferred from the input to the output. The general RTL representation of this process is

Output <= Input;

where "Input" can be a complex combination of various types of functions, and the "<=" indicates the transfer of values.

Since RTL permits a fairly high level design style, designing ASIC logic systems using RTL level VHDL can provide the greatest gains in design productivity. Designs described at the RTL level are also technology and ASIC vendor independent prior to synthesis, which provides for flexibility and design reusability. The major commercial VHDL logic synthesis systems normally accept RTL level descriptions as input to the design process, and the design techniques used to synthesize logic will be covered in great detail in succeeding chapters of this book.

Structural Level

This syntax of VHDL coding is similar to the use of a netlist, which shows the design as a structure consisting of "components" interconnected by signals. The components can be as simple as gates and flip flops or can be larger blocks described as behavioral or RTL level code, and VHDL allows any mix of these styles to be used simultaneously in an overall design. For instance, in a top down design approach, you could start with your top level blocks described abstractly while you evaluate and simulate different architectural and algorithm choices. In time, as each block proceeds through various stages of the design process, you could then replace that block with its RTL or gate level equivalent and confirm that the overall system still functions correctly. Eventually the entire ASIC with its I/O, synthesized blocks, and compiled or megafunction blocks could be simulated in VHDL at the gate level, possibly combined with other ASICs and discrete parts in a larger system simulation.

Throughout all of these phases of the design process, the structural capabilities of VHDL allow interconnection of the blocks with each other, and with the "test bench" that is used to generate the simulation stimulus signals and to monitor the output results. The use of structural VHDL will be covered in Chapter 9.

Describing and Simulating Hardware

Since VHDL was developed to have the flexibility to describe very general design problems, it is not surprising that there is a corresponding relationship between standard design methods and the equivalent VHDL operations. For example, if you look at the general method of how digital systems are developed from Chapter 2, a block diagram is created identifying the inputs and outputs, the main functional blocks, and the interconnection between those blocks. The design of the major functions may in turn require a more detailed block diagram to assist both the final gate level logic design and the planning for logic simulation and test pattern generation. The VHDL language has been designed to support all of these types of operations, including partitioning and design hierarchy, in a straightforward manner. In addition to describing the behavior of the design, VHDL also has capabilities for the generation of the test bench used to accurately simulate and verify the results of the simulation, which will be covered in Chapter 9.

Physical Entities

When you want to describe a physical electronic system, you may think of the overall system itself, a chassis, a group of circuit boards, Multi-Chip Modules, ASICs, or even parts of logic within any of these other pieces. Whatever level of complexity you choose, there is some physical entity involved that communicates with the rest of the environment by means of signal connections. VHDL appropriately names this an **ENTITY**, and its connections are called **PORTS** that may be inputs, outputs, or bidirects, as shown in Figure 3-1. An entity is given a name, in this case CPU, and as far as synthesis goes, is simply a way to identify a functional piece of a system and its I/O connections.

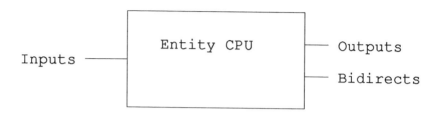

Figure 3-1 : A VHDL Entity

The syntax for the Entity simply defines its name, the names of its ports, the direction of the ports (eg, input, output,etc), and a VHDL "logic type" designator (called Std_Ulogic in this example) that will be explained below :

```
Entity CPU is

  Port ( CLOCK : In    Std_Ulogic;
         ENAB  : In    Std_Ulogic;
         RESET : In    Std_Ulogic;
         Gate  : Out   Std_Ulogic);

  End CPU;
```

Figure 3-2: Entity Syntax

Notice the following general VHDL syntax features:

- VHDL words and names can be upper case or lower case or both; for example, the name RESET, reset, or Reset are all accepted as being the same name. It is recommended however that a consistent style be used, both for neatness and for compatibility with the requirements of other CAE tools.

An important section on allowable names and other "identifiers" will be found at the end of this chapter.

- The use of semicolons, parentheses, and colons is important to the syntax and should appear as shown. In particular, semicolons are used to terminate lines, as shown in Figure 3-2. Like most other compilers, if the punctuation is incorrect you can get some confusing error messages, so it pays to be aware of their usage.

   - The PORT statements define the inputs and outputs of the Entity and follow the above syntax of

   Port (**Signal name : Direction   Logic_type** ;
            Signal name :   Direction   Logic_type ;
               etc ...
            Signal name :   Direction   Logic_type);

where **Direction** is usually In, Out, or Inout for bidirectional ports, and where **Logic_type** is explained below under Signals. An additional output type called **Buffer** is also useful and will be covered below.

Note that each line of signals in the PORT list ends in a semicolon except the last line which ends with a closed parenthesis and semicolon.

You could also group several signals on one line such as

   Clock, reset, enab : In Std_Ulogic;

but while this is perfectly legal VHDL, the previously shown Figure 3-2 with one signal per line is much easier to read. The extra lines of code can be easily handled by the most text editors or by one of the commercial VHDL "shell" generators (see Chapter 9).

- The order of the signals in the PORT list is your choice but should be selected to make subsequent testing and documentation easier. A good choice might be to group all inputs together in alphabetical order, then all outputs, in order to simplify making later changes.

Comments can be added to the code for documentation and ease of review and should be used liberally. VHDL comments are always preceded by two dashes (--) and are in effect until the end of that line. This means that they can be added on the same line as the actual code for convenience, such as

RESET : In   Std_Ulogic;   **-- Master reset, 2 usec wide**

**--   This is a comment on its own line. If you need more than one
-- line, each must be preceded by the two dashes.**

If you need more than one line for a VHDL statement, you simply continue onto as many lines as you need. The termination character, usually a semicolon, tells the VHDL compiler where the end of the statement is regardless of how many lines of text it requires. It is good practice to enter any continued lines neatly as shown with the signals in the entity above, since it is easier to read and understand. This practice will be illustrated throughout the book.

Signals have been mentioned here without having been formally defined because basically the idea is already familiar to designers. Signals are used in VHDL the same way they are used in an ASIC or on a board or system. In a typical design, you will normally have multiple entities, each of which defines some portion of the overall system. The signals form the connections between entities by means of the ports, analogous to the connector pins on a board or other module. Additional signals are also used inside of entities to perform the function of that entity, just like the internal signals on a PC board that do not go to the board I/O connector.

Signals

In a physical hardware system, signals can be single bits, such as a clock or reset, or they can be busses of some specified width. Both of these signal types can be easily described in VHDL once you understand the underlying principles. The VHDL language by itself does not predefine the characteristics of signals such as their logic states or driving strengths, in order to provide flexibility for many different types of applications. Instead it makes provision for doing that by means of VHDL files grouped into "packages" and "libraries", which are normally included as part of a VHDL simulator or synthesis tool, or can be developed by the user for special applications.

Originally, the simplest signal types were named **BIT** and were defined to have only the logic states 0 and 1, while a later logic type named **MVL4** (Multi-Valued Logic 4) allowed for three-state logic and unknowns during simulations by using four states (0,1,Z,X). The MVL4 signal type was sufficient for describing basic logic operations but had shortcomings in its ability to model more complex cases, such as open-collector logic and "don't care" states.

In order to provide even greater flexibility, the IEEE fostered the industry development of a logic state system that included driving strength and other capabilities for synthesis and simulation. This definition has become the accepted standard logic library as of the early 1990s. Known as the **IEEE Standard_Logic_1164** library, it defines a 9 state logic value system given the nomenclature **Std_Ulogic** and a similar value system called **Std_Logic**. This logic system is also referred to as MVL9, for Multi-Valued Logic, 9 states. Using this library, signals can have the normal logic states 0 and 1, as well as

U   for Uninitialized
X   for Unknown
Z   for Tri-state
W   for Weak strength
H   for High (resistive)   - used for open collector outputs
L   for Low (resistive)   - used for open emitter outputs
-    for Don't Care.

In terms of general simulations, these nine states are sufficient to describe most common logic conditions. They also provide other useful capabilities, such as the U for a signal that has never attained a legal value during the simulation run, and the "-" to indicate synthesis "don't care" conditions.

Libraries are specified by an overall library statement and a "use" statement that provides the name of the specific library data. The following example is typical of the library specification used with most VHDL simulators and synthesis tools :

Library IEEE;
  Use ieee.std_logic_1164.all;

A library is a convenient mechanism for storing commonly used VHDL functions and for defining data types. The synthesis support library commonly consists of two main groups of VHDL code. The first, called the **Package**, generally defines the names and the inputs and outputs of the functions in the library. The second part is called the **Package Body** and contains the VHDL code that makes up the functions in the library. In many cases, the only libraries you will need will be included with the simulation or synthesis tools and merely need to be invoked as shown above. Ways to use your own packages will be covered in later chapters, but for now it is sufficient to know that they exist.

In your VHDL file, the library statements precede the Entity section. For synthesis, most vendors also provide a second library with special functions for arithmetic and logical operations, plus additional logic types. The format of the second library is similar to the example, but with a different name, and also precedes the Entity statement.

Signals are declared with both a Name and a Type as in

SIGNAL CLOCK : Std_Ulogic;

This particular signal has the name CLOCK and is of type Std_Ulogic. As we shall see below, the signal name and its type are defined once and thereafter can be simply referred to in the rest of the VHDL code as CLOCK.

Using the IEEE MVL9 system, busses or registers or any multibit signals are referred to as "**vectors**", of type **Std_Ulogic_Vector,** whose bits can be defined two different ways.

a. If you normally think of binary data values where the most significant binary value is at the leftmost end of the word, then the "downto" convention should be used :

Signal Name : STD_ULOGIC_VECTOR (M downto N);   for busses where the most significant bit is on the left. For example

> **Databus : STD_ULOGIC_VECTOR (7 downto 0);** -- an 8 bit bus whose MSB (Bit 7) is at the left end of the bus, and has the binary weight 2^7 or 128 :

Databus

7	6	5	4	3	2	1	0

b. If you prefer the reverse bit ordering, then the "to" convention should be used, where n < m :

Signal Name : STD_ULOGIC_VECTOR (n to m);   for busses where the bits are numbered from left to right such as

> **Instr_Reg : Std_Ulogic_Vector (0 to 9);** -- for a 10 bit buss.

Instr_Reg

0	1	2	3	4	5	6	7	8	9

The **individual bits** in a vector are simply identified by their bit number, as for example :

> **Databus (4); -- for bit 4**

The implication of defining an 8 bit bus or register as either REG_A (7 downto 0) or REG_B (0 to 7) is that, if the contents of the register is the binary number one (00000001), then for the individual bits

REG_A (0) is logic 1 and all other bits are 0, whereas
REG_B (7) is a 1 and all other bits are zero.

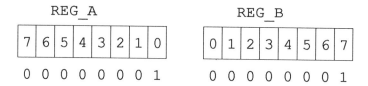

The convention used affects how you view the value of the individual bits in the word. In REG_A, bit 5 is the value of 2^5 or 32, whereas in REG_B bit 5 is the binary value 4. There is no problem with signal assignments using either format, such as REG_A <= REG_B, and the result would be exactly as shown above.

Although most applications and binary arithmetic texts use the format where the MSB (largest positive binary value) corresponds to the bit number, VHDL provides the flexibility to handle either system. Most of the examples in the book however will keep the MSB on the left, as in REG_A, and will use the "downto" format.

You will probably need to use part of a bus in your designs, especially in data processing applications. A portion of a bus or register is referred to as a **SLICE**, and is written as

Databus **(5 downto 2)**; - A 4 bit "Slice" of Databus

Databus : Std_Ulogic_Vector (7 downto 0) ;

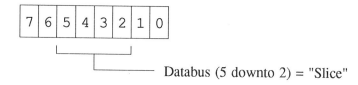

Databus (5 downto 2) = "Slice"

Note that the "to" or "downto" of the slice must agree with the "to" or "downto" of the vector declaration.

This has been a brief overview of some of the types of signals and how they are defined. Signals can also be of other types than Std_Ulogic which, along with additional ways to use signals, will be shown below and in the design examples in later chapters. At this point, a general understanding of the basics of defining entities and signals for synthesis is sufficient to continue with a discussion of VHDL ideas. Before we discuss how to model the behavior of the entity however, let's look at a typical system or ASIC situation and see how to plan the effort.

Estimating and Sizing Entities

As covered in Chapter 2, early in the design cycle you will have generated a block diagram and will have defined functional blocks of logic that belong together. Now there is no fundamental reason why all of the logic blocks could not be included in a single large Entity, but that would be just like drawing all of the logic design on a single schematic page. It is more logical to keep the same sense of organization shown in the block diagram by using multiple Entities to represent the VHDL ASIC. Therefore, each of the major blocks, as shown in Figure 3-3, should become one of the entities for the design, with one additional practical constraint for synthesis purposes. The relative size of the logic within an entity should be limited to a "reasonable" value, dependent on your workstation's performance and the speed of your synthesis tool.

Estimating the size or rough gate count of any entity is typically something that you have probably done during the block diagram phase so that the overall ASIC type and size can be selected. This is also one of the reasons why the block should show sufficient detail to make estimating possible. This estimate is fairly rough and is based on your previous design experience that, for example, a four bit binary counter is around 50 gates, or a 16 bit register is 16 flip flops at 8 gates per flop or 128 gates, etc. The point is that it is a good idea to be aware of how big a design "should" be, for initial estimating, for identifying entities, and

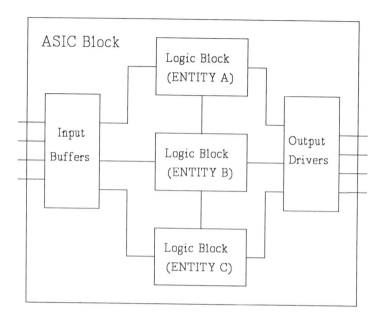

Figure 3-3 : A Typical Partitioned ASIC Block Diagram

as a general "sanity check" on the size of synthesized designs.

A "reasonable" size for a synthesized entity in the early 1990s is in the thousands of gates range, rather than tens of thousands, based on both workstation performance and time of synthesis. Most synthesis tools can actually handle tens of thousands of gates, but at the expense of long synthesis runs. In general, in day to day design activity involving making changes to a design or optimizing the logic for desired performance, you should prefer to keep synthesis runs to under an hour, hence the limit to several thousand gates. The reasonableness factor is simply a practical guideline that will surely improve in the future as workstations get faster, but the rule that your computer can never be too fast will probably continue to be true with this type of technology.

Architectures - Defining the Function

The **Architecture** is the VHDL name used for the section of code that defines how the entity behaves or what it is composed of. The VHDL description of the Architecture can be abstract, RTL, or structural as covered earlier, although for synthesizable functions the RTL behavioral style will be used most often. In any particular simulation or synthesis run, there will be one Architecture associated with each entity. This is mentioned here because in a top down design methodology you may actually have several alternative Architectures for an Entity within your file directory, and VHDL provides a mechanism for selecting one of them by means of an optional CONFIGURATION statement. In any event, an Architecture has a "type" and an association with the Entity such as :

Architecture RTL of CPU is

    ... many other lines ...

End RTL;

The type, in this case called RTL, can be anything you choose and has no significance to the VHDL compiler, but should be given a name meaningful to the activity. Other useful names could be BEHAVIORAL, STRUCTURAL, SCHEMATIC, TEST_BENCH, etc, to indicate the type of VHDL design being modelled. Capital letters are used here simply for clarity and emphasis.

Recall that an ASIC will usually consist of several Entities, each representing a functional block of logic, and signals that connect the Entities via their Ports. Each Architecture has access to all of its Port signals by definition. In addition you will normally define a number of other signals that are used only within the Architecture and which will be synthesized into the registers, counters, and gates of the Architecture.

Examples of the syntax of the signal specification for a clock, a signal named Load, a 16 bit instruction register, and an 8 bit counter are :

Architecture RTL of CPU is

Signal CLOCK	: Std_Ulogic;
Signal Load	: Std_Ulogic;
Signal Instr_Reg	: Std_Ulogic_Vector (15 downto 0);
Signal Counter	: Std_Ulogic_Vector (7 downto 0);

These statements are called **Signal Declarations**. The signals are always defined at the beginning of the architecture, or in the "declaration section", as shown here and throughout the rest of the book.

It is also good practice to add comments for documentation and for review purposes, such as :

Signal CLOCK : Std_Ulogic; **-- 10 MHz system clock**

The general rule for synthesis is that every clocked storage device on your detailed block diagram must be given a name and assigned to a logic type. Good practice is to use the same names for the VHDL signals as were used on the block diagram, since this simplifies reviewing and understanding the design. We will see later on that for combinational logic it is not always necessary to explicitly name signals if they are only used as inputs to clocked devices.

Similar to signals are definitions of data **Constants** that can be given names for convenience, such as :

Constant ALL_ONES: Std_Ulogic_Vector := "11111";

The **Constant Declarations** are also placed in the declaration section at the beginning of the architecture, and a good location would be just following the signal declaration statements. Note that a Constant Declaration uses a similar syntax to the Signal Declaration, with the addition of the ":=" and the value (in this case five binary ones). Binary values of vectors are always enclosed in double quotes as shown, and are referred to as bit string literals. Notice that unlike signals, you may but you need not specify how many bits are in the constant, such as with the "(4 downto 0)" syntax.

Following the Signal and Constant Declarations, there may be a few other lines that will be discussed later. Basically however the **BEGIN** statement comes next and shows where the function of the Architecture starts. A simple Entity and its Architecture might now look like this :

-- The Libraries

Library IEEE;
　Use ieee.std_logic_1164.all;

Library Synthesis;   -- This name depends on synthesis tool used
　Use synthesis.synth.all;

-- The Entity

Entity CPU is

```
Port  ( CLK33M    : In    Std_Ulogic;
        ENAB      : In    Std_Ulogic;
        RESET     : In    Std_Ulogic;
        Datain    : In    Std_Ulogic_Vector (15 downto 0);
        Dataout   : Out   Std_Ulogic_Vector (15 downto 0);
        GATE      : Out   Std_Ulogic);
End CPU;
```

-- The Architecture

Architecture RTL of CPU is

```
Signal CLOCK       : Std_Ulogic;
Signal Load        : Std_Ulogic;
Signal Instr_Reg   : Std_Ulogic_Vector (15 downto 0);
Signal Counter     : Std_Ulogic_Vector (5 downto 0);
```

Constant All_Ones : Std_Ulogic_Vector := "111111";

Begin

-- The rest of the Architecture goes here

End RTL;

Figure 3-4 : General Form of an Entity and Architecture

The "rest of the Architecture" is the part of the VHDL code that defines how this block of logic performs in response to its inputs. Before we continue however, there are two concepts that must be understood. In general, in any digital system you can describe the behavior of the logic by the states of its signals, and by the times that it reaches those states. It is important therefore to see how VHDL is used to set the state of a signal, and how the language handles the idea of time.

Setting the State of a Signal

Having declared the I/O signal ports of the Entity and the internal signals within the Architecture, the next step is to see how to set signals to their desired values. The VHDL term for this is **Signal Assignment** and the syntax is

SIGNAL <= VALUE;

where VALUE can essentially be either a constant, a signal, or the result of a logical or arithmetic operation performed on signals or constants. Expressed in words, the "<=" symbol indicates that the state of SIGNAL becomes the state of VALUE.

As a simple example, consider

    Counter <= ALL_ONES;

which means "Set the value of the signal Counter to ALL_ONES", eg, the binary value 111111 as defined in the Constant Declaration above. You could also have gotten the same result without using a declared "constant" by

    Counter <= "111111";

The " <= " **symbol is used with signals** to indicate that the value on the right is transferred to the named signal on the left. In this case, the signal **Counter** has been defined earlier as a 6 bit signal (Std_Ulogic_Vector (5 downto 0)), and the righthand value must also be the same type and length, otherwise the VHDL compiler will produce an error message. As you can see, the Constant All_Ones agrees, and the "111111" syntax shows 6 bits, with the double quotes used to denote multi-bit values, or "string literals". The topic of type agreement and "overloading" will be covered in a later section of this chapter.

Signal Assignments can also be done with individual signals such as

    Clock <= CLK33M;
or
    Load <= '1';
or
    Load <= Enab AND Instr_Reg (3);

Note that for individual bits the binary value is enclosed in single quotes (a "character literal"), whereas bus or vector values are enclosed in double quotes. As we shall see below, the righthand side of the expression can be quite complex compared with these simple examples, and can include arithmetic and logical operations using numerous signals. But for now, understanding the basic signal assignment idea is sufficient to continue.

The second part of signal behavior is determining **when** it reaches its logic state. The features of VHDL that allow it to accurately describe general logic operations are its built-in concepts of time and time delay, and its ability to handle simultaneous, or concurrent, operations.

Time and Concurrent Operations

Consider a typical digital design consisting of various registers, counters, multiplexers, and control logic, all synchronously clocked by a system clock signal. In a physical system the logic may consist of separate independent functions or may have functions that interact, but all of the logic essentially operates simultaneously, or in other words concurrently, in each section of the design. In addition, the output of every gate or flip flop solely depends on its response to its own inputs, in terms of both the logic state and response time. A hardware description/simulation system must therefore be able to account for both concurrent operation of the logic and the response time or time delay of its outputs. The built-in **concepts of time and concurrency** are two of the major differences between VHDL and conventional software programming languages and make it possible to accurately describe complex logic as well as more abstract systems.

In a standard programming language such as PASCAL, C, or Fortran, all operations happen sequentially as the code is executed step by step. Since these languages were not developed to model real time processes, there is no concept of time at all, and handling of concurrent operations requires application specific programming to "calculate" time and to compute the next state of all variables from the previous state. These factors tend to make it cumbersome to do simulation with a general purpose language.

With VHDL however, time and concurrent operations are part of the language and therefore allow modelling of actual logic behavior in a system. In this regard, the concept of time is key to VHDL's capabilities and is analogous to the way that real hardware works. Previously we showed a Signal Assignment of

    Load <= Enab AND Instr_Reg (3);

where the Load signal is the output of a logical AND of the signals Enab and Instr_Reg (3). The VHDL syntax to add a 5 nsec propagation delay to the gate is

Load <= Enab AND Instr_Reg (3) **after 5 nsec**;

This is fairly intuitive syntax that will be expanded to show further possibilities later on. But if this code results in a 5 nsec delay in the output state, then what happens when no delay is specified? Well, you can rest assured that there is no such thing as a true zero delay, but there **is** a concept of an infinitely small "delta" delay until the output changes. This technique allows the VHDL simulator to **evaluate** all of the logic states in the system at one point in time T, compute any changes, and then let those changes take effect an instant later at (T + delta T). The actual evaluation of the logic states is only done when there is a change on one or more of the inputs to the system, or when a time delay process ends, such as the 5 nsec delay of the Load signal above.

With the system in a quiescent or steady state condition, at any instant of time all of the inputs and outputs of the gates and storage elements are at a particular logic state, which we could call the Present State of the system. The VHDL simulator essentially waits at the beginning of the Architecture for some change to occur. If a change occurs on one or more of the system input signals at some time T0, then the simulator must compute whether the outputs of the gates or storage devices will change and what their output time delays will be. It does this by proceeding through the code of the Architecture from the Begin statement to the End, examining all signals at their present states as of time T0. As a result of the evaluation there will be a new state of the system, which may or may not change from the present state, and the new state will be "scheduled" to change at T0 + delta T. The delta time, which is also referred to as a "time tick", is the small increment of time at which signals are updated if they have no delay, and at which point the delays are scheduled for those signals that do have delays specified. After the time tick, the "present state" of the system is updated and the process waits again for the next change to occur.

Note that this mechanism is important, since without it signals would be updated in the order in which they appear in the code. That would not be the way actual hardware works and in addition would make it extremely difficult to code a design.

The way VHDL signals are evaluated should be familiar to most logic designers, since it is analogous to how storage devices are updated in a synchronous logic system. At the instant of the clock edge, all of the inputs to the flip flops, counters, registers, and latches are "evaluated" and the results transferred to the outputs just after the clock edge. Once the clock edge has occurred, and after their propagation delays, the changed outputs become the new state of the system. Some additional examples of how signals are handled will be given below in the section on VHDL "processes".

There are other important time related operations that are needed in addition to concurrent evaluations and gate delay specifications. The **WAIT** statement is key to many synthesis and simulation issues, and has several variations. The simplest form is

Wait for {Time value}; such as

Wait for 12.3 usec;

This type of operation is unique to languages such as VHDL and can be used to model many types of time delays. Other variations of wait statements are

WAIT UNTIL condition;
and
WAIT ON signals;

In the IEEE 1164 standard library, functions are defined to detect the rising edge or falling edge of a signal. Using these functions along with the WAIT UNTIL statements will allow us to describe and synthesize clocked logic. With clocked hardware devices, the concept that nothing changes until the clock edge is familiar. The equivalent VHDL operation is the statement

WAIT UNTIL Rising_Edge (CLOCK);

followed by Signal Assignments to the flip flops or storage registers. The Rising_Edge function from the 1164 library operates on the signal CLOCK and causes the simulator to wait until the clock edge occurs. The Rising_Edge function is equivalent to the VHDL expression

(CLOCK'event and CLOCK = '1')

which is used by some synthesis tools instead of Rising_Edge. The Rising_Edge will be used throughout the book because it is simpler, but depending on which synthesis system you use will determine the appropriate syntax.

The WAIT statement above does just what it says, and after the clock edge statement you will add the code to set the clocked signals to their desired states. The result is that in the logic synthesizer, all of these clocked signals will eventually become the flip flops and registers of the logic circuit. This subject will be expanded below.

But now we have a dilemma. If the simulator is waiting for a clock edge or some other WAIT condition, how do all of the other conditions that have nothing to do with clock edges get evaluated? The answer is that operations that depend on WAIT conditions get placed in VHDL "**processes**", while those that have no WAIT conditions are evaluated concurrently outside of processes. The simulator then keeps track of the conditions that each process is waiting for, and until those conditions occur it continues to update the other signals as required.

Just as an ASIC consists of a number of Entities, a VHDL Architecture typically consists of one or more processes that implement the logic of your block diagram. We will see below that **all** of the signals that are **assigned to values** in a "**Clocked Process**" will become flip flops or other storage devices. Any other signals that you want to be purely combinational can be located either between processes or in combinational processes. The question now is what is a process, what makes it a clocked process, and how do processes relate to the digital logic on the block diagram.

Processes

From the discussion above, processes wait for some condition to occur and, in response, cause some other action. For example, in the case of sequential, or clocked, logic, the process waits for a clock edge or a master reset signal and then sets the signals representing flip flops to their desired states. The signals that the process waits for are included in that process's "**Sensitivity List**", which is equivalent to the "WAIT ON signal list" syntax used above.

Processes are given names, or **Labels**, and have the general format :

LABEL : Process (Sensitivity List)

Begin

... Statements ...

End Process LABEL;

As an example, the process Sig_Gen has the signals Clock and Reset in its sensitivity list :

Sig_Gen : Process (Clock, Reset)
Begin

... Statements ...

End Process Sig_Gen;

During the normal flow of a VHDL simulation, the process waits for a change to occur on one of the signals in the sensitivity list, at which point it executes the statements within the process once from the Begin to the End. In the case of Sig_Gen, nothing happens in the process unless there is a change on the Clock or the Reset signal. The statements within the process must determine which signal has changed and what to do in response to that change.

In addition to the signals included in the sensitivity list, processes are implicitly "connected" to all the other signals in the Architecture and to the Ports in the Entity. This means that without needing to specifically list those signals, the process can respond to their states and can control signals as outputs. The only restrictions on signals in processes is that processes cannot use the Entity's output ports as inputs, or, expressed in VHDL terms, the process cannot read output ports. In practical terms this is not really a problem however, and suitable techniques will be shown below.

At this point the basic structure for a process has been described but without any actions following the Begin statement. The actions can be of two forms, a "clocked" process that will be synthesized into clocked or sequential logic, and an unclocked process that will produce combinational logic.

Clocked Processes

Clocked processes are used to synthesize circuits with flip flops, registers, latches, or any other type of clocked logic. There will be combinational logic generated by clocked processes, but the logic functions will be internal to that circuit's process, as inputs to the flip flops. The point is that clocked processes will be used to synthesize both combinational and sequential logic, with the distinction as stated above that any signal that is explicitly assigned to a value will become a clocked device. This point will be illustrated in the examples in later chapters.

In a clocked process, the sensitivity list always includes the clock signal and usually includes an asynchronous reset signal (as an ASIC design rule, not a VHDL rule). There can only be one clock and one reset in a process. If you have more than one reset signal, they should be OR'ed together outside of the process.

First let's consider the simplest case where no reset is used and assume our ASIC library uses positive edge triggered flip flops. The process will test for the rising edge of the clock and then load the contents of a databus into a register, both of which we'll assume have been defined earlier as 8 bit vectors. The Process will look like :

```
Loadreg : Process ( Clock )

Begin

   If Rising_Edge ( CLOCK ) then
           REG <= Databus;
   End if;

   End Process Loadreg;
```

The use of the simple If-then statement shown here is fairly intuitive and says "If the rising edge of the Clock signal occurs then transfer the value of the databus into the register REG, otherwise don't do anything". This code will synthesize into a simple register composed of D flip flops, without a direct clear.

Recall that a Process is activated whenever there is a change on a signal in the sensitivity list. In the case of our clock, the only changes are on the rising, or leading, edge and the falling, or trailing, edge. In the code above you can see that only the Rising_Edge is tested and results in action. On the falling edge, the IF condition would not be true and the Process would then be completed or "suspended", until the next change on its sensitive signals.

In this clocked process, note the simple expression

            REG <= Databus.

The synthesizer will automatically produce a register from this statement, where the number of bits had been previously specified in the signal definition, eg REG : Std_Ulogic_Vector (7 downto 0). If you change REG to be 23 bits, you will then get a 23 bit register ! **This single line of code can therefore be used to generate a register, whether it is one** bit or N bits wide, without requiring any more work on your part! This fact is one of the contributors to the greatly increased productivity gained from the use of logic synthesis, and applies to both sequential and combinational logic.

Now let's see how to add an asynchronous, active high clear to the register. The process will need to have two signals in its sensitivity list, the clock and the reset. Since there are two signals, we must also decide which one should have priority if they should both happen simultaneously. In normal design practice, the reset would override any other action by the circuit, resulting in the following Process :

Loadreg : Process (**Clock, Reset**)
Begin

If Reset = '1' then

REG <= "00000000"; -- Comment - Clear REG

Elsif Rising_Edge (CLOCK) then

REG <= Databus;

End if;
End Process Loadreg;

The ELSIF statement, short for ELSE IF, is active only if the first IF condition is false, ie, if RESET is low, or 0. Notice that **priority** is automatically included in the IF-ELSIF statements. The Elsif part really means "Only if there is no reset, then check for the clock edge actions". In other words, if both conditions occur simultaneously, the reset has priority over the clock. You can also see that, except for the direct reset, action takes place only at the clock edge, which is why all signals that are set within a clocked process become flip flops or registers in the synthesized circuit. If you need to combinationally decode some state of a register or counter, you will add those statements outside of the process, as will be shown later.

This VHDL syntax is the standard format for clocked logic with a direct clear capability :

a. Both clock and reset signals are included in the sensitivity list, with only be one clock and reset in a process.

b. The "If Reset ... " statement is followed by the action taken when the reset occurs, and

c. The "Elsif Rising_Edge ..." statement is followed by the clocked action.

Logic design examples given later in the book will show that the clocked actions can be much more powerful than simply loading a register. The basic structure of the Clocked Process will remain the same however. Deciding how much logic to include in a process and how many processes to create will be discussed following the section on unclocked processes below.

A word of caution. You may have some sections in your design that have an asynchronous reset and others that do not. For instance, although most ASIC suppliers prefer a reset on every flip flop for improved testability and initialization of both the simulation and test, you may have pipelined data registers or other logic that will be clocked to a known state early in the simulation or test. If you make the choice not to provide a reset for this logic, be careful to put those registers into a separate process from those with resets. The reason is that in the simulation, if the reset is on, the clocked (elsif Rising_edge ...) logic will not be processed and none of the flip flops or registers in that process will change state. In the actual ASIC hardware however, if the clock is running and the reset occurs, those flip flops that do not have a reset will be clocked and could change state. **This fact would result in the physical ASIC working differently from the simulation, and could be a serious problem in some cases.** This situation can and should be avoided by being careful to separate the logic without resets into a separate process, or by using an asynchronous reset on all clocked logic.

Unclocked, or Combinational, Processes

Combinational logic can be described either outside of a process or by means of an unclocked process. The primary difference is in the types of syntax available in each situation, which will be covered below. An unclocked process otherwise looks very similar to a clocked process in terms of its format.

To illustrate the point, consider a simple multiplexer with inputs A, B, and SEL and output Y. The Y output will be equal to A when SEL is low and equal to B when SEL is high. As a process, we'll want to have A, B, and SEL in the sensitivity list so that a change on any of those signals will activate the process and evaluate the state of the output Y.

```
MUX2_1 : Process ( A, B, SEL )  -- a 2 to 1 Multiplexer

Begin
    If SEL = '0' then  Y <= A ;

    Else  Y <= B ;

    End If;
End Process MUX2_1;
```

One observation is that there is no Rising_Edge (or Falling_Edge) operation used as in the clocked process. The result is that this process will produce a combinational logic circuit rather than a clocked circuit.

Also note that the signals A, B, and Y could be single bits or they could be 32 bit words, depending on how they were defined in the Architecture. The code for the process would remain the same, but the circuit would of course be much larger in the 32 bit case. This is another good example that shows that the number of VHDL lines of code does not necessarily relate to the number of gates in the synthesized circuit. The 32 bit circuit would obviously be 32 times as large as the single bit case without requiring any additional changes to the VHDL code other than the signal definitions.

This process example used another variation of the IF syntax, the IF-THEN-ELSE, which can be generalized to :

IF Condition is true **THEN**
Some action ...

ELSE Some action ...

End IF;

The ELSE implies that the first condition is not true. The IF-THEN-ELSE syntax avoids having to use separate IFs to test for the condition being true and false, such as

IF Condition is true THEN
    some action;
End IF;

IF Condition is false THEN
    some action;
End IF;

These IF type statements are used extensively when modelling logic systems and will be illustrated throughout the text examples. **Each IF "clause" needs an END IF statement, and the "IF" syntax can only be used within processes.** Other syntax used to do the same function outside of a process will be given later.

Latched Processes

A third variation on processes is used to create latches, ie, level sensitive storage devices. Also referred to as transparent or flow through latches, they are similar to flip flops except that their output follows the input while the "clock" or latch enable is in the active state, then retains the last state of the input when the enable becomes inactive. The form of a latched process is similar to the clocked process except for the handling of the enable signal and the inclusion of the input data in the sensitivity list. Assume the latch uses an active high enable and reset:

```
Load_latch : Process ( Enable, Reset, Databus)
Begin

    If Reset = '1' then

        LATCH <= "00000000"; -- Comment : Clear LATCH

    Elsif Enable = '1' then     -- Comment : Latch enable

        LATCH <= Databus;

    End if;
End Process Load_latch;
```

Aside from the reset condition, this process is activated on any change to the enable or databus inputs. Adding the databus to the sensitivity list ensures that even if the enable stays high, a change on the databus input will cause the latch to be updated.

At this point, you may be wondering how the synthesizer knows whether to create a latch or simply a gate for the above process. After all, the signal names Reset and Enable are not specialized or "reserved" words that carry any special significance to VHDL or to most synthesis tools. In light of that fact, why doesn't the above process behave as an and-or gate, where the output is "00000000" when reset is '1' and is equal to databus when Enable is '1'? The answer is that some synthesis tools use an additional mechanism, **such as an "attribute" or a synthesis control command,** that identifies certain signal names as being clocks, resets, or latch enables. This mechanism then tells the synthesizer that these signal names are special and are to be made into the proper type of storage device. Other synthesizers analyze the reset and clock from the syntax, which must follow the strict format given above.

Signals Revisited

Now that we have covered processes and have learned how to create a flip flop in a clocked process, let's look at an example of the signal updating mechanism. Consider 3 flip flops connected as a simple shift register in Figure 3-5.

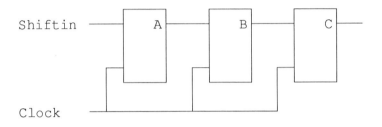

Figure 3-5 : A Simple 3 Stage Shift Register

This could be coded within a clocked process as

Shiftreg : Process (**Clock, Reset**)

Begin
  If Reset = '1' then

            A <= '0';
            B <= '0';
            C <= '0';

   Elsif Rising_Edge (CLOCK) then

            A <= Shiftin;
            B <= A;
            C <= B;
   End if;
End Process Shiftreg;

At the time of the clock edge, the state of this circuit is determined by the flip flops A, B, and C and the input Shiftin. Assume the system has been stable with all four signals at logic 0. If a change now occurs to make Shiftin a logic 1, then at the next clock edge at time T the VHDL simulator will compute the new states of A, B, and C based on the conditions at time T, namely Shiftin = 1 and A, B, and C = 0. The new state of the system will thus be A = 1, B and C = 0, at time T + delta T. Having reached the end of the process, the simulator advances time to (T + delta T) and waits for the next clock or reset, and the state of the system is now A = 1, B = 0 and C = 0. It should be noted that the **order** of the signal assignments does not affect the results - the following code would work exactly the same :

```
B  <=  A ;
C  <=  B ;
A  <=  Shiftin ;
```

since only the conditions at time T (Shiftin = 1 and A, B, and C = 0) are used to compute the new state. Figure 3-6 shows the states of the flip flops after several clocks.

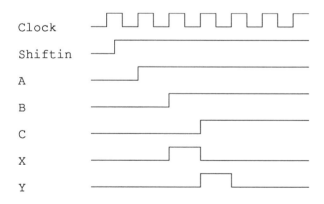

Figure 3-6 : Waveforms of Shift Register

Compare that operation with a conventional programming language which would not only be order dependent but would compute the new values as they occurred. Thus in a language such as PASCAL the code

A <= Shiftin;
B <= A;
C <= B;

would set A to 1, then B = A (ie, 1), then C = B (ie, 1). It is clear that
the ideas of time and concurrent signal assignments are vital to describing
the operation of logic functions.

Clocked vs. Unclocked Action

To illustrate the difference between setting a signal in a clocked
process and setting it outside of a process, let's generate the signal

B and (not C);

which should be a simple AND gate, assuming that both polarities of flip
flop C are available. Let signal **Y** be this function inside of the clocked
process and signal **X** be assigned to the same function outside of the
process :

Shiftreg : Process (**Clock, Reset**)

Begin
 If Reset = '1' then

        A <= '0';
        B <= '0';
        C <= '0';
        Y <= '0';

 Elsif Rising_Edge (CLOCK) then

        A <= Shiftin;
        B <= A;
        C <= B;

Y <= B and not(C); -- Within the process

End if;
End Process Shiftreg;

X <= B and Not(C); -- Outside of the process

As can be seen in the signal waveforms in Figure 3-6, the signal X becomes its value as soon as B or C changes, whereas Y must wait for the next clock edge to assume its new value. As you can see, X is directly produced as B and not(C) whereas Y is a clock period delayed. The Y output will be a flip flop preceded by the gate B and not(C), since it is part of a clocked process. The rule is again :

All signals that are assigned new values within a clocked process will be synthesized as flip flops or other storage devices.

Signals that are assigned new values outside of processes or in unclocked processes will be synthesized as combinational logic.

Processes within an Architecture

We have now seen that a typical Architecture in an ASIC will consist of some clocked logic and some combinational logic. Figure 3-7 graphically shows the general VHDL organization of such an Architecture with several different clocked and combinational processes, and with other combinational logic shown between processes.

Architecture RTL of Entity_Top is

Signals and Constants ...

Begin

Combinational logic Signal Assignments...

```
Clocked Process #1
  Begin
   .. Statements ...
  End;
```
(A Sequential
logic Function)

Combinational logic Signal Assignments ...

```
Clocked Process #2
  Begin
   .. Statements ...
  End;
```
(Another Sequential
logic function)

Combinational logic Signal Assignments ...

```
UnClocked Process #1
  Begin
   .. Statements ...
  End;
```
(A Combinational
logic function)

Combinational logic Signal Assignments ...

End RTL; -- End of the Architecture

Figure 3-7 : A VHDL Architecture with Multiple Processes

The **order** that the processes appear does not affect the operation or the resulting synthesized logic, nor does it matter where the combinational signal assignments are placed. It does make sense, however, to locate the combinational logic expressions near other functions to which they are related. For example, the signal X of the previous code should be located near the process that generated B and C, or else near a process where it will be used.

How Many Processes?

There are several things to consider when assigning blocks of logic from your block diagram to VHDL processes. Since a particular clocked process can only have a single clock and reset input, if you need to have separate clocks in the system, then each must be in a separate process. In general, most ASIC vendors recommend the use of synchronously clocked design practices, with preferably a single clock, which not only makes the logic easier to control but also makes testing and testability design simpler. Nevertheless, there are usually applications where multiple clocks are required and, as a result, multiple processes will be needed.

Even in the case where you have a single clock, you will probably want to break the logic into several processes from an organizational viewpoint, as was done earlier with Entities. Again there is no VHDL reason why all of the logic could not be in one large process, but depending on the application you should try to keep independent logic circuits together in their own process. On the other hand, don't make the processes so small that you lose the relationship to the block diagram and to the overall system functions. For instance, making a process for just a 4 bit counter, separate from its overall function, is not a productive means of designing with synthesis. Rather the emphasis should be on the overall circuit function and not the individual components within that circuit. In other words, **THINK FUNCTION, NOT GATES!** The design examples shown in later chapters will emphasize this point in more detail.

Defining the Circuit Behavior

Now that you have an Entity and an Architecture defined with its signals and I/O ports, and various clocked processes blocked out, the next task is to transfer your block diagram approach into synthesizable VHDL code. In preparation for the design examples beginning in Chapter 4, we need to know the types of VHDL operations that are available for defining circuit behavior. In this chapter the different features will be simply shown for reference, since details of their usage will be covered in the examples.

Signal Assignment Operations : The standard logical and arithmetic operations that can be synthesized directly are shown in Table 3-1. The logical and relational operators are straightforward, but you should refer to Chapter 7 for a discussion of additional data format considerations and restrictions that apply to the arithmetic functions. Note that some of the standard VHDL operations, such as Multiply and Divide, are not included because they cannot be **directly** synthesized as simply as Y = A * B, or Y = A/B. They can be synthesized, but the synthesis tools need more information, such as which of the many algorithms to use and whether the function should be serial or parallel. This is a case where your detailed block diagram should show the desired hardware architecture. These issues will be discussed further in Chapter 7.

Although the later chapters will fully illustrate operator usage, the following brief examples show some of the possibilities :

a.   Y <= (A and B) or Not (C);   -- Logical operations.

Parentheses must be used where the intended function may be ambiguous. The expression "A and B or Not(C)" could be interpreted as "A and (B or Not(C))" or "(A and B) or Not(C)", which are not equivalent.

b.   If D_BUS >= "101101" then ...
     (Note that >= is "greater than or equal", not a backwards arrow!)

Type	Symbol	Operation
Arithmetic	+	Addition
	-	Subtraction
Sign	-	Negation
Logical	and	Logical And
	or	Logical Or
	nand	Logical Nand
	nor	Logical Nor
	xor	Logical Exclusive-or
	not	Logical Complement
Relational	=	Equal
	/=	Not Equal
	<	Less Than
	>	Greater Than
	<=	Less Than or Equal
	>=	Greater Than or Equal
Concatenation	&	Concatenation

Table 3-1 : Major Synthesizable Operations

c. **If Reg_A <= Reg_B then ... -- ***!**

(*****! : Without the IF, this would be a Signal Assignment, but within the IF statement it means "less than or equal"!**)

Note that this distinction applies as well to relational or comparison operations.

    d.   Dataout <= Data_A + Bus_B + "0101"; -- Addition

    e.   Y_Bus <= - B_Bus;   -- Negation

Concatenation is used to **join** two or more bits or vectors together into a longer data word, such as :

If Output is a 10 bit register and Input is a 6 bit register whose value is "000101", then the operation

    Output <= "0100" & Input;

results in Output having the value "0100000101". Similarly

    Output <= Input & "0100" results in "0001010100"

Testing for Conditions in the Logic

The various means of testing for logic conditions include the IF, WHILE, and WHEN statements, and the CASE and SELECT statements. IF and CASE statements are used only in Processes, whereas SELECT and WHEN are only used outside of processes and thus will produce combinational circuits.

Be careful when testing for comparison conditions, such as equal, less than, and greater than, etc, that both sides of the comparison agree in size. If you compare an 8 bit vector with something other than 8 bits, you will not get the expected result. For example, if Reg_A is a 6 bit vector, then the comparison statement

    If Reg_A = "000101"   will always be true when Reg_A equals 5,

but the statement

    If Reg_A = "0101"

where Reg_A is 6 bits and "0101" is 4 bits **may never be true**. To avoid any problems, just make sure that the number of bits agree for any comparisons.

1. **IF:**

   a. If condition then
           Some action;
       End If;

   b. If condition 1 then
           Some action;
       Elsif condition 2
           Some other action;
       End If;

2. **WHEN** : (A "Conditional Signal Assignment" that synthesizes as combinational logic)

   a. SIGNAL <= '1' WHEN Databus > Register Else '0';
       (Synthesizes as an N bit "greater than" comparator, depending on the number of bits in "Databus")

   b. DBus <= REG_A When (CONTROL = '1') or (XYZ = '0') or (BUS_C = "0101") Else REG_B;

       (Synthesizes as a multiplexer controlled by logic that tests for the three separate conditions. See Chapter 8)

   c. Sum <= A + B When Select = '1' Else C + D;
       (Synthesizes into a multiplexed Adder - see Chapter 7)

3. **SELECT** : (A "Selected Signal Assignment" that synthesizes into combinational logic)

Syntax : With signal SELECT
           Target output <=
           Waveform 1 when Condition 1,

Waveform 2 when Condition 2,

...

Waveform N when Condition N;

a.   With SigA SELECT
      Output <=
         DataA when "00",
         DataB when "01",
         DataC when "10",
         DataD when "11";

(Synthesizes into a 4 to 1 multiplexer controlled by SigA, a two bit control signal, whereby the Output is switched to one of the four Data busses)

The "I" or vertical bar symbol can be used to indicate multiple conditions or choices in a list that will have the same action. For example, instead of

DataA when "00",
DataA when "01",

you could simplify the code by using

DataA when "00" I "01",

This shorthand can be used for Select and Case operations.

4. **CASE** : (A "Selected Signal Assignment" within a Process)

Case is similar to the SELECT above, but for each of the tested conditions, there can be multiple actions.

a.   CASE SigA is
         When "00" => Output <= DataA;
         When "01" => Output <= DataB;

         ...

         When "11" => Output <= DataD;
      End Case;

(Within an unclocked process, this creates the same logic as the SELECT code above)

b. CASE State is
       When S0 => Output <= A + B;
                        SIG1   <= '1';
                        State   <= S1;
       When S1 => Pulse   <= '0';
                        State   <= S0;
    End CASE;

(CASE tests the signal named State and depending on whether it has the value S0 or S1, appropriate actions are performed)

The CASE statement must check for all possible values of the CASE condition. However, if not all values are of interest, the syntax

   When **Others** => Desired Action or Null;

where null indicates no action to be taken. In addition, you can use the vertical bar symbol " I " to test for more than one condition that has the same action :

   When "00" I "10" => Output <= Dataval;

ie, when the expression is "00" or "01" then perform the action.

Repetitive Operations

VHDL supports repetitive operations by using loops, similar to standard programming languages. The loop types include unconditional loops, and WHILE and FOR loops, but some of these may not be supported in all synthesis tools. Related statements used with loops are the NEXT and EXIT statements.

1. **Unconditional Loop** :

    Loop_label : LOOP
      Statements
    End loop Loop_label;

The identifier shown as Loop_Label is optional, but is a good documentation practice and should be used.

2. **FOR Loop** :

    Loop_Label :
    For Loop_Variable in RANGE LOOP
      Statements
    End loop Loop_Label;

Examples will be given in Chapters 7, 8 and 9.

3. **WHILE Loop** :

    Loop_Label :
    WHILE condition LOOP
      Statements
    End Loop loop_Label;

4. **NEXT** : Causes the loop to begin again at the top when the condition is true.

    NEXT Loop_Label when condition;

5. **EXIT** : Terminates the loop, in two variations.

    a. EXIT Loop_Label;

    b. EXIT Loop_Label when condition;

These are the basic VHDL operations used in defining the function of an Architecture or a circuit. The use of these constructs for design and synthesis of typical logic functions will be shown in the examples beginning in Chapter 4.

Signal types and Variables

Up until now, signals have been mostly defined as being of type Std_Ulogic from the IEEE 1164 standard library, and the operation of signals within VHDL Entities and Architectures has been illustrated. Some synthesis tools use the Std_Logic type rather than Std_Ulogic, and it is also possible to define signals as other types, such as BIT or MVL4 mentioned earlier. Likewise, some tools allow you to define vectors as Integers, or as "sub-types" of Std_Logic_Vector called Signed and Unsigned. All of the above fall into the class called **Enumerated Types**, which, along with Integer, Floating Point, and Time Types, comprise the basic VHDL type classes.

VHDL requires that types be assigned to all ports, signals, variables, and constants. Built in to the language are a small group of "pre-defined" types, including Integer and Real for specifying numerical values, and Boolean, Bit, and Character for describing logic values. The latter types are called enumerated types, since the allowable values for these types are defined by listing them.

The language defines type Boolean to include only the states true and false. The type description of Boolean is

   Type Boolean is (False, True);

Similarly, type Bit only includes the states '0' and '1'.

Enumerated types can also be user created, and can include any values except for certain keywords reserved by the VHDL language, given in Table 3-1. The IEEE accepted Std_Ulogic type is an example of an enumerated type that has been defined as a standard type to allow portability and reusability of designs. Std_Ulogic is defined to include the values 'U', 'X', '0', '1', 'Z', 'W', 'L', 'H', and '-' :

Type Std_Ulogic is ('U','X','0','1','Z','W','L','H','-');

For general behavioral and synthesizable design work, it is very convenient to be able to define your own enumerated types. For example, a user defined type named Primary_Colors could be defined to include just three values : Red, Yellow, and Blue.

Type PRIMARY_COLORS is (RED,YELLOW,BLUE);

while a State Machine might have states named State0 and State1, or S0 and S1, or Initial, Ready, and Go. From a design viewpoint, you need not always describe logic operations in terms of 0s and 1s, and user defined types can be very descriptive for an application. Examples showing how to use your own enumerated types are given beginning in Chapter 6.

A related issue concerning signals of an enumerated type is their initial state for simulation. VHDL specifies that the values of all signals are deterministic at all times, which guarantees that each time a simulation is performed on a circuit, the same result is generated. The language requires that all uninitialized signals be set to the leftmost value of the defined type. For instance, for signals of type Primary_Colors, all uninitialized signals will be set to the value red, whereas for Std_Ulogic the uninitialized state is 'U'.

Technically, since you can only synthesize circuits that have logic states of '0' and '1', for synthesis the BIT type would be sufficient by itself. However, for multi-bit signals the Integer type, which is just like an integer in a programming language, is also very convenient for expressing and manipulating busses. A bus signal of type integer could be defined as

Databus : Integer Range 0 to 255;

and then used in expressions such as

If Databus = 27 then ...

or

Databus <= 125;

Defined as a Std_Ulogic_Vector, you would have to use the syntax :

> If Databus = "00011011" then ...

or

> Databus <= "01111001";

or use a Type Conversion routine as covered below.

As far as synthesis goes, either type of signal will result in the same circuit. If all we were doing was describing logic, it certainly looks easier to use Integer rather than Std_Ulogic. However, there are significant differences to the overall logic simulation process of ASICs that make the latter a preferred choice. First, if a signal or a bus is unknown (ie, X) or tri-state (Z), the Std_Ulogic type will respond to those states correctly whereas they would be illegal for the Integer type. A Std_Ulogic register whose input is unknown will produce unknowns on its output, but the Integer version, used with a type conversion, will produce all zeroes, as will the BIT type.

The significance of this behavior to the ASIC design process is that you should be using the same test patterns for simulating at the RTL level as at the gate level for final verification, and you should expect to get the exact same results! Typically, gate level logic simulators produce unknowns when devices are not initialized, or when timing glitches occur, or when inputs become tri-state or exhibit bus contention. If the RTL and gate level results don't agree, due to the response of Integer or BIT types of signals, you will have to do a considerable amount of additional work to resolve the differences.

Second, when using VHDL for logic design and synthesis, the use of Std_Ulogic and other types of vector notation retains an awareness of what the logic is doing. For instance, if a signal is assigned the value "01001100" you immediately know that it is an 8 bit function, contrasted with the integer value of 72 which gives you no clue as to the word length. This may not be a major issue, but it does remind you that by using synthesis you are designing hardware circuits, not simply coding VHDL.

On the other hand, when using VHDL for higher level behavioral simulation such as in a top down design process, the more abstract nature of the system description lends itself to the use of integer and real types for signals and provides greater flexibility. The VHDL concept of "overloading" also allows the mixed use of Std_Ulogic, integers, and other types for comparisons and relational operations, which will be illustrated below.

The Std_Ulogic and Std_Logic types have been mentioned as being part of the IEEE 1164 library. Both types describe a 9 state logic system, with the difference that a **Std_Ulogic** signal can only be driven from a single process or architecture while a **Std_Logic** signal can be driven from multiple processes. Allowing multiple drivers is similar to using a wired-or in hardware and is made possible by the VHDL mechanism called a **Resolution function.** In other words, if one source is driving a logic 0 and another source a logic 1, the Std_Logic type would define how to "resolve" the state of that signal, in this case as an unknown or "X".

The use of resolved signal types brings up several technical issues. First, many ASIC vendors recommend not using wired-or or even tri-state logic internal to the ASIC, because of potential problems with bus contention and resulting noise and power dissipation problems. Second, the resolved logic type requires an additional computation by the VHDL simulator and consequently can cause the simulator to run slower. And third, you may inadvertantly drive a signal from more than one source and be unaware of the problem due to the signal resolution, whereas Std_Ulogic would immediately produce an error message indicating multiple drivers for a signal. You should be aware of these concerns since some synthesis systems only support the Std_Logic type (or related types) for vectors.

It is possible to convert one VHDL Type to another by means of **Type Conversion** functions. Common type conversions, such as Integer to Std_Ulogic_Vector and vice versa, are usually provided with VHDL simulation and synthesis tools, but can also be written as needed. Examples are provided in Chapter 5, and in Chapter 9 under Test Benches.

Despite having stressed the use and operation of signals throughout this chapter, VHDL does have another mechanism called **Variables** that can also be assigned to values. Variables have three major differences compared with signals :

a. Variables assume their values as soon as they are assigned, exactly the way software variables are set in PASCAL and other languages. This means that, unlike signals, the order that variables are assigned is critical to the operation of the process.

b. Variables can only be used within Processes, Functions, and Procedures, and are strictly local to that block. The value of a variable cannot be passed out of the process for use by another process or concurrent operation.

c. The variable assignment uses the symbol ":=" instead of the "<=" used for signal assignments.

For synthesis design, there is no reason why signals can not be used exclusively, except for the use of variables in VHDL Functions and Procedures (covered below). For completeness, the use of variables will be shown in the examples, but in general you must exercise care in using them, especially in large processes. Since the order of statements matters for variables, making changes to the design requires that you keep track of and analyze how the variables will be affected by the change. This is somewhat analogous to designing logic using both edges of the clock - it can be done but it is usually tricky to maintain and modify. For this reason, this book will emphasize the use of signals over variables for most synthesis design work, although variables certainly can be used effectively in test benches and in functions and procedures.

Arrays

Signals and data can be organized as tables of values, or arrays, which can be most easily represented in hardware as RAM or ROM memories. Arrays are part of the VHDL class of Composite Types, where each element of the array is of the same type. The specification of an array is analogous to that of a memory in that the array is composed of a number of rows of data each of which has a width. For example, a 32 word by 8 bit memory array can be described in VHDL as

> Type Data_File is Array (0 to 31)
> > of Std_Ulogic_Vector (7 downto 0);

or in general

> Type NAME is Array (Number of values) of Type of values;

The number of values in the array is the **Range** of the array, which can consist of fixed values, such as (31 downto 0), or an unspecified number of values. The first type is called a "constrained" array, while the latter is an "unconstrained" array whose size is determined by its subsequent usage in the VHDL description. For example, an 8 word x 4 bit ROM lookup table could be defined and used as follows :

> Type Lookup is Array (0 to 7) of Integer Range 0 to 15;

> Constant ROM_Table_1 : Lookup := (12, 4, 6, 14, 1, 7, 12, 0);

> Constant ROM_Table_2 : Lookup := (7, 5, 3, 1, 6, 4, 2, 0);

The constrained array type "Lookup" is defined once as 8 Integer values, each of which can be between 0 and 15, and that type is used twice as ROM_Table_1 and ROM_Table_2, each of which contains different values. The number of values in the table must match the number in the array definition or a compiler error will occur, thus providing a check that the specification of the array size is correct.

The same array could be defined as an unconstrained array using the syntax "(natural range <>)", which means that the range of the array consists of the natural numbers with no specific upper or lower limit, or to be precise, the left and right bounds of the range are not defined. For example,

> Type Lookup is Array (natural range <>) of Integer (range 0 to 15);

Constant ROM_Table : Lookup := (1, 3, 5, 7, 9, 11, 13, 15, 17);

where the size or range of ROM_Table is determined by the number of values listed, or in this case, 9. One advantage of using unconstrained arrays is that the number of elements in the array can be changed without having to edit the array definition statement.

Examples of how arrays can be used in logic design applications are given in Chapters 5, 8, and 9.

Functions and Procedures

VHDL Functions and Procedures are VHDL subprograms used to simplify the coding of repetitive or commonly used circuit operations. If a certain block of code is used repeatedly in a design to implement a particular logic function, it can instead be placed in a Function or Procedure once and then referred to by name as needed.

The two types of subprograms differ in that Functions can only have one output while Procedures can have multiple outputs. Signals and variables can be passed into and out of the Procedure or Function by means of a list. For example, a Function named MULT to multiply two numbers A and B would be called as :

Product <= MULT (A, B) ;   -- Using a function

whereas a Procedure would use the syntax :

MULT (A, B, Product) ; -- Using a Procedure

Internal to the subprograms, variables can be used as needed to perform the desired operation and as always are only valid within that Function or Procedure.

The **Function** for MULT is defined as follows:

Function MULT (List of Inputs) RETURN Output IS

[Variable or Constant Declarations;]

Begin
        Statements;

        Return statement;
End MULT;

A **Procedure** for MULT would be defined as :

Procedure MULT (List of inputs and outputs) IS
    [Variable or Constant Declarations;]

Begin
    Statements
    Output signal assignments
End MULT;

Detailed examples of the use of Functions and Procedures for synthesis are given in Chapters 5 through 9. One interesting point about the use of these subprograms is that the synthesized circuit could actually be different depending on the type of inputs to the subprogram. For example, if a digital filter application is multiplying the contents of a register by another register, you would expect to get a normal multiplier circuit. If however the filter is multiplying the contents of a register by a constant, the synthesizer would delete the gates where the constant contains logic 0s, and similarly would alter the gating for the logic 1 inputs, since they are also constant. Thus the same Multiply subprogram would produce two correct, but different, synthesized circuits. In contrast to this behavior, a software language's subprogram or subroutine would be expected to produce the same results regardless of the input values.

Type Agreement and Overloading

VHDL is similar to PASCAL and other modern programming languages in that it is a "strongly typed" language, ie, the "type" of expressions on both sides of a signal assignment or other operation must agree. For instance, if the left side is a 12 bit vector, the right side must be a 12 bit vector. This characteristic of a language allows for error checking by the compiler and can assist in the process of debugging code. On the other hand, strict type agreement can also make coding more difficult. For example, if you want to add 1 to an eight bit bus or vector, you would have to use the syntax

      Result <= Databus + "00000001";

It would be convenient in this case to be able to use

      Result <= Databus + 1; (Std_Ulogic plus integer)
or   Result <= Databus + '1'; (Std_Ulogic plus a bit value)

     VHDL uses a technique called **Overloading** to define operations between the same or different types of data, which makes it possible to mix Std_Ulogic, Integer, Bit, and other types. The overloading definitions are normally included in the VHDL libraries and packages that are provided with your simulation and synthesis tools, and define the operations allowed by the tools. Most common overloads are included in logic synthesis systems and permit such operations as adding bits to vectors or comparing vectors with integers (IF vector = 27 then ...). The examples in succeeding chapters will use the syntax available with most commercial synthesis tools, assuming that the overloaded functions are available.

     Type Conversion functions can also be overloaded, in the sense that you may have several conversions to Integer format that use the same name for the Type Conversion function. For instance, functions that convert Bit_Vectors to Integer and Std_Ulogic_Vectors to Integer can both be named "To_Int" and used with the appropriate input data. The function "To_Int" is then called overloaded because it applies to as many data types as the functions are defined.

VHDL Predefined Attributes

VHDL has a number of built in or Predefined Attribute functions that can be used to determine the size and position of information in vectors and arrays. These attributes will be found useful in writing general purpose functions such as Type Conversions, in handling variable sized arrays, and in various loop operations. The syntax of the attributes is the apostrophe followed by the attribute name, such as **'HIGH**. As an example, for a Std_Ulogic_Vector named DATABUS defined as (15 downto 0), the highest bit number can be found by using

DATABUS'High, which returns the value 15,

and the lowest bit number by

DATABUS'Low, which returns 0.

The size of the vector, or its length, can be determined by using

DATABUS'Length, or the value 16.

You might use this information in a FOR loop such as :

For i in Databus'Low to Databus'High loop ....

Using the attributes can provide great flexibility in generating behavioral VHDL code, test benches, and in synthesis applications. Some of the attributes used in the examples later on in this text are 'HIGH, 'LOW, 'LEFT, 'RIGHT, and 'LENGTH, plus the 'EVENT attributes shown earlier in the clock edge syntax. A complete list and definition of the predefined attributes can be found in one of the VHDL texts in Appendix A or the IEEE Std 1076 Language Reference manual.

Reserved Words and Naming Conventions

VHDL has a list of reserved words (Table 3-2) that are part of the language and cannot be used for Signal or Entity names, or any other identifiers. In addition, the ASCII characters that can be used for names or identifiers are limited to alphabetic and numeric characters, plus the underscore, with a few restrictions :

a. Names must begin with an alphabetic letter.

b. The underscore (_) may be used to improve readability, but you may not use two in a row, nor may it be the last character in the name.
    Sig_A and BUS_07 are correct, but SIG_ _ A and BUS_ are not allowed.

c. Spaces are not allowed within names.

Most of the other keyboard characters are used for operations (+, -, *, /, &) or various types of delimiters (; , () <> = #) and therefore cannot be included in name identifiers.

The length of names used for signals, processes, entities, etc, is relevant to how they affect other design tools external to the VHDL synthesis system. The normal output of synthesis is a netlist that may be used with an ASIC verification system or gate level simulator, and the names used in the VHDL are reflected in the netlist. Many commercial systems however limit the length of signal names, including instance and hierarchy names, which could result in truncation of the names used in the netlist. Since this is a bad if not disastrous situation, it is prudent to be aware of your system's restrictions and, if need be, limit the length of names and identifiers used in your VHDL code.

abs	generate	procedure
access	generic	process
after	guarded	
alias		range
all	if	record
and	in	register
architecture	inout	rem
array	is	report
assert		return
attribute	label	
	library	select
begin	linkage	severity
block	loop	signal
body		subtype
buffer	map	
bus	mod	then
		to
case	nand	transport
component	new	type
configuration	next	
constant	nor	units
	not	until
disconnect	null	use
downto		
	of	variable
else	on	
elsif	open	wait
end	or	when
entity	others	while
exit	out	with
file		
for	package	xor
function	port	

Table 3-2 : VHDL Reserved Words

VHDL Literals

Literals are symbols used to represent values in VHDL, and include six types : integer, floating point, character, string, bit string, and physical. The VHDL references provide greater detail, but some examples are shown here :

Integer : A number without a decimal point. Different number bases such as Decimal, Hex, Octal, and Binary may be used to represent integers.

Floating Point : Not synthesizable. Used for numbers such as 26.32, 1.00, or 45.2E3 that contain decimal points.

Character : A single ASCII character enclosed in single quotes, such as '0', 'Z', '1'.

String : Several ASCII characters enclosed in double quotes, such as "ABCD", "State_0", "Mode_1".

Bit String : Several ASCII characters used to represent numbers in Hex (X), Octal (O), Binary (B). Only characters 0-9, A-F, and a-f are valid. Examples are :

```
B"11010001"    -- Binary
O"77320"       -- Octal
X"FC23"        -- Hexadecimal
```

Physical : For synthesis, the physical units represent Time values such as :

```
6 ns    - 6 nanoseconds
27 us   - 27 microseconds
```

Summary

This chapter has introduced most of the concepts of VHDL that apply to logic synthesis and ASIC design. The examples beginning in Chapter 4 will illustrate the use of VHDL in designing logic functions, and will present many ASIC related ideas. It is recommended that a VHDL language textbook also be used for additional information about the language and its formal syntax rules. A number of excellent texts that cover the entire VHDL syntax are given in the list of references in Appendix A.

Chapter 4

Synthesis of Sequential Circuits

Chapter 3 covered the basic synthesis rules for clocked processes and the fact that they could be used to design sequential logic functions. This chapter presents the first discussions of how to design common sequential circuit functions using VHDL synthesis, and uses various types of counters as the vehicle to illustrate the techniques. Design for testability will also be addressed as it applies to general counting applications.

Counters are the basic building blocks of a number of different kinds of sequential logic functions, and because of their importance will be discussed in detail in this section. Design techniques will be developed gradually through a series of examples, beginning with the basic counter syntax and proceeding to add the capabilities normally needed in typical real world systems. Following the general discussion of counters, examples of how they can be synthesized into other types of applications will be demonstrated in chapter 5.

It will become clear in the next few chapters how synthesis can help create good logic designs while still giving the designer control over the basic design decisions and techniques used. In fact, it will be seen that designer experience is essential to the synthesis process in most cases and allows design work to move to a higher level while using the synthesis tool to handle the gate level details.

Basics of counters

In the current industry practice there are many different ways to implement counters, such as those that count up, down, or either up and down depending on a control signal. In most applications, counters have separate control signals to enable counting as well as to reset or load the counter to a desired state. Although binary counters are the most common form of counter design, other variations such as BCD (Binary Coded Decimal), Johnson, Gray or pseudo-random counters may be preferred depending on the application. Fortunately all of these counter types are readily synthesizable from VHDL constructs.

As explained in chapter 3, for any particular synthesis tool various VHDL packages are usually provided by the tool vendor that handle operations such as type conversions, operator overloading, and synthesis specific attributes and which make the VHDL coding task simpler and easier to understand. These capabilities will be pointed out when they are first used in the examples below in order to clarify their usage. In addition, although synthesizers can often handle various VHDL signal types such as Bit and Std_Logic, this book will use Std_Ulogic most often since this IEEE-adopted 9 state logic definition yields better flexibility during VHDL simulation.

The basic binary counter

In order to set the foundation for the design of more complex functions in succeeding chapters, let's consider the VHDL code that can be used to synthesize a basic counter, as shown in Figure 4-1.

Figure 4-1 : A Basic Binary Counter

In a VHDL synthesizable description, a counter of N bits is usually assigned to a VHDL signal, commonly of type Std_Ulogic_Vector or of type Integer, whose operation is defined in a process within a design's architecture. As an example, if you need an 8 bit binary up counter, you might define it as

   Signal Upcount : Std_Ulogic_Vector (7 downto 0);

or Signal Upcount : Integer range 0 to 255;

or a similar data type that may be synthesis tool dependent.

   To implement the counter's operation within a process, the basic counter syntax is simply

         Upcount <= Upcount + '1';   -- Where Upcount is a vector
or
         Upcount <= Upcount + 1;   -- Where Upcount is an Integer

where this operation will be performed at a clock edge, or in response to some other signal. Naturally there will be more code required to handle the reset conditions and other features which will be covered below, but the VHDL code to implement the fundamental counting operation is straightforward. Notice that by describing a counter at this RTL level and using logic synthesis tools you need only to specify the function of the counter, not any of the details of the flip flops and gates required to implement the function. Notice the VHDL "overloading" that permits adding a Std_Ulogic signal to the literal '1' or the Integer value 1. Without that capability you would have to use

Upcount <= Upcount + "00000001";

in order to produce correct VHDL. This is one of the conveniences of having complete VHDL overloading packages as part of the synthesis tool and makes the VHDL coding task clearer and easier to use for logic design.

Understanding What Synthesis Produces

When this basic counter is synthesized, it is interesting to see that without any timing constraints the circuit will be implemented as a register and a simple adder which adds 1 to the register contents on every clock, as shown in Figure 4-2.

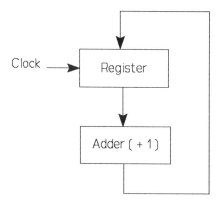

Figure 4-2 : Register = Register + 1

In terms of the function this synthesized circuit technique seems very logical, but if you are a designer with TTL experience or have been trained to use formal counter design techniques, the result is surprising, since you would not normally implement a counter this way. However, if you examine the logic equations for each flip flop in the adder-based counter, you will find that they are exactly the same as those of a conventional counter design.

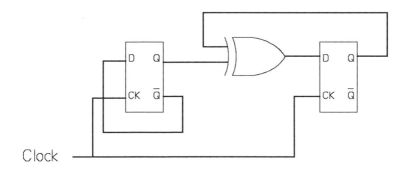

Figure 4-3 : Basic Binary Counter Logic

Synthesis designs are usually based on D flip flops rather than JK or other types of flops. The actual gate level circuit produced by the synthesis tool also depends on the timing constraints imposed for the design; if there are no constraints the circuit will normally be optimized for minimum gate count and will produce a serial chain of gates. As the synthesis constraint on clock frequency is increased, the simple gate network will be automatically replaced by carry look-ahead logic where needed. The significant point regarding a good synthesis tool is that it continues to try to minimize the gate count while meeting the timing constraints, and in fact will not just put carry look-ahead logic everywhere but only where needed!

A second point regarding the circuit produced is that the above exclusive-or operation is likely to be implemented differently depending on the cell devices available in the ASIC vendor's library. For example, the LSI Logic 100k series has a number of Ex-Or functions and also several combinations of And-Or gates, all of which have different equivalent gate counts and propagation delays. The synthesizer will usually pick an AO2, which is two 2-input And gates whose outputs are NORed together but which is only 2 equivalent gates, instead of the exclusive-or (EO2) gate which is 3 equivalent gates and is slightly slower. A further sophistication, as shown in Figure 4-4, is that rather

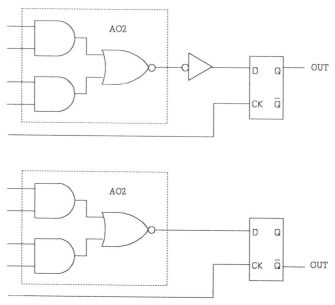

Figure 4-4 : Synthesis Minimization Details

than invert the output at a cost of an inverter (and its delay), the synthesizer knows that flip flops are symmetrical and if the D input is inverted, the Q output is inverted and therefore the Q-not output is the correct polarity. It is good to be aware of these tradeoffs being made by the tools, both for understanding as well as having an appreciation of the quality design details produced by the synthesizer.

VHDL Counter Design

Now let's consider the complete VHDL process syntax needed to implement our counter, and at the same time add an asynchronous reset. For this example assume that the system design requirements call for a 32 MHz clock. At some point, the clock and reset will be defined either as input ports to the entity or as signals within an architecture. For this example, we'll name the 32 MHz clock CLK32M and the master reset will be called MRESET.

The general principles for VHDL synthesis of sequential logic are reiterated in the code of Example 4-1 :

a. VHDL processes for designing sequential logic are referred to as "clocked" processes. In terms of hardware this simply means that the circuit uses a clock, whereas in VHDL terms it means that the process will wait for a clock edge in order to execute. This is handled by use of the "sensitivity list" in defining the process.

The process for our counter includes both clock and reset in its sensitivity list, which causes the process to execute whenever either signal changes.

b. The IF-THEN-ELSE syntax is used to define operations in clocked processes. The code for the reset and clock conditions are placed in separate IF clauses as shown, and the given VHDL structure should be considered as the normal technique for sequential logic design.

c. The Rising_Edge function is used for the clock (in ASIC families in which flip flops are clocked on the leading edge of the clock. Otherwise use the Falling_Edge function). Both functions are included in the IEEE 1164 standard logic library for 9 state logic. The equivalent of the Rising_Edge operation required for some synthesis systems is

(CLK32M'event and CLK32M = '1'); -- Same as Rising_Edge

The code for this simple binary counter with asynchronous reset is :

```
-- Signal Definitions

Signal CLK32M, MRESET : Std_Ulogic;
Signal Upcount : Std_Ulogic_Vector (5 downto 0);
```

Begin

-- Process definition

Upcounter : Process (CLK32M, MRESET)

Begin

   If MRESET = '1' then

     Upcount <= "000000";

   Elsif Rising_Edge (CLK32M) then

     Upcount <= Upcount + '1';

   End if;

End Process Upcounter;

Example 4-1 : VHDL Code for the Basic Binary Counter

There are several concepts that should be recalled relative to the VHDL process. First, a process executes from the Begin statement to the End statement whenever there is a change on any of the signals in the sensitivity list, ie MRESET or CLK32M in this case. Signals that get assigned to new values during the process, such as Upcount, are only updated at the end of the process. For a logic designer, this situation is exactly analogous to the operation of clocked flip flops, where the D inputs may change prior to the clock but the flip flop output does not change until the next clock edge. In the code above if the state of the counter Upcount is "000100" when the process begins, it will retain that value during the process and will only become Upcount + 1, or "000101", at the end of the process. In this simple example there may not appear to be much significance to this fact, but in more complex designs where values are tested during the process it is important to remember this behavior of signals and signal assignments.

The counter synthesized from this code will have a direct clear on every flip flop connected to the MRESET signal, but otherwise is the same basic circuit as our simpler counter. If the clock runs continuously, when the 6 bit counter reaches all ones (ie, 63 decimal) the next count becomes all zeroes and the cycle repeats itself.

Adding enable and load logic

While this circuit illustrates some of the basic requirements for VHDL synthesis of sequential logic, in most cases counters need an enable signal to control the count process and often a capability of being synchronously loaded to some desired value. The counterpart to this type of counter in standard TTL logic would be the 54163 family of 4 bit counters, although synthesis is much more flexible and allows any number of bits to be easily designed using the same technique. Let's extend the previous example by adding a count enable signal ENABLE and a LOAD signal which will force the count to the value on a databus named DATABUS. For this example, assume that the enable must be high in order for loading or counting to occur:

```
-- Signal Definitions

Signal CLK32M, MRESET, ENABLE, LOAD : Std_Ulogic;
Signal Upcount, DATABUS : Std_Ulogic_Vector (5 downto 0);

Begin

-- Process definition

Upcounter : Process (CLK32M, MRESET)

Begin
```

If MRESET = '1' then

  Upcount <= "000000";

Elsif Rising_Edge (CLK32M) then

  If ENABLE = '1' then

    If LOAD = '1' then

      Upcount <= DATABUS;

    Else
      -- Normal counting

      Upcount <= Upcount + '1';

    End if;

  End if; -- End of If enable ...

  End if;

End Process Upcounter;

Example 4-2 : The binary counter with enable and load

The synthesized circuit shown in Figure 4-5 is similar to the basic counter of Example 4-1 except for additional multiplexing into each flip flop for the data to be loaded and the enable controls. The equations for the D input to each flip flop are exactly what a designer would do manually, for example :

$$D0 = (\overline{ENAB})(Q0) + ENAB*((Load)(Databus0)$$

$$+ (\overline{Load})(\overline{Q0})).$$

In words, if ENABLE is low stay at the present state, if ENABLE and LOAD are both high go to the databus bit value, and if ENABLE is high and LOAD is low then toggle the flip flop (ie, count). Instead of the gating shown, some synthesis tools may use 2 to 1 multiplexer cells if they exist in the ASIC cell library.

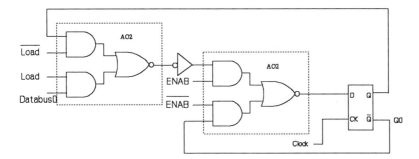

Figure 4-5 : A Synthesized Counter Stage with Load and Enable

Notice that the load operation is implemented as an IF statement within another IF statement that checks for the presence of the enable signal. In this case the load operation will only be performed if the enable is high, otherwise it is ignored. In the case that you want the load to always be available, you would simply modify the VHDL code as follows:

If LOAD = '1' then
    Upcount <= DATABUS;

Elsif ENABLE = '1' then
    Upcount <= Upcount + '1';

End if;

The use of the IF-ELSE and IF-ELSIF syntax is important as far as synthesis is concerned because it implies a **priority** of one operation over the other. Consider the situation in which you code the conditions independently such as :

```
If LOAD = '1' then
    Upcount <= DATABUS;
End if;

If ENABLE = '1' then
    Upcount <= Upcount + '1';
End if;
```

VHDL rules state that if a signal is assigned to multiple values, such as

Upcount <= DATABUS followed by Upcount <= Upcount + '1',

the last signal assignment determines the value of the signal, and the previous assignment is ignored. A synthesis tool will use this fact in the previous section to give counting priority over loading, which may not be what you intended, and reversing the order of those statements would result in a different circuit.

In order to avoid ambiguity and/or design errors, determine how the circuit should respond to multiple inputs, which is a design decision that you should make before coding the VHDL. The basic rule is to place the operations that take precedence first in the IF-THEN-ELSE clause, followed by the remaining functions. As shown in the case of Example 4-2, if the LOAD signal is active then that operation would override the normal counting function.

Modifying the Length of the Counter

Up until this point the counters have been allowed to count through all of their binary states. In many applications it is desirable to count to a specific value and then return to zero and repeat that sequence. For Example 4-3, assume that the 32 Mhz clock needs to be divided down by 50 to yield an output frequency of 32/50 MHZ or 640 Khz. Let's choose to have the counter count from 0 to 49 and repeat, for a total of 50 clocks. In order to do this, the counter will require 6 bits, and we will

have the circuit detect the value of 49 and then return to 0 on the next
clock :

```
-- Signal Definitions

Signal CLK32M, MRESET, ENABLE, LOAD : Std_Ulogic;
Signal Upcount, DATABUS : Std_Ulogic_Vector (5 downto 0);

Constant LAST_VALUE : Std_Ulogic_Vector := "110001"; -- Value 49
Constant ZERO : Std_Ulogic_Vector := "000000";

Begin

-- Process definition

Upcounter : Process (CLK32M, MRESET)

Begin

   If MRESET = '1' then

      Upcount <= "000000";
                  -- (or alternatively Upcount <= ZERO;)

   Elsif Rising_Edge (CLK32M) then

      If ENABLE = '1' then

         If LOAD = '1' then

            Upcount <= DATABUS;

         Elsif Upcount = LAST_VALUE then

            Upcount <= ZERO;              -- Reset count to zero
         Else
```

```
        Upcount <= Upcount + '1';          -- Normal counting

     End if;
   End if;
 End if;
End Process Upcounter;
```

Example 4-3 : Modified binary counter with enable and load

Notice the use of named constants for the final count value (LAST_VALUE) and for the beginning count of (ZERO). This style of coding can be used at the discretion of the designer and has been used here to improve the readability of the VHDL. This approach also is beneficial when making changes to the design parameters of the circuit. Rather than hunting for and changing every instance of a value such as 49, the defined constant can be found at the beginning of the process and need only be changed once.

In this code, when the count reaches LAST_VALUE (or 49) it will go to all 0s on the next clock edge. The technique shown for controlling the length of a count sequence is common to many applications and will be seen often in later design examples. Note the simplicity and ease of understanding of the VHDL code, as well as the straightforward ability to change the count length by modifying one constant.

The synthesized circuit is similar to that of the basic counter from Example 4-2 except that gates are added to detect the value 49, and when that occurs, gating logic into the D inputs of the flip flops forces the count back to 0 on the next clock edge. Since both polarities of the flip flops are available in most ASIC families, the last_count detection gate is an N bit And gate, where N is the number of bits in the counter.

This is simple for fixed length counters, but what if you need to vary the count length based on a system parameter that is not a constant? This would be that case where you need to generate a time delay whose duration depends on a system value, or you need to count to a specified value that is varied as the system operates, as shown in Figure 4-8.

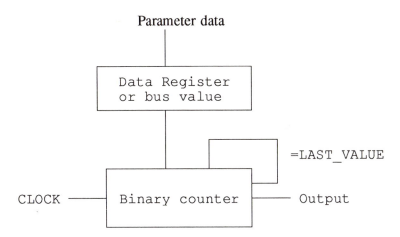

Figure 4-6 : Counter with Variable Count Length

Counting by a variable amount

This is a typical situation where logic design experience is necessary in order to produce an efficient circuit. The desired function can be implemented either as a counter that counts from 0 up to the parameter value and then synchronously resets to 0, or as a counter that starts at the parameter value and counts down to zero and is synchronously set back to the value. From design experience, you know that to detect a count being equal to a data value requires an N bit comparator, whereas to detect a fixed value requires only an N bit gate. On the other hand, depending on the ASIC library cells used, it can require more gates to load a counter than to clear it to zero. The significant point is that these are architecture choices based on design experience - a VHDL synthesis tool would not be able to make this judgement by itself.

Two other variations on this theme also exist - you could count up from the complement of the value to all 1's, a common TTL logic technique, or you could count down from all 1's to the complement value, and similar arguments apply as in the above discussion. Often the variable parameter value is loaded into a latch or register in which case both polarities would be available, ie the value and its complement. This type of planning can help direct the most efficient approach to take in many cases. Armed with this knowledge, let's look at the variable count length logic.

Counting down

Up to this point we have discussed simple up counters, ie, those whose binary value increases at each clock. It should be intuitive by now that to make a down counter the syntax simply changes to :

Downcount <= Downcount - '1' ;

For basic down counting, the VHDL code would be essentially the same as in Example 4-1 with subtraction replacing the add operation. To illustrate this usage, let's assume that you need a counter whose length is determined by the value on a databus and you choose to implement the function as a down counter that begins at that value and counts down to zero. Example 4-4 uses a process named Dncounter and a signal Dncount to implement the circuit in VHDL, and maintains the same enable and load capability as in previous examples.

```
-- Signal Definitions

Signal CLK32M, MRESET, ENABLE, LOAD : Std_Ulogic;
Signal Dncount, DATABUS : Std_Ulogic_Vector (5 downto 0);

Constant ZERO : Std_ULOGIC_Vector := "000000";

Begin
          -- Process definition

Dncounter : Process (CLK32M, MRESET)

Begin
```

```
If MRESET = '1' then
   Dncount <= ZERO;

Elsif Rising_Edge (CLK32M) then

   If ENABLE = '1' then

      If LOAD = '1' or Dncount = ZERO then

         Dncount <= DATABUS;
      Else

         Dncount <= Dncount - '1';    -- Normal counting
      End if;
   End if;
End if;

End Process Dncounter;
```

Example 4-4 : Variable length binary down counter

Counting up and down

Now that we've studied up and down counters, it should be a straightforward procedure to design an up/down counter. This type of counter is used in many applications ranging from error counters in control systems, duty cycle measurement systems and display systems, and variations include circuits with a common up/down control signal as well as separate up and down controls.

Consider the case where a common control (Up_Down) is used to determine the count direction :

-- Signal Definitions

```
Signal Clk32M, Mreset, Up_Down, Enable, Load : Std_Ulogic;
Signal Count, DATABUS : Std_Ulogic_Vector (5 downto 0);

Constant ZERO : Std_ULOGIC_Vector := "000000";

Begin
                -- Process definition

UPDOWNCNT : Process (CLK32M, MRESET)

Begin

   If MRESET = '1' then
      Count <= ZERO;

   Elsif Rising_Edge (CLK32M) then

      If ENABLE = '1' then

         If LOAD = '1' then
            Count <= DATABUS;

         Elsif UP_DOWN = 1 then

            Count <= Count + '1';          -- Counting up
         Else
            Count <= Count - '1';          -- Counting down
         End if;
      End if;
   End if;

End Process UPDOWNCNT;
```

Example 4-5 : Up/down binary counter with common control

Notice the priority given to the loading of the counter over the counting operation by use of the IF-Elsif-Else syntax.

BCD Counters

Binary coded decimal or BCD counters are often used when the count value must be displayed for people to view on an alphanumeric display or when transferred as data to a computer. Digital watches are a primary example of the use of BCD displays, although there are countless other applications. The major difference between BCD counters and binary counters is that BCD counters only count from 0 to 9 and are therefore 4 bits in length, and succeeding BCD stages advance when the previous stages go from all 9s to 0.

As an example, consider a 3 digit BCD application that will permit counting from 0 to 999 decimal. Name the signals representing the digits LSD (for least significant digit), LSD2, and MSD. The technique used in the example is to check the value of the LSD and if it equals 9, set it back to 0 (on the next clock) and check the LSD2. If it equals 9 do the same, otherwise let the LSD2 count up, and so on for the MSD. The VHDL code for this process is a series of If-then-else statements as shown in Example 4-6.

```
-- Signal Definitions

Signal CLK, MRESET : Std_Ulogic;
Signal LSD         : Std_Ulogic_Vector (3 downto 0);
Signal LSD2        : Std_Ulogic_Vector (3 downto 0);
Signal MSD         : Std_Ulogic_Vector (3 downto 0);

Constant Zero      : Std_Ulogic_Vector (3 downto 0) := "0000";
Constant Nine      : Std_Ulogic_Vector (3 downto 0) := "1001";

Begin

-- Process definition

BCDcount : Process (CLK, MRESET)

Begin
```

If MRESET = '1' then

  LSD <= Zero;
  LSD2 <= Zero;
  MSD <= Zero;

Elsif Rising_Edge (CLK) then

  If LSD = Nine then
    LSD <= Zero;

    If LSD2 = Nine then
      LSD2 <= Zero;

      If MSD = Nine then
        MSD <= Zero;
      Else
        MSD <= MSD + '1';
      End if;

    Else
      LSD2 <= LSD2 + '1';
    End if;

  Else
    LSD <= LSD + '1';
  End if;
End if;

End Process BCDcount;

Example 4-6 : A 3 digit BCD counter

This code synthesizes into 3 separate BCD counters with the appropriate enable to each counter determined by decoding the state 9 of each previous stage.

A More Complex Problem - the Gray Counter

The binary sequence known as the Gray code has the characteristic that only one bit changes at a time in the count sequence. There are numerous binary sequences that satisfy this rule, and two variations are shown in Figure 4-7 for a 3 bit Gray code counter :

```
000            000
001            010
011            110
010            111
110            011
111            001
101            101
100            100

000            000    ..repeats sequence
```

Figure 4-7 : Two valid 3 bit Gray code sequences

The Gray coding scheme is one example of a class of sequential counters that do not generate spikes or glitches when gates are used to decode particular states of the counter, unlike the case with binary or BCD counters. Gray codes have been used for many years with mechanical switches and shaft encoders, for the same reason that no false decodes would occur as the switch was moved from one position to another. The significance to our synthesis discussion is that this is the first example in which there is no obvious algorithm that defines a Gray code counter, such as Count <= Count + '1' as used in the binary counters. This is a good time to consider the general logic design problem as well as the various VHDL synthesis alternatives available.

If you were designing a Gray code counter by conventional means, you would first have to decide how to implement the logic to generate that sequence. For this example, let's use the first code sequence from

Figure 4-7. The classical approach to this type of problem is to list the code sequence for the state prior to the clock edge, and then the next state which is to occur after the clock edge, and then write the equation for the D input to each flop from that process. Figure 4-8 illustrates this technique for the 3 bit example, where the 3 bits are named A, B, and C.

	Present state		Next state
(Decimal)	(Binary) C B A	====>	C B A
	─────		─────
0	0 0 0		0 0 1
1	0 0 1		0 1 1
3	0 1 1		0 1 0
2	0 1 0		1 1 0
6	1 1 0		1 1 1
7	1 1 1		1 0 1
5	1 0 1		1 0 0
4	1 0 0		0 0 0

Figure 4-8 : Gray code design equations

There are several methods to write and minimize the equations for each flip flop, including the use of Karnaugh maps, Quine-McCluskey, etc. A brute force approach for this simple case for flip flop C would be:

Next state of C = 1 when the present state is 2, 6, 7, and 5. The logic equations become

$$C(next) = \overline{C}B\overline{A} + CB\overline{A} + CBA + C\overline{B}A$$

$$= \overline{A}B + AC.$$

This is the equation for the D input to flip flop C, and could be implemented as a 2 to 1 multiplexer. Similarly

B(next) = $\overline{A}$B + A$\overline{C}$ (2 to 1 mpx)

A(next) = CB + $\overline{C}$ $\overline{B}$ (Exclusive nor gate)

Having gotten the equations, the resulting logic circuit would consist of 3 flip flops, 2 2-1 multiplexers, and an ex-nor gate, or the equivalent gating logic. At this point we've done the design manually, with the logic minimized by conventional means. Without synthesis you would proceed to draw schematics for the resulting circuit, while with synthesis, the same equations can be used directly to specify the circuit.

There are several possible approaches to designing this type of circuit with VHDL that typify general techniques usable in a wide range of applications. The following example will show both a state machine method and a technique similar to the process shown above, both which yield efficient synthesized circuits.

A State Machine technique

Since there is no simple algorithm for this type of circuit, the VHDL code must be fairly explicit in what you expect it to do. The technique is to use the table of desired states (Figure 4-7) to drive a simple state machine within a process, using the Case statement syntax. Let's define a 3 bit signal named Graycnt, and use the typical Clock and Reset signals as inputs :

Architecture RTL of Graycounter is

Signal Clock, Reset   : Std_Ulogic;
Signal Graycnt   : Std_Ulogic_Vector (2 downto 0);

  Begin

  Gray : Process (Clock,Reset)
  Begin

```
If Reset = '1' then
          Graycnt <= "000";

Elsif Rising_Edge (Clock) then

-- For each state of Graycnt, explicitly define the next state ...

     Case Graycnt is

          When "000"    => Graycnt <= "001";
          When "001"    => Graycnt <= "011";
          When "010"    => Graycnt <= "110";
          When "011"    => Graycnt <= "010";
          When "100"    => Graycnt <= "000";
          When "101"    => Graycnt <= "100";
          When "110"    => Graycnt <= "111";
          When "111"    => Graycnt <= "101";
          When others   => null;
     End case;
   End if;
 End process Gray;

End RTL;--(End of Architecture Graycounter)
```

Example 4-7 : Gray Code Counter Using State Method

The synthesized circuit is very efficient and is identical to the manually derived equations presented earlier. This design technique is very general and will be seen in other examples of different types of logic, as well as in true state machine sequencer design. One particular advantage of this style is that it is very easy to understand what is intended by the VHDL code, which simplifies both design reviews as well as making any future changes to the sequence. Notice that other than specifying the desired state sequence, you can allow the synthesizer to fully determine the logic design. On the negative side it is obvious that for larger numbers of bits and states the code can get lengthy, but that is due to the nature of this type of sequencer design. We will see in succeeding chapters that for general state machines it is not always necessary to define the binary states of the machine as we have done here, but instead the synthesis tool can be given the freedom to use the optimum coding scheme to produce the most efficient logic.

Logic design or "Dataflow" method

A second technique takes advantage of your logic design knowledge and makes use of the equations derived above for the Gray code counter. This style, also referred to as "dataflow" style, represents a lower level of VHDL abstraction than all of the previous examples, but also gives a finer control over the logic produced while still not resorting to a true gate level structure. Referring back to the earlier example, the logic equations were :

$$C(next) = \overline{C}\,\overline{B}\,\overline{A} + C\overline{B}\,\overline{A} + CBA + C\overline{B}A$$

$$= \overline{A}B + AC.$$

$$B(next) = \overline{A}B + A\overline{C}$$

$$A(next) = CB + \overline{C}\ \overline{B}$$

For the VHDL implementation, let's take our 3 bit Graycnt vector from Example 4-7 and for convenience assign the VHDL "Alias" name C to bit 2, B to bit 1, and A to bit 0, the least significant bit. For synthesis tools that cannot handle the Alias syntax, use signals A, B, and C instead of the 3 bit vector. The VHDL process becomes a single clocked process where new values for A, B, and C are assigned based on the equations above.

Architecture RTL of Graycnt is

Signal Grayff :Std_Ulogic_Vector (2 downto 0);

Alias C : Std_Ulogic is grayff (2);
Alias B : Std_Ulogic is grayff (1);
Alias A : Std_Ulogic is grayff (0);

Begin

Grff : Process (Clock, Reset)
Begin

    If Reset = '1' then

        C <= '0';
        B <= '0';
        A <= '0';

    Elsif Rising_Edge (Clock) then   -- Use the Boolean Equations

        C   <= (not (A) and B) or (A and C);
        B   <= (not (A) and B) or (A and not (C));
        A   <= (C and B) or (not (C) and not (B));

 End if;
End process Grff;

End RTL;

Example 4-8 : Dataflow coded Gray counter

You can see that the logic equations for flip flops A, B, and C are implemented directly in the VHDL code, and as a result the synthesizer's task is made much simpler. The resulting circuit is the same as the previous state machine design, which means that the two approaches are indeed equivalent.

It should be noted that in the dataflow style, the example above used minimized equations in the VHDL code, but this is not really necessary when using a good synthesis tool. An advantage of synthesis in this case is that if you gave it the raw logic equations before any minimization, the synthesizer tool would proceed to optimize the logic to produce the most efficient design. As a designer you have the option of choosing either technique depending on the information available or on your individual preference.

The dataflow style can be used in wide variety of design applications, but is particularly useful in those cases where you want to have more precise control of the synthesis process. Typical situations include circuits that have demanding timing requirements, or cases in which you want to use a specific circuit technique from a previous design. The next section shows a different application implemented in both the state machine and dataflow styles, showing that the dataflow method can often be a very efficient way to describe and synthesize a circuit.

A Johnson Counter Design

Johnson counters are a common form of shift register based circuits that are convenient for timing and multi-phase clock generation and allow for simple decoding using two input gates. No gates are normally needed to implement the design, but for an N state sequence N/2 flip flops are needed. The Johnson counter can have an even or odd number of states, which in the classical design is implemented by connecting the D input of the first shift register stage to the inverted output of the last stage or a combination of the last two stages. Figure 4-9 shows the waveforms of a very simple 3 bit Johnson counter, where JC0 is the least significant bit.

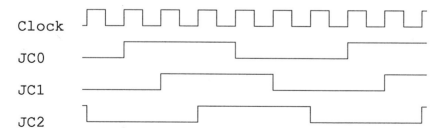

Figure 4-9 : Johnson Counter Waveforms

There are two different ways to think about this circuit. In the first case you might simply write out the states as a binary code and use a VHDL case statement in a clocked process as shown above for the Gray counter:

000
001
011
111
110
100

000   ... repeats

In this sequence, states 010 and 101 never occur and would be considered "don't cares" in a classical design. Good design practice in this case however is to ensure that if those states ever did occur that the sequence would get back into a legal state. The VHDL code should include those states in order to make the synthesis task easier, and as shown in the example below state 101 goes to 010 and 010 goes to 100, which is a legal state in the sequence.

The second approach to this design is the logic design or "dataflow" approach, in which the shift register of the Johnson counter is implemented as a data operation shown in Figure 4-10. The VHDL code for this circuit will use the vector slice syntax and the concatenation operation (&) to implement a shift register, such as the following for an 8 bit signal REG :

To shift right with zeros :

(In a clocked process)

REG (7 downto 0) <= '0' & REG (7 downto 1);

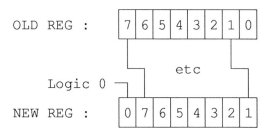

Figure 4-10 : Shifting by means of VHDL concatenation

To rotate right:

REG (7 downto 0) <= REG (0) & REG (7 downto 1);

To shift left with zeros:

REG (7 downto 0) <= REG (6 downto 0) & '0';

Given those two basic techniques, let's implement each one as a separate process within the Example 4-9 architecture. The state machine variable has been named JC (for Johnson Counter) and the dataflow version has been named JCreg, both using 3 bit vectors :

Architecture RTL of JOHNSON is

Signal JC : Std_Ulogic_Vector (2 downto 0);
Signal JCreg : Std_Ulogic_Vector (2 downto 0);

Begin

-- **** A State Machine Version ****

```vhdl
Johnscnt : Process (Clock, Reset)

Begin
    If Reset = '1' then

        JC <= "000";

    Elsif Rising_Edge (Clock) then

        Case JC is

            When "000" => JC <= "001";
            When "001" => JC <= "011";
            When "011" => JC <= "111";
            When "100" => JC <= "000";
            When "110" => JC <= "100";
            When "111" => JC <= "110";

            When "010" => JC <= "100"; -- The extra states
            When "101" => JC <= "010";

            when others => null;    -- For example, when "XXX"

        End case;
    End if;
End process Johnscnt;

-- **** The DATAFLOW Version *******

Jcnt : Process (Clock, Reset)

Begin
    If Reset = '1' then

        JCreg <= "000";

    Elsif Rising_Edge (Clock) then

        JCreg <= JCreg (1 downto 0) & not (JCreg (2) );
```

End if;

End Process Jcnt;

End RTL; -- (End of Architecture JOHNSON)

Example 4-9 : Two Versions of the Johnson Counter Design

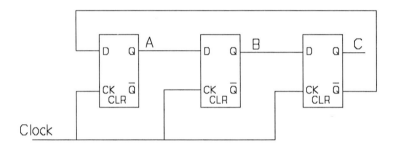

Figure 4-11 : Synthesized Johnson Counter

It should be noted that for larger numbers of bits, the dataflow technique is much simpler and will remain as one line of code. The state technique grows by two times the number of bits, which can become cumbersome to code for larger designs. It is however a useful technique and has been shown here mainly to illustrate the VHDL methods available to solve a problem. Both circuits synthesize to the expected result of a shift register with feedback, with the difference that the dataflow code shown did not try to account for the missing states (010 and 101) whereas the state machine explicitly defined what to do in those cases. If desired, the logic could also be added to the dataflow version.

It is worthwhile to consider what would happen to the state machine version if the unused states were not specified as

```
When "010" => JC <= "100"; -- The extra states
When "101" => JC <= "010";
```

but were left out to be included in the "others" statement:

when others => null;

In this case, the synthesizer might try to decode the unused states and ensure that under those conditions nothing happens to the circuit. This could result in the addition of extra logic with no purpose (ie, if in State 010 stay in 010), and for that reason it is prudent to define explicitly what should happen in those states as shown in the example code. The states you choose will depend on your logic design experience and in this case it is probably wise to examine the resulting logic created by the synthesizer.

Design for Testability Issues

The testability of counters and complete ASICs is a serious concern in simulation and in achieving high quality fault coverage for physical testing of ASIC devices. While testability in general can be the subject of an entire book, it is worthwhile to briefly discuss the problems, and to demonstrate some of the possibilities available through VHDL synthesis.

The basic goals of design for testability are to ensure that a given logic design can be completely simulated and tested, using a finite set of test patterns. During test of a system or ASIC, the test patterns must be sufficient to detect most or all of the physical defects that might be present. In an ASIC, the defects can be caused by opens and shorts in the metal layers, various silicon defects that can result in incorrect operation of the transistors or gates, and problems due to physical handling and processing. Once packaged, the effects of any type of chip failure must be detectable during the test phase by using a set of input test patterns to drive the chip and by observing the state of the output signals. As designs become larger, it becomes increasingly more difficult to exercise all of the logic using a reasonable number of test patterns, and by simply observing the output pins. For this reason, the issues of circuit controllability and observability are the focus of design for testability methods.

Counters can easily illustrate some of the problems involved in testability design. In the simplest case, if a design contains a 24 bit binary counter, it would take 2^{24}, or over 16 million clocks (ie test patterns), to test a 300 gate circuit. If the rest of the design also acted in response to the states of the counter, additional tests would be needed to test for decodes and control logic. The generation of that many test patterns would be unacceptable in terms of the capability of physical test equipment and the duration of logic and fault simulation runs. In addition, if the outputs of the counter are not observable at the pins of the ASIC, then the effect of any minor problem in that counter's operation would have to be detected by observing some other output signal or signals. It is easy to see how the problem is compounded as designs become larger relative to the number of I/O pins. Without consideration of the testability problem, the ability to detect faults is greatly reduced, resulting in "tested" chips that in fact have undetected problems.

Testability techniques can be formal in nature, such as serial scan based approaches, or informal or "ad hoc" techniques in which additional logic is added to improve the circuit's controllability or observability as needed. For instance, a parallel data processing chip consisting of ALUs and combinational logic may be easily testable by generating precise input data patterns and observing the calculated results at the output. In the more typical case, a complex signal processor may need multiplexing logic added to the output pins to be able to observe internal circuit nodes more easily during test, or additional input pins added to be able to preset logic to a known state. More rigorous approaches use some form of scan logic to serially shift data into most or all of the flip flops in an ASIC, then clock the system and shift out the results to an output pin. While this technique requires even more logic to be added for test purposes, it has the advantage that it can provide complete fault detection, and the generation of the serial test patterns can be automated. For this reason, many synthesis and ASIC vendors support and can provide scan synthesis tools and automatic test pattern generation (ATPG).

Some ad hoc techniques for testability can be added in VHDL using multiplexers and other combinational logic, covered in Chapter 8. Others can use the ability to preset registers and counters to known states via input data busses, as illustrated earlier in this chapter. Or the counter can be split into smaller sections, with each section counted separately by means of a test control signal. For instance, if the 24 bit counter were broken up into three 8 bit sections, then the second group of eight would normally count only when the previous counter reached all 1s, or on every 256th clock. In test mode it could be made to advance on every clock edge, and would only require 256 clocks to count through its entire range. In VHDL, the enable logic for the counter would be

> If (First_Count = All_Ones) or (Test_Mode = '1') then
>
>> Count <= Count + '1';
> End if;

Although this type of testability logic is pretty simple, these approaches are commonly recommended by ASIC suppliers and can easily be added as needed in a design.

Scan logic can also be added using VHDL if a specialized scan synthesis tool is not available. The technique is to add a serial shift capability to all sequential logic so that data can be shifted into and out of the flip flops. Every register and counter would therefore have a separate test signal, plus a serial in and out line. For a parallel loaded 8 bit data register, the code would typically be

> S_out <= Reg (0);
>
> If Test_Mode = '1' then
>
>> Reg <= S_in & Reg (7 downto 1); -- A shift Register
>
> Else if Load = '1' then
>
>> Reg <= Databus;
> End if;

The counter would be handled in the same fashion, except that the condition for normal counting would be added.

In a full scan design, each of the separate serial paths would be connected together to form longer serial chains, with the input and final output connected to I/O pins. Therefore the register serial output Reg (0) could be the serial input to the counter :

If Test_Mode = '1' then

Count <= Reg (0) & Count (7 downto 1);

etc.

The main point regarding testability is that it can be directly designed in using the logic synthesizer if desired, and can provide the same capability as a conventional design approach.

Summary

This chapter covered techniques for synthesis of a number of basic counter types used in most electronic systems. Design approaches using simple processes, state machines, and dataflow styles were introduced, along with common terminology. The next chapter will build on these techniques in order to demonstrate how to design more complex functions such as pulse generators, waveform and timing generators, and memory and bus controllers.

Chapter 5

Sequential Counter Applications

This chapter will build on the previous counter synthesis techniques to develop procedures for designing the more common logic building blocks used in most systems. The functions covered here are more typical of those required in signal processing systems as compared to data processing systems, but the general VHDL synthesis techniques are common to both types of systems.

Logic functions designed in this chapter will include the following:

- Pulse or gate generators
    Fixed width, Variable width
    Retriggerable, not retriggerable
- Leading edge detector, trailing edge, both.

- Square wave generators
- Lookup Tables and ROMs
- Event counters - synchronous/asynch
- Error counters - up/down
- Time measurement functions

It may appear that there are many variations of some functions such as the pulse generators, but upon inspection you will find that there are basic concepts to be learned from each of the examples that can be applied to other design problems. In many cases a simple VHDL coding of a circuit may have to use a state machine approach because of a change to the input signals, or you may find that a dataflow technique may be simpler than a higher level style.

The Basic Pulse Generator

A common application required in many logic designs is a pulse generator that produces a waveform with a desired pulse width when a trigger signal occurs, as shown in Figure 5-1. This function is sometimes referred to as a one-shot from its early days as an analog circuit, but in its digital form can be designed to have either fixed or variable pulse width and can be made retriggerable during the pulse or not as desired.

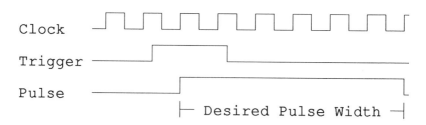

Figure 5-1 : Basic Pulse Generator Waveforms

As a first example, let's consider a simple non-retriggerable pulse generator whose trigger input is shorter in duration than the generated pulse output. Considerations of retriggering and handling of different trigger widths will be discussed in succeeding sections. Let's assume that the spec for this design will be to generate a 1.5 usec wide pulse synchronous to the clock whenever the input trigger goes high, using a 16 Mhz Clock. The technique will use a flip flop to enable and reset a counter, in which the flip flop is clocked high after the trigger and clocked low when the counter times out the desired time interval. The flip flop thus is the desired output of the pulse generator. Figure 5-2 shows the basic circuit function that will be synthesized.

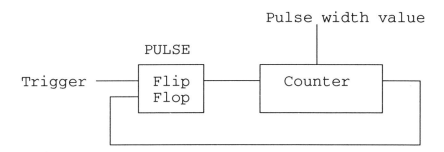

Figure 5-2 : Pulse Generator Block Diagram

The VHDL code uses signals TRIG, Clock, and Reset as inputs and PULSE as the output signal. In Example 5-1, note the use of a counter with an enable (the Std_Ulogic_Vector signal COUNT) as covered in Chapter 4. Since the clock is given as 16 MHz, the counter must be able to count for 24 clocks to yield 1.5 usec, and therefore the signal COUNT requires 5 bits. As in previous examples, we will let the counter count from 0 to 23 and when it equals 23 (ie, the constant named WIDTH) we will end the pulse output.

Architecture RTL of PGEN is

Signal CLOCK, RESET, TRIG, PULSE : Std_Ulogic;
Signal COUNT   : Std_Ulogic_Vector (4 downto 0);

```
Constant WIDTH   : Std_Ulogic_Vector := "10111"; -- (23)

Begin

PULSEGEN : Process (CLOCK, RESET)
Begin
    If RESET = '1' then
        PULSE    <= '0';
        COUNT    <= "00000";

    Elsif Rising_Edge (CLOCK) then

        If TRIG = '1' then              -- Set Pulse on
            PULSE <= '1';
        End if;

        If PULSE = '1' then
            If COUNT = WIDTH then   -- Done? Reset PULSE
                PULSE    <= '0';
            Else                        -- Otherwise count up
                COUNT   <= COUNT + '1';
            End if;
        Else
            COUNT <= "00000"; -- Clear counter when PULSE is off
        End if;
    End If;
End Process Pulsegen;
End RTL;
```

Example 5-1 : Fixed width pulse generator

In this coding, if the trigger is high then the Pulse gets set high, on the next clock edge. The Pulse signal itself then enables the counter to count until it reaches the value WIDTH, at which point Pulse is reset and the counter is cleared. This is a common technique that produces an efficient logic design and can be used in many applications.

Up to this point, all of the counters have been VHDL signals of type Std_Ulogic_Vector. It is also possible to code the counters as VHDL **Variables** instead of signals, where the counter variable is not an output of the process but is only used internally. It must be kept in mind that variables are different from signals in that they assume their new values as soon as they are assigned, in contrast to signals, which do not get updated until after the next clock edge. In a simple process such as the pulse generator, the variable only gets assigned a value once and is easy to follow. In more complex processes however, care must be taken if variables are updated at more than one place within the process and then are tested elsewhere. This obviously can lead to problems in debugging the code during simulation and should be handled carefully.

The other difference between signals and variables is that variables are defined and used only within a process, so that they cannot drive ports or other processes. As long as variables are used internal to the process, as in the example below, then there is no problem. If it is necessary to drive another process or port, then you will have to create a signal for that purpose and set the signal to the state of the variable. In general it is easier to just use signals in the first place.

Architecture RTL of PGEN is

```
    Signal CLOCK, RESET, TRIG, PULSE : Std_Ulogic;
    Constant WIDTH : Integer := 24;

Begin

PULSEGEN : Process (CLOCK, RESET)

    Variable COUNT : Integer Range 0 to 31; -- Five bits wide

    Begin

If RESET = '1' then
    PULSE      <= '0';
    COUNT      := 0;        -- Note syntax 0 for integer
```

```
    Elsif Rising_Edge (CLOCK) then

  If Trig = '1' then
     PULSE <= '1';
  End if;

  If PULSE = '1' then
     If COUNT = WIDTH then
       PULSE      <= '0';
     Else
       COUNT    := COUNT + 1;
     End if;
  Else
     COUNT := 0;
  End if;
  End If;
End Process Pulsegen;

End RTL;
```

Example 5-2 : Fixed width pulse generator using variables

In both of these examples a down counter could have been used and would have produced similar results - the choice is personal preference in this case. When a variable pulse width is required however, a down counter approach can be more efficient as explained in Chapter 4. Consider, for example, a system where a pulse must be generated whose width is determined by a value in a register. Let's name the register PW for Pulse Width and assume it is an 8 bit value, as is the counter.

```
Architecture RTL of VAR_PGEN is

Signal CLOCK, RESET, TRIG, PULSE : Std_Ulogic;
Signal COUNT, PW : Std_Ulogic_Vector (7 downto 0);
```

Constant ZERO : Std_Ulogic_Vector := "00000000";

Begin

PULSEGEN : Process (CLOCK, RESET)
   Begin

   If RESET = '1' then
     PULSE   <= '0';
     COUNT   <= ZERO;

   Elsif Rising_Edge (CLOCK) then

     If Trig = '1' then   -- Set Pulse on
       PULSE   <= '1';
     End if;

     If PULSE = '1' then
       If COUNT = ZERO then   -- Reset Pulse
         PULSE   <= '0';
       Else
         COUNT   <= COUNT - '1'; -- Count Down
       End if;
     Else
       COUNT <= PW;   -- Preset counter to Pulse Width
     End if;
   End If;
End Process Pulsegen;
End RTL;

Example 5-3 : Variable width pulse generator

There should be no surprises in this example since this is simply an application of the variable down counter designed in chapter 4.

Trigger Width Problems

At this point it is time to consider what would happen if the trigger signal were longer than the pulse duration. Inspection of the code of Examples 5-1 to 5-3 shows that if the trigger is still high (logic 1) when the pulse ends, it will turn on again after one clock time. This may be a good technique for designing signal generators that are gated on and off by an enable signal rather than the short trigger of Examples 5-1 to 5-3. But how do you handle the case where you only want to generate a single pulse when the trigger goes high, as shown in Figure 5-3?

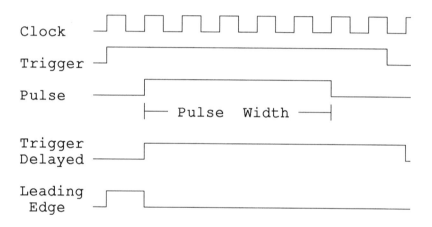

Figure 5-3 : Pulse Generator With Long Trigger

There are several possible implementations of this function. In terms of VHDL techniques, two will be considered, one based on logic design experience and a second based on treating the circuit as a state machine. Discussion of the state machine version will be deferred until Chapter 6.

An experienced logic designer can deduce from the waveforms of Figure 5-3 that if the leading edge of the trigger signal is detected, then that signal could be used to trigger the simple pulse generators shown earlier. This approach is illustrated in Figure 5-4 below.

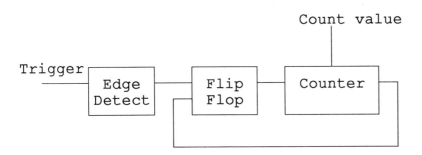

Figure 5-4 : Edge Detecting the Pulse Trigger

A Leading Edge Detector

A leading edge detector circuit, whose output is shown in Figure 5-5, can be simply designed using a flip flop and an AND gate when the trigger is synchronous to the clock, or two flip flops and an AND gate when the trigger is asynchronous, as shown in the schematic of Figure 5-6 :

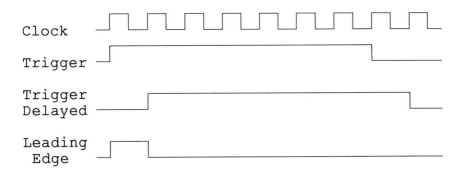

Figure 5-5 : Leading Edge Detector Waveforms

For this example of the pulse generator, let's assume that the trigger is synchronous to the clock. We will add the leading edge detector to the VHDL code of Example 5-1 using a dataflow style of coding as follows:

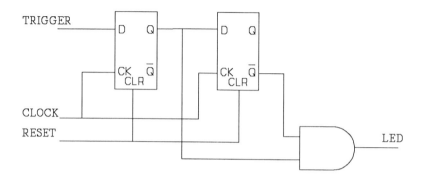

Figure 5-6 : A Leading Edge Detector

After the Rising_Edge (Clock) code, generate a signal Trig_Delayed which will follow the Trigger signal but delayed by one clock period, as shown in Figure 5-6:

    Trig_Delayed <= Trig;

    Leading_Edge_Pulse <= Trig AND not (Trig_Delayed);

or
    If Trig = '1' AND Trig_Delayed = '0' then
       Leading_Edge <= '1';
    Else
       Leading_Edge <= '0';
    End if;

In the example code below, notice that instead of generating a **separate** leading edge pulse, the same conditions will be simply used in an IF statement to set the PULSE signal high.

Architecture RTL of PGEN is

Signal CLOCK, RESET, TRIG, Trig_Delayed, PULSE : Std_Ulogic;
Signal COUNT   : Std_Ulogic_Vector (4 downto 0);
Constant WIDTH : Std_Ulogic_Vector := "10111"; -- (23)

Begin

```
PULSEGEN : Process (CLOCK, RESET)
    Begin
    If RESET = '1' then
       PULSE    <= '0';
       COUNT    <= "00000";

    Elsif Rising_Edge (CLOCK) then

          Trig_Delayed <= TRIG;

       If TRIG = '1' and Trig_Delayed = '0' then -- "Leading Edge"
          PULSE <= '1';
       End if;

       If PULSE = '1' then
          If COUNT = WIDTH then -- Done? Reset PULSE
             PULSE <= '0';
          Else                              -- Otherwise count up
             COUNT <= COUNT + '1';
          End if;
       Else
          COUNT <= "00000";  -- Clear counter when PULSE is off
       End if;
    End If;
End Process Pulsegen;
End RTL;
```

Example 5-4 : Pulse Generator Independent of Trigger Width

This technique is very simple and uses minimal hardware for the leading edge detector. Using this approach also makes it fairly simple to add a **retriggering** feature to the circuit.

Retriggering the Pulse Generator

Some applications may require that the pulse generator be retriggerable such that if the trigger goes low then back high during the pulse, the pulse will be restarted from that point as shown in Figure 5-7.

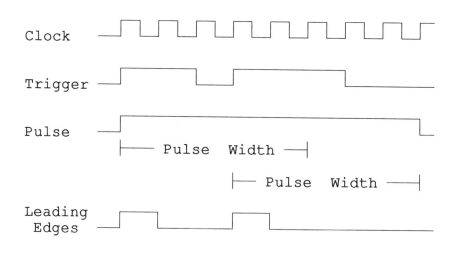

Figure 5-7 : Retriggering the Pulse Generator

As shown in Figure 5-7, the pulse width is retimed from the occurrence of the second trigger (or from any subsequent triggers during the pulse). As a logic design requirement, by using the leading edge detector we can see that we simply have to reset the counter with the leading edge pulse and otherwise keep the design the same as in Example 5-4. In Chapter 6 however we will see that describing this fairly simple retriggering requirement as a state machine gets surprisingly complicated and again points out how design experience should be used effectively in the synthesis process.

The code below will use the leading edge conditions to both start the Pulse signal as well as reset the counter :

~~~~~~

```
Architecture RTL of PGEN is

Signal CLOCK, RESET, TRIG, Trig_Delayed, PULSE : Std_Ulogic;
Signal COUNT    : Std_Ulogic_Vector (4 downto 0);
Constant WIDTH : Std_Ulogic_Vector := "10111"; -- (23)

Begin

PULSEGEN : Process (CLOCK, RESET)
  Begin

    If RESET = '1' then
        PULSE    <= '0';
        COUNT    <= "00000";

    Elsif Rising_Edge (CLOCK) then

        Trig_Delayed <= TRIG;

      If TRIG = '1' and Trig_Delayed = '0' then -- "Leading Edge"

        PULSE      <= '1';            -- Start the Pulse
        COUNT      <= "00000";        -- and reset the counter
      End if;

    If PULSE = '1' then
        If TRIG = '1' and Trig_Delayed = '0' then

          COUNT <= "00000";  -- Retrigger on each leading edge

        Elsif COUNT = WIDTH then    -- Done? Reset PULSE
          PULSE    <= '0';
        Else                        -- Otherwise count up
```

```
        COUNT    <= COUNT + '1';
      End if;
   End if;


   End If;
End Process Pulsegen;
End RTL;
```

Example 5-5 : Code for the Retriggerable Pulse Generator

As expected, this circuit only requires a few extra gates to add the retriggering capability using the leading edge detection logic.

Before leaving the subject of edge detectors, the implementation of a trailing edge detector and both edge detector is straightforward and is given here for completeness, with the waveforms shown in Figure 5-8 :

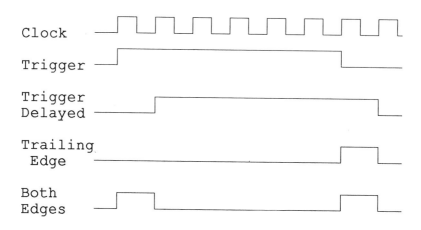

Figure 5-8 : Trailing Edge and Both Edge Detection

Trailing Edge Detector :

Trailing_edge <= Not (Trigger) and Trig_Delayed;

Both Edge Detector :

Both_Edges <= Trigger EX-OR Trig_Delayed;

Waveform Generators

Many applications require signals of various frequencies to be generated from a system clock, either as outputs or to be used internal to the system. Sometimes these waveforms can be outputs from binary counters or other counters as shown in chapter 4, but often the desired signal characteristics require a more complex circuit design.

Let's begin with a design requirement to generate a repetitive waveform with a period of 25 usec that is to be low for 7.7 usec and high for 17.3 usec, using a 10 Mhz clock, as shown in Figure 5-9.

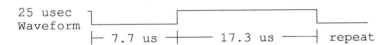

Figure 5-9 : An Asymmetrical Repetitive Waveform

A logic designer would realize from this requirement that some form of counter should be used to count for 7.7 usec, then count for 17.3 usec, and toggle a flip flop at the end of each count interval. Alternatively the counter could count up to 25 usec repetitively and decoders could be used to detect 7.7 usec and 25 usec. In either case, with a 10 Mhz clock, an 8 bit counter would be required for the longest count interval of 173 (17.3 usec at 10 MHz) or 250, and the design approach chosen would be made based on other factors such as the ease of decoding vs. resetting the counter twice. For this example, let's choose the second alternative of an

8 bit binary counter counting from 0 to 249 (ie, 250 clocks), with the waveform flip flop to be set after a count of 76 and reset after 249 by decoding logic. Naturally the decode of 249 will also be used to reset the counter synchronously back to zero. Notice that these design choices must be made based on logic design techniques and are not driven by any particular VHDL requirements.

For the code of Example 5-6, we will use the signals CLOCK, RESET, and COUNT and the output waveform WAVE. Let's make COUNT an integer for ease of specifying the decoding values and define its Range to be 0 to 255, or 8 bits. Without a defined range limit, a synthesis tool may default to 32 bits and thus create undesirable additional logic. You should therefore always control the synthesizer by specifying the range of integer signals and variables.

```
Architecture RTL of Sqwave is

Signal CLOCK, RESET, WAVE : Std_Ulogic;
Signal     COUNT     : Integer Range 0 to 255;
Constant   SetWave    : Integer := 76;
Constant   ResetWave : Integer := 249;
Constant   Period     : Integer := 249; -- For clarity of code

Begin

WAVEGEN : Process (CLOCK, RESET)
   Begin

      If RESET = '1' then
         WAVE      <= '0';
         COUNT     <= 0;

      Elsif Rising_Edge (CLOCK) then

         If COUNT = Period then        -- Counter controls
            COUNT <= 0;
         Else
```

```
        COUNT <= Count + 1;
     End if;

     If COUNT = SetWave then        -- WAVE controls
        WAVE   <= '1';
     Elsif COUNT = ResetWave then
        WAVE   <= '0';
     End if;
   End If;
End Process WAVEGEN;

End RTL;
```

Example 5-6 : 25 usec Waveform Generator

In this example, the decoding logic has been simply defined as

```
If COUNT = value then ...
```

which is very clear to understand and review, and if the values are defined as VHDL constants then any design changes to the parameters can be made easily. By extension of this example if additional output waveforms are required with the same period but different duty cycles or phases, additional signals and decode values can be simply added to the WAVEGEN process code and resynthesized.

ROM and Lookup Table Applications

There are many applications where it is convenient to store data patterns in a ROM and then use a counter or similar address scheme to access the patterns. Depending on how the data is arranged in the ROM or lookup table, the table can function as an encoder, a decoder, a data converter from one format to another, a list of parameters or numerical constants, microcode instructions, or simply timing waveform patterns. Compared with using a true ROM macrocell or discrete device, which has a fixed number of words and bits, a synthesized table will normally be minimized depending on the actual data contained in the table. For example, if one bit in a 256 word ROM table is mostly 0s with only a few 1s, the logic would be reduced to a few gates to decode those states. For large ROMs, it will usually be more efficient to use a ROM cell provided by the ASIC supplier, but there are many small applications where synthesis can be used effectively.

ROMs and lookup tables are described in VHDL by means of Arrays. Although you can address the table by various means, let's consider a case where a counter is used to cycle through each address of the table. For the first example, the table will contain 32 8-bit words defined as integers. The array definition would be, as shown earlier in Chapter 3 :

a. A constrained array :

Type Rom_Table is Array (0 to 31) of Integer Range 0 to 255;

Constant Lookup : Rom_Table :=
(1, 3, 5, 7, 9, 11, 13, 15, 17, 19, 21, 23, 25, 27, 29, 31,
 255, 127, 96, 64, 48, 32, 24, 16, 14, 12, 10, 8, 7, 6, 5, 0);

This array is defined to have 32 values that will be in addresses 0 to 31, with each address containing an 8 bit value as given in the Constant statement. The name of the table is Lookup, whose type is Rom_Table as defined above, and the data format is each value separated by a comma, and parentheses around the entire data section. There can be multiple tables of type Rom_Table contained within a design, each of which would have a unique name such as Lookup, ASCII_Table, COST, etc.

To use the lookup table in a design, first define the address counter ADDR and a data register (if needed) to store the data :

```
Signal ADDR        : Integer Range 0 to 31;
Signal Datareg     : Integer Range 0 to 255;

Begin
Int_Table : Process (Clock, Reset)
Begin

If Reset = '1' then
    ADDR      <= 0;
    Datareg   <= 0;

Elsif Rising_Edge (Clock) then

    Datareg <= Lookup(ADDR);

    ADDR <= ADDR + 1;

End if;
End Process Int_Table;
```

Example 5-7 : A Counter Driven Lookup Table

The synthesized table can be thought of as encoding logic that uses the address inputs to generate the output bit pattern given in the table, and usually terms can be combined to minimize the logic.

b. An Unconstrained Array (see Chapter 3) :

```
Type Rom_Table2 is Array (Natural range <>) of
                        Integer Range 0 to 255;

Constant Lookup2 : Rom_Table2 :=
(1, 3, 5, 7, 9, 11, 13, 15, 17, 19, 21, 23, 25, 27, 29, 31,
```

255, 127, 96, 64, 48, 32, 24, 0);

The unconstrained array removes the requirement to specify how many entries are in the table, which makes it easier to manage adding or deleting entries in the table. On the other hand, you lose the checking that the VHDL compiler does to ensure that if you specified 32 entries, that you do have 32 entries in the table. The choice depends on the application and on your preference for data checking versus easy table expansion. The use of either type of array in the code of Example 5-7 will produce the same results.

At times, it is more meaningful to express the data in a different format, such as Hexadecimal, or binary. The various ways of expressing VHDL literals makes this straightforward. Consider a table of Hex values, where the address and data are Std_Ulogic_Vectors :

```
Signal Addr        : Std_Ulogic_Vector (4 downto 0);
Signal DataReg     : Std_Ulogic_Vector (7 downto 0);


Type Rom_Table3 is Array (Natural range <>) of
                         Std_Ulogic_Vector (7 downto 0);

  Constant Hex_Val : Rom_Table3 :=
  (X"3C", X"25", ... etc ... ,X"4D");
```

The address to the array is an Integer, and yet we have the signal Addr as a Std_Ulogic_Vector. The conversion to integer can be handled easily by a Type Conversion provided by the VHDL tool supplier, or by means of a conversion as shown below. Other type conversions are discussed in Chapters 3 and 9.

This function accepts a Std_Ulogic_Vector INPUT whose length is determined by using the VHDL predefined attributes 'LOW and 'HIGH. For a vector of (7 downto 0), the 'LOW value is 0 and 'HIGH is 7. This allows you to make a general purpose conversion function that can be used for any size vector :

```
Function To_Int (INPUT : Std_Ulogic_Vector) return Integer is
      Variable RESULT : integer := 0;
Begin
      For i in INPUT'high downto INPUT'low loop

      RESULT := RESULT * 2;
         If (INPUT (i) = '1') then
              RESULT := RESULT + 1;
         End if;
      End loop;
   Return RESULT;
End To_Int;
```

Note that these conversion functions are often placed in a library package so they can be used in general by many users.

The code for the Hexadecimal data table would now be similar to that of Example 5-7, except for the use of the Type Conversion function. The process would be modified as follows:

```
Signal Addr         : Std_Ulogic_Vector (4 downto 0);
Signal DataReg      : Std_Ulogic_Vector (7 downto 0);

Type Rom_Table3 is Array (Natural range <>) of
                          Std_Ulogic_Vector (7 downto 0);

Constant Hex_Val : Rom_Table3 :=
      (X"3C", X"25", ... etc ... ,X"4D");

Begin
Hex_Table : Process (Clock, Reset)
Begin
```

```
If Reset = '1' then
    ADDR         <= "00000";
    Datareg      <= "00000000";

Elsif Rising_Edge (Clock) then

    Datareg <= Hex_Val (To_Int (ADDR));
    ADDR <= ADDR + '1';
End if;
End Process Hex_Table;
```

Example 5-8 : A Hexadecimal Table Lookup

In this example, the assumption is made that the number of entries in the table and the largest value of ADDR agree so that you won't address non-existent locations. This may not be the case however, since ADDR is a 5 bit vector and the Hex_Val table may be shorter than 32 words. The VHDL 'HIGH attribute can again be used here to detect when the ADDR counter reaches the maximum length of the table. The highest value in the table Hex_Val is Hex_Val'HIGH, an integer. In order to do the comparison below, your synthesis library package will need to have comparison functions (eg, $>$, $<$, $=$, etc) for vectors and integers. Otherwise simply make the address ADDR an integer.

In order to detect the largest address in the table, instead of

```
ADDR <= ADDR + '1';
```

substitute the code

```
If ADDR < Hex_Val'HIGH then
    ADDR <= ADDR + '1';
Else ADDR <= '00000';
End If;
```

This tests whether the address counter is within the highest value of table Hex_Val and resets it back to 0, on the next clock edge.

The design applications that use tables of values are extensive, and for reasonable size tables the synthesis techniques can be very effective. The tradeoff of whether to use a ROM megacell over a synthesized ROM should be examined on a case by case basis and depends on considerations of the size of the ROM and its area, and on the characteristics of the data patterns used.

Other Counting Applications

Many applications involve counting the number of events that have occurred, where the events may be signals received from an antenna, or parts moving down a conveyor, or the number of scan lines on a CRT monitor. Assuming that this function will be designed using a synchronous system clock, it can be easily designed as a counter with an enable input. The enable would be a one clock period wide pulse occurring for each event, so that the counter advances on each enable. An edge detector technique can be used to make the enable signal, and the VHDL description would be simply

```
If enable = '1' then Count <= Count + '1';
End if;
```

Rather than continuously counting events, you may want to detect that a threshold value was exceeded, and you may want to stop counting at a certain maximum value. Both of these types of variations can be coded as :

```
Signal     Count      : Integer Range 0 to 255;
Signal     Detect     : Std_Ulogic;
Constant   Threshold  : Integer := 27;
Constant   MaxValue   : Integer := 160;
```

Then, in a clocked process :

```
If Enable = '1' then
      If Count < MaxValue then   -- Only count if less than max.
```

```
            Count <= Count + '1';
      End if;
   End if;

   If Count = Threshold then
        Detect <= '1';  -- Set a Detect Flip Flop
      End if;
   End Process;
```

This is a simple technique for counting to a value and holding that value until a reset occurs, or until a computer or other circuitry reads the count value. You would use a similar approach if the counter was used as a memory address counter and you wanted to store data into addresses 0 to 1000 and then stop.

Summary

This chapter has covered some of the practical issues that must be dealt with in designing and synthesizing counting and signal processing functions. It should have also demonstrated that logic design experience is essential to planning a good design approach, and that VHDL synthesis is more than just a software process. Chapter 6 continues with a discussion of the design of control logic functions and state machines.

Chapter 6

Control Logic and State Machines

Nearly all systems require some form of control logic to direct the flow of operations. Designing this type of logic, which is also referred to as a sequencer or a finite state machine, uses a different type of design approach from that previously discussed. This chapter will discuss the basic types of control logic, the analysis methods used for their design, and the practical VHDL techniques for their synthesis. The commonly used VHDL syntax used for designing control logic, such as enumerated types and case statements, will also be elaborated here.

Basics of Control Logic Design

The operation of the basic logic functions in a system is managed by various types of control logic. Previous chapters have covered counters, signal generators, and registers as separate building blocks, but in a complete system it is the control logic that usually directs when the counters should count and when the registers should be loaded. The term control logic can be applied equally to both the central control of several system functions or to the control of a simple logic function within a larger block. The fundamental idea of control logic is that it is sequential in nature and its operation depends on both its inputs as well as its internal state.

The control function is performed by a synchronous form of logic named a "sequencer". Sequencers allow actions to be performed based on the current state of the system and the values of certain inputs to the sequencer. This combination is completely deterministic, since for each combination of inputs and current state, there is one and only one specified set of responses.

A **state machine** is a form of sequencer that is organized as a finite set of states. Each state represents one set of actions, such as enabling a counter to increment or generating an acknowledge output. Almost all states also contain a method of transferring control to another state based on certain conditions. Any state that does not have a means of going to another state would have to be the last state of a state machine, and the system would remain in this state forever (or until the sequencer is asynchronously reset). Transferring between states can be conditional (based on the values of other signals in the system), or non-conditional.

The design and analysis of sequential control logic often uses graphical means of showing the control flow and operations performed. **Flowcharts** and **State Diagrams** are two common tools that can be used to represent this process. A flowchart is the type of diagram used to emulate the execution of a process, in which conditions and actions are combined to solve a logical problem. The flowchart is not limited to logic design applications but can be used to analyze the flow of any generalized process. The logic designer's equivalent of the flowchart is called a state diagram, which shows a box or circle for each "state" of the process or sequencer. Within the diagram, each state describes both the actions to be performed during that state and the branching condition to allow progression to another state. The assignment of the next state can be conditional (based on the current state and the conditions of the inputs), or non-conditional (based only on the current state). A "reset" oval is also used to show an initialization state which, in hardware, is usually implemented as an asynchronous reset input. During simulations, initialization sets all unknown outputs and signals to known values. In a complete system design, initialization allows the simulation of the circuit to be completely deterministic throughout all levels of representation.

Sequencers and state machines are used in a wide variety of applications. In general they can be grouped into two classes, one in which a sequence of operations is performed unconditionally, and one in which the operation flow is dependent on input conditions. Figures 6-1 and 6-2 show these groups represented in flow chart form and as an equivalent state diagram.

An Unconditional Sequencer

The simplest type of sequencer is one that steps through its operations non-conditionally and does no testing of conditions that would alter its flow, as represented in the leftmost charts in Figures 6-1 and 6-2. The most common logic application is a timing sequencer or controller that generates a fixed series of signals that perform some system action. As an example, let's consider a simple bus controller for a memory system and assume that the data and address for the memory come from several sources in the system. In order to control writing into the memory, the controller must disable the input busses, then enable the bus from the selected source, write the data into the memory, and then release the bus for the next operation. The desired signal waveforms, shown in Figure 6-3, are spaced in time to avoid bus contention and write timing problems, and the entire sequence takes 6 clock periods.

For a simple controller as this, there are many ways to generate the desired signals. For instance, you could simply use a binary, Johnson, or

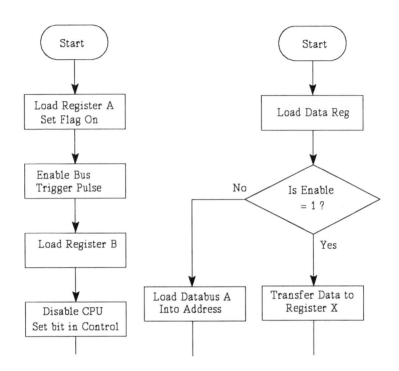

Figure 6-1 : Sequencers with and without Branching

Gray counter and decode the signals from the appropriate states of the counter. However, let's look at this problem as a sequencer design, in which we will code the desired action in VHDL and allow the synthesizer to select the best circuit implementation.

Since sequencers and state machines are clocked functions, they will be generated in clocked processes, following the rules from Chapter 3. Each state of the machine will perform some specific action, in this case generation of the memory controller signals, and then proceed to the next state. A flow chart would show a straight line sequence of operations similar to Figure 6-1 above, except with different actions.

The sequencer has six states, as shown in Figure 6-3, which can either be "explicitly" defined as 0 through 5, or "implicitly" defined by a VHDL Type statement as states S0 through S5. For this example, we'll assume that we want the sequence to be binary 0 to 5 and therefore will explicitly

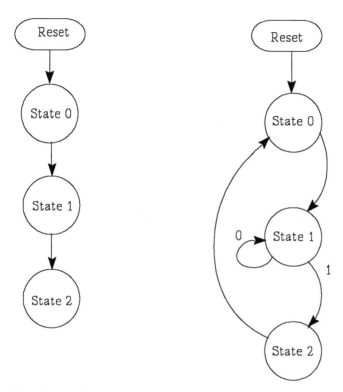

Figure 6-2 : State Diagrams of Two Types of Sequencers

use those states.

The basic idea below in Example 6-1 is that a signal State is defined, and, using a Case statement, every value of State is tested for the action to be performed, and then the next state is explicitly given.

Signal Clock, Reset : Std_Ulogic;
Signal Trigger, Bus_en, Data_en, Wr_data : Std_Ulogic;
Signal State : Integer Range 0 to 5 ;

Begin

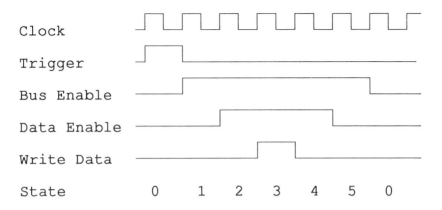

Figure 6-3 : Waveforms for Memory Bus Controller

Bus_Cont : Process (Clock, Reset)
    Begin
        If Reset = '1' then
            State   <= 0;
            Bus_en   <= '0';
            Data_en   <= '0';
            Wr_data   <= '0';

        Elsif Rising_Edge (Clock) then

            Case State is   -- What State are we in?
                When 0 =>
                    Bus_en   <= '1' after 5 ns;
                    State   <= 1;
                When 1 =>
                    Data_en   <= '1' after 5 ns;
                    State   <= 2;
                When 2 =>
                    Wr_data   <= '1' after 5 ns;
                    State   <= 3;
                  When 3 =>

```
                    Wr_data      <= '0' after 5 ns;
                    State        <= 4;
              When 4 =>
                    Data_en      <= '0' after 5 ns;
                    State        <= 5;
              When 5 =>
                    Bus_en       <= '0' after 5 ns;
                    State        <= 0;
          End Case;
       End if;
End Process;
```

Example 6-1 : Code for a Sequencer without Branching

This is a pretty simple case of a non-branching sequencer that essentially runs continuously after being reset. The basic technique however is that you explicitly control every state and what happens in that state. There can be multiple actions in each state, and the states need not be in order as they are here.

Some synthesis tools allow a simpler technique to be used for this type of sequencer, making use of a "multiple clock" process where the states are not explicitly defined.

The Multiple Clocked Process

The major difference from the standard clocked process syntax is that the process has no sensitivity list and instead makes use of multiple waits for clock edges. A loop is used for the process and tests are made at every clock for the presence of the reset signal. If the reset occurs, the Next statement returns to the top of the loop and resets the logic. Using this approach, the code for the sequencer would become :

Signal Clock, Reset : Std_Ulogic;
Signal Trigger, Bus_en, Data_en, Wr_data : Std_Ulogic;
Signal State : Integer Range 0 to 5 ;

Begin

Bus_Cont : Process
Begin

Reset_loop : Loop

    If Reset = '1' then
        State   <= 0;
        Bus_en   <= '0';
        Data_en   <= '0';
        Wr_data   <= '0';
    End if;

Wait until Rising_Edge (Clock);
   Next Reset_loop when Reset = '1';
       Bus_en   <= '1' after 5 ns;

Wait until Rising_Edge (Clock); Next Reset_loop when Reset ='1';
       Data_en   <= '1' after 5 ns;

Wait until Rising_Edge (Clock); Next Reset_loop when Reset = '1';
       Wr_data   <= '1' after 5 ns;

Wait until Rising_Edge (Clock); Next Reset_loop when Reset = '1';
       Wr_data   <= '0' after 5 ns;

Wait until Rising_Edge (Clock); Next Reset_loop when Reset = '1';
       Data_en   <= '0' after 5 ns;

Wait until Rising_Edge (Clock); Next Reset_loop when Reset = '1';
       Bus_en   <= '0' after 5 ns;

Wait until Rising_Edge (Clock); Next Reset_loop when Reset = '1';

    End Loop Reset_loop;
End Process Bus_cont;

▬▬

Example 6-2 : The Sequencer Using a Multiple Clocked Process

Note that the sequencer has now become a repetitive series of Wait Until clock and Next statements and actions, but that no state machine has been identified. This allows the synthesizer to pick the implementation of the sequencer based on a number of algorithms and use the minimum circuit approach. This code creates a synchronous reset since the reset test follows the clock edge. A variation available on at least one synthesis system allows for an asynchronous reset by using :

    Wait until Rising_Edge (Clock) or Reset = '1';
      Next Reset_loop when Reset = '1';

The multiple clock capability is not available with all synthesis tools, but provides a convenient means of implementing state machines where you choose not to specify it yourself. This feature is likely to be added to other systems in the future.

The Leading Edge Detector

Before showing examples of branching sequencers and state machines, let's illustrate the use of flow charts and state diagrams with a simple example that was used in Chapter 5, the Leading Edge Detector, shown in Figure 6-4. The desired output is to generate a one clock wide pulse just after the leading edge of the Trigger input. This function can be thought of as having two separate states, the first in which it waits for the Trigger to go high and then sets the Leading Edge (LED) high, and the second in which it sets the Leading Edge low and then waits for the Trigger to go low.

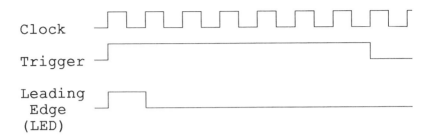

Clock

Trigger

Leading
  Edge
(LED)

Figure 6-4 : Waveforms for Leading Edge Detector

The operation of this two state function can be represented by a flowchart as shown in Figure 6-5. Note the two conditional test nodes that control the flow of the function and the two blocks that indicate action. Implicit in each of the blocks is a separate clock edge, since this is a synchronous design, and therefore the LED goes high on one clock and returns low on the next clock.

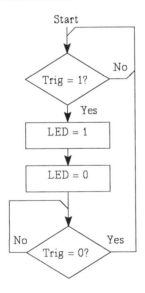

Figure 6-5 : A Flow Chart for the Leading Edge Detector

Many designers prefer to use state diagrams to describe functions, which show the operation in a different form. In this case, the designer explicitly identifies the states of the logic, the conditions that cause a change of state, and the outputs at each state. A state diagram of the Leading Edge Detector is shown in Figure 6-6.

TRIG Input / LED Output :

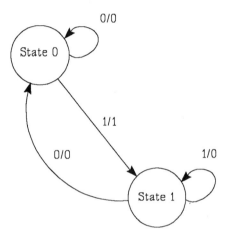

Figure 6-6 : State Diagram for Leading Edge Detector

In this case, the two states are identified with names as State 0 and State 1, and the operation is assumed to be synchronous with the clock. When in State 0, if the Trigger stays low, then the LED stays low and the state does not change. When Trigger goes high, the LED goes high and a transition is made to State 1. The actions that take place in State 1 are similarly shown.

Working from the flowchart of the leading edge detector, it will be an easy step to convert each of the flow chart steps into its VHDL equivalent below. A **sequencer is a clocked process** and thus follows the normal rules from Chapter 3 of having the clock and reset in the process's sensitivity list, and separate code sections for the reset and clock edge conditions. Second, a **Case statement** will be used to test for the state of the machine and to direct the actions. In this example there are only two states and the testing could be done with simple IF statements, but in larger designs the Case syntax is the easiest and most straightforward technique. Also the signal State has been "**explicitly**" given Std_Ulogic states '0' and '1' in Example 6-3, while in Example 6-4 it will be assigned to enumerated states S0 and S1.

Entity LEDDET is
        Port (Clock   :   In   Std_Ulogic;
                   Reset   :   In   Std_Ulogic;
                   Trigger   :   In   Std_ULogic;
                   LED   :   Out   Std_Ulogic);
End LEDDET;

Architecture RTL of LEDDET is
    Signal State : Std_Ulogic;

Begin

LED_EX : Process (Clock, Reset)
Begin

  If Reset = '1' then
       Led   <= '0';
       State   <= '0';

  Elsif Rising_Edge (Clock) then

    Case State is   -- What State are we in?

      When '0' =>
        If Trigger = '1' then
           Led   <= '1' after 5 ns;
           State   <= '1';
        End if;

```
        When '1' =>
             Led <= '0' after 5 ns;
             If Trigger = '0' then state <= '0';
             End if;

        When others => null;
   End Case;
  End If;
End Process LED_EX;
End RTL;
```

Example 6-3 : VHDL Code for Leading Edge State Machine

The LED signal has been given delays of 5 nsec for use during simulation only and this delay is ignored by the synthesizer.

In a more complex sequencer, explicitly assigning the states may restrict the synthesizer and may not produce the optimum design. A more flexible approach is to define the states "implicitly" by using an enumerated Type as :

   Type Seq_State is (S0, S1);

In other words, the new type Seq_State has only two values, named S0 and S1 in this case. We can then define **signal** State to use that type, as in :

   Signal State : Seq_State;

The resulting Architecture is shown in Example 6-4. Note that the signal State now tests for S0 and S1, which the synthesizer can assign to specific logic values for the most efficient logic.

▒▒▒

```
Architecture RTL of LEDDET is

    Type Seq_State is (S0, S1);

    Signal State : Seq_State;

Begin

LED_EX : Process (Clock,Reset)
  Begin

        If Reset = '1' then
            Led      <= '0';
            State    <= S0;   -- Forces asynchronous reset

        Elsif Rising_Edge (Clock) then

            Case State is    -- What State are we in?

                When S0 =>
                    If Trigger = '1' then
                        Led     <= '1' after 5 ns;
                        State   <= S1;
                    End If;
                When S1 =>
                    Led <= '0' after 4 ns;
                    If Trigger = '0' then
                        State   <= S0;
                    End If;
            End Case;
        End If;
End Process LED_EX;
End RTL;
```

▒▒▒

Example 6-4 : Implicit States in Leading Edge State Machine

In general, conditional branching is implemented in VHDL via IF and CASE constructs, while non-conditional branching is performed with the assignment statement (such as State <= S0).

The Retriggerable Pulse Generator

The next example will create a state-machine to generate a retriggerable pulse generator, whose desired operation is shown in Figure 6-7. This function was designed in Chapter 5 using a leading edge detector and pulse generator logic, which required some degree of design knowledge. The basic problem can also be approached as a state machine whose inputs are the clock and trigger, internal signals that decode the end of the pulse, and the counter and pulse output. Assume that if a second trigger occurs at the same time as End would have been, retrigger the pulse (ie, give retriggering priority).

A conditional flip-flop and counter circuit was designed in Chapter 5 to generate a retriggerable leading edge detector. The equivalent state machine version requires separate states to detect transitions on the trigger input and separate states to reset the counter and to transition the output pulse. The state diagram for the design is shown in Figure 6-8 and shows that 5 separate states have been defined. The logic synthesizer will therefore require at least 3 flip flops to decode the 5 required states.

Referring to the state diagram, the reset condition places the state machine into the idle state. At this point, the counter is cleared, and the output pulse is reset to inactive (='0'). The trig input is monitored. No activity occurs until trig becomes active (='1'). When trig='1', the sequencer advances to the gen_pulse_a state. This sets the pulse output active, and allows the counter to advance. This state is waiting for one of two events: a change in the value of the trig input, or an indication that the counter value (count) has reached the predefined pulse width value (width). Until either of these events occurs, the sequencer will remain in gen_pulse_a state, and the counter is incremented. This is indicated in the state diagram by an arrow both starting and ending at the gen_pulse_a state.

Case A. (Short trigger-width)

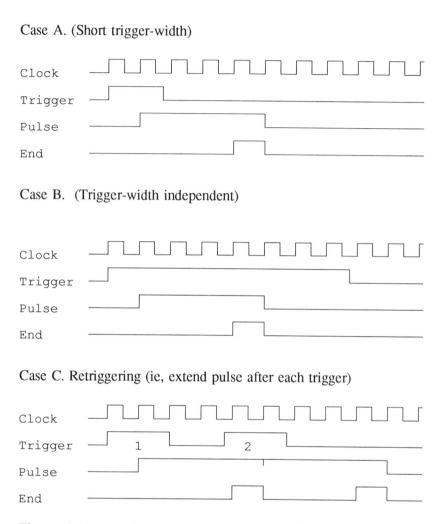

Case B. (Trigger-width independent)

Case C. Retriggering (ie, extend pulse after each trigger)

Figure 6-7 : Waveforms for a Retriggerable Pulse Generator

First, let's see what happens when count equals width. The state machine will advance to the end_pulse state. At this point, the output pulse is reset to inactive. The sequencer remains in the end_pulse state until trig goes inactive (='0'). This guarantees that the sequencer will not retrigger due to a trigger width longer than the predetermined pulse width. When trig finally goes to '0', the sequencer returns to the idle state and waits to be triggered.

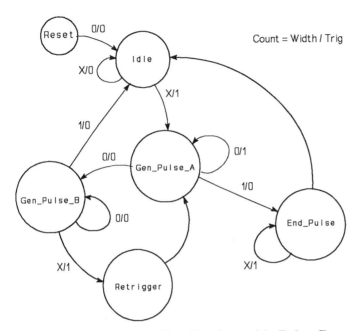

Figure 6-8 : State Diagram For Retriggerable Pulse Generator

From the gen_pulse_a state, if count < width and the trig input goes inactive (='0'), then the sequencer advances to the gen_pulse_b state. During this state, the state machine is still incrementing the counter value, and is waiting for one of two events: a retriggering of the trig input, or an indication that the counter value has reached the predetermined pulse width value. If no retriggering occurs within the pulse width window, count will increment until the count value equals width. At this point, since the trig input is already inactive, the state machine advances directly to the idle state. The output pulse is reset to inactive, and the sequencer waits for a leading edge on the trig input.

But what if the trig input retriggers before the count value equals width? From state gen_pulse_b, the sequencer will advance to the retrig state, which clears the counter and returns the state machine directly to the gen_state_a state. The sequencer again waits for activity on the trig input or a count equals width indication.

The VHDL code for the state machine version of the retriggerable pulse generator is shown in Example 6-5. The entity is named PULSEGEN, containing three inputs CLK, RESET, and TRIG, and one output named PULSE. The TRIG input is used to initiate an event on the output PULSE. Within the architecture, a type is defined to contain all of the required states. This type, PULSEGEN_STATE_TYPE, is an enumerated type containing only the values IDLE, GEN_PULSE_A, GEN_PULSE_B, END_PULSE, and RETRIGGER, which are the states that have been defined in the state diagram of Figure 6-8. Two additional signals are defined to contain the values of the states throughout the execution of the design and are named NEXT_STATE and CURRENT_STATE, both declared as type PULSEGEN_STATE_TYPE.

```
-- Retriggerable Pulse Width Generator

library IEEE;
 use IEEE.std_logic_1164.all;

Entity PULSEGEN is
   Port ( CLK      : In      Std_Ulogic;
          RESET    : In      Std_Ulogic;
          TRIG     : In      Std_Ulogic;
          PULSE    : Out     Std_Ulogic);
End PULSEGEN;

Architecture STATE_MACHINE of PULSEGEN is

-- Define type

Type PULSEGEN_STATE_TYPE is (Idle, Gen_Pulse_A,
    Gen_Pulse_B, End_Pulse, Retrigger);

Signal CURRENT_STATE, Next_State :
       PULSEGEN_STATE_TYPE;

Signal    COUNT   : integer range 0 to 31;
Constant  WIDTH   : integer range 0 to 31 := 5;
```

Begin

State_Mach_Proc : Process (CURRENT_STATE, TRIG, COUNT)

```
    Begin
                    -- State Machine
        Case CURRENT_STATE is

            when Idle           =>  If TRIG='1' then
                                        Next_State <= Gen_Pulse_A;
                                    End If;
            when Gen_Pulse_A    =>  If COUNT = WIDTH then
                                        Next_State <= End_Pulse;
                                    Elsif TRIG= '0' then
                                        Next_State <= Gen_Pulse_B;
                                    End If;
            when End_Pulse      =>  If TRIG= '0' then
                                        Next_State <= Idle;
                                    End If;
            when Gen_Pulse_B    =>  If TRIG= '1' then
                                        Next_State <= Retrigger;
                                    Elsif Count = Width then
                                        Next_State <= Idle;
                                    End If;
            when Retrigger      =>  Next_State <= Gen_Pulse_A;

    End case;

End Process State_Mach_Proc;
```

PULSE_PROC: Process (CLK, RESET)

```
Begin

    If RESET = '1' then
        PULSE               <= '0';
        COUNT               <= 0;
        CURRENT_STATE       <= Idle;
```

Elsif Rising_Edge (CLK) then

```
-- Update Current State
   CURRENT_STATE <= Next_State;

-- Set Outputs
   Case Next_State is

       when Idle          =>   PULSE   <= '0';
                               COUNT   <= 0;
       when Gen_Pulse_A   =>   PULSE   <= '1';
                               COUNT   <= COUNT + 1;
       when End_Pulse     =>   PULSE   <= '0';
                               COUNT   <= 0;
       when Gen_Pulse_B   =>   PULSE   <= '1';
                               COUNT   <= COUNT + 1;
       when Retrigger     =>   COUNT   <= 0;

   End case;
 End If;
End Process PULSE_PROC;
```
End STATE_MACHINE;

Example 6-5 : The Retriggerable Pulse Generator State Machine

Note that a signal COUNT and constant WIDTH have also been declared in Example 6-5. Both are of type integer, which allows a direct comparison of the two values. Both are constrained to the range of 0 to 31 and therefore Count and WIDTH can only be set to values inclusively between 0 and 31. Since this range contains 32 values, a five bit storage element will be created by synthesis to contain the value of COUNT. Since WIDTH is a constant, and will never change throughout the design, no storage elements are necessary to contain the WIDTH value.

The body of the STATE_MACHINE architecture contains two

processes. STATE_MACHINE_PROC is a non-clocked process. This process monitors the CURRENT_STATE, TRIG, and COUNT values to determine the next state to be performed. The VHDL code includes a case statement with embedded if statements. The synthesized logic will generate no storage elements from this process.

The PULSE_PROC process is a clocked process. Reset condition initializes the PULSE output to '0', COUNT to 0, and CURRENT_STATE to Idle. Two concurrent operations are defined by PULSE_PROC. The assignment statement

CURRENT_STATE <= NEXT_STATE

will load the NEXT_STATE value generated by the STATE_MACHINE_PROC process into CURRENT_STATE on the rising edge of CLK. The other concurrent operation is a Case statement used to control the signal COUNT and the output PULSE. All actions on COUNT and PULSE will occur when either RESET goes active , or upon a rising edge of CLK. While RESET is inactive (='0'), each rising edge of CLK will execute the actions determined by the present value of CURRENT_STATE. For example, if CURRENT_STATE equals GEN_PULSE_A, the PULSE output is set to '1' and the COUNT value is incremented. Synthesis will generate storage elements for signals CURRENT_STATE and COUNT, and for the output PULSE.

The Traffic Light Controller State Machine

One of the classic textbook examples of a simple state machine is a traffic light controller. The traffic light is placed at the intersection of a busy highway and a seldom used farm road. A sequencer could be defined to control the traffic light's action in response to cars arriving at this intersection. Preferably, the light should always allow cars on the highway to travel through the intersection, unless a car appears on the farm road. At this time, the light should change to allow the car to cross, and then return to allow the highway traffic to flow freely.

There are two basic graphical representations to describe this sequencer. The flowchart version is shown in Figure 6-9.

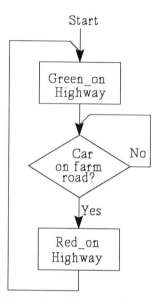

Figure 6-9 : Flow Chart for Car on Highway

The traffic light example begins with an action: the light is set to green for cars on the highway (this implies a red-light for cars on the farm road). This is followed by a decision: is there a car on the farm road? If the answer is "yes", then the light is set to red for cars on the highway (green for cars on the farm road). This execution is followed by returning to the beginning of the flowchart, and setting the light back to green for cars on the highway. If the answer to the decision is "no", then the flowchart keeps asking this question until the answer is true.

The state diagram for the simple traffic light controller is shown in Figure 6-10.

The sequencer is asynchronously initialized to the state "Green_on_highway". The action performed during this state is to set the light to green for cars on the highway (red on farm road). The next state is conditionally based on whether there is a car on the farm road. An If statement is used to check the condition. If this condition is false, then

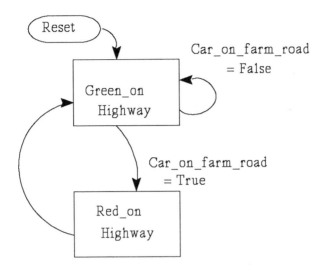

Figure 6-10 : State Diagram for Car on Highway

the sequencer stays in the "Green_on_highway" state. If this condition is true, then the sequencer advances to the state "Red_on_highway". The action performed during this state is to set the light to red for cars on the highway (green on farm road). The next state is non-conditional: always return to the "Green_on_highway" state. An assignment statement sets Next_State to Green_on_highway, which returns the state machine to its initialization state, and the sequence repeats continuously.

This example shows again that you need not specify a design in terms of logic states 0 and 1, but by using VHDL's Enumerated Types you can describe the action in terms such as "Green_on_highway". A VHDL description for the traffic light sequencer is shown in Example 6-6 and is extracted directly from the state diagram. The only additional information needed is the clock rate at which to execute the sequence, which is specified by the design requirements and will be implemented in the test bench (see Chapter 9). For this example, a fast clock rate might cause problems. If the clock is allowed to change every second, then cars might never be able to cross through the intersection! A more realistic clock rate for this sequence might be in the range of 10 to 30 seconds. In addition to the clock, the other inputs are Reset, which asynchronously initializes the sequencer, and Car_on_farm_road. This is an indicator that is externally set to '1' when there is a car waiting on the farm road. The output from the traffic_light_controller Entity is the signal Set_green_on_highway, which sets the light to green for highway traffic when equal to '1'. When this output is set to '0', the light is reversed, setting the light to red for highway traffic.

```
-- Simple Traffic Light Controller Sequencer

library IEEE;
 use IEEE.std_logic_1164.all;

Entity TRAFFIC_LIGHT_CONTROLLER is
  Port ( CLK                      :  In    Std_Ulogic;
         RESET                    :  In    Std_Ulogic;
         CAR_ON_FARM_ROAD         :  In    Std_Ulogic;
         SET_GREEN_ON_HIGHWAY :  Out   Std_Ulogic);
End TRAFFIC_LIGHT_CONTROLLER;

Architecture SYNTHESIZABLE of TRAFFIC_LIGHT_CONTROLLER
is

-- Define type
  Type TRAFFIC_LIGHT_STATE is
    (GREEN_ON_HIGHWAY, RED_ON_HIGHWAY);

  Signal NEXT_STATE : TRAFFIC_LIGHT_STATE;

Begin

  SEQUENCER : Process (CLK, RESET)
  Begin

  If RESET = '1' then
      NEXT_STATE <= GREEN_ON_HIGHWAY;
      SET_GREEN_ON_HIGHWAY <= '1';
```

```
Elsif Rising_Edge (CLK) then

Case NEXT_STATE is
    when GREEN_ON_HIGHWAY =>
        SET_GREEN_ON_HIGHWAY <= '1';
        If CAR_ON_FARM_ROAD = '1' then
            NEXT_STATE <= RED_ON_HIGHWAY;
        Else
            NEXT_STATE <= GREEN_ON_HIGHWAY;
        End If;

    when RED_ON_HIGHWAY =>
        SET_GREEN_ON_HIGHWAY <= '0';
        NEXT_STATE <= GREEN_ON_HIGHWAY;
    End Case;
    End If;
End process SEQUENCER;
End SYNTHESIZABLE;
```

Example 6-6 : State Machine for the Car on Highway

In this example, the VHDL architecture for the simple traffic light controller defines a type (traffic_light_state) to include all possible states of the traffic light sequencer. In this case there are only two possible states, Green_on_highway and Red_on_highway. An internal signal named Next_State is defined to be of type "traffic_light_state", so it can only be set to the states Green_on_highway or Red_on_highway. The body of the architecture contains one synchronous (clocked) process that monitors the value of the car_on_farm_road input. This input, along with the value of the signal Next_State, determines the execution to be performed and the state to follow.

The Next_State signal is initialized by the reset input to the state Green_on_highway. This directly affects the logic generated by synthesis. The storage elements that will be used to decode the state will have a direct asynchronous clear or preset. The number of flip flops required depends on the possible number of states. For this example, there are only two possible states and therefore only one flop will be required. The minimum number of storage elements is equal to the closest factor of two that is greater than or equal to the possible number of states. For example, if 5,6,7 or 8 states exist, then 3 storage elements are needed ($2**3$). If 30 states exist, then 5 storage elements are necessary ($2**5$).

The output Set_Green_on_highway also is initialized to '1' by reset. Since this occurs inside of a clocked process, the resulting synthesis creates a storage element to contain the value of this output. The storage element also will contain a direct asynchronous reset or set. If the reset clause had been omitted, then the resulting storage element would not be asynchronously initializable. Many gate array and standard cell libraries would use a physically smaller storage element in this case, but difficulties could arise in comparing the VHDL simulations with the resulting gate-level simulations, since these elements need to be initialized through other circuitry.

The actual state machine logic is performed by a VHDL Case statement. All values of Next_State are checked in the case statement. If Next_State is equal to Green_on_highway, then the Set_Green_on_highway output is set to '1' (no change if the output was already set to '1'). The input car_on_farm_road is monitored to determine if the state should be changed. If the input equals '1', then the Next_State signal is changed to the Red_on_highway state. Since the process is synchronous, this execution does not occur until the next active edge of the clock, which signifies the beginning of the next cycle. If the Next_State equals Red_on_highway, then the actions are non-conditional. The Set_Green_on_highway output is set to '0', and the Next_State is changed to Green_on_highway (again to execute with the next active clock edge).

Summary

This chapter has shown the basic techniques of state machine and control logic design using single and multiple clocked processes, explicit and implicit state definition, and the use of Enumerated Types for defining the states. The use of some of these VHDL features will be further illustrated in Chapter 7 as applied to data processing functions.

Chapter 7

Data Processing Functions

The types of circuits used in data processing differ significantly from the signal processing functions presented in the previous chapters. Data processing functions typically perform calculations on numerical data, and considerations must be given to how the data is represented and to the speed at which it can be processed. For more complex functions, such as multipliers, there are many different algorithms that can be chosen depending on the application, and the synthesizer must therefore be closely directed to produce the desired result. This chapter discusses how common computing functions can be synthesized and optimized, and how the various data formats can be controlled during that process.

Background

In addition to traditional computer systems, many other types of systems today make use of data processing functions. High speed, pipelined architectures are widely used in digital filtering processors, image and video processing, real time military applications, and television systems. From a logic circuit viewpoint, these different system applications are implemented by basic functional units such as registers, shifters, stacks, accumulators, multipliers, arithmetic logic units, and digital filters.

The major difference between data processing functions and signal processing functions is that, compared with sequencers, pulse generators, and counters that can be designed using a limited set of register and incrementer/decrementer combinations, data processing functions can be implemented in logic in many varying ways. Part of the reason for this is that numbers can be represented in so many different formats, such as unsigned or two's complement, real or integer, floating point, and fractional. In addition, functions such as multiply, divide, and square root can be implemented by many different algorithms, each of which has advantages depending on the specific application. Finally, data processing functions generally have speed or throughput requirements that can be greatly affected by the choice of the circuit approach. For example, even a simple function such as an adder can be synthesized to create a minimum size design, or it can be built as a maximum speed design, and the resulting logic from these two operations can differ greatly. These factors should be taken into account when planning the block diagram, the VHDL coding, and the synthesis constraints.

Data processing circuits can also be generated from macrocell libraries or datapath compilers, if available. You may still want to synthesize these functions however, to gain more control over the area and speed parameters or because of specialized circuit requirements. For example, in video processing with digital filters, a common requirement is for scaling, ie, to multiply a data value by a constant. A synthesized multiplier under these conditions would be greatly minimized when compared with a fixed multiplier macro.

Logic Functions

Most data processing functions operate on data contained in some form of register or storage device, and transfer the results into a similar register structure. These functions are true Register Transfer Level (RTL) operations, and this means that the designer must explicitly define each stage of registers required on the block diagram, and similarly in the VHDL signal definitions. The data processing functions then operate between these groups of registers, which computer technology and digital filter designers commonly refer to as levels of pipeline. The design data flows from register to register, or from pipeline to pipeline. The logic must also be able to propagate logic from register to register within a specified time and this maximum allowed propagation time, the system's cycle time, is a key design specification. The amount of data processing that can be implemented between two sets of registers depends on both the speed of the ASIC fabrication technology, as specified by the manufacturer, and the cycle time. In most cases the maximum gate delay through the function must not exceed the system's cycle time.

Several of the data processing functions have already been shown briefly in Chapter 3, such as

a. Adder :   Sum <= A + B ;

b. Subtracter : Diff <= C - D - 4 ;

The VHDL code for these operations is simple, but what does it really mean in terms of your specific application? In the adder, are the numbers unsigned or two's complement? Are they the same word length? A logic synthesizer has no inherent means for knowing exactly what you need for a particular application, and the technique used must make it clear.

Consider the case of adding two unsigned numbers. If both numbers are 16 bits wide, then the sum will be 17 bits wide. The impact that this has on the VHDL code is that the SUM must be a vector of length 17, otherwise the most significant bit of the sum (bit 17) will be truncated and therefore will not produce the correct sum. If the SUM is 17 bits, then the input terms A and B must also be made 17 bits wide. Since the values of A and B are 16 bits wide, an extra 0 should be concatenated to the MSB of those signals, such as

'0' & A

The resulting VHDL expression would then be

Sum <= ('0' & A) + ('0' & B);

which is the 17 bit unsigned result of the addition. This will be further illustrated in examples to follow.

A second point about complex data functions is how to organize the VHDL expression to attain the desired speed performance. As an example, if you want to add 4 databusses together in a parallel adder, you might code the expression as :

Sum <= BusA + BusB + BusC + BusD;

This code is correct functionally, but it leaves a lot of assumptions for the synthesizer to determine. The synthesizer might design the adder in series, which would result in a long propagation delay through some of the adder paths. By using parentheses however, you can direct the synthesizer as to how to organize the adder logic :

Sum <= (BusA + BusB) + (BusC + BusD);

which would produce a more balanced circuit, as shown in Figure 7-1. The general rule is to use parentheses in this manner to force the structure of ambiguous combinational functions such as the adder, and whenever an expression might not be exactly clear from its syntax. For instance :
Out <= A + B AND C + 4;

which could also be better described as

Out <= (A + B) AND (C + 4);
As can be seen from these simple cases, the VHDL syntax for arithmetic operations is straightforward but the other issues of data

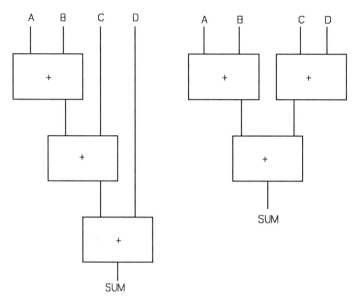

Figure 7-1 : Controlling Adder Architectures

formatting and controlling the synthesized architecture must be addressed early in the design process.

A Simple Unsigned Adder

As a first example, let's consider a simple pipelined adder that will add the contents of a databus to the contents of a register and store the result in another register. The first step in designing for synthesis is to draw a block diagram of the system, where each register is defined, and the functions between registers are described. Figure 7-2 shows a simple block diagram of the adder placed between two registers, from which we can identify two VHDL signals as Reg_A and Reg_B.

Let's make the input data 8 bits wide and assume that the data is represented as positive integers, requiring no sign bit. As a designer, you know that when two 8 bit numbers are added, the result requires 9 bits. Synthesis tools generally need your help in identifying that fact, since they have no means of automatically determining whether the data is

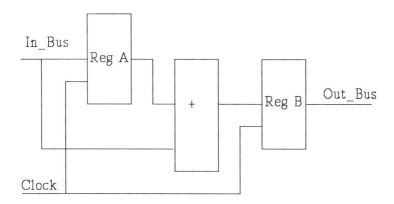

Figure 7-2 : Block Diagram of a Pipelined Adder

represented as signed, unsigned, integer or two's complement values. In this example therefore, Reg_B will be explicitly specified as a 9 bit vector. In handling various data processing functions, you will directly control factors such as word lengths, sign bits, and overflow conditions that vary depending on the specific system design.

The input bus, IN_BUS, is defined to be an 8 bit positive unsigned number. This allows values between the range "00000000" and "11111111" (0 to 255 integer). The internal signal Reg_A must also be an 8 bit value, again containing the range integer range 0 to 255. The result of the adder function is assigned to the signal B_Reg. This value must be between the integer values of 0 and 510 ("000000000" and "111111110"), since the minimum and maximum results of the addition are:

```
Decimal    Binary              Decimal    Binary

   0       "00000000"             255    "11111111"
 + 0     + "00000000"           + 255  + "11111111"
 ---     ------------           -----  ------------
   0       "000000000"            510    "111111110"
```

Reg_B has been defined to be 9 bits wide to allow the full adder range. In order to create the registers, a clocked assignment must be generated for each signal, using a standard clocked process from Chapter 3. This is shown inside the process CLOCKED_PROC, which is triggered on the rising-edge of CLK. The assignment

    Reg_A <= IN_BUS;

notifies the synthesizer of the actions: that Reg_A is a register clocked by the rising-edge of CLK, and that IN_BUS is the input to this register. The assignment

    Reg_B <= Reg_A + IN_BUS;

specifies three actions to the synthesizer: Reg_B is a register also clocked by the rising-edge of CLK, an adder must be created to perform the function Reg_A + IN_BUS, and the result from this adder is the input to register Reg_B. The width of each register is specified by the signal declaration, and determines the size of the adder required. In this example, IN_BUS and Reg_A are 8 bits wide and Reg_B is 9 bits wide, so the adder generated will add two 8 bit numbers to create a 9 bit result. Note that each of the inputs to Reg_B has a logic 0 concatenated to the MSB so that it becomes a 9 bit value consistent with Reg_B. If both sides of the expression do not agree in length, the VHDL compiler will generate an error message.

The VHDL code to represent this block diagram is listed in Example 7-1.

```
library IEEE;
 use IEEE.std_logic_1164.all;

Entity PIPELINE is

  Port ( CLK         :  In   Std_Ulogic;
         IN_BUS      :  In   Std_Ulogic_Vector (7 downto 0);
         OUT_BUS     :  Out  Std_Ulogic_Vector (8 downto 0));

End PIPELINE;
```

Architecture RTL of PIPELINE is

-- Define signals

Signal Reg_A: Std_Ulogic_Vector (7 downto 0);
Signal Reg_B: Std_Ulogic_Vector (8 downto 0);

Begin

OUT_BUS <= Reg_B;

CLOCKED_PROC: Process (CLK)

Begin

If Rising_Edge (CLK) then

Reg_A <= IN_BUS;
Reg_B <= ('0' & Reg_A) + ('0' & IN_BUS);

End if;

End Process CLOCKED_PROC;

End RTL;

Example 7-1 : VHDL Code for a Simple Pipelined Adder

This is a simple case where it is clear to the synthesizer how the logic should be generated. In the general case, a more complex arithmetic logic unit (ALU) capable of addition, subtraction, and various logical operations may be required. In a system the data for the ALU may come from numerous sources, which raises the issues of whether multiple ALUs are needed or whether the input data should be simply multiplexed into a single ALU. The answer of course depends on whether the operations can be done in series or whether they must be done simultaneously, and the cycle time also becomes a factor. While you must consider these issues as you architect the system, several logic synthesis tools also analyze whether logic functions can be shared efficiently. In the language of logic synthesis, this is called **Resource Sharing.**

As an example, consider the following VHDL code:

Y <= A + B When Select = '1' Else Y <= C + D;

This seems pretty straightforward, that the output Y is the sum of A and B when Select is high, otherwise it is the sum of C and D. But this function could be designed as two separate adders followed by a multiplexer, or a single adder whose inputs are multiplexed, as shown in Figure 7-3. The code above is very simple and straightforward, but does not indicate which design approach to take.

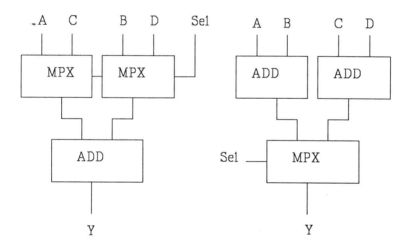

Figure 7-3 : Alternative Adder Architectures

One possibility would be to write the code in a different form to force the two designs shown:

a. Two multiplexers, one adder :

```
Mpx1  <= A when Select = 1 Else C;
Mpx2  <= B when Select = 1 Else D;
Y     <= Mpx1 + Mpx2;
```

b. Two adders, one multiplexer :

```
Sum1  <= A + B;
Sum2  <= C + D;
Y     <= Sum1 when Select = '1' Else Sum2;
```

As you can see, each of these description is a lot more code than the original single line. A synthesis tool that can detect the possibilities for resource sharing will analyze the code, compare both alternates and select the most efficient approach. This can produce a significant reduction in the number of gates needed to implement a function, while allowing you to code that operation efficiently.

As a more complete example of specifying the desired architecture, consider the block diagram of Figure 7-4. In this case, it is desired to add Reg_A to the IN_BUS if the signal B_SELECT is 0 and add Reg_A to itself if B_SELECT is high. Example 7-2 shows the VHDL code to replace the assignment to Reg_B with a conditional assignment statement based on a new input signal: B_SELECT. This description allows the synthesizer much flexibility in creating the logic design. The most straight forward approach is diagrammed in Figure 7-4. This design consists of two registers, two adders, and one multiplexer (generated from the "if" statement).

-- Modified Pipeline Example --

```
library IEEE;
 use IEEE.std_logic_1164.all;

entity PIPELINE is

    Port ( CLK          : In  Std_Ulogic;
           B_SELECT     : In  Std_Ulogic;
```

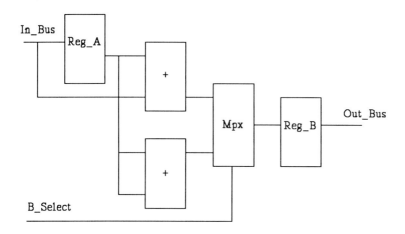

Figure 7-4 : A Dual Adder Implementation

```
        IN_BUS           : In  Std_Ulogic_Vector (7 downto 0);
        OUT_BUS          : Out Std_Ulogic_Vector (8 downto 0));
End PIPELINE;
```

Architecture RTL of PIPELINE is

```
-- Define signals
Signal Reg_A             : Std_Ulogic_Vector (7 downto 0);
Signal Reg_B             : Std_Ulogic_Vector (8 downto 0);
```

Begin

OUT_BUS <= Reg_B;

CLOCKED_PROC: Process (CLK)

Begin

 If Rising_Edge (CLK) then

        Reg_A <= IN_BUS;
```
```

```
        If B_SELECT = '0' then
            Reg_B  <= ('0' & Reg_A) + ('0' & IN_BUS);
        Else   -- B_SELECT = '1'
            Reg_B  <= ('0' & Reg_A) + ('0' & Reg_A);
        End if;
     End if;
   End Process CLOCKED_PROC;
End RTL;
```

Example 7-2 : A Multiplexed Pipelined Adder

One consideration with this particular design is that both of the adder paths can never be selected to be loaded into Reg_B during the same cycle, because the multiplexer guarantees that the data from only one of the adders can be used during any cycle.

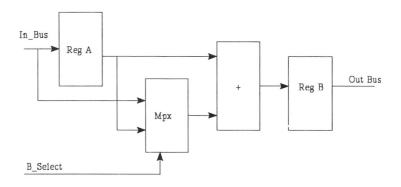

Figure 7-5 : Single Adder Implementation

If the architecture is changed as in Figure 7-5 so that the multiplexer is placed before the adder, with inputs to the adder based on the value of the B_SELECT input, then only one adder is required. This will create a design requiring two registers, a multiplexer, and only one adder, and will synthesize to less logic gates as a result of the explicit resource sharing, since adders require more logic than multiplexers. Example 7-3 shows a VHDL description to explicitly share the adder. Notice that this code divides the logic into two sections: the multiplexer is defined concurrently (non-clocked), and the adder is defined inside of the clocked process in order to create the pipeline register. This description requires an additional signal, ADDER_INPUT, resulting in slightly more complicated coding.

```
-- Explicit Resource Sharing --

Architecture RTL of PIPELINE is

-- Define signals
Signal Reg_A            : Std_Ulogic_Vector (7 downto 0);
Signal Reg_B            : Std_Ulogic_Vector (8 downto 0);
Signal ADDER_INPUT   : Std_Ulogic_Vector (7 downto 0);

Begin

  OUT_BUS  <= Reg_B;

-- The multiplexer - Combinational

  ADDER_INPUT  <= IN_BUS when B_SELECT = '0' Else Reg_A;

  CLOCKED_PROC: Process (CLK, Reset)

Begin
  If Reset = '1' then
        Reg_A <= "00000000";
        Reg_B <= "000000000";

  ElsIf Rising_Edge (CLK) then

        Reg_A  <= IN_BUS;
        Reg_B  <= ('0' & Reg_A) + ('0' & ADDER_INPUT);
```

End if;
End Process CLOCKED_PROC;
End RTL;

Example 7-3 : Multiplexing Data into the Adder

The designer's choice of coding style can greatly affect the resulting synthesized logic. In the previous example, when the B_SELECT input equals '1', then Reg_A is added to itself and then loaded into Reg_B. Since the IN_BUS value is a positive unsigned number, an equivalent function would be to left shift Reg_A by one place, and fill the least significant bit with a '0'. In VHDL, for an 8 bit vector, this assignment is written:

   Reg_B <= Reg_A (7 downto 0) & '0';

where the '&' character represents the concatenation. The resulting block diagram of this shifting representation is shown in Figure 7-6. Notice that once again the design uses 2 registers, 1 multiplexer, and 1 adder.

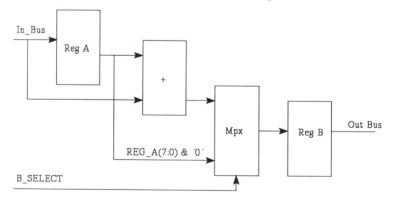

Figure 7-6 : Adder replaced by Left Shift

Other shifting functions can be described just as easily in VHDL.

Again, for a positive unsigned value, a right shift and fill the most significant bit with '0' (equivalent to a divide by 2) is written:

Reg_B <= '0' & Reg_A (7 downto 1);

Still another equivalent function to Reg_A + Reg_A is:

Reg_B <= 2 * Reg_A;

This is a VHDL multiply operation, and if a true multiplier were synthesized to perform this function, then the resulting logic would be much larger than necessary. Due to the size of logic multipliers and dividers, caution should be exercised when synthesizing these functions. Some vendors will not allow general multiplication or division in synthesizable VHDL code and, in these cases, the designer must instantiate the operation (an example is shown later this chapter). One exception to this rule is the multiplication or division by a factor of two. Some advanced logic synthesis tools recognize these operations as simple shift functions.

A Two's Complement Adder

What if the value on IN_BUS can be either a positive or negative number? Now the inputs to the adder can be positive or negative, and the result of the adder could also be positive or negative. Let's define the IN_BUS value to be in two's complement notation. In this number system, if all of the bits equal '0', then the value equals zero. Else, if the most significant bit is '0', then the value is considered positive. Else, the value is negative. For an eight bit vector, the range for a two's complement number is:

```
     Binary        Decimal

   "01111111"       +127
   "01111110"       +126
   . . . . . . . .   . . . .
   "00000100"        +4
   "00000011"        +3
```

```
"00000010"              +2
"00000001"              +1
"00000000"               0
"11111111"              -1
"11111110"              -2
"11111101"              -3
"11111100"              -4
..........            ....
"10000001"            -127
"10000000"            -128
```

The most positive number is "01111111" (integer +127) and the most negative value is "10000000" (integer -128). The minimum and maximum results of the adder are:

```
Decimal          Binary            Decimal       Binary

   (-128)        "10000000"          127        "01111111"
+  (-128)      + "10000000"       +  127    +   "01111111"
---------      -------------       -----    -------------
   (-256)        "100000000"         254        "011111110"
```

The range of Reg_B is therefore between integer -256 and +254, which is contained in a 9 bit two's complement number. Notice that the maximum positive range has been inherently cut by approximately half to allow negative values.

By the definition of two's complement notation, when converting an 8 bit number to a 9 bit number, a positive value must remain positive and a negative number must remain negative. To accomplish this, the value must be sign-extended. The msb of the 8 bit value must be appended as the new msb, thus maintaining the sign bit. This is shown in Example 7-4, and again, the adder requires a 9 bit result. This time, though, it is a two's complement number within the range -256 and +254.

-- Two's Complement Adder Example --

Architecture RTL of PIPELINE is

-- Define signals
 Signal Reg_A : Std_Ulogic_Vector (7 downto 0);
 Signal Reg_B : Std_Ulogic_Vector (8 downto 0);

Begin

 OUT_BUS <= Reg_B;

CLOCKED_PROC: Process (CLK, Reset)

Begin
  If Reset = '1' then
        Reg_A <= "00000000";
        Reg_B <= "000000000";

  ElsIf Rising_Edge (CLK) then

        Reg_A <= IN_BUS;

        If B_SELECT = '0' then
           Reg_B <= Reg_A (7) & Reg_A + IN_BUS (7) & IN_BUS;
        Else -- B_SELECT = '1'
           Reg_B <= Reg_A (7) & Reg_A + Reg_A (7) & Reg_A;
        End if;
  End if;
 End Process CLOCKED_PROC;
End RTL;

Example 7-4 : A Two's Complement Adder

A Sign-Magnitude Adder

The value on IN_BUS could also be a positive or negative number represented by sign-magnitude notation. In this case, the most significant bit indicates whether the value is positive or negative, and all other bits determine the magnitude. If the msb = '0', then the number is positive, and if the msb = '1', the number is negative. For an 8 bit number, both "00000000" and "10000000" equal zero. The range for a sign-magnitude 8 bit number is:

Binary	Decimal
"01111111"	+127
"01111110"	+126
....	
"00000100"	+4
"00000011"	+3
"00000010"	+2
"00000001"	+1
"00000000"	0
"10000000"	0
"10000001"	-1
"10000010"	-2
"10000011"	-3
"10000100"	-4
....	
"11111110"	-126
"11111111"	-127

The most positive number is "01111111" (integer +127) and the most negative value is "11111111" (integer -127). The minimum and maximum results of the adder are:

Decimal	Binary	Decimal	Binary
(-127)	"11111111"	127	"01111111"
+ (-127)	+ "11111111"	+ 127	+ "01111111"
(-254)	"111111110"	254	"011111110"

So the range of Reg_B is between integer -254 and +254, which is contained in a 9 bit sign-magnitude number. Once again, the maximum positive range has been cut by approximately half to allow negative values.

Addition of sign-magnitude numbers is not as straight-forward as the two's complement case. If the input is negative, then a conversion has to be performed. This extra process needs to be included in both the input and output stages of the adder. Figure 7-7 shows a modified block diagram to include conversions for negative inputs. Notice each input to

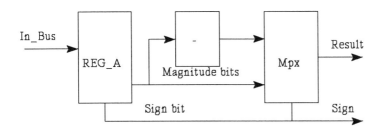

Figure 7-7 : Sign-Magnitude Addition Logic

the adder block. The most significant bit is evaluated to determine if the value is negative. If the number is zero or positive, then the msb is set to '0' (to account for the "10000000" case), and the least significant bits pass unchanged. If the number is negative, the msb is set to '1', and the two's complement of the value of the least significant bits is sent to the adder block. The adder block sums both two's complement numbers, and returns a two's complement result. Now the adder output must be converted back to sign-magnitude notation. If the result is negative (msb='1'), then the least significant bits are two's complemented to create a positive magnitude. The sign bit is maintained and passed directly to the output register. The VHDL code for the sign-magnitude addition is shown in Example 7-5.

Note the use of the negation (-) or "unary subtraction" in the code for

Reg_A below.

━━━

-- Sign-magnitude Adder --

Architecture RTL of PIPELINE is

```
-- Define signals
Signal Reg_A              : Std_Ulogic_Vector (7 downto 0);
Signal Reg_B              : Std_Ulogic_Vector (8 downto 0);
Signal ADDER_INPUT   : Std_Ulogic_Vector (7 downto 0);
Signal ADD_A, ADD_B : Std_Ulogic_Vector (7 downto 0);
Signal ADD_C             : Std_Ulogic_Vector (8 downto 0);

Begin

 OUT_BUS  <= Reg_B;

 ADDER_INPUT  <= IN_BUS when B_SELECT = '0'
              Else Reg_A;

 ADD_A <= '0' & Reg_A (6 downto 0) when
    (Reg_A (7)='0' or Reg_A (6 downto 0) = "0000000")
     Else '1' & ( - Reg_A (6 downto 0));

 ADD_B <= '0' & IN_BUS (6 downto 0) when
    (IN_BUS (7) = '0' or IN_BUS (6 downto 0) = "0000000")
      Else '1' & ( - IN_BUS (6 downto 0));

 ADD_C <= ADD_A + ADD_B;
```

CLOCKED_PROC: Process (CLK, Reset)

```
Begin
 If Reset = '1' then
       Reg_A  <= "00000000";
       Reg_B  <= "000000000";
```

ElsIf Rising_Edge (CLK) then

    Reg_A <= IN_BUS;

    If ADD_C (8) = '1' then   -- Negate ADD_C :
      Reg_B <= ADD_C (8) & (- ADD_C (7 downto 0));
    Else
      Reg_B <= ADD_C;
    End if;
  End if;
End Process CLOCKED_PROC;
End RTL;

Example 7-5 : A Sign-Magnitude Adder

The magnitude conversions from positive to negative, and from negative to positive, are performed by unary subtraction. Notice the assignment:

  Reg_B <= ADD_C(8) & (- ADD_C (7 downto 0));

This is equivalent to the assignment:

  Reg_B <= ADD_C(8) & (0 - ADD_C (7 downto 0));

The vectors ADD_A, ADD_B, and ADD_C have been defined as intermediate signals. These vectors determine the direct inputs and output of the combinational adder. They do not create registers, since the assignment to each register is performed outside of the clocked process.

Multipliers

Many current logic synthesis tools do not directly create multiplication logic from VHDL code, using syntax such as Y <= A * B. And because of the size of the resulting logic, even synthesizers that have this capability should be monitored closely to confirm that extra logic is not created. If multipliers are needed in a design, there are many common algorithms to perform binary multiplication, depending on the data format. The techniques range from serial to parallel, the simple shift and add, or more complex techniques such as Booth's algorithm.

For a parallel, unsigned multiplication, the shift and add algorithm is the binary equivalent of long-hand decimal multiplication. Starting with the least significant bit of the second multiplicand, it is multiplied by the first multiplicand to form the first product term. Then you move to the next column or bit and do the same for all of the bits. The final product is simply the sum of the product terms.

As an example, if the first number is "0110" (integer 6), and the second number is "1010" (integer 10), then the expected result is an 8 bit number equal to "00111100" (integer 60). This is shown below:

```
                        Binary                    Decimal

        M1              "0110"                          6
        M2            x "1010"                       x 10
                     ----------------               ----
Product terms          "0000"                          0
                       "0110"                          6
                      "0000"
                     "0110"
                     ------------                   ----
Product              "00111100"                       60
```

Binary bit multiplication is much easier than decimal however since each product term will either be 0 or the first multiplicand.

The maximum result from multiplying two 4 bit numbers is:

```
                Binary              Decimal

                "1111"                15
              x "1111"             x  15
              ---------            ----
              "11100001"             225
```

This result is contained in an 8 bit vector, and in the general case the width of the result is equal to the sum of the widths of the two multiplicands.

This algorithm forms the basis for designing combinational parallel multipliers. In VHDL, the same approach can be done using a direct technique, and then a more flexible technique using a VHDL Procedure.

In the direct method, the partial product terms will be formed and added directly to produce the final result. Consider a 4 bit multiplier to multiply inputs A and B :

```
Signal P0, P1, P2, P3    : Std_Ulogic_Vector (7 downto 0);
Signal A, B              : Std_Ulogic_Vector (3 downto 0);
Signal Product           : Std_Ulogic_Vector (7 downto 0);

-- A combinational, concurrent design

Begin

P0   <= ("0000" & A      )      when B(0) = '1' Else "00000000";
P1   <= ("000"  & A & '0' )     when B(1) = '1' Else "00000000";
P2   <= ("00"   & A & "00" )    when B(2) = '1' Else "00000000";
P3   <= ('0'    & A & "000")    when B(3) = '1' Else "00000000";

Product <= (P3 + P2) + (P1 + P0);
```

Example 7-6 : A Simple Brute Force Parallel Multiplier

As you can see, each partial product is formed depending on the state of the corresponding bit in the B data word, and is "shifted over" by being concatenated with 0s. The product is formed by two sets of adders grouped with parentheses in order to balance the delay and to indicate how the synthesizer should arrange the adder logic.

This approach is quite simple and produces very controllable results. The synthesized circuit consists of the adders for the partial products P0 to P3, and each adder is driven by its A input ANDed with the appropriate B bit for that term. Note that with this technique the length of the code is proportional to the number of bits in the data, so that an 8 bit multiplier would thus require 8 partial product terms. This is not really that difficult, but let's do the multiplier two additional ways, one that does the brute force method but in a Procedure, and the last a loop algorithm that can be used with any word length.

-- A VHDL Procedure for Multiply

**Procedure MPY   (Signal A, B : In Std_Ulogic_Vector (3 downto 0);
                     Signal Prod : Out Std_Ulogic_Vector (7 downto 0))
            Is**

Variable P0, P1, P2, P3 : Std_Ulogic_Vector (7 downto 0);
Constant Zero : Std_Ulogic_Vector := "00000000";

Begin

If B(0) = '1' then P0 := ("0000" & A);
Else   P0 := Zero;
End If;

If B(1) = '1' then P1 := ("000" & A & '0');
Else   P1 := Zero;
End If;

If B(2) = '1' then P2 := ("00" & A & "00");
Else   P2 := Zero;
End If;

If B(3) = '1' then P3 := ('0' & A & "000");
Else   P3 := Zero;
End If;

Prod <= (P3 + P2) + (P1 + P0);

End MPY;

Example 7-7 : Parallel Multiplier using a Procedure

The Procedure can be included directly in the Architecture, and a convenient place to locate procedures is following the signal and constant declarations. Note that the Procedure has two inputs and one output, whose names only have meaning within the procedure. The multiplier can then be used within the architecture by passing in the appropriate signals. Also note that any internal values must be declared as Variables, such as the partial product terms P0 to P3, or as Constants.

Within the architecture for the overall circuit, the multiplier procedure calls for three separate multipliers could look like :

Begin -- of Architecture

    MPY (X, Y, ProductZ);   -- One multiplier circuit

    MPY (M, N, Prod_L);   -- Another

    Five <= "0101";
    MPY (SigA, Five, SigB); -- Multiply by Constant Five ("0101")

End RTL; - of Architecture

Procedures make it easier to generate a particular logic circuit many times within a design, and also allows for design re-use between different designs. The interesting thing about a synthesizable procedure such as the MPY example above is that the circuit resulting from the procedure may be different if one of the inputs is a constant value rather than both inputs being signal values. Logic minimization in the synthesizer would remove gates connected to the constant values and create only the required logic for the multiplication. Chapter 8 further discusses the minimization issues and shows additional examples.

The two examples above are both very controllable and produce efficient results, but they are dedicated to particular word lengths because the vectors and constants are defined as (7 downto 0) for instance. It would be nice to be able to create a multiplier that could adapt to a given word length specified once in the code. This could be done by using constants in the Entity and Architecture, or alternatively VHDL "Generics" could be used to pass the values into the code. Generics are the mechanism by which parameters, such as the width of ports and vector signals, can be passed from the Entity into Architectures. The following example will use the Generic statement to pass the word length into a multiply routine and has the format :

Generic (Constant Width : Integer := 4);

If this statement is used in the Entity, then the Ports and Signals in the Entity and Architecture can use this value as the parameter Width.

Now we have the problem of how to define values such as ZERO, which normally have a format of "00000000", depending on the word length. Of course these values could also be Generics, but a more flexible technique would be to use a Type Conversion to convert the integer value 0 into a vector of WIDTH. Common Type Conversions are normally provided by the supplier of the simulation and synthesis tools, and examples of their coding are also given in Chapter 9.

Assume for now that a Type Conversion from Integer to Std_Ulogic_Vector might be called I_to_SUV and would be given inputs of the integer value and length of the vector, such as

    Result := I_to_SUV (0, 8); -- To make an 8 bit vector "00000000"
or
    Result := I_to_SUV (0, WIDTH); -- Width determines the size

The latter form would provide the ability to create variable length constants based on the WIDTH parameter, which would suit our multiplier requirement perfectly.

Now we need an algorithm that would avoid the concatenation of many terms as used above. A "For" loop can be used that would perform the add and shift procedure for the number of bits in WIDTH. Consider forming each partial product by steering the data A into the appropriate bit position by using

If B(i) = '1' then
PP((i + Width -1) downto i) := IN_A;

where PP was initially zero and i is the loop index. In other words, if i = 0, then the A data will be in PP bits WIDTH downto 0, which for 4 bits would be 3 downto 0. If i is 1, then the data will be in 4 downto 1, and so on. If each of these partial products is added to the result each time through the loop, the final product will be the sum of all of the partial products.

Since this design is to be a parallel multiplier, the result should be done as a combinational process, and variables will be used within the loop so that each partial product is formed at that instant within the loop. The following example illustrates this process. Notice the arithmetic that can be done using Width to specify vector lengths.

```
library IEEE;
 use IEEE.std_logic_1164.all;

-- Unsigned 4 x 4 Multiply

Entity MULTIPLY is
    Generic (Constant Width : Integer := 4);

    Port (  IN_A : In  Std_Ulogic_Vector (Width - 1 downto 0);
            IN_B : In  Std_Ulogic_Vector (Width - 1 downto 0);
            RESULT : Out Std_Ulogic_Vector (2*Width - 1 downto 0));
End MULTIPLY;
```

Architecture ALGORITHM of MULTIPLY is

Begin

MPY_Proc : Process (IN_A, IN_B)
   Variable PP, Res : Std_Ulogic_Vector (2*Width -1 downto 0);

  Begin

   Res := I_to_SUV(0, 2*width); -- The Type Conversion
--   Same as Res := "00000000"; for a 4 x 4 multiply

   Loop1:
   For i in 0 to Width-1 loop

--   Set PP to zero each time through loop

         PP := I_to_SUV(0, 2*width);   -- = "00000000"

         If IN_B(i) = '1' then

            PP((i + Width - 1) downto i) := IN_A;
         End if;

        Res := Res + PP;
      End loop Loop1;

    Result <= Res;
End Process;
End ALGORITHM;

Example 7-8 : A Flexible Multiplier Algorithm

This code produces a series of adders preceded by AND gates depending on the state of the B bus, but the adders are in series rather than being grouped as in Examples 7-6 and 7-7. Variations can also be coded to help split the adders into more balanced delay chains as shown in Figure 7-1. The advantage of this approach is in the reusability of the code for many word lengths, and the general technique of using Generics, variable width data busses and loops can be applied to many different applications.

A variation on the previous example uses a technique where the sum is shifted rather than the partial products. This algorithm may not be possible on all synthesis tools unless arrays can be handled.

If the shift and add routine is used again, with M1 being the first multiplicand, M2(3) representing the msb of the second multiplicand, and M2(0) as the lsb of the second multiplicand, then the result of the multiply operation is:

```
M2(3) = '1' :    result = "0110"

M2(2) = '0' :    result = "0110" & '0'
                        = "01100"

M2(1) = '1' :    result = ("01100" & '0') + M1
                        = "011000" + "0110"
                        =      "011000"
                        +        "0110"
                               --------
                        =      "011110"


M2(0) = '0' :       result = "011110" & '0'
                           = "0111100"
```

This algorithm will be implemented by shifting the sum of the partial products inside of a loop, as shown in Example 7-9. The entity MULTIPLY defines two 8 bit inputs, IN_A and IN_B. The output is a 16 bit vector named RESULT. The architecture contains only one non-clocked process, which is needed to be able to declare the variable SUM. Since the assignment to a variable occurs immediately, the variable SUM can be updated many times before any simulation time advances. This repetitive assignment occurs inside of the loop named SHIFT_ADD_LOOP. Since the multiply output RESULT is an output port, which executes as a signal, the assignment to RESULT will wait until the loop has completed.

-- Unsigned 8 x 8 Multiply Algorithm

```
library IEEE;
  use IEEE.std_logic_1164.all;

entity MULTIPLY is
    Port (  IN_A       :  In   Std_Ulogic_Vector (7 downto 0);
            IN_B       :  In   Std_Ulogic_Vector (7 downto 0);
            RESULT     :  Out  Std_Ulogic_Vector (15 downto 0));
End MULTIPLY;

Architecture ALGORITHM of MULTIPLY is
Begin

    MULT_PROC: Process (IN_A, IN_B)

    variable SUM: Std_Ulogic_Vector (15 downto 0);

    Begin

    SUM := "0000000000000000";

    SHIFT_ADD_LOOP: for i in 7 downto 0 loop

        If IN_B(i) = '1' then    -- Compute Partial Product
            SUM := SUM + ("00000000" & IN_A);
        End if;

        If i /= 0 then           -- Shift Sum left
            SUM := SUM (14 downto 0) & '0';
        End if;
    End loop SHIFT_ADD_LOOP;

    RESULT <= SUM;
```

End Process MULT_PROC;

End ALGORITHM;

Example 7-9 : An Unsigned 8 x 8 Multiplier Algorithm

When the process is triggered, the variable SUM is first cleared (set to "0000000000000000"). Next, the shift and add loop is entered. The loop SHIFT_ADD_LOOP is a "For" loop that is executed 8 times as defined by the range of 7 downto 0 of the index "i". The loop starts with "i" equal to 7, and evaluates IN_B(7). If this equals '1', then the value of IN_A is added to SUM. The index "i" is evaluated to determine if it is not equal to 0, and in this case since "i" is not equal to 0 then SUM is left shifted by one bit. When the "end loop SHIFT_ADD_LOOP" statement is reached, "i" is decremented, and is checked against the range defined for the index "i". Since "i" now equals 6, and this value is inside of the range 7 downto 0, the loop is repeated. The last execution of the loop occurs when "i" equals 0. If IN_B(0) equals '1', then the value of IN_A is added to SUM, but SUM is not shifted. The loop is exited, and the present value of SUM is assigned to the output RESULT.

By using similar techniques with adders and subtracters, algorithms for two's complement multiplication can also be designed.

Register Stacks

This section shows two different versions of synchronous register stacks. Stacks are used for buffering data streams, either within one system or between different systems. Typically, the data write and data read operations are independent, and can occur overlapping at differing frequencies.

Last-In First-Out (LIFO)

The first stack example is a Last-In First-Out (LIFO) stack. This type of storage device is used for many priority handling applications, where the last data loaded is the most important. Write operations always load data to the "top" of the stack. The top is defined as the next word in line to be read. All other words currently loaded into the stack are "pushed" further down into the stack, and the bottom location in the stack is lost. Figure 7-8 shows an example of a 4 deep LIFO stack. The top of the stack contains the data "1000". The bottom of the stack contains the data "0001". When a write operation loads new data "1111" into the stack, all currently stored data words are "pushed" down in the stack.

Last-In First-Out Stack : Write Operation

```
                 Before Write          After Write

DATA_IN    "1111"                    ------

  TOP    | "1000" |               | "1111" |

         | "0100" |               | "1000" |

         | "0010" |               | "0100" |

BOTTOM   | "0001" |               | "0010" |
```

Figure 7-8 : Writing to a LIFO Stack

When a read request is generated to the LIFO stack, the last data word that was loaded is read back from the stack. All locations are "popped" up in the stack, with the last word read being removed. The next data word becomes the top of the stack, as shown in Figure 7-9. Before the read operation, the top of stack data word is "1111". After the read, "1000" becomes the top. An additional read operation would read "1000" from the stack. When the stack is empty, meaning that all words that were written to the stack have been read, then no "pop" operation occurs.

Last-In First-Out Stack : Read Operation

        Before Read   After Read

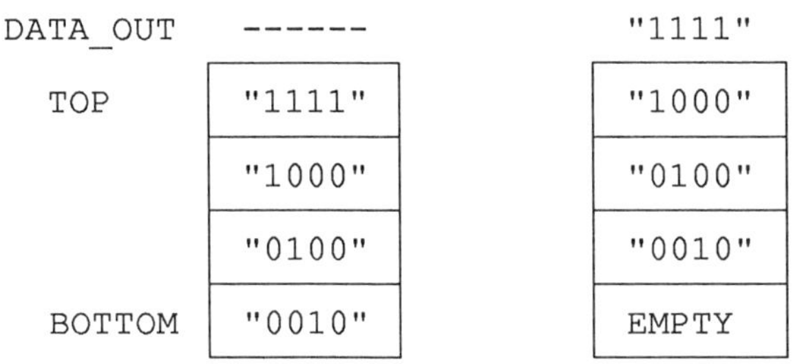

```
DATA_OUT       ------              "1111"
    TOP    ┌──────────┐        ┌──────────┐
           │ "1111"   │        │ "1000"   │
           ├──────────┤        ├──────────┤
           │ "1000"   │        │ "0100"   │
           ├──────────┤        ├──────────┤
           │ "0100"   │        │ "0010"   │
           ├──────────┤        ├──────────┤
 BOTTOM    │ "0010"   │        │ EMPTY    │
           └──────────┘        └──────────┘
```

Figure 7-9 : Reading from a LIFO Stack

Example 7-10 lists the VHDL code for a LIFO stack. The ports of the entity include a clock (CLK), reset (RESET), read strobe (RD), write strobe (WR), 8 bit DATA_IN, and 8 bit DATA_OUT. Inside the architecture, an assortment of constants, signals, types, and functions are declared. First, the constant DEPTH is defined to be an integer equal to 16, which is used to specify the depth of the internal stack array. All references to the size of the array, along with specifying checks of the last location of the array, are performed with this constant. If the size of the array needs to be changed later, then all that changes is the value of the constant DEPTH.

The internal array named DATA is defined by declaring a type and a signal. The type named DATA_ARRAY defines the width of the array as 7 downto 0 (8 bits), and the data type as integer. The length of the array is defined with the actual signal declaration to DATA and for this example the length is defined as 0 to DEPTH-1 (16 words long). Another type, named STATUS_TYPE, is also declared as an enumerated type with just two values, EMPTY and NOT_EMPTY. The signal STATUS is declared to be of type STATUS_TYPE, and will monitor whether the LIFO is empty or not.

The last part of the architecture declaration section defines a Function named INCR_PTR. Functions are a convenient way of separating code that is to be executed multiple times, or might be universal enough to be used by multiple programs. When a Function is declared at the top of an architecture, then it can be called from any place within the architecture body. The Function INCR_PTR reads two values, PTR and DEPTH, and returns an updated PTR value. The Function name, input values and the return type are declared with the Function declaration statement:

Function INCR_PTR (PTR : Integer; DEPTH : Integer)
Return Integer is

Inside of the Function, the "return" statement outputs the updated value of PTR, and exits the Function.

The actual execution of the LIFO is split into one clocked process (LIFO_PROC), and one concurrent assertion statement. Based on the conditions of the inputs, the clocked process loads the internal data stack, updates the pointer, and/or outputs data from the internal stack. The concurrent assertion to the enumerated type STATUS continuously monitors the PTR value, and sets the STATUS value appropriately.

--LAST-IN FIRST-OUT STACK EXAMPLE

```
library IEEE;
 use IEEE.std_logic_1164.all;

entity LIFO_STACK is
  Port ( CLK        : In   Std_Ulogic;
         RESET      : In   Std_Ulogic;
         DATA_IN     : In   Std_Ulogic_Vector (7 downto 0);
         RD         : In   Std_Ulogic;
         WR         : In   Std_Ulogic;
         DATA_OUT   : Out  Std_Ulogic_Vector (7 downto 0));
End LIFO_STACK;

Architecture SYNTHESIZABLE of LIFO_STACK is
```

```
-- Define Depth of Stack
    constant DEPTH: Integer := 16;

-- Define Data Array
    type DATA_ARRAY is array ( Integer range <> )
            of Std_Ulogic_Vector ( 7 downto 0 );
    Signal DATA: DATA_ARRAY ( 0 to DEPTH-1 );

-- Define type for Lifo Status
    type STATUS_TYPE is ( EMPTY, NOT_EMPTY );
    Signal STATUS: STATUS_TYPE;

-- Define pointer index
    Signal PTR: Integer range 0 to DEPTH-1;

-- Define Function INCR_PTR
--  Increments in range ( 0 to DEPTH - 1 )
    Function INCR_PTR (PTR : Integer; DEPTH : Integer)
        Return Integer is

    Begin
        If PTR = (DEPTH - 1)
            then  Return DEPTH - 1;
        Else
            Return PTR + 1;
        End if;
    End INCR_PTR;
--

Begin

LIFO_PROC: Process (CLK, RESET)

 Begin

 If RESET = '1' then
  PTR   <= 0;
  DATA_OUT <= "00000000";
```

```
Elsif Rising_Edge (CLK) then

  If WR = '1' then
     DATA (0) <= DATA_IN;
      If RD = '0' or STATUS = EMPTY then
         for i in 1 to DEPTH - 1 loop
             DATA (i) <= DATA (i-1);
         End loop;
         PTR <= INCR_PTR ( PTR, DEPTH ); -- Function Call
      End if;
  End if;

  If RD = '1' and STATUS = NOT_EMPTY then
     DATA_OUT <= DATA (0);
     If WR = '0' then
        for i in 1 to DEPTH - 1 loop
            DATA (i-1) <= DATA (i);
        End loop;
        PTR <= PTR - 1;
     End if;
  End if;
 End if;
End Process LIFO_PROC;

-- Monitor Lifo Status
STATUS <= EMPTY when PTR=0  Else NOT_EMPTY;

End SYNTHESIZABLE;
```

Example 7-10 : A LIFO Stack

First-In First-Out (FIFO)

The second stack example is a First-In First-Out (FIFO) stack. FIFO's
are mainly used to buffer streams of data between systems with different
data handling capacities. This might be an interface between a high-speed

system such as a computer chip and a slower-speed peripheral. In a FIFO, read operations always read data from the "bottom" of the stack. All other words are pushed down, with the next word in line now becoming the bottom of the stack. Figure 7-10 shows an example of a 4 deep FIFO. The bottom of the stack contains the data "1000". The top contains the data "0001". A read operation sets the value "1000" onto the output data bus (DATA_OUT). All currently stored words are pushed down in the stack. If the stack is empty, then no word is read.

Fast-In First-Out Stack : Read Operation

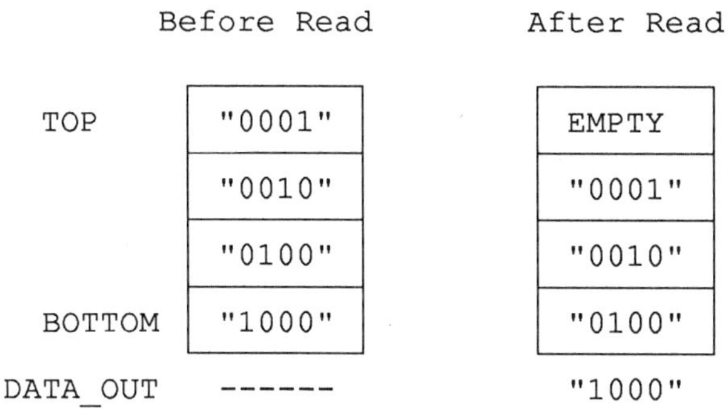

```
              Before Read            After Read

                  _____            _____
   TOP           | "0001"    |          |  EMPTY    |
                 |_____|          |_____|
                 | "0010"    |          | "0001"    |
                 |_____|          |_____|
                 | "0100"    |          | "0010"    |
                 |_____|          |_____|
   BOTTOM        | "1000"    |          | "0100"    |
                 |_____|          |_____|
   DATA_OUT         ------                 "1000"
```

Figure 7-10 : Reading from a FIFO

When a write request is generated to the stack, if the stack is not full, then the data is pushed onto the stack. If the FIFO is empty, then this data becomes the bottom of the stack, and is the next word to be read. If the FIFO is not empty, then this data is placed after the last loaded location. Figure 7-11 shows an example of writing to an empty FIFO.

Fast-In First-Out Stack : Write Operation
  Empty Stack

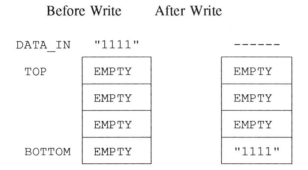

Figure 7-11 : Writing to an Empty FIFO

Figure 7-12 shows the before and after data values when writing to a stack with only one open location.

Fast-In First-Out Stack : Write Operation
  One Empty Location

	Before Write	After Write
DATA_IN	"1111"	------
TOP	EMPTY	"1111"
	"0010"	"0010"
	"0100"	"0100"
BOTTOM	"1000"	"1000"

Figure 7-12 : FIFO Write Operation

A FIFO VHDL description is shown in Example 7-11. The ports of the entity are the same as for the LIFO example. Inside the architecture, the type STATUS_TYPE includes an additional state: FULL. The FIFO stack does not write to a full stack, and therefore this condition must be monitored. Also, there are now three PTR signals instead of one. The RD_PTR indicates the location of the next word to be read from the internal array, the WR_PTR points to the next location to be written to, and the DIFF_PTR signal contains the difference between the two pointers. The pointers are actually a revolving index, incrementing from 0 to DEPTH minus 1, and then back to 0 again. This indexing is performed by the Function CHANGE_PTR.

FIRST-IN FIRST-OUT STACK EXAMPLE

```
-- First-In First-Out Stack
--

library IEEE;
  use IEEE.std_logic_1164.all;

entity FIFO_STACK is
  Port ( CLK          : In    Std_Ulogic;
         RESET        : In    Std_Ulogic;
         DATA_IN      : In    Std_Ulogic_Vector (7 downto 0);
         RD           : In    Std_Ulogic;
         WR           : In    Std_Ulogic;
         DATA_OUT     : Out   Std_Ulogic_Vector (7 downto 0));
End FIFO_STACK;

Architecture SYNTHESIZABLE of FIFO_STACK is

-- Define Depth of Stack
        Constant DEPTH: Integer := 16;

-- Define Data Array
        Type DATA_ARRAY is array ( Integer range <> )
                of Std_Ulogic_Vector ( 7 downto 0 );
        Signal DATA: DATA_ARRAY ( 0 to DEPTH - 1 );
```

```
-- Define type for Fifo Status
        Type STATUS_TYPE is (EMPTY,NOT_EMPTY,FULL);
        Signal STATUS: STATUS_TYPE;

-- Define pointer indexes
        Signal Rd_Ptr, Wr_Ptr: Integer range 0 to DEPTH-1;

-- Define pointer difference
        Signal Diff_Ptr: Integer range 0 to DEPTH;


-- Define Function CHANGE_PTR
--  Counts in range ( 0 to DEPTH-1 )

    Function Change_PTR (PTR : Integer; DEPTH : Integer)
          Return Integer is
     Begin
          If PTR = ( DEPTH - 1 )
                then  Return 0;
          Else  Return PTR + 1;
          End if;

    End CHANGE_PTR;
--

Begin

FIFO_PROC: Process (CLK, RESET)

 Begin

 If RESET = '1' then
     RD_PTR   <= 0;
     WR_PTR   <= 0;
     DIFF_PTR <= 0;
     DATA_OUT <= "00000000";

 Elsif Rising_Edge (CLK) then
```

```
If WR ='1' and
   (STATUS = NOT_EMPTY or STATUS = EMPTY) then
         DATA ( Wr_Ptr ) <= DATA_IN;
         Wr_ptr <= Change_ptr ( Wr_ptr, depth ) ; -- Function call
             If rd= '0' then
                 diff_ptr <= diff_ptr + 1;
             End if;
End if;

If rd= '1' and (status = not_empty or status = full) then

   data_out <= data ( Rd_ptr );
       Rd_ptr <= Change_ptr ( Rd_ptr, depth ); -- Function call
       If wr = '0' then
           diff_ptr <= diff_ptr - 1;
       End if;
   End if;
 End if; -- End of overall If
End Process FIFO_PROC;

-- monitor fifo status
   Status <=   Empty when diff_ptr = 0
               Else Full when diff_ptr = depth
               Else Not_empty;

End Synthesizable;
```

Example 7-11 : A FIFO Stack

Summary

This chapter has looked at some of the problems in formatting numerical data and methods to control the synthesis of data processing circuits. Multiple adder configurations were used to focus on controlling the resulting architectures, both with and without resource sharing, and multipliers were used to illustrate alternative VHDL techniques. Some of the VHDL language elements covered included the use of Procedures and Functions, loops, arrays, and generics. The next chapter will focus on the design and synthesis of combinational logic and related synthesis issues such as technology independence, logic minimization and optimization.

Chapter 8

Combinational Logic and Optimization

This chapter discusses the synthesis of purely combinational circuits, and how the logic optimization capabilities of synthesis tools can be used to control gate count and path delays through the logic. In the examples, the interaction of the ASIC library with the optimization tools will be illustrated along with the effect of the library's rules for loading and wire delays.

Combinational Logic

In earlier chapters, we have seen that combinational functions can be often generated within a larger sequential function and need not be explicitly defined. However, there are also times that you may desire to generate a specific block of combinational logic. In either case, the resulting logic can greatly depend on the way in which the source VHDL is written, since inefficient code can create extra gates, slower circuitry, and untestable logic. It is important to understand the effects that a specific piece of VHDL code will have on the resulting synthesized logic.

Logic designers regularly make use of different types of common combinational blocks. Functions such as multiplexers, decoders, encoders, and comparators are functions that can be identified and partitioned on a block diagram, and then coded directly at that level. Other types of combinational circuits such as adders, ALUs, and multipliers are also directly synthesizable and were discussed in Chapter 7.

General Synthesis Goals

Before considering combinational circuit designs, it is worthwhile to understand the concept of synthesis and optimization goals. The goal of logic synthesis is to create a functionally equivalent logic circuit from the VHDL description that meets or exceeds the specified design and target technology requirements. These requirements vary depending on the individual design and the resulting target technology. For instance, a common set of synthesis requirements for a CMOS gate array might be area, speed, and fanout. For different target technologies, other goals could be specified : you might have layout goals such as minimizing the number of nets, or power goals for ECL arrays, or CLB (Configurable Logic Block) goals for FPGA arrays.

Synthesis goals can generally be grouped by priority. Consider a CMOS gate array with these goals :

1. Implement a functionally equivalent design.

   This is always the highest priority, and is absolutely necessary or the synthesis build is considered unsuccessful. This goal is not mentioned for the rest of the chapter, but is always assumed.

2. Ensure that the synthesized circuit meets the target technology's fanout and design rule requirements.

>   This is the next highest priority. If not satisfied, the build is completed, but the resulting design might not meet target technology goals. This will usually result in an error in the ASIC vendor's signoff requirements and must be corrected before layout.

3. Ensure that the circuit speed is greater than or equal to the design's speed requirements.

>   Circuit speed is essential but while it can be attained by means of logic optimization, it can also be controlled by the VHDL design architecture used.

4. Ensure that the circuit area is less than or equal to the design's area, or gate-count, requirement.

>   This is the lowest priority of the four goals. It is certainly more important to have a correctly working design that meets all of the design constraints before worrying about the gate count.

Random Gate Logic

The simplest types of combinational logic consist of the random gate functions needed in most systems. A typical situation could be an output signal needed to detect a particular state of a counter, register, databus, or state machine. This type of low level logic is often implemented as a PLD or simple boolean gates in board designs, and in an ASIC can be synthesized easily using various styles of VHDL. For instance, at the lowest level you could code the boolean equations directly such as :

Out <= (Sig1 AND Sig2) OR Sig3 OR Not(Sig4);

Any of the VHDL logical operators can be used, and, as with any boolean equations, parentheses may be needed to avoid ambiguity. The VHDL packages that handle logical operations for the various signal types (eg, Std_Ulogic, Std_Logic) are normally supplied with the synthesis and simulation tools.

The same signal can be generated outside of a process by using the When syntax :

Out <= '1'   When (Sig1 = '1' AND Sig2 = '1') OR Sig3 = '1'
              OR Sig4 = '0' Else '0';

Detecting the state of a databus or register can also be done using the When format :

Sigout <= '1' when RegA = "1101000011110101" else '0';

You can extend this technique to multiple AND-OR terms in order to detect multiple values :

Constant Val1 : Std_Ulogic := "11110000";
Constant Val2 : Std_Ulogic := "00110100";

Sigout <= '1' when RegA = Val1 OR RegA = Val2 else '0';

In addition to these gating functions, more complex functions such as multiplexed adders were shown in Chapter 7, using a similar style of VHDL code. The next section discusses multiplexers in general.

Multiplexers

Multiplexers are typical of the type of combinational circuits that can be described simply in VHDL. For example, the description of a multiplexer :

" Select input bus A when S is inactive or select bus B when S
  is active "

represents a 2 input to 1 output multiplexer. Assuming that '1' is the active state, and '0' is the inactive state, this is easily written as a VHDL architecture at this level, as shown in Example 8-1. This example shows two VHDL alternates to implement the multiplexer, one using an unclocked process and the other outside of a process. Both techniques are equivalent and will synthesize into similar circuits.

Architecture SYNTH of Multiplexer is

Constant WIDTH : Integer := 2;   -- Define vector width = 2 bits
Signal A,B,Y,Z   : Std_Ulogic_Vector (WIDTH - 1 downto 0);
Signal S   : Std_Ulogic;

Begin

   -- **Outside of process : A conditional assignment statement**

      Y <= A when S = '0' else B;

   -- Inside of process : Using "If" statement

      Multiplexer_PROCESS: Process (A, B, S)

      Begin

         If S = '0' then   Z <= A;
         Else   Z <= B;
         End if;

      End process Multiplexer_PROCESS;

End SYNTH;

Example 8-1 : A 2 to 1 Multiplexer

The constant named WIDTH is defined and used in the vector size specification to allow the width of the vectors and thus the multiplexer to be varied easily. In this case, WIDTH = 2, resulting in a 2 bit wide 2 to 1 multiplexer. By changing just one constant declaration, the multiplexer can be changed to any desired width. In this example, the results on signals Y and Z are identical. The synthesis process will generate similar gating to generate both Y and Z. It should be noted that the resulting gate-level circuits might be identical, or might be slightly different from each other. This is an important distinction - the synthesized circuits must be functionally equivalent, but need not necessarily be identical in terms of their gates.

In general, if the design is not speed critical, then there is no reason to spend effort and gates on speeding up the paths. The emphasis should be on generating a minimum gate design that meets the target library requirements. For now, let's concentrate on minimum area only constraints.

The VHDL descriptions for the above multiplexer can be synthesized into many varied gate-level designs. Two examples for a 2 bit 2 to 1 multiplexer are shown in Figures 8-1 and 8-2. The first schematic uses

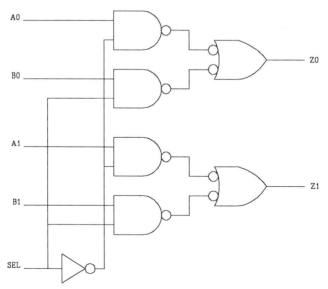

Figure 8-1 : Nand Implementation of 2 : 1 Multiplexer

only "nand" gates to implement the and-or functionality of the mux. The second schematic uses "and" and "or" gates to implement the same function. The synthesis tools must decide which design better meets the design constraints. In this case, there are no speed requirements, so the only goal is to generate a minimum gate multiplexer. For example, if the "nand" gate design uses fewer gates, then this implementation is chosen.

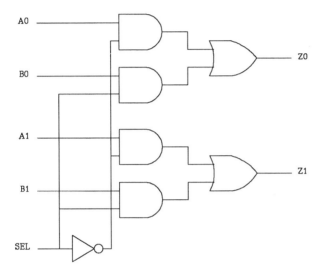

Figure 8-2 : And-Or Implementation of 2 : 1 Multiplexer

In the schematics of Figure 8-1 and 8-2, all signals of type "Std_Ulogic_Vector" have been expanded into bits. The vector A in the VHDL code now becomes signals named A1 and A0, although the exact signal names generated by synthesis for vectors depend on the specific synthesis tool. Similarly, the vectors B and Z have been expanded into 2 bits each. If the width of the multiplexer is changed (by changing the value of the constant WIDTH), then the number of bits generated from each vector would change accordingly. If WIDTH is set to 32, a 32 bit 2-1 multiplexer would be generated, and each vector would be expanded to 32 bits. This would generate considerably more logic from a simple change of one constant parameter, and again illustrates that the ability to make changes easily and rapidly is one of the great advantages of logic synthesis.

Example 8-1 was a very simple case of a 2-1 multiplexer. For larger multiplexers, such as 4 to 1 or more, the SELECT or CASE syntax is easier to implement and would again produce similar logic.

Architecture SYNTH of Multiplexer is

```
Constant WIDTH      : Integer := 2;   -- Define vector width = 2 bits
Signal A, B, C, D   : Std_Ulogic_Vector (WIDTH - 1 downto 0);
Signal Y, Z         : Std_Ulogic_Vector (WIDTH - 1 downto 0);
Signal S            : Std_Ulogic_Vector (1 downto 0);

Begin
    -- Outside of process : The Select statement

        With S SELECT
        Y < =
            A when "00",
            B when "01",
            C when "10",
            D when "11",
            A when others;

    -- Inside of a process : Using "CASE" statement

    Multiplexer_PROCESS: Process (A, B, C, D, S)
    Begin
        CASE S is

            When "00"      = > Z < = A;
            When "01"      = > Z < = B;
            When "10"      = > Z < = C;
            When "11"      = > Z < = D;
            When Others    = > null;

        End CASE;
```

End process Multiplexer_PROCESS;
End SYNTH;

<hr>

Example 8-1a : A 4 to 1 Multiplexer

The use of the SELECT and CASE syntax is intuitively simple and can easily be adapted to suit different applications. The "null" statement indicates to do nothing, a logic "don't care", for the other conditions.

Comparators

Combinational comparator functions can also be easily described in words, and subsequently can be described easily in VHDL. An "equal-to" comparator in words is:

" If A equals B, then the result is TRUE, else the result is FALSE "

A and B can be individual signals or data busses whose values are being compared. The VHDL code for this comparator (two bits wide) is shown in Example 8-2.

<hr>

Architecture SYNTH of EQUAL_COMPARATOR is

Constant WIDTH : Integer := 2;
Signal A,B   : Std_Ulogic_Vector (WIDTH-1 downto 0);
Signal Y,Z   : Boolean;

Begin

    -- outside of process
  Y <= TRUE when A = B Else FALSE;

```
      -- inside of process
 EQUAL_COMP_PROC: process(A,B)

Begin
   If A = B then
       Z <= TRUE;
   Else
       Z <= FALSE;
   End if;
End process EQUAL_COMP_PROC;

End SYNTH;
```

Example 8-2 : VHDL Code for an Equal To Comparator

Once again, based on the target library, one of many gate implementations could be created to produce the identical functions generating signals Y and Z. The given speed and area constraints will determine which gate level design is chosen. Figure 8-3 shows a schematic for a 2 bit "equal" comparator that uses exclusive-or gates.

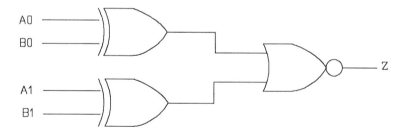

Figure 8-3 : A Synthesized Equal-To Comparator

Other types of comparators that can also be synthesized easily include the set of VHDL comparison operators :

/=   not equal to
<   less-than
<=   less-than or equal to
>   greater-than
>=   greater-than or equal to

The VHDL code for a less-than comparator is shown in Example 8-3.

Architecture SYNTH of LESS_THAN_COMPARATOR is

Constant WIDTH : Integer := 2;
Signal A,B   : Std_Ulogic_Vector (WIDTH-1 downto 0);
Signal Y,Z   : Boolean;

Begin

-- outside of process
Y <= TRUE when A < B Else FALSE;

-- inside of process
LESS_THAN_COMP_PROC: process (A,B)

Begin
    If A < B then
          Z <= TRUE;
    Else
          Z <= FALSE;
    End if;

End process LESS_THAN_COMP_PROC;
End SYNTH;

Example 8-3 : A "Less Than" Comparator

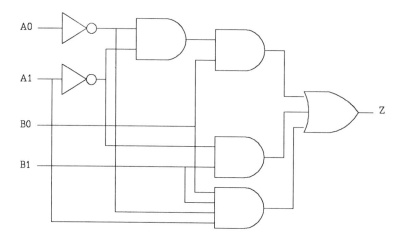

Figure 8-4 : Synthesized Less-than Comparator

Figure 8-4 shows a gate-level implementation of a 2 bit "less-than" comparator. The VHDL descriptions for the "equal" and "less-than" comparators are almost identical, but notice that the resulting gate-level designs look very different.

Decoders

Many common combinational components can be described with VHDL "select" and "case" syntax. As discussed in Chapter 3, the "select" statement is used outside of a process, while the "case" statement can only be used inside of a process. A 2 to 4 bit VHDL decoder could be described as shown in Example 8-4, with its synthesized gate-level schematic shown in Figure 8-5.

Architecture SYNTH of DECODER is

  Signal A   :   Std_Ulogic_Vector (1 downto 0);

Signal Y,Z : Std_Ulogic_Vector (3 downto 0);

Begin

-- Outside of process, uses a Selected Signal Assignment statement

With A Select
    Y <=   "0001" when "00",
             "0010" when "01",
             "0100" when "10",
             "1000" when others; -- when "11"

-- inside of process
-- "case" statement

DECODER_PROC: process(A)
Begin
 Case A is
    when "00"   => Z <= "0001";
    when "01"   => Z <= "0010";
    when "10"   => Z <= "0100";
    when others => Z <= "1000"; -- when "11"
 End case;

End process DECODER_PROC;
End SYNTH;

Example 8-4 : A 2 to 4 bit Decoder

A Priority Encoder

Similar to the decoder, a priority encoder could be described with either a "select" or "case" statement. In this example, bit 0 has priority. If bit 0 is set, then the encoder will output a result of zero, and set the ACTIVE output true. Else, if bit 1 is set, the output goes to one and again the ACTIVE output is set to true. This checking continues up to the

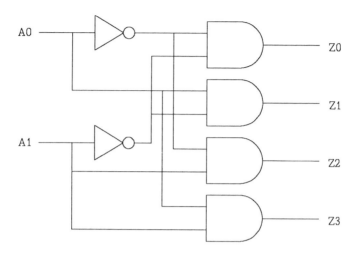

Figure 8-5 : A Synthesized 2 to 4 Decoder

highest input bit number. If no input bits are set, the output is invalid and the ACTIVE output is reset to false. An example of VHDL code for a 4 to 2 bit priority encoder is shown in Example 8-5. For the encoder description outside of a process, a concurrent assignment is used. Inside of the process, two examples are shown : a "case" version and an "if" version. Figure 8-6 shows one synthesized schematic representation of the 4 to 2 bit priority encoder. There are many possible gate-level designs that could be built from the same VHDL description. Remember, the actual built design varies due to the specific target gate-array library and design goals specified.

Architecture SYNTH of ENCODER is

Signal A   : Std_Ulogic_Vector (3 downto 0);
Signal X,Y,Z   : Std_Ulogic_Vector (1 downto 0);
Signal ACTIVE   : Boolean;

Begin

```
--   ***   outside of process  ***
   Y <=  "00"     when A (0)   = '1'
    Else "01"     when A (1)   = '1'
    Else "10"     when A (2)   = '1'
    Else "11"; -- ie, when A(3) ='1' or A="0000"

ACTIVE <= TRUE  when A > "0000"  Else FALSE;

--   ***   inside of process   ***

ENCODER_PROC: Process ( A )
Begin
   If A > "0000" then
        ACTIVE    <= TRUE;
   Else
        ACTIVE    <= FALSE;
   End if;

   -- "If" version
   If        A (0)   = '1'    then  X <= "00";
   ElsIf     A (1)   = '1'    then  X <= "01";
   ElsIf     A (2)   = '1'    then  X <= "10";
   Else                             X <= "11";
            -- ie, when A(3)='1' or A="0000"
   End If;

   -- "Case" version
   Case A is
   When    "0001" | "0011" |
           "0101" | "0111" |
           "1001" | "1011" |
           "1101" | "1111"   => Z <= "00";

   When    "0010" | "0110" |
           "1010" | "1110"   => Z <= "01";
```

```
When     "0100" | "1100"    => Z <= "10";

When    others              => Z <= "11";
        -- ie, when "1000" or "0000"
    End case;
  End process ENCODER_PROC;
End SYNTH;
```

Example 8-5 : VHDL Code of a Priority Encoder

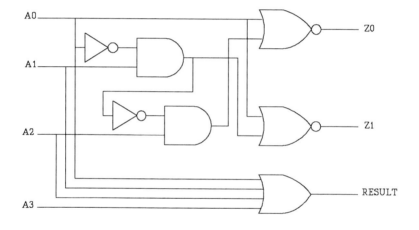

Figure 8-6 : Synthesized Priority Encoder

Logic Optimization Between Blocks

The functions discussed in this chapter have been solely combinational. No sequential flip-flops or latches are used to build the gate level designs. When these functions are connected together in a design, the logic flows through all paths within the same clock cycle. The synthesis tool has the ability to optimize logic gating to produce either a smaller or faster design, based on the area and timing goals. Much of this performance gain results from optimizing logic within each combinational block. But some performance and area improvement also results from combining the logic between blocks. Removing the dividing lines between the gate-level circuits creates a better design by allowing optimization of the entire circuit. The increase in performance comes with a price however : The direct correlation between the VHDL source and the gate-level result becomes much more difficult after minimization since each VHDL statement would no longer match directly to a section of gates. Instead, the optimized result of many VHDL statements is combined into a much larger section of gating. You should be aware of this behavior when debugging timing problems, since a problem in a section of gates will have to be identified with the source VHDL description in order to make changes.

As an example of what happens when combinational circuits are optimized, Figure 8-7 shows a block diagram of a multiplexer driving a priority encoder, followed by an equal-to comparison.

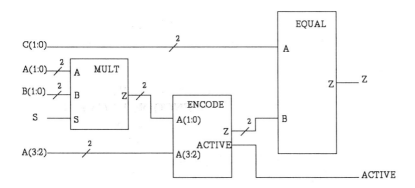

Figure 8-7 : Multiple Combinational Blocks

The VHDL code is shown in Example 8-6.

```
Entity COMPARISON is
  port (   A        : In      Std_Ulogic_Vector (3 downto 0);
           B        : In      Std_Ulogic_Vector (1 downto 0);
           C        : In      Std_Ulogic_Vector (1 downto 0);
           S        : In      Std_Ulogic;
           Z        : Out     Boolean;
           ACTIVE   : Out     Boolean);
End COMPARISON;

Architecture SYNTH of COMPARISON is
  Signal MULT_OUT      : Std_Ulogic_Vector (1 downto 0);
  Signal ENCODE_IN     : Std_Ulogic_Vector (3 downto 0);
  Signal ENCODE_OUT    : Std_Ulogic_Vector (1 downto 0);

Begin
 -- Multiplexer
  MULT_OUT <= A (1 downto 0) when S='0'
       Else B (1 downto 0); -- when S='1'

 -- Combine A (3:2) and MULT_OUT into vector
  ENCODE_IN <= A (3 downto 2) & MULT_OUT (1 downto 0);

 -- Encoder
  ENCODE_OUT <=        "00" when ENCODE_IN (0) = '1'
        Else           "01" when ENCODE_IN (1) = '1'
        Else           "10" when ENCODE_IN (2) = '1'
        Else           "11"; -- when ENCODE_IN(3) = '1'
              --     or ENCODE_IN = "0000"
  ACTIVE <= TRUE when (ENCODE_IN > "0000")  Else FALSE;
 -- Equal-to
  Z <= TRUE when (C = ENCODE_OUT)  Else FALSE;

End SYNTH;
```

Example 8-6 : VHDL Code for Multiple Combinational Functions

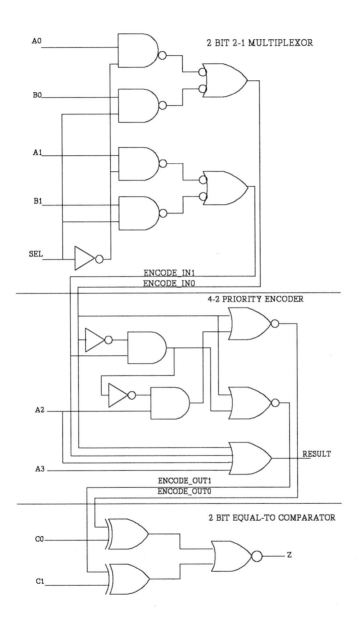

Figure 8-8 : Multiple Blocks Before Optimization

If each function is synthesized independently and left untouched, the schematic design could be directly correlated to the VHDL code, as illustrated in Figure 8-8.

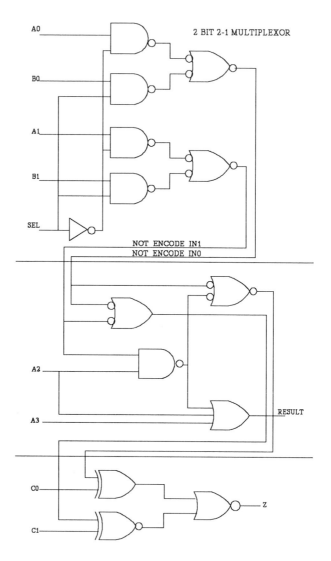

Figure 8-9 : Multiple Blocks After Logic Minimization

It can be seen that this design requires 17 gates, and the longest path contains 10 levels of gating (ie, the longest path from any input to any output passes through 10 gates).

If the synthesis optimizer is given the ability to optimize both internal to the blocks and the interconnections between blocks, then the resulting synthesized design could look much different. A schematic of a design optimized without boundary restrictions is shown in Figure 8-9. After optimization, this design contains 14 gates, and the longest path passes through only 7 levels of gating.

Logic Minimization

There are some basic boolean logic minimization steps that most logic synthesizers are able to perform. Some of the simple methods are reduction of constants, inverter deletion, and removal of redundant gating. More complicated techniques include Karnough mapping, Rand-McCluskey minimizations, and other boolean logic techniques that are outside of the scope of this text. Realize however that the ability to minimize logic is a key consideration in choosing a logic synthesis tool. Figure 8-10 shows two examples of simple logic minimization.

Figure 8-11 shows four ways to reduce logic when an input is a constant. This situation is quite common in most applications, occurring for instance when you add a constant to a signal : (RegA + "00101100") or if you test for a constant value (If RegA = "00111100" ...). The logic synthesized in these cases is greatly reduced from that of a full adder or comparator due to the minimization process.

Redundant gates can appear when logic minimization has not occurred fully, or when multiple blocks of gating are concatenated (such as in a hierarchical design). An optimal logic synthesis tool should have the capability to remove all redundant gates. Consider the following example of lines of VHDL code that are inadvertantly redundant, but not obviously wrong by inspection :

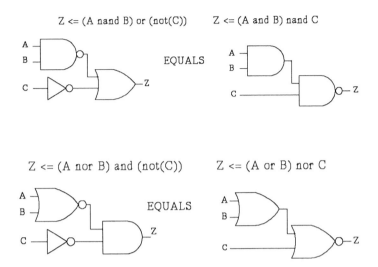

Figure 8-10 : Simple Logic Minimization

C <= A or (B and Not (SEL));

... perhaps more code in between ...

Z <= A when SEL='0' Else C;

This situation could arise in a long section of code where the two lines in question are not adjacent. In this case the logic is totally redundant and could be written equivalently as: "Z <= A". If the result of synthesis leaves any gating (other than buffers or inverters if necessary), then the circuit will contain redundant gates. Figure 8-12 shows one possible circuit that would be functionally equivalent to the above VHDL code. All six gates in this schematic are logically redundant and would be minimized to a simple connection of Z to A.

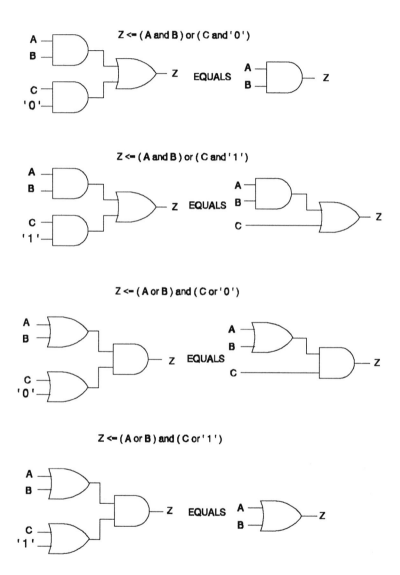

Figure 8-11 : Minimization of Constant Inputs

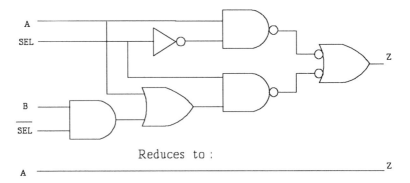

Reduces to :

Figure 8-12 : Redundant Gates

Optimization for Speed

In addition to minimization, speed optimization is a necessary feature of a logic synthesis tool. All critical speed paths need to be specified, and a minimum speed goal must be set. The synthesis process must invoke an internal timing analyzer to determine the delay of each path in the synthesized circuit. If speed is not a concern, then the resulting circuitry will be minimized for area, which in the case of gate array technology directly equates with gate count. In this case, logic optimization techniques will be employed with no regard for speed. But if certain logic paths are given speed requirements, or if specific clock timing constraints are specified, then other techniques must be used. After logic minimization, the timing analyzer is activated to determine the delays of the critical paths. If the timing of a path is too slow, then the path is isolated to try to improve the speed to meet the specified goal. This portion of the synthesis process is greatly dependent on the selected target technology. Consider a simplified example as shown in Figure 8-13. This type of gate circuit could occur if you are doing a wide decode such as

SigA <= '1' when Count = "1101000011001010" Else '0';

Since most ASIC libraries cannot directly implement a 16 bit AND gate, the synthesizer would have to create the equivalent from smaller gates, similar to the case in Figure 8-13.

Z <= (((((A and B) and C) and D) and E) and F) or G

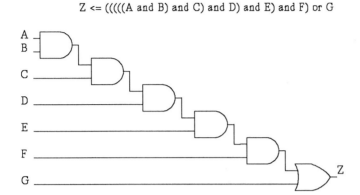

Figure 8-13 : Example of a Slow Delay Path

In this circuit, if the propagation delay through each gate is 2 nanoseconds (ns.), then the total A to Z path flows through 6 gates, creating a delay of 12 ns. If A to Z is the critical path, and a speed goal of 4 ns. maximum is required, then the circuit must be modified. Synthesis optimization for speed can reach this goal by isolating the "and" portion of the circuit as shown in Figure 8-14.

Y <= (((((A and B) and C) and D) and E) and F)
Z <= Y or G

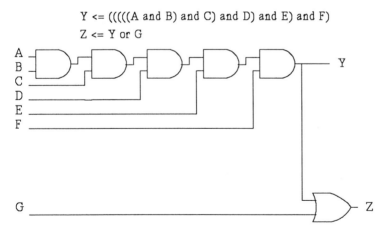

Figure 8-14 : Separating the And/Or Terms

The internal signal Y has been inserted to separate the "and" and "or" functions. The A to Z path can be shortened by reducing the A to Y path. This can be accomplished by swapping the A and F inputs, reducing the path to a total of 2 gates, with a delay of 4 ns. This circuit containing a faster A to Z path is shown in Figure 8-15.

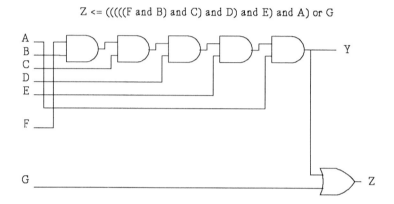

Figure 8-15 : Circuit Optimized for A to Z Delay

CMOS gate array speed is also greatly dependent on gate fanout (specified as loading). In general, the more places a gate must drive, the longer the total propagation delay will be for that network. As an example, Figure 8-16 shows a decoder in which the "and" gate driving the signal SELECT fans out to 8 places. Consider a particular gate-array library in which a gate that drives 1 load takes 2 nsec to propagate, but for the same gate to drive 8 loads requires 9 nsec. A table specifying the load characteristics of this gate is shown in Figure 8-17.

In Figure 8-16, the path from A_1 to OUT_1 takes 13 ns. If this is the critical speed path, and a maximum speed goal of 11 ns is required, then the synthesis tool must try to reduce the path delay. Simple swapping of input pins will not increase the path speed, since all delays from the inputs to the OUT_1 output are equal. Two other techniques can be used to reduce the delay. The first technique will reduce the propagation delay through the gate by inserting buffering. Buffer gates have the ability of driving higher fanouts with less propagation delay. A sample speed vs. loading characteristic table for a buffer gate is also shown in Figure 8-17.

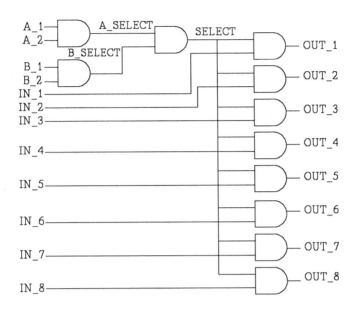

Figure 8-16 : Decoder gate driving 8 loads

GATE SPEED vs. OUTPUT LOADING

	NUMBER OF LOADS							
	1	2	3	4	5	6	7	8
SPEED (ns)	2	3	4	5	6	7	8	9

BUFFER SPEED vs. OUTPUT LOADING

	NUMBER OF LOADS							
	1	2	3	4	5	6	7	8
SPEED (ns)	2	2	3	3	4	4	5	5

Figure 8-17 : Typical gate and buffer speed vs. loading

Consider a buffer inserted between the gate generating SELECT and the 8 places that SELECT drives. The buffer can drive the 8 loads with a

propagation delay of only 5 ns. Now the path from A_1 to OUT_1 flows
through 3 "and" gates and 1 buffer, with a delay of 11 ns.

If the speed goal from A_1 to OUT_1 was 7 ns, then inserting
buffering would not be enough to meet the requirement. The second
synthesis technique that could be employed is to reduce the loading on
SELECT by adding parallel "and" gates. If a parallel "and" gate is added
as shown in Figure 8-18, then the loading on A_SELECT and
B_SELECT doubles, increasing the propagation delay on each gate from
2 ns. to 3 ns. But the "and" gate generating SELECT_1 now only drives
1 load, requiring only 2 ns, and the propagation delay from A_1 to
OUT_1 through the internal signal SELECT_1 is reduced to only 7 ns.

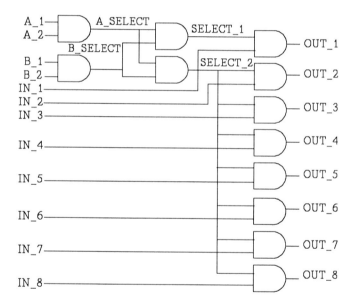

Figure 8-18 : The speed optimized decoder circuit

The synthesizer has added only a single gate to the minimal area
design, but the speed goal has been met, by trading size (number of
gates) for speed (propagation time). The implementation details of how
logic is optimized for speed and area is greatly dependent on the target
technology and the particular ASIC family used, and also on the

sophistication of the synthesis tools themselves.

Adders Revisited

Simple speed optimization techniques can effectively decrease the delay through specific logic paths. These methods work on a portion of the circuit without a knowledge of functionality, since only the connectivity of the gates is known to the speed optimizer. However, many times the choices that are made before the gates are created can greatly affect the speed of the resulting circuit.

Consider a simple description for an adder from Chapter 7 containing the assignment

RESULT <= A + B;

There are many equivalent ways to design an add function, ranging from a serial connection of full adder cells to a sopisticated multilevel carry look ahead scheme. The synthesized results of a minimal area design can greatly differ from the maximum speed design, and therefore choices must be made depending on your application. If the minimal code approach is used, such as A + B, and yet you have a stringent speed requirement, the synthesizer may or may not be able to implement the fastest look ahead scheme directly. When using functions such as adders and other ALUs, two different approaches are commonly taken.

The first approach is very technology and ASIC library specific. During synthesis, a predesigned adder could be selected from a macro library which has been tailored to create different speed grades of adders, with the faster circuits usually requiring the most area. A different macro could be selected for each application, based on the speed goal desired. This macro would be instantiated in-place, and left untouched by the synthesis tool.

The second approach is to insert carry look-ahead circuitry, either by specialized algorithms built into the synthesizer, or by adding the specific logic equations to the VHDL description. Carry look ahead can be implemented in numerous ways, depending on the number of bits used and the speed required. With this technique, the carry path through the adder is designed to minimize the delay through that particular path. This speeds up the slowest paths through the adder but results in a much larger circuit, often requiring 25 to 50 percent more area.

Summary

As with other logic functions, the code for combinational logic can be straightforward, but considerations of speed and logic minimization are often more demanding and require good design planning. By using good engineering judgement, logic synthesis tools can provide a powerful means to control and optimize a design in order to meet the requirements of many applications. The next chapter will discuss how the various design entities that comprise a complete ASIC can be put together and combined with test benches for functional logic simulation and test.

Chapter 9

Putting the Pieces Together

Up to this point, the discussions have focussed on VHDL and logic synthesis design issues. This chapter discusses how to combine the various design entities and related pieces together into a complete ASIC, and how to generate VHDL test benches for simulation.

Creating a Complete ASIC

The previous chapters have dealt with design of logic circuits that most likely would be part of a single VHDL entity or functional block of logic. As discussed in Chapter 2, these blocks should generally be limited in size to a few thousand gates so that they can be synthesized and simulated relatively quickly, and to help in floorplanning the ASIC. Larger functions such as wide ALUs or data processors can often be broken into sections by doing the calculations in smaller bit slices, or by reducing the width of the datapaths temporarily until the basic logic functionality is verified. In cases where that is not possible, larger blocks can still be synthesized but at some cost of time and schedule if design changes or problems occur.

In any event, most larger ASICs consist of multiple entities, plus I/O cells, high current drivers for clocks and resets, nand trees for input current testing, and perhaps functions such as boundary scan test structures. These pieces are interconnected by means of VHDL structural syntax, as shown below. Depending on your ASIC supplier, there may be individual preferences as to how these pieces are grouped into a complete netlist, how the use of hierarchy affects layout and floorplanning, and how the entity and netlist structures are generated.

The basic structure of larger designs can be generated in several ways. Up to this point, the VHDL code for Entities and Architectures has been shown as text throughout the examples, but that does not mean that the text must necessarily be typed into a text file. Most of the commercially available design tool systems provide a graphical schematic means of entering a design entity, from which the basic VHDL Entity and Architecture "shell" can be automatically generated. The "schematic" for an Entity is simply a block showing the input and output signals that are the ports of the Entity, and is simply a pictorial view of the Entity. Files normally generated from this block include the Entity with its ports, an empty Architecture structure, and usually a test bench structure, covered later in this chapter. The empty Architecture file is where the RTL code for the entity's functionality is entered.

As the various Entities are created, the task of combining them together into a larger design module or even the complete ASIC can be done schematically or by entering the text directly. At the schematic level, the various Entities are handled as components and are interconnected graphically as in any other schematic system, that is, each component "part" has a designator and has specific signals connected to its pins. When the VHDL code for this schematic is generated, it creates a VHDL structural netlist containing the Components, designators, and signal names on the ports of the components. The following section will present the format of this VHDL structure.

VHDL Structural Descriptions

Consider the schematic of a larger design entity that consists of multiple blocks, as shown in Figure 9-1.

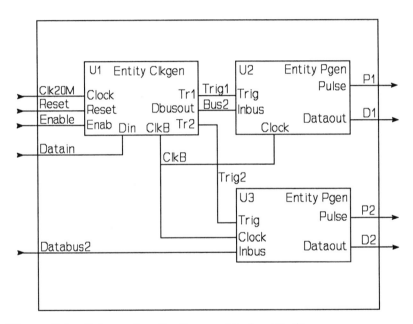

Figure 9-1 : Schematic of a Larger Design Entity

This design uses two different Entities, named CLKGEN and PGEN, and there are two separate instances of PGEN, designated as components U2 and U3. Note that like any other schematic, the actual signals connected to the component pins may have different names than the individual Entity's port names. Without this capability, you would not be able to use an Entity more than once in a design, and the signal naming flexibility would be greatly restricted and difficult to manage.

Assuming the two Entity types had been previously designed, their Entity descriptions would be as follows :

```
Entity Clkgen is
    Port (
        Clock       : In    Std_Ulogic;
        Reset       : In    Std_Ulogic;
        Enab        : In    Std_Ulogic;
        Din         : In    Std_Ulogic_Vector(15 downto 0);
        ClkB        : Out   Std_Ulogic;
        Tr1         : Out   Std_Ulogic;
        Tr2         : Out   Std_Ulogic;
        Dbusout     : Out   Std_Ulogic_Vector(15 downto 0)
    ); -- (end of port)

End Clkgen;

Entity Pgen is
    Port (
        Clock       : In    Std_Ulogic;
        Reset       : In    Std_Ulogic;
        Trig        : In    Std_Ulogic;
        Inbus       : In    Std_Ulogic_Vector(15 downto 0);
        Pulse       : Out   Std_Ulogic;
        Dataout     : Out   Std_Ulogic_Vector(7 downto 0)
    ); -- (end of port)

End Pgen;
```

The overall top level design consists of another VHDL entity and architecture pair named Top_gen, whose I/O ports are as shown in Figure 9-1:

Entity Top_gen is

```
Port (
    Clk20M    : In    Std_Ulogic;
    Reset     : In    Std_Ulogic;
    Enable    : In    Std_Ulogic;
    Datain    : In    Std_Ulogic_Vector(15 downto 0);
    Databus2  : In    Std_Ulogic_Vector(15  downto  0);
    P1        : Out   Std_Ulogic;
    D1        : Out   Std_Ulogic_Vector(7 downto 0);
    P2        : Out   Std_Ulogic;
    D2        : Out   Std_Ulogic_Vector(7 downto 0)
    ); -- (end of port)
End Top_gen;
```

The Architecture of Top_gen will be a structural VHDL interconnection of the entity components and signals. The architecture will consist of two parts. The **Declaration** section includes

Signal Declarations for signals such as Trig2 that are internal to the Top_gen entity,

Component Declarations for the "parts" used, and

Constant Declarations, if needed.

The **Instantiation** section includes **Component Instances**, such as U1, U2, etc, with their specific signal connections defined in a **Port Map declaration**.

The Architecture for Top_Gen therefore looks like this :

Architecture Schematic of Top_gen is

```
Signal Trig1   : Std_Ulogic;
Signal Trig2   : Std_Ulogic;
Signal ClkB    : Std_Ulogic;
Signal Bus2    : Std_Ulogic_Vector ( 15 downto 0);
```

A **Component Declaration** for each entity type being used within Top_gen. The Component Declaration simply references the Entity and the names and types of its I/O ports.

Component Clkgen
    Port (

Clock	: In	Std_Ulogic;
Reset	: In	Std_Ulogic;
Enab	: In	Std_Ulogic;
Din	: In	Std_Ulogic_Vector(15 downto 0);
ClkB	: Out	Std_Ulogic;
Tr1	: Out	Std_Ulogic;
Tr2	: Out	Std_Ulogic;
Dbusout	: Out	Std_Ulogic_Vector(15 downto 0)

    ); -- (end of port)
End Component; -- Clkgen

Component Pgen
    Port (

Clock	: In	Std_Ulogic;
Reset	: In	Std_Ulogic;
Trig	: In	Std_Ulogic;
Inbus	: In	Std_Ulogic_Vector(15 downto 0);
Pulse	: Out	Std_Ulogic;
Dataout	: Out	Std_Ulogic_Vector(7 downto 0)

    ); -- (end of port)
End Component; -- Pgen

An **Instance name, or instantiation label,** for the device such as U1, U2, etc. The instance of the Entity is similar to having an occurrence of a part on a schematic, in which you provide an identifier such as U1 and the specific signals that connect to its I/O ports in a list called a **Port Map.** This ability to map signal names to the ports allows you to have multiple copies of a component, each with different I/O signals, operating together within the higher level entity. The syntax for the signals in the port map list is :

Entity port name => Schematic signal name

The Begin statement indicates where this part of the architecture starts.

Begin

U1 : Clkgen

```
Port Map (
        Clock       => CLK20M, -- A "Named" Port Map
        Reset       => Reset,
        Enab        => Enable,
        Din         => Datain,
        ClkB        => ClkB,
        Tr1         => Trig1,
        Tr2         => Trig2,
        Dbusout     => Bus2
        ); -- End Port Map
```

U2 : Pgen

```
Port Map (
        Clock   => ClkB,
        Reset   => Reset,
        Trig    => Trig1,
        Inbus   => Bus2,
        Pulse   => P1,
        Dataout => D1
        ); -- end port map
```

U3 : Pgen

```
Port Map (
        Clock    => ClkB,
        Reset    => Reset,
        Trig     => Trig2,
        Inbus    => Databus2,
        Pulse    => P2,
        Dataout  => D2
        ); -- end port map
```

The Port Map syntax of

Entity port signal name => Schematic signal name

is referred to as a Named Port Mapping and has the advantage that the signal list is order-independent and, as shown, different signals can be connected to the entity ports. You can also use an implicit or Positional Port Map :

U3 : Pgen

        Port Map (
                Clock,
                Reset,
                Trig,
                Inbus,
                Pulse,
                Dataout
                ); -- end port map

    where the order of the signals must be exactly the same as in the entity port list. The use of this form requires more care in maintaining the port name order, but in the case where the same signal names are used as the port names, the list can be easily copied from the entity.

    Gathering all of these sections together, the complete syntax for the design Top_gen is shown in Example 9-1 :

Entity Top_gen is
        Port (
                Clk20M   : In   Std_Ulogic;
                Reset   : In   Std_Ulogic;
                Enable   : In   Std_Ulogic;
                Datain   : In   Std_Ulogic_Vector(15 downto 0);
                Databus2 : In   Std_Ulogic_Vector(15 downto 0);
                P1   : Out   Std_Ulogic;

```
        D1        : Out   Std_Ulogic_Vector(7 downto 0);
        P2        : Out   Std_Ulogic;
        D2        : Out   Std_Ulogic_Vector(7 downto 0)
    ); -- (end of port)
End Top_gen;
```

Architecture Schematic of Top_gen is

```
    Signal Trig1      : Std_Ulogic;
    Signal Trig2      : Std_Ulogic;
    Signal ClkB       : Std_Ulogic;
    Signal Bus2       : Std_Ulogic_Vector ( 15 downto 0);
```

Component Clkgen
```
    Port (
        Clock      : In    Std_Ulogic;
        Reset      : In    Std_Ulogic;
        Enab       : In    Std_Ulogic;
        Din        : In    Std_Ulogic_Vector (15 downto 0);
        ClkB       : Out   Std_Ulogic;
        Tr1        : Out   Std_Ulogic;
        Tr2        : Out   Std_Ulogic;
        Dbusout    : Out   Std_Ulogic_Vector (15 downto 0)
    ); -- (end of port)
End Component; -- Clkgen
```

Component Pgen
```
    Port (
        Clock      : In    Std_Ulogic;
        Reset      : In    Std_Ulogic;
        Trig       : In    Std_Ulogic;
        Inbus      : In    Std_Ulogic_Vector (15 downto 0);
        Pulse      : Out   Std_Ulogic;
        Dataout    : Out   Std_Ulogic_Vector (7 downto 0)
    ); -- (end of port)
End Component; -- Pgen
```

Begin

U1 : Clkgen
   Port Map (
          Clock   => CLK20M,
          Reset   => Reset,
          Enab   => Enable,
          Din   => Datain,
          ClkB   => ClkB,
          Tr1   => Trig1,
          Tr2   => Trig2,
          Dbusout => Bus2
          ); -- End Port Map

U2 : Pgen
   Port Map (
          Clock   => ClkB,
          Reset   => Reset,
          Trig   => Trig1,
          Inbus   => Bus2,
          Pulse   => P1,
          Dataout => D1
          ); -- end port map

U3 : Pgen
   Port Map (
          Clock   => ClkB,
          Reset   => Reset,
          Trig   => Trig2,
          Inbus   => Databus2,
          Pulse   => P2,
          Dataout => D2
          ); -- end port map

End Schematic; -- Architecture of Top_gen

Example 9-1 : A Structural Description of a Multiple Entity Design

Design Hierarchy

A Design Entity such as Top_gen could represent an entire ASIC or a functional sub_block of a design. During the course of the design phase, you may choose to simulate individual entities as well as functional groups of entities, and eventually the entire synthesizable design. The structural syntax will be used to connect entities together, and the entities to the test benches, as covered below. As you can see from Example 9-1, there is some overhead in having too many entities in a design in that the structure will then consist of many components, instances, and signal-to-port assignments. On the other hand, as explained in Chapter 2, the entity sizes should be limited to a moderate gate count so that the synthesis process can be done in a reasonable length of time. This tradeoff is a practical guideline that will change as workstations become faster and synthesis software becomes more efficient, and the use of larger synthesizable blocks will be more feasible in the future.

VHDL Simulation and Test Benches

Throughout the design process, each Design Entity is normally tested and debugged at the RTL level using the VHDL simulator. As with any other simulation system, you need both the data representing the design entity being simulated and a means of generating the input signals or stimulus to the design, along with a way to monitor the simulation results. In the VHDL environment, the input signals to the Entity Under Test are generated by means of another VHDL Entity referred to as a **Test Bench**. The concept of a test bench is an accepted VHDL standard means of simulating, testing, and documenting an entity, and the entity/test bench pair can form the basis for executable specifications and documentation in a top-down design methodology. A separate test bench can be created for each Entity being simulated, as well as for multiple Entities and finally for the complete ASIC.

Figure 9-2 shows a test bench that generates the input signals for the design being tested and monitors the results at the output ports or at points internal to the entity. Note that the test bench is a complete module that encompasses and communicates with the entity under test but has no I/O ports to the outside world.

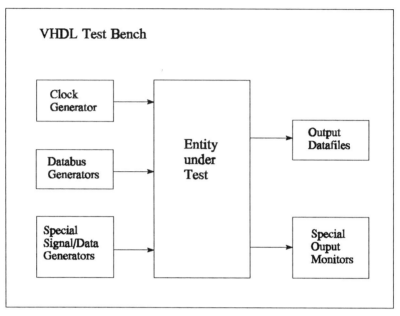

Figure 9-2 : A Basic VHDL Test Bench

From the VHDL viewpoint, the test bench includes a component declaration for the entity being tested and an instantiation of that component, just as in the structural entity Top_gen of example 9-1. This declaration includes a port map identical to the ports of the design's entity. In addition, the test bench declares all of the signals connecting to the component, and also the code to generate the input signals that drive the component. The test bench itself is normally organized as a separate file from the entity being tested because it does not get synthesized and because several different test benches may be necessary to completely test a design. For instance, you may have a simple test that checks basic functions and a more complex set that exhaustively checks all combinations in an ALU or function generator, and keeping them separate allows more control during the simulation process.

The difference between a test bench entity and a synthesizable design entity is that the test bench can take advantage of using the full VHDL language. It is not restricted by synthesis guidelines and design rules. This fact allows for great flexibility throughout the design cycle in generating different types of stimulus signals, examples of which will be shown below. In addition, output results from the test bench can consist of logic states as a function of time, messages to announce the presence of some desired conditions, and file outputs that can be used to drive other design tools such as the gate level simulator and the final ASIC verification tools. Different VHDL simulators also provide graphical waveforms and analysis capabilities specific to the tools.

Let's simulate the Top_gen entity, as shown in Example 9-1. The input signals consist of a 20 MHz clock, a master reset, an enable signal, and two separate data busses. Using this data, the test bench can be generated.

Test Bench Structure

The Entity of the test bench contains no external ports, and therefore the entity declaration is simply :

Entity Tb_Top_gen is   -- There are no ports in the test bench
End Tb_Top_gen;

A good practice is to give each of your test bench entities a name that helps identify its use, such as the prefix **Tb_** and the name of the entity being tested. The same convention should be also be applied to the names of the test bench files.

The Architecture of the test bench usually includes signals, component declarations, component instances, and signal stimulus generators:

1. A declaration list of the signals used to connect with the entity or entities being tested. This list is a normal signal declaration list, and can optionally specify an initial state of a signal for simulation. In the list below, note that the signal Clk20M is initialized to logic 1 for the simulation by using the syntax (:= **value**) :

```
Signal  Clk20M   : Std_Ulogic := '1'; -- Initial value ;
Signal  Reset    : Std_Ulogic := '0';
Signal  Enable   : Std_Ulogic;
Signal  Datain   : Std_Ulogic_Vector (15 downto 0);
Signal  Databus2 : Std_Ulogic_Vector (15 downto 0);
Signal  P1       : Std_Ulogic;
Signal  D1       : Std_Ulogic_Vector (7 downto 0);
Signal  P2       : Std_Ulogic;
Signal  D2       : Std_Ulogic_Vector (7 downto 0)
```

In a test bench where the only component is the entity being tested, the signal names are usually made to agree with the entity port names for simplicity.

During simulation, signals that are not initialized are unknown at the start of the simulation. For Std_Ulogic signals and variables, a separate uninitialized state 'U' is defined for this purpose. It is a good idea to initialize the signals generated in the test bench to some known value, since the design entity itself will also contain unknowns until it is reset or clocked to a known state, and the unknown conditions that exist in the simulation can impede debugging.

2. **The Component Declaration** for the Top_gen entity being driven by the test bench :

Component Top_gen
```
    Port (
        Clk20M   : In    Std_Ulogic;
        Reset    : In    Std_Ulogic;
        Enable   : In    Std_Ulogic;
        Datain   : In    Std_Ulogic_Vector (15 downto 0);
        Databus2 : In    Std_Ulogic_Vector (15 downto 0);
        P1       : Out   Std_Ulogic;
        D1       : Out   Std_Ulogic_Vector (7 downto 0);
        P2       : Out   Std_Ulogic;
        D2       : Out   Std_Ulogic_Vector (7 downto 0)
    ); -- (end of port)
```
End Component;

3. The Begin and Component Instantiation of Top_gen :

Begin

U1 : Top_gen
 Port Map (
  Clk20M => Clk20M, -- Explicit signal map
  Reset   => Reset,
  Enable,   -- Implicit signal map
  Datain,
  Databus2,
  P1,
  D1,
  P2,
  D2
  ); -- end port map

4. The VHDL code that generates the signals driving the entity under test.

As shown in Figure 9-1, signals can be either single discrete waveforms or data busses. Discrete signals can include repetitive waveforms such as clocks, single pulses used to reset logic or load registers, asynchronous timing waveforms and pulse trains, and serial data streams. These types of signals can be independent or can be related to each other, such as in a group of signals used to control, read, and write from a memory. Data busses can consist of a few data words in simple simulations to complex data patterns or values. In many cases, these signals can be generated from simple VHDL code, whereas some types of signals can best be generated by modelling a hardware process or by reading signal values from a data file.

Signal Generation

As an example, let's consider how to generate a few simple signals.

A. DC Signals :

These signals are constant levels that are simple to create, and are used at some point in most simulations:

```
SigA <= '1';
SigB <= '0';
Data <= "00101111";
```

B. Repetitive clock signals :

One of the simplest types of signals to generate is a symmetrical clock or square wave. The code for a 10 MHz symmetrical clock called Clk10M could be as simple as :

Clk20M <= **Not** (Clk20M) **after 25 ns**;

Assuming that CLK20M was initialized to 0 or 1, the signal is simply inverted every 25 nsec. A better way to set the clock rate however is to define a constant of type TIME, which is a VHDL predefined type, as follows:

Constant Clk_period : Time := 50 ns;

and then generate the clock as

Clk20M <= Not (Clk20M) **after Clk_period/2;**

The use of constants such as Clk_period makes it easier to make changes as the design progresses.

C. Asymmetrical repetitive waveforms or pulse trains

In this case, a simple process using Waits can be used. Let's make a pulse train whose period is 100 usec and is active high for 1 usec :

Constant Clk_period : Time := 99 us;

```
Pulse1 : Process
Begin
  Pulse <= '1';
  Wait for 1 us;
  Pulse <= '0';
  Wait for Clk_period;
End Pulse1;
```

Pulse

The time specified in the Wait statement need not be an integer, but can be a real number such as 12.36 us. Each independent pulse train with a unique period must be in its own process and can be totally asynchronous to any other signal or clock if desired. You can however generate several signals within one process if they are related in time and have the same period. For example, consider signal SIGA that is 8 usec wide and is delayed from signal Pulse by 12 usec:

```
Pulse   <= '1';
Wait for 1 us;
Pulse   <= '0';
Wait for 11 us;
SIGA    <= '1';
Wait for 8 us;
SIGA    <= '0';
Wait for 80 us;  -- ie, Sum of waits is 100 usec
```

This technique can be extended for multiple signals but obviously can become lengthy for complex timing delays.

D. Simple non-repetitive signals :

Signals that occur once, such as a master reset signal, can be simply generated as :

Constant Reset_Width : Time := 140 ns;

Reset <= '0', '1' after Reset_Width; - Active low reset signal

In this example, the first assignment to logic 0 occurs at time 0 of the simulation. For more complex but manageably simple signals, you can extend the technique :

Enable <= '1', '0' after 27 us, '1' after 78 us, 'Z' after 112 us,
        '0' after 189 ns;

Constants for the time intervals can also be used here. Note that the 'Z' is the high impedance state of Std_Ulogic.

E. Regular Data Patterns

It is often necessary to drive a data bus with some type of regular data pattern, such as an incrementing or linear value, an arithmetic function, or a binary walking bit pattern. In previous chapters we synthesized similar functions, but in the test bench the technique is simpler. Consider the linear ramp, otherwise known as a counter :

Signal Databus : Std_Ulogic_Vector (7 downto 0) := "00000000";

Databus <= Databus + '1' after 75 ns; --

Notice that this function can be easily implemented in one line rather than the multiple lines of a synthesizable clocked process. Better yet would be to define a constant such as

Constant Data_delay : Time := 75 ns;

and then   Databus <= Databus + '1' after Data_delay;

A pattern of walking ones is often used in test and can be implemented by setting a vector to the value 1 and either adding the value to itself in a loop or alternatively multiplying the value by 2 :

```
Walk1 : Process
Begin
     DBUS <= "00000001";

     Loop1:
          For i in 0 to DBUS'length loop
               Wait for Data_delay;
               DBUS <= DBUS + DBUS;
          End loop Loop1:
End Process Walk1;
```

In this example, rather that using "For i in 0 to 7 loop", the VHDL predefined attribute " 'Length " is used to get the length of DBUS. If the size of DBUS is changed, the process would automatically be accommodated.

Other arithmetic functions such as multiply and divide can be used directly in test benches to implement more complex data patterns or time varying patterns.

F. Table driven Data Patterns

Many applications have data inputs that assume specific values, such as the data used to load control parameters into registers or processors. A simple means of handling this situation is to create a table of the desired values and then cycle through the table to generate the databus signals. The following shows the basic use of a VHDL array to hold a table of integer values for frequencies, which will then be outputted as signals DBUS (an integer) and DBUS_V (a Std_Ulogic_Vector of 16 bits). The Type and Constant definition can appear either in the Entity or in the Architecture, or in an external Package :

```
Type freq is array (1 to 6) of Integer; -- An array "constrained" to 6
                                         -- values
Constant Freqval : freq := (   6820,
                            10230,
                            10231,
                             9507,
                              235,
                             1238  );
```

```
Signal DBUS : Integer;
Signal DBUS_V : Std_Ulogic_Vector (15 downto 0);
Constant Bus_delay : Time := 75 ns;
```

```
Begin
```

```
Table_gen : Process
 Begin
     Loop1: for i in 1 to 6 loop

         DBUS <= Freqval(i);
         DBUS_V <= To_SUV (Freqval(i),16);

         Wait for Bus_delay;
     End Loop Loop1;
 End process;
```

Notice that the integer values in the Constant table Freqval get converted to Std_Ulogic_Vector format by a Type Conversion routine named "To_SUV". The code for this function will be shown in this chapter under File Input, although similar type conversion functions are often provided by the supplier of your VHDL simulation or synthesis tools. Without this function, the table would have been generated in vector format rather than integers, which might be less meaningful in some cases.

The table above also is limited in that it uses fixed values for the size of the array and for the loop. If you need to increase the size of the table, you would have to edit the constants (1 to 6 ...), and could be prone to error. A better technique would be to use an "unconstrained" array, where the array and loop size is determined by the VHDL compiler based on the amount of data provided. The two lines of fixed constants would be replaced as follows :

--   Type freq is array (1 to 6) of Integer;
   Type freq is array (natural range <>) of Integer;

---   Loop1: for i in 1 to 6 loop
   Loop1: for i in Freqval'LOW to freqval'HIGH loop

where the 'LOW and 'HIGH VHDL predefined attributes are used to compute the array lower and upper bounds. Now you can add values to the table freqval and the test bench will automatically adjust to the new count of values. This technique can be extended to a wide range of applications.

G. Design Oriented Signal Generation

Some types of signals should be generated by techniques similar to those used for logic design. For instance, it would be very cumbersome to generate a serial data stream for a communications system if you had to write each bit individually. In this case, the data words or messages are normally thought of in terms of entire words, and a logic designer would typically use a parallel to serial converter to create the signal.

Generation of this type of signal in VHDL should use a similar technique, in which the parallel data words are outputted bit by bit using a loop :

Assuming a 16 bit signal named Databus in Std_Ulogic_Vector format, and a shift rate of 4 MHz (or 250 nsec/bit) :

```
Serial : Process
Begin
  Loop1 :
    For i in 0 to Databus'High loop
      Serial <= Databus(i);
      Wait for 250 ns;
    End Loop Loop1;
End Process Serial;
```

By extension, you could use a second loop to retrieve different data
values for Databus from a table, and then convert each to serial form as
above.

In situations where a signal like the serial data stream is an output of
the entity, the problem arises as to how to monitor and verify the results.
You could visually inspect the state of the signal as the simulation runs,
but checking bit-wise results is tedious at best. A better approach is to
think about building "hardware" to convert the serial data back to parallel,
as part of the test bench. This technique is applicable to a wide range of
applications, although the example below will only show serial data
conversion.

Assume there is a serial data output named Serial, whose bit rate is 20
MHz. Issues such as synching to the data have been determined and a
signal Valid indicates that data is shifting in, least significant bit first. A
simple process can handle the task of shifting the data into the register
Outreg, every 50 nsec :

```
Shreg : Process
  Begin
    If valid = '1' then
      Outreg <= Serial & Outreg(15 downto 1);
    end if;
  Wait for 50 ns;
  End process;
```

In a more complex test bench, you would also take care of the problem
of synching to the data, oversampling, etc. But the point is that the test
bench can be made to mimic the actual external hardware interface to the

design being tested, allowing for an accurate test of the outputs.

It is also possible to generate complex time varying signals, such as used in some DSP application, by waiting for a time value that is a result of a calculation. As a simple example, consider a signal of type Time

    Signal Delay_time : Time := 3 us;

Then, in a loop :

    SigA <= '1';
    Wait for 100 ns;
    SigA <= '0';
    Wait for Delay_time;   -- Wait for a variable time value
    Delay_time <= Delay_time + 50 ns;   -- Compute new time value
    If Delay_time = 12 us then
       Delay_time <= 3 us;
    End if;

This example shows a simple time varying function for a pulse train whose period increases linearly from 3.1 usec to 12.1 usec and then repeats. Considerably more complex functions can be created as needed for specific applications.

Miscellaneous Test Bench Operations

Assertion Statements

The VHDL **Assertion Statement** can be used to generate a text message when some event occurs during a simulation. The assertion consists of 3 parts, the condition, the reported message, and the type or severity of the message. The message and message severity are sent to the terminal when the assertion condition is False :

Assert pulse = '0'   -- If false then report message
  Report " Pulse has gone high "
    Severity Warning;

In words, this means to check for the signal Pulse to be not 0, then write the message out with severity Warning. Types of severity can be **Failure, Error, Warning, and Note,** and can be applied to fit your own circumstances as needed. Severity type Error need not really be an error, but is simply a designator you can use to indicate the status of a condition. The same applies for the other Severity types.

File Input and Output

The VHDL standard package TEXTIO includes routines that allow you to read and write files containing data or text to and from your computer system. You can read a table of values and store them in an array in your test bench, or you can write out files containing signal names and logic states that can be used in an external gate level simulator or some other system. This type of coding generally requires more VHDL skill and knowledge of how to handle various Type Conversions and the use of the TEXTIO Read and Write routines, which are somewhat primitive. An example of a Type Conversion from Std_Ulogic to Bit will be shown below.

File Output

A brief example will help illustrate the technique, and will show how to write numbers, text, signals, and simulation data and time values to a file. Suppose you want to create a file named SIM_DATA.OUT for use with a gate level simulator, and the file needs :

a. The number of signals, in this example, 2.
b. The signal names, one per line.
c. A list containing the simulation time in nsec, right justified to column 10, followed by 2 spaces and then the state of the signals at that time.

An example of this file would look like :

```
2
Clk
Mreset
           0    01
         100    11
         200    01
         300    11
         400    01
         500    11
         600    01
         700    11
         800    01
         900    11
        1000    00
        1100    10
        1200    00
```

The file is created below in a process named Write_Output that is part of the test bench. Within the process, the name of the output file is specified in the "File outfile" line as Simout.dat. The VHDL TEXTIO routines are defined so that you write all of the data that goes on a line to a temporary variable, named "ptr" in this case, and then when the line is complete it is written to the file using the Writeline function. The TEXTIO functions can only handle data in the predefined VHDL formats of BIT, Boolean, Bit_Vector, Character, String, Integer, Real, and Time format, which means that for Std_Ulogic, Type Conversion functions must be used where needed, as shown below.

When reading and writing to files from a test bench, you will need to convert signals from Std_Ulogic to a format such as Characters in order to use the VHDL WRITE functions in the standard TEXTIO package. A simple Type Conversion function is shown here to illustrate the technique. For this conversion the Std_Ulogic 9 states will be converted to 4 (0,1,Z, and X) for output.

```
FUNCTION To_Char ( A : Std_Ulogic; ) RETURN Character IS
BEGIN
    CASE A IS
        WHEN '0' | 'L' =>    RETURN '0';
        WHEN '1' | 'H' =>    RETURN '1';
        WHEN 'Z' | 'H' =>    RETURN 'Z';
        WHEN OTHERS =>       RETURN 'X';
    END CASE;
END;
```

This function will be used below in writing the states of the signals to the file.

In addition, VHDL variables are defined for :

```
Ptr          : to hold the text for one complete line
Now_integer  : to convert simulation time into integer format
Init_flag  :   used to write the header data once at the top of the file
```

Constants are defined for the signal names (Clk and Mreset).

The process begins by defining the variables and constants, then the number 2 is written to the file, followed by the signal names one per line. Then the process waits for a change on any of the signals during the simulation run. When that happens, the time NOW is converted to integer for output to the file, and the states of the signals are converted to Character and stored in Ptr. (The VHDL function NOW returns the current simulation time and is a built-in feature of the language). The entire process then repeats for the duration of the simulation run and thereby stores all of the signal states and times in the output file.

Write_output : Process

```
Variable init_flag     : Boolean := False;
File outfile           : Text is Out "SIMOUT.DAT";
Variable Ptr           : Line;
Variable Now_integer   : Integer;
```

```
        Constant n_Clk          :  String := "Clk";
        Constant n_Mreset       :  String := "Mreset";

Begin
    If init_flag = false then       -- Write the header once only
        Write (Ptr,2);              -- Write the number 2
        Writeline (outfile, Ptr);
        Write (Ptr, n_Clk );
        Writeline (outfile, Ptr);   -- Now write the signal names to the file
        Write (Ptr, n_Mreset );
        Writeline (outfile, Ptr);
        init_flag := true;          -- set the flag true
    End If;

    Wait until       -- Wait for a change on either signal during simulation

        Clk /= Clk'last_value or
        Mreset /= Mreset'last_value;

    Now_integer := Now / 1 ns;   -- Get the current time, in Integer

    Write (Ptr, Now_integer, FIELD=> 10); -- Put the time value into Ptr
    Write (Ptr,' ');
    Write (Ptr,' ');

        -- and now the signal states
        -- with some type conversions

    Write (Ptr, To_Char (Clk));    -- Convert Std_Ulogic to Char
    Write (Ptr, To_Char (Mreset));

    Writeline (Outfile, Ptr);          -- Write the line Ptr to the file

End Process Write_output;
```

Variations of this type of process can be used to generate files for other applications and is a very powerful means of passing data directly from

VHDL to CAE tools and other design systems.

Preceding the Entity declaration, add the library declaration :

Use Std.textio.all;

in order to provide access to the standard TEXTIO package.

File Input

In a similar manner, you may want to read time and data values from a file in order to drive signals in the test bench. To illustrate the technique, let's read a file containing Time values, a databus (DATAIN) and a single signal ENABLE. Type Conversion functions are needed to convert the file numerical value to Std_Ulogic format, and the functions named To_SU and To_SUV are shown below.

```
Library ieee;
   use ieee.std_logic_1164.all;

Use std.textio.all;

 Signal DATAIN: Std_Ulogic_Vector (15 downto 0);
 Signal ENABLE: Std_Ulogic;

Function To_SU (B: Bit) return Std_Ulogic is
   Begin
       Case B is
         When '0'=> return '0';
         When '1'=> return '1';
       End case;
End To_SU;
```

Function To_SUV (B: Bit_Vector) return Std_Ulogic_Vector is

Variable RESULT: Std_Ulogic_Vector (B'range);

Begin

```
  For i in B'range loop
    Case B(i) is
      When '0'=> RESULT(i) := '0';
      When '1'=> RESULT(i) := '1';
    End case;
  End loop;

 Return RESULT;
```
End To_SUV;

Begin

READ_INPUT: Process

```
Variable LAST_TIME : Integer := 0;
```
File INFILE: Text is IN "datain.in";
```
Variable Ptr : line;
Variable NOW_INTEGER : Integer;
Variable DATAIN_VAR : bit_Vector (15 downto 0)
                         := "0000000000000000";
Variable ENABLE_VAR : bit := '0';
```

Begin

  While Not (endfile (INFILE)) loop

```
    Readline (INFILE, Ptr);
    Read (Ptr, NOW_INTEGER);      -- read time value
    Read (Ptr, DATAIN_VAR);       -- read bit_vector
    Read (Ptr, ENABLE_VAR)        -- read bit
```

    Wait for (NOW_INTEGER - LAST_TIME) * 1 ns;

```
    DATAIN <= To_SUV (DATAIN_VAR);
    ENABLE <= To_SU (ENABLE_VAR);
    LAST_TIME := Now / 1 ns;

  end loop;
Wait;
```

End process READ_INPUT;

As with the Write_Output code, you simply add the code for the Read_Input process or a similar process to the test bench.

A Complete Test Bench

To summarize, an example of a complete test bench used to test the Entity Top_gen is shown below:

```
Library ieee;  -- IEEE standard packages
  use ieee.std_logic_1164.all;

Library Synth;   -- Vendor dependent packages
  use Synth.packages.all;
```

Entity Tb_Top_gen is   **-- There are no ports in the test bench**
End Tb_Top_gen;

Architecture Test_Bench of Tb_Top_gen is

-- Define all of the signals going to the Top_gen Entity

```
    Signal   Clk20M   : Std_Ulogic := '1'; -- Initial value ;
    Signal   Reset    : Std_Ulogic := '0';
    Signal   Enable   : Std_Ulogic;
```

```
Signal  Datain    : Std_Ulogic_Vector (15 downto 0);
Signal  Databus2  : Std_Ulogic_Vector (15 downto 0);
Signal  P1        : Std_Ulogic;
Signal  D1        : Std_Ulogic_Vector (7 downto 0);
Signal  P2        : Std_Ulogic;
Signal  D2        : Std_Ulogic_Vector (7 downto 0);

Constant Clk_Period    : Time := 50 ns;
Constant Reset_Width   : Time := 140 ns;
```

-- Declare the Entity Top_gen as a component

Component Top_gen
```
    Port (
        Clk20M    : In    Std_Ulogic;
        Reset     : In    Std_Ulogic;
        Enable    : In    Std_Ulogic;
        Datain    : In    Std_Ulogic_Vector (15 downto 0);
        Databus2  : In    Std_Ulogic_Vector (15 downto 0);
        P1        : Out   Std_Ulogic;
        D1        : Out   Std_Ulogic_Vector (7 downto 0);
        P2        : Out   Std_Ulogic;
        D2        : Out   Std_Ulogic_Vector (7 downto 0)
    ); -- (end of port)
```
End Component;

Begin

-- Assign one instance of Top_gen as U1 and assign signals to ports

U1 : Top_gen

```
    Port Map (
        Clk20M,      - Positional Port Map
        Reset,
        Enable,
        Datain,
        Databus2,
        P1,
```

```
            D1,
            P2,
            D2
            ); -- end port map
```

-- **** Add your test bench VHDL here *****

TB : Block

Begin

```
    Clk20M    <= Not (Clk20M) after Clk_period / 2; -- A 20 MHz clock

    Reset     <= '1' after 10 ns,'0' after Reset_Width;

    Enable    <= '1' after 520 ns, '0' after 840 ns, '1' after 3010 ns,
                   '0' after 6010 ns, '1' after 7010 ns;
```

-- Note that Signal "Enable" has not been initialized and therefore will
-- have the value 'U' until 520 nsec;

```
    Datain    <= "0000000000010101";  -- Value 21

    Databus2  <= "1111000000010101", "1010101011110000" after
                   2 us;
```

-- "File Read" Process if desired ...

-- "File Write" Process if desired ...

 End Block TB;

End Test_Bench;

Summary

This chapter has covered the use of structural VHDL for combining entities into a hierarchical structure and for connecting to test benches for simulation. The wide variety of code possibilities that can be used in test benches provides for great flexibility in both simulation and in converting VHDL data into formats for use with other design tools. In that regard, a more complete VHDL syntax textbook, such as given in the references, would be valuable for more elaborate test bench development.

Chapter 10

Evaluating A Synthesis System

This chapter will discuss the issues involved in evaluating a VHDL synthesis tool. The goal of the evaluation is to be able to make an accurate assessment of the capabilities of a system prior to purchase, and to understand how the tool can be integrated into your own design environment.

The task of evaluating synthesis tools is one of the major steps in making the transition to this technology, primarily due to the relative high cost of these tools and their general complexity. It is also important to realize that the quality and accuracy of the synthesis and other ASIC design tools are absolutely essential to successful ASIC development. Since budgets for design tools are a serious concern to most businesses, it is important to make an accurate decision prior to purchasing the tools. The criteria used for judging various tools include a number of technical issues, business issues such as the overall cost and expected return, and supplier relationships. This chapter will focus primarily on the technical aspects of synthesis tool assessment, since the other factors will depend more heavily on specific company management processes.

Technical Evaluation Criteria

Ideally, you should expect a logic synthesis tool to be able to produce logic designs that are as good as your own designs, to operate quickly and efficiently, and to be closely integrated with all of your other design tools. These criteria would thereby allow you to create high quality designs while improving productivity and reducing the overall ASIC development schedule and cost. Synthesis tool suppliers are working to solve these requirements, but, since this is a fairly new technology, differences exist among the available tools that can have an effect on their overall performance and effectiveness.

Some of the major technical aspects that affect a tool's overall capability, not necessarily in order, include :

> Speed
> VHDL language support
> Design quality and efficiency
> Optimization capability
> ASIC Libraries
> Gate capacity
> Integration with other tools
> Testability

Speed

The speed of the synthesis tool is an obvious but important factor in assessing tools. Just as there are major speed differences between VHDL simulators and other design tools, synthesis tools vary widely in the time it takes to synthesize a given block of logic. Related tasks such as optimization of logic delay paths and schematic generation also follow the same pattern and should be measured when comparing systems. The important issue is to be able to synthesize or optimize the typical several thousand gate block in a reasonable amount of time, since you may have to iterate this step at a critical point in the ASIC design schedule.

Language Support

As covered in Chapter 3, logic synthesis tools utilize a subset of the VHDL language. This does not necessarily imply a shortcoming, since many capabilities of VHDL pertain to simulation and modelling areas not applicable to synthesis. In reality however, some synthesis tools restrict the available VHDL syntax to an even smaller subset that can affect the ease and flexibility of designing logic. In light of the examples given in earlier chapters, an assessment should be made of the VHDL syntax supported and that which is not supported by a tool, and how that would affect the design process. Some features, such as the use of Alias or Generic parameters, are convenient and efficient but not essential to designing logic, whereas having a complete set of arithmetic and logical operations available is an important capability.

Design Quality and Efficiency

The quality of the logic produced and its gate count should be a major requirement of any synthesis system. You should expect that, when the VHDL is coded correctly, the synthesized logic should be as good as a good designer would produce. It may be different in the arrangement of the gates, but it should be efficient and above all, correct. The tools should also automatically minimize the gate level logic for the specific ASIC library chosen and should use the most efficient basic cells available. For example, AND-OR gate cells should be used rather than individual AND and OR gates since they result in less total gates, less area, less wiring, and faster speed. This book has emphasized being aware of the gate count and controlling the synthesis process to do what you want it to do, and you should always expect high quality results. In this regard, you should also verify that the synthesized designs conform to the ASIC supplier's rules for loading and drive capability, and that the gate delays take into account estimated wiring delay and capacitance effects. These rules are critically important for CMOS ASIC families using 1 um or sub-micron technology, and close cooperation is required between the ASIC supplier and synthesis tool developers in order to produce accurate and efficient results.

Optimization

Along the same lines, the ability to optimize a design for speed should be tested by using some of the large counter or data processing examples as test vehicles. The synthesizer should be able to optimize critical paths in the design based on either the clock rate, the delay between input or output ports and clocked logic, and the combinational delay from input to output. The use of a long counter chain for this assessment is convenient, since the VHDL code is fairly simple but the logic can be considerable. You should be able to see how the logic design changes as the speed constraints are made more difficult, ranging from minimum gate count at slow speeds to more complex look-ahead carry chains as the speed increases. How does the tool respond when it cannot meet the speed constraints? Can you get a report of the critical path delays and the timing margin? The analysis of critical timing problems is one area where the synthesized schematics can be helpful, and some tools also provide a schematic view of the critical path logic that traverses the design hierarchy.

ASIC Libraries

As mentioned above, the ASIC libraries are a critical issue. Factors to consider are the availability of libraries for your current and future ASIC suppliers, how often they are updated, and how closely the ASIC and synthesis developers work together in refining the design rules for synthesis. The reason that this is important is that, near the end of the ASIC development cycle, the final ASIC verification simulations and timing analysis are done, either using the ASIC supplier's "golden" tools or another gate level simulator. If timing or design rule problems show up at this point, you will have to analyze the gate level source of the problem and resynthesize or re-optimize the logic with tighter constraints. An even worse situation would be if the synthesizer's design rules or libraries are in error, which could result in longer delays in correcting the problem. It should be clear that the success of ASIC development depends heavily on the success of the tools and libraries, and on a close working relationship with the suppliers.

Size Limitations

Another tool issue that should be assessed is the limit, if any, on the maximum synthesizable gate count. This is a concern for the synthesis and optimization of large logic functions that either cannot be split into smaller entities or where there is a need for optimization across several entities. As mentioned earlier, because of the length of time involved to iterate large blocks, you should try to avoid synthesizing very large designs wherever possible. However, assuming you have a fast workstation and sufficient time for the synthesis, any limit on the maximum synthesizable gate size should be a factor in comparing different tools.

Tool Integration

Integration of the synthesis tools with other design and verification tools is often handled by means of netlist or file transfers using EDIF, VHDL, or other formats. Issues such as schematic and VHDL shell generation have been discussed in Chapter 9 and should be easily accommodated by the design tools, if not fully integrated. Keep in mind that at certain times in the design cycle you will have to iterate the loop from VHDL simulation to synthesis, gate level simulation, and design verification, and you will want this loop to be as seamless and simple as possible.

Testability

A final criteria to apply to synthesis tools is their ability to handle testability structures and test patterns for ASIC test. This is a fairly complex issue and involves tradeoffs between the size of testability logic and the ability to fully test the ASIC in production. Regardless of the tools used, you must decide on the overall scheme to be used, such as full or partial scan, ad hoc approaches, JTAG boundary scan, or full built-in-self-test (BIST). Structured testability schemes such as scan lend themselves nicely to automated synthesis of the serial scan chains and to subsequent automatic test generation, at a cost of roughly 15% additional logic to replace all flip flops with scan flops. Criteria for selecting test synthesis tools include how easily the scan chains get generated, how effective the automatic test patterns are in terms of achieving 100% fault coverage, how fault coverage is measured, and whether the results are compatible with your ASIC supplier's test equipment and software.

Summary

There are many considerations in selecting design tools, and this chapter has tried to provide some technical guidelines to use in evaluating commercially available competing tools. The next chapter looks at some of the capabilities that can be expected from synthesis and ASIC design tools in the future.

Chapter 11

Future Prospects for ASIC Synthesis

In general, the major forces driving the development of logic synthesis and other design tools today and in the future are fairly obvious. First, the pace of technology and new product development shows no signs of slowing down, and second, the need to remain competitive demands that the cost of the products and their development costs and schedules must be continually improved. Since ASICs are such an integral part of new product development, their continued growth in size and performance requires ever more capable design and logic synthesis tools.

Although today's logic synthesis tools are being used to design large and complex ASICs, there are improvements in several areas that can be expected and indeed are necessary for the future. These issues can be loosely grouped into three categories - tighter integration of synthesis into the ASIC design process, tool performance, and higher level synthesis and design tools.

Integration

One of the main problems with current point synthesis tools is that they are not as closely integrated with an ASIC supplier's layout and verification tools as is desired. Technically, this involves issues of the synthesis libraries and library accuracy, the loading and delay rules used, and the timing problems resulting from layout effects.

The accuracy of the timing delay values used in a synthesis library affects the optimization portion of the process. Synthesis tools typically use static timing analysis algorithms during optimization to analyze whether different logic configurations will satisfy the speed or clock rate constraints. The minimum gate configuration that meets those constraints is then selected as the best circuit choice. The key to the accuracy of the timing analysis depends on the accuracy of the gate delay data and the computation of the effects of loading or fanout and the effects of wirelength and capacitance.

One of the problems faced by synthesis developers is that the intrinsic gate delay data and fanout rules are continually being updated by the ASIC suppliers as they characterize and refine their fab processes. Updating the synthesis libraries for gate delay should be a straightforward technical matter between the ASIC supplier and the library developer, assuming that they have a mutual business interest. The computation of estimated delay due to loading, wirelength, and capacitance can be more difficult however. For instance, the simple RC equations that were sufficient for slower technologies are not accurate enough for sub-micron designs and require different algorithms to be used. Depending on how the computations and algorithms are handled in the synthesizer, the updating of the tools and their ultimate accuracy can be affected and could lag behind the ASIC vendor's timing rules by an extended period.

The result to the designer is that a synthesized circuit that is passed through the ASIC supplier's simulation and verification tools may fail due to timing errors, glitches, or violations of internal fanout or rise and fall time rules. This can be a fairly serious problem that hopefully can be corrected by resynthesizing the logic, since editing the gate level netlist is a tedious and error-prone alternative that should be avoided. Long range, the difficulty may have to be resolved by the synthesis tool vendor, resulting in schedule delays. It is clear that the accuracy of the synthesis timing data is crucial to using synthesis efficiently and must remain an industry focus in the future.

The pre-layout timing data derived in this manner is however only an estimate, based on the ASIC supplier's experience with the technology. Estimates of the wirelength delays, which are a great portion of the total delay for a signal, are based on knowledge of the physical area and layout of the individual gates relative to each other. Without information of the layout at the time of synthesis, the tools make an estimate based on the number of gates in an entity and use a factor based on that size to compute the delay effects. The fact that this is only a rough estimate leads to design problems between the pre-layout synthesis and the post-layout design verification.

Since the ASIC gate delays are affected by the actual chip floorplanning and layout, this information must be made available to the synthesis tools in an integrated fashion in the future. In other words, as the synthesis process is working to optimize a design, a "layout" process must be automatically running in the background, providing timing data back to the synthesizer. This sounds easy to say, but based on today's workstation capability and design and layout software, it is a very formidable problem indeed. Both synthesis and layout are extremely CPU intensive processes, and the difficulty is compounded by the fact that overall design sizes are growing steadily. This problem is the subject of significant industry attention at this point and will require a combination of increased computer capability and cooperation between ASIC suppliers and design tools developers to resolve. In the short term, estimates will become more refined as these processes mature and as designers push for closer integration between ASIC design tools.

Tool Performance and Accuracy

As discussed in earlier chapters, the speed of the synthesis process is a factor of the workstation performance and the efficiency of the software operating on that platform. Current early 1990's performance of these tools effectively limits the size of a synthesized entity to the several thousands of gates range. For designs that are becoming larger and more complex each year, this size barrier must be raised by an order of magnitude in order to allow for both synthesis of larger blocks and optimization between blocks of a large ASIC. Coupled with the eventual need for layout based timing data, the need for speed and efficiency of synthesis tools will be an essential requirement for future design work.

Another factor that will be improved in the future is the ability to synthesize logic unambiguously from a VHDL description. As stressed throughout this book, the style and correct implementation of the VHDL description has a great deal to do with attaining correct and efficient synthesized logic. However, the current state of synthesis technology is not yet at the point where synthesis can be blindly trusted to produce a correct circuit design. Even discounting the delay effects covered above, design correctness is an issue that is being constantly refined and will continue to do so in the future. In the interim, gate level logic simulation and other design verification tools serve to catch any errors in the synthesized design, assuming of course that the simulation effort is well planned and executed. It is hard to overemphasize the importance of the simulation and verification efforts, and the planning that goes into them, to the ultimate first pass success of an ASIC development.

Higher Level Tools

Helping in solving some of the earlier problems are a variety of higher level design tools. Some of these tools include formal verification systems, specialized datapath and function generators, and higher level synthesis techniques.

Currently, the correctness of a design, as it moves from behavioral VHDL description to RTL level to gate level, is confirmed by means of comparing logic simulation results. While this process is effective, it can often be time consuming due to the length of the simulation activity. Formal design verification tools are being developed that will be able to confirm the equivalence of two designs by analyzing their descriptions or netlists, independent of the simulation process and without requiring extensive test patterns. This type of tool will help reduce errors throughout the overall design process and will help to assure the correctness of synthesized designs as well as edited netlists. This is a difficult technology that will require extensive testing and development before it becomes a trusted design tool, but can provide a much needed capability in the future.

Other forms of "synthesis" tools such as datapath compilers and specialized function generators may also be a part of the ASIC design toolkit of the future. While not universally usable like a general synthesis system, these tools primarily have application for very regular structures in DSP and data processing systems, such as adders, multipliers, and pipelined processors. Their benefit is in speed, since they use specialized rules based on pre-defined circuit configurations to produce the required number of bits in the function. In addition, a higher level VHDL model of the function can be produced for use during the simulation of the entity or ASIC, in somewhat of a reverse synthesis operation.

Conclusion

These are exciting times for designers of electronic systems. The rapid changes in technology and design methods means that the task of learning and adapting new ideas is a continuous part of an engineer's career. In that sense, using VHDL and logic synthesis is simply one more step along the path to using the technology of the future. Based on the progress made over the past decade, it should be a very interesting journey!

Appendix A

Reference Materials

1. R. Lipsett, C. Schaefer, C. Ussery - VHDL : Hardware Description and Design, Kluwer Academic Publishers, 1989

2. D. Coehlo - The VHDL Handbook, Kluwer Academic Publishers, 1989

3. D. Perry - VHDL, McGraw Hill, 1989

4. IEEE Standard VHDL Language Reference Manual, IEEE Std 1076-1987

5. S. Mazor, P. Langstraat - A Guide to VHDL, Kluwer Academic Publishers, 1992

Index

TÁIN BÓ CÚALNGE

from the Book of Leinster

EDITED BY

CECILE O'RAHILLY

SCHOOL OF CELTIC STUDIES
DUBLIN INSTITUTE FOR ADVANCED STUDIES

Reprinted by
Dundalgan Press, Ltd., Dundalk, Co. Louth

CONTENTS

FOREWORD

It is now more than sixty years since Windisch published his edition of Táin Bó Cúalnge from the Book of Leinster. Besides the LL-text with translation into German Windisch's monumental work contained a long Introduction, a critical apparatus with readings from all available manuscripts, grammatical notes, a full glossary and indexes, forming in all a huge volume of 92 + 1120 pages.

It might well be thought that such a thorough piece of work could never be superseded. There are, however, two reasons why a new edition of Táin Bó Cúalnge is now called for. First there is the difficulty of procuring a copy of Windisch's book, long since out of print. Secondly, since 1905 an immense amount of work on Irish palaeography, grammar and lexicography, has been done by Celtic scholars. All the known versions of TBC have been published, and Thurneysen's researches have thrown further light on the development and relationship of the recensions of the saga.

For the LL-text Windisch relied on the RIA Facsimile, a lithographic reproduction of O'Longan's transcript (1880) and on a copy made for him directly from the MS. by Standish H. O'Grady. So too for his quotations from LU Windisch again used the RIA Facsimile, also a reproduction of O'Longan's transcript (1870), and for YBL the facsimile published in 1896. Since Windisch's day diplomatic editions of the Book of Leinster (vol. II 1956) and of Lebor na hUidre (1929) have appeared, and the text in these shows many corrections of the facsimiles. The full text of YBL-TBC was published in 1912. In the same year Dr. R. I. Best in his Notes on the Script of Lebor na hUidre (*Ériu* vi) published his account of the various hands in LU which has been of such value for the study of Recension I. In addition the full text of the Stowe Version of TBC from which Windisch quoted freely has appeared (1961), and a late copy of Recension I has come to light and been published (1966). The most significant step in the study of TBC was made by Thurneysen who dealt in many articles with the tradition and transmission of the tale and the relationship of the different recensions, and showed that the hypotheses of Zimmer, Nettlau and Windisch were erroneous. Finally in *Die irische Helden- und Königsage* (1921) Thurneysen published his invaluable analysis of Táin Bó Cúalnge, giving a comprehensive view of the saga and comparing and evaluating the three recensions in the light of his own researches.

Any editor of a new edition of TBC must take those later studies and researches into account, but must at the same time warmly acknowledge a great debt to the pioneering work of that great scholar Windisch and build on the foundation of his monumental edition.

For the present edition a photostat of the manuscript has been used together with the printed text of the Diplomatic Edition. Some passages of the text were much stained and difficult to decipher; for these the reading of the Diplomatic Edition has been taken. Emendations suggested in footnotes in the Diplomatic Edition have been adopted and some others have been suggested by the editor. Where useful for comparison, readings from other manuscripts (LU, YBL, C, St) have been given in the footnotes. In a few cases the division into words of the Diplomatic Edition has not been followed, but where such changes have been made, they have always been indicated in footnotes or in Notes to Text. In general the paragraphing of the Diplomatic Edition has been adopted. Punctuation and capitals have been introduced and marks of length supplied. Contractions have been, with a very few exceptions, silently expanded; any doubt as to the manuscript reading can easily be cleared by reference to the Diplomatic Edition in which italics are more freely used. Words or letters added by the editor are indicated by square brackets. In the translation of the text elegance of style has been sacrificed to the aim of providing as literal a rendering as possible of the Irish, in the belief that such a version will be of most use to the student.

My thanks are due to the Board of Trinity College for permission to edit the text from a manuscript in their possession. I should like to express my gratitude to Professor D. A. Binchy and to Professor Myles Dillon who have always given unstintingly of their broad learning whenever I was faced with doubts and difficulties in the editing of this text. For whatever errors and misunderstandings remain in the work I alone am responsible.

CECILE O'RAHILLY

INTRODUCTION

§1 Irish Heroic Saga

Táin Bó Cúalnge, the Cattle-raid of Cooley, is the longest and most important heroic tale of the Ulster cycle. Plundering raids, especially cattle-raids, are a characteristic feature of Irish heroic saga. Táin Bó Cúalnge, the most famous of these raids, tells of a foray made by Medb of Connacht into the territory of the Ulaid for the purpose of carrying off the bull Donn Cúalnge from the district of Cúalnge, present-day Cooley, Co. Louth.

The earliest recension of the tale, a conflation of two 9th-century versions, is found in a manuscript written about 1100. The story existed in the first half of the eighth century, and Thurneysen held that it may have been recorded as early as the middle of the seventh century[1]. But the general conditions, the political structure and the methods of warfare described in TBC are not those which prevailed in seventh-century Ireland. Tradition assigns the events of the Ulster tales to the beginning of the Christian era, and it was at one time uncritically accepted that the civilisation depicted in the tales accorded with this tradition. Some scholars believe that the Ulidian tales are wholly mythical in origin and that the leading characters are euhemerized deities. Others, including Windisch, maintain that the sagas contain little that can safely be called myth, though mythological elements are found in them. In the 19th century many scholars held a firm belief in the historicity of Cú Chulainn, Medb and Conchobor, but nowadays it is generally acknowledged that the characters and events of early Irish sagas are purely imaginary and have no connection with history proper. Nevertheless the sagas are of historic importance in that they contain genuine traditional material and the picture they give of early Irish life and civilisation tallies in many points with the account of those of Gauls and Britons before the Roman invasion.

According to T. F. O'Rahilly[2], the Ulidian tales preserve a genuine tradition of a state of warfare between the Ulaid and the Connachta. But these Connachta, he holds, were originally the Midland Goidels, descendants of Conn, who later made themselves masters of the land

---

[1] Heldensage 112, ZCP xix. 209.
[2] EIHM 175, 180.

westwards of the Shannon. By the time the Ulidian cycle of tales
took shape, the name Connachta was narrowed in meaning and the
convention was adopted by the saga writers that the enemies of the
Ulaid lived not in Tara but in the province of Connacht, where Medb
and Ailill had their capital in Crúachu. The five provinces of
Ireland were invented with the object of depicting the province of
Connacht and all the rest of Ireland as leagued together against the
Ulaid. By such adaptations of early traditions the tale of Táin Bó
Cúalnge evolved and gave us the account of Cú Chulainn's defence
of the Ulster marches against the four provinces.

In his Cours de Littérature Celtique[1], H. D'Arbois de Jubainville
compared the civilisation of the Celts with that of Homer's Greeks,
and discussed the agreement between the picture of Irish heroic
life in the Ulidian tales and the accounts given by various classical
writers of life among the Gauls and Britons before the Roman
invasion. In the long and masterly Introduction to his edition of
Táin Bó Cúalnge from the Book of Leinster, Windisch developed
and expanded the ideas of de Jubainville[2]. The passages he quotes
from Diodorus Siculus, Strabo and Caesar, all deriving directly or
indirectly from the lost work of the historian Posidonius, deal with
the structure of society in Gaul, with the characteristics and customs
of Gaulish warriors and with their methods of warfare. In these
points, as he shows, the background and setting of the Ulster cycle
and the characteristics and weapons of the Ulster warriors agree to
a remarkable extent with the descriptions of the classical authors.

Diodorus Siculus tells us of the Gauls: 'They have also lyric poets
whom they call Bards. They sing to the accompaniment of instru-
ments resembling lyres, sometimes a eulogy and sometimes a satire.
They have also certain philosophers and theologians who are treated
with special honour whom they call Druids. They further make use
of seers, thinking them worthy of high praise'[3]. Corresponding to
the bard, druid and *vatis* among the learned classes in Gaul we have
in the Irish tales the *fili*, *druí* and *fáith*. Both Diodorus Siculus and

---

[1] Vol. vi, La Civilisation des Celtes et celle de l'Épopée Homérique
(Paris 1899).

[2] Part II of Introduction, Die irische Heldensage und die Nachrichten
der Alten über die Celten pp. xi–xxxix.

[3] In a recent valuable article on The Celtic Ethnography of Posidonius
(PRIA ix C 189–275, 1959–60) Professor J. J. Tierney has dealt with the
three later Greek authors, Athenaeus, Diodorus Siculus and Strabo, in
whose work material from Posidonius is reproduced or summarised. It is
from Professor Tierney's translation of the relevant passages that I quote
here and in the following notes.

Strabo speak of the warlike qualities of the Celts, of their boastfulness and courage. 'The whole race . . . is madly fond of war, highspirited and quick to battle'[1]. 'Some of them so far despise death that they descend to battle unclothed except for a girdle'[2]. Examples of such 'heroic nudity' are found in TBC. The Gaulish weapons which are described by these authors, the shield, the long sword, the spear and the sling[3] are those of the Irish warriors in TBC and other tales. The use of the war-chariot had disappeared in Gaul in Caesar's time though it was still found in Britain. Diodorus Siculus tells us that for their journeys and in battle the Gauls used chariots drawn by two horses and carrying both charioteer and chieftain[4]. 'When they meet with cavalry in the battle', he says, 'they cast their javelins at the enemy and then descending from the chariot join battle with their swords'. So in TBC we have a description of the noise and tumult of the Ulster warriors advancing to battle in their chariots with clatter of wheels and loud hoof-beats of horses. But in the ensuing battle no mention is made of chariots; the warriors arm themselves with shield and spear and sword and fall to hacking and slaughtering their enemies[5]. The Gauls cut off the heads of their enemies slain in battle and attached them to their horses' necks[6]. So Cú Chulainn beheads the enemies he has slain, impales the heads on a branch or brandishes them before the enemy as a sign of victory. Diodorus tells us that at feasts they honour brave men with the finest portion of the meat[7]. Athenaeus, quoting directly from Posidonius, writes: 'And in former times when the hindquarters were served up, the bravest hero took the thighpiece, and if another man claimed it, they stood up and fought in single combat to the death'[8]. In the sagas the custom is represented as still surviving among the Irish, for to this corresponds the *curadmír*, the champion's portion, for which the heroes contended in Fled Bricrend and in Scéla Mucce Meic Dathó.

In his Introduction Windisch quoted most of these passages from classical writers and gave parallels from Irish heroic literature

---

[1] Strabo IV, iv. 2.

[2] Diodorus Siculus V. 29.

[3] Strabo IV. iv. 3.

[4] Diodorus Siculus V. 29.

[5] Exceptionally in the Fer Diad episode Cú Chulainn and Fer Diad fight from their chariots (3141-2), perhaps a late invention of the compiler of that passage.

[6] Diodorus V. 29, Strabo IV. iv. 5.

[7] Diodorus V. 28.

[8] Athenaeus IV. 40.

to show that the background, characteristics and battle-customs of
the Ulster heroes as a whole were similar to those of the Gauls
and Britons in the second and first centuries B.C. He points out
that this heroic life lasted longer in an Ireland free from Roman
influence. He accepts the traditional date assigned to the events
of the sagas and suggests that the main body of the oldest Irish
tales must be dated no later than the last pre-Christian or the first
post-Christian centuries.

In a recent lecture[1] Professor Kenneth H. Jackson has again
dealt with the relevant passages from classical writers. He admits
the extraordinary similarities between the traditional background
of the Ulster cycle and the La Tène Iron Age civilisation of the
Gauls and Britons in the second and first centuries B.C. He rejects
the traditional dating of the events told in the Ulster stories. He
notes that historical documentation cannot go back earlier than
the 5th century and that, to date the Ulster cycle of tales, we must
rely on internal evidence alone. These tales represent a state of
affairs older than the 5th century with a wholly pagan background,
and, according to Professor Jackson, depict the civilisation of the
Early Iron Age which in Ireland, untouched by Roman influence,
lasted much longer than in Gaul or Britain. He suggests that one
way in which La Tène culture reached N. Ireland was via Northern
Britain from Gaul in the second century B.C. or earlier, and that
these immigrants brought with them such characteristics as their
iron swords and decorated scabbards, and, in particular, their use
of the war-chariot. The heroic life in the Ulster tales is then, he
says, 'a. picture of life among the descendants of those La Tène
Celts of Northern British and ultimately of Gaulish origin'. Therefore, he concludes, the formation of the tradition of these tales falls
somewhere between the second century B.C. and the fourth century
A.D. Professor Jackson accepts the view that these stories had
been handed on by oral tradition for a very long time before being
written down in the middle of the seventh century. The question
is how long an oral tradition may be assumed. The traditional
dating suggests a period of 650 years, but Professor Jackson thinks
that to postulate so long an oral life strains our credulity. He
believes that 'the picture of the background against which it was
felt such stories should be composed had been formulated not later
than the fourth century'. This would allow for a period of some

---

[1] The Oldest Irish Tradition: A Window on the Iron Age, Rede Lecture
1964.

300 years during which heroic traditions were gradually moulded into these stories and handed on by professional *filid* and storytellers.

This view, that Irish heroic tales had a long oral existence before they took literary shape, is that accepted by the majority of Celtic scholars. It has, of course, been generally recognised that the tales, though they embody older pre-Christian heroic traditions, have come down to us transcribed by Christian monks. Hence some of their pagan characteristics have been eliminated or softened, and they show the influence of Christianity or of foreign literatures. Thus, in a deliberate attempt to add Christian flavour, the death of Conchobor is made to synchronise with the death of Christ (Aided Chonchobair LL 14296 ff.), and Cú Chulainn appears to King Laegaire in later times and calls upon him to believe in God and St. Patrick (Siaburcharpat Con Culaind LU 9221–9548). The compiler of the LU-Táin was acquainted with the Aeneid and equates the Morrígan with the Fury Allecto (*Allechtu* LU 5320). In the LL-Táin and in Recension III an interpolated poem attributes one of the deeds of Hercules to Cú Chulainn; he is said to have slaughtered the Amazons (infra 1292). In all versions of TBC the narrative is interrupted while the banished Ulstermen recount to Medb and Ailill the noteworthy exploits of Cú Chulainn's boyhood. This has been compared with Aeneid, Books ii and iii. Some scholars would even go so far as to see in Táin Bó Cúalnge a deliberate attempt to provide Ireland with an epic comparable in scope to the Aeneid.

Professor James Carney does not subscribe to the view that Irish heroic saga had a long oral tradition. In the chapter 'The External Element in Irish Saga' in his Studies in Irish Literature and History, he contends that early Irish saga is in part traditional, in part an imaginative reconstruction of a remote pagan past in a form which belongs to the mixed culture of early Christian Ireland. All scholars are in agreement with him on one point: heroic traditions preserved by scribes of a later age are inevitably influenced by the literary attainments of the compilers. But as Professor David Greene has noted: 'Certain elements belonging to the coherent society portrayed in the Táin—totem and tabu, head-hunting, fighting from chariots—are unknown in early Christian Ireland and cannot, therefore be inventions of literary men influenced by Latin learning'[1]. Professor Carney, however, seems to deny an oral origin to TBC. He believes that 'it is against all likelihood

1 Thomas Davis Lectures, Irish Sagas (1959) p. 97.

that an epic produced under monastic influence and inspired to
some extent by the Aeneid should more or less in its present form
join the general stream of Irish oral tradition'. The main point of
Professor Carney's thesis is that the sagas were deliberate works
of art, based to a small extent on oral tradition but making considerable use of written sources. Such a view would lessen the value
of the sagas as repositories of genuine tradition while making
them more impressive as works of art.

If, however, we may postulate, as most scholars believe we can,
a highly developed oral art of saga-recitation before the stories
took literary form, then despite any influence on the saga-writer's
technique or matter from written sources, we may conclude that
the inspiration of the sagas is ultimately oral.

§2 Recensions of TBC

Táin Bó Cúalnge has been preserved in three recensions. As
reference must frequently be made to these recensions, it will be
well to deal with them briefly.

(1) Recension 1 is generally called the LU-Version, as Lebor na
hUidre (LU) is the oldest manuscript which contains part of it[1].
This recension is found also in Egerton 1782 (W) and in the Yellow
Book of Lecan (YBL)[2]. None of the three manuscripts offers a
complete text, and none is copied from the other. LU contains
ll. 1–2181 of the Strachan-O'Keeffe edition of TBC; W contains
ll. 1–829, 914–1423, 1498–1581; YBL lacks the opening part,
beginning l. 206, and there are other lacunae in the text, but this
manuscript carries the tale to its end. Thus common to all three
manuscripts are only ll. 206–829, 914–1361, 1498–1581.

In the text of LU-TBC two main scribes have been distinguished,
A who wrote 55a–b[33] and M who wrote the remainder of the text,
while a third, denoted as H, has erased some passages in the text,
and, partly in the space so gained, partly in intercalated parchment
sheets, added many interpolations. In the part of TBC common
to all three manuscripts, we find that the H-interpolations of LU
are lacking in YBL but are found incorporated in the text of W.
We can verify the interpolations in Recension I only in the part
contained in LU, but we can take it as certain, as they are not found

[1] Lebor na Huidre, ed. Best and Bergin, ll. 4480–6720.
[2] TBC from Eg. 1782, ed. Windisch ZCP ix. 121–58; Táin Bó Cúailnge
from the Yellow Book of Lecan, ed. Strachan and O'Keeffe.

in the corresponding passages of YBL, that they did not continue in the remainder of the text.

In addition to the three manuscripts mentioned, Recension I is found also in the recently discovered O'Curry MS. I (C), a late sixteenth-century paper manuscript[1]. In this the text is incomplete. It is acephalous, opening with the arrival of the boy Cú Chulainn at Emain (= 1. 711 Strachan-O'Keeffe edition of TBC[2]) and breaking off after the description of Eogan mac Durthacht in *Toichim na mBuiden* (= 1. 3191 Strachan-O'Keeffe). There are lacunae in the narrative owing to loss of leaves or illegibility.

The main body of the text in C is that of Recension I, that is, the interpolated Recension I (U^H) as far as the LU version extends, and the YBL text (with some influence of Recension II) for the remainder. The order of events is that of Recension I, and C contains the same reduplications and contradictions. C, however, has not been copied from any of the three existing manuscripts of Recension I. Occasionally it offers a better reading than any of these manuscripts, and it contains a few additional lines of verse or rhetoric. The most interesting point about C is that it also contains material which is not found in Recension I but which occurs in Recension II, and in addition it contains two poems which are interpolated in the modern version of LL, the so-called Stowe version.

(2) Recension II is the version contained in the so-called Book of Leinster (LL)[2]. The text is complete except for the loss of one page (between 74^b and 75^a). Recension II is represented also by the modernisation and expansion of that version in RIA MS. C vi 3, formerly in the Stowe Collection, which is known as the Stowe version[3]. (The version in C vi 3 and in later manuscripts based on that version was called Version IIb by Thurneysen). Recension II is a carefully unified narrative. Much of Recension I is lacking, in particular some of the long rhetorical passages. No references are made to variant versions; contradictions are eliminated; many episodes are expanded and elaborated and some poems and prose passages are added. In this recension alone we get an introductory section giving the reason for Medb's foray into Ulster.

---

[1] Táin Bó Cuailnge, ed. Pádraig Ó Fiannachta, Dublin 1966.

[2] Die altirische Heldensage Táin Bó Cúalnge nach dem Buch von Leinster, ed. Ernst Windisch, Leipzig 1905; The Book of Leinster II, ed. Best and O'Brien, Dublin 1956.

[3] The Stowe Version of Táin Bó Cuailnge, ed. C. O'Rahilly, Dublin 1961.

(3) Recension III is preserved in fragmentary form in two late manuscripts, Egerton 93 and H 2. 17.[1] It is acephalous and opens with the return of the boy Cú Chulainn to Emain at the end of the *Macgnímrada*, and ends with the death of Ferchú Loingsech. This recension agrees in many points with the LL-version, and, significantly, often where LL differs from Recension I. But it contains also passages which agree verbatim with some of the H-interpolations of LU, and occasionally elsewhere the wording of this recension is nearer to Recension I than to Recension II. In addition it offers a small amount of material not found in any other recension[2].

§3 GROWTH AND STRUCTURE OF TBC

Like all heroic tales which have had a long oral existence, Táin Bó Cúalnge presents a central incident around which have been built up various themes—the theme of single combat or of a fight against odds, the theme of the arming of a warrior or of an army advancing into battle[3]. The story of the Táin, told countless times in oral recitation, must have varied continuously with the additions and improvisations of each teller elaborating and developing a traditional theme.

The central themes of Táin Bó Cúalnge may be summarised thus: Medb and Ailill of Connacht make a foray into Ulster to capture the bull Donn Cúalnge. Among Medb's forces is a band of exiled Ulstermen led by Fergus mac Róig and Cormac Conn Longas. The Ulstermen in Ulster are at this time suffering from a debility, *ces naíden*, and the defence of the province falls to Cú Chulainn, a boy of seventeen.

To this outline various episodes were gradually added. Thurneysen held that by the ninth century the two old parallel versions of TBC which he postulated contained all the essential episodes of the tale. These he enumerated as the prophecy of Fedelm, the carrying off of the bull Donn Cúalnge, some of the more important

---

[1] H.2.17, ed. Thurneysen, ZCP viii. 538–554; Eg. 93, ed. Nettlau, RC xiv. 254–6, xv. 62–78, 198–208.

[2] In the discussion of TBC which follows I have not dealt with the Fer Diad episode which calls for separate treatment. It is doubtful if *Comrac Fir Diad* formed a part of Recension I; the only reference to Fer Diad in LU occurs in an H-interpolation (5896–7). See Heldensage 219–235.

[3] Here I may refer to the brilliant researches of the American scholar A. B. Lord on composition by theme and formula in oral epic poetry in his book The Singer of Tales (Harvard University Press 1960).

duels, the final fight at Gáirech and Ilgáirech and the death of the bulls, Findbennach and Donn Cúalnge. In the course of time still other additions were made. These include the catalogues of heroes and of the places traversed in Medb's journey from Crúachu to Cúalnge, and the elaborate descriptions of the Ulster warriors in *Toichim na mBuiden*. Some themes received fuller treatment and were developed in great detail, in particular the long segment entitled *Breslech Maige Murthemne*. This long descriptive passage is later in style and language than the rest of the tale in LU, the oldest MS. version, and Thurneysen suggested that it may have been a separate piece later incorporated into the text. This is demonstrably true of *Comrac Fir Diad* which must have been originally an independent tale. It has also been suggested that the *Macgnimrada* passages did not form part of the original tale but were worked in at some period by a skilful compiler.

The duplication of incidents, the occurrence of what Thurneysen calls 'doublets', is a marked feature of TBC. In Recension I, as distinct from Recension II and Recension III, some of these doublets, or what appear to be doublets, are due to the introduction of variants. Thus in Recension I the explanation of the name Bernas Bó Cúalnge is given twice, once in the main narrative (where it also occurs in Recension II and Recension III)[1], and again in what the LU-scribe calls *córugud aile* where the name is changed to Bernas Bó nUlad[2]. In YBL a short notice with the words *iar slicht aile seo* tells how Cú Chulainn killed at Áth Meislir six men .i. *Meis Lir* ⁊rl[3]. This is a doublet of the passage *Aided Trí Mac nGárach*[4]. The killing of Medb's herd Lóthar by Donn Cúalnge is given in a passage noted as a variant in YBL (*slicht sain so sís* TBC[2] 872). It is a doublet of the killing of the herd Forgemen which is told in all three recensions[5]. Again, the warning which, in Recension I, Cú Chulainn takes to Conchobor: *Mná brataitir, éti agatair, fir gonaitir* (LU 5526-30) has a doublet in the later passage in which Sualtaim goes on Cú Chulainn's behalf to warn the Ulstermen and speaks the same words[6]. The first occurrence is in the long passage noted by the scribe as *córugud aile*.

Still other doublets have arisen from the manner in which the original nucleus of the tale grew by the accretion of incidents, and

---

[1] LU 5369-72, LL 1366-71, ZCP viii. 543.
[2] LU 5395-9.
[3] TBC[2] 762-4.
[4] LU 5270-84, LL 1247-57.
[5] LU 5826-30, LL 1795-1802, RC xiv. §15 257-8.
[6] TBC[2] 2985, LL 4011-2.

these doublets are found in all recensions of TBC. Here the same motif or narrative element is repeated in a different context or repeated with variation of detail. For example, Cú Chulainn seeks to hinder the progress of Medb's army into Ulster. He casts on to a pillarstone a withe bearing an ogam inscription forbidding the Connachtmen to continue their march until one of their number has cast a withe in like manner[1]. The same motif is repeated in the cutting of the four-pronged branch on which Cú Chulainn impales the heads of four Connachtmen[2]. In Recension I alone it occurs a third time in a later passage described as a variant, *slicht sain*: Cú Chulainn cuts down an oak-tree in the path of the army, with an ogam inscription defying them to pass until one of them has leapt across the tree in his chariot[3]. Again, when a Connacht warrior scorns to fight with a beardless youth, Cú Chulainn smears a false beard on his chin to deceive his opponent. This is found in *Comrac Nath Crantail*, in Recension I only[4], and again in the fight with Lóch in all recensions[5]. Cúr mac Da Lóth refuses to go against a beardless lad and suggests sending a youth of Cú Chulainn's own age from his people to encounter Cú Chulainn[6]. This point is elaborated in the later incident when Lóch sends his young brother to fight with Cú Chulainn since he himself scorned to fight with a mere boy[7]. Again, in *Aided Tamuin Drúith*, when Ailill's golden diadem is put on the jester's head, Cú Chulainn kills him, supposing him to be Ailill[8]. The same incident seems to be told in the brief mention of the death of Maenén Drúth (in Recension I only), though there the wearing of Ailill's crown is omitted[9]. In Recension II and Recension III the same disguise-motif occurs in the killing of Medb's handmaid[10]. The equivalent passage in Recension I has merely the words *Indar la Coin Culaind ba sí Medb* (LU 5341), but explicitly in the other recensions we are told that she is wearing Medb's crown. Again when Cethern mac Fintain attacks the Connacht camp. Ailill's diadem is put, 'through fear of Cethern', upon a pillar-stone. Cethern is tricked and thrusts his

---

[1] LU 4697–8, 4736–66, LL 456–60, 471–507.
[2] LU 4796–8, 4807–10, LL 560–5, 598–636.
[3] LU 5213–8.
[4] LU 5757–8.
[5] LU 6131–2, LL 1974–9 (Stowe), RC xv. 63 §§69–70.
[6] LU 5959–61, LL 1822–5.
[7] LU 6106–6111, LL 1964–7, RC xiv. 266 §§61–62.
[8] LU 6670–73, LL 2460–72.
[9] LU 5300–2.
[10] LL 1341–8, ZCP viii. 542.

sword into the stone[1]. In Recension III alone we find a repetition of this motif, in the section *Breslech Maige Murthemne*, when Follomain leads the youths of Ulster to Cú Chulainn's help. Ailill's diadem is placed on a pillar-stone between the two sons of Bethe mac Báin. Follomain had sworn never to return to Emain until he had obtained Ailill's head with its golden crown. When, as he thinks, he attacks Ailill, he is killed by the sons of Bethe[2].

Such repetition of themes or motifs in the development and expansion of the original tale, as represented now by LU, is merely an indication that the story had existed for a long period in tradition. As the central theme was elaborated and the tale grew by the accretion of episodes, the same theme was introduced more than once, with variation of context or with additional detail[3]. This is what we find throughout in the theme of a duel between the hero and a Connachtman. But Thurneysen's view of the origin of doublets is different. He seems to have held that a doublet of this type cannot occur within one version of a tale. To him the repetition of a motif denotes a different version. For example, Lugaid is greeted with a set form of words by Cú Chulainn (LU 5526–30) and Thurneysen assigns the passage to Source B. When the same formula is repeated in Cú Chulainn's greeting of Fergus (LU 5645–9), Thurneysen ascribes it to A[4]. When Cú Chulainn assumes a false beard to deceive Nath Crantail (LU 5757–8), Thurneysen assigns it to Source B. But the repetition of the same motif in the fight with Lóch (LU 6128–30) he assigns to A[5]. There are three passages which describe Cú Chulainn's attempt to halt the progress of the army. One of these, the ogam-inscribed oak-tree at Belach Áne, occurs in a variant (LU 5213–5) and Thurneysen attributes it to B, because he has already assigned the other two incidents to A. These two passages, the ogam-inscribed withe forbidding the army to pass and ogam-inscribed *gabal*, occur together in the section preceding the *Macgnímrada*. So tenacious is Thurneysen of his belief that one motif denotes one version that

---

[1] TBC² 2882–5, LL 3794–3800.

[2] RC xv. 75 §§127, 129.

[3] Cp. G. S. Kirk, The Songs of Homer (pp. 72–80), who notes the prominence of repeated formulas, passages and themes in the Homeric poems.

[4] 'Als nicht zu B gehörig erweist sich namentlich . . . die Begrüssung Fergus' durch CuChulainn, eine Dublette zu §35, der Begrüssung Lugaids' (Heldensage 152 n. 3.).

[5] 'Anderseits hat B den falschen Bart CuChulainns schon in §39 verwendet; dieses Motiv stammt also hier aus A' (ib. 170 n. 4).

he suggests for the second incident here that the sentences telling of the ogam inscription on the *gabal*[1] are a later addition, perhaps added by the compiler of LU[2].

We must, however, agree with Thurneysen when he attributes the inconsistencies and contradictions of Recension I to the conflation of two or more earlier versions. This would explain the uncertainty about Conall Cernach who is represented several times as being among the *dubloinges*, but twice is mentioned as being with the Ulstermen in Ulster. When Bricriu is said to be with the Connacht forces, as in one passage of the *Macgnimrada* and in the fight with Lóch, Thurneysen attributes these references to B. When, at the end of TBC we are told (in an obvious reference to *Echtra Nerai*) that he had lain ill in Connacht during the whole course of the Táin and rose from his sick-bed to witness the fight of the two bulls, Thurneysen attributes the passage to A.[3] So too the death of Findabair is assigned to A, her survival, where she is mentioned in the concluding paragraph of the YBL version, to B.[4] In both sources Cú Chulainn is said to have had a friend and confidant in the Connacht forces. In one source he is Fiacha mac Fir Febe (A), in the other Lugaid mac Nóis (B), and where both Lugaid and Fiacha are mentioned, as in the Fer Báeth episode, Thurneysen suggests that the compiler has worked both sources together.

Many more variants and doublets occur in interpolations in LU. That part of Recension I which is found in the manuscript LU contains many passages written in a different hand (H) from that of the two main scribes of LU (A and M). These passages occur also in the fragmentary version in Eg. 1782 where it coincides with the LU text[5], but they do not occur at all in YBL. These H-interpolations, which must have been taken by H from *w*, the forerunner of W (as Thurneysen names Eg. 1782), are found throughout the text of LU. Sometimes they consist of short glosses or of

---

[1] *gabal* . . . ⁊ *ainm ogaim iarna scríbend ina tóeb* . . . *Ardléga fer dib in n-ogum ro boi i tóeb na gabla .i. óenfer rod lá in gabail cona óenláim ⁊ ní théssid secce conda rala nech uaib co h-áenláim* (LU 4804-9).

[2] Heldensage 128 n. 3. When Thurneysen gives as an argument that the injunction on the *gabal* is not carried out, he overlooks the fact that this is true also of the injunction on the *id*-inscription.

[3] Heldensage 216 n. 2.

[4] ib. 219 n. 1.

[5] Eg. 1782 contains the poem *Atchiu fer find*, many but not all of the marginal or interlinear glosses, and ll. 5834–5890 of the long interpolation LU 70b–71b (the rest of this passage, no doubt, also on a lost page of Eg. 1782).

passages *in rasura*. Two long sections have been written on intercalated parchment leaves (71–72, 75–76). The language of these
passages is more modern, and, as Thurneysen noted, they contain
constructions not found before the 11th century[1]. The H-interpolations are definitely additions and form no part of the original
epic. They may first have been inserted into an incomplete version
like that contained in the MS. LU by some compiler acquainted
with the later part of TBC as told in the YBL version. At any rate
some of the interpolations are a re-telling of incidents which occur
in a later context but which the interpolator tries to fit into the
earlier narrative. Indeed some compiler of the text may have
'composed' these passages, relying on memory or on an oral
development of the tale rather than on a written version.

In the uninterpolated Recension I (but not in Recensions II and
III) Lugaid mac Nóis goes twice to parley with Cú Chulainn. This
occurs in the long passage noted by the scribe as *córugud aile*.
Then three times in Recension I (twice in Recensions II and III)
Mac Roth goes to offer terms to Cú Chulainn, and Fergus names
the conditions and goes to make a final settlement with Cú Chulainn.
In the H-interpolation, LU 5834–5868, are several variant accounts
of these incidents: (1) Lugaid is sent to ask for a truce and tells
Cú Chulainn that his terms, named in detail, have been rejected
(and this although the passage comes after that in which Fergus
has made final terms with Cú Chulainn); (2) Fergus is sent to ask
permission for them to move camp; (3) an unspecified messenger
goes to offer half the captured cows to Cú Chulainn who agrees to
these terms; (4) Maine Aithremail goes and offers Findabair to
Cú Chulainn; (5) Lugaid goes again to him with the same offer.
Here are given again the two missions of Lugaid which had already
been described in great detail in the uninterpolated narrative.
The mission of Maine Aithremail in the H-interpolation is close in
wording to that of Mac Roth, already told, and in H the description
of the snow melting around Cú Chulainn because of his ardour
corresponds to a passage in Recension II which precedes the first
mission of Mac Roth[2].

In this same H-interpolation we get a variant account of the
disguise-motif, common in all recensions. Here the jester ac-
companying Findabair wears Ailill's crown, but Cú Chulainn is

---

[1] Heldensage 235, ZCP ix. 421.
[2] In Recension III this passage occurs in Mac Roth's account of Cú
Chulainn on his return from his first mission.

not deceived and kills the jester by thrusting a pillar-stone through him.

In this H-interpolation also, while Ailill and Medb are still at the stage of offering terms to Cú Chulainn and before the Ulstermen have recovered from their debility[1], there is inserted a passage entitled *Comlond Munremair ⁊ Con Roí* which has been derived and adapted from the later passage where, in Recension I and Recension II, Amargin and Cú Roí cast stones which collide in the air overhead. The interpolator tries to justify Munremar's intervention at this early stage by adding *Ro scáich noínnin Ulad fo sodain*. The passage which follows, *Aided na Macraide*, is also out of context. It is a variant of the incident which occurs later in all recensions in the section *Breslech Maige Murthemne*. There it is in the correct place: only women and boys were exempt from *ces naiden*; the Ulstermen were as yet unable to come to Cú Chulainn's aid, so the youths of Ulster came. It was to avenge the killing of these youths that Cú Chulainn fought *Sesrech Breslige*. After *Aided na Macraide* in the same H-interpolation comes *Bánchath Rochada*, a variant of the later *Bángleo Rochada*. Here again the interpolator feels called upon to explain how the Ulsterman Rochad could fight at this stage. Cú Chulainn sends his charioteer to ask for Rochad's aid: *Téit in gilla . . . ⁊ asbert fris techt do fórit[h]in Con Culaind ma dodeochaid asa noennin*. There follows an account of the killing of six of Medb's royal mercenaries (*sesiur rígamus*). This parallels the later passage where the seven under-kings of Munster in Medb's army rise to take revenge for the trick that has been played on them in promising Findabair as wife to each man.

The second long H-interpolation occurs LU 6131–6206. The insertion of an interpolation at this point may have been suggested by the preceding passage (in LU only, not in YBL), in which Medb says *Is sáeth dam nách accim in gilla imma n-ágar sund . . . Cía fer sucut, a Ferguis?* (6116–20). This could lead to the introduction of the meeting between Medb and Cú Chulainn at Ard Aignech which seems to be a variant not of any incident in the uninterpolated Recension I but of the meeting of Medb and Cú Chulainn at Glenn Fochaíne which occurs in Recension II and Recension III[2]. In

---

[1] In all uninterpolated versions the first Ulsterman to recover is Óengus mac Óenláime Gábe. Note in a poem in Comrac Fir Diad the line *Munbud Chonchobor 'na chess* (infra 2775, TBC² 2407).

[2] The meeting at Ard Aignech occurs also in Recension III which contains this H-interpolation verbatim.

Recension III this meeting is introduced by Medb's complaining that she does not know Cú Chulainn and wishes to see him: *Maith a Fherghais, bar Medb, nachan aithnidh dam fein gid etir Cú Chulainn* (ZCP viii. 548). In the H-interpolation the incident opens with the sentence *Conniacht Medb comarli dús cid dogenad fri Coin Culaind ar ba aincis mór lei an ro bíth leis dia slógaib* (6134–5). which suggests that it should come earlier in the narrative as does the meeting at Glenn Fochaíne. Compare Medb's complaint, *Nibat buana ar slúaig fón samlaid seo dia mmarbad Cú Chulaind cét láech cach n-aidche úan* (LL 1467–70), in the passage which introduces the missions of Mac Roth. It is of course possible that the Ard Aignech meeting occurred in the variant version drawn on by the LL-compiler and the compiler of Recension III. and that the LL-compiler suppressed the incident. being unwilling to depict Medb's treachery.

In that part of the tale which precedes the section *Breslech Maige Murthemne*, the theme of a duel between the hero and an enemy warrior occurs several times. Cú Chulainn fights with Nath Crantail. Cúr, Fer Báeth and Lóch, and the skill with which these encounters are varied in circumstance and detail is remarkable. Other encounters are more briefly told. These episodes. if we except the Fer Diad story, an obvious interpolation, end with Cú Chulainn's meeting with Fergus, his last encounter with one man. which in Recension III is aptly called *Bánchomrac Fergusa*, the bloodless encounter with Fergus.

The theme of a fight against odds is particularly common. Cú Chulainn is attacked by the two sons of Nera and their charioteers, and by the three sons of Gárach and their charioteers. Sometimes such fights are only briefly mentioned. as when we are told that Cú Chulainn killed three druids and their wives (sent by Medb to attack him, LL) or that Medb sent a hundred men of her household against him. In more detail is described the attack of Gaile Dána (Calatín Dána, LL), and his twenty-seven sons and his nephew (grandson, LL). In the original tale this seems to have been the culminating fight of Cú Chulainn. In Recension I. in the section *Sírrabad Sualtaim*, no account is taken of the long *Comrac Fir Diad* interpolation. Sualtaim hears of the attack by Gaile Dána and his sons[1] and speaks of *gair mo maicse re n-ecomlann*[2].

---

[1] *buadrugud a meic Con Culaind fri da mac dec Gaile Dana ⁊ mac a sethar* TBC² 2974–5, an obvious confusion with the fight against Ferchú and his twelve followers.

[2] Again, in TBC² 3387–8 reference is made to Cú Chulainn's fight against odds, *Cú Chulaind . . . iarna chrechtnugud in n-ecomlund.*

Similarly, and surprisingly, this is the attack referred to here in LL[1], although mention has been made in the previous section of Cú Chulainn's having been wounded by Fer Diad.

In the later part of the tale the fights described are those of Ulster warriors coming to the aid of Cú Chulainn as they recover from their debility. These Ulstermen fight against the men of Ireland, Medb's forces, some single-handed like Cethern mac Fintain and Iliach, others supported by their troops like Fintan, Mend and Reochaid. Cethern's attack is described in a few words, but there follows a long passage describing how Fíngin Fáthlíaig, Conchobor's physician, examines Cethern's wounds and Cú Chulainn identifies his attackers. This elaborate passage, which must surely belong to a later stratum of the tale, is of just the kind which allowed freedom to expand to scribe or compiler who could always insert an additional description of wound and attacker to his taste. Some of Cethern's opponents are well known figures from the enemy camp, Medb, Illann mac Fergusa, Ailill and his sons, but others bear names which do not occur elsewhere in the tale[2].

The last part of TBC contains *Tochostul Ulad*, a catalogue of the warriors summoned to the final battle by Conchobor's messenger. We may note here that, in a later passage in YBL, there is another long catalogue entitled *Tochostul Fer nÉrend*. Ailill sends a messenger Traigthrén to summon the *ferchuitredaig* to his aid. Thurneysen takes this second catalogue to have been inserted by the compiler as a Connacht counterblast to the earlier list of Ulstermen[3]. He notes that the title of this section does not occur in the list of episodes which in YBL precedes the final part of the tale. The second catalogue, however, occurs also in LL in shortened form and without the title *Tochostul Fer nÉrend*[4]. After *Tochostul Ulad* comes the long description of the Ulster heroes as they advance into battle with their troops. Twenty companies of Ulstermen arrive at Slemain Mide, each band headed by its lord whose physical appearance, clothing and armour are given in a recognised formulaic pattern. Each band is described by Mac

---

[1] Only in Recension IIb is added here a reference to Cú Chulainn's fight with Fer Diad (Stowe 3947–8), but note in a later passage in LL *ní lécgat Ulaid ind é ar bith a chned ⁊ a chréchta dáig ní hiomchomlaind ⁊ ní hiochomraic aithle chomraic Fir Diad* (4591–3).

[2] Note, for instance, that in YBL there are three more sets of attackers than in LL, none of them well known figures.

[3] Heldensage 105.

[4] See Stowe TBC notes 4795, 4798.

Roth and identified for Ailill by the Ulsterman, Fergus. It might be thought that in this section of TBC so disproportionate an amount of the narrative has been devoted to these elaborate descriptions that the battle itself, which is hardly described, comes as an anti-climax[1]. This, however, is to judge TBC by the canons of modern literary taste. We can be sure that far different standards of dramatic effect existed for the 12th-century storyteller.

It must, however, be conceded that by any standards the final passages of the tale in Recension I are disappointing. It is as if the storyteller, having concentrated so long on the doings of heroic men, has little time to spare for the bulls and deals with their death in the casual manner of one tying up loose ends[2]. The story flags sadly and the narrative dwindles to a mere *dindshenchus*, giving eight place-names derived from the encounter and death of the bulls and bearing the stamp of pure invention.

The episodic nature of TBC, the result of continual accretions, is precisely what we should expect in an orally preserved tale. Further the saga is uneven and lopsided, some parts having been elaborated and expanded and stylistically embellished. It has been suggested that the native genius of the Irish writer is better suited to the short story than to a work of long and complicated structure. As Windisch noted, the genius of a great poet was lacking which might have welded all the disparate parts of the tale into a great whole: 'Die Táin ist auf der Vorstufe zu einem grossen Epos stehen geblieben' (Introd. p. lx).

§4 Sources of LL-Version

A. LL and Recension I

Thurneysen held that the interpolated LU-version of TBC was the sole source of the LL-version. In support of this theory he quotes three passages from the H-interpolations and from LL[3], and his argument is as follows:

(1) The killing of Fir Focherda is told twice in LU, once in the same place in the narrative as in LL (LU 5817–20), where the number

---

[1] Compare Thurneysen's criticism of *Togail Bruidne Da Derga*, Heldensage 627.

[2] Recension II gives a short account of the bulls' encounter (4872–4880).

[3] ZCP ix. 426–431. 'Der Verfasser der LL-Version hat eine erweiterte LU-Version gekannt, wie sie in der Hs. W vorliegen würde wenn diese vollständig wäre' (ib. 429).

of the slain is twenty, and again in an interpolated passage (6159–68) where it is fourteen[1]. In LL the names of Fir Focherd are given (but only seventeen names for the twenty men); in LU the names are given only in the interpolation. This, Thurneysen argues, shows that the LL-compiler knew and drew on the H-interpolations of LU. He suppressed the second episode and transferred the names to the first.

(2) In the H-interpolated *Comlond Munremair ⁊ Con Roí* (LU 5884–97), when the stones cast by the two warriors collide in the air overhead, the Connachtmen protect themselves with their shields *dia sáerad for barnib na cloch* (5888). In the later incident in LL and YBL the same encounter is related of Amargin and Cú Roí, and the collision of the stones in the air is described in YBL with the words *co 'mmafrecraidis na clocha isinn aer* (TBC[2] 2963–4), and in LL *co 'mbofrecraitis na bairendlecca bodba i nnélaib ⁊ i n-áeraib úasu* (3960–1). The word *bairendlecca* shows, says Thurneysen, that the LL-compiler knew and drew on the passage in hand H which occurs earlier in the text with the phrase *dia sáerad for barnib na cloch*[2].

(3) The poem *Atchíu fer find firfes chless* is written on an erased passage in LU by hand H (4546–4591). It occurs in the same place in LL in a form close to that of LU.

Only the last passage, the poem, can be adduced as a proof of Thurneysen's theory. The other two passages do not seem to be of sufficient weight to uphold it. There is, perhaps a negative argument on Thurneysen's side. The two long interpolations on intercalated leaves in LU (5834–5952, 6131–6206) represent in the main variant versions of incidents related later in the uninterpolated Recension I. If the LL-compiler had before him the interpolated LU-version, he would naturally suppress all of these variants. So it would not be surprising to find in LL nothing more of these long passages than the names of the Fir Focherda (which in LL, however, do not agree exactly with those of the H-passage[3]) and the word

---

[1] Still another variant of the Fir Chronige episode occurs in an earlier H-interpolation. Here the interpolator splits the incident in two: twenty men are sent against Cú Chulainn and he kills them all; then he kills forty-eight men *hi Crónig* whose names are given (LU 5839–44). Cp. Heldensage p. 237.

[2] For this point one might equally well argue that the H-interpolator knew or had a vague memory of a version like LL. The words *bairne na cloch* occur TTr[2] 1866.

[3] See infra note to text 1767–71.

bairendlecca[1]. On the other side we might argue that the H-interpolator and the LL-compiler took the poem, *Atchíu fer*, from the
same source. A poem, being of fixed form, remains practically
unchanged whatever its provenance. Where Thurneysen argues
that the LL-compiler drew on the H-interpolations in LU, I should
be inclined to conclude only that he had access to the same version
of the poem as the H-interpolator[2]. Neither theory is capable of
definite proof.

Throughout the part of Recension I contained in LU we get
variant versions of incidents[3]. Sometimes the scribe notes that a
certain episode occurs in a different place in the narrative *iar
n-araili slicht*. For instance, after the incident when Cú Chulainn
puts an ogam-inscribed withe on a pillar-stone, all three manuscripts state that, according to another version, it was then that
Fedelm Banfháith met Medb, and not earlier as had already been
told (LU 4770–76). So too for the killing of the pet bird and the
pet marten, which, according to YBL, another version places after
the killing of Medb's pet hound, Baiscne (TBC[2] 766–9). In all three
manuscripts the death of Medb's herd Forgemen occurs after
Medb's destruction of Dún Sobairche, but in an earlier passage
(denoted as *slicht sain* in YBL) a doublet of this incident is found
in the death of Medb's herd Lóthar (LU 5348–5355). A long variant
passage is given LU 5389–5582 introduced by the words *Dogniat
immorro augtair ⁊ libair aile corugud aile fora n-imthechtaib a
Findabair co Conaille*. In this occurs a variant account of the
incident where Fergus's sword is taken from its scabbard. In the
remainder of Recension I, that is, in the continuation of the tale in
YBL (and in C as far as it goes), it is remarkable that the only
variant version referred to is that of Sualtaim's death (TBC[2]
3003–5). The other instances in this part of Recension I of what
seem to be variants are due to mere uncertainty on the part of

---

[1] The only other agreement with the H-interpolations is the reference
in LL to the presence of Flidais Fholtchaín in the Connacht camp (306–7).
In LU the name Flidais is inserted here by H in an interlinear note. I
believe, however, that the reference to Flidais here in LL comes from a
version of Táin Bó Flidais II. See note to text. Here again we might
argue that the H-interpolator was acquainted with a version like Recension
II. Significantly this H-note is not in Eg. 1782.

[2] The LL- poem is nearer to LU than to Eg. 1782, but there is not
exact agreement between the H-interpolated poem and the LL-poem.
Note that the poem, unlike some of the other H-interpolations, does not
represent a doublet or variant.

[3] These are denoted as variants by the scribe and are to be distinguished
from the variants or doublets of the H-interpolations.

xxviii TÁIN BÓ CÚALNGE

the scribe as to who uttered a certain rhetoric, c.g. *Is and sin asbert Celtchair . . . Nó comad he Cuscraid Mend Macha mac Concoba[i]r rochanad in laig sea* (TBC² 3073. 3078–9. Similarly 3093, **3443**, **3454**. 3466).

It has been generally assumed that the compiler of LL had before him a version like Recension I with all these variants, and that he discarded the variants in the interest of unification[1]. This would imply a definite process of selection. It is a curious fact that of all the alternatives offered in Recension I prefixed by the word *nó*, only four are found in the LL-compilation.

(1) In the passage describing the four men who went ahead of Medb's army. these men are called in LU *Eirr ⁊ Inell, Foich ⁊ Fochlam a nda ara. Ceithri meic Iraird meic Ánchinne* (4728–30), with a marginal note in hand M: *nó cethri meic Nera meic Núado meic Taccain ut in alis libris inuenitur.* (YBL has merely *Eirr ⁊ Indell. Foich ⁊ Fochlam a nda ara*, omitting the father's name TBC² 243–4). In LL they are *dá mac Nera meic Nuatair meic Tacáin .i. dá mac rechtairi na Crúachna, Err ⁊ Inell. Fráech ⁊ Fochnam anmand a n-arad* (468–70)[2]. It seems as if the marginal note is due simply to the scribe's knowledge of a version like Recension II.

(2) In LU the boy Cú Chulainn kills Culann's hound by dashing it against a pillar-stone. *Mad iar n-arailiu immorro is a líathróit ro lásom inna beolu co rruc a inathar thrit* (5014–6). In LL the two versions are combined; the boy casts his ball into the hound's mouth and then dashes the animal to pieces against the stone (881–885).

(3) In LU Cú Chulainn casts a sling-stone and kills the pet marten on Medb's shoulder. Then he kills the bird on Ailill's shoulder. *Nó dano is for gúalaind Medba batár immalle eter togán ⁊ én* (5292–6). The alternative account is that of LL; both marten and bird are killed on Medb's shoulder (1273–7)[3]. It is significant

---

[1] But note Nettlau (RC x. 332–3): 'We must, I should think, assume that when these texts [TBC] were first written down the MSS. assumed soon the character of an accumulation of different accounts of the single episodes, doublets, confused and contradictive reports, the natural attributes of every collection of materials in these credulous and naive times. We cannot believe that a uniform version like the LL text has been collected *then* by people competent to judge authentically what belonged to the same "version" and what not'.

[2] This passage is not contained in the fragmentary Recension III.

[3] Similarly in Recension III (ZCP viii. 542). Note that in the YBL passage inserted after the killing of Medb's hound Baiscne no mention

that the variant is omitted in Eg. 1782 (ZCP ix. 142). Here again I suggest the addition of a scribe who knew of a version like Recension II.

(4) At the end of the encounter with Fer Báeth we read in LU: *Focherd sin ém, or Fer Báeth. Is de atá Focherd Murthemne. Nó iss é Fiacha asrubart, Is beóda do feocherd* [sic] *indiu, a Cu Chulaind,* or se (6027-9). In LL it is Fiachu who says at this point *Maith trá in focherd, a Chúcúc* (1903)[1].

Apart from these four instances which seem of no particular significance, we have no proof that the LL-compiler had before him the many variants of Recension I. The only variant quoted as such in LL is a marginal note at the end of *Breslech Maige Murthemne*: *Iss ed atberat araile ro fich Lug mac Eithlend la Coin Culaind Sesrig mBresslige* (2322-3)—a sentence which occurs in the text of LU, but not in YBL nor in Recension III[2].

One of these variant passages in LU, the incident at Belach Áne and *Aided Fraich*, has been studied in great detail by Professor James Carney. He was the first to point out that *Aided Fraich* is an interpolation in LU-TBC.

All versions of TBC go back, according to Thurneysen, to Recension I which was a conflation of two 9th-century written versions now lost. These 9th-century versions, which he names Source A and Source B, derive from oral tradition. Thurneysen enumerates certain criteria by which we can decide from which source a particular passage or episode has been taken, and in his analysis of Recension I he attributes accordingly some episodes to Source A, others to Source B. But he himself notes that some passages have been attributed to B only because they cannot definitely be assigned to A, and vice versa, and that possibly a third source, not a complete version of TBC, may be in question[3]. Where doublets occur, Thurneysen, in default of other criteria, generally assigns one to Source A, the other to Source B. The incident *Aided Fraich* is inserted clumsily into one such doublet which Thurneysen assigns to Source B. The passage is a variant

---

is made of Ailill and the marten is not on Medb's shoulder: *Mad iar naraile slicht immorro is and romarbad in togan boi hi carput la Meidb ⁊ in peatac eoin da urchoraib* (TBC[2] 766-8).

[1] In Recension III it is Cormac Conn Longas who says *Is beodha in fhoithcheird gaisgidecht, a Chugan* (RC xiv. 261 §40).

[2] Thurneysen suggested that this was a marginal note incorporated in the text by the LU scribe. Compare the marginal note on *cethri meic Nera* mentioned above.

[3] Heldensage 119.

(*slicht sain*)[1] and recounts one of Cú Chulainn's attempts to impede the progress of Medb's army. He fells an oak-tree in their path, cuts on it an ogam inscription defying anyone to pass until a single warrior has driven across it in his chariot. In trying to accomplish this, thirty horses fall and thirty chariots are broken. Eventually Fergus drives across the tree in his chariot. The encounter of Cú Chulainn and Fróech and Fróech's death by drowning at Cú Chulainn's hands are inserted awkwardly before the final sentence, *Lingid Fergus darsin n-omnai ina charput* (LU 5238).

Professor Carney has argued that the tale *Táin Bó Fraích* is a composition of the 8th century which had no oral existence prior to that date. He shows that the Fróech incident in Recension I TBC, *Aided Fraích*, is based on *Táin Bó Fraích* and forms no part of the original epic. He further maintains that *Aided Fraích* was written in the 8th century by the author of *Táin Bó Fraích* and grafted on to a copy of TBC used by the compiler of Recension I. Professor Carney attributes the incident of the felled oak-tree to the same author who, in an attempt to make his interpolation appear genuine, 'modelled his work upon an incident in the original', (that is, the Áth Gabla incident). Here then, he holds, we have a divergence between Thurneysen's Sources A and B which cannot be attributed to oral differentiation. Hence Professor Carney rejects Thurneysen's theory of the origin of doublets and of oral tradition as the basis of TBC.

There are two points to note here. First, Thurneysen did not recognise that *Aided Fraích* was an interpolation, but when he referred to this passage (LU 5211–5238) as a doublet, he must have meant only the incident of the felled oak-tree. That this incident is not an interpolation but a variant known to the compiler is shown by the sentence in the next section, *Aided Órláim*, which refers back to it: *Ro brisisem ar carpait oc tofund na ailite ucut Con Culaind* (5257)[2]. Into this variant a scribe or compiler acquainted

---

[1] As such it is not found in LL.

[2] Also in LL, though the variant does not occur: *dáig ar bith ro mebtatar* (sic) *ar carpait inné ic tafund na hailite urdairce út .i. Con Culaind.* This sentence is not in Recension III. Note that just before *Aided Órláim* the H-interpolator erased a variant passage referring to the killing of Medb's pet bird and marten (TBC[2] 766–9) and substituted for it: *Mór in cuitbiud dúib, ol Medb, can tophund na erri angceóil ucut fil co for nguin. Doberatsom iarom topund fair iar sin* (iar sini MS.) *coro brisiset fertsi a carpat oca* (LU 5243–5), a repetition of the variant but based on l. 5257, which was probably intended as an introduction to *Aided Órláim*, *Aided Fraích* having been interpolated after the breaking of the chariots at Belach Áne.

with the first part of *Táin Bó Fraích* inserted an invented account of the drowning of Fróech by Cú Chulainn[1]. *Aided Fraích* is here completely out of context. Not until terms have been arranged with Cú Chulainn (LU 5585–5622) do individual Connacht warriors come forth at Medb's request to fight against him. What we should have expected here, if anything intervened before the final sentence, is an account of how Fergus tried to drive across the tree but broke a number of chariots before finally succeeding in his own chariot. When *Aided Fraích* reads *Tonfóir, a Fraích, ol Medb. Discart dín in n-écin fil fornd* (5220–1), we should expect *Tonfóir, a Ferguis* etc. As Professor Carney points out, the wording here is almost identical with that of the Áth Gabla incident where it is Fergus whom Medb calls upon to help them in this strait. In his study of *Táin Bó Fraích* Professor Carney had noted that the author had based the character of Fróech in part on that of Fergus, and he says that the most significant feature of *Aided Fraích* is that Fróech plays in it the part played by Fergus in the Áth Gabla incident of which it is a doublet. But this is not so. Except for the similarity in wording of Medb's appeal, there is no sign here of a doublet of Fergus's role in the Áth Gabla incident. Fróech makes no attempt to drive a chariot across the tree. It is perhaps possible that the Fróech interpolator took over here some of the wording of a suppressed passage of the kind I have suggested. More likely, however, he borrowed the phrases from the Áth Gabla passage. Further, we may note that in *Aided Fraích* Cú Chulainn fights against an unarmed opponent. The only other instance of such a fight occurs in the H-interpolation *Comrac Maind* (6706–6722). In *Aided Fraích* Cú Chulainn wrestles with Fróech in water (*Atnagait co céin móir oc imtrascrad forsind usci*); in *Comrac Maind* he wrestles with Mand on land (*Gabait iarom for imtrascrad fri ré cian*)[2].

The second point is that *Aided Fraích* is not the only interpolation of this kind in Recension I. Thurneysen noted that the passage in LU entitled *Imacallaim na Mór[r]igna fri Coin Culaind* (6081–6100) is an anticipation of events to come rather awkwardly inserted in the narrative. As long ago as 1887 Windisch suggested

---

[1] The death of Fróech *a comrac usci ar Táin Bó Cuailgne* is told in the prose Dindshenchus (RC xvi. 137 §132) and in the verse (MD III. 362–4). The account is clearly based on the TBC passage.

[2] Contrast the Nath Crantail episode where Cú Chulainn says *Ní gonaimse nech cen arm* (5745). Nath Crantail is the first Connachtman called on by Medb to fight Cú Chulainn after terms have been arranged.

that this *Imacallam* was a later interpolation in TBC[1]. Thurneysen assigns the whole passage to Source B, but when, in the fight with Lóch, we find the sentence *Is and sin trá dogéni Cú Chulaind frisin Mórrigain a trede dorairngert di hi Táin Bó Regamna* (6232–3), he suggests that these words have been inserted by the LU-compiler. The *Imacallam* is based on *Táin Bó Regamna*, and I believe that we may go farther than Thurneysen and attribute that whole passage also to the LU-compiler.

There is still another passage in Recension I which seems to me to bear the marks of an interpolation, namely, the second version of *Imroll Belaig Eóin*. The first version (6647–54) is clearly a story of casts which missed their mark (*imroll*), but in it the place-name Belach Eóin is left unexplained. It seems as if the compiler tried to remedy this by giving a variant version which opens with the words *Tiagait na sloig do Beluch Eúin* (6655). But the tale which follows contains no reference to a mis-cast. On the contrary each man kills his opponent. There are other points here suggestive of an interpolation. The name Diarmait son of Conchobar does not occur elsewhere in TBC. The only reference to him seems to be that of *Cóir Anmann*: ⁊ *is fri Furbaidhe mac Conchubair aderthí Diarmaid mac Conchubhair íarsin* (§255). But Furbaide mac Conchobuir appears later in *Toichim na mBuiden*, and he survived to kill Medb in later times (LL 14390–14434). His death is certainly out of place here. Again, at this stage of the narrative, a message from Conchobor to Ailill and Medb is unexpected. The description ·of the fight between Diarmait and Maine (which of the seven Maines is unspecified) is a doublet of the Recension III fight between Cú Chulainn and Buide mac Báin Blai, with the difference that here both are killed in the exchange of missile spears. We may note too the use of the word *marcach*, apparently with the meaning of 'messenger'; it occurs in a similar context in an H-interpolation (LU 5936). Finally, the wording of Conchobor's message suggests that the compiler was thinking of *Echtra Nerai*: ⁊ *tabár in tarb aníar cosin tarb ille co comairset uair ro báge Medb* (6661). In *Echtra Nerai*, which is connected with *Táin Bó Regamna*, Medb swears that she will not lie down nor sleep nor eat and drink until she sees the two bulls fighting: *conam rabat na da tharb sin ar mo uelaib a comrac* (RC x. 226. 188–9)[2].

---

[1] Ir. T. II² p. 240.

[2] Another reference to *Echtra Nerai* which I should attribute to some compiler occurs at the end of TBC: *Bai Bricriu Nemtenga thiar ina thor iar mbrisiud a chind do Fergus cusna feraib fitchilli* (TBC² 3653–4).

Scribes and compilers must have been completely familiar with the various tales of the Ulster cycle. Scribe M of LU-TBC refers in a marginal note (4792) to *Togail Bruidne Da Derga* (and also to *Cath Maigi Tured*), and in a note between columns (4835) to *Tochmarc Emire*, while the interpolator H shows his acquaintance with *Táin Bó Flidais* (4623). *Aided Fraích* occurs in a section entitled *Slicht sain so co aidid nÓrláim* (5211). Órlám is the hero in the *remscél Táin Bó Dartada* as Fróech is in the *remscél Táin Bó Fraích*. The account of Órlám's death in the next section may well have suggested the interpolation of *Aided Fraích* at this point. I suggest that the LU-compiler also drew on the *remscél Táin Bó Regamna* for the passage *Imacallam na Mórrígna fri Coin Culaind*.

None of these three passages, *Aided Fraích*, *Imacallam na Mórrígna* and the second version of *Imroll Belaig Eóin*, is found in the LL-compilation. This seems to me an additional argument in favour of attributing them to the compiler of Recension I or to the compiler of one of the manuscripts, now lost, which must have intervened between the earliest written version of TBC and that of LU. Their interpolation by a compiler would not, however, invalidate the theory that the earliest written version of TBC was based on oral tradition.

Besides these three passages, there is also in Recension I a good deal of matter which is not found in LL. The following five passages contain this extra material:

(1) TBC² 198–214 (Thurneysen's §5).
(2) TBC² 416–483 (§§12–15).
(3) TBC² 1575–1584 (§48).
(4) LU 6115–6125 (§51b).
(5) TBC² 3680–3684 (§94).

(1) §5 Dubthach in a trance prophesies the arrival of Cú Chulainn and the subsequent bloodshed, in a poem of five quatrains ('in einem unvollständigen und nicht fehlerlos erhaltenen Gedicht', Thurneysen). This is the only poem in LU which is not in LL. The opening words of the passage, *Asberatsom is and sin ro gab Dubthach in laíd seo*, are almost the equivalent of *Asberat araili . . .* One line of the poem seems to be a reference to the tale *Cophur in Dá Muccado*. But note too that one quatrain (205–6) appears later in a variant passage (1033–4) where it fits the context better, and that two lines (211) are but slightly different from those of a quatrain which occurs in all recensions of TBC after the death of Nath Crantail (1313), while the last two lines are almost identical with the opening lines of the verses spoken by Dubthach in a later

passage (2056). It seems as if a compiler had borrowed from other verses in the tale to fill out his poem. The reference to Cormac's son (207), not mentioned elsewhere, is obscure. The verses are followed by an account of an attack by the war-goddess, *ind Nemain*, and the phrase *Nip si sin adaig ba sámam dóib* is one which is twice repeated in the last part of the narrative (3478–80, 3568–70). Thurneysen takes the whole passage to be a doublet of the later *Aislinge Dubthaich* (§80, 3082–3092) where Dubthach in his sleep recites a rhetoric, *Amra maitne* etc. There too we get the sentence *Cotmesca ind Emain forsin slog* (3091). Thurneysen attributes §5 to A, §80 to B. I am inclined to take the whole of passage §5 as an addition by a compiler. Passage §80 occurs in LL. Passage §5 is omitted in LL, yet in the same place in that recension, that is, next to the account of the killing of eight score deer at Móin Coltna, there is in LL a passage where Fergus warns of Cú Chulainn's arrival in a poetic dialogue (9 quatrains) with Medb: *Fó dúib faitchius 7 fót* (393–435). Did the compiler of Recension I or of one of its forerunners know that here, 'according to another version', there occurred a passage warning of Cú Chulainn's arrival, and, not having that version at his disposal, invent a warning by Dubthach based on the later *Aislinge Dubthaich*? Alternatively we might assume that here the LL-compiler substituted a warning by Fergus for that of Dubthach. But elsewhere in LL we do not find such a drastic change.

(2) §§12–15 All this extra material is an addition to Fergus's narration in the *Macgnímrada* section. Conall Cernach's account which follows is of one exploit only, the killing of Culann's hound. So too, in the last section, Fiacha describes how the boy takes up arms and first wields them. But in Recension I Fergus describes not one but several exploits. The words *Fecht n-and dano, or Fergus, in tan bá gilla . . .*, introducing this extra material, are superfluous here and suggest the addition of a compiler. There is repetition too in the account of Cú Chulainn's ball-playing with the youths in *Aided na Macraide*. In Fergus's opening story §11 Cú Chulainn merely knocks down the boys in Emain. In Conall's account §16 he kills a hound. But Fiacha's story §17 reaches a climax when the boy kills three of the Ulstermen's enemies and bears their bloody heads in triumph to Emain, while a childish touch is added by representing him as carrying in equal triumph the birds and deer that he has captured alive. With the extra matter §§12–15 there is no sense of climax, for in that part Cú Chulainn is said to have killed again and again. It has sometimes

been suggested that the whole of the *Macgnímrada* section, containing neither verse nor rhetorics, is an interpolation in TBC. Here, however, I suggest only that the extra material in Fergus's narration was added at some time by some compiler of Recension I who could not leave well alone. Without the passages §§12–15 the *Macgnímrada* section is more consistent and more artistically satisfying. This perhaps might be made an argument in favour of a skilful compiler of the LL-version who condensed to the advantage of the narrative. But I think that this is to attribute too great a power of constructive imagination to the LL-compiler[1].

(3) §48 This passage prefaces the fight with Láiríne mac Nóis. Thurneysen finds in it no criteria by which it may be assigned to Source A or Source B. Here Ailill suggests that, since no one can be found to go against Cú Chulainn, each man shall be plied with wine and promised Findabair if he brings back Cú Chulainn's head. But Cú Chulainn kills each man in turn until none can be found to oppose him. Yet, in contradiction to this statement, immediately after this passage, Láiríne is summoned and consents to fight. The passage is an obvious interpolation with repetition of phrases from the Fer Báeth episode. The sentence which precedes the passage, *Neach uaib ambaroch ar cend bar cele, or Lugaid*, would lead naturally to the sentence which follows it, *Congairther doib Lairene mac Nois*, if we omitted the interpolation.

(4) §51b This passage occurs in LU (and C), but not in YBL. It is obviously a doublet of the later passage which introduces the poem *Masu hé in riastarthe* (2037–2054). Here Cú Chulainn appears in ceremonial garb and the women climb on men's shoulders to catch a glimpse of him. Medb complains that she cannot see him, and—a curious touch—asks who he is. Fergus answers with a rhetoric in praise of Cú Chulainn, *Gilla araclich claideb* etc. Then Medb too climbs up to see him. The passage intervenes between a sentence telling us that Lóch would have killed the slayer of his brother Long had he known that he had a beard, and a sentence describing how the women induced Cú Chulainn to smear a false beard on his chin. It is an obvious interpolation, or possibly it is a variant version clumsily introduced. The mention of *Lethrend echaire Ailella* is to be noted; he does not occur elsewhere.

---

[1] Note Zimmer's remark concerning the passages §§12–15 : 'Dies Stück kann aus innern Gründen nicht der LL- Recension angehört haben' (KZ 28. 483). Zimmer looked upon LL as one of the two sources of Recension I, but that does not affect his argument here which is in the main my argument above.

(5) §94 This forms the concluding paragraph in YBL. The whole passage seems to be an addition, in particular the sentence *Anaid Findabair la Coin Culaind*, inserted by a compiler who liked a neat and happy ending. Thurneysen ascribed the passage to B because Findabair's death had been already told in a passage which he had assigned to A.

According to Thurneysen, the LL-compiler, whose sole source was the LU-version with all its variants and interpolations, worked the tale, 'mit kräftiger Hand', into a unified whole, discarding all contradictions and omitting many episodes[1]. Thus, we may assume, Thurneysen would explain the omission in LL of all the passages which I have enumerated. The LL-compiler discarded a large part of the material at his disposal[2]. Against this theory, I suggest that there existed a version of TBC like Recension I as it would appear if stripped of its variants and interpolations (and here I include as interpolations many passages other than the H-interpolations). With such a recension the LL-version would agree closely in content and sequence of incidents. Accepting this view, we must alter our idea of the manner in which the LL-compiler worked. The LL-version has been taken as a conscious literary attempt to give a complete and consistent narrative. When some passages of Recension I are not found in LL, as e.g. part of the *Macgnimrada*, it has generally been assumed that the compiler of LL is skilfully condensing his material. Elsewhere when he gives a more detailed account of an incident, he is supposed to have expanded and elaborated the barer text of Recension I. Where variants are quoted in Recension I, the LL-compiler suppressed them. In addition, the matter which LL contains in excess of Recension I is attributed by Thurneysen to the LL-compiler's own invention. I shall deal with the last point in a comparison of the LL-version with Recension III.

[1] ZCP ix. 432. That the LU- version was the basis of all later versions, as Thurneysen pointed out (ib. 436), is of course true. But Recension I as it now stands in the MS.LU represents an attempt to gather together all the known variants of the Táin.

[2] Nettlau (RC x. 332–3) was the first to stress what he called 'the uniform character of the LL- text' and to realise that such uniformity and unity of narrative did not necessarily imply unity of source. But Nettlau, if I understand him aright, implied that the LL- compiler was moved by purely aesthetic considerations when he chose a particular version of a particular episode, which assumes of course that he had a choice to make. The point in this theory which I find inexplicable is why invariably his choice should fall on that part of LU which forms the main narrative and not on those parts in LU which are variants nor on any of the numerous interpolations.

B. LL and Recension III

When editing the Stowe version of TBC, I had occasion to deal with the relationship of Recension III to Recensions I and II. Thurneysen had reached the conclusion that Recension III represented a more modern form of an LL-version which was in some points closer to Recension I[1]. After a comparison of the three recensions I ventured to disagree with Thurneysen and suggested that Recension III was largely based on a version of the LL-type, that it had drawn some episodes and passages from a version like LU and that it contained a small amount of material found neither in LL nor in LU[2]. A closer study of LL-TBC has caused me to modify this view in one point. I no longer believe it necessary to assume that the compiler of eh, the forerunner of Recension III drew on l, the archetype of the LL-version[3]. Hitherto all explanations of the agreement between the 12th-century LL and the late Recension III (found only in 15th-16th-century MSS.) seem to have been unconsciously influenced by the question of date. But the tradition to which Recension III goes back may well be older than the date of the compilation of LL, and agreement between LL and Recension III may be explained by assuming an early variant version or versions of the Táin (x)[4] on which the compiler of LL drew for the passages wherein LL differs from LU. From this early version would be descended also the late Recension III, and where Recension III agrees with LL against LU, it is not necessary to conclude that the compiler of eh was using a copy of l; rather we may assume that he drew independently on x for

---

[1] ZCP viii. 530, ix. 426.

[2] Stowe TBC Introduction p. xvii. In a footnote I suggested that for some passages the H-interpolator and the compiler of Recension III drew on the same source (p. xvii n. 2).

[3] I here use l for the original of the LL- version, unlike Thurneysen who gives l as the original of ls (from which he derives LL and St) and also as the original of eh (ZCP viii. 530).

[4] Nettlau (RC x. 333) pointed out that we have no proof of the existence of different versions of TBC: 'Versions of every episode existed of course but no versions of the whole text'. So too Thurneysen ZCP ix. 422. 'Einheitliche Sammlungen der ganzen Táin hat es wahrscheinlich in älterer Zeit nicht gegeben, sondern nur verschiedene uneinheitliche Sammlungen einzelner Episoden.' For my x, however, I assume a version of TBC which in the main outline or 'plot' conformed to the LU-version but which differed in its version of certain episodes and included some extra material. I see no reason to deny the existence of a complete version of this type. Note that some of the extra material in the Stowe version occurs in the late section Toichim na mBuiden. See Stowe TBC p. xii–xiii.

these passages and incidents. Where in a few points Recension
III shows agreement with Recension I against LL, it can sometimes
be shown that the compiler of Recension II may have had reason
to disregard these points. Where such agreement is purely verbal,
there is no difficulty in assuming that the LL-compiler changed
the wording of his source. Further, for some passages, the compiler
of Recension III drew on the same source as the H-interpolator
of LU.

Thurneysen held that the only important additional matter in
LL is the Introduction (1–146)[1] and the passage where Cú Chulainn
meets Medb and Fergus at Glenn Fochaíne (1398–1463)[2], and this
extra material he attributes to the redactor of the LL-version[3].
This, however, is to oversimplify the question. In many other
passages LL contains matter in excess of Recension I. Thus we
have in LL several poems spoken by Fergus which are not found
in Recension I (400–435, 604–627, 665–684, 1285–1300). After the
death of Lóch LL inserts a long poem which is obviously at variance
with the preceding prose (2016–2091). In the fight with Calatín
Dána (Gaile Dána, Recension I) two passages, the grief of Fergus
and his request for a witness (2546–2555), the flight and death of
Glas mac Delga (2588–2598), have no equivalent in Recension I.
In the later part of the tale there is a long passage of extra material
in LL, inserted between Mac Roth's two accounts of the Ulster
army (4228–4278). His first account is a description of noise and
tumult, of mist and flashing fire etc., interpreted for Ailill by Fergus
as the approach of a great army. Then comes the LL passage
describing how Conchobor and Celtchair go to Slemain Mide ahead
of the army, in the hope of being the first to shed blood. They are
seen by Mac Roth who describes them as a small band of three
thousand chariot-warriors. Ailill and Medb speak contemptuously
of this small force and Cormac Cond Longas rises to take revenge
for their boastful words. Then Conchobor attacks the Connachtmen,
killing 800 men, and retires unwounded to Slemain Mide[4]. In the

---

[1] The Introduction may well be the invention of Thurneysen's
Bearbeiter C. See infra note to text 1–146.

[2] See Stowe TBC p. xx n. 1. The passage is clumsily inserted into
LL but seems to be introduced into its correct place in Recension III.

[3] ZCP ix. 435.

[4] Thurneysen makes no comment in his Analysis of TBC on this extra
matter in LL here. Note the repetition (4278–83) of the passage in which
Ailill summons Mac Roth to reconnoitre the plain of Meath (4160–65)
immediately after this extra material in LL, which suggests that the
compiler himself realised that this was an interpolation. Recension I
does not contain the second summons.

section which follows, *Toichim na mBuiden*, in Recension I Mac Roth explains that Cú Chulainn did not come with the Ulstermen to Slemain Mide because he had been wounded, but, inconsistently, he goes on to describe Cú Chulainn and his charioteer who are then identified for Ailill by Fergus (TBC² 3387–3404). In place of this passage LL has a description of Cú Chulainn's followers, *tricha cét Murthemne*, sorrowing for the absence of their lord (4560–4586). In addition to all this extra matter, we find that in LL some incidents related in a few lines in the earlier version have been expanded into a full and detailed narrative, for example, the death of Loche (1341–1348), *Tuige im Thamon* (2460–2472), *Cinnit Ferchon* (2510–2531). All these differences in LL, not merely the last three, Thurneysen seems to attribute to a process of expansion and elaboration undertaken by the LL-compiler, Bearbeiter C.

Now the fragmentary Recension III coincides with LL only in a comparatively small portion of the whole tale, equal to some 1350 lines of the present edition. The portion common to LL and Recension III includes the section *Breslech Maige Murthemne* (215 lines in this edition) which is practically identical in all three recensions. So that finally we have about 1140 lines of this edition of LL to compare with Recension III. Yet in this small amount we find an astonishingly large agreement between LL and Recension III.

Leaving aside the many smaller points of agreement[1], we find in these lines of LL four passages in excess of Recension I which occur also in Recension III.

(1) Fergus, visited by a premonition of the arrival of Cú Chulainn, warns the men of Ireland in a poem: *Damb ró Cú Chulaind Cúalnge* (1281–1300). Similarly in Recension III where there is an extra quatrain (ZCP viii. 542).

(2) In both LL and Recension III the Morrígu warns Donn Cúalnge in a speech before the rhetoric which is common to all recensions, and the rhetoric is followed by a list of the attributes of Donn Cúalngo (1306–1309, 1320–1333. ZCP viii. 541).

(3) In LL Fiachu mac Fir Aba goes on a mission to Cú Chulainn, and Medb and Fergus meet Cú Chulainn at Glend Fochaíne. Two poems occur here: *Más é ucain in Cú cain* and *A Chú Chulaind cardda raind* (1398–1463). Similarly in Recension III, but one

---

[1] For these see notes to text 1177–8, 1182, 1189, 1199, 1248, 1341–48, 1368, 1389, 1394–7, 1478, 1555, 1589, 1605, 1698, 1758, 1994–6, 2361–4, 2469, 2494.

poem only, *A Chú Chulainn caraid raind*, the first being summarised briefly in the prose (ZCP viii. 548.)

(4) The poem beginning *Airg uaim, a Laíg, laíder slúaig* occurs in both LL and Recension III (2016–2091, RC xv. 69–70). This poem has obviously been taken into LL from the version of TBC which was the ancestor of Recension III. The details of the fight with Lóch which are given in the poem are inconsistent with those of the prose account given in the LL-version, but are those of the prose version in Recension III. In LU here we have a short poem of four stanzas, *M'oénurán dam ar étib* (6216–6231), corresponding to the nineteen stanzas of LL, and these four stanzas come before Lóch is killed by the *ga bulga*. In LU the poem is not addressed to Láeg; Cú Chulainn complains of having to fight single-handed and asks that Conchobor be summoned quickly to his aid (*Aprad nech fri Conchobor | cía domissed níbo rom* 6220–1). In LL and Recension III the poem comes after the death of Lóch; it is addressed to Láeg; it describes Cú Chulainn's wounding by Lóch and his encounter with the Morrígu as well as the death of Lóch. Thurneysen's suggestion that the compiler of Recension III remodelled his prose account of the fight with Lóch to suit the verse account in LL will not stand, for it fails to explain why LL should give in the verse an account of the fight at variance with the preceding prose account. On the other hand, if we assume that the LL-compiler, for once regardless of inconsistency, took over the verses from a variant version of TBC, the discrepancy is easily explained.

Finally, and significantly, in one passage in LL the compiler's inadvertence shows that he had before him, in addition to the LU-version, a manuscript which resembled the ancestor of Recension III. He takes over a sentence from a variant version which differed from the Recension I account which he has chosen to give at that point, and the sentence can refer only to the version given in Recension III. In LU and LL Buide mac Báin Blai is killed by a spear (*certgae*) cast by Cú Chulainn. There is no mention of any attack by Buide on Cú Chulainn. But LL continues: *Cian gar ro bátar forinn uropair sin ic clóechlód na dá chertgae . . . rucad in Dond Cúalnge . . . úadib* (1784–6). This sentence is obviously based on the account of the fight in Recension III: *Maith, a Bhuidhi, bhar Cu Chulainn, tairr romaind bharsin áth sa sís go cláechlomais da urchar dha chele. Tangadar forsin ath ⁊ da chlaechlodar dha urchur ann* (RC xiv. 258 §16). Corresponding to the LL sentence we find in Recension III: *Cidh tra acht gérbha ghairid do bhadar in*

*daña churaidh sin agus in dana chathmhilidh ag cláechlodh in
dana urchur sin etarru, rugadar . . . in Donn Cúailghni* (ib. §17).

Side by side with these agreements between LL and Recension
III, there is also in Recension III a long passage (RC xv. 63–66.
§§71–76) which agrees word for word with ll. 6133–6192 in LU.
Here Lóch, for no obvious reason and despite the fact that Cú
Chulainn has assumed a false beard, postpones his fight with Cú
Chulainn for a whole week. During this time Cú Chulainn fights
against others who are named. Then Medb treacherously arranges
a meeting with Cú Chulainn at Ard Aignech. and fourteen of her
men lie in wait there for Cú Chulainn who kills them all. The
important point about this passage is that in LU it is a interpolation
in hand H and does not occur in YBL. (YBL omits LU 6131–6190).
The opening lines 6131–2, which are not taken over by Recension
III, give a variant account of Cú Chulainn's assuming a false beard.
already told ll. 6128–30 in the part of the text written by M. Again
the killing of several men during Cú Chulainn's week at Áth Grencha
ll. 6137–6142 is a repetition of ll. 6100–6102 in hand M. The meeting
at Ard Aignech may be taken as a variant or doublet of that at
Glenn Fochaíne, found only in Recension II and Recension III.
The derivation of the name Focherd is a repetition of an earlier
passage at the end of the Fer Báeth episode.

Because this passage §§71–76 in Recension III agrees verbatim
with the H-interpolation, Thurneysen concludes that it was taken
over into Recension III from a manuscript of the interpolated
LU-version[1]. But if that were so. we should have expected many
more such borrowed passages elsewhere in Recension III[2]. All
we can conclude is that for this passage the Recension III compiler
or scribe had access to the same version on which the H-interpolator
drew in the 12th century.

In that part of the same H-interpolation which was written to
coalesce with the LU text on page 77a. there is a passage (ll.
6200–6206) which is found in the same wording in Recension III
(RC xv. 67 §82). When Cú Chulainn has been wounded by Lóch.
Fergus calls for someone to incite Cú Chulainn to further efforts
and Bricriu responds with a taunting speech. This passage is not
in Recension II which, owing to the loss of a page in LL. is here
represented by the Stowe version, Recension IIb. It is just possible
to conjecture that it was omitted by the Stowe compiler who had

---

[1] Heldensage p. 240 n. 1; ZCP ix. 427 n. 2.

[2] The long H-interpolation (LU 5834–5952) is not represented in
Recension III.

in mind the later reference to Bricriu's having lain ill in Connacht during the whole course of the Táin. But Stowe follows the LL-version so closely elsewhere in the text that we are justified in concluding that the reference to Bricriu did not occur in LL. What is of most interest here is that the wording of the Recension III passage is that of the H-interpolation, not that of YBL. Thus YBL reads *Gresid in fer, or Fergus fria muinter. Roscaich do bruth, or Bricriu, brattan bec dotrascair intan dofil Ultu asa ces* (TBC² 1715–6), which is expanded by the H-interpolator into: *Olc ón om, for Fergus, a ngním sin hi fiadnaisi námat. Gressed nech úaib, a firu, for se fria muintir, in fer nár tháeth i n-ascid. Atraig Bricriu Nemthenga mac Carbatha ⁊ gabais for gressacht Con Culaind. Ro scáich do nert, ol se, in tan is bratan bec dottrascair in tan dofil Ultu asa ces chucut. Dolig duit gním n-erred do gabail fort hi fiadnaisi fer nÉrend ⁊ laech ansa do dingbail a gasciud fon samail* [*sin*] (LU 6200–6206). It is this expanded version which is found word for word in Recension III §82.

Again in *Éle Loga* in the section *Breslech Maige Murthemne*, the last four lines are written by H on an erased passage in LU, and the Recension III reading corresponds exactly. What stood here originally we cannot tell for there is loss of text here in YBL.

There is agreement between Recension I and Recension III against LL in a few passages. Thus after the death of Lóch Cú Chulainn kills five men (RC xv. 71 §96–98 = LU 6239–41) and in the section *Slánugud na Mórrígna* we find the lines *Maith a Chu Chulainn, bar in Mhórríghu, bharairngertais damsa nach bh[f]aighind furtacht ná fóirithen dhod lamaibh. Ma da fesaind gomadh tú do bheth ann, níris faighbhithea* (ib. 72 §111). Note, however, that the wording of the equivalent passage in LU is not the same: *Atbirt frim trá, or in Mórrigain, nim bíad íc lat co bráth. Acht rofessin combad tú, ol Cu Chulaind, nit icfaind tria bith sír* (LU 6255–7). The passage does not occur in LL.

For some small points in which Recension III agrees with LU against LL, Thurneysen assumes that they were also in the fore-runner of LL but were omitted, partly through carelessness[1]. This would hold for the mention of the herd Forgemen (LU 5330, ZCP viii. 541), for the place-name Botha which is mentioned and explained in Recension I and Recension III (LU 5379–80, ZCP viii. 544). For other passages there may be a different explanation. The LL-compiler seems inclined to suppress or minimise any reference to magic help given to Cú Chulainn. He omits, for instance, the

[1] ZCP viii. 532.

naming of Lug mac Eithlend when he comes to aid Cú Chulainn
and the singing of *Éle Loga* over him, a passage which occurs
in Recension I and Recension III. So too he may well have
deliberately omitted the reference to the rivers which rose in flood
to impede Medb's progress (LU 5362, 5380–81, 5385–6, ZCP viii.
543).

As Thurneysen pointed out[1], Recension III sometimes agrees in
wording with Recension I against LL. But it is noteworthy that
these agreements occur in passages which are otherwise nearer to
the LL-version than to that of Recension I. Thurneysen quoted,
for instance, from *Comrac Fergusa* the phrase *Teilg traigid dam*
which is common to Recension I and Recension III (LU 6693,
RC xv. 206 §210) where LL has *Teich romum-sa* (2494). But the
rest of *Comrac Fergusa* in Recension III is closer to LL than to
LU[2]. Again Thurneysen quoted the passage where the boy Cú
Chulainn returns to Emain at the end of the Macgnímrada. In
LU we get the sentence *Ardailfe fuil laiss cach dune fil isind lis
mani foichlither* (LU 5191–2) and in Recension III *Mana faigligther
lib-si e, ferfaidh* [leg. *ferfaidher*?] *cru dar cuiged Concobair anocht
aigi* (ZCP viii. 538), while in LL the sentence reads *Ocus meni
frithálter innocht é, dosfaíthsat óic Ulad leis* (1182). But note that
LL and Recension III put the speech in the mouth of Leborcham
while in LU the sentence is spoken by *in dercaid i nEmain*. In
Aided Órláim Recension I and Recension III describe Cú Chulainn
as drawing the chariot-poles *tria ladra a glac . . . etir rúsc ⁊ udba*
(LU 5260–1, ZCP viii. 540. 7) while LL has *tria ladraib a choss ⁊
a lám i n-agid a fiar ⁊ a fadb* (1233). But in other details *Aided
Órláim* in Recension III agrees with LL against LU.

These agreements in wording between Recension I and Recension
III led Thurneysen to the conclusion that there existed a version
of TBC which in essentials corresponded to the forerunner of LL
but which in some points was nearer to the original of Recension I.
From this version (his *l*), he derives on the one hand LL (and Stowe),
on the other Recension III. But occasional similarity of phrasing
does not lead inevitably to such a conclusion. Whether Recension
III as it now stands represents only the variant *x* or whether it is
a confused compilation which drew on *uy* as well as on the variant
cannot now be determined. The variant *x* may have agreed
occasionally in wording with Recension I, and where Recension
III agrees with Recension I and LL disagrees with the wording of

---

[1] ib. 528–9.

[2] See note to text 2494.

both, there seems no reason why we cannot assume that in these passages the LL-compiler altered his original.

Finally, Recension III differs occasionally from the other two recensions and contains a small amount of extra material[1] which may be the addition of the compiler or scribe or may be attributed to the variant version or versions represented by Recension III. That such variants existed is proved by the extra material in the Stowe version of TBC. It is improbable that the variant drawn on by the compiler of Recension III was a complete version of the tale. The concluding words in Eg. 93 (RC xv. 208) suggest that his original did not go beyond the *Cinnit Ferchon* episode. He had probably merely a disparate collection of incidents from the Táin on which to base his recension (if indeed Recension III merits the name of recension at all). In Recension III, for instance, the order of the episodes which follow the killing of the sons of Gárach is neither that of Recension I nor that of Recension II. Again, in Recension III Redg Cáinte's death comes late in the narrative, between *Comrac Fergusa* and *Cinnit Ferchon*, whereas in the other versions it comes before the fight with Cúr mac Da Lóth. Nor is the style of Recension III homogeneous; it varies from the adjective-laden, verbose description of the boy Cú Chulainn returning in triumph to Emain, obviously the work of a late scribe expanding the text, to the simpler narrative of the passage equivalent to the H-interpolation of LU which must have been taken unchanged from an early version[2].

I have already suggested that the LL-compilation is based solidly on a version of Recension I without variants and without interpolations. I further suggest that the LL-compiler drew also on a variant version x. The greater part of Recension III represents the later development of this variant[3].

In my view that LL and Recension III both drew on an early variant version or versions of parts of TBC, my disagreement with Thurneysen is, however, more apparent than real. Thurneysen's belief was that Recension III was mainly based on a version like

---

[1] See Stowe TBC Introduction pp. xviii–xix.

[2] That variant versions of some passages common to LL and Recension III existed in different wording seems to be shown by the late MS. C. In the Glenn Fochaíne meeting and in the list of Donn Cúalnge's attributes, C differs sharply in wording though not in content, and the wording is not of the kind that we should expect a 16th century scribe to use if he were merely paraphrasing the LL- text.

[3] As already noted, the Recension III compiler had also access to the source of some H-interpolations.

LL with resemblances here and there in wording to the LU-version[1]. From this version which he named *l* he derived *ls*, the ancestor of LL, and *eh*, the ancestor of Recension III. But Thurneysen's *l* is directly and solely descended from the expanded LU-version[2], and this is where I differ. To my mind *l* should represent a version of Recension I without variants and interpolations, greatly influenced by a variant version *x*. I disagree with Thurneysen about the extra material in LL. Thus, for instance, he ascribes the meeting at Glenn Fochaíne to the invention of the LL-compiler. I should assume that the incident occurred in the variant version *x*. In particular two passages in LL which I have already mentioned strengthen this theory, viz., that the LL-compiler based some of his material on a variant: the poem, *Airg uaim, a Laíg*, which gives the version of Lóch's fight which is told in the prose version of Recension III; a sentence in the account of Buide mac Báin Blai's fight which shows the LL-compiler's knowledge of the variant version given in Recension III. I am at one, then, with Thurneysen when he holds that agreement between LL and Recension III may be traced to a common source, but I disagree with him when he derives that source solely from the expanded Recension I.

§5 THE LL-VERSION

I assume then that the LL-version was based on a version like Recension I shorn of its variants and of its H-excrescences and other interpolations, and that the LL-compiler also knew and made use of a variant version *x*. Such an account of the origin of the LL-version is of course grossly oversimplified. There must have been many other complex factors at work—the influence of inter-related tales well known to the compiler, the unconscious adaptation of oral developments of TBC, the re-shaping of older material to the compiler's own taste to produce a new and personal version. For it is above all clear that Thurneysen's Bearbeiter C intended to provide a full, accurate and definitive narrative to be the pattern for all future scribes. In his colophon he calls down a blessing on those who leave his version unchanged and uninterpolated.

The text of TBC is found in the manuscript 53b to 104b (xl–lxvi mediaeval foliation). In a recent article (Celtica vii pp. 1–31),

---

[1] ZCP viii. 530.

[2] See Thurneysen's stemma ZCP ix. 441.

'Notes on the Script and Make-up of the Book of Leinster', William O'Sullivan, Keeper of Manuscripts, T.C.D., has made a valuable palaeographical study of the Book of Leinster, correcting the views of Best, restoring the mediaeval foliation of the manuscript, and showing that LL is not a lay patron's book but rather a scholar's book written by ecclesiastics filled with a desire to preserve Irish secular learning. Mr. O'Sullivan notes that all the styles of writing in LL are of the same family. Hence Best's suggestion that LL was the work of a single scribe. But Mr. O'Sullivan distinguishes four main hands, A (that of Aed mac meic Crimthaind), F (probably the hand of Bishop Find's scribe), T (a hand found throughout the manuscript) and U. The Táin is written in hand T, but F is responsible for slightly more than three pages in the middle of the tale, ll. 2122–2360, that is, from the opening of the *Breslech Maige Murthemne* section to midway in the description of Cú Chulainn in ceremonial garb. Note that this section of TBC is one which Thurneysen suggested may have been a separate piece later incorporated in the tale.

The language of the LL-text is later than that of Recension I. On internal evidence Thurneysen assigned the composition of the LL-version to the first third or even the first quarter of the 12th century. LL-TBC shows the spread of the *s*-preterite and *f*-future, the *-it* ending of the 3 pl. perfect passive, the loss of the deponent, the increasing use of the independent pronoun. All these points and many others have already been ably dealt with elsewhere in detail[1]. The language of the LL-text also shows some characteristics of its own, e.g. the interchange of preverbs and the frequent use of *con-* in relative construction[2].

The style of the LL-text is more diffuse and more detailed than that of Recension I. In LU the narrative is generally bare, the style terse. Sometimes a passage is so condensed that the full meaning is learnt only by a comparison with the equivalent passage in LL. Compare, for instance:

Teilg traigid dom, or Lóch. Doléiciseom Cú Chulaind combo thairis docer (LU 6236–7).

Teilg traigid dam corop ar m'ugid sair tóethus ⁊ nárap dar m'aiss siar co firu Hérend, arná rádea nech dib is roi madma nó techid

---

[1] Máirín O Daly, The Verbal System of the LL Táin, Ériu xiv. 31–139. See also M. Dillon, ZCP xvi, 313–356, xvii. 307–346 and Gearóid S. Mac Eoin, ZCP xxviii, 76–136, 149–223. Professor Mac Eoin would place the composition of TTr[1]. earlier than that of LL-TBC.

[2] These points are dealt with in the notes to text. Cp. also Ériu xx. 104–111:

dam remut-sa . . . Ocus teilgis Cú Chulaind traigid ar cúl dó (LL 2006–10).

In the derivations of place-names, we sometimes find in Recension I merely a short sentence, *Is de atá Imroll Belaig Eúin* or *Is de atá Áth Tamuin ┐ Tuga im Thamun.* In LL the full explanation is given by quoting a remark made by 'the men of Ireland'. Thus *And sin atrubratar fir Hérend: Is imroll díbairgthi, bar íat-som, a tarla dona feraib, cách díb do guin a charat ┐ a choibnesaim badessin. Conid Imroll Belaig Eóin and sin* (2456–8) and *Is tuige im thamon duit-siu ám, a Thamuin drúith, bar íat-som, étgud nAilella ┐ a imscim n-órda immut* (2466–7). This is particularly noticeable in the episodes *Fiacalgleó Fintain, Rúadrucce Mind* and *Bángleó Rochada*, where there is no explanation of the names in Recension I, whereas here again in LL they are explained clearly in a remark made by the men of Ireland (3831–5, 3843–5, 3889–93).

Again and again we get examples in LL of additional details added as the context demands, e.g.

Tiagait íarom eich in cethrair i n-agid in tslóig ┐ a fortchai forderga foraib (LU 4801–2).

Ocus léicis Cú Chulaind eocho in fiallaig sin i n-agid fer ṅHérend i frithdruing na sliged cétna ┐ a n-ésse airslaicthi ┐ a méide forderga ┐ cuirp na curad ic tindsaitin a fola sell sís for crettaib na carpat. Dáig nibá miad nó níba maiss leiss echrad nó fuidb nó airm do brith óna corpaib no marbad (LL 570–7).

In the passage in LU where Ailill asks Fergus if it is Conchobor who has come to Áth Gabla, Fergus answers *Ní tergadside co hor críche cen lín catha immi* (4824). Similarly in answer to the question if it is Celtchar mac Uthidir or Eogan mac Durthacht (4825–8). But the LL equivalent passage is *Dámbad ésin tísad and, ticfaitis slúaig ┐ sochaide ┐ forgla fer nUlad filet maróen ris ┐ giano betis fora chind i n-óenbaile ┐ i n-óendáil ┐ i n-óentochim ┐ i n-óenlongphurt ┐ i n-óentulaig fir Hérend ┐ Alban, Bretain ┐ Saxain, cath dobérad dóib, reme no maissed ┐ ní fair no raínfithé* (696–700). A similar exaggerated answer is given in LL for Celtchair, Eogan and Cuscraid Mend Machae (700–716).

Such elaborations and additions explain why the LL-version is longer than Recension I. The matter has been spun out in this way. In addition the compiler's style is repetitive. Not alone does he describe similar incidents in similar words, but when, for instance, a message is given to a messenger, it is delivered in a repetition of the same words. Note the exact repetition of Fergus's message to Cú Chulainn (1719–21) given by Fiachu (1724–6), of

Lugaid's message to Cú Chulainn (1874–80) given by Láeg (1887–93), and the repetition by Sualtaim of Cú Chulainn's complaint (4023–32). This does not happen in Recension I. Note, for instance, in the account of Mac Roth's first mission: *Adfét dó in n-imarchor n-ule amal asrubartmár* (LU 5604–5), where repetition is avoided. In the passage *Fuile Cethirn* the LL-compiler's delight in repetition is clearly seen where the account of every examination by Fíngin is prefaced with the words *Féga latt dam in fuil seo, a mmo phopa Fíngin* and every identification of the attackers by Cú Chulainn ends with the words *Ba búaid ⁊ coscur ⁊ commaídium leó gea dofáethaiste-sa dá lámaib*[1].

The tendency to pile up nouns and adjectives, synonymous or almost synonymous, which is a dominant feature of the style in late texts, is very marked in the LL-version. In Recension I this is noticeable only in passages which are obviously of later composition, e.g. in *Breslech Maige Murthemne* and in *Comrac Fir Diad*. In LL: *Luid iarum Láeg ar cul do saigid Con Culaind co cendtromm n-imthursech n-anfálid n-osnadach. Is cendtromm n-imthursech n-anfálid n-osnadach dotháet mo phopa Láeg dom indsaigid-se, bar Cú Chulaind* (1881–4). Contrast the equivalent passage in LU: *Soid Láeg afrithisi co airm a mboi Cú Chulaind. Ni forbáelid mo popa Laeg dia athiusc, or Cú Chulaind* (5998–9). In Recension I Ailill asks Medb what shall be done with the Gailioin if they neither go with Medb's forces nor yet stay at home, and Medb grimly answers *A nguin* (LU 4634–5). In LL Medb says *Bás ⁊ aided ⁊ airlech is accobor lem-sa dóib* (331). Again *a bás ind fir* (TBC² 2996) is in LL *a bás ⁊ a éc ⁊ a aided ind fir* (4032). Of the harpists of Caínbile we read in LU *roptar druid co móreolas* (5311) but in LL *batir fir co mórfiss ⁊ go mórfáistine ⁊ druídecht iat* (1271). The final battle is hardly described in Recension I but in LL we are twice told how the men of Ireland fought: *Da gabsat fir Hérend cách díb bar slaide ⁊ bar slechtad, for tóchtad ⁊ bar tinmi, for airlech ⁊ for essargain araile ri ré cían ⁊ ra reimes fata* (4668–70, 4683–5).

There are in LL occasional omissions of small points which are found in Recension I. Some of these may be due to scribal carelessness. Thus when Culand bewails the loss of his hound and says *Is bethu immudu ⁊ is trebad immaig mo trebad i ndegaid mo chon*

---

[1] Cp. A. B. Lord's study of the formula in the technique of oral verse-making (The Singer of Tales, chap. III). The term 'formula' covers not only 'tags' and 'clichés' but also applies to any group of words constructed on the same syntactical pattern to express a given idea. We may assume the same technique is saga-recitation.

(LU 5021–2), in the equivalent sentence in LL the words *i ndegaid mo chon* are omitted, probably through oversight[1]. So when the Morrígu warns Donn Cúalnge, she sits on a pillar-stone *i ndeilb eúin* (LU 5321), but again the phrase is omitted in the LL-version[2]. In Recension I we are told that Iliach came to fight naked (*osse tarrnocht . . . contibset in fear tarnocht* TBC[2] 2937, 2939), but this is not mentioned in LL, although the remark of the men of Ireland (3911–2) seems to imply his nakedness.

Notable in LL is the omission of Medb's reasons for objecting to the Gailioin. In LU she says that it is useless for the rest to go on the expedition for the credit of victory will go to the Gailioin (*is foraib biaid búaid in tslóig* 4631) and on their return they will seize Medb's land (*Ficfit fornd iar tíachtain ⁊ gébtait ar tír frind* 4633–4)[3]. Again the LL-version omits the description of Cú Chulainn's distortion in the *Macgnímrada* which is found in Recension I but which is surely out of place there[4]. Omitted also in Recension II is what might be to Cú Chulainn's discredit, e.g. the wailing of Nechta Scéne after the killing of her three sons, and the killing by Cú Chulainn of Órlám's charioteer[5]. This last incident was no doubt omitted in LL because it was in direct contradiction to what Cú Chulainn has just said: *Ní gonaim aradu nó echlacha nó áes gan armu* (1240), a sentence repeated in the fight with Nath Crantail (1728).

There are many inconsistencies in Recension I, generally the result of the conflation of two or more versions. The unified LL-version is as a rule free from contradictions. In both LU and LL there is, however, one notably inconsistent passage where the LL-compiler has eliminated some but not all of the contradictions. In both versions Cú Chulainn writes an inscription on a withe and casts it around a pillar-stone. In LU the inscription warns them that they must not pass until one of them, but not Fergus, has cast a withe in like manner (4739–41). In LL it warns them that

---

[1] The phrase occurs in Stowe (932).

[2] Omitted also in Stowe and Recension III.

[3] Possibly the LL-compiler felt that this reason was a reflection on Medb's courage.

[4] Zimmer (KZ 28. 485) suggested that ll. 4879–84 should be omitted in this passage of LU. But note, in the encounter with Nath Crantail in Recension I: *Sia[ba]rtha im Choin Culaind amal dorigni frisna maccu i nEmain* (LU 5778).

[5] The apologetic passage referring to this in LU and YBL, *Ní fír trá amlaid sin na marbad Cú aradu. Ní marbad ém cen chinaid cip innus,* is not in Eg. 1782. Note that in LU also Cú Chulainn assures the charioteer *ni gonaimsea aradu etir* (5267).

they must stay by it for a night in camp until one of them has cast a withe in the same way (502–7). LL omits the stipulation about Fergus (⁊ *friscuriur mo phopa Fergus*). In neither version is any attempt made to carry out the injunction. All that we are told is that the army takes refuge in a wood until morning. (This of course fulfils part of the LL-injunction). Next day Cú Chulainn cuts a four-pronged branch with one blow of his sword and casts it into the ford with one hand. On the branch is an ogam inscription which, in LU, defies them to pass until one of them, but not Fergus (*cenmothá Fergus*), has cast a branch in the same manner (4807–10). In LL there is no mention of Fergus, the inscription merely says that one of them must draw the branch from the ford with one hand even as it had been cast there (634–6). In the event this is what is done by Fergus in *both* versions. Here the LL-compiler has eliminated the contradictions in the LU-version of the *gabal* incident.

The compiler of the LU-version drew on the *remscél Táin Bó Fraích* for the passage *Aided Fraích*, and probably, as I have suggested, on the *remscéla Táin Bó Regamain* and *Echtra Nerai* for other passages. The compiler of the LL-version drew on *Echtra Nerai* for his account of Bricriu's illness in Crúachu which is given in greater detail than in Recensïon I. I believe too that he drew on some older version of *Táin Bó Flidais II* for some passages[1], in particular for his account of the loss of Fergus's sword. Note how he prefaces that passage with the words *Bliadain riasin sceol sa* (2487) just as he prefaces his account of Bricriu's illness with *Bliadain résin scél sa Tánad Bó Cúalnge* (4862), both being references to a tale outside the Táin. In the long variant passage in LU, *córugud aile*, is described how Fergus's sword was taken from its scabbard and replaced by a wooden sword. The incident takes place not in the preceding year in Crúachu (as in LL) but in Ulster when Medb and Fergus part from the rest of the army on the journey to Conaille Murthemne. Cuillius, the charioteer of Ailill, comes upon Medb and Fergus, takes the sword and gives it to Ailill. Fergus, perceiving his loss, goes into the wood and cuts a wooden sword to replace it. This incident is referred to in a later passage in Recension I: *Is and sin asbert Ailill re araid, Domiceb* (leg. *domiced*) *in claideb cuilleis toind*[2]. *Tongu do dia*

---

[1] See infra notes to text 299, 306–7, 307–8, 4824–32. Alternatively we might assume that TBFlid II drew on *x*, the variant version of TBC. See infra note 1–146.

[2] I suggest that the name *Cuillius* in the LU variant may have come from a misreading of *cuilleis* (= *choilles* LL) in this passage by some

*toingeas mo thuath mad meso a blath lat indiu olldas a llaithe
dondmbiurtsa duit isin liter i crich nUlad, cia no beidis fir hErind
ocot anocol airimsa nit ansitis* (TBC² 3554–7). Note the words
i ccrích nUlad which show that the incident referred to is that
of the variant. In the LL-version, where the story of the loss of
the sword occurs later in the text in *Comrac Fergusa*, it is Ailill
himself who comes upon the guilty pair in Crúachu, a year before
the events of the Táin. He takes the sword from its sheath, gives
it to his charioteer (unnamed) and puts a wooden one in its place
(LL 2487–91). In the later reference in LL (= TBC² 3554–7) we get
Is and atbert Ailill rá araid badessin .i. ra Fer Loga[1]: *Domraiched
craum claideb choilles toind, a gillai, bar Ailill. Nátiur-sa bréthir
mad messu a bláth ná 'lessugud lett indiu é andá in lá tucus barin
lettir i Crúachnaib Aí, dá mbet fir Hérend ⁊ Alban acott anacul forom
indiu, nit ainset uile* (4712–8). The account in LL agrees exactly
with that given in *Táin Bó Flidais II*[2]. Here the incident is said
to have taken place in Crúachu and it was Ailill who removed the
sword and put a wooden blade in its stead. Note further that, in
the reference to the incident in *Aided Etarcomail* in LU, Ailill is
given as the one who removed the sword: *Atchoas dam dano . . . ro
gab Ailill a mbáegul inna cotlud héseom ⁊ Medb, ⁊ dorétlaistir a
claidiub ar Fergus ⁊ dorat dia araid dia toscaid ⁊ doratad claideb
craind ina intech* (5641–4). This is like the LL account. Again in
an abbreviated reference to the loss of the sword in LU we find the
words *Ar gatsai Ailill ass ut praediximus* (6691)[3]. The account of
the episode in the variant LU 5400–5418 seems to have been an
elaboration of the more usual account, with Cuillius substituted for
Ailill and various details added, as, for example, Ailill's excuse
for Medb's conduct (*Is dethbir disi, or Ailill. Is ar chobair ocon
Táin dorigni* 5410). At the end of this long variant passage in
LU the killing of Cuillius, the charioteer, by Cú Chulainn is told
(5578–81). Note the contradiction here with the later passage in
Recension I where Ailill calls upon the charioteer to whom he had

compiler. If so, it is an exact parallel to the invention of the name *Íthall*
for one of the Connacht physicians in LL. See infra note 3653. In the
late MS. C the only instance of the name written in full is *Cuillesc*, perhaps
a reminiscence of the name *Cú Cuillesc Cáinte* in the late version of *Oided
Con Culaind*, itself a deformation from *cainte co culluaisc* (LL 13940).

[1] In ScMMD Ailill's charioteer is Fer Loga. Thurneysen suggested that
the LL- compiler borrowed the name from ScMMD (Heldensage 212).

[2] Glenmasan MS., Celtic Review I. 228; repeated II 312.

[3] Thurneysen does not distinguish between the two accounts in
Recension I but attributes all references to the incident to Source B
(Heldensage 149, 156 n. 1, 186, 212 n. 2).

entrusted the sword. There is no contradiction in LL for the whole long variant passage is wanting in that version.

Windisch, comparing Recension I and Recension II, concluded that the LL-version followed more closely that part of Recension I which is the continuation in YBL[1]. But his remark follows a passage in which he noted that the long rhetorical dialogues and other rhetorics in LU (most of which are in what the scribe notes as variants) are not represented in LL. Also, in comparing LL with the second half of the tale in YBL, he takes into account the Fer Diad interpolation. In fact, however, LL follows the main tale in LU more closely and is nearer in wording to LU than it is in the latter half to YBL. In particular the section from the end of *Toichim na mBuiden* to the end of the tale shows divergence in LL[2]. The order of passages is not the same. In YBL the rhetorical prophecies are more numerous and are attributed to different men. When Fergus's sword is restored to him. Ailill chants a long rhetoric in YBL: in LL Fergus welcomes the sword in a few words (taken from the rhetoric). In YBL the list of *ferchuitredaig* is entitled *Tóchostul Fer nÉrend*, a catalogue of a hundred and one triads whom Ailill summons by the messenger Traigthrén. In LL we are merely told that the *ferchuitredaig* arrived and the names of thirty-three triads are given. Their function in LL (as in *Tochmarc Ferbe*) is to act as bodyguard to Ailill and Medb; in YBL this function is assigned to the chariot-warriors from Irúad. In YBL Fergus is said to have killed a hundred of the Ulstermen before being turned aside by Conall Cernach who repeats Cormac's advice to Fergus to work off his wrath by shearing off the tops of the three hills, Máela Mide. In LL it is Cormac alone who gives this advice, Conall Cernach does not appear and Fergus kills no one. There is additional matter in this section in LL, the attack made by Conchobor on Medb's forces (4228–4277), the passage describing Cú Chulainn lying wounded in Fert Sciach, restrained by the Ulstermen from joining the battle (4587–95), the incident which explains the place-name Fual Medba (4821–32), together with additional details in the reference to Bricriu's illness. In LL too the battle of the bulls is described in some detail (4872–80) and Donn Cúalnge is said to have slaughtered women and boys in Cúalnge before his end. In YBL Donn Cúalnge is depicted as drinking from streams and fords on his way back to Cúalnge, a detail omitted in LL. The final

[1] Introduction p. lxxxiii.
[2] It has to be remembered that the text of the late MS. YBL must have suffered some change and corruption in the course of time.

paragraph of the YBL version, which I take to be a scribal interpolation, is altogether omitted in LL which ends with the death of
Donn Cúalnge.

In the Introduction which is peculiar to the LL-version Medb's
character is defined once and for all. She it is who takes the
initiative and makes decisions. Later in the tale when Fergus
reads the ogam inscription and its threats on the withe, in LU
Ailill says *Ní háil dúinni ém guin dúne dín fó chétóir* (4766–7) but
in LL it is Medb who speaks and her words are more emphatic:
*Ní hed sain ba háil linni ém, nech d'fuligud nó d'fordergad foirn ar
tíchtain isin cóiced n-aneóil se .i. i cúiced Ulad. Alcu dún fuligud ⁊
fordergad for nech* (507–9). Even such a small point as *Redg cáinte
Ailella* (LU 5805) is in LL *Redg cáinte Medbi* (1807). But the
passage in LL which most clearly shows Medb's dominant role is
that in which Mac Roth is sent to make terms with Cú Chulainn.
In LL it is Medb who complains that her army will not survive
Cú Chulainn's attacks (*Níbat búana ar slúaig* etc. 1468–71) and
suggests what terms shall be offered to him. Ailill plays a sub-
servient role and asks *Ciarso choma sain?* (1471) and *Cia ragas
frisin coma sin?* (1473), while it is Medb who decides the terms and
says that Mac Roth shall take them to Cú Chulainn. Again when
Mac Roth returns, it is Medb who questions him (1512–15, 1539–43).
Finally Medb contradicts Ailill, who disapproves of the terms, and
says *Is assu lind óenláech úaind cach laí dó-som oldás cét láech cach
n-aidchi* (1558–9). In the LU-version Ailill plays the part that
Medb takes in LL. It is he who says *Bid dimbuan ar slóg la Coin
Culaind in cruth sa* and offers terms. It is Ailill who says, when
finally accepting terms: *Is assu ém duinni in fer cech laí andás a
cét cach n-aidchi* (5621–2). Still more striking is the fact that in
this passage in Recension III (which is close in wording to LL and
agrees with LL in the number of Mac Roth's missions), it is also
Ailill who takes the leading role. Contrast with LL: *Berar
chomaidh uainne do, bar Oilill. Carsad comhadha sin aile? bar
Medhbh,* and *Cia rachas risna comadhaibh sin aile? bar Medhbh.
Cia dho rachadh ann acht Mac Roth in rimheachlach dano? ar Oilill*
(ZCP viii, 544), and so on throughout the passage. I believe that
we have here a deliberate alteration of his exemplar by the LL-
redactor who wished to show Medb as the stronger character[1].

In the closing part of the tale, that is, the part contained in
YBL, Medb rarely appears but her presence is felt more in the
LL-version than in the YBL-text. In LL Medb rebukes Cú Roí

---

[1] Stowe agrees with LL throughout this passage.

for his stone-fight with Amargin: *Ar fír do gascid frill, a Chú Rui, scuir dún don diburgun dáig ní furtacht ná fórithin tic dún de acht is mifurtacht tic dún de* (3962–4). There is no reference to Medb in this passage in YBL. When Fergus has identified the phenomena described by Mac Roth as the onrush of a great army, Ailill says *Artanesamar. Itat oic lind doib* (TBC² 3132). But in LL it is Medb who says *Ní dénam robríg de. Atethatar daglaích ⁊ degóic acainni dá n-acallaim* (4224–5)[1]. In a long passage in LL which has no equivalent in Recension I Medb is represented as speaking scornfully of Conchobor and his *tricha cét* who come to attack them. She suggests that Conchobor and his men can be taken unwounded, that is, without resistance, by her forces, and is attacked by Cormac (4250–59). In still another passage, not represented in Recension I, Medb boasts of all the ravages that she has wrought in Cú Chulainn's district and is rebuked by Fergus (4573–86). In LL but not in Recension I, when his sword has been restored to Fergus he asks *Cia fara n-immér-sa so?* and Medb answers *Arna slúagaib immut immácúaird* (4721–2). Where YBL reads *Is iarum ro indis a ara do Choin Chulaind bith do Ailill ⁊ do Meidb og guidhi Fergusa im thecht isin cath ⁊ asbertadar fris narbo chol do ar doradsad mor do maith do fora lonnges* (3540–2), the LL passage has no reference to Ailill and we get direct speech: *Is and sain atbert Medb ri Fergus: Ba bág ám dait-siu ga dobertha do greimm catha gan díchill lind indiu* etc. (4704–7). We are told in LL, but not in Recension I, that Medb had sent Donn Cúalnge to Crúachu, 'as she had promised'[2], that there he might meet the bull Findbennach (4822–4), and that she had gathered all the men of Ireland to Crúachu that they might see the bulls' encounter there (4852–3).

To sum up: the impression felt by the reader of LL-TBC is that this version is a unity, that the various elements of the tale have been brought together to form a new coherent whole by a single main 'editor', Thurneysen's Bearbeiter C[3]. The style is homogeneous throughout. Considering the length of the tale the standard of consistency attained is remarkable. The few interpolations it contains may be easily detected, and the only doublet, the repetition

---

[1] The same passage is repeated in TBC² 3182 but there it is, as in LL, Medb who says *Fogebad a n-acallaim sund* (= LL 4361–2).

[2] Probably a reminiscence of *Echtra Nerai*. Cp. LU 6661–2.

[3] On the basis of style and language Thurneysen attributed also to Bearbeitor C the compilation of LL-*Mesca Ulad* and the invention of LL-*Cath Ruis na Ríg*. See N. Power, The Common Authorship of Some Book of Leinster Texts, Ériu ix. 118–146.

of a warning by Fergus, though it differs in wording from the first occurrence, is noted by the compiler himself with the words *amal ra scríbsam remaind* (1284)[1]. Above all the unity of character is observed throughout in the depiction of Medb in all her pride and arrogance.

---

[1] See infra note to text 1280–1300.

FEc[h]t n-óen do Ailill ⁊ do Meidb íar ndérgud a rígleptha dóib i Crúachanráith Chonnacht, arrecaim comrád chind cherchailli eturru. 'Fírbriathar, a ingen,' bar Ailill, 'is maith ben ben dagḟir.' ⁻'Maith omm,' bar ind ingen, 'cid dia tá lat-su ón?' 'Is ⁵ de atá lim,' bar Ailill, 'ar it ferr-su indiu indá in lá thucus-sa thú.' 'Ba maith-se remut,' ar Medb. 'Is maith nach cúalammar ⁊ nach fetammar,' ar Ailill, 'acht do bith-siu ar bantincur mnáa ⁊ bidba na crích ba nessom duit oc breith do ṡlait ⁊ do chrech i fúatach úait.' 'Ní samlaid bá-sa,' ar Medb, 'acht m'athair i n-ardrígi ¹⁰ Hérend .i. Eocho Feidlech mac Find meic Findomain meic Findeoin meic Findguill meic Rotha meic Rigéoin meic Blathachta meic Beothechta meic Enna Agnig meic Óengusa Turbig. Bátar aice sé ingena d'ingenaib: Derbriu, Ethi¹ ⁊ Éle, Clothru, Mugain, Medb. Messi ba úasliu ⁊ ba urraitiu díb. Bam-sa ferr im rath ⁊ tidnacul ¹⁵ díb. Bam-sa ferr im chath ⁊ comrac ⁊ comlund díb. Is acum bátar cóic cét déc rígamus do maccaib deórad dar tír ⁊ a chomméit n-aill do maccaib aurrad ar medón, ⁊ dechenbor cach amuis díbside,² ⁊ ochtur ri cach n-amus, ⁊ mórfessiur cach amuis, ⁊ sessiur cach amais, ⁊ cóicḟiur cach amuis³, ⁊ triur ri cach n-amus, ⁊ días cach ²⁰ amuis, amus cach amuis. Bátar ocom sain ri gnáthteglach,' ar Medb, 'conid aire dobert m'athair cúiced de chóicedaib Hérend dam .i. cóiced Crúachna. Conid de asberar Medb Chrúachna frim. Táncas ó Ḟind mac Rosa Rúaid ríg Lagen dom chungid-sa ⁊ ó Chairpriu Nia Fer mac Rosa ríg Temrach, ⁊ táncas ó Chonchobur ²⁵ mac Ḟachtna ríg Ulad, [⁊] táncas ó Eochaid Bic. Ocus ní dechad-sa, dáig is mé ra chunnig in coibchi n-ingnaid nára chunnig ben ríam remom ar fer d'ḟeraib Hérend .i. fer cen neóit, cen ét, cen omon. Diambad neóit in fer 'gá mbeind, níbad chomadas dún beith

maróen fo bíth am maith-se | im rath ⁊ tidnacul, ⁊ bad cháined ³⁰

---

¹ Ethne *St, sic leg.*   ² ⁊ nōnmar la cec[h] n-amhus *add. St*
³ ⁊ cethrar cecha hamuis *add. St*

1

dom fir combadim ferr-sa im rath secha, ⁊ níbad cháined immorro
combar commaithe acht combadar maithe díb línaib. Diambad
úamain m fer, ní mó bad chomdas dún beith maróen, úair brissim-sea
catha ⁊ cumleṅga ⁊ congala m'óenur, ⁊ bad cháined dom fir
35 combad beódu a ben indá ⁊ ní cáined a mbeith combeóda acht
combat beóda díb línaib. Dámbad étaid in fer 'cá mbeind, níbad
chomdas béus, dáig ní raba-sa ríam can fer ar scáth araile ocum.
Fuarusa dano in fer sain .i. tussu .i. Ailill mac Rosa Rúaid do
Lagnib. Nírsat neóit, nírsat étaid, nírsat déaith. Tucusa cor ⁊
40 coibchi duit amal as dech téit do mnaí .i. timthach dá fer déc
d'étuch, carpat trí secht cumal, comlethet t'aigthi do dergór,
comthrom do riged clí do finddruini. Cipé imress méla ⁊ mertain
⁊ meraigecht fort, ní fuil díri *nó* eneclann duit-siu ind acht na fil
dam-sa.' ar Medb. 'dáig fer ar tincur mná atatchomnaic.' 'Ní
45 amlaid sin bá-sa.' ar Ailill, 'acht dá bráthair limm, fer díb for
Temraig ⁊ fer for Lagnib .i. Find for Lagnib ⁊ Carpre for Temraig.
Léicsius rígi dóib ara sinsirecht ⁊ níp[1] ferra im rath *nó* thidnacul
andú-sa. ⁊ ní chúala chúiced i nHérind ar bantinchur acht in cúiced
sa a óenur. Tánac-sa dano, gabsus rígi sund [2]i tunachus[2] mo
50 máthar. dáig ar bíth Máta Murisc ingen Mágach mo máthair, & gia
ferr dam-sa rígan no biad ocum andaí-siu, dáig ingen ardríg Hérend
atatchomnaic.' 'Atá dano,' ar Medb, 'is lia mo maith-sea indá do
maith-siu.' 'Is ingnad linni anísin,' ar Ailill, 'ar ní fil nech is lia
seóit ⁊ moíne ⁊ indmassa andú-sa, ⁊ rafetar ná fail.'
55   Tucad dóib anba táriu dá sétaib co festais cia díb dámbad lia
seóit ⁊ moíne ⁊ indmassa. Tucad chucu a n-éna ⁊ a ndabcha ⁊ a
n-iarnlestair, a mílain ⁊ a lóthommair ⁊ a ndrolmacha. Tucait dano
cucu a fánne ⁊ a falge ⁊ a fornasca ⁊ a n-órdúse ⁊ a n-étguda, eter
chorcair ⁊ gorm ⁊ dub ⁊ úaine, buide ⁊ brecc ⁊ lachtna, odor, alad
60 ⁊ riabach. Tucait a murthréta caírech d'aicthib[3] ⁊ d'urlannaib ⁊
rédib. Ra rímit ⁊ ra hármit ⁊ ra achnít corbatar cutrumma comméti
comlínmair. Acht baí raithi sainemail for caírchaib Medba ⁊ ba
gabálta i cumail é, ⁊ boí rethi a [f]recartha for caírchaib Ailella.
Tucait a n-eich ⁊ a n-echrada ⁊ a ṅgrega d'férgeltaib ⁊ scoraib.
65 Baí ech sainemail ar graig Medba ⁊ ba gabálta i cumail. Baí ech
a [f]recartha oc Ailill. Tucait dano a murthréta mucc a fedaib
⁊ fánglentaib ⁊ díamairib. Ra rímit ⁊ ra hármit ⁊ ra hachnít. Boí
torc sainemail oc Meidb ⁊ araile dano la hAilill. Tucait dano a
mbótháinte bó ⁊ a n-alma ⁊ a n-immirge dóib a fedaib ⁊ fásaigib in
70 chúicid. Ra rímit ⁊ ra hármit ⁊ ra hac[h]nít, ⁊ roptar cutrumma comméti
comlínmair dóib. Acht boí tarb sainemail ar búaib Ailella ⁊ ba

---

[1] *sic. read* níptar   [2-2] a dualgus *St*   [3] = d'faithchib

lóeg bó do Meidb atacomnaic ⁊ Findbennach a ainm. Acht nírbo
54ᵇ miad leis beith for bantinchur, | acht dochúaid co mboí for búaib
in ríg, & ba samalta re Meidb ná beth penning a selba lé ar ná baí
tarb a chomméit lé fora búaib. Is and sain conacrad Mac Roth ⁷⁵
ind echlach co Meidb ⁊ conscomarc Medb¹ de ar co fessed Mac Roth
airm i mbiad tarb a samla sút i cúiciud de chúicedaib Hérend.
'Rofetar omm,' bar Mac Roth, 'airm i fail tarb as dech ⁊ is ḟerr
dorísi i cúiciud Ulad i tríchait cét Cúalnge i tig Dáre meic Ḟachtnai²
.i. Dond Cúalnge a ainm.' 'Tó duit-siu connici sain, a Meic Roth, ⁸⁰
⁊ cunnig dam-sa for Dáre íasacht ṁbliadna do Dund Cúalnge, ⁊
ragaid lóg a íasachta dó i cind bliadna .i. coíca samaisci ⁊ Dond
Cúalnge fadessin. Ocus ber-siu comaid aile latt, a Meic Roth: mad
olc ra lucht na críchi ⁊ ind ḟeraind in sét sainemail sin do thabairt
.i. Dond Cúalnge, taít-sum féin ra tharb. Ragaid comméit a ḟeraind ⁸⁵
féin do mín Maige Aí dó ⁊ carpat trí secht cumal, ⁊ ragaid cardes
mo [ṡ]liasta-sa fessin.'

Lotar iar sain na echlacha dó co tech Dáre meic Ḟiachnai. Is é
lín luid Mac Roth nónbor echlach. Ra ferad fálti iar tain fri Mac
Roth i tig Dáre. Deithbir sin, prímechlach uile Mac Roth. Ra ⁹⁰
iarfacht Dáre do Mac Roth cid dobretha imthecht fair ⁊ cid 'ma
tánic. Innisis ind echlach inní imma tánic ⁊ innisid immarbáig
eter Meidb ⁊ Ailill. 'Ocus is do chungid íasachta don Dund Cúalnge
i n-agid ind Ḟindbennaig tánac,' arse, '⁊ atetha lóg a íasachta .i.
coíca samasci ⁊ Dond Cúalnge fessin, & araill aile dano béus, tair-siu ⁹⁵
féin lat tarb ⁊ fogéba comméit th'ḟeraind féin de mín Maige Aí ⁊
carpat trí secht cumal ⁊ cardes sliasta Medba air sin anechtair.'
Ba aitt la Dáre aní sin ⁊ ra mbertaig co raímdetar úammand a
cholcthech faí & atrubairt: 'Dar fír ar cubais, ³cid an ní ra Ultaib³,
bérthair in sét sa in cur sa do Ailill ⁊ do Meidb .i. Dond Cúalnge, ¹⁰⁰
i crích Connacht.' Ba maith dano la Mac Roth ra ráde [Mac]
Fiachna.

Ra frithálit iar sain ⁊ ra hecrait aíne ⁊ úrlúachra fóthib. Tucad
caíne bíd dóib ⁊ ra fordáled fled forro ⁴co mbátar búadirmesca⁴,
& dorécaim comrád eter dá echlaig díb. 'Fírbriathar,' ar indara ¹⁰⁵
echlach, 'is maith fer in taige i tám.' 'Maith omm,' bar araile,
'In fuil cid [d']Ultaib nech is ḟerr andás?' ar ind echlach taísech
béus. 'Atá omm,' bar ind echlach tánaise. 'Ferr Conchobor 'cá
tá, ⁊ cid immi gabtais Ulaid uile ane, níbad nár dóib.' 'Mór in
maith dó aní i mbiad opair cethri n-ollchóiced ṅHérend do brith ¹¹⁰
a crích Ulad .i. Dond Cúalnge do thabairt dúnni nónbur echlach.'

---

¹ Meidb *MS*.   ² *sic, for* Fiachnai
³⁻³ *sic MS.*, read cid [olc] an ní [sin] ra Ultaib   ⁴⁻⁴ *sic*

4   TÁIN BÓ CÚALNGE

And sain dano conarraid in tres echlach comrád forru. 'Ocus cid
ráter lib-si?' ar sí. 'Ind echlach út atbeir is maith fer fer in taige
i táam. Maith omm, bar araile. In fail cid d'Ultaib nech is ferr
¹¹⁵ andá? ar ind echlach thaísech béus. Atá omm, ar ind echlach
tánaise. Ferr Conchobor 'cá tá, ⁊ gid imme gabtais Ulaid uili ane,
níbad nár dóib. Mór in maith dó aní i mbiad opair cethri n-ollchóiced ṅHérend do brith a crích Ulad do thabhairt dúnni nónbor
echlach.' 'Nírb uráil limm sceith cró ⁊ fola 'sin mbél assa tic sain,
55ᵃ dáig cenco tucthá | ar áis, dobértha ar écin.'
   Is and sin doruacht fer uird rainne Dáre meic Fiachnai 'sin tech ⁊
fer fo lind leis ⁊ fer fo bíud, ⁊ atchúala anra chansat, ⁊ táncatar
fergga dó ⁊ turnaid a bíad ⁊ a lind dóib, ⁊ ní ebairt riu a chathim ⁊
ní ebairt a nemchathim. Dochúaid assa aithle issin tech i rrabi
¹²⁵ Dáre mac Fiachnai & ra rádi: 'In tú thuc in sét suachnid út dona
hechlachaib .i. Dond Cúalnge?' 'Is mé omm,' for Dáre. 'Ní raib
rígi airm i tucad, ar is fír aní rádit, ar [cen]co tuca-su ar áis,
dombéra ar écin fri sochraiti Ailella ⁊ Medba ⁊ ra móreólas Fergusa
meic Róig.' 'Dothuṅg mo deo dá n-adraim ná co mberat¹ ar écin
¹³⁰ samlaid nacha mbérat ar áis.' Fessit samlaid co matin. Atragat
na echlacha co moch arnabárach ⁊ dochúatar i tech i mbaí Dáre.
'Eólas dún, a úasail, co rísem bail a tá in Dond Cúalnge.' 'Nithó
omm,' ar Dáre, 'acht diambad bés dam-sa fell for echlacha nó for
aes n-imthechta nó tastil sliged, ní ragad nech úaib i mbethaid.' 'Cid
¹³⁵ són?' ar Mac Roth. 'Fail a mórabba,' ar Dáire. 'Ra ráidsebair cenco
tucaind ar áis dobéraind ar écin ra sochraiti Ailella ⁊ Medba ⁊ ra
móreólas Fergusa.' 'Aile,' ar Mac Roth, 'giped no ráditís echlacha
dot lind-su ⁊ dot bíud, ní hed ba tabartha do aíg *nó* do aire *nó*
d'airbire do Ailill ⁊ do Meidb.' 'Ní thibér-sa trá, a Meic Roth,
¹⁴⁰ mo tharb din chur sa dianetur.'
   Lotar na echlacha ar cúl dó samlaid ⁊ ráncatar Crúachanráith
Connacht. Conscodarc Medb scéla díb. Adféta Mac Roth scéla,
ná tucsat a tharb ó Dáre. 'Cid fotera són?' ar Medb. Rádis Mac
Roth aní dia mbaí. 'Ní hécen féth dar fudbu de, a Meic Roth, ar
¹⁴⁵ rafess,' ar Medb, 'ná tibértha ar áis [cen]co tuctha ar écin, ⁊
dobérthar ón.'

U Rthatar techta ó Meidb cosna Manib arco tístaís co Crúachain,
na secht Mani cona secht tríchtaib cét .i. Mane Máthremail,
Mane Athremail, ⁊ Mane Condagaib Uili, Mane Míngor ⁊
¹⁵⁰ Mane Mórgor ⁊ Mane Conda Mó Epert. Urthatar techta aile co

¹ mbérat *MS.*

maccaib Mágach .i. Cet mac Mágach ⁊ Anlúan mac Mágach ⁊ Mac
Corb mac Mágach ⁊ Bascell mac Mágach ⁊ Én mac [Mágach ⁊]
Dóche mac Mágach, Scandal mac Mágach. Táncatar sain ⁊ ba sed
a llín deich cét ar fichit cét fer n-armach. Urthatar techta aile
úathib[1] co Cormac Cond Longas mac Conchobuir ⁊ co Fergus mac 155
Róig, ⁊ táncatar deich cét ar fichit cét a llín.

In cétna lorg cétamus forthí berrtha forro. Bruit úanidi impu.
Delggi argait intib. Lénti órsnáith fria cnessaib ba tórniud[2] do
dergór. Claidib gelduirn léo co n-imdurnaib argit. [3]'Inn é Cormac
sút?' for cách. 'Nád é om,' for Medb[3]. 160

In lorg tánaise berrtha núa léo. Bruitt forglassa uli impu.
55ᵇ Lénti glégela | fria cnessaib. Claidib co muleltaib óir ⁊ co
n-imdurnib argait léo. 'Inn é Cormac sút?' for cách. 'Nád é
omm,' bar Medb.

In lorg dédenach berrtha lethna léo. Monga findbuide forórda 165
forscaílti forru. Bruitt chorcra cumtaichthi impu. Delgi órdai
ecorthi ós ochtaib dóib. Lénti sémi setai sítaidi co tendmedón
traiged dóib. I nn-óenfecht dostor[g]baitis[4] a cossa ⁊ dofairnitis
arís. 'Inn é Cormac sút?' ar cách. 'Is é ón ém,' ar Medb.

Ra gabsatar dúnad ⁊ longphort in n-aidchi sin corba dlúim diad ⁊ 170
tened eter chethri áthaib Aí .i. Áth Moga ⁊ Áth mBercna, Áth Slissen
⁊ Áth Coltna, & tarrassatar ed cían cóicthigis i Crúachanráith
Connacht ic ól ⁊ ic ánius ⁊ ic aíbnius combad esaiti léo a fecht ⁊ a slógad.

Ocus is and sain rádis Medb fria haraid ar co ngabad a echraid di
co ndigsed d'acallaim a druad d'iarfaigid fessa ⁊ fástini de. 175

A ránic Medb airm i mbaí a druí, ra iarfacht fiss ⁊ fástini de.
'Sochaide scaras fria chóemu ⁊ fria chairdiu sund indiu,' ar Medb,
'⁊ fria chrích ⁊ fria ferand, fria athair ⁊ fria máthair, ⁊ meni thíset
uli i n-imslánti, forom-sa combenfat a n-osnaid ⁊ a mmallachtain.
Araí sin ní théit immach ⁊ ní anand i fus as diliu lind oldámmit 180
fadessin. Ocus finta-ssu dún in tecam fo ná tecam.' Ocus ra ráid
in druí: 'Cipé [tic] nó ná tic, ticfa-su fessin.'

Impáis in t-ara in carpat ⁊ dotháet Medb for cúlu. Co n-accai ní
rap ingnad lé .i. in n-áenmnaí for fertais in charpait 'na farrad ina
dochum. Is amlaid boí ind ingen ic figi chorrthairi ⁊ claideb 185
findruini ina láim deiss cona secht n-aslib do dergór ina déssaib[5].
Bratt balla breccúani impi. Bretnas torrach trénchend 'sin brutt
ósa brunni. Gnúis chorcra chrúmaínech[6] lé. Rosc glass gáirectach
lé. Beóil derga thanaide. Dét níamda némanda. Andar let

---

[1] b *add. above line, also in marg. later*   [2] *sic: read* dergindliud (*Windisch*)
[3]⁻[3] *add. marg. inf.*   [4] no turgbadis (no tarccbhadais) *St*
[5] dessaib *MS.*   [6] chaoimhoineach *St*

¹⁹⁰ batar frossa findnémand erctais ina cend. Cosmail do núapartaing
a beóil. Binnidir téta mendchrot acá seinm a llámaib śírśúad
bindfogur a gotha ⁊ a caínurlabra. Gilidir snechta sniged fri
óenaidchi taídlech a cniss ⁊ a colla secha timthach sechtair.
Traigthi seta sithgela, ingni corcra córi crundgéra lé. Folt findbudi
¹⁹⁵ fata forórda furri. Teóra trillsi dá fult imma cend. Trilis aile
combenad foscad fri[a] colptha.
      Forrécacha Medb furri. 'Ocus cid dogní-siu and sain innossa, a
56ª ingen?' for Medb. 'Ic tairdeilb | do lessa-su ⁊ do lítha. Ic teclaim
⁊ ic tinól cethri n-ollchóiced ṅHérend lat-su i crích ṅUlad ar cend
²⁰⁰ Tána Bó Cúalnge.' 'Cid 'má ndénai-siu dam-sa sain?' ar Medb.
'Fail a mórabba dam. Banchumal dit muntir atamchomnaic.'
'Cóich dom muntir-sea tussu?' ar Medb. 'Ni handsa ém. Feidelm
banfáid a Síd Chrúachna atamchomnaic-se.' 'Maith and sin,

a Feidelm banfáid, cia facci ar slúag?'

²⁰⁵ 'Atchíu forderg forro, atchíu rúad.'

'Atá Conchobor 'na chess noínden i nEmain ém,' ar Medb.
'Ráncatar m'echl[ach]a-sa connice. Ní fail ní itágammar-ne la
Ultu. Acht abbair a fír, a Feidelm.

Feidelm banfáid, cia facci ar slúag?'

²¹⁰ 'Atchíu forderg forro, atchíu rúad.'

'Atá Cuscraid Mend Macha mac Conchobuir i nInis Cuscraid ina
chess. Ráncatar m'echlacha [connice]. Ní fail ní itágammar-ne
la Ultu. Acht abbair-siu fír, a Feidelm.

Feidelm banfáid, cia facci ar slúag?'

²¹⁵ 'Atchíu forderg forro, atchíu rúad.'

'Atá Eogan mac Durthacht ic Ráith Airthir 'na chess. Ráncatar
m'echlacha connice. Ní fuil ní itágammar-ne la Ultu. Acht
abbair-siu fír rind, a Feidelm.

Feidelm banfáid, cia facci ar slúag?'

²²⁰ 'Atchíu forderg forro, atchíu rúad.'

'Atá Celtchair mac Cuthechair ina dún 'na chess. Ráncatar
m'echlacha connice. Ní fuil ní itágammar-ne la Ultu. Act abbair
fír, a Feidelm.

Feidelm banfáid, cia facci ar slúag?'

²²⁵ 'Atchíu forderg forro, atchíu rúad.'

'Ní bá lim-sa aní dá tá lat-su sain, dáig ó condricfat fir Hérend óenbaile, betit debtha ⁊ irgala ⁊ scandlacha ⁊ scandrecha eturru im chomríchtain tossaig nó derid nó átha nó aband. im chétguine muicce nó aige nó fíada nó fiadmíla. Acht abbair fír rind, a Feidelm.

Feidelm banḟáid, cia ḟacci ar slúag?' 230
'Atchíu forderg forro, atchíu rúad¹.'

Ocus ro gab ic tairṅgiri ⁊ remḟástine Con Culaind d'ḟeraib Hérend. ⁊ doringni laíd:

'Atchíu fer find firfes chless
co lín chret² ima cháemcnes, 235
lond láith i n-airthiur a chind,
óenach búada ina thilchind.

'Fail secht [n]gemma láth ṅgaile
ar lár a dá imcaisne,
fail fuidrech fora rinne, 240
fail leind deirg drolaig imme.

'Ro fail gnúis is grátam dó,
dober mod do banchuireo.
gilla óc is delbdu dath
tadbait delb drecoin don chath. 245

'Cosmail ³a ḟind sa³ gaile
fri Coin Culaind Murthemne,
nocon ḟetar cóich in Cú
Culaind asa Murthemniu,
acht rafetar-sa trá imne 250
bid forderg in slúag-sa de.

'Cethri claidbíni cless n-án
⁴ra fail cechtar⁴ a dá lám,
condricfa a n-imbirt for slúag,
i[s] sain gním ris téit cech n-aí úad. 255

¹ Atchiu fer find firfes chless *written here in stain; misplaced, omitt.*
² crechta *LU,* crecht *St* ³⁻³ = innas a *LU, sic leg.*
⁴⁻⁴ fil í cechtar *LU, St*

'A gae bulgae mar domber
cenmothá a chlaideb 's a śleg,
fer¹ i furchrus¹ bruitt deirg,
dobeir a choiss for cach leirg.

260 'A dá śleig dar fonnad nglé,
ard ás gail in ríastarde,
cruth domarfáit air co se
derb limm no chlóemchlaífed gnée.

'Ro gab tascugud don chath,
265 meni faichlither bid brath,
don chomlund is é farsaig,
Cú Chulaind mac Sualtaim.

'Slaidfid for slúaga slána,
concurfe far tiugára,
270 faicébthai leis óg for cend;
ní cheil in banfáid Feidelm.

'Silfid crú a cnessaib curad,
bud fata bas chianchuman; |
56ᵇ beit cuirp cerbtha, caínfit mná,
275 ó Choin na Cerdda atchíu-sa.' A.

Tairngire ⁊ remfástini ⁊ cendphairt in sceóil ⁊ fotha a fagbála
⁊ a dénma, ⁊ comrád chind cherchaille doringni Ailill ⁊ Medb i
Crúachain connice sain.

280 SLigi na Tána in so ⁊ tossach in tślúagid ⁊ anmand na sliged
dochúatar cethri ollchóiced Hérend i crích Ulad .i.²
I Mag Cruinn³, for Tóm Móna, for Turloch Teóra Crích, for
Cúl Sílinni, for Dubfid⁴, for Badbna⁵, for Coltain, for Sinaind, for
Glúine Gabur, for Mag Trega, for Tethba túascirt, for Tethba in
descirt, for Cúil, for Ochain, for Uata fothúaid, for Tiarthechta
285 sair, for Ord, for Slaiss, for Inndeóin, ⁶for Carn⁶, for Mide, for Ortrach,

---

¹⁻¹ *read* i cathfochrus (i cathfochrus *LU*, i cathocras *St*)   ² *names arranged*
in two columns MS.   ³ *originally* Cruimm *MS.* m *stroke over* i *and finial of*
m *effaced;* Muicc Cruimb *LU,* for Muccrumib *Eg.,* for Mucdhruim *St*
⁴ Dubloch *MS.* och *erased,* fid *written in marg. sup.* (for Fid *LU* 4594, for
Fidh Dubh *St* 287)   ⁵ Badbgna *MS. with* g *expunged;* Badgna *St*
⁶⁻⁶ for Carn *bis*

for Findglassa Asail, for Druiṅg, for Delt, for Duelt, for Delaind,
for Selaig, for Slabra, for Slechta conselgatar claidib ria Meidb
⁊ Ailill, for Cúil Siblinni, for Dub, for Ochun, for Catha, for Cromma,
for Tromma, for Fódromma, for Sláne, for Gort Sláne, for Druim
Licci, for Áth ṅGabla, for Ardachad, for Feoraind, for Findabair, 290
for Assi, for Airne, for Aurthuile, for Druim Salaind. for Druim
Caín, for Druim Caímthechta, for Druim mac ṅDega. for Eódond
ṁBec, for Eódond Mór, for Méide in Tog*maill*, for Méide ind Eóin.
for Baile, for Aile, for Dall Scena, for Ball Scena, for Ros Mór. for
Scúaip, for Timscúaib, for Cend Ferna, for Ammag. for Fid Mór 295
i Crannaig Cúalṅgi, for Druim Caín i Sligid Midlúachra.

Cétna uidi dochúatar na slóig co faítar in n-aidchi sin for Cúil
Siblinni & focress a phupall do Ailill mac Rosa in n-aidchi sin.
Pupall Fergusa meic Róich dia láim deiss. Cormac Cond Longas
mac Conchobuir fora lám-ide[1]. Íth mac Étgaíth fora lám-ide[1]. 300
Fiachu mac Fir Aba fora lám-ide[1]. Gobnend mac Lurgnig fora
lám-ide[1]. Suidigud pupaill Ailella dia láim deiss arin tṡlúagud in
sin ⁊ trícha cét fer nUlad dia láim [deiss] in sin, & ba aire dobered
trícha cét fer nUlad dia láim deiss combad faesiti in cocur ‑ in comrád
⁊ na hairigthi bíd ⁊ lenna dóib-sium. Medb Chrúachan immorro 305
do chlí Ailella. Findabair fora lám-ide[1]. Flidais Foltcháin ben-side
Ailella Find arna feis la Fergus ar Táin Bó C*úalnge* & is sí no bered
in sechtmad n-aidchi iṅgalad[2] d'feraib Hérend forin tṡlúagad do
lacht eter ríg ⁊ rígain ⁊ rígdomna ⁊ filid ‑ fogl*aimthid*. Med[b] ba
dédenach[3] dona slúagaib in lá sain ic iarfaigid fessa ‑ fástini | 310
⁊ eólais ar co fessed cia lasbad lesc ‑ lasbad laind in slúagad do
thecht. Ní arlacair Medb ara turnta a carp*at* *nó* ara scorthea a
eich co rálad cor di 'sin dúnad.

And sain ra díchurit eich Medba ⁊ ra turnait a carp*ait* ⁊ dessid
ar láim Ailella meic Mágach. Ocus confóchta Ailill scéla di Meidb 315
ar co fessed Medb cia lasmad laind nó nemlaind nó lasmad lesc in
slúagad. 'Hespach do neoch a thríall acht dond óenfialluch[4].' ar
Medb. 'Cia maith fogníat in tan moltair sech cách?' ar Ailill.
'Fail dlug molta forro,' ar Medb. 'Tráth ro gab cách dúnad ⁊
longphort do dénam, ro scáich dóib-sium botha ‑ bélscaláin do 320
dénam. Tráth ro scáich ra cách botha ⁊ bélscaláin, ro scáich
dóib-sium urgnam bíd ⁊ lenna. Tráth tarnaic do chách urgnom
bíd ⁊ lenna, scáich dóib-sium praind ⁊ tomailt. Tráth ro scáich
praind ⁊ tomailt [do chách][5], bátar-som 'na cotlud andsaide. Feib

57ᵃ

[1-1] láim sidi *LU;* laim sén *St*   [2] *read* a ṅgalad   [3] deidhencha *St*
[4] .i. do trichait cet na nGailian *add. St*   [5] *om. MS., add. from St*

³²⁵ ra deligetar a ndáer¹ ┐ a mogaid de dóeraib ┐ mogadaib fer ṅHérend,
deligfit a ndeglaích ┐ a ndegóic do deglaíchaib ┐ de degócaib fer
ṅHérend in chur sa forin tṡlúagud.' 'Ferrde linni sain,' bar Ailill,
'dáig is lind imthiagat ┐ is erund conbágat.' 'Níba lind ragait ┐
níba erund conbágfat.' 'Anat i fus didiu,' bar Ailill. 'Ní anfat,'
³³⁰ bar Medb. 'Cid dogénat didiu,' bar Findabair, 'meni digset ammach
┐ nád anat i fos?' 'Bás ┐ aided ┐ airlech is accobor lem-sa dóib,' bar
Medb. 'Mairg atber ón omm,' bar Ailill, 'ar abba dúnad ┐ long*phort*
do gabáil dóib co hellom ┐ co héscaid.' 'Dar fír ar cubais,' ar Fergus,
'ní diṅgnea bás dóib-siút acht intí dogéna bás dam-sa.' 'Ní rim-sa
³³⁵ is ráite duit-sin sain, a Ḟerguis,' ar Medb, 'dáig itó-sa lín do gona ┐
t'airlig co tríchait cét Galían immut, dáig atát na secht Mani cona
secht tríchtaib cét ┐ meic Mágach cona tríchait cét ┐ Ailill cona
tríchait cét ┐ atú-sa com thegluch 'no. Atám and sain lín do
gona-su ┐ t'airlig cot tríchait cét Galían immut.' 'Ní comadas a
³⁴⁰ rád frim-sa sain,' ar Fergus. 'Dáig atát lim-sa sund na secht
n-airríg do Mumnechaib cona secht tríc[h]taib cét. Failet sund
trícha cét a n-as dech di ócaib Ulad². Fail sund a n-as dech dagóc
fer ṅHérend, trícha cét Galían. Messi dano as chor ┐ as glinni ┐
trebairi friu ó tháncatar ó críchaib dílsib fadesin, ┐ lim congébat
³⁴⁵ 'sind ló bága sa. Atá ní,' ar Fergus, 'níbat ecra ind ḟir sin—is inund
ón ┐ ní deceltar dam. Díscaílfet-sa úaim in tríchait cét Galian
út fo ḟiru Hérend conná bia cóiciur díb i n-óenbale.' 'Fó lim-sa
ón,' ar Medb, 'case chruth i mbet acht nád bet 'sin chaír chomraic
i tát ammáin.' Is and sain ra díscaíl Fergus in tríchait cét sain fo
³⁵⁰ feraib Hérend nád boí cóiciur díb i n-óenbaile. |

57ᵇ   Lotar na slóig iar sodain i cend séta ┐ imthechta. Fa dolig dóib
fri[th]airle in tṡlúaig romóir lotar forsin fecht frisna iltúathaib ┐
frisna ilmaicnib ┐ frisna ilmílib dosbertatar leo co 'mmanactís ┐ co
'mmafessaitis ³combad chách³ cona cháemaib ┐ cona chairdib
³⁵⁵ ┐ cona chomdúalas forin tṡlúagud. Atbertatar dano ba samlaid
bad chóir a thecht. Atbertatar dano béus cinnas bad chomdas
in slúagad do thecht. ⁴Atbertatar dano ba samlaid bad chóir a
thecht⁴: cach droṅg imma ríg, cach réim imma muirech ┐ cach
buiden imma tuísech, cach rí ┐ cach rígdomna d'ḟeraib Hérend ina
³⁶⁰ thulaig fo leith. Ro ráidset béus cia bad chóir do eólus rempu
eterna dá chúiced, & atbertsat combad é Fergus, ar bíth ba slúagad
bága dó in slúagad, dáig is é boí secht ṁblíadna i rrígu Ulad ┐ iar
marbad mac nUsnig fora ḟaísam ┐ fora chommairgi, tánic estib, '┐
atá sec[h]t ṁblíadna déc fri Ultu ammuig ar loṅgais ┐ bidbanas.'

---

¹ ndaoir *St*   ² .i. tricha cet na dubloingsi *add. St*
³⁻³ *sic*   ⁴⁻⁴ *sic, omitt., dittogr.*

Is aire sin bad chomadas a dul ria cách do eólas. Luid iarum assa [365]
aithli sin Fergus ria cách do eólas, & tánic ell condailbi im Ultaib
dó ⁊ dobretha cor n-imruill fothúaid ⊣ fodess dona slúagaib &
urthatar techta úad co rrobthaib do Ultaib & ro gab ic fostud ⊣ ic
imfuiriuch in tslúaig. Rathaigis Medb aní sin ⊣ benais Medb
béim n-aisce fair ⊣ ro chan in laíd: [370]

'A Ferguis, ca rádem de?
Cinnas conaire amse[1]?
Fordul fodess is fothúaid
berma dar cech n-ailethúaith.'

F.   'A Medb cid not medra-su? [375]
Ní cosmail fri brath in so.
Is for[2] Ultu, a ben trá,
in tír darsa tiagu-sa.'

M.   'Ardattágadar [3]co ngail
Ailill án cona slúagaib[3]. [380]
Ní tharddais menmain—[4]comal nglé[4]
fri imthús na conaire.'

F.   'Ní for amlesaib in tslúaig
biurt-sa cach fordul ar n-úair,
dús in n-imgabaind iar tain [385]
Coin Culaind mac Sualtaim.'

M.   'Écóir duit amles ar slúaig,
a Ferguis meic Rosa Rúaid,
mór de maith fuarais i fus
ar do longais, a Fergus.' [390]

'Nád bíu-sa resna slúagaib [5]níbas íriu[5],' bar Fergus, 'acht iarra-su
nech n-aill ríam remán rempo.' Dessid Fergus riasna slúagaib araí.
Bátar cethri ollchóicid Hérend bar Cúil Sílinni in n-aidchi sin.
Tánic gérmenma géribrach Con Culaind do Fergus & ra ráid ra firu
Hérend fatchius do dénam. Dáig dasficfad in leom[6] letarthach ⊣ in [395]
bráth bidbad ⊣ in bidba sochaide ⊣ in cend costuda ⊣ in cirriud
mórslúaig ⊣ in lám tidnaicthi ⊣ in chaindel adanta .i. Cú Chulaind mac

---

[1] cingme *LU;* an ní se *St*   [2] la *LU, St, sic leg.*
[3-3] dia mbrath | Ailill Aie lía slúagad *LU,* dia brath ... *St*
[4-4] co se *LU, St, sic leg.*   [5-5] ní bus faidi *St*   [6] leoman *St*

Sualtaim. Ocus ro boí icá thairṅgire samlaid ⁊ doringni in laíd ⁊ ro
[f]recair Medb:

400
'Fó dúib fatchius ⁊ fót
co n-ilur arm ⁊ óc.
ticfa intí 'tágammar de,
mórglonnach [mór]¹ Murthemne.'

405
'Condalb sain, condelg n-ága,
duit-siu, a meic Róig rodána. |
58ᵇ Óic ⁊ airm ²limm for laind²
do [f]rithálim Con Culaind.'

'Ro chatir, a Medb don Maig,
óic is airm don immargail,
410
for cind marcaig Léith Mache
cach aidche is cach óenlathe.'

'Atát acum sund for leith
curaid ri cath is ri creich,
trícha cét di chodnaib gíall
415
do láthaib gaile Galían.

Na curaid a Crúachain chain,
na láeich a Lúachair lendglain,
cethri chóicid Gáedel ṅgel
dingébait dím in óenḟer.'

420
'Buidnech Bairrche ⁊ Banna
consreṅgfa crú dar cranna,
concicher ar múr is ar ṅgrian
in trícha sin fer Galían.

I n-athlaime na fandle,
425
i llúais³ na gaíthe gairbe,
is⁴ amlaid bís mo Chú cháem chain
oc imguin ós análaib.'

'A Ḟerguis doroichet raind,
roiched úait do Choin Chulaind,
430
rab é a thúachail beith 'na thost,
atetha a Crúa[c]hain crúadchosc.'

'Bid ferda firfitir fuidb
i n-airiur ingine Buidb,
Cú na Cerda, crithrib cró,
snigfid fairne ferga fó.'   435

A haithle na laíde sin: Táncatar cethri ollchóiceda Hérend dar
Móin Coltna sair in lá sain ⁊ da 'marallsatar dóib ocht fichtiu oss
n-all*aid*¹. Sernsat ⁊ immsit na slóig impu ⁊ ros gonsat conná bátar
élódaig díb. Acht atá ní, garbat díscaíltig trícha cét Galían, cóic
aige in amáin ba hé cuit fer [n]Hérend díb. Rodasfuc in t-óentrícha   440
cét uli na hocht fichtiu oss.

Is hé in lá cétna tánic Cú Chulaind mac Sual*taim* ⁊ Sualtach
sídech a athair co ngeltatar a n-eich geilt immon corrthe ic Ard
Chuillend. Congeltat eich Sual*taim* fri coirthi atúaid fér co húir.
Fogeltat eich Con Culaind fri corthi aness fér co húir ⁊ connici na   445
lecga lomma. 'Maith a phopa Sual*taim*,' ar Cú Chulaind, ²·gérmenma
in tṡlóig form-sa², ⁊ urtha-su dún co rrobtha do Ultaib arnád bet
ar maigib rédib co ndigset i fedaib ⁊ fásaigib ⁊ fánglentaib in
chúicid ar imgabáil fer ṅHérend.' 'Ocus tussu, a daltáin, ³cid
amgéna³ dano?' 'Am écen-sa tocht i n-herus inalta Feidilmthi   450
Noíchruthaige fodess co Temraig ram glinni fodessin co matin.'
'Mairgg théit ón ám,' ar Sual*taim*, '⁊ Ulaid do lécud fo chossaib
a nnámat ⁊ a n-echtrand ar thecht i comdáil n-óenmná.' 'Amm
écen-sa trá techt, dáig meni digius, gúigfitir dála fer ⁊ fírfaitir briathra
ban.'   455

Urtha Sual*taim* co robthaib do Ultaib. Luid Cú Chulaind fón
fid ⁊ tópacht and cétbunni darach d'óenbéim bun barr ⁊ ro sníastar
ar óenchois ⁊ ar óenláim ⁊ óensúil ⁊ dorir.gni id de. Ocus tuc ainm
n-oguim 'na mennuc inn eda ⁊ tuc in n-id im cháel in chorthe ic
Ard Chuillend. Scibis in n-id co ránic remur in chorthe. Luid   460
Cú Chulaind 'na bandáil asa aithle sin.

Imthúsa fer ṅHérend imráter sund iar tain. Táncatar-saide
connici in corthe ic Árd Cullend. Ocus gabsat oc fégad in chúicid
aneóil úadib, cúicid Ulad. Ocus dias de muntir Medba no bíd
rempo do grés i tossuch cach dúnaid ⁊ cach slúagaid, cach átha ⁊   465
58ᵇ cacha aband | ⁊ cacha bernad. 'S aire dognítis-[s]ium sain arná
tísad díamrugud dia timthaigib na rígmac i n-eturturtugud ná [i]
n-imchumung slúag ná sochaide .i. dá mac Nera⁴ meic Nuatair
meic Tacáin .i. dá mac rechtairi na Crúachna, Err ⁊ Innell. Fráech
⁊ Fochnam anmand a n-arad.   470

---

¹ nall–*MS.*, (n-all*ta Dipl. Ed.*) ocht fichit oss n-allaid *LU*, ocht ficit oss
all*aidh St*   ²⁻² *supply* atá? *Cp.* atta menma in tṡlóig ocum *LU*, ata
germenma an tsluaig agam-sa *St*   ³⁻³ *sic;* cret do-genair-si *St*   ⁴ Nena *MS*

Táncatar mathi Hérend connici in corthi ⁊ gabsat oc fégad ¹[na]
ingelta¹ ro geltsat na eich immon corthi ⁊ gabsat oc fégad ind idi
barbarda forácaib in rígnia immun corthi. Ocus gebid Ailill in n-id
inna láim² ⁊ tuc i lláim Fergusa ⁊ airlégais Fergus in n-ainm n-oguim
475 baí 'na menuc ind eda ⁊ innisis Fergus d'feraib Hérend inní ro chan
in t-ainm oguim baí 'sin menuc. Ocus is amlaid ro gab icá innisin
⁊ doringni laíd:

'Id in so, cid sluinnes dún?
In t-id cid imma tá a rún?
480   Is cá lín ³ra lád³ co se,
inn úathad nó in scchaide?

'Mad dia tístai sccha innocht
can anad aice⁴ i llongphort,
dabarró⁵ in Cú cirres cach n-om;
485   táir foraib a sár[u]god.

'Co tabair irchóit don tslúag
mad dia mberthai uide úad;
finnaid, a druíde, and sain
cid imma ndernad in t-id.

490   'Crefnas curad cur ro lá,
lánaircess fri ecrata.
Costud ruirech, fer co ndáil,
ras cuir oenfer dá óenláim.

'Furópair fír ra feirg fúair
495   Con na Cerdda 'sin Chráebrúaid.
Iss naidm níad, ní nasc fir mir,
is é ainm fil isinn id.

'Do chur chesta, cétaib drend,
for cethri cóiceda Hérend,
500   nocon fetar-sa ac[h]t mad sin
cid imma ndernad in t-id.'   Id.

Aithli na laíde sin: 'Atbiur-sa mo bréthir frib,' ar Fergus, 'mad
dia sárgid in n-id sin ⁊ in rígnía doringni can aidchi ndúnaid ⁊
longphuirt sund ná co nderna fer úaib id a mac samla sút ar óenchois
505 ⁊ óensúil ⁊ óenláim feib doringni-sium, cid airchind bes-sum fó
thalmain nó i tig fó dúnud conáirgife guin ⁊ fuligud dúib ria tráth

¹⁻¹ na hingealta St (in gelta Dipl. Edn.)   ² in n-id repeated after láim MS.
³⁻³ ro lá LU   ⁴ oidhci St   ⁵ dobro St

éirge immbárach diana sárgid.' 'Ní hed sain ba háil linni ém,' ar
Medb, 'nech d'fuligud *nó* d'fordergad foirn ar tíchtain isin cóiced
n-aneóil se .i. i cúiced Ulad. Alcu dún fuligud ꝛ fordergad for nech.'
¹'Ní sáríagum¹ itir in n-id se,' bar Ailill, 'ꝛ ¹ní sáríagum¹ in rígnía ⁵¹⁰
danringni, acht ragma i mmunigin ind feda móir se tess co matin.
Gebthar dúnad ꝛ longphort acund and.' Lotar na slóig iarum ꝛ
baslechtat rempu in fid dia claidbib riana carptib. Conid Slechta
comainm ind inaid sin béus airm i táat Partraigi Beca ra Cenannas
na Ríg aniardess amne bar Cúil Sibrilli. ⁵¹⁵

Ferais tromsnechta dóib in n-aidchi sin ꝛ baí dá met co roiched
co formnaib fer ꝛ co slessaib ech ꝛ co fertsib carpat combatar
clárenig uile cóiceda Hérend din tsnechtu. Acht níra sádit sosta
59 *nó* botha *nó* pupla in n-aidchi sin. Ní dernad | ² urgnam bíd *nó*
lenna. Ní dernad praind *nó* tomaltus. Ní fitir nech d'feraib ⁵²⁰
'Hérend in cara fa náma ba nessam dó co solustráth éirge arnabárach.
Is comtig·conná fúaratar fir Hérend aidchi ndúnaid *nó* longphuirt
bad mó dód *nó* doccair dóib indá inn aidchi seo bar Cúil Sibrilli.
Táncatar cethri ollchóiceda Hérend co moch arnabárach la turcbáil
ngréne dar taitnem in tsnec[h]tai ꝛ lotar rempo assin chrích i ⁵²⁵
n-araile.

Imthúsa Con Culaind immorro, ní érracht-saide mochthrád etir co
tormalt feiss ꝛ díthait, co foilc ꝛ co³ fothraic in lá sain. Rádis fria
araid ar co ragbad in n-echrad ꝛ ar co n-indled in carpat. Gebis in
t-ara in n-echraid ꝛ indlis in carpat, & luid Cú Chulaind ina charpat ⁵³⁰
ꝛ táncatar for slichtlorg in tslúaig. Fúaratar lorgfuilliucht fer
nHérend seccu assin chrích i n-araile. 'Amae, a phopa Láeig,' ar
Cú Chulaind, 'ní ma lodmar dar mbandáil arráir. Iss ed is lugu
condric ó neoch bís i cocrích égim nó iachtad nó urfócra nó a rád cia
thic 'sin sligid ní tharnic úan do rád. Lodatar fir Hérend sechund ⁵³⁵
i crích nUlad.' 'Forairngert-sa duit-siu, a Chú Chulaind, sain' ar
Láag, 'cia dochúadais it bandáil co ragad méla a mac samla⁴ fort.'
'Meith a Láeig, eirgg dún for slichtlurg in tslúaig ꝛ tabair airdmes
forro, ꝛ finta dún cá lín dochúatar sechund fir Hérend.' Tánic
Láeg i slicht in tslúaig ꝛ baluid dia lethagid in luirg ꝛ tánic dia ⁵⁴⁰
lettaíbꝛ luid dara ési. 'Mesc fri⁵ árim⁶ fort, a mo phopa Laíg,' ar Cú
Chulaind. 'Is mesc écin,' for Láeg. 'Tair isin carpat didiu ꝛ
dobér-sa ardmes forro'. Tánic in t-ara 'sin carpat. Luid Cú
Chulaind i slichtlurg in tslúaig ꝛ dobretha airdmes forro, & tánic
dia lettáeb ꝛ luid dar' éisi. 'Is mesc fri⁵ árim fort, a Chúcúc,' bar ⁵⁴⁵

---

¹⁻¹ ní shaireocham *St*   ² *pp.* 59, 60 *are written in long lines*
³ coro *MS.,* ro *add. incorrectly with caret markes above line*   ⁴ sút *add. St*
⁵ = fria; f *with overhead stroke* = for *MS.*   ⁶ airm (air *comp.*) *MS.*

Láeg. 'Ní mesc,' bar Cú Chulaind, 'dáig forḟetar-sa in lín dochúatar
sechund na slúaig .i. ocht tríchait chét déc, ⁊ ra ḟodlad dano in
t-ochtmad trícha chét déc fo ḟiru Hérend.' —Ra bátar trá ilbúada
ilarda imda for Coin Chulaind: búaid crotha, búaid delba, búaid
550 ṅdénma, búaid snáma, búaid marcachais, búaid fidchilli ⁊ branduib,
búaid catha, búaid comraic, búaid comluind, búaid farcsena, búaid
n-urlabra, búaid comairle, ¹búaid foraim¹, búaid ṁbánaig, búaid
crichi² a crích comaithig.

'Maith a mo phopa Laíg, innill dún in carpat ⁊ saig brot dún arin
555 n-echraid. Cuindscle dún in carpat ⁊ tabair do chlár clé frisna
slúagaib dús in tairsimmís tossach nó deired nó medón dona slúagaib,
dáig níbam béo meni thaeth cara nó náma limm d'feraib Hérend
innocht.' 'S and saigthis in t-ara brot forin n-echraid. Dobretha
a chlár clé frisna slúagaib co tarlaic i Taurloch Caille Móre fri
560 Cnogba na rRíg atúaid frisi ráter Áth ṅGabla. Luid Cú Chulaind
fón fid and sain ⁊ tarmlaiṅg asa charput ⁊ tópacht gabail cethri
ṁbend bun barr d'óenbéim. Ros fúacha ⁊ ros fallsce ⁊ dobreth
ainm n-oguim 'na táeb, ⁊ dobretha rót n-urchair di a iarthur a
charpait do ind a óenláime co ndechaid dá trían i talmain connach
565 boí acht óentrían úasa. Is and sin tarthatar in dá gilla chétna
.i. dá mac Nera meic Uatair³ meic Tacáin é icond auropair sin &
ba tetarrachtain dóib cia díb no bérad a chétguine fair ⁊ no benfad
a chend de ar tús. Impádar Cú Chulaind friu ⁊ tópacht a cethri
cinnu díb⁴ colléic ⁊ tuc cend cech ḟir díb ara beind do bennaib na
570 gabla. Ocus léicis Cú Chulaind eocho in ḟiallaig sin i n-agid fer
ṅHérend i frithdruing na sliged cétna ⁊ a n-ésse airslaicthi ⁊ a méide
forderga ⁊ cuirp na curad ic tindsaitin a fola sell sís for crettaib
na carpat. Dáig níbá miad *nó* níba maiss leiss echrad *nó* fuidb
nó airm do brith óna corpaib no marbad. Atchondcatar na slúaig
575 dano eich⁵ ind ḟiallaig bátar ríam remán rempu ⁊ na cuirp i n-écmais
a cend ⁊ cuirp na curad ic tindsaitin a fola sell sís for crettaib na
carpat. Anais tossach in tslúaig fria deired ⁊ focuirethar i n-artbe,
cumma ⁊ i n-armgrith uile.

Doroacht Medb ⁊ Fergus ⁊ na Mane ⁊ Meic Mágach. Dáig ar bíth
580 is amlaid no imthiged Medb ⁊ noí carpait fóthi a óenur. Dá charpat
rempe díb ⁊ dá charpat 'na diaid ⁊ dá charpat cechtar a dá táeb ⁊
carpat eturru ar medón cadessin. Is aire fogníd Medb sin arná
rístais fótbaige a crúib greg *nó* uanfad a glomraib srían *nó* dendgur
mórslúaig *nó* mórbuiden arná tísad díamrugud don mind óir na
rígna. 'Cid in so?' for Medb. 'Ní handsa,' ar cách. | 'Eich ind

60

---

¹⁻¹ *read* búaid foraime? (*om. St*)   ² ccuarta *St*   ³ *sic, for* Nuatair
⁴ diob fen ⁊ da n-aradaibh *St*   ⁵ *sic*

fiallaig ro bátar ríam remán sund ⁊ a cuirp ina carptib i n-écmais
a cend.' Ra cruthaiged comairle occu ⁊ ba demin leó combad
slicht sochaide sút ⁊ gomba tadall mórslúaig ⁊ gondat Ulaid
ardastánic. Ocus ba sed a comairle ra cruthaiged leó, Cormac
Cond Longes mac Conchobuir do lécud úadib dá fiss cid baí 'sind 590
áth, ar bíth ciano betis Ulaid and, ní gonfaitis mac a rríg dílis
fodessin. Tánic assa aithli Cormac Cond Longes mac Conchobuir
⁊ bá sed a lín tánic deich cét ar fichit chét fer n-armach dia fis cid
baí issind áth, ⁊ á ránic, ní facca ní ac[h]t in ṅgabail ar lár ind átha
co cethri cennaib furri ic tinnsaitin a fola sell sís fri creitt na gabla i 595
sruthair na haband & glethe¹ na dá n-ech ⁊ fuilliucht ind óencharptig
⁊ slicht inn óenlaích assind áth sair sechtair.

Táncatar mathi Hérend connici in n-áth ⁊ gabsat oc fégad
na gabla uili. Ba machtad ⁊ ba iṅgnad leó cia ro chuir in
coscur. 'Cá ainm ind átha sa acaib-si gustráthsa, a Ḟerguis?' 600
ar Ailill. ²'Áth ṅGrena'² bar Fergus, '⁊ bid Áth ṅGabla a
ainm co bráth din gabuil se ifec[h]tsa,' ar Fergus, & ra ráid in
laíd:

'Áth ṅGrena, claímchlaífid ainm
do gním Chon rúanaid rogairb. 605
Fail sund gabuil cethri ṁbend
do cheist for feraib Hérend.

'Fail ar dá mbeind, mana n-áig,
cend Fráech ⁊ cend Fochnáim.
Fail araile ar dá mbeind 610
cend Eirre ⁊ cend Innill.

'Gá ogum sút ina táeb,
finnaid, a druíde co n-áeb,
is cia dorat inti sain,
gía lín ros cland i talmain ? ' 615

'In gabul út co ṅgráin guiss
atchí-siu sund, a Ḟerguis,
ros tesc óenfer, as³ mo chin,
de builli chrichid chlaidib.

---

¹ sliocht *St*
²⁻² Áth Grena LU 4956 *but* Áth Grencha *ib.* 6101, Áth Grencha *Eg. St*
³ ar *MS.*, as *St*

620   'Ros fúach is ras fuc fria aiss,
        gid ed ní heṅgnam imthaiss,
        is dassarlaic sís ar sain
        dá gait d'fir úaib-si as talmain.

        'Áth ṅGrena a ainm mad co se,
625   méraid ra cách a chumne.
        Bid Áth ṅGabla a ainm co bráth
        din gabail atchí 'sind áth.'   A.

Aithli na laídi: 'Machdath ⁊ iṅgnad lim-sa, a Ḟerguis,' bar Ailill,
'cia no thescfad in ṅgabail ⁊ bífed in cethrur buí remoind i traiti se.'
630 'Ba córu machtad ⁊ iṅgantas dontí ro tesc in ṅgabail atchí d'óenbéim
bun barr, ros fuach ⁊ ros faillsce ⁊ tarlaic rofut n-urchair di a íarthur
a charpait d'ind a óenláime co ndechaid dara dá trían i talmain
connach fil acht a óentrían úasom ⁊ nach class cona chlaideb rempe
acht is tria glaslec clochi conindsmad ⁊ conid geiss d'ḟeraib Hérend
635 techt do lár ind átha sa ná co tuca nech díb hí anís do ind a óenláime feib dosfarlaic-sium sís ó chianaib.' 'Is dár slúagaib duit-siu,
a Ḟerguis,' ar Medb, '⁊ tabair in ṅgabail dún do lár ind átha.' 'Domroched carpat', ar Fergus, & dobretha carpat do Ḟergus, & dobretha
61ᵃ Fergus frosse forsin ṅgabail ⁊ doriṅgni minbrúan | ⁊ minscomart
640 din charput. 'Domroiched carpat,' ar Fergus. Dobretha carpat do
Ḟergus & dobretha Fergus feirtche forsin ṅgabail ⁊ doriṅgni briscbruan
⁊ minscomartach din charput. 'Domroched carpat,' ar Fergus.
Ocus dobretha Fergus nertad forsin ṅgabail ⁊ doriṅgni briscbruan
⁊ minscomartach din charput. Airm i mbátar na secht carpait
645 déc do charptib Connacht, doringni Fergus díb uile briscbruan
⁊ minscomartach, ⁊ ní chaemnaic in ṅgabail do gait do lár ind átha.
'Ale léic ass, a Ḟerguis,' ar Medb, 'na bris dúin cairptiu ar túath ní
as siriu, dáig ar bíth meni bethe arin tṡlúagud sa a chur sa doraismis
Ultu co n-airnelaib braiti ⁊ bóthánti lind. Rafetamar-ni aní dia
650 ndénai-siu sain d'ḟostud ⁊ d'immḟuiriuch in tṡlúaig co n-érsat
Ulaid assa cess ⁊ co tucat cath dún, cath na Tána.' 'Domroiched
crim carpat,' ar Fergus. Ocus dobretha a charpat fadessin do
Ḟergus & dobretha Fergus tepe forsin ṅgabail, ⁊ níro gnuisistar ⁊
níro gésistar roth *nó* fonnud *nó* fertas d'ḟertsib in charpait. Cia
655 baí dia chalmacht ⁊ dia churatacht dosfarlaic intí dosfarlaic sís,
baí dia nertmaire ⁊ dá óclachas dasfucastar in cathmílid ⁊ in
chliathbern chét ⁊ [in t-]ord essorgni ⁊ in bráthlec bidbad ⁊ in cend
costuda ⁊ in bidba sochaide ⁊ in cirriud mórṡlúaig ⁊ in chaindel
adantai ⁊ in toísech mórchatha. **Dosfuc anís do ind a óenláime co**

ránic aidlend a gúaland ⁊ dobretha in ṅgabail i lláim Ailella. Ocus ⁶⁶⁰
tincais Ailill furri .i. nos fégand. 'Crichiditi lim-sa in gabul,' ar
Ailill, 'dáig is óentescad atchíu-sa bun barr furri.' 'Crichiditi
omm,' bar Fergus. Ocus ro gab Fergus ¹ar[a] admolad¹ ⁊ dobretha
laíd furri:

'Atá sund in gabul glúair   ⁶⁶⁵
'coa rabi Cú Chulaind chrúaid,
'gá farcgaib ar ulc ri nech
cethri cinnu comaithech.

'Is derb ní theichfed úadi
ria n-óenḟer calma crúadi,   ⁶⁷⁰
giaras fácaib Cú gan chess,
arthá crú 'ma chaladchness.

'Mairg ragas in slúagad sair
ar cend Duind Chúailgne chrúadaig.
Betit curaid arna raind   ⁶⁷⁵
fa neim claidib Con Culaind.

'Níba hascid a tharb trén
'moa mbia comrac arm ṅgér,
ar [to]crád² chloicgne cach cind,
gol cacha aicmi i nHérind.   ⁶⁸⁰

'Nuchum thá ní rádim de
im dála meicc Deictire,
concechlafat fir is mná
din gabuil sea mar atá.'   A.

Aithli na laídi sin: 'Sáditer sosta ⁊ pupaill lind,' ³ar Ailill³, '⁊ ⁶⁸⁵
déntar urgnam bíd ⁊ lenna lind ⁊ cantar ceóil ⁊ airfiti lind ⁊ [déntar]⁴
praind ⁊ tomaltus. Dáig is⁵ comtig ara fagbaitis fir Hérend ríam
nó iarum aidchi ṅdúnaid nó longphuirt mad⁶ mó dód nó doccair
dóib andás ind aidchi se arraír.' Ra sádit a sosta ⁊ ra suidigit a
61ᵇ pupla. Darónad urgnam | bíd ⁊ lenna leó ⁊ ra canait ceóil ⁊ airfiti ⁶⁹⁰
leó ⁊ darónad praind ⁊ tomaltus. Ocus ra iarfacht Ailill do Ḟergus:
'Is machtad ⁊ iss iṅgantus lim cia ticfad cucaind co hor críchi ⁊

¹⁻¹ ar admolad Con Culainn St   ² tocrád St
³⁻³ nó Fergus *written overhead MS.;* ar Fergus St   ⁴ om. *MS.,* dēntar St
⁵ *read* ní   ⁶ = bad

no bífed in cethrur buí remaind i traiti se. In dóig inar tísed
Conchobor mac Fachtna Fáthaig ardrí Ulad?' 'Nád dóig ém,'
695 ar Fergus, 'dáig líach a écnach 'na écmais. Ni fil ní nád gellfad dar
cend a enig. Dáig dámbad ésin tísad and, ticfaitis slúaig ⁊ sochaide
⁊ forgla fer ṅHérend¹ filet maróen ris, ⁊ giada² betis fora c[h]ind i
n-óenbaile ⁊ i n-óendáil ⁊ i n-óentochim ⁊ i n-óenlongphurt ⁊ i
n-óentulaig fir Hérend ⁊ Alban, Bretain ⁊ Saxain, cath dobérad
700 dóib, reme no maissed ⁊ ní fair no raínfithé.' 'Ceist didiu, cia bad
dóig diar tíachtain? Dóig innar tísed Cúscraid Mend Machae mac
Conchobuir ó Inis Cuscraid?' 'Nír dóig,' ar Fergus mac ind ardríg.
'Ní fil ni nád gellfad dar cend a enig, dáig dámbad é ³no thiasad³
and, ticfaitis meic ríg ⁊ rígthuísech failet maróen riss ic reicc a
705 n-amsa, ⁊ gana betís ara c[h]ind i n-óenbaile ⁊ i n-óendáil ⁊ i n-óentochim ⁊ i n-óenlongphurt ⁊ i n-óentulaig fir Hérend ⁊ fir Alban,
Bretain ⁊ Saxain, cath dobérad dóib, reime no maidsed ⁊ ní fair
no⁴ raínfide.' 'Ceist didiu inar tísad Eogan mac Durthacht rí
Fernmaige?' 'Nár dóig omm, ár diambud ésin tísad and, ticfaitis
710 fosta fer Fernmaige leiss, ⁊ cath dobérad ut ante.' 'Ceist didiu,
cia bad dóig dar tíchtain? Dóig ;arnar⁵ tísad Celtchair mac
Uthechair?' 'Nár dóig omm. Líach a écnach 'na écmais. Bráthlecc
bidbad in chóicid ⁊ cend a costuda uili ⁊ comla chatha Ulad⁶, ⁊
gana betís fora chind i n-óenbaile ut ante ⁊ fir Hérend uile ó iarthur
715 co airthiur, ó desciurt co túascert, cath dobérad dóib, reme no
maidsed ⁊ ní fair no raínfide.'
'Ceist didiu, cia bad dóig diar tíachtain?' 'Iṅge ém meni thísed
in gilla bec, mo dalta-sa ⁊ dalta Conchobuir, Cú Chulaind na Cerdda
atberar friss.' 'Ia ómm ale,' bar Ailill. 'Atchúala lib in mac
720 ṁbec sain fecht n-aill i Crúachain. Ced ón cinnas a áesa-sum in
meic bic sin innossa?' 'Ní hí a áes is dulgium dó etir,' ar Fergus,
'dáig ba ferda a gníma in meic sin inbaid ba só andás in inbaid
inan fail.' 'Ced ón,' ar Medb, 'in fail cid d'Ultaib innossa com-
lonnaid a áesa is duilgium andásum?' 'Ní airgem and ⁷fá[e]l bad
725 fuilchuiriu⁷ nó láth bad luinniu nó comlannaid a áesa raseised co
62ᵃ trian nó go cethramad | comluind Con Culaind. Ní airge and,' ar
Fergus, 'caur a chomluind nó ord essorgni nó bráth for borrbuidni
nó combág urgaile basad inraicciu andá Cú Chulaind. Ní airge
and conmessed a áes ⁊ a ás ⁊ a forbairt ⁊ a ánius ⁊ a urfúath ⁊ a
730 urlabra, a chrúas ⁊ a chless ⁊ a gasced, a forom ⁊ a ammus ⁊ a
ammsigi, a brath ⁊ a búadri ⁊ a búadirsi, a déini ⁊ a dechrad ⁊ a
tharpige ⁊ a díanchoscur co cliss nónbair ar cach find⁸ úasu mar

¹ nUlad St, sic leg.   ² = giano   ³⁻³ sic for tísad   ⁴ ro MS.
⁵ sic ↘   ⁶ an fer sin add. St   ⁷⁻⁷ faol bus feolchairi St   ⁸ rind LU, St

Choin Culaind.' 'Ní dénam robríg de,' ar Medb. 'I n-óenchurp
atá. Imgeib guin immoamgeib gabáil. Is áes ingini macdacht
ármthir leis & ní géba fri féta in serrite óc amulchach atberthe.' 735
'Ní focclam-ne ón,' ar Fergus, 'dáig ba ferda a gníma in meicc sin
inbaid bad śóó andás inbaid inad fail.

Incipiunt macgnímrada Con Culaind

'Dáig alta in mac sin i tig a athar ⁊ a máthar ¹icon Air[g]dig¹ i
mMaig Muirthemne, & adféta dó scéla na maccáem² i 740
nEmain. Dáig is amlaid domeil Conchobor in rígi óro gab
rígi in rí .i. mar atraig fó chétóir cesta ⁊ cangni in chóicid d'ordugud;
in lá do raind i trí asa athli: cétna trian de fó chétóir ic fégad na
maccáem ic imbirt chless cluchi ⁊ immánae, in trian tánaise dond
ló ic imbirt brandub ⁊ fidchell, ⁊ in trian dédenach ic tochathim bíd ⁊ 745
lenna conda geib cotlud for cách. Áes cíuil ⁊ airfitid dia thálgud fri
sodain. Cia 'táim ane ar longais riam reme, dabiur bréthir,' ar
Fergus, 'ná fuil i nHérind nó i nAlbain óclach mac samla Conchobuir.
'Ocus adféta don mac sin scéla na maccáem ⁊ na maccraide i
nEmain, ⁊ rádis in mac bec ria máthair ar co ndigsed dá chluchi 750
do chluchemaig na Emna. "Romoch duit-siu sain, a meic bic," ar
a máthair, "co ndeoch ánruth do ánrothaib Ulad lett nó choímthecht
écin do chaímthechtaib Conchobuir do chor th' [f]aesma ⁊ t'imdegla
forin maccraid." "Cían lim-sa di śodain, a máthair," ar in mac
bec, "⁊ ní bíu-sa ocá idnaide, acht tecoisc-siu dam-sa cia airm i tá 755
Emain." "Is cían úait," ar a máthair, "airm indas fil. Slíab
Fúait etrut ⁊ Emain." "Dobér-sa ardmes furri amne," ar ésium.
'Luid in mac remi ⁊ gebid a adbena ániusa. Gebid a chammán
créduma ⁊ a liathróit n-argdide, ⁊ gebid a chlettíni díburgthi, ⁊
gebid a bunsaig mbaísi mbunloscthi, ⁊ fogab³ ic athgardigud a 760
62ᵇ śliged díb. Dobered béim | din chammán dá liathróit co mbered
band fota úad. ⁴No t[h]eilg[ed]⁴ dano a chammán arís d'athbéim
cona berad⁵ níba lugu andá in cétband. No thelged a chlettín ⁊
no sneded a bunsaig ⁊ no bered rith baíse 'na ndíaid. No gebed dano
a chammán ⁊ no geibed a liathróit ⁊ no geibed a chlettíne ⁊ ní 765
roiched bun a bunsaige lár tráth congebed a barr etarla etarbúas.
'Luid reme co forodmag na hEmna airm i mbátar in maccrad.
Trí cóicait maccáem im Follomain mac Conchobuir icá clessaib for

---

¹⁻¹ ocond Dairggdig LU: icon Airgthic St   ² maccaemi *Dipl. Edn.;*
what looks like i *with a stroke through it after* maccaem
³ *sic, for* ro gab (do gab St)   ⁴⁻⁴ do telccedh St   ⁵ band *add.* St

faidche na Emna. Luid in mac bec issin cluchimag eturru ar medón
770 ⁊ ecrais cid in liathróit i ndíb cossaib úadib ⁊ nís arlaic sech ard
a glúne súas ⁊ nís arlaic secha adbrond sís ⁊ ris eturturthig ⁊ ros
comdlúthaig i ndíb cossaib ⁊ ní rocht nech díb bir *nó* bulle *nó* béim
nó fargum furri. Ocus rosfuc dar brúach mbáire úadib.
'Nad fégat uili i n-óenfecht amaide. Ba machtad ⁊ ba ingantus
775 leó. "Maith a maccu", ar Follomain mac Conchobuir, "nobar
benaid uili fóe sút ⁊ táet a bás lim dáig is geiss dúib maccáem do thíchtain infar cluchi can chur a faísma foraib, ⁊ nobar benaid uile fóe i
nn-óenfecht, ar rofetammar is do maccaib ánroth Ulad sút ⁊ ná
dernat bés tuidecht infar cluchi can chur a faísma foraib *nó* a
780 commairge."
'Is and sin ros bensat uile fóe i n-óenfecht. Tarlaicset a trí
coíctu cammán ar ammus a chendmullaig in meicc. Turcbaid-
sium a óenluirg n-ániusa ⁊ díc[h]uris na trí coícait lorg. Tarlacait
dano na trí coícait liathróiti¹ ar ammus in meic bic. Turcbaid-sium
785 a dóti ⁊ a rigthi ⁊ a dernanna ⁊ díchuris na trí coíctu líathróiti¹.
Tarlacit dó na trí coícait bunsach baísi bunloscthi. Turcbais in
mac a scéthíni slissen ⁊ díchuris na trí coícait bunsach. Is and
sain imsaí-sium fóthib-sium. Scarais coíca rígmac im thalmain
díb fóe. Luid cóiciur díb,' ar Fergus, 'etrum-sa is Chonchobor 'sin
790 magin i mbámmar ic imbirt fidchilli .i. na Cendcháeme, for forodmaig
na hEmna. Luid in mac bec 'na ndíaid dia nn-imdibe. Gebid
Conchobor a rígláma in meic bic. "Ale atchíu ní fóil amberai-siu,
a meic bic, in maccrad." "Fail a mór*damnae* dam-sa," ar in mac
bec. "Ní fúarusa fíad n-oíged ga² thánac a tírib imciana ican
795 maccraid iar torachtain." "Ced són, cia tussu?' for Conchobor.
"Sétanta bec missi mac Sualtaim, mac-sa Dechtiri do derbsethar-su, ⁊
ní lat-su fo dóig lim-sa mo chrád d'fagbáil samlaid." "Ced ón, a
meic bic," for Conchobor, 'nád fetar armirt fil do[n] macraid conid
geiss dóib mac dar tír cuccu can chur a faísma forro?" "Ní fetar,"
800 bar in mac bec. "Dia fessaind, combeind 'na fatchius." "Maith
a maccu," bar Conchobor, "geibid foraib faísam in meic bic."
"Ataimem omm," bar siat.
63ᵃ 'Luid in mac bec | for faísam na maccraidi. 'S and sain scaílset
láma de-sium, ⁊ amsoí fóthu arís. Scarais *coíca* rígmac i talmain
805 díb fóe. Fa dóig la n-athrechaib is bás dobretha dóib. Níba sed
ón acht uathbás bretha impaib do thulbémmennaib ⁊ múadbémmen-
naib ⁊ fotalbémmennaib móra. "Aile," for Conchobor, "cid ataí dóib-
sin béus?" "Dothongu-sa mo dee dia n-adraim co ndigset-som uili ar
m'[f]ó[e]sam-sa ⁊ ar m'imdegail feib dochúadusa ara fáesam-sun ⁊ ara

---

¹ *sic*   ² = ce, cia

n-imdegail, conná gét-sa láma díb conas tarddur ̣uile fo thalmain.'' ⁸¹⁰
"Maith a meic bic, geib-siu fort fáesam na maccraide." "Ataimim
omm," ar in mac bec. And sain dochúatar in maccrad fora
[f̣]áesam ⁊ fora imdegail.

'Mac bec doriṅgni in gním sain,' ar Fergus, 'i cind chóic ṁblíadan
iarna brith coro scart maccu na curad ⁊ na cathmíled ar dorus a ⁸¹⁵
llis ⁊ a ndúnaid fadessin, nocorb éicen machta *nó* iṅgantus de ciano
thísed co hor cocríchi, gana thescad gabail cethri ṁbend, gana
marbad fer nó dís nó triur nó cethrur in am indat slána .xvii.
ṁbliadna de for Táin Bó Cúalnge.'

Is and sin atubairt Cormac Cond Longas mac Conchobuir: ⁸²⁰
'Doringni in mac bec sin gním tánaise 'sin bliadain ar cind doridisi.'
'Ciaso gním?' bar Ailill. 'Culand cerd buí i crích Ulad. Ro urg-
nastar fleid do Chonchobur ⁊ dochúaid dá thoc[h]uriud co Emain.
Rádis friss ara tísed úathad leis meni thucad fíraígid leiss ár nach
crích *nó* ferand baí aice acht a uird ⁊ a indeóna ⁊ a duirn ⁊ a thendchore. ⁸²⁵
Atbert Conchobor co ticfad úathad a dóchum.

'Tánic Culand connice a dún reme do frestul ⁊ frithálim lenna ⁊
bíd. Dessid Conchobor i nEmain corbo amm scaílti co tánic
deired dond ló. Gebid in rí a f̣iallgud¹ n-imétrom n-imthechta
immi ⁊ luid do chelebrad don maccraid. Luid Conchobor arin ⁸³⁰
faidchi co n-acca ní ba iṅgnad leiss: trí coícait mac 'sindara chind
dind f̣aithchi² ⁊ óenmac barin chind aile di. Dobered in t-óenmac
búaid ṁbáire ⁊ immána óna trí coíctaib maccáem. Tráth ba
cluchi puill dóib—cluichi puill fogníthi for faithchi³ na Emna—
ocus tráth ba leó-som díburgun ⁊ ba lesium imdegail, congeibed ⁸³⁵
na trí coícait liathróit fri poll immuich ⁊ ní roiched ní secha 'sin
poll. Tráth ba leó-som imdegail ⁊ ba leisium díburgun, no chuired na
trí coícait liathróit 'sin poll can imroll. Tráth fo imtharruṅg n-étaig
dóib, no benad-som a trí choícait ṅdechelt díb ⁊ ní chumgaitis uili a
delg do béim ⁴assa brut-som [nammá]⁴. Tráth ba imtrascrad dóib, ⁸⁴⁰
concured-som na trí coícait cétna i talmain foí ⁊ ní roichtis-[s]ium
uili immi-sium lín a urgabála. Arrópart Conchobor ic forcsin in
63ᵇ meic bic. "Amae a ócu," bar | Conchobor, "mo chin tír asa tánic
in mac bec atchíd dá mbetis na gníma óclachais aice feib atát na
macgníma." "Ní comdas⁵ a rád," ar Fergus. "Feib atré in mac ⁸⁴⁵
bec atrésat a gníma óclachais leis." "Congarar in mac bec dún co
ndig lind do ól na fledi dia tíagam." Conágart in mac bec do

---

¹ edgadh *St*   ² f̣aichthi *MS.*   ³ faichthi *MS.*
⁴⁻⁴ brutsom *add. in marg. In an erased space in text* brutsom *written in
with a fine pen, prob. over erased* nammá. (asa brat-somh amhain *St,* asa
brotsom nammá *LU*)   ⁵ comadas *St*

Chonchobur. "Maith a meic bic," ar Conchobor, "tair-siu linni d'ól na fledi dia tíagum." "Ní rag omm," bar in 850 mac bec. "Ced són?" bar Conchobor. "Ar ní dóethanaig in maccrad do chlessaib cluchi *nó* ániusa, ⁊ ní rag-sa úadib corbat doíthanaig cluchi." "Is cían dúni beith acot irnaidi ri sin, a meic bic ⁊ nicon bíam itir." "Táit-si round," ar in mac bec, "ocus rag-sa far ṅdiaid." "Nídat eólach etir, a meic bic," bar Conchobor. 855 "Géb[at]-sa¹ slichtlorg in tṡlúaig ⁊ na n-ech ⁊ na carpat."

'Ocus tánic Conchobor iar sin co tech Culaind cerdda. Ro fritháiled in rí, ⁊ ro fíadaiged ar grádaib ⁊ dánaib ⁊ dligedaib ⁊ úaslecht ⁊ caínbésaib. Ro hecrait aíne ⁊ úrlúachair fóthu. Gabsat for ól ⁊ for aíbnius. Ro iarfacht Culand do Chonchobur: "Maith 860 a rí, inra dális nech innocht it [d]íaid don dún sa?" "Níra dálius omm,' bar Conchobor, dáig níba cuman dó in mac bec dálastar 'na díaid. "Cid són?" bar Conchobor. "Árchú maith fil ocum. Á fúaslaicthir a chonarach de, ní laimthanoch² tasciud do óentríchait chét fris do ḟir chúardda *nó* imthechta, ⁊ ní aichne 865 nech acht missi fodessin. Feidm cét and do nirt." And sin atbert Conchobor: "Oslaicther dún dond árchoin coro imdegla in tríchait cét." Ra fúaslaiced dind árchoin a chonarach ⁊ fochuir³ lúathchúaird in tríchait cét, & tánic connice in forud i mbíd ic comét na cathrach ⁊ baí and sain ⁊ a chend ara mácaib. Ocus ba borb 870 barbarda bruthmar bachlachda múcna matnamail cách baí and sain.

'Imthúsa na maccraide bátar i nEmain corbo amm scaílti dóib. Luid cách díb da thig a athar ⁊ a máthar, a mumme ⁊ a aite. Luid dano in mac bec i slichtlurg na slúag co ránic tech Culaind cerda. 875 Gab[ais]⁴ icc athgarddigud na sliged reme dá adbenaib ániusa. Ó ránic co faidche in dúnaid i mbaí Culand ⁊ Conchobor, focheird a adbena uile riam acht a liathróit nammá. Rathaigid in t-árchú in mac ṁbec ocus glomais fair co clos fosnaib túathaib uili gloimm inn árchon. Ocus ní raind fri fes ba háil dó acht a slucud i n-óenḟecht 64ᵃ dar compur | a chléib ⁊ dar farsiuṅg a brágat ⁊ dar loiṅg a ochta. Ocus ní baí lasin mac cóir n-imdegla reme acht focheird róut n-urchair din liathróit conas tarla dar gincráes a brágat dond árchoin co ruc a mboí di ḟobaig inathair and dar' iarcomlai, ⁊ gebis i ndíb cossaib é ⁊ tuc béim de immun corthe co tarla 'na gabtib 885 rointi im thalmain. Atchúala Conchobor gloimm inn árchon. "Amae a ócu," bar Conchobor, "ní ma táncamar d'ól na fledi se." "Cid són?" bar cách. "In gilla bec ra dál im díaid, mac mo ṡethar,

¹ Gebatsa *St*   ² second *a* on erasure, *o* formed on *e*.   ³ *sic;* ro cuir *St*
⁴ *abbrev. stroke om.;* gebi*d*h *St*

Sétanta mac Sualtaim, dorochair lasin coin.'" Atragatar i
nn-óenḟecht uli Ulaid ollbladacha. Ciarbo óebéla oslaicthi dorus
na cathrach, dochúaid cách 'na irchomair ¹dar sondaib in dúnaid¹ 890
immach. Cid ellom condránic cách, lúaithium conarnic Fergus ⁊
gebis in mac ṁbec do lár thalman fri aidleind a gúaland - dobretha i
fiadnaisi Conchobuir. Ocus tánic Culand immach - atchondaire a
árchoin 'na gabtib rointi. Ba béim cride fri cliab leis. Dochúaid
innund isin dún asa aithle. "Mo chen do thíc[h]tu, a meic bic." bar 895
Culand, "ar bíth do máthar ⁊ t'athar, ⁊ ní mo chen do thíc[h]tu fort
féin." "Cid taí-siu don mac?" ar Conchobor. "Ní ma tánac-su
dam-sa do chostud mo lenna ⁊ do chathim mo bíd, dáig is maith
immudu ifec[h]tsa mo maith-se ⁊ is bethu immuig mo bethu².
Maith in fer muntiri rucais úaim. Concométad éite - alma - indili 900
dam." "Nádbad³ lond-so etir, a mo phopa Culand." ar in mac bec,
"dáig bérat-sa a ḟírbreth sin." "Cá breth no bértha-su fair, a
meic?" for Conchobor. "Má tá culén do ṡíl in chon út i nHérind,
ailēbthair lim-sa gorop inengnama mar a athair. Bam cú-sa
imdegla a almai ⁊ a indili ⁊ a ḟeraind ⁴in n-ed sain⁴." "Maith rucais 905
do breth, a meic bic," for Conchobor. "Nís bérmais ém." ar
Cathbath, "ní bad ḟerr. Cid arnach Cú Chulaind bias fort-su de
suidiu?" "Nithó," bar in mac bec. "Ferr lim mo ainm fodéin,
Sétanta mac Sualtaim." "Nád ráid-siu sin, a meic bic," ar
Cathbath, "dáig concechlabat fir Hérend ⁊ Alban in n-ainm sin ⁊ 910
bat lána beóil fer ṅHérend ⁊ Alban din anmum sin." "Fó limm
did*iu* cid sed bess form," ar in mac bec. Conid de ṡódain ro lil in
t-ainm aurdairc fair .i. Cú Chulaind, ó ro marb in coin boí ic Culaind⁵
cherd.

64ᵇ 'Mac bec doriṅgni in gním sin,' | ar Cormac Cond Longas mac 915
Conchobuir, 'i cind sé ṁbliadan arna brith, ro marb in n-árchoin
ná laimtís slúaig *nó* sochaide tascud i n-óentríchait cét fris, nírb écen
machtad *nó* iṅgantus de gana thísed co hor cocríchi, giano t[h]escad
gabail cethri ṁbend, gana marbad fer nó dís nó triur nó chethrur
in am inat ṡlána .xvii. ṁbliadna de for Táin Bó Cúalnge.' 920
'Doriṅgni in mac bec in tres gním isin bliadain ar cind dorís,' ar
Fiachu mac Fir Aba. 'Gá gním doringni?' bar Ailill. 'Cathbad
druí buí ⁶oc tabairt [tecaisc?]⁶ dá daltaib fri hEmain anairtúaith ⁊
ocht ṅdalta do áes in dána druídechta 'na farrad. Iarfacht ⁷[fer
díb]⁷ dia aiti ciaso ṡén ⁊ solud buí forin ló i mbátar, in ba maith 925

---

¹⁻¹ dar sond abdain in dunaid *MS.;* tar sondaigibh sitharda an dunaidh *St*
² i ndegaid mo chon *add. LU,* a ndiaig mo chon *add. St* ³ *sic;* ná badat
l. 956; na bat *St* ⁴⁻⁴ inn edsam *MS* ⁵ *sic;* Culann *St* ⁶⁻⁶ *a word
dropped here;* ic denamh foghlama *St* ⁷⁻⁷ *om. MS., supplied from St*

fá in ba saich. And atbert Cathbad mac bec congébad gasced, bad án ⁊ rabad irdairc, rabad duthain ⁊ dimbúan. Rachúala-som anísin ⁊ sé fria chlessaib chluchi fri hEmain aniardes, ⁊ focheird a adbena ániusa uli úad ⁊ dochúaid i cotultech Conchobuir. "Cach
930 maith duit, a rí féne," bar in mac bec.—Aithesc dano cungeda neich ó neoch in t-athesc sain.—"Cid connaige, a meic bic?" ar Conchobor. "Airm do gabáil," ar in mac bec. "Cia dotrecoisc, a meic bic?" bar Conchobor. "Cathbad druí," ar in mac bec. "Nít mérad-su ¹.i. nít mairnfed¹ sain, a meic bic," ar Conchobor. Tobert
935 Conchobor dá śleig ⁊ claideb ⁊ scíath dó. Bocgais ⁊ bertnaigis in mac bec na harmu ²[co nderna]² minbruan ⁊ minscomairt díb. Tuc Conchobor dá śleig aile dó ⁊ scíath ⁊ claideb. Bocgais ⁊ bertnaigis, crothais ⁊ certaigis co nderna minbruan ⁊ minscomairt [díb]³. Airm i mbátar na cethri airm déc bátar ic Conchobur i nEmain ic frithálim na
940 maccáem ⁊ na maccraide—ciped mac díb no gabad gasced combad Conchobor doberad trelam fúaparta dó, búaid n-eṅgnama leis assa aithle—cid trá acht⁴ doriṅgni in mac bec sin minbruan ⁊ minscomairt díb uili.

"'Ní maith ám and na airm se, a ṁo phopa Conchobuir," ar in
945 mac bec. "Ní thic mo diṅgbáil-se di śodain." Tuc Conchobor a dá śleig fodessin ⁊ a sciath ⁊ a chlaideb dó. Bocgais ⁊ bertnaigis, crothais ⁊ certaigis conarnic a fográin aice fria n-irlaind, ⁊ níras robris na harmu ⁊ ros fulgetar dó. "Maithi na ha[i]rm se omm," bar in mac bec. "Is é so mo chomadas. Mo chin in rí asa gasced ⁊
950 trelam so. Mo chin tír asa tánic." 'S and sin tánic Cathbad druí 'sin pupull ⁊ atbert: "Airm conagab sút?" ar Cathbad. "'S ed écin omm," bar Conchobor. "Ní do mac do⁵ máthar bad áil dam a ṅgabáil 'sind ló sa," ar Cathbad. "Cid són? Nach tussu darrecoisc?" ar Conchobor. "Nád mé omm," bar Cathbad. "Cid lat, a śiriti
955 śíabairthi," ar Conchobor, "in bréc dobertais immund?" |
65ª "Ná badat lond-su immorro, a mmo phopa Conchobuir," ar in mac bec, "dáig ar bíth is ésium domrecuisc-se araí ár iarfoacht a dalta dó ciaso śén baí forin ló ⁊ atbert-som mac bec no gébad gasced and, bad án ⁊ bad urdairc, ba[d] duthain ⁊ dimbúan immorro." "Fír
960 dam-sa ón," bar Cathbad. "Bat án-su ⁊ bat urdairc, ba[t] duthain ⁊ dimbúan." "Amra bríg canco rabur acht óenlá ⁊ óenadaig ar bith acht co marat m'airscéla ⁊ m'imthechta dimm ési." "Maith a meic bic, airg i carpat ⁶ar iss ed na cétna dait⁶."

---

¹⁻¹ marginal note   ²⁻² om. MS., cp. infra l. 938 ; ⁊ do-gní St
³ om. MS., cp. infra l. 936   ⁴ om. MS. suppl. from St   ⁵ a LU, St
⁶⁻⁶ arin sēn cētna St

'Dotháet i carpat, & in cétna carpat i tánic béus dano bocgais ┐
bertnaigis imme co nderna minbruan ┐ minscomairt de. Luid issin ⁹⁶⁵
carpat tánaise co nderna minbruan ┐ minscomairt de fón cumma
cétna. Doriṅgni minbruar don tres carput béus. Airm i mbátar
na sec[h]t carpait déc bátar oc frithálim na maccraide ┐ na maccáem
ic Conchobur i nEmain, doringni in mac bec minbruan ┐ minscomairt díb uile ┐ níro fulṅgetar dó. "Nít maithe and na carpait ⁹⁷⁰
so, a phopa Chonchobuir," ar in mac bec. "Ní tháet mo diṅgbáil-se
díb-so." "Cia airm i tá Ibar mac Riangabra?" ar Conchobor.
"Sund ém," ar Ibar. "Geib lat mo dá ech féin [dó]¹ sút ┐ inill
mo charpat." Gebid iarum in t-ara in n-echraid ┐ indliss in carpat.
Luid in mac bec 'sin carpat iarum. Bocais in carpat imme ┐ ro ⁹⁷⁵
fulṅgestar dó ┐ níro briss. "Maith in carpat sa omm," ar in mac
bec, "┐ iss ed and so mo charpat comadas."

"'Maith a meic bic," bar Ibar, "léic na eocho ara férgeilt ifechtsa."
"Romoch sin béus, a Ibair," ar in mac bec. "Tair round timchull
na Emna indiu. Indiu mo chétlá-sa do gabáil arm, coro[b] búaid ⁹⁸⁰
eṅgnama dam." Táncatar fo thrí timchull na Emna. "Léic na
eocho ar férgeilt ifec[h]tsa, a meic bic," ar Ibar. "Romoch sin
béus, a Ibair," ar in mac bec. "Tair round ar co mbennachat in
maccrad dam-sa, indiu mo chétlá do gabáil arm." Lotar rempu
don magin i mbátar in maccrad. "Airm congab sút?" ar cách. ⁹⁸⁵
"'S ed écin són." "Rob do búaid ┐ cétguine ┐ choscur sin, acht ba
romoch lind congabais armu fo bíth do deligthi ruind ocna clessaib
cluchi." "Ní scér-sa frib-si etir, acht do seón congabsa² armu indiu."
"Léic, a meic bic, na eocho ar férgeilt ifec[h]tsa," ar Ibar. "Romoch
sin béus, a Ibair," bar in mac bec. "Ocus in tsligi mór sa imthéit ⁹⁹⁰
sechond, gia leth imthéit?" ar in mac bec. "Cid taí-siu di?" ar
Ibar. "Aile it fer saignéch-su atchíu, a meic bic," bar Ibar. "Maith
lim, a maccáin, prímsligeda in chóicid d'iarfaigid. Cia airet
imthéit?" "Téit co Áth na Foraire i Sléib Fúait," ar Ibar. "Cid
'ma n-apar Áth na Foraire fris in fetar-su?" "Rafetar-sa omm," ⁹⁹⁵
bar Ibar. "Dagláech de Ultaib bís ic foraire ┐ ic forcomét and
arná tíset óic *nó* echtranna i nUltu do fúacra comraic forru, corop
é in láech [sin]¹ conairr comrac dar cend in chóicid uli. Dá ndig
dano áes dána fo dímaig a Ultaib | ┐ assin chóiciud, corop é conairr
séta ┐ maíne dar cend aenig in chóicid dóib. Dá tí dano áes dána ¹⁰⁰⁰
'sin crích, corop é in fer [sin]¹ bas chommairge dóib co rrosset
colbo Conchobuir, corop siat a dúana-sain ┐ a dréchta gabtair ar
tús i nEmain ar ríchtain." "In fetar-su cia fil icond áth sain
indiu?" "Rofetar omm," bar Ibar. "Conall Cernach curata

65ᵇ

---

¹ *supplied from St*   ² *read* congabus-sa; do gabus *St*

¹⁰⁰⁵ comramach mac Amargin, rí láech Hérend," bar Íbar. "Tó rouind
duit-siu, a maccáin, ar co rísem in n-áth." Lotar rempu co dreich
inn átha i mbaí Conall. "Airm congab sut?" ar Conall. "'S ed
écin," bar Ibar. "Rop da búaid ⁊ choscur ⁊ cétguine sin," ar Conall,
"acht bad romoch lind ra gabais armu, dáig ar bíth nít ingníma-su
¹⁰¹⁰ béus dámbad chommairgi ricfad a less intí ticfad sund, ar badat
slánchommairgi-siu bar Ultaib uli n-óg ⁊ atréstaís maithi in
chóicid rit báig." "Cid dogní and sin, a phopa Chonaill?" ar in
mac bec. "Foraire ⁊ forcomét in chóicid sund, a meic bic," bar
Conall. "Eirgg-siu dot tig ifechtsa, a phopa Conaill," ar in mac
¹⁰¹⁵ bec. "⁊ ¹no léicfe¹ dam-sa foraire ⁊ forcomét in chóicid do dénam
sund." "Xithó, a meic bic, ar Conall. "Nídat túalaing comrac ri
degláech co se." "Ragat-sa sechum fodes di*diu*," ar in mac bec,
"co ²Fertais Locha Echtrand² colléic dús in fagbaind mo láma do
fuligud for caraigt nó námait indiu." "Rag-sa a meic bic," ar
¹⁰²⁰ Conall, "dot imdegail arná tiasair th'óenur ³['s]in cocrích."³
"Xithó," ar in mac bec. "Rachat omm," bar Conall, "dáig benfait
Ulaid form do lécud th'óenur 'sin cocrích."

'Gabtair a eich do Chonall ⁊ ro indled a charpat & dochúaid
d'imdegail in meic bic. Ó rasiacht Conall ard fri aird fris, demin
¹⁰²⁵ leis giano thachrad écht dó, ná lécfad Conall dó a dénam. Gebid
lámchloich do lár thalman dárbo lán a glacc. Focheird róut
n-urchoir úad ar ammus cuṅgi carpait Conaill coro bris cuiṅg in
charpait ar dó co torchair Conall tríít go talmain co ndechaid a
máel asa gúalaind. "Cid and so, a meic?" ar Conall. "Messi
¹⁰³⁰ tarlaic dia fis dús in díriuch m'urchor nó cinnas díbargim etir nó
amm⁴ adbar gascedaig atamchomnaic." "Neim ar th'urchur ⁊
neim fort féin. Cid do chend fácba lat námtiu ifesta, nicon tías
dot imdegail níba siriu." "'S ed sin conattecht-sa foraib," ar
ésium, "dáig is geis dúib infar nUltaib techt dar éclind infar
¹⁰³⁵ carptib." Tánic Conall fothúaid arís co Áth na Foraire ar cúlu.

'Imthúsa in meic bic dochúaid-se⁵ fodes co ²Fertais Locha Echtrand². Baí and co tánic deired dond ló. "Dá lammais a rád frit, a meic
66ᵃ bic." ar Ibar, "ropa mithig lind techt co hEmain | ifechtsa, dáig ro
gabad dáil ⁊ raind ⁊ fodail i nEmain á chíanaib ⁊ fail inad urdalta
¹⁰⁴⁰ lat-su and di cach lóu rodicfa bith eter dá choiss Conchobuir, ⁊ ní
fail lim-sa acht bith eter echlachu ⁊ oblóire tigi Conchobuir. Mithig
lim-sa techt do imscrípgail friu." "Geib lat dún ind echrad didiu."
Gebid in t-ara in n-echraid ⁊ luid in mac issin carpat. "Aile a Ibair,
gá tulach and in tulach sa thúas innossa?" ar in mac bec. "Sliab

---

¹⁻¹ *sic*   ²⁻² Fertais Locha Echtra *LU, St*   ³⁻³ isin choiccrích *St*
⁴ = imba (*Windisch*) (an adbar gaiscedaigh me *St*)   ⁵ *sic*

Moduirn sin innossa," ar Ibar. "Ocus gia findcharn sút i mmullaig 1045
in tslébe?" "Findcharn dano Slébe Moduirn," ar Ibar. "Aile is aíbind
in carn út," ar in mac bec. "Óebind omm." bar Ibar. "Tair roind, a
maccáin, co rrísam in carn út". "Aile at fer ¹saignesach-su lista¹
atchíu," for Ibar, "acht is é seo mo chétfecht-sa lat-su. Bud é mo fecht
dédenach co brunni mbrátha mad dá ríus Emain óenfec[h]t." Lotar 1050
co mullach na taulcha araí. "Maith and, a Ibair," ar in mac bec,
"tecoisc-siu dam-sa Ulaid ar cach leth dáig ním eólach-sa i crích
mo phopa Conchobuir etir." Tecoscis in gilla dó Ulaid ar cach
leth úad. Tecoiscis dó cnuicc ⁊ céti ⁊ tulcha in chóicid ar cach
leth. Tecoscis dó maigi ⁊ dúne ⁊ dindgnai in chóicid. "Maith 1055
and sin, a Ibair," ar in mac bec, "gia mag and in cúlach cernach
ochrach glennach sa ruind aness?" "Mag mBreg," bar Ibar.
"Tecoisc-siu dam-sa déntai ⁊ dindgnai in maige sin." Tecuscais in
gilla dó Temair ⁊ Taltiu, Cleittech ⁊ Cnogba ⁊ Brug Meic inn Óóc
& Dún mac Nechtain Scéne. "Aile nach siat na meic Nechtain sin 1060
maídes nach mó fail 'na mbethaid d'Ultaib andá a torchair leó-som
díb?" "Siat ómm," bar in gilla. "Tair romuind co dún mac
Nec[h]tain," ar in gilla bec. "Mairg atbir ón omm!" bar Ibar. "Is
fis dún conid mór in bert baísi a rád. Gibé dig," bar Ibar, "níba
missi." "Ragaid do beo nó do marb," ar in mac bec. "Is mo beo 1065
ragas fades," ar Ibar, "⁊ mo marb fócebthar² icon dún rofetar .i.
oc dún mac Nechtain."

Lotar rempo connice in dún & tarmlaing in mac assin charput
forind faithche³. Amlaid boí faithchi⁴ in dúnaid ⁊ corthi furri ⁊ id
iarnaidi 'na thimchiull ⁊ id niachais éside ⁊ ainm n-oguim 'na menoc. 1070
Ocus is é ainm boí and: Gipé tísed in faidche, diamba gascedach,
geis fair ar thecht dind faidchi cen chomrac n-óenfir do fúacra.
Airlégais in mac bec in n-ainm ⁊ tuc a dá rigid 'mun coirthi, mar
boí in coirthi cona id. Tarlaic sin linnid⁵ co toracht tond taris.
"Andar lind," ar Ibar, "ní ferr sin ná a bith i fail i rraba,& rofetamar 1075
fogéba forin faidchi se aní 'coa taí iarair don chur sa .i. airdena báis
⁊ éca ⁊ aideda." "Maith a Ibair, córaig fortcha in charpait ⁊ a

66ᵇ fortgemni dam | coro thurthaind cotlud bicán." "Mairg atbeir ón
ám," ar in gilla, "dáig is crích bidbad so ⁊ ní faidchi airurais."
Córaigis in gilla fortcha in charpait ⁊ a fortgemne. Taurthais in 1080
gilla bec cotlud forind faidche.

'And sain tánic mac do maccaib Nechtain forin faidchi .i. Fóill
mac Nechtain, "Ná scuir na eochu itir, a gillai," ar Fóill. "Ní

---

¹⁻¹ *read* saignesach-su .i. lista? (As fer liosta tú *St*) ² faicfither *St*
³ faichthe *MS*. ⁴ faichthi *MS*. ⁵ *second* i *subscr.* (*phrase om. St*, isin
linn *H'*, isin linnidh *P*)

triallaim itir," ar Ibar. "Atát a n-ési ⁊ a n-aradna im láim béus."
1085 "Cóichi¹ na eich sin etir?" for Fóill. "Dá ech Conchobuir," ar in
gilla, "na dá chendbricc." "'S í sin aichni dobiur-sa forru. Ocus
cid tuc na eocho sund co hor cocríchi?" "Máethmaccáem congab
armu lind," ar in gilla. "Tánic co hor cocríchi do thasselbad a
delba." "Nírop do búaid nó choscur ón," ar Fóill. "Dia fessaind
1090 combad ingníma, is a marb ricfad fathúaid arís co hEmain ⁊ níbad
a béo." "Ní ingníma omm," bar Ibar, "ní comad[as]² gid a rád
ris etir. Isin tšechtmad bliadain arna breith don fail." Conúargaib
in mac bec a gnúis ó thalmain ⁊ tuc a láim dara gnúis ⁊ doringni
rothmol corcarda de ó mulluch co talmain. "Isam ingníma omm,"
1095 ar in mac bec. "Docho lim ná 'ráda duit nídat ingníma." "Bid
docho duit acht condrísem forsind áth, acht eirg-siu ar cend t'arm
dáig atchíu is midlachda tánac, ar ní gonaim aradu nó echlacha
nó aes cen armu." Bidcais in fer sain ar cend a airm. "Cóir duit
arechus dúin fris sút, a meic bic," ar Ibar. "Ced ón écin?" ar in
1100 mac bec. "Fóill mac Nechtain in fer atchí. Ní ngabat renna nó
airm nó faebair itir." "Ní rum-sa is chóir duit-siu sain do rád, a
Ibair," ar in mac bec. "Dobér-sa mo láim fón deil cliss dó .i. fón
n-ubull n-athlegtha n-íarnaide, ⁊ tecéma i llaind a scéith ⁊ i llaind a
étain ⁊ béraid comthrom inn ubaill dá inchind ³tria chúladaig³ co
1105 ndingne retherderg de fria chend anechtair combat léiri lésbaire
aeóir triana chend." Tánic immach Fóill mac Nechtain. Tuc-som
a láim fón deil cliss dó ⁊ focheird róut n-urchair úad co tarla i llaind
a scéith ⁊ i llaind a étain ⁊ berid comthrom inn ubaill dá inchind
³tria chúladaig³ co nderna retherderg⁴ de fria chend anechtair
1110 comba léir lésbaire aeóir triana chend, & tópacht-som a chend dia
méde.

'Tánic in mac tánaise immach arin faidchi, Túachall mac Nechtain.
"Aile atchíu commaídfide lat sain," ar Túachall. "Ní maídim limm
chétus óenláech ⁵do marbad.⁵" "Ní maídfe-su ón afechtsa dáig
1115 dofaíthaisiu limm-sa." "Tó duit-siu ar cend t'arm dáig is midlachda
tánac." Bidgais in fer sain ar cend a arm. "Cóir duit arechus dúin
risiút, a meic bic," bar Ibar. "Cid són?" ar in mec bec. "Túachail
mac Nechtain in fer atchí. Meni arrais din chétbulli nó din
67ᵃ chéturchur nó din chéttadall | ní arrais etir chaidche a⁶
1120 amansi ⁊ a⁶ airgigi non imrend im rennaib na n-arm." "Ní
rim-sa is rátti sin, a Ibair," ar in mac bec. "Dobér-sa mo láim fón
manaís murnig Conchobuir, fón cruísig neme. Tecéma 'sin sciath

---

¹ *sic for* cóichit   ² comadhas *St*   ³⁻³ *sic for* triana chúladaib
⁴ rechderg *MS.*   ⁵⁻⁵ dom marbad *MS.*, do t[h]uitim lem *St*
⁶ ar *St* (= ara)

ósa broind ⁊ ¹brúifet tria asna a tháeib¹ bas siriu úaim ar tregdad a chridi 'na chliab. Bud aurchor deóraid sin ⁊ níba hicht urraid. Níba teg legis *nó* othrais úaim-se dó co bruinne mbrátha." Tánic ¹¹²⁵ Túachall mac Nechtain immach arin faidchi ⁊ focheird in mac bec a láim fón manaís Conchobuir dó ⁊ dorecgmaing 'sin scíath ósa broind [⁊]² ³brúis trí asna isin taíb ba siriu úad³ ar tregdad a chridi 'na chlíab. Benaid-sium a chend [de]⁴ riasiu sessed dochum talman. 'And sin tánic immach sósar na clainde forsin faidche .i. Faindle mac ¹¹³⁰ Nechtain. "Is báeth in lucht condránic frit and sin," ar Fandle. "Cid ón?" ar in mac bec. "Tair sechut sís arin lind bail ná ró do choss lár." Bidgais Fandle reme forin lind. "Cóir duit arechus dúin risiút, a meic bic," bar Ibar. "Cid ón écin?" ar in mac bec. "Fandle mac Nechtain in fer atchí. Is de dia tá in t-ainm fair, mar fandaill ¹¹³⁵ nó mar íaraind imthéit muir. Ní chumgat snámaigi in talman ní dó." "Ní rim-sa is chóir sin do rád, a Ibair," ar in mac bec. "'S aichnid duit-siu ind aband fil ocuind i nEmain, Kalland⁵. Tráth nos immet in maccrad do chlessaib cluchi furri ⁊ úair nach foísam in lind, berim-se maccáem cechtar mo dá dernand tarsi and sin ⁊ ¹¹⁴⁰ maccáem cechtar mo dá gúaland, ⁊ ní fliuchaim fadesin gid mo adbrunnu fóthu." Condránic dóib forind lind ⁊ furmid in mac bec a rigthi tharis co tarla in muir aird fri aird fris ⁊ dobretha tathulbéim do chlaidiub Conchobuir dó ⁊ tópacht a chend dá médiu. Ocus léicis in colaind lasin sruth ⁊ dobretha a cend leis. ¹¹⁴⁵

Lotar isin dún iar tain ⁊ ra airg[set]⁶ in cathraig ⁊ ra loiscset connárbdar airdiu a déntai andát a immélaig. Ocus imsóiset rempu i Slíab Fúait ⁊ dobrethsat trí cind mac Nechtain leó.

Confaccatar in n-alma do aigib alta rempu. "Cóchit na innili imda imdíscire⁷, a Ibair," ar in mac bec. "Pettai sút nó inn aigi ¹¹⁵⁰ chena?" "Aige chena omm," bar Ibar. "Almai d'aigib alta sain bít i ndiamraib Sléibi Fúait." "Saig brot dún forsin n-echraid ⁸dús ar co nn-ársimmís⁸ ní díb." Saigis in t-ara brot forin n-echraid. Ní chaemnactar eich roremra ind ríg in damrad do chomaitecht. Luid in mac bec assin charput ⁊ gebis dá n-ag lúatha látiri díb. ¹¹⁵⁵ Ceṅglais d'fertsib ⁊ d'[f]ithisib ⁊ d'iallaib in charpait.

Lotar rempu co forodmag na hEmna co 'mafaccatar in n-elta do gésib gela seccu. "Cóichi⁹ and na eóin sin, a Ibair?" ar in mac bec.

---

¹⁻¹ *read* brúifid trí asna isin táeb; (brisfidh tri asna isin taob *St*). *Cp. infra* coro brúi tri asna 'sin táeb ba siriu úad (1781) ² ⁊ *expunged after* broind *MS.; ⊦ St* ³⁻³ bruis sin asna ina thaíb ba siriu úad *MS.;* brisis tri asna isin taobh ba síriu uadh *St* ⁴ *om. MS., supplied from St* ⁵ .i. Callann a hainm *St* ⁶ airg- *end of line, hyphen denoting accidental omission;* ro airgsit *St* ⁷ imdaiscaire *MS. infl. of preceding word;* imdisgire *St* ⁸⁻⁸ *omit* dús (da fhios in ngepmaois *St*) ⁹ *sic, for* cóichit

67ᵇ "Indat pettai | sút nó indat eóin chena?" "Eóin chena omm," bar
1160 Ibar. "Elta do gésib sin tecait di chlochaib ⁊ carrgib ⁊ ailénaib in
mara móir immuich do geilt for maigib ⁊ rédib Hérend." "Cia bad
irdarcu a mbeó sút do rochtain Emna nó a mmarb, a Ibair?" ar
in mac bec. "Airdarcu a mbeó omm," bar Ibar, "dáig ní cách
conairg na eóin beóa do gabáil." And sain dobretha in mac ceird
1165 ṁbic forru. Fostaid ocht n-eóno díb. Ocus dobretha ceird máir
iar sain ⁊ fastaid sé eóin déc díb. ¹Ceṅglais do ḟertsib ⁊ d'ḟithisib
⁊ iallaib ⁊ d'ḟolomnaib ⁊ tétaib in c[h]arpait.¹ "Tuc lat na eónu,
a Ibair," ar in mac bec. "'Tú-sa i ndulig," ar Ibar. "Cid són écin?"
ar in mac bec. "Fail a mórabba dam. Dianom glúasiur itir assin
1170 magin i tú, nom thescfat roith iarnaidc in charpait [re]² feramla ⁊
fertsigi ⁊ fortressi céimmi inna hechraide. Dánam luur itir dano,
nom thollfat ⁊ nom thregtaifet benna na n-aigi." "Aile nít fírlaec[h]-
su béus, a Ibair," ³ar in mac bec³, "dáig in fégad fégfat-sa forna
echaib, ní ragat assa certimthecht. In tincud tincfat forsna haigib,
1175 cromfait a cinnu ar m'ecla ⁊ ar m'úamain, ⁊ fó duit-siu gid dia
mbendaib no chiṅgthe⁴."

Lotar rempo co ráncatar Emain. Is and sin rathaigis in Leborcham íat. Ingen-saide Aí ⁊ Adairce. "Óencharptech sund," for
Leborcham, "⁊ is úathmar thic. Cind a bidbad fordergga 'sin
1180 charput aice. Eoin áille óengela ic imuarad aice 'sin charput. Aige
altamla anríata i ceṅgul ⁊ chrapull ⁊ chuibrech ⁊ charcair aice.
Ocus meni frithálter innocht é, dosfaíthsat óic Ulad leis." "Rodafetammar in carptech sin," ar Conchobor, "in gilla bec, mac mo
ṡethar, dochóid co hor cocríche, ro derg a láma, ⁊ ní doíthanach
1185 comraic, ⁊ meni frithálter dano, dofaíthsat óic Emna uili leis."
Ocus ba sed in chomairle ra cruthaiged leó in bantrocht da lécud
immach do ṡaigid in meic .i. trí coícait ban .i. deich mnáa ⁊ secht
fichit díscir derglomnocht i n-óenḟecht uili ⁊ a mbantóesech rempo,
Scandlach, do thócbáil a nnochta ⁊ a nnáre dó. Táncatar immach
1190 in banmaccrad uile ⁊ túargbatar a nnochta ⁊ a nnáre uile dó.
Foilgid in mac a gnúis forru ⁊ dobretha a dreich frisin carpat arná
acced nochta *nó* náre na mban. And sain ro irgabad in mac bec
isin charput. Tucad i trí dabchaib úaruscib é do díbdud a ḟerge.
Ocus in chétna dabach i tucad in mac bec, ro díscaíl dá cláraib ⁊
1195 dá circlaib amal chnómaidm imbi. In dabach tánaise configfed
durnu di. In tres dabach fer fos foilṅged ⁊ fer ní foilṅged etir.

---

¹⁻¹ *this sentence misplaced. To be read at end of par. as in LU, St*
² *om. MS., supplied from St* ³⁻³ iarum *MS., reading supplied from St*
⁴ chiṅgthé *MS.*

And sain ¹[tíagait] fergga¹ in meic for cúlu² ⁊ conácbad a thimthach
68ª immi². Táncatar a delba dó | & doriṅgni rothmól corcra de ó
mulluch co talmain. Secht meóir cechtar a dá choss ⁊ secht meóir
cechtar a dá lám, ⁊ secht meic imlessan cechtar a dá rígrosc iarum ¹²⁰⁰
⁊ secht ṅgemma de ruthin ruisc fo leith cech mac imlesan díb.
Cethri tibri cechtar a dá grúad : tibri gorm, tibri corcra, tibri úane,
tibri buide. Coíca urla fégbuide ón chlúais go 'cheile dó amal chír
ṁbethi nó amal bretnasa bánóir fri taul ṅgréne. Máel glé find fair
mar bó ataslilad. Brat úanide imme, delg n-argait indi.³ Léni ¹²⁰⁵
órṡnáith immi. Ocus ra sudiged in mac eter dá choiss Conchobuir
& ro gab in rí ic slíachtad a maíle.

'Mac bec doringni na gníma sin i cind a ṡecht ṁbliadan arna
breith, barroscart na curaid ⁊ na cathmílid ris torcratar dá trian
fer nUlad ⁊ ná fúaratar a dígail forro ná co n-érracht in gein sin ¹²¹⁰
chucu, nocorb éicen machtad nó iṅgantus de giano thísed co hor
críche, gana marbad fer nó dís nó triur nó chethrur in aim inat
ṡlána secht ṁbliadna déc [de] for Táin Bó Cúalnge.'

Conid innisin do macgnímaib Con Culaind sin for Táin Bó Cúalnge,
ocus remthús in sceóil ⁊ na sliged ⁊ imthechta in tṡlúaig a Crúachain ¹²¹⁵
connici sin.

In scél fodessin is ní and fodechtsa.

TÁncatar cethri ollchóicid Hérend arnabárach dar Cruind ⁴.i.
Sliab⁴ sair. Luid Cú Chulaind ríam remáin rempu conarnaic fri
araid Órláim meic Ailella ⁊ Medba ro baí oc Tamlac[h]tain Órláib ¹²²⁰
fri Dísert Lochad atúaid. Buí ic buiṅg na fertas carpait culind
issin ḟid. 'Amae, a Láeig, ar Cú Chulaind, 'is tarpech in mod do
Ultaib más iat benas in fid fón samlaid se ar cind fer ṅHérend, &
airis-[s]iu sund bic co fessur-sa cia benas in fid fón samlaid se.'
Luid Cú Chulaind iarum conarnaic frisin n-araid. 'Cid dogní-siu ¹²²⁵
sund, a gillai ?' ar Cú Chulaind. 'Itú-sa ém,' ar in gilla, 'oc boiṅg na
fertas carpait culind sund dáig ar bíth ro mebtatar⁵ ar carpait
inné ic taffund na hailiti urdairce út .i. Con Culaind, & ar bíth
t'óclachais-[s]iu, a óclaích, coṅgain lim-sa nacham thair in Cú
Chulaind urdairc sin.' 'Roga duit, a gillai,' ar Cú Chulaind, 'a ¹²³⁰
n-imtheclamad nó a n-imscothad nechtar de.' 'Dogén a n-imtheclamad dáig is assu.' Forrópart Cú Chulaind fora n-imscothad, ⁊
nos tairṅged tria ladraib a choss ⁊ a lám i n-agid a fíar ⁊ a fadb co

---

¹⁻¹ teid fercc *St*   ²⁻² ⁊ do cuiredh a edach aonaigh uime *St;* do gabad
faedaran [*sic leg.*] gormchorcra uime *Rec.* III; *read* co ngabad, *omitting* ⁊
³ *sic*   ⁴⁻⁴ *gloss written overhead*   ⁵ *sic for* mebdatar

ndénad a féth ⁊ a snass ⁊ a slemnugud ⁊ a cermad. Nos bláthiged
[1235] conná tairised cuil forru tráth nos léiced úad. And sin nod fégand
dano in gilla. 'Dar lim ám ale, ní h'opair chóir dombiurt-sa fort-su
itir. Cóich thussu itir?' bar in gilla. 'Is missi in Cú Chulaind
airdairc atbertais-[s]iu imbuarach.' 'Ro mairc-se ón ém,' ar in
68[b] gilla, 'darochar deside co bruinni mbrátha.' | 'Nád bia etir, a gillai,'
[1240] ar Cú Chulaind, 'ar ní gonaim aradu *nó* echlachu *nó* áes gan armu.
Is cia airm i tá do thigerna-su chena ale?' 'Aracút tall forin
fertai,' ar in gilla. 'Dó duit-siu connice ⁊ urtha [co] robud dó ⁊
¹ar dogné¹ fatchius, dáig dia condrísam, dofáeth lem-sa.' Luid iar
sain in t-ara do ṡaigid a thigerna, ⁊ cid lúath condránic in gilla,
[1245] lúathiu conarnic Cú Chulaind ⁊ tópacht a chend de Órláb, & turcbais
⁊ tasbénais do feraib Hérend in cend.

'S and sin táncatar trí meic Árach barsin n-áth ic Ard Chiannacht
i n-herus Con Culaind, Lon ⁊ Úal[u] ⁊ Díliu. MasLir ⁊ MasLaig ⁊
MasLethair anmand a n-arad. Is aire condeochatar sin i comdáil
[1250] Con Culaind dáig ba immarcraid gním leó doriṅgni in lathe reme
forro .i. dá mac Nera meic Nuatair meic Thacáin do marbad ic
Áth Gabla, Órláb mac Ailella ⁊ Medba do guin dano ⁊ a chend do
thaisselbad d'feraib Hérend. Coro gontais-[s]ium Coin Culaind
fón samlaid sin ⁊ go ructaís a chend leó i taisselbad. Lotar fón fid ⁊
[1255] ro bensat trí fidṡlatta findchuill i llámaib a n-arad condrístais a sessiur
i n-óenfecht gliaid fri Coin Culaind. Impádar Cú Chulaind friu ⁊
benais a sé cinnu díb. Torchratar meic Árach samlaid la Coin Culaind.

Tánic dano Lethan fora áth for Níth i crích Conailli Murthemne
do chomruc fri Coin Culaind. Barrópart forsin n-áth. Áth Carpait
[1260] a chomainm inn átha áit mal connairnechtatar² dáig conmebdatar
a carpait ic imthrutt isinn áth. ³Is and sin focera Mulchi forsin
taulaig eter na dá n-áth.³ Conid de dá tá Gúalu Mulchi dano béus.
Is and sin dano condránic Cú Chulaind ⁊ Letha[n] & dofuit Lethan
fri Coin Culaind ⁊ tópacht a chen[d] dia méde forsin áth ⁊ nad
[1265] n-ácaib leis .i. fácbais a chend la cholaind. Conid de atá in t-ainm
forsinn áth ó ṡain .i. Áth Lethan i crích Conailli Murthemne.

And sain ⁴táncatar crutti⁴ Caínbili ó Ess Rúaid dá n-airfitiud.
Indar leó-som rapo da t[h]ascélad [forro] ó Ultaib ⁊ rucsat in
tṡlúaig tafond dírecra i fat forro co ndechatar i ndelbaib oss n-alta
[1270] úadib icna corthib ic Líic Móir, ar giarsa Chruitti Caínbili atberthea
friu, batir fir co mórfiss ⁊ go mórfástine ⁊ druídec[h]t iat.

---

¹⁻¹ *sic, ar omitt.*   ² *first* ta- *added above line*
³⁻³ Is and sin focera c̄c̄ Mulchi ... *MS.,* c̄c̄ *omitt.;* Docer Mulcha ara Lethain
isin gúala fil etarro *LU;* Docher ... *YBL;* Is ann sin ro tuit Mulcha forsan
tulaigh itir na da ath *St*   ⁴⁻⁴ tancatar .i. crutti *MS.,* .i. *omitt.*

And sain bágais Cú Chulaind port i faicfed Meidb dobérad
chloich furri ⁊ níbad chían ó lethchind.   Fír dó-som.   Port indas
facca Meidb, focheird chloich assa thabaill furri coros ort in petta
ṅ-eóin buí fora gúalaind fri áth aniar.   Luid Medb dar áth sair ⁊ ¹²⁷⁵
dobretha cloich assa thabaill furri béus goro ort in petta togmalláin
baí fora gúalaind fri áth anair.   Conid Méide in Togmaill ⁊ Méde ind
Eóin a n-anmand na n-inad sin béus, ⁊ conid Áth Srethe comainm
ind átha dara sredestar Cú Chulaind in cloich assa thabaill.

TÁncatar cethri ollchóiceda Hérend arnabárach. Gabsat ar argain ¹²⁸⁰
Maigi Breg ⁊ Maigi Murthemne. Ocus tánic gérmenma géribrach
69ᵃ | Con Culaind dá aiti, do Fergus.   Ocus atrubairt fri firu
Hérend faitchius in n-aidchi sin dáig ar bíth dosficfad Cú Chulaind,
⁊ a formolad i fus doridisi amal ra scríbsam remaind, ⁊ dorigni laíd:

'Damb ró Cú Chulaind Cúalṅge   ¹²⁸⁵
ria curadaib Cráebrúade.
Beti fir i fuilib de
d'argain Maige Murthemne.

'Dochúaid-sium turus ba¹ sía
go ránic slébi Armenia.   ¹²⁹⁰
Ralá ág dar[a]² aiste,
ra chuir ár³ na Cíchloiste.

'Ba handsu dó meic Nec[h]tain
do chur assa prímlepthaib,
cú na cerda—ba moɔl n-áig—   ¹²⁹⁵
⁴do marbad cona óenláim⁴

'Nochom thá ní rádim de
im dála meic Deic[h]tere,
is sí mo chobais, ní gó,
⁵co corrossid dob anró.'⁵   ¹³⁰⁰

Aithle na laíde sin: Is hé in lá cétna tánic in Dond Cúalnge co
crích Margín⁶ & coíca samséisci immi de samascib, & foclassa
búrach dó—is inund són ⁊ focheird úir dá lúib taris. Is é in lá cétna

---

¹ bad *MS*   ² dara *St*   ³ ár ar *MS.*   ⁴⁻⁴ do marp iet co n-aonlaim *St*
⁵⁻⁵ ɔ corrossid *MS.; read* cen corrossid do-bar-ró *(Pedersen)*   ⁶⁻⁶ go cric[h]
Mairgine *St,* i crích Mhairgi *Rec III* (cp. hi Tír Marccéni LU 5332)

tánic in Mórrigu ingen Ernmais a sídib co mboí forin chorthi i
1305 Temair Chúalṅge ic brith robuid don Dund Chúalnge ria feraib
Hérend, ⁊ ro gab acá acallaim ⁊ ¹[is ed asbert]¹: 'Maith, a thrúaig,
a Duind Chúalnge,' ar in Mórrigu, 'déni fatchius dáig ar [bíth]
dotroset fir Hérend ⁊ not bérat dochum loṅgphoirt meni déna
faitchius.' Ocus ro gab ic breith robuid dó samlaid ⁊ dosbert na
1310 briathra sa ar aird:
    ²'Nach fitir dub duṡáim. dal na inderb. esnad fiacht. fiacht.
fiacht nad cheil.   cuardait námait do thuaithbregaib binde ar
tánaib tathaib rún.   rafiastar dúib danis murthonna fer forglas
forláib lilasta áeb agesta in mag meldait.   slúaig scothníam buidb
1315 bógei[m]nech febdair fiach fir nairm³ rád n-iṅguir crúas Cúalṅge
⁴có icat⁴ do bás mórmaicne féc muintir ar n-éc muntire do námait⁵
écaib.'
    Tánic iarum Dond Cúalnge.   Urtha reme go Glend na Samaisce i
Sléib Chulind ⁊ coíca samaisce leis dia ṡamascib.
1320   Aill do búadaib Duind Chúalnge and so .i. coíca samaisce no daired
cach laí. Bertís láegu riasin trá[t]h arṅabárach & do neoch ná⁶
bered lóegu díb, no scaíltis imma lóegu, dáig ní ḟulṅgitis compert
Duind Chúalnge accu.   Ba do búadaib Duind Chúalnge coíca do
maccáemaib no bítis ic clessaib cluchi cacha nóna ara cháemdruim.
1325 Ba do búadaib Duind Chúalnge cét láech no dítned ar thess ⁊ ar
uacht ba⁷ foscud ⁊ ba⁷ imdegail. Ba do búadaib Duind Chúalnge ná
laimed bánanach nó bocánach nó genit glinni tascud d'óentríchait
chét friss. Ba do búadaib Duind Chúalnge crandord dogníd cacha nóna
ic tiachtain ar ammus a liss ⁊ a léis ⁊ a machaid, ba leór ceóil ⁊
1330 airfiti dond ḟir i túasciurd ⁊ i ndesciurd ⁊ in n-etermedón tríchait⁸
chét Cúalnge uili in crandord dogníd ⁹cacha nóna ic tiachtain dó ar
ammus a liss ⁊ a léis ⁊ a machaid⁹. Conid ní de búadaib Duind
Chúalnge in sin.

    Dollotar na slúaig iarum im ailib ⁊ im airtraigib críchi Conaille
1335 Murthemne arṅabárach, & rádis Medb ara tarta amdabach do
scíathaib ósa cind nachas díburged Cú Chulaind de chnoccaib nó
chétib nó thulcaib. Ocus trá ní roacht Cú Chulaind guin nó ath-
ḟorgab for feraib Hérend ¹⁰im chríchaib¹⁰ im ailib im artraigib Conaille
Murthemne in lá sain.

---

¹⁻¹ om. MS.; is edh adubairt St   ² .r. (retoiric) prefixed in marg.
³ máirm LU; mairb St   ⁴⁻⁴ coigde LU; coicti St   ⁵ om. LU, St
⁶ no MS. (do neoch nach beredh St, mina berdais H.2.17)   ⁷ = fo a (ima
fhosccad ⁊ imdegail C)   ⁸ sic   ⁹⁻⁹ sic MS. with marks of reference to
same passage earlier   ¹⁰⁻¹⁰ after chríchaib, Mur̄ cancelled; im chríchaib omitt.

Co foítar cethri ollchóiceda Hérend i rRéde Loche i Cúalnge ⁊ co ¹³⁴⁰
ragbatar dúnad ⁊ longphort and in n-aidchi sin. Rádis Medb fria
cáeminailt comaitechta dá muntir tec[h]t ar cend usci oóil ⁊ innalta
dochum na haba di. Loche comainm na hingene, & dotháet iarum
Loche ⁊ coíca ban impi ⁊ mind n-óir na rígna ósa cind. Ocus foceird Cú
69ᵇ Chulaind | cloich assa thabaill furri co rróebriss in mind n-óir í trí ¹³⁴⁵
⁊ coro marb in n-ingin inna réid. Conid de atá Réde Loche i Cúalngiu.
Ba dóig trá la Coin Culaind i n-écmais a fessa ⁊ a eólusa ba hí Medb
boí and.

L Otar na slúaig arnabárach go ráncatar Glaiss Cru[i]nd & barrró-
bratar in nGlassi ⁊ forfémdetar a techt. Ocus Clúain Carpat ¹³⁵⁰
comainm in chétinaid áit mal connarnectar. Is de dia tá Clúain
Carpat forsin dú sin ar bíth cét carpat ruc in Glassi díb co muir. Rádis
Medb fria muntir ar co ndigsed láech díb do fromad na haba. Ocus
atraacht óenláech prósta mór di muntir Medba, h Úalu a chomainm,
⁊ gebis nertlía cloche fria ais ⁊ dotháet-aide dia fromad na Glassi ¹³⁵⁵
⁊ focheird in Glaiss for cúlu é, marb cen anmain, a lía for[a] druim.
Rádis Medb ar co tucthá anís ⁊ ara claitte a fert ⁊ ara túargabtha a
lía. Conid de atá Lía Úaland i crích Cúalnge.
Lilis Cú Chulaind co mór dena slúagaib in lá sain ic iarair comraic
⁊ comluind forru ⁊ marbais cét láech díb im Róen ⁊ im Roí, im dá ¹³⁶⁰
senchaid na Tána.
Rádis Medb fria muntir ara tíastais i comruc ⁊ i comlund fri Coin
Culaind. 'Níba messi,' ⁊ 'níba mé,' ar cách assa magin. 'Ní dlegar
cimmid dom muntir. Giano dlestea, ní mé no ragad i n-agid Con
Culaind dáig ní réid comrac ris.' ¹³⁶⁵
Táncatar na slúaig fri táeb na Glassi dáig fosrémdetar a techt go
ráncatar airm i táet in Glassi assin tslíab, ⁊ dámbad áil dóib,
bacóistís eter in Glassi ⁊ in slíab acht ní arlacair Medb, acht in
slíab do chlaidi ⁊ do letrad rempi combad ail ⁊ combad athis for
Ultaib, & conid Bernais Tána Bó Cúalnge ainm in inaid ó sain dáig ¹³⁷⁰
taris rucad in táin iar tain.
Gabsat cethri ollchóiceda Hérend dúnad ⁊ longphort in n-aidchi
sin ic Bélut Aileáin. Bélat Aileáin a ainm connici sain, Glend Táil
immorro a ainm ó sain ara mét ra thálsat na halma ⁊ na immirgl
a loim ⁊ a lacht and do feraib Hérend. Ocus Líasa Líac ainm aile ¹³⁷⁵
dó, ar is de atá in t-ainm fair ar is and ro sáidset fir Hérend léss ⁊
machad dia n-almaib ⁊ dia n-immirgib.
Táncatar cethri ollchóiceda Hérend co ráncatar inní co Sechair.
Sechair a ainm na aband co sin. Glass Gatlaig a ainm ó sain. Is

1380 de dano atá in t-ainm furri ¹[úair ba]¹ i ṅgataib ⁊ róoib tucsat fir
Hérend a n-alma ⁊ a n-immirgi tarsi ⁊ léicset na slúaig uile a ngait ⁊
a róe lasin ṅglais iar tíachtain tarsi. Is as sain ainm Glassi Gatlaig.
Táncatar cethri ollchóiceda Hérend co ragbatar dúnad ⁊ longphort i
nDruim Én i crích Conaille Murthemne in n-aidchi sin & gabais
1385 Cú Chulaind acond Ferta i lLergga 'na fírḟocus in n-aidchi sin.
Ocus cutlaigis² ⁊ bertaigis ⁊ crothais Cú Chulaind a armu in n-aidchi
sin co n-ebailt cét láech din tṡlúag ar gráin ⁊ ar ecla ⁊ ar úamun Con
Culaind. Rádis Medb fri Fiachu mac Fir Aba di Ultaib ar co
ndigsed d'acallaim Con Culaind do brith choma dó. 'Ciarso choma
1390 no bértha dó?' for Fiachu mac Fir Aba. 'Ní handsa.' ar Medb.
'Imdéntar leis do neoch ro milled d'Ultaib coro ícthar friss feib
as dech atbérat fir Hérend. Feiss i Crúachain do grés dó, fín ⁊
70ª mid do dáil | fair ⁊ tíchtain im géilsine-se ⁊ i ṅgélsine Ailella dáig is
sochru dó andás beith i ngélsine ind ócthigerna icá tá.'—Conid sí
1395 bríathar is mó gén ⁊ tarcassul ro ráided for Táin Bó Cúalnge .i.
ócthigern do dénam din chúicedach is dech buí i nHérind .i. di
Chonchobur.
Tánic iarum Fiachu mac Fir Aba do acallaim Con Culaind.
Ferais Cú Chulaind fáilti fris. 'Tarissi limm.' 'Tairisi duit-siu
1400 ón.' 'Dot acallaim tánac ó Meidb.' 'Cid dobertais latt?' 'Im-
dénta[r]³ latt anro milled d'Ultaib coro ícthar frit feib as dech
atbérat fir Hérend. Feiss i Crúachain dait, fín ⁊ mid do dáil fort,
& tíchtain i ngélsine Ailella ⁊ Medba dáig i[s] sochru dait andás
bith i ngélsine ind ócthigerna⁴ icá taí.' 'Nithó omm,' ar Cú Chulaind.
1405 'Ní recfaind-se bráthair mo máthar bar ríg n-aile. 'Ocus ar co tís
co moch imbárach i n-erus Medba ⁊ Fergusa co Glend Fochaíne.'
Luid iarum Cú Chulaind co moch arnabárach co Glend Fochaíne.
Dotháet dano Medb ⁊ Fergus 'na chomdáil. Ocus tincais Medb ar
Coin Culaind ⁊ cessis a menma fair in lá sain dáig ar bíth ní mó ná
1410 mod maccaím lée atacaemnaic. 'Inn é sút in Cú Chulaind airdairc
atberi-siu, a Ḟerguis?' ar Medb. Ocus ra gab Medb ar acallaim
Fergusa ⁊ dorigni laíd:

> 'Más é ucain in Cú cain
> itirid-si⁵ infar nUltaib,
1415   ní thabair a thraig⁶ fri tend
> ná diṅgaib d'ḟeraib Hérend.'

---

¹⁻¹ om. MS.; úair ba i ngataib . . . LU, St; doigh i ngadaib . . . Rec. III
² read cuclaigis (Windisch)   ³ imdhentar St   ⁴ octhigirn MS.
⁵ aderti-si St, atbeirthi-si C   ⁶ thraigid MS., thraig St, C

'Cid óc in Cú sin atchí
imriada Mag Murthemni,
ní tabhair fri talmain traig
ná dingba ar galaib óenḟir.'   1420

'Berar coma úan don láech,
mad dia tí taris is báeth,
leth a bó dó is leth a ban
⁊ clóechládh sé a gasced.'

'Fó lim gana[1] chlóthar úaib   1425
in Cú din Murthemni múaid,
ní hecal ria ṅgním ṅgarb ṅglé
rafetar más éside.'   M.

'Acaltar úait Cú Chulaind, a Ḟerguis,' ar Medb. 'Nithó,' ar
Fergus, 'acht acall-su fessin é,' for Fergus, 'dáig ní cían etruib 1430
immun glend sund, im Glend Fochaíne.' Ocus forḟópart Medb for
acallaim Con Culaind ⁊ dorigni laíd:

'A Chú Chulaind cardda raind,
dingaib dín do chrantabaill,
[2]ron báid-ne[2] do gleó garb glé   1435
ror[3] briss is ror[3] búaidre.'

'A Medb do Múr mac Mágach,
nídam drochláech dimbágach,
noco tréciub duit frim ré
immáin Tána Bó Cúalṅge'.   1440

'Mad dia ṅgabtha-su úanni,
a Chú chomramach Cúalṅgi,
leth do[4] bó ⁊ leth do ban
[5]rot biad dáig is ecengal[5]'.

'Dáig is mi re recht rubad   1445
ársid imdegla Ulad,
noco géb co tartar dam
cach bó blicht, cach ben Gáedel.'

---

[1] *sic, for* cia no; cen co *St, C; sic leg.*   [2-2] ron baidne nor forraig
MS., nor forraig *being a gloss incorporated* (*Windisch*)   [3] ron *St*
[4] da *MS.*   [5-5] rot bia duit uaind ar h'oman *St. C*

'Is romór a nad-maíde
1450 ar cur á[i]r¹ ar ṅdegdaíne,
formna ar n-ech is formna ar sét
²araí óenḟir² d'imchomét.'

'A ingen Echach Find Fáil,
nídam maith-se oc immarbáig,
1455 [acht]³ cidam láech-sa—líth ṅglé—
att úaitte mo chomairle.'

'Ní athis duit na 'tbere,
a meic droṅgaig Dechtere,
is robladach duit-siu in raind,
1460 a Chú chomramach Culaind.'   A.

Aithle na laíde sin: Níra gab Cú Chulaind nach comai conaittecht
fair. Ra díscaílset immo[n] ṅglend fón samlaid sin & balotar ass tria
chomḟeirg di leith for leth.

Gabsat cethri ollchóiceda Hérend dúnad ⁊ longphort trí lá ⁊ trí
1465 aidche ic Druim Én i Conaille Murthemne. Acht níro sádit sosta
nó pupla ⁊ ní dernad praind nó tomaltus leó ⁊ níra canait ceóil ná
arfiti leó na trí aidchi sin, & no marbad Cú Chulaind cét láech cach
70ᵇ n-aidchi díb co solustráth n-érgi arnabárach. | 'Níbat búana ar
slúaig fón samlaid seo,' ar Medb, 'dia mmarbad Cú Chulaind cét
1470 láech cach n-aidche úan. Cid ná berar coma do ⁊ nach acaltar
úaind é?' 'Ciarso choma sain?' ar Ailill. 'Berar 'n-as blicht dond
alaid dó ⁊ 'n-as dáer na braiti & cosced a chranntabaill d'ḟeraib
Hérend & léiced écin cotlud dona slúagaib.' 'Cia ragas frisin coma
sin?' ar Ailill. 'Cia,' bar Medb, 'acht Mac Roth ind echlach.' 'Ní
1475 rag omm,' bar Mac Roth, 'dáig nírsa eólach etir ⁊ ní ḟetar gia airm
inda fil Cú Chulaind.' 'Iarfaig do Ḟergus,' ar Medb, 'is dóig a ḟiss
lais.' 'Ní ḟetar-sa ém,' ar Fergus, '⁴ac[h]t óen⁴ ba dóig lem a bith
eter Ḟochaín ⁊ muir ic lécud gaíthe ⁊ gréne fóe ar nemchotlud na
aidchi arráir ic slaide ⁊ ic áirdbe in tslúaig a óenur.' Fír dó-som sain.
1480 Ferais tromṡnechta in n-aidchi sin corbo chlárḟind ⁵nó corbo
clárenech⁵ uili cóiceda Hérend don tṡnechtu. Ocus focheird Cú
Chulaind de na sec[h]t cneslénti fichet cíardai clárda bítis fo thétaib
⁊ rifetaib fria chnes arnacha ndechrad a chond céille tráth doficfad
a lúth láthair. Ocus legais in snechta trícha traiged ar cach leth
1485 úad ra méit brotha in míled ⁊ ra tessaidecht cuirp Con Culaind, &

---

¹ áir St, C   ²⁻² araí oenḟer MS., araí n-aoinḟir St, C   ³ om. MS.; sic St, C
⁴⁻⁴ acht aonní chena St; sic leg.   ⁵⁻⁵ gloss added in marg.

ní chaemnaic in gilla bith i comḟocus dó itir ra mét na feirge ⁊
bruthmaire in míled ⁊ ra tessaidecht in chuirp.
'Óenláech cucaind, a Chúcúcáin,' for Láeg: 'Cinnas láech?' ar
Cú Chulaind. 'Gilla dond drechlethan álaind. Bratt dond
derscaigthech immi, bruthgae umaidi 'na brut. Tarbléni trebraid ¹⁴⁹⁰
fria chness. Dá bernbróic etera dá choiss is talmain. Mátadlorg
ḟindchuill issindara láim. Claideb lethḟáebair co n-eltaib dét 'sind
láim anaill dó.' 'Aile a gillai,' ar Cú Chulaind, 'comartha n-echlaige
sin. Cia d'echlachaib Hérend sin ¹do imlúad¹ athisc ⁊ irlabra frim-sa.'
Doroacht Mac Roth iarum co ránic airm i mbáe Láeg. '²Ciarsat ¹⁴⁹⁵
comainm céli-siu², a gillai?' ar Mac Roth. 'Am chéli-se ind óclaíg
út túas,' ar in gilla. Tánic Mac Roth cosin magin i mbaí Cú Chulaind.
'²Ciarso comainm céli-siu², a óclaíg?' ar Mac Roth. 'Am céle-se
Conchobuir meic Fachtnai Fáthaig.' 'In fail ní as derbiu latt ná
sain?' 'Lór sain itráthsa,' ar Cú Chulaind. 'Ar co festa-su dam-sa ¹⁵⁰⁰
cia airm i faigbind in Coin Culaind airdirc seo immna³ n-egat fir
Hérend in cur sa barin tṡlúagud sa.' 'Cid atbértha-su friss nad
ebertha frim-sa?' ar Cú Chulaind. 'Dá acallaim tánac ó Ailill ⁊ ó
Meidb ra coma ⁊ ra caínc[h]omrac dó.' 'Cid dobertaisiu latt dó?'
'A n-as blicht dond alaid dó ⁊ a n-as dáer don brait, ⁊ coisced a chrann- ¹⁵⁰⁵
tabaill dona slúagaib, dáig ní súairc in torandchless dogní-sium
forro cacha nóna.' 'Cid airchind beth intí connaigi-siu i comḟocus,
ní gébad na comai conattgisiu, dáig mairfit Ulaid a mblechtach do
gressaib ⁊ glámmaib ⁊ gríssaib dar cend a n-aenig mani bé sescach
occu, & dano dobérat a mná dáera bar lepthaib dóib ⁊ ásfaid ¹⁵¹⁰
dáermaicne i crích Ulad a lleth ó máthrechaib samlaid.' Luid
Mac Roth ar cúl. 'Nád fúarais aile?' ar Medb. 'Fuar-sa ém gilla
grúamda ferggach n-úathmar n-anniaraid eter Ḟocháin ⁊ muir. Ní
ḟetar ém inn é in Cú Chulaind.' 'Inra gaib na comai sin?' 'Nád
ragaib écin.' Ocus innisis Mac Roth inní donára gaib. 'Is ésium ra ¹⁵¹⁵
acallais,' ar Fergus.
'Berar coma aile dó,' ar Medb. 'Ciarso choma?' bar Ailill. 'Berar
'n-a[s] seisc ind alaid dó ⁊ 'n-a[s] sáer na braiti, ⁊ [coisced]⁴ a
chranntabaill dona slúagaib dáig ní súairc in torandchless dogní
forro cacha nóna.' 'Cia ragas frisin coma sin?' 'Cia acht Mac Roth.' ¹⁵²⁰
'Ragad omm,' ar Mac Roth, 'dáig amm eólach don chur sa.' Tánic
Mac Roth d'acallaim Con Culaind. 'Dot acallaim tánac don chur
sa dáig rafetar is tú in Cú Chulaind airdairc.' 'Cid dobertais latt
samlaid?' ''N-a[s] seisc ind alaid ⁊ 'n-a[s] sáer na braiti, ⁊ coisc

¹⁻¹ tic d'imluad St; read tic do imlúad   ²⁻² sic, read Cia dianda chéli-siu?
(Cia dian celi-si St)   ³ = imma
⁴ illegible at line end MS.; coisced St

71ª do cranntabaill | do feraib Herend ┐ léic cotlud d'feraib Hérend ¹nó
dona slúagaib¹, dáig ní súairc in torandchless dogní-siu forro cacha
nóna.' 'Nád géb-sa na coma sain, dáig mairfit Ulaid a sescach dar
cend a n-aenig, ar it fíala Ulaid, & beit Ulaid can sescach ┐ can
blechtach itir. Dobérat a mnáa sáera ar bróntib ┐ lostib ┐ mugsaine
1530 ┐ dáeropair dóib. Ní maith lim-sa ind ail sin d'ácbáil i nUltaib dar
m'éis, cumala ┐ banmogaid do dénam d'ingenaib ríg ┐ rígthoísech
Ulad.' 'In fail coma gaba-su itir ifechtsa?' 'Fail écin,' ar Cú
Chulaind. 'In n-epir-ssu frim-sa in coma amlaid?' ar Mac Roth.
'²Dar [mo] bréthir²,' ar Cú Chulaind, 'ní mé adféta dúib.' 'Ceist
1535 didiu?' ar Mac Roth. 'Má tá ocaib 'sin dún ar medón,' ar Cú
Chulaind, 'rofessad na coma fail ocum-sa, ráded frib. Ocus máni
fail, ná tecar dom innaigid-se ní bas mó im choma nó im cháen-
chomrac, ar cipé tí, bid sé fot a sáeguil.' Luid Mac Roth ar cúl &
imfacht Medb scéla de. 'I[n] fúarais?' ar Medb. 'Fúar omm écin,'
1540 ar Mac Roth. 'Inra gab?' ar Medb. 'Nád ragab,' ar Mac Roth.
'I[n] fail coma gabas?' 'Fail dano atbeir.' 'In n-ébairt-sium frit-su
in choma sain?' 'Is hí ém a bríathar,' ar Mac Roth, 'nába é dosféta
dúib.' 'Ceist didiu?' ar Medb. 'Acht má tá lind ar medón rofessed
na coma fail lais-[s]ium, asberad frim, ┐ meni fail, ná tecar dá
1545 indsaigid níbad siriu nó bas mó. Acht is é seo óenní móedim-se
chena,' ar Mac Roth, 'cid rígi Hérend ³do[bertha dam]³, ná rag-sa
fessin dá maídib friss.'
  Is and sin tincais Medb for Fergus. 'Ciarso choma connaig sút,
a Ferguis;' ar Medb. 'Ní accim maith dúib itir din chomai connaig,'
1550 ar Fergus. 'Ciarso choma sin? ar Medb. 'Óenfer do feraib Hérend
do chomruc fris cach dia. I[n] fat bethir icá marbad ind fir sin.
imthecht do lécud don tslúag frissin. Mar thairc dano in fer sin
do marbad, láech aile for áth dó-som nó nechtar de longphort ┐
dúnad do gabáil d'feraib Hérend and sin co solustráth érge arna-
1555 bárach, ┐ a biathad ┐ a étiud Con Culaind forin tánaid se béus
úaib-si.'
  'Is í ar cubus,' ar Ailill, 'is coma dímaig.' 'Is maith an condnaig,'
ar Medb, '& atetha-som na comai sin, dáig ar bíth iss assu lind
óenláech úaind cach laí dó-som oldás cét láech cach n-aidchi.'
1560 'Cia ragas frisnaib comai sin dia innisin do Choin Chulaind?' 'Cia
dano acht Fergus,' ar Medb. 'Nithó,' for Fergus. 'Cid són?' for
Ailill. 'Co tartar cuir ┐ glinni, rátha ┐ trebairi imm airisium arna
comai sin ┐ 'ma tabairt di Choin Chulaind.' 'Ataimim-si ém,' ar
Medb, ┐ aurnaidmis Fergus fón samlaid cétna foraib.

---

¹⁻¹ *added overhead*   ²⁻² Dar mo breit[h]ir am *St*   ³⁻³ *sic St*

RO gabad echrad Fergusa ⁊ ra indled a charpat. Ocus ro gabait a **1565**
dá ech do Etarcumul mac Feda ⁊ Lethrinni, máethmaccáem di
muntir Medba ⊣ Ailella. 'Cid imluid-siu?' ar Fergus. 'Lodma
lat-su,' ar Etarcumul, 'd'fégad chrotha ⊣ delba Con Culaind ⊣ do
thaidbriud fair.' 'Día n[d]ernta-su form-sa,' ar Fergus, 'ní targtha
'manetir.' 'Cid són amai?' 'Do sobcha ⊣ do saisillecht, a lunni **1570**
immorro ⊣ a ágmairi ⊣ a ainserci in meic dá tégi innaigid, & is dóig
lim-sa debaid dúib ria n-imscar.' 'Nach fétfa-su ar n-etráin?' ar
Etarcumul. 'Rafétad,' for Fergus, '¹[acht] nád chunnis fodessin¹.'
'Nad chunnius ón co brunni mbrátha.'

Lotar iarum rempu iar tain do saigid Con Culaind. A mboí **1575**
Cú Chulaind eter Fochaín ⊣ muir oc imbirt búanbaig fria araid, ⊣
²ní théiged² isin mag can arigud do Láeg ⊣ no bered cach ra cluchi
for Coin Culaind asin búanbaig béus arapa. 'Óenláech cucund, a
Chúcúc,' ar Láeg. 'Cinnus láeich?' ar Cú Chulaind. 'Métithir lim
71ᵇ óen na prímslíab is mó bís for mórmachairi | in carpat fil fónd **1580**
óclaig. Métithir lim óen na prímbili bís for faidchi prímdúni in folt
craíbach dúalach findbudi forórda forscaílti fail immo chend.
Fúan corcra corrtharach inaithi³ immi. Delg n-órda n-ecortha 'sin
brut. Manaís lethanglas ar derglassad 'na láim. Scíath cobradach
condúalach co cobraid óir deirg úasu. ⁴Claideb fata sithlaí⁴ ⁵co **1585**
n-ecrasaib serrda⁵ for díb sliastaib sudigthi dond óclaíg móir
borrfaid fail isin charput ar medón. 'Ale mo chen a thíchtu inar
ndochum-ni ind oíged sin,' ar Cú Chulaind. 'Rafetamar-ni in fer sin.
Mo phopa-sa Fergus dotháet and sin.' 'Atchíu-sa óencharpdech
aile 'nar ndochum-ne béus. Is lór n-árgigi ⊣ n-óebinniusa ⊣ n-ániusa **1590**
amthiagat a eich.' 'Cia do maccáemaib fer [n]Hérend sin, a mo
phopa Laíg,' ar Cú Chulaind. 'D'fégad mo chrotha-sa ⊣ mo delba
dotháet in fer sain dáig am urdairc-sea leó-som 'na ndún ar medón.'
Doriacht Fergus ⊣ tarblaing assin charpat, ⊣ ferais Cú Chulaind
fáilti fris. 'Tarisi lim,' ar Fergus. 'Tarisi duit-siu ón om,' ar **1595**
Cú Chulaind, 'dáig dia tóichle inn iall én 'sin mag, rot bia cadan
co lleth araile. Dia tuinne iasc i n-inberaib, rot bia éicni co leith
anaill. Rot bia dorn bilair ⊣ dorn femmaig ⊣ dorn fothlochta.
Dámsat éicen⁶ comrac nó chomlond, missi ragas dit ráith for áth &
rot bia foraire ⊣ forcomét co táthais do suan ⊣ do chotlud.' 'Maith **1600**

---

¹⁻¹ acht nat cuinge fen ugra no imreasain *St*   ²⁻² *read* ní théiged ní *or*
ní théiged nech? (ni tigedh ni *St;* ni theighed bethadhach *H*.2.17)
³ *sic; read* i faithi?   ⁴⁻⁴ *read* claideb fata sithithir laí (claideb **sithidir**
loí churaig *LU;* cloidem foda comfada fri raim curaig *St*)
⁵⁻⁵ *corrupt for* i n-ecrus sesta (*Windisch*); cp. claideb orduirnd i n-**ecrus sesta**
fora dib sliastaib *LU* 10226-7, *Tochm. Em.*   ⁶ éicni *MS.*

amin rofetamar mar atá th'óegedchaire in chur sa for Táin Bó
Cúalnge. Acht in cor sa conattecht for firu Hérend, comlund
óenḟir, atetha. Is dó thánac-sa dia naidm fort, & geib-siu fort.'
'Ataimim omm,' bar Cú Chulaind, 'a mo phopa Ferguis.' Ocus
1605 ní baí ní ba siriu ná sain ac comlabra arná ráditis fir Hérend a mbrath
nó a trécun do Ḟergus fria dalta. Ro gabtha a dá ech do Ḟergus
⁊ ro indled a charpat, ⁊ luid for cúlu.

Dessid Etarcumul dia éis ic fégad Con Culaind fri ed cían. 'Cid
fégai-siu, a gillai?' for Cú Chulaind. 'Fégaim-se tussu,' for
1610 Etarcumul. 'Ní fota in rodarc ém duit-siu ón,' ar Cú Chulaind,
'immonderca súil i sodain duit. Acht¹ dia festa-su, is andíaraid in
míl bec ḟégai-siu [.i.]² missi. Ocus cinnas atú-sa acut frim ḟégad
didiu?' 'Is maith lim ataí immorro. Maccáem tucta³ amra álaind
tú co clessaib ána imḟacsi ilarda. Mad t'árim immorro bail i mbiat
1615 daglaích nó dagoíc nó láith gaile nó ord⁴ essoirgne, nít áirmem itir ⁊ nít
imrádem.' 'Rofetar-sa⁵ is commairgi dait immar thánac assin
longphort ar einech mo phopa Fergusa. Toṅg-sa mo dee dá
n-adraim chena, menbad [ar] bíth einig⁶ Fergusa, ní ricfad acht do
chnámi mintai ⁊ t'áigi fodailti arís dochum longphuirt.' 'Aile
1620 nacham thoma-sa itir ní bas íriu de sodain dáig in cor sa ra chungis
for firu Hérend, comlund óenḟir, ní ḟil d'ḟeraib Hérend tí imbárach
dit ḟópairt acht missi.' 'Tair-siu ass ón, ⁊ gid moch thís, fogéba-su
missi sund. Ní thechiub-sa riam remut.' Luid Etarcumul ar cúlu
⁊ ro gab ar chomrád fria araid. 'Isam écen-sa trá imbárach comrac
1625 fri Coin Culaind, a gillai,' bar Etarcumul. 'Ra gellais trá,' ar in
72ᵃ t-ara. | 'Ní ḟetar-sa chena in comella.' 'Ocus cia ferr a dénam
imbárach nó innocht fo chétóir?' 'Is í ar cubus,' ar in gilla, 'acht
ní búaid a dénam imbárach, is mó is dimbúaid a dénam innocht
dáig⁷ is nessu do urgail.' 'Impá dún in carpat, a gillai, arís for
1630 cúlu dáig ar bíth toṅgu-sa na dé dá n-adraim, ní rag-sa ar cúl co
brunni ṁbrátha co rucur cend na herre út lim i tasselbad, cend
Con Culaind.'

Imsoí in t-ara in carpat arís dochum inn átha. Tucsat a clár
clé fri airecht ar amus ind átha. Rathaigis Láeg. 'In carpdech
1635 dédenach baí sund ó chíanaib, a Chúcúc,' ar Láeg. 'Cid de-side?'
ar Cú Chulaind. 'Dobretha a chlár clé riund ar ammus ind átha.'
'Etarcumul sain, a gillai, condaig comrac cucum-sa, & ní ramaith
lim-sa dó ar bíth ainig m'aiti ar[a] tánic assind longphurt ⁊ ní
[a]r bíth a imdegla-som atú-s[a] itir. Tuc-su latt, a gillai, m'arm

---

¹ acht is MS., is omitt.   ² sic St, om. LL   ³ sic; tuchtach LU, sic leg.
⁴ uird St, sic leg.   ⁵ sic, for rofetar-su (ro-fhetraisi St)   ⁶ éinig MS.
⁷ ar is dáig MS. ar is omitt., an anticipation of aris in foll. line

dam-sa connici in n-áth. Ní miad lim-sa diam túscu dó icond áth ná [1640]
dam-sa.' Ocus luid iarum Cú Chulaind connicè in n-áth ⁊ nochtais
a chlaideb ósa gelgúalandchor ⁊ baí urlam forsinn áth for cind
Etarcomla. Dorocht dano Etarcumul. 'Cid iarrai, a gillai?' ar
Cú Chulaind. 'Comrac frit-su iarraim-se,' bar Etarcumul. '¹Na
dernta form¹, ní thargtha itir,' ar Cú Chulaind, 'ar bíth ainig [1645]
Fergusa ara tánac assin longphurt ⁊ ní ar bíth t'imdegla-su itir
itú-sa.' Tuc trá Cú Chulaind fotalbéim dó goro thesc in fót boí
fo bund a chossi conid tarla bolgfáen is a fót fora broind. Dámbad
áil dó is dá orddain dogénad de. 'Dó duit ifechtsa ar dob*iurt*-sa
robud dait.' 'Ní rag-sa condrísam béus,' bar Etarcumul. Tuc Cú [1650]
Chulaind fáebarbéim co commus dó. Tópach[t] a folt ó chúl có
étan de ón chlúais co araile marbad do altain áith étruim nad
berrtha. Níro fulig trác[h]tad fola fair. 'Dó duit ifechtsa.' for
Cú Chulaind, 'ar dob*iurt*-sa gén fort.' 'Ní rag-sa condrísam béus
ón co rucur-sa do chend-su ⁊ do choscur [⁊] do chommaídim nó co [1655]
ruccu-ssu mo chend-sa ⁊ mo choscur ⁊ mo chommaídim.' ''S ed
trá bias de a n-atberi-siu fo deóid, missi béras do chend-su ⁊ do
choscur ⁊ do chommaídim.' Tucastar Cú Chulaind múadalbéim
dó i comard a chind co rocht a imlind. Tucastar béim tánaise dó
urtharsna conid i n-óenfecht rángatar a trí gáibti rainti co talmain [1660]
úad. Dorochair Etarcomul mac Feda ⁊ Lethrinne samlaid.

Ocus ní fitir Fergus in comrac do dénam, dáig ba deithbir són ar
níro fég Fergus dara ais ríam ic suidi *nó* ic érgi *nó* ic astar *nó* ic
imthecht ⁊ chléith ⁊ chath *nó* chomlund ar nád ráided nech ba
fatchius dó fégad dara éiss, acht na mbíd ríam remi ⁊ aird fri haird [1665]
friss. Rasiacht gilla Etarcomla aird i n-aird fri Fergus. 'Cá airm
inda fil do thigerna-su immanitir, a gillai?' ar Fergus. 'Dorochair
ó chíanaib forsinn áth la Coin Culaind,' ar in gilla. 'Nírbu chóir
ém,' ar Fergus, 'don serriti síabarda mo sárgud immontí thánic
for m'[f]oísam. Impá dún in carpat, a gillai,' ar Fergus, 'ar [1670]
condrísam immacallaim fri Coin Culaind.'

Imsoí iarum in t-ara in carpat. Lotar dó rempo a dochum ind
átha. 'Cid latt mo sárgud, a serriti síabarda,' ar Fergus, 'immontí

72ᵇ tánic for m'[f]oísam ⁊ for mo chommairgi?' | 'Dond altram ⁊ dond
iarfaigid dobertaisiu form-sa, ráid dam cia de bad ferr lat-su, mo [1675]
choscur-sa ⁊ mo chommaídim-se dó-som oldás a choscur-som ⁊ a
chommaídim-sium dam-sa. Ocus anaill béus, ²iarfaig-siu a gilla-
som² cia bad chintach úan fri araili.' 'Ferr lemm na ndernais.
Bendacht for[in] láim dofárraill.'

---

¹⁻¹ *sic; read* Ma dernta form (*Windisch*); *cp. supra* dia ndernta-su form-sa
²⁻² iarfaig día araid *LU;* fiarfaigh-si dia giolla fen *St, H.*2.17.

1680 And sin trá ra ceṅglait dá n-id im chaílaib choss Etarcomla ⁊ ra sreṅgad i ndegaid a ech ⁊ a charpait. Cach all ba amréid dó no fácbaitis a scaim ⁊ a thrommai im ailib ⁊ im airtdrochib. Cach bali ba réid dó na chomraictis a gabti rainti 'mon echraid. Ra sreṅgad samlaid dar fiartharsna longphuirt co dorus pupla Ailella 1685 ⁊ Medba. 'Fail and sain trá,' bar Fergus, 'far maccaím dúib, ar cach assec cona thassec is téchta.' Dotháet Medb immach co dorus a pupla ⁊ dobreth a hardguth for aird. 'Dar lind ém,' bar Medb, 'ba mór bruth ⁊ barand in chuliúin-se ¹tús laí¹ dia ndechaid assin longphurt. Andar lind ní ainech athfir in t-ainech forsa ndechaid, 1690 ainech Fergusa.' 'Cid ra mer in cali ⁊ in banaccaid?' bar Fergus. 'Cid ón, ciarso dúal don athiuchmatud saigid forsin n-árchoin ná lamat cethri ollchóiceda Hérend tascud nó tairisin dó? Cid mi fadéin, ba maith limm tíchtain imṡlán úad.' Torchair trá Etarcumul fón samlaid sin.
1695 Conid Comrac Etarcomla fri Coin Culaind sin.

ANd sin atraacht láech prósta mór do muntir Medba, Nath Crantail a chomainm, & tánic do fúapairt Chon Culaind. Nír fiú leis airm do thabairt leis itir acht trí noí [m]bera culind at é fúachda follscaide forloiscthi. Ocus is and boí Cú Chulaind 1700 forsin lind fora chind.—²Ocus níba fáesam cid sí & bátar noí ṁbera trethi. Ní bíd esbaid Con Culaind for ³nach óenbir³ díb².—And sain focheird-sium bir for Coin Culaind. Cingis Cú Chulaind co mbaí for ind úachtarach in bera contarlaic, & tarlaic Nath Crantail béus in bir tánaise. Tarlaic Nath Crantail in tres ṁbir & ciṅgis 1705 Cú Chulaind do ind in bera tánaise co mbaí for ind in bera dédenaig.
Is and sin [techis]⁴ inn íall én 'sin mag. Luid Cú Chulaind 'na ndíaid mar cach n-én conná ragtaís úad co fargdais cuit na aidchi dádaig⁵. Dáig iss ed arfurad ⁊ arfognad Coin Culaind iascach ⁊ énach ⁊ osfeóil for Táin Bó Cúalnge. Ac[h]t atá ní, fo glé ra Nath 1710 Crantail iss i roí madma ⁊ techid dochúaid Cú Chulaind úad, & luid reme co dorus pupla Ailella ⁊ Medba ⁊ dobreth[a] a ardguth ar aird. 'In Cú Chulaind airdairc-se atberthai-si,' ar Nath Crantail, 'dochúaid i rroí madma ⁊ techid riam reme⁶ ambuarach.' 'Rafetam-mar,' ar Medb, 'rapad fír, acht con[d]arístaís daglaích ⁊ dagóic, 1715 ní gébad fri féta in serriti óc amulchach sain, ár in am dosfarraid dagláech, ní riss ra gabastar acht is riam reme ro madmastar.' Ocus rachúala Fergus anísin ⁊ ba níth mór la Fergus óen do maídim

¹⁻¹ i ttús laoí St   ²⁻² this parenthesis om. LU, St   ³⁻³ óen nach bir MS.
⁴ om. MS., supplied from St   ⁵ aice add. St   ⁶ read remom (reomsa LU 5736, romamsa St 1755)

thechid fri Coin Culaind, & rádis Fergus fri Fiachu mac Fir Aba

73ª ar co ndigsed do acallaim Con Culaind. 'Ocus ráid-siu | friss fíal
dó bith forsna slúagaib cian gar dorigéni gnímrada gaile forro & ba ¹⁷²⁰
féile dó a immḟolach oldás teched ria n-óenláech díb.' Dotháet
iarum Fiachu do acallaim Con Culaind. Ferais Cú Chulaind fáilte
fris. 'Tarissi lim-sa ind ḟálti sin, ac[h]t dot acallaim tánac ót aiti
ó Ḟergus, & atbert fíal duit bith forsna slúagaib cian gar doringnis
gnímrada gaile ⁊ ba féliu duit th'immḟoluch oldás teiched ria ¹⁷²⁵
n-óenláech díb.' 'Cid ón, cia nod maíd acaib-si sin?' bar Cú
Chulaind. 'Nath Crantail ém,' bar Fiachu. 'Cid ón, ná fetar-su ⁊
Fergus ⁊ mathi Ulad ná gonaim-se aradu *nó* echlacha *nó* áes gan
armu, & ní airm baí lais-[s]ium acht bir craind, & ní gonfaind-se
Nath Crantail co mbeth arm leiss. Ocus ráid-siu friss arcom thé ¹⁷³⁰
co moch imbárach sund & ní thechiub-sa riam reme.' Ocus ba fata
ra Nath Crantail corbo lá cona ṡollsi dó do ḟúapairt Con Culaind.
Tánic co moch arnabárach do ḟópairt Con Culaind. Atraig Cú
Chulaiṅd co moch ⁊ dofáncatar a ḟerga laiss in lá sain ⁊ focheird
fáthi ferge¹ dia brutt taris co tarla darin corthi clochi ⁊ co tópacht ¹⁷³⁵
in corthe clochi a talmain eturru 's a bratt, & ní fitir sin itir ar méit
na ferggi dofánic & ra síabrad immi. And sain trá dotháet Nath
Crantail & atbert: 'Cia airm i tá in Cú Chulaind se?' for Nath
Crantail. 'Aracút tall aile,' ar Cormac Cond Longas mac Conchobuir.
'²Ní hé sút cruth ardomḟarfaid-se² indé,' ar Nath Crantail. 'Diṅgaib- ¹⁷⁴⁰
siu trá in láech út,' bar Cormac, '⁊ is samalta duit ⁊ feib no diṅgébtha
Coin Culaind.'

Tánic iarum Nath Crantail ⁊ focheird rót n-urchair dia chlaidiub
úad for Coin Culaind conda tarla immun corthi boí eter Coin
Culaind ⁊ a bratt co róebriss in claideb immon corthi. Ciṅgid Cú ¹⁷⁴⁵
Chulaind do lár thalman co mbaí for úachtar cobraidi scéith Nath
Crantail ⁊ dobretha táthbéim dó sech barrúachtur in scéith co
tópacht a chend dia médi. Túargab a lám³ co immathlam darís
⁊ dothuc bulli n-aill i mmulluch in mét*h*i co ndergeni dá gabait
rainti⁴ co talmain. Torchair Nath Crantail fón samlaid sin la ¹⁷⁵⁰
Coin Culaind. Atbert Cú Chulaind assa aithle:

'Má dorochair Nath Crantail,
⁵[bid formach dond imargail.]⁵
Apraind can chath ⁵[isind úair]⁵
do Meidb co tríun in tṡlúaig.'   **¹⁷⁵⁵**

---

¹ *omitt.*? (focheird fáthi n-imbi *LU*)
²⁻² nípu samlaid domarfás *LU*, ni hé sut cruth ro taidbredh dam-sa fair *St*
³ *sic, for* láim   ⁴ de *add. St*   ⁵⁻⁵ om. *MS., supplied from LU (St, C)*

DOlluid iarum Medb co tríun in tṡlúaig [1]fer ṅHérend impi co
ránic inní co Dún Sobairc[h]i fathúaid. [2]Ocus lilis Cú Chulaind
co mór do Meidb in lá sain co luid Medb i nGuiph ria Coin
Culaind[2] comdar[3] techt fathúaith marbais Cú Chulaind Fer Taidle
1760 dia tát Taidle, & marbais Maccu Buachalla dia tá Carn Mac
ṁBuachalla, & marbais Lúasce i lLettri dia tát Lettre Lúasce, &
marbais Bó Bulge ina grellaig dia tá Grellach Bó Bulge, & Murthemne
fora dind dia tá Delga Murthemne.

Conid iar sain da tarraid Cú Chulaind atúaid dorísi do imdegail
1765 ⁊ do imdítin a c[h]rích ⁊ a f[h]eraind fodessin, dáig ba handsa lais
andá crích ⁊ ferand neich n-aile.

Is and sin trá forecmangaid Firu Crandce .i. dá Artinne ⁊ dá
mac Licce, dá mac Durcridi, dá mac Gabla ⁊ Drúcht ⁊ Delt
⁊ Dathen, Te ⁊ Tualaṅg ⁊ Turscur ⁊ Torc Glaisse ⁊ Glass ⁊
1770 Glassne, inund sain ⁊ fiche Fer Fochard. Basnetarraid Cú Chulaind
73ᵇ ic gabáil | longphuirt ria cách co torchratar lais.

Is and sin dorecmaiṅg do Choin Chulaind Buide mac Báin Blai
de chrích Ailella ⁊ Medba ⁊ do ṡainmuntir Medba. Cethror ar ḟichet
láech. Bratt i filliud im cach fer. Dond Cúalnge i rrithur ⁊ i
1775 fúatuch rempu iarna thabairt a Glind na Samaisci i Sléib Chulind,
⁊ coíca samaisci dia ṡamascib imme. 'Can doberid in n-alaid?'
for Cú Chulaind. 'As[in] tṡléib út amne,' ar Buide. 'Cá do chomainm-
siu badessin?' bar Cú Chulaind. 'Nít charadar, nítt ágedar,' ar
Buide, 'Buide mac Báin Blai missi do chrích Ailella ⁊ Medba.'
1780 'Asso fort in certgae so di*diu*,' bar Cú Chulaind, ⁊ focheird in sleig
fair. Forecmaing 'sin scíath ósa broind coro brúi trí asna 'sin táeb
ba siriu úad iar tregtad a chridi 'na chlíab. Ocus dorochair Buidi
mac Báin Blai. Conid de atá Áth mBuide i crích Ross ó ṡain.

Cian gar ro bátar forinn uropair sin ic clóechlód na dá chertgae,
1785 dáig ní fo chétóir conarnic úadib, rucad in Dond Cúalnge i rrithur ⁊ i
fúatach úadib dochum longphuirt uadib[4] amal as dech berair mart
longphuirt. Conid ésin méla ⁊ mertain ⁊ meraigecht is mó tucad for
Coin Culaind forsin tṡlúagud sa.

Imthúsa Medbi cach áth forsa mbaí, Áth Medbi a chomainm.
1790 Cach bail ro ṡáid a pupaill, is Pupall Medba a ainm. Cach bail
ro ṡáid a echlaisc, is Bili Medba a chomainm.

Ra chuir trá Medb din chúaird sin cath fri Findmóir mnaí Celtchair for dorus Dúni Sobairchi ⁊ ro marb Findmóir ⁊ ra airg Dún
Sobairchi.

---

[1] *remainder of col. much stained and difficult to decipher* [2-2] *Cp.* Is and sin
luid Medb co tríun in tsloig le hi Cuib dó chungid in tairb ⁊ luid Cu Chulaind ina
ndiaid *LU* 5787-8; Lenais Cu Chulainn go mor diob in la sin. Luid Medb
i cCúib d'iarraid in tairb *Stowe* 1796-7 [3] *sic; read* conid ar [4] *sic, omitt.*

TÁncatar trá cethri ollchóiceda Hérend i cind chían chóic- ¹⁷⁹⁵
thigis dochum dúnaid ⁊ longphuirt eter Meidb ⁊ Ailill ⁊ fiallach
tabartha in tairb, ⁊ ní arlaic a búachaill dóibide Dond Cúalnge,
co ndasrimmartatar co cruind for scíath fair conda mbertatar i
mbernaid cumaiṅg, co ndaralastar¹ na halma i talmain a chorp
tríchait traiged co ndernsat minscomartaig ⁊ minbruan dia churp. ¹⁸⁰⁰
Forgemen a chomainm.
Conid Bás Forgaimin in sin for Táin Bó Cúalnge.

Odariachtatar fir Hérend go óenbaile, eter Meidb ⁊ Ailill ⁊
fiallach tabartha in tairb, dochum in dúnaid ⁊ longphuirt,
atbertatar uili nábud chalmu chách Cú Chulaind menibeth ¹⁸⁰⁵
in clessín iṅgantach baí aice, clettín Con Culaind. Conid and sin
foídset fir Hérend úadib Redg cánti Medbi do chungid in chlettín.
Conattecht Redg in clettín & nád tard dait[h] Cú Chulaind in clettín
dó. Ní sain ⁊ nád éscaid laiss a thabairt. Rádis Redg no bérad
ainech Con Culaind. And sin tarlaic Cú Chulaind in clettín dó 'na ¹⁸¹⁰
díaid conid tharlathar i classaib a dá chúlad co ndechaid dara bél
a dochum talman. Ocus ní tharnaic úad acht a rád: 'Is sólom dún
in sét sa,' tráth conroscar a anim fria chorp forsin áth. Conid de
asberar in t-áth sin ó ṡin Áth Sólomṡét. Ocus fochuridar a uma²
don chlettín forsin sruth, conid de atá Umanṡruth ó ṡin. ¹⁸¹⁵

74ᵃ A Trubratur fir Hérend cia bad chóir d'ḟúapairt Con Culaind |
accu, & atbertatar uile combad é Cúr mac Dá Lóth bad chóir
dá ḟúapairt, dáig amlaid buí Cúr níba súair[c] comlepaid nó
comáentu friss. Ocus atbertatar cid sé Cúr táetsad, ba diṅgbáil
trommad³ dona slúagaib; diambad é Cú Chulaind, bá ferr són. ¹⁸²⁰
Conácart Cúr i pupaill Medba. 'Cid táthar dam-sa?' ar Cúr. 'Do
ḟúapairt Con Culaind,' ar Medb. 'Is cert ár⁴ [m]búaid lib, is amra
lib tráth is ro máethmaccáem a ṡamla sain nom ṡamlaid. Dia
fessaind-se fessin, ní thicfaind la sodain. Bad [leór]⁵ lim gilla a
chomaís dim muntir do thecht 'na agid for áth.' 'Ale is acca a rád ¹⁸²⁵
samlaid sin,' ar Cormac Cond Longas mac Conchobuir. 'Rabad
amra bríg duit fadessin mad dia tóetsad latt Cú Chulaind.' 'Dénaid-
si arrgraige n-imthechta frí úare⁶ na matne i mmucha imbárach,
dáig suba sliged dogníu-sa de. Ní hed nobar furgfe sib guin na
hailiti út, Con Culaind.' And sin atraacht co moch arnabárach ¹⁸³⁰

---

¹ *read* ndaralatar   ² úma *MS.*   ³ = tromma (trumma)   ⁴ *sic MS., and*
diminutive ar-*compendium in marg.*   ⁵ *om. MS., supplied from LU, St* (robadh
lor lim C *with* lor *added overhead with caret mark*)   ⁶ *sic; read* úair (húair *St*)

Cúr mac Da Lóth. Tucad aire feóin leis do threlam gascid do
fópairt Con Culaind, & barópairt ac folmasi a gona. Dochúaid
Cú Chulaind trá fora chlessaib co mmoch in lá sain. At eat a n-uli
anmand .i. ubul[l]chless ᚄ fóenchless ᚄ cless clettínech ᚄ tétchless ᚄ
¹⁸³⁵ corpchless ᚄ cless caitt ᚄ ích n-errid ᚄ cor ṅdelend ᚄ léim dar néim
ᚄ filliud eirred náir ᚄ gai bulgga ᚄ baí brassi ᚄ rothchless ᚄ gless¹
for análaib ᚄ brúud gine ᚄ sían curad ᚄ béim co fommus & táthbéim
ᚄ réim fri fogaist co ndírgud chretti fora rind co fornaidm níad.
Ar is aire dogníd Cú Chulainn cacha maitne ar mucha cach cless
¹⁸⁴⁰ díb ²ar lus na lethláim³ amal as dech téit catt cróich² ná digsitís
ar dermat nó díchumni úad. Ocus tarrasair mac Da Lóth co trían
in laí i túaim a scéith ic folmaisse gona Con Culaind. Is and sain
rádestar Láeg fri Coin Culaind: 'Maith a Chúcúc, frithálti in láech
fail ic folmaisi do gona.' Is and sin tincais Cú Chulaind fair—is
¹⁸⁴⁵ inund ón ᚄ no fégand—& is and sain tórgaib ᚄ tarlaic na hocht
n-ubla i n-airddi. ⁴Dolléci in nómad uball⁴ róot n-urchair úad do
Chúr mac Da Lóth co tarla i llaind a scéith ᚄ a étain, co ruc com-
thromm inn ubaill dia inchind triana chúladaib. Co torchair dano
Cúr mac Da Lóth fón samlaid sin ra Čoin Culaind.
¹⁸⁵⁰ 'Dán fargabat far cuir ᚄ far rátha ifechtsa.' bar Fergus, 'láech
aile for áth dó sút, nó gabaid dúnad ᚄ longphort sund co solustráth
n-éirge imbárach, dáig darochair Cúr mac Da Lóth.' 'Arapa a fáth
táncammar,' ar Medb, 'is cubés dún cid isna puplaib cétnaib
bemmit.' Dessid dóib issin longphort sain co torchair Cúr mac Da
¹⁸⁵⁵ Lóth & Lath mac Da Bró & Srub Dare mac Fedaig .i.⁵ mac teóra
Maignech⁶. Torc[h]ratar sain trá ra Coin Culaind ar galaib óenḟir.
Acht is emilt engnam cach ḟir fo leith díb d'innisin.

IS and sin rádis Cú Chulaind fria araid fri Láeg: 'Dó duit, a
phopa Laíg,' ar Cú Chulaind, 'i llongphort fer ṅÉrend ᚄ beir a
74^b n-imc[h]omarc | úaim-se dom áes chomtha ᚄ dom chomaltaib
ᚄ dom chomdínib. Beir a imchomarc do Ḟir Diad mac Damáin &
do Ḟir Dét mac Damáin & do Bress mac Ḟirb, do Lugaid mac Nóis
& do Lugaid mac Solamaig, do Ḟir Báeth mac Baetáin & do Ḟir
Báeth mac Ḟir Bend, & a imchomarc féin béus dom derbchomalta,
¹⁸⁶⁵ do Lugaid mac Nóis, dáig is é óenḟer coṅgeib commond ᚄ caratrad
frim-sa don chur sa forin tslúagud, & beir bennachtain ⁷ar co
n-eperta-som⁷ frit-su cia dotháet dom ḟúapairt-se imbárach.'

---

¹ cleas *St*   ²⁻² om. *Rec. I, St, C, Rec. III*   ³ *sic, read* lethláime?
⁴⁻⁴ *add. in marg., read by Facsimilist*   ⁵ *sic; read* ᚄ   ⁶ *sic;* nAignech
LU etc., Rec. III   ⁷⁻⁷ *sic, intended for* ar co n-eprea-som; *scribe probably*
began to write ar co n-epertha-su fris

Luid iarum Láeg reme i llongphort fer ṅHérend ⁊ ruc a n-imchom-
arc d'áes chumtha ⁊ do chomaltaib Con Culaind & dano dochúaid
i pupaill[1] Lug*dach* meic Nóis. Ferais Lugaid fálte fris. 'Tarissi lim,' [1870]
ar Lóeg. 'Tarissi duit-siu ón,' bar Lugaid. 'Dot acallaim tánac
ó Choin Chulaind,' ar Láeg, '& tucad t'imchomarc do glaine ⁊ do
léire úad duit, ⁊ ar co n-epertha-su frim-sa cia dotháet dá fúapairt
Con Culaind indiu[2].' 'Mallach[t] a chommaind ⁊ a chomaltais ⁊
a charatraid ⁊ a chardessa fair, a derbchomalta díless dúthaig [1875]
fadessin .i. Fer Báeth mac Fir Bend. Rucad i pupaill Medba ó
chíanaib. Tucad ind ingen Findabair ara lethláim. Is í dortes
curnu fair. Is í dobeir phóic la cech n-óendig dó. Is í gaibes láim fora
chuit. [3]Ní do chách[3] la Meidb in lind dálter for Fer ṁBáeth. Ní
thucad acht aire coícat fén de dochum longphuirt.' [1880]
  Luid iarum Láeg ar cúl do ṡaigid Con Culaind co cendtromm
n-imthursech n-anfálid n-osnadach. 'Is cendtromm n-imthursech
n-anfálid n-osnadach dotháet mo phopa Láeg dom indsaigid-se,'
bar Cú Chulaind. 'Is nech trá écin dom chomaltaib dotháet dom
fúapairt.'—Ar ba messu lais-sium fer a chomgascid andá láech [1885]
anaill.—'Maith and, a mo phopa Laíg.' ar Cú Chulaind, 'cia dotháet
dom fúapairt-se indiu[4]? 'Mallacht a chommaind ⁊ a chomaltais ⁊
a charatraid ⁊ a chardessa fair, do chomalta díles dúthaig fadessin
.i. Fer Báeth mac Fir Bend. Rucad i pupaill Medba ó chíanaib.
Tucad ind ingen fora lethláim. Is sí dortes curnu fair. 'S í dobeir [1890]
phóic la cech n-óendig dó. Is í geibes láim fora chuit. [3]Ní do
chách[3] la Meidb in lind dáilter for Fer ṁBáeth. Ní tucad acht aire
coícat fén de dochum longphuirt.'
  Ní tharrasair Fer Báeth co mmatin itir acht luid fo chétóir
d'athchur a charatraid for Coin Culaind. Ocus conattecht Cú [1895]
Chulaind in caratrad[5] ⁊ in commund ⁊ in comaltus friss, ⁊ nír
fáemastar Fer Báeth [cen][6] in comrac do dénam. Luid Cú Chulaind
tria ḟeirg úad ⁊ fosnessa sleig culind ina bond traiged coras fothraic
eter feóil ⁊ chnám ⁊ chroicend. Tarngid Cú Chulaind in sleig arís ar
cúlu assa frémaib ⁊ dosfarlaic dara gúalaind i ndegaid Fir Baíth, ⁊ [1900]
fó leis gid no ríssed ⁊ ba fó leis ginco ríssed. Dotarlai*c* in sleg i
classaib a chúlaid co ndechaid trina bél dochum talman co torchair
Fer Báeth amlaid. 'Maith trá in f[h]ocherd, a Chúcúc,' bar Fiacha
mac Fir Aba, ar ba focherd lais in cathmílid do marbad den bir
culind. Conid de atá Focherd Murthemne béus arin inad i mbátar. [1905]

---

[1] pupl- MS. (pupaill *St*)   [2] amarach *St*   [3-3] ní do chách berar *St*
[4] amarach *St*   [5] charatrad *Facs. and Dipl. Edn.; MS. stained and*
difficult to decipher here   [6] *om. MS.* (nir fhoem F. B. cen in comrac
do denam *St* 1930)

Irg-siu dam-sa, a mo phopa Láeig, da acallaim Lugdach i
llongphort fer ṅHérend ⁊ finta latt in ránic ní Fer Báeth fo ná
ránic¹[⁊ fiarfaid de cia tig im aghaidh-si amárach. Tét Láog
roime go pupaill Lugdach. Ferais Lugaidh fáilte fris. 'Tairisi liom in
1910 fáilte sin,' ar Láog. 'Tairisi duit ón,' ar Lugaidh. 'Dot hagallaimh
táncus-[s]a ót c[h]omalta co nn-innisi dam in ránic Fer Báoth (an
longphort).' 'Ráinic ón,' ar Lughaidh, '⁊ bendacht arin láim
dusfaraill úair torchair marb isin gleand ó chianaibh.' 'Indis dam-sa
cia tic imárach i n-aghaidh Con Culainn do c[h]omrac (fris).'
1915 'Atáthar agá rádha fri brát[h]air fil agam-sa toidhecht ina aghaidh,
drút[h]óglach sotal soisil ⁊ é bailcbémnech buanaisech, ⁊ is uime
curtha[r] do c[h]omrac fris é da t[h]uitim les (⁊) co ndechaind-si
dá dhíogail fair-siomh, ⁊ ní rach-sa ann go bruinne mbrátha.
Láiríne mac Í Blaitmic in bráthair sin. Rachat-sa d'agallaimh Con
1920 Culainn uime sin,' ar Lugaidh. Ro gapadh a dhá ech do Lugaidh
⁊ ro hindledh a c[h]arpat forra. Táinic i ccomdáil Con Culainn co
ránic imagallamh eatorra. Is ann sin itbert Lughaidh: 'Atáthar
agá rádha re bráthair fil agam-sa tec[h]t do c[h]omrac frit-sa .i.
drúthóclach borb barbarda bailc búadnasach é ⁊ as uime curthar
1925 do chomrac frit-sa é dá thuitiom-siom leat ⁊ dá f[h]échain an
rac[h]ainn-si dá díogail fort, ⁊ ní rac[h]-sa ann sin go bruinne
mbrát[h]a. Ocus arin ccompántas fil edrainn aróen, ná marb-sa mo
brát[h]air-si. Dar ar ccubus ámh,' ar sé, 'cid tánaisde báis dusbéra
dó, as ced liom, úair dar mo s[h]árugadh téid it haghaidh-si.' Luid
1930 iaramh Cú Chulainn ar cúl ⁊ tét Lughaidh don longphort.
    Is ann sin do goiredh Láiríne mac Nóis i pupaill Oilella ⁊ Medba,
⁊ tucadh Fionnabair fora láim. As í no dáiledh corna fair ⁊ doberedh
póic la gac[h] n-áondig dhó ⁊ do gabhadh lám fora c[h]uid. 'Ní do
c[h]ác[h] berar la Meidb an lionn dáilter for Láiríne,' ar Fionnabair.
1935 'Ní tucc ac[h]t eri cáoga[t] fén de doc[h]um an loncphuirt.' 'Cia
ráidhi?' ar Oilill. 'An fer út thall.' ar sí. 'Cidh ésidhe?' ar Oilill.
'Minic lat h'aire do t[h]abairt do ní nábadh coimdigh. Ba córa
duit h'aire do t[h]abairt don láneamain as mó mait[h], mied ⁊
maisi dá bfuil ar énshlige i nÉrinn .i. Fionnabair ⁊ Láiríne mac
1940 Nóis.' 'Docím-si iat mar sin', ar Oilill. Is ann sin tuc Láiríne bogadh
⁊ bertnugadh fair gur maídedar úamanna na ccoilcech bátar faoí
curbo brec faithc[h]i an longphuirt dia cclúmaibh.
    Fada les gurbo lá cona lánsoillsi ann doc[h]um Con Culainn

---

¹ *lacuna here in LL, an entire page; supplied from Stowe* 1943-2040; *opening
words of Stowe for this passage read:* Erg-si, a Laoig, ar Cu Chulainn, d'agallaimh Lugdach mic Nóis i loncphort bfer nErenn aris ⁊ fionta in rainic Fer
Baoth an longphort . . . (1941–3)

d'fúabairt. Táinic i mucha na maïdne arnamárach ⁊ tuc eri feóin do t[h]realmaibh gaiscidh les ⁊ táinic forsan áth i ccomdáil Con 1945 Culainn. Níba fiú la dagláochaibh in dúnaidh nó an longphuirt tec[h]t d'féchain c[h]omraic Láiríni ac[h]t mná ⁊ giollanradh ⁊ ingena d'fochaidbedh ⁊ d'fanamad ima comrac. Táinic Cú Chulainn ina c[h]omdáil conici in áth ⁊ níorb fiú les airm do t[h]abairt les acht táinic diairm ina dháil.   Benais Cú Chulainn a airm uile asa 1950 láimh mar benas neach a aidme áineasa a láimh mhic bhic. Ro mel ⁊ ro cumail Cú Chulainn itir a lámhaibh é, non cúrond ⁊ non ceanclonn, non carcrann ⁊ nos crot[h]onn co seabaind a c[h]aindebar uile as gurbo ceó aéerda an ceat[h]araird i mboí. Telgis úada é iar sin do lár in át[h]a fíart[h]arsna in longphuirt go dorus puible 1955 a brát[h]ar. Cid trá acht níor érig riam gan éccaíne ⁊ níro loing gan airc[h]iseacht ⁊ ní raibi riamh ó sin amac[h] gan maíthe medóin ⁊ gan cumga cléb⁊ gan bronngalar ⁊ gan tat[h]aige amach ar mince. Is é sin trá aoinfer térna iar ccomrac fri Coin Culainn ar Táin Bó Cúailnge,⁊ táinic ris fós iersma an galair sin conadh é bás ruc ier sin. 1960 Conad comrac Láiríne ann sin for Táin Bó Cúailnge.

IS ann sin do goiredh Lóch Mór mac Mo Febhis i bpubaill Oilella ⁊ Medpa. 'Cid fa bfuilter dam-sa lib?' ar Lóch. 'Do c[h]omrac duit fri Coin Culainn,' ar Medhb. 'Ní rac[h]-sa don turas sin úair ní miedh nó maisi liom móethmaccóemh óg gan ulchain gan 1965 fhésóig d'ionnsaige ⁊ ní do bém aisge fair, acht atá agam fer a ionnsaighthe .i. Long mac Emónis, ⁊ gébaidh coma uaib-si.' Do gairedh Loncc i bpubaill Oilella ⁊ Medba, ⁊ geallais Medb mórc[h]omadha dhó .i. timtacht dá fher déc d'éudgadh gacha datha ⁊ carpat ceit[h]ri sheacht cumal ⁊ Finnabair d'óenmnaoi, ⁊ fes i 1970 cCrúachain do grés ⁊ fíon do dháil fair. Táinic ieram Long d'ionnsaige Con Culainn ⁊ marbais Cú Chulainn é.

Ráidis Medb fria banchuire teacht do agallaimh Con Culainn da rádha fris ulc[h]a smérthain do dénam fair. Tángatar in bantrac[h]t rempa ar amus Con Culainn co n-ebertitar fris ulcha 1975 smérthain do gabáil fair: 'Úair ní fiú la dagláoch isin loncphort techt do c[h]omrac frit ⁊ tú gan ulchain.' Do c[h]uir ieramh Cú Chulainn ulcha smérthain fair ⁊ táinic arin tulaigh ós cionn bfer nÉrenn ⁊ taisbénais in ulcha sin dóib uile i ccoitc[h]inne.

Atchonnairc Lóch mac Mo Febhis sin ⁊ is edh adubairt: 'Ulc[h]a 1980 sút ar Coin Culainn.' 'As edh ón atchíu,' ar Medp. Geallais Medb na mórc[h]omadha cétna do Lóch ar coscc Con Culainn. 'Rachat-sa dá ionnsaige,' ar Lóch.

Tic Lóch d'ionnsaighe Con Culainn go dtarla dá c[h]éle iet icin
1985 áth inar thuit Long. 'Tair romainn arin áth n-úac[h]tarach[h],'
ar Lóch, 'úair ní comraicfem arin áth so.' Ar ba háth heascoman
les-[s]iom in t-áth fora dtorc[h]air a bhráthair. Iar sin ro
comraicsit forsan áth úachtarach.

Is ann sin táinic in Morrígan ingen Ernmais a síodaibh do
1990 admilledh Con Culainn, ar ro gellastair for Táin Bó Regamna go
dtiocfad do aidhmilledh Con Culainn in tráth do beith ig comrac
fri degláoch for Táin Bó Cúailnge. Táinic ieramh in Morrígan ann
sin i rriocht samhaisci finne óderge co coícait samasc uimpi ⁊ ronn
fiondruine itir gach dá samaisc díoph. Dobertsat in banntracht gesa ⁊
1995 airmberta for Coin cCulainn dá ttísadh úadh gan fhosdadh gan
aidmilledh fuirre. Dobert Cú Chulainn rót n-urc[h]uir di gur bris
let[h]rosc na Morrígna. Táinic dano in Morríghan ann sin i rriocht
escuinge slemne duibi lasan srut[h]. Tét ieramh forsan lind gurrus
iomnaisg fo chosaibh Con Culainn. An fad boí Cú Chulainn agá
2000 díchur de, ro ghon Lóch urt[h]arsna é tre c[h]ompar a c[h]léb.
Táinic ieramh in Morrígan i riocht saidhi gairbi glasrúaidhi. Cien
goirit boí Cú Chulainn igá díchur dhe, ro ghon Lóch é. Iar sin ro
75ᵃ érigh fercc Con Culainn ris] (⁊) | ¹ gonais Cú Chulaind Lóch din gai
bulga coro thregda a chridi 'na chlíab. 'Ascid dam ifechtsa, a
2005 Chú Chulaind,' bar Lóch. 'Gia ascid connaige?' 'Ní ascid anacail
nó midlachais iarraim-se fort,' bar Lóch. 'Teilg traigid dam corop
ar m'agid sair tóethus ⁊ nárap dar m'aiss síar co firu Hérend, arná
rádea nech díb is roí madma nó techid dam remut-sa, dáig torchar
din gae bulga.' 'Teilcfet,' bar Cú Chulaind, 'dáig is láechda ind ascid
2010 connaigi.' Ocus teilgis Cú Chulaind traig[i]d ar cúl dó. Conid de
fil in t-ainm forsind áth ó sin .i. Áth Traiged i cind Tíri Móir.

Ocus gebis athrechus mór Coin Culaind in lá sin .i. bith forin
Táin itir a óenur, & rádis Cú Chulaind fria araid fri Láeg techt
do innaigid Ulad ar con tístaís do chosnam a tána. Ocus ²ro
2015 gab-sum merten ⁊ athscís forru² & doringni rand:

'Airg úaim, a Laíg, laíder³ slúaig.
Cain dam i nEmain adrúaid⁴
am tursech cach dia 'sin chath
condam créchtach crólinnech.

---

¹ *LL resumes; in Stowe after the word* ris *the sentence continues as:* gurrus
gon don gae bulga é gurro tregd a c[h]roidi ina c[h]liab

²⁻² do gab mertin ⁊ aithsgis mor C.C. *St:* ros gab mifri ocus merten é *C*

³ laitter *St*, laiter *C, Rec. III*   ⁴ armruaidh *St, C, Rec. III*

'Mo tháeb dess is mo tháeb clé,   2020
andsa mess for cechtar de.
Ní lám Fíngin roda slaid,
¹dirgid fola fidfaebraib.¹

'Apair fri Conchobor cáem
atú tursech tiachairtháeb.   2025
Trén ra chlóechla chruth amne
mac dil droṅgach Dechtire.

'M'óenurán dam ar éitib
acht nís léicim, nís étaim.
²Atú im ulc, ním fuil im maith²,   2030
m'óenur dam ar iláthaib.

'Feraid bróen fola for m'arm
go ndamrala créchtach ṅgarb.
Ním thic cara ar báig *nó* ar blait
acht mad ara óencharpait.   2035

'Mad úathad dochanat form,
ní airfitiud nach n-óenchorn.
Mad ilar ceól a cornaib,
iss ed is binniu din choblaig.

'Senḟocal so, srethaib cland,   2040
ní lassamain cech³ n-óenchrand.
Dá mbetis a dó nó a trí,
lasfaitis a n-athinni.

'In t-óenchrand ní hassu a chlód
meni fagba a [ḟ]rithadód;   2045
ar úathad imrither⁴ gó
noco modmar cach³ n-óenbró.

'Nach cúala tú in cach than
clóentar gó ar úathad, fír dam.
Iss ed ná fulaṅgar de   2050
turscolbad na sochaide.

---

¹⁻¹ dirge*d*h fola fadfhaobra*i*bh *St;* dirgadh folae fidfaobra*i*bh *C*
²⁻² atu i n-ulc nim fuil i maith *St*   ³ nach *St, C*
⁴ imerther *St,* imberthar *C,*

'Gid úathad lín in chaire
dochaitter menma aire.
Cuit in tslóig, is é a samail,
2055 ní berbther é ar óengabail.

'M'óenur dam i cind in tslóig
'gund áth i cind Tíri Móir.
Ba lía Lóch co lleith Bodba
go remfoclaib Regomna.

2060 'Ra lettair Lóch mo dá lón,
rom tesc in tsód garb glasród;
ro geguin Lóch mo thromma,
rom tresgair in esconga.

'Is rem chlettín-se a cosc,
2065 an tsód [1]ó ro[1] mill a rosc,
ro brisses a gerr gara
do thosuch na hégrada.

'[2]Ó indill[2] Láeg in gae Aífe
risin sruth, ba seól faethe.
2070 Ro theilgesa in ngae ngér guis
dar thóeth Lóch mac E[mon]is.

'Cid d'Ultaib nach fegat cath
d'Ailill is d'ingin Echach?
Tráth atú-sa sund i n-ach
2075 is mé créchtach crólinnech.

75ᵇ 'Apair ri Ultu ána
tecat i ndiaid a tána.
Rucsat meic Mágach a mbu
⁊ ros raindset eturru.

2080 'Bágim-se báig airdgella[3]
⁊ [4]ra comall[4] chena,
bágim-se a heniuch caem chon
nacham tair-se óen m'óenor.

[1-1] coro *St, C*   [2-2] ro innill *St,* innlis *C leg.* ro indill
[3] rod gealla *St, C*   [4-1] ra comallad *St;* ro comailled *C*

'Ac[h]t is fálid[1] brain berna
i llong*phurt*[2] Ailella ⁊ Medba.   2085
Tursig núallana reme
rena ngáir i mMaig Murthemne.

'Conchobor ní thic immach
náco raib a lín 'sin chath.
tráth nach fálid[3] é amne.   2090
ansu árim a fergge.'
                                              Airg.

Conid Comrac Lóich Móir meic Ma Femis fri Coin Culaind sin
[for] Táin Bó Cúalnge.

Nd sain faítti Medb in sessiur úadi i n-óenfecht do fúapairt [2095]
Con Culaind .i. Traig ⁊ Dorn ⁊ Dernu. Col ⁊ Accuis ⁊ Eraísi.
Tri ferdruíd ⁊ trí bandruíd. Basrópart Cú Chulaind síat co
torchratar lais. Ára brissed fír breth for Coin Culaind ⁊ comlund
óenfir, gebis Cú Chulaind a chrantabaill ⁊ basrópart in slúag do
díburgun a Delggain andess in lá sain. Giambtar liri fir Hérend [2100]
in lá sin, barémid nech díb a aged do soud fodess in lá sin do
choin nó ech nó duine.
     And sin tánic in Mórrígu ingen Ernmais[4] a sídib i rricht sentainne
co rrabi ic blegu[n][5] bó trí sine [6]'na fiadnaisse[6]. Is immi tánic-si
[7][mar] sin[7] ar bíth a fórithen[8] do Choin Chulaind. Dáig ní gonad [2105]
Cú Chulaind nech ara térnád co mbeith cuit dó féin 'na legius.
Conattech[9] Cú Chulaind blegon furri iarna dechrad d'íttaid. Dobretha-
si blegon sini dó. 'Rop slán a n-éim dam-sa so.' Ba slán a lethrosc
na rígna. Conattech[9]-som blegon sini furri. Dobreth[a]-si dó.
'I n-éim rop slán intí doridnacht'. Conaittecht-som in tres ndig [2110]
⁊ dobretha-si blegon sine dó. 'Bendacht dee ⁊ andee fort. a ingen.'—
Batar é a ndee in t-áes cumachta. & andee in t-áes trebaire.—Ocus
ba slán ind rígan.
     And sain faítte Medb in cét láech i n-óenfecht do fúapairt Con
Culaind. Basrópart Cú Chulaind siat uili co torchratar leiss. 'Is [2115]
cuillend dúin guin ar muntiri samlaid,' ar Medb. 'Níp sé sút a
chétchuillend dúin ind fir chétna,' bar Ailill. Conid Cuillend Cind
Dúni comainm béus ind inaid i mbátar ó sin & conid Áth Cró ainm

---

[1] faoil*idh* St, C   [2] sgor St, C, *sic leg.*   [3] faoilidh C
[4] Ernnais MS.   [5] n-*stroke om.*   [6–6] i fiadnaise Con Culainn St
[7–7] mar sin St   [8] fóirithne St. *sic leg.*,   [9] *sic*

ind átha fors mbátar. Dethbir ara méit dá crú ⁊ dá fuil dochúaid
2120 fo sruthair na haband.

Breslech Maige Murthemne so sís

RO gabsat cethri chóicid Hérend dúnad ⁊ longphort isin Breslig
Móir hi Maig Murthemne. Ro láiset a n-ernail búair ⁊ braite
76ᵃ   secco fodes i Clithar Bó Ulad. | Gabais Cú Chulaind icon
2125 Fert i lLerggaib i comfocus ⁊ i comfochraib dóib & atáis a ara
tenid dó tráth nóna na haidchi sin .i. Lóeg mac Riangabra.
Atchonnairc-seom úad grístaitnem na n-arm nglanórda úas chind
chethri n-ollchóiced nÉrend re fuiniud néll na nóna. Dofánic ferg ⁊
luinni mór icá n-aiscin re ilar a bidbad, re immad a námat. Ro
2130 gab a dá sleig ⁊ a scíath ⁊ a chlaideb. Crothais a scíath ⁊ cressaigis
a slega ⁊ bertnaigis a chlaidem, ⁊ dobert rém curad asa brágit coro
[f]recratar bánanaig ⁊ boccánaig ⁊ geniti glinni ⁊ demna aeóir re
úathgráin na gáre ¹dosbertatar ar aird¹, coro mesc ind Neamain²
forsin tslóg. Dollotar i n-armgrith cethri chóicid Hérend im
2135 rennaib a sleg ⁊ a n-arm fadessin co n-erbaltatar cét láech díb di úathbás
⁊ chridemnas ar lár in dúnaid ⁊ in loṅgphairt in n-aidchi sin.
     Dia mbaí Lóeg and co n-acca ní, in n-óenfer dar fiartharsna in
dúnaid cethri n-ollchóiced nHérend anairtúaith cach ndíriuch ina
dochum. 'Óenfer sund chucund innossa, a Chúcúcán,' ar Lóeg.
2140 'Cinnas fir and sin ale?' or Cú Chulaind. 'Ní handsa. Fer caín
mór dano. Berrad lethan lais. Folt casbuide fair. Bratt úanide
i forcipul imme. Cassán gelargit isin brutt úasa bruinne. Léne
de sról ríg fo derggindliud do derggór i custul fri gelchness co
glúnib dó. Dubscíath co calathbúali finndruini fair. Sleg cóicrind
2145 ina láim. Foga fogablaigi ina farrad. Iṅgnad ém reb ⁊ ábairt ⁊
adabair dogní. Acht ní saig nech [fair]³ ⁊ ní saig-som dano for nech
feib nacha n-aicced nech issin dúnud chethri n-ollchóiced [n]Hérend.'
'Is fír anísin, a daltán,' for sé, 'cia dom chardib sídchaire-sa sein⁴
dom airchisecht-sa dáig ar bíth foretatar-som in t-imned mór
2150 anam [f]uil-sea m'óenurán i n-agid chethri n-ollchóiced nHérend
ar Táin Bó Cúalṅgi don chur sa.' Ba fír ém do Choin Culaind
anísin. A nad-ránic in t-óclách airm i mboí Cú Chulaind, argladais
⁊ airchisis de. 'Cotail-siu ém bic, a Chú Chulaind,' or in t-óclách,

---

¹⁻¹ do-rinde *St*; doberadh ar aird *Rec. III*; *read* dobert ar aird?
² *in marg.* .i. in Badb   ³ *om. MS., supplied from LU (C)*   ⁴ tic *add. St*

'do thromthoirthim chotulta icon Ḟerta Lergga co cend trí láa ⁊
teóra n-aidchi, ⁊ firbat-sa forna slógaib in n-airet sin.'   2155
Is and sin cotlais Cú Chulaind a thromthairthim cotulta icond
Ḟerta i lLergaib co cend teóra láa ⁊ [teóra n-]¹ aidche. Bá deithbir
són céro boí do mét in chotulta boí do mét na athscísi. Ón lúan
re samain sainriuth cossin cétaín iar n-imbulc níra chotail Cú
Chulaind risin ré sin acht mani chotlad fithisin ṁbic fria gaí iar 2160
medón midlaí ⁊ a chend ara dorn [⁊ a dorn]¹ imma gaí ⁊ a gaí ara
76ᵇ glún | [acht]¹ ic slaidi ⁊ ic slechtad, ic airlech ⁊ ic essorgain chethri
n-ollchóiced ṅHérend frisin re [sin]¹. Is and sin focheird in láech
lossa síde ⁊ lubi ícci ⁊ slánsén i cnedaib ⁊ i créchtaib, i n-áladaib
⁊ i n-ilgonaib Chon Culaind co térno Cú Chulaind ina chotlud cen 2165
rathugud dó etir.

Is hí sin amser dollotar in maccrad atúaid ó hEmain Macha,
tri choícait mac do maccaib ríg Ulad im Ḟollomain mac Conchobuir,
⁊ dosbertsat teóra catha dona slúagaib co torchratar a trí comlín
[leó]¹ ⁊ torchratar in macrad dano acht Follomain mac Conchobuir. 2170
Bágais Follomain ná ragad ar cúlu co hEmain co brunni ṁbrátha
⁊ betha co mberad cend Ailella leis cosin mind óir boí úaso. Nírbo
réid dó-sum anísin úair dofárthetar dá mac Beithe meic Báin, dá
mac mumme ⁊ aite do Ailill, ⁊ rod gonat co torchair leo. Conid
Aided na Maccraide Ulad in sin ⁊ Ḟollom*na* meic Conchobuir.   2175
Cú Chulaind immorro buí ina ṡúanthairthim cotulta co cend teóra
láa ⁊ t[e]óra [n-]aidche icon Ḟerta i lLerggaib. Itraacht Cú
Chulaind iar sin assa chotlud ⁊ dobert lám² dara agid, ⁊ doriṅgni
rothnúall corcra [de]³ ó mulluch co talmain, ⁊ ba nert leis a menma
⁊ tíasad i n-óenách nó i toichim nó i mbandáil nó i coirmthech 2180
nó i prímóenach do prímóenaigib Hérend. 'Cia ḟot itú-sa isin
chotlud sa innossi, a ócláich?' ar Cú Chulaind. 'Trí láa ⁊ trí
aidche,' for in t-óclách. '⁴Ron marg-sa⁴ de-side,' or Cú Chulaind.
'Cide⁵ ón?' or in t-ócláech. 'Na slóig cen ḟópairt frisin ré sin,' ar Cú
Chulaind. 'Ní ḟilet-som ón etir,' or in t-óclách. 'Ceist cia rodas 2185
fópair?' ar Cú Chulaind. 'Lotar in macrad atúaid ó Emain Macha
trí choícait mac im Ḟollomain mac Conchobuir do macaib ríg
Ulad ⁊ dobertsat teóra catha dona slúagaib ri hed na trí láa ⁊ na trí
n-aidche i taí-siu it chotlud innossa, ⁊ torchratar a trí comlín leó ⁊
torchratar in macrad acht Follomain mac Conchobuir. Bágais 2190
Follomain ⁊rl.' 'Apraind ná bá-sa for mo nirt de-side úair dia mbend-
sea for mo nirt, ní thóethsaitis in macrad feib dorochratar ⁊ ní
thóethsad Fallomain.'

---

¹ *om. MS., supplied from LU (St, C)*   ² *sic, for* láim   ³ *om. MS., om.*
LU, *supplied from St, C*   ⁴⁻⁴ ⱳom maircsi *C*   ⁵ = cid de *LU*

'Cosain archena, a Chúcán, ní haisc dot inchaib ⁊ ní táir dot
²¹⁹⁵ gaisciud.' 'Airis-[s]iu sund inocht dún, a ócláig,' ar Cú Chulaind,
'ar co ndíglom 'malle in maccraid forsna slúagaib.' 'Nád anaeb
ém ale,' ol in t-óclách, 'úair cid mór do chomramaib gaile ⁊ gaiscid
dogné nech it arrad-su, ní fair bias a nós nach a allud nach a aird-
ercus acht is fort-so. Is aire-sin nád anub-sa, acht [imbir]¹ féin
²²⁰⁰ do gním gaiscid th'óenur forsna slúagaib úair ní leó atá commus
t'anma don chur so.'
'Ocus in carpat serda, a mo phopa Lóeg,' ar Cú Chulaind, 'in
coemnacar a innell? Má cotnici a innell ⁊ má dotá a t[h]relom,
non inill, ⁊ mani fil a t[h]relom, nacha inill.'

²²⁰⁵ IS and sin atracht in t-ara ⁊ ro gab a fíaneirred araidechta immi.
⁷⁷ᵃ Ba dond | [f]íaneirriud aradachta sin ro gabastar-som imbi a inar
bláith bíannaide, is é étrom aérda, is é súata srebnaide, is é úaigthe
osslethair, conná gebeth² ar lúamairecht lám dó anechtair. Ro
gabastar-som forbratt faiṅg taris-sein 'anechtair doriṅgni Simón
²²¹⁰ Druí do ríg Rómán conda tarad³ Dair do Chonchobur conda tarat
Conchobor do Choin Culaind co tarat Cú Chulaind dá araid. Ro
gabastar in t-ara cétna a chathbarr círach clárach cetharchoir⁴
co n-ilur cech datha ⁊ cacha delba dara midgúallib sechtair. Ba
somaissi dó-som sin ⁊ nírbo thortrommad. Tarraill a lám leis in
²²¹⁵ gipni ṅdergbuide marbad land dergóir do bronnór bruthi dar or
n-indeóna re étan do indchomartha⁵ araidechta⁶ secha thigerna.
Ro gab idata aurslaicthi a ech ⁊ a del intlaissi ina desra⁷. Ro
gabastar a éssi astuda ech ina thúasri .i. aradna a ech ina láim
chlí, re imchommus a araidec[h]ta.
²²²⁰ Is and sin focheird a lúrecha iarnaidi intlaissi immo echraid
congebethar dóib ó thaul co aurdorn ⁸do gaínib⁸ ⁊ birínib ⁊ slegínib
- birchrúadib corbo birfocus cach fonnud issin charput sin corbo
chonair letartha cach n-uill[e]⁹ ⁊ cach n-ind ⁊ cach n-aird ⁊ cach
n-airchind don charput sin. Is and sin focheird bricht comga
²²²⁵ tara echraid ⁊ tara chomalta connárbo léir do neoch issin dúnud
iat - corbo léir dóib-sium cách issin dúnud sin. Ba deithbir ém cé
focheirded-som innísin dáig ar bíth bátar teóra búada araidechta

¹ MS. obscure, faire suggested tentatively Dipl. Edn.; imbir LU, St, C
² gebtar MS., corr. to gebeth: gebethar LU, gabadh St, gebeth C
³ = tarat (final tenuis > media bef. initial voiced cons.)   ⁴ cethrochair
LU, ceithireochair St   ⁵ indochomartha MS.   ⁶ airaidechta MS.
⁷ dessa MS., desra LU   ⁸⁻⁸ omission before do gaínib? Cp. da nditen ar
gainibh St, co n-egar di gainib C, lan do ghainibh Rec. III   ⁹ n-ulind LU

forind araid in lá sin .i. léim dar boilg ⁊ foscul ṅdíriuch ⁊ immorchor
ṅdelind.

Is and sin ro gab in caur ⁊ in cathmílid ⁊ in t-innellchró Bodba ²²³⁰
fer talman, Cú Chulaind mac Sualtaim, ro gab a chatheirred catha
⁊ comraic ⁊ comlaind imbi. Ba don chatheirred catha sin ⁊ comraic
⁊ comlaind ro gab-som imbi secht cneslénti fichet cíartha clárda com-
dlúta bítís ba thétaib ⁊ rothaib ⁊ refedaib i custul ri gelchnes dó arna cha
ndechrad a chond nach a chiall ó doficed¹ a lúth láthair. Ro ²²³⁵
gabastar a chathchriss curad taris anechtair do chotutlethar
crúaid coirtchide do fhormna secht ṅdamseiched ṅdartada congabad
dó ó thana [a] thaíb co tiug a oxaille. Ro bíth imbi ic díchur gaí
⁊ rend ⁊ iaernn ⁊ sleg ⁊ saiget, dáig is cumma focherdditis de ⁊
marbad de chloich nó charraic nó choṅgna ²ro chiulaitis². Is and ²²⁴⁰
sin ro gabastar a [fʼ]úathbróic srebnaide sróil cona cimais de bánór
bricc friá fri móethíchtur a medóin. Ro gabastar a dond[fʼ]úath-
bróic ṅdondlethair ṅdegsúata do formna cethri ṅdamseiched
ṅdartada cona chathchris do cholomnaib ferb fua dara fúathróic
srebnaide sróil sechtair. Iss and sin ro gabastar in rígníath ¦ a ²²⁴⁵
chatharm catha ⁊ comraic ⁊ comlaind. Ba don chatharm chatha sin
ro gabastar a ocht claidbíni im[a] cholg ṅdét ṅdrechsolus. Ro
gabastar a ocht sleigíni imma sleig cóicrind. Ro gabastar a
ocht [n]gothnata ʼma gothnait ṅdét. Ro gabastar a ocht
clettíni ʼma deil chliss³. Ro gabastar a ocht scíathu cliss imma ²²⁵⁰
chrommscíath ṅdubderg ina téiged torc taisselbtha ina thaul tárla
cona bil aithgéir ailtnidi imgéir ina urthimchiull contescfad finna
i n-agid srotha ar áithi ⁊ ailtnidecht ⁊ imgéiri. Inbaid fogníth in
t-óclách fáeborchless di, is cumma imthescad dia scíath ⁊ dia sleig
⁊ dá chlaideb. Is and sin ro gab a chírchathbarr catha ⁊ comlai[n]d ²²⁵⁵
⁊ comraic imma chend as[a] ṅgáired gáir cét n-óclách do sí́réigim
cecha cúli ⁊ cecha cerna de, dáig is cumma congáiritis de bánanaig ⁊
bocánaig ⁊ geiniti glinne ⁊ demna aeóir ríam ⁊ úasu ⁊ ina thimchuill cach
ed immatéiged re tesitin fola na míled ⁊ na n-aṅglond sechtair. Ro
chres a cheltar chomga tharis don tlachtdíllat Tíre Tairṅgire ²²⁶⁰
dobretha dó ó Manannán mac Lír ó ríg Thíre na Sorcha.

Is and sin cétríastarda im Choin Culaind co nderna úathbásach
n-ilrechtach n-iṅgantach n-anachnid de. Crithnaigset a chairíni
imbi immar chrand re sruth nó immar bocsimind ri sruth cach ball
⁊ cach n-alt ⁊ cach n-inn ⁊ cach n-áge de ó mulluch co talmain. Ro ²²⁶⁵
lá sáebchless díbirge dia churp i mmedón a chracaind. Táncatar a
thraigthe ⁊ a luirgne ⁊ a glúne co mbátar dá éis. Táncatar a sála ⁊

77ᵇ

---

¹ doficed MS., doficed LU   ² ⁻² do chiuchlaidis Rec. III (RC 15.77)
³ chniss MS.; cliss LU (St)

a orccni ⁊ a escata co mbátar riam remi. Táncatar tullféthi a orcan
co mbátar for tul a lurggan comba méitithir muldorn míled cech
2270 meccon dermár díbide. Sreṅgtha tollféithe a mullaig co mbátar
for cóich a muneóil combá mé[t]ithir cend meic mís cach
mulchnoc dímór dírím dírecra dímesraigthe díbide.
And sin doriṅgni cúach cera dia gnúis ⁊ dá agaid fair. Imsloic
indara súil dó ina chend; iss ed mod dánas tarsed fíadchorr tagraim
2275 do lár a grúade a iarthor a chlocaind. Sesceiṅg a séitig co mboí
fora grúad sechtair. Ríastarda a bél co urthrachda¹. Sreṅgais
in n-ól don fidba chnáma comtar inécnaig a inchróes. Táncatar a
scoim ⁊ a thromma co mbátar ar eittelaig ina bél ⁊ ina brágit.
Benais béim n-ulgaib leómain don charput úachtarach fora forcli
2280 comba métithir moltcracand teóra ṁblíadan cech slamsrúam
teined doniged ina bél asa brágit. Roclos bloscbéimnech a chride
re chlíab immar glimnaig árchon i fotha nó mar leóman ic techta
78ᵃ fo mathgamnaib. | Atchessa ²na coinnle Bodba ⁊² na cidnélla³
nime ⁊ na haíble teined trichemrúaid i nnéllaib ⁊ i n-áeraib úasa
2285 chind re fiuchud na fergge fírgairbe itrácht úaso. ⁴Ra chasnig⁴
a folt imma chend imar craíbred ṅdercscíath i mbernaid athálta.
Céro craiteá rígaball fo rígthorud immi, is ed mod dá rísad ubull
díb dochum talman taris acht ro sesed ubull for cach n-óenfinna
and re frithchassad na ferge atracht da folt úaso. Atracht in lónd
2290 láith asa étun comba sithethir⁵ remithir áirnem⁶ n-ócláig. Aird-
dithir remithir tailcithir tressithir sithithir seólchrand prímluṅgi
móre in bunne díriuch dondfola atracht a fírchléithe a chendmullaig
i certairddi co nderna dubchíaich ṅdruídechta de amal chíaich de
rígbruidin in tan tic rí dia tenecur hi fescur lathi gemreta.
2295 Iarsin ríastrad sin ríastarda im Choin Culaind iss and sin
dorroeblaiṅg ind err gaiscid ina chathcharpat serda ⁷cona erraib⁷
iarnaidib, cona fáebraib tanaidib, cona baccánaib ⁊ cona birchrúadib,
cona thairbirib níath, cona glés aursloicthi, cona tharṅgib gaíthe
bítis ar fertsib ⁊ iallaib ⁊ fithisib ⁊ folomnaib dun charput sin.
2300 Is and sin focheirt torandchless cét ⁊ torandchles dá cét ⁊ torandchless trí cét ⁊ torandchless cethri cét, ⁊ tarassair aice for torandchless
cóic cét úair nírbo furáil leis in comlín sin do thuitim leis ina chétchumscli ⁊ ina chétchomliṅg catha for cethri chóicedaib Hérend.
Ocus dotháet ass fón cumma sin do innsaigid a námat ⁊ dobreth a

---

¹ úrtrachta LU   ²⁻² na klne Bodba MS., so also LU; om. St; na
coindli Bodhbha ⁊ Rec. III (RC 15.199)   ³ cithnella LU   ⁴⁻⁴ ra canig
MS., ra chasnaig LU (St)   ⁵ sic LU; sithe MS. (sithremigthir St)
⁶ so also LU, but airtimh St; read airtem   ⁷⁻⁷ co n-erraib LU; read
cona serraib (cona searraib St)

charpat mórthimchell cethri n-ollchóiced ṅHérend amaig anechtair, [2305]
&dosbert seól trom fora charpat. Dollotar rotha iarnaidi in c[h]arpait
hi talmain corbo leór do dún ⁊ do daiṅgen feib dollotar rotha
iarnaide in charpait i talmain, úair is cumma atraachtatar cluid ⁊
coirthe ⁊ carrge ⁊ táthlecca ⁊ murgrian in talman aird i n-aird
frisna rothaib iarnaidib súas sell sechtair. Is airi focheird in circul [2310]
ṁBodba sin mórthimchell chethri n-ollchóiced ṅHérend ammaig
anechtair arná teichtis úad ⁊ arná scaíltís immi co tórsed forro
[1]re tenta fritharggain[1] na maccraide forro. Ocus dotháet issin
cath innond ar medón ⁊ fálgis fálbaigi móra de chollaib a bidbad
mórthimchell in tṡlóig ammaig annechtair, & dobert fóbairt bidbad [2315]
fo bidbadaib forro co torcratar bond fri bond ⁊ méide fri méide,
ba sé tiget a colla. Dosrimchell aridisi fa thrí in chruth sin co
farggaib cossair sessir impu fá mórthimchell .i. bonn tríir fri méide
tríir fócúairt timchell immon dúnad. Conid Seisrech Bresslige a
ainm issin Táin, ⁊ iss ed tres ṅdírime na Tána .i. Sesrech Breslige [2320]
⁊ Imṡlige Glennamnach ⁊ iṅ cath for Gárich | ⁊ Irgáirich, acht ba
cumma cú ⁊ ech ⁊ duine and. [2]Iss ed atberat araile ro fích Lug
mac Eithlend la Coin Culaind Sesrig ṁBresslige[2].

Nicon fess a árim ⁊ ní chumaṅgar a rím cia lín dorochair and do
dáescarṡlóg acht ro rímthé a tigerna nammá. It é in so sís a n-anm- [2325]
and-side .i.[3] Dá Chrúaid, dá Chalad, dá· Chir, dá Chíar, dá Éicell,
trí Cruimm, trí Curaid, tri Combirgi, cethri Feochair, cethri
Furachair, cet[h]ri Caiss, cet[h]ri Fotai, cóic Caurith, cóic Cermain,
cóic Cobthaig, sé Saxain, sé Dauich, sé Dáiri, secht Rochaid, secht
Rónáin, secht Raurthig, ocht Rochlaid, ocht Rochthauid, ocht [2330]
Rinnaich, ocht Mulaig, noí ṅDaigith, noí ṅDáiri, noí ṅDamaich,
deich Féic, deich Fiachaig, deich Feidlimid. Deich ríg ar sé fichtib
ríg ro bí Cú Chulaind issin Bresslig Móir Maige Murthemne.
Díríme imorro archena di chonaib ⁊ echaib ⁊ mnáib ⁊ maccaib ⁊
mindóenib ⁊ drabarṡlóg, ar nír érna in tres fer do feraib Hérend [2335]
cen chnáim leissi nó lethchind nó lethṡúil do brissiud nó cen bithanim
tria bithu betha.

Dotháet Cú Chulaind arnabárach do thaidbriud in tslóig ⁊ do
thasbénad a chrotha álgin álaind do mnáib ⁊ bantrochtaib
⁊ andrib ⁊ ingenaib ⁊ filedaib ⁊ áes dána, uair nír miad ná [2340]
mais leis in dúaburdelb druídechta tárfás dóib [fair][4] in adaig sin
riam reme. Is aire sin dano tánic do thasselbad a chrotha álgin
álaind in lá sin.

Álaind ém in mac tánic and sin do thaisselbad a chrotha dona
2345 slúagaib .i. Cú Chulaind mac Sualtaim. Trí fuilt bátar fair: dond
fri toinn. cróderg ar medón. mind órbuide ardatuigethar. Caín
cocáirisi ind fuilt sin *concuirend* [1]teóra imsrotha[1] im chlaiss a chúlaid
combo samalta ꝉ snáth órsnáith cach finna faithmainech forscaílte
forórda dígrais dúalfota derscaigthech dathálaind [2]dara formna[2]
2350 síar sell sechtair. Cét cairches corcorglan do derggór órlasrach
imma bráigit. Cét snáthéicne do charrmocul chummascda i
timthacht fria chend. Cethri tibri cechtar a dá grúad .i. tibre
buide ꝉ tibre úane ꝉ tibre gorm ꝉ tibre corcra. Secht ṅgemma
de ruithin ruisc cechtar a dá rígrosc. Secht meóir cechtar a dá
2355 choss. Secht meóir cechtar a dá lám co ṅgabáil iṅgni sebaicc, co
forgabáil iṅgne griúin ar cach n-aí fo leith díb.
Gabaid-som dano a díllut óenaig immi in láa sin. Baí dá étgud
immi .i. fúan caín cóir corcra cortharach cóicdíabuil. Delgg find
findarggait arna ecor d'ór intlaisse úasa bánbruinni gel immar
79[a] bad lócharnn lánsolusta nád chumgaitis | súli dóeni déscin ar
gleórdacht ꝉ ar glainidecht. Cliabinar siric fri[a] chness [3]arna
imthacmaṅg massi[3] de chimsaib ꝉ chressaib ꝉ chorrtharaib óir
ꝉ argit ꝉ findruni, condriced go barrúachtur a dondfúathbróci
donderggi míleta imme de sról ríg. Scíath dígrais dondchorcra
2365 fair co mbil argit óengil ina imthimchiull. Claideb órduirn intlassi
bara chlíu. Gae fata fáeburglass re faga féig fóbarta co súanemnaib
loga, 'co semmannaib findruine issin charput ina farrad. Noí cind
isindara láim dó ꝉ deich cind isin láim anaill, & ros croth úad risna
slúagaib do chomartha a gascid ꝉ a eṅgnama. Laigis Medb a heinech
2370 fa damdabaich scíath arnáras díbairged Cú Chulaind in lá sin.
Is and sin ra attchetar in ingenrad firu Hérend 'ma tócbáil bar
lébennaib scíath ás gúallib feróclách do thaidbriud chrotha Con
Culaind. Ár rap iṅgnad leó-som in delb álaind álgen atchondcatar
in lá sin fair ic athféscain na duabordelbi dóescairi druídechta ra
2375 condcas fair inn adaig ríam reme.
Is and sin ra gab ét ꝉ elcmaire ꝉ immfarmat Dubthach Dáel Ulad
imma mnaí ꝉ dabert comairle braith ꝉ trécthi Con Culaind dona
slúagaib .i. cathetarnaid imme far cach leth ar co táetsad leó.
Ocus rabert na briathra sa:

1-1 teorae himsrethae *C*, teóra sretha *Rec. III* (*RC* 15.203)
2-2 dara as formna *MS.*; dara a formna *YBL*, dara formna *LU*, *C*; tara
ais *St*
3 3 *sic MS.*, imthacmaṅg *to be altered to* imthacmassi, *but scribe neglected to*
delete-maṅg (arna imthacmaisi *Rec. III RC* 15.203)

'Másu é in riastarde,   2380
betit colla dóene de,
betit éigme de¹ im lissu,
betit buind ri harissu,
²betit brain ri brainessu.²

'Betit corrthe de ³im lechta³,   2385
bud fórmach ⁴do rigmartra⁴.
Ní maith fararlith⁵ in cath
ar leirg risin fóendelach⁶.

'Atchíu chruth inn fóendelaich,
nóe⁷ cind leis ⁸i foendelaib⁸,   2390
atchíu fadb leis 'na brétaig,
deich cind ina rosétaib.

'Atchíu forthócbat far mná
a n-aidche ósna urgalá,
atchíu-s[a] far rígain máir   2395
ná hérig dond imforráin.

'Dámbad mé bad chomarlid
da betís óic di cach leith,
⁹coro gartigtis a ré,
mása é in riastarde⁹.'   M.   2400

Atchúala Fergus mac Róig aní sein ┐ ba dimbág leis comairle braith
Con Culaind do thabairt do Dubthach dona slúagaib, & rabretha
trénlúa tarpech dá choiss úad riss ¹⁰co tarla darráib¹⁰ ra budin
anechtair. Ocus ¹¹ra faismis¹¹ fair na huli ulcu ┐ écóra ┐ fell ┐ mebol
doriṅgni ríam ┐ íarum ra Ultaib & rabert na briathra and: 2405

'Más é Dubthach Dóelteṅga
ar cúl na slúag bosreṅga,
nocho dergena nach maith
ó geguin in n-ingenraid.

---

¹ om. LU, YBL, C   ²⁻² om. LU, YBL, C   ³⁻³ i llechtaib LU, YBL, C
⁴⁻⁴ do rígmartaib LU, YBL, di righmartraib C   ⁵ no fichid LU, YBL,
C; confighed Rec. III   ⁶ n-oennenach LU, YBL, C   ⁷ ocht LU, YBL, C
⁸⁻⁸ inna chuillsennaib LU, YBL, C   ⁹⁻⁹ this half-quatrain is in C and
in Rec. III (RC 15.204) but is lacking in LU, YBL   ¹⁰⁻¹⁰ read co tarla dá
srúib? (co n-arrasair di(a) sruib LU, YBL, C; co tarla a srubh i freslige St)
¹¹⁻¹¹ ro faisneid C, ro foillsigh St

2410
'Ferais écht ṅdochla ṅdogair,
guin Fiachach meic Conchobuir,
nocho cáeme rachlas dó,
guin Charpri meic Fedilmtheo.

2415
'Ní flaith Ulad nod chosna
mac Lugdach meic Casruba,
iss ed ragní ¹ra dóenib¹,
cách nas ruba ¹ris faidib¹

'Ní maith ra loṅgis Ulad
guin a meic nach allulach,
2420
costud Ulad danfor tí,
consaífet far n-immirgi.

'Scérdait ²far n-óendili² i fat
re nUltaib acht co n-éirset,
betit échta sceoil mára,
2425
betit rígna ḋérmára.

'³[Biait collai fó chossaib]³,
Betit buind fri brannusa⁴,
Betit fáenscéith ⁵fri lerga⁵,
Bid tórmach ⁶na ndíberga⁶.

2430
'Atchíu ras furcbat far mná
a ṅgnúis ásna hirgala,
Atchíu bar rígain in máir,
ní érig don immḟorráin | .

79ᵇ
'⁷Ní dergían gaisced *nó* gart
2435
a⁸ *meic* Lugdach ⁹gan nach láechdacht⁹,
ría ríg ní rúamnat renna
más é Dubthach Dóelteṅga.⁷'

Carpat Serda connice sin.

---

¹⁻¹ *read* fri doíni *and* cotsoídi? (iss ed dogní fri doíni | a nad rubad cosaídi LU 6599–60)   ²⁻² *in marg.* .i. far nindili   ³⁻³ *om. MS., supplied from* LU, YBL, C   ⁴ branfossaib *LU,* branḟesaib *YBL, C*   ⁵⁻⁵ hi lergaib LU, YBL, C   ⁶⁻⁶ do díbergaib *LU, YBL, C*   ⁷⁻⁷ *this quatrain not in* LU, YBL, *but in* C *and Rec. III*   ⁸ *add. below line*   ⁹⁻⁹ nach *omitt.* (cen laochdacht *C,* cona laechdhacht *Rec. III) or better delete* a *at beg. of line and read* mac Lugdach gan nach láechdacht

IS and sin ras fárraid óclách rodána do Ultu na slúagu dárbo chomainm Óengus mac Óenláime Gábe, & imsóe reme na slúagu a Modaib[1] Loga risi ráter Lugmud in tan sa co Áth Da Ḟert i Sléib Ḟúait. Iss ed marímat eólaig dámmad ar galaib óenḟir dosfístá Óengus mac Óenláime Gaibe [2]ar co[2] táetsaitis leis riam remáin reme ar galaib óenḟir[3]. Ní hed ón dogníset-som itir acht dogníth cathetarnaid imme bar cach leth go torchair accu ac Áth Da Ḟert i Sléib Ḟúait.

Imrol[l] Belaig Eóin and so innossa

IS and sin radechaid chucu-som Fiacha Fíaldána do Ultaib d'acallaim meic sethar a máthar, Mane Andóe de Chonnachtaib. Ocus is amlaid tánic-som ⁊ Dubthach Dóel Ulad maróen riss. Is amlaid tánic in Mane Andóe 'no ⁊ Dóche mac Mágach aróen riss. [4]Á danaccaig[4] in Dóche mac Mágach in Fiacha Fíaldána, tarlaic sleig fair fá chétóir co mboí tríana charait fadessin, tri Dubthach Dáel Ulad. Tarlaic in Fiacha Fíaldána sleig for Dóche mac Mágach co mbaí tríana[5] charait badessin, tri Mane Andóe de Chonnachtaib. And sin atrubratar fir Hérend: 'Is imroll díbairgthi,' bar íat-som, 'a tarla dona feraib, cách díb do guin a charat ⁊ a choibnesaim badessin.' Conid Imroll Belaig Eóin and sin. Ocus Imroll aile Belaig Eóin ainm aile dó 'no.

Tuige im Thamon and so innossa

AND sin ra ráidsetar fir Hérend ri Tamun drúth étgud Ail*ill* ⁊ a imscimm n-órda do gabáil immi ⁊ techt farin n-áth bad ḟiadnaissi dóib. Ra gabastar-som 'no étgud nAilella ⁊ a immscimm órda immi ⁊ tánic barin n-áth bad ḟiadnaisi dóib. Ra gabsat fir Hérend ac cluchi ⁊ ac gredan is ac fochuitbiud imme. 'Is tuige im thamon duit-siu ám, a Thamuin drúith,' bar íat-som, 'étgud nAilella ⁊ a imscim n-órda immut.' Corop Tuigi im Thamon and sain. Dachonnaic Cú Chulaind é ⁊ indar leis i n-écmais a ḟessa ⁊ a eólais ba sé Ailill baí and fadessin, & bosreóthi cloich assa c[h]rantabaill úad fair co n-art Tamun drúth can anmain barsinn áth i rrabi.

Corop[6] Áth Tamuin and sin ⁊ Tugi im Thamon.

---

1 Moda bí *MS.* (oc Modaib Loga LU, as Muighibh Loga *St*)
2–2 *sic*; *read* co
3 oenḟer *MS.*; *from* riam remáin *to* oenḟir *omitt.*
4–4 O't-connairc *St*
5 trí | na *MS.*
6 conadh *St*

RA gabsat cethri ollchóicid Hérend dúnad ⁊ longphort acon chorthe i Crích Ross in n-aidchi sin. And sin conattec[h]t
²⁴⁷⁵ Medb firu Hérend im nech díb do chomlond ⁊ do chomrac ra Coin Culaind arnabárach. Iss ed atdeired cach fer: 'Níba missi,' ⁊ 'Níba mé as mo magin. Ní dlegar cimbid dom chenéol.'

And sin conattecht Medb Fergus do chomlund ⁊ do chomrac ra Coin Culaind ar ros fémmid firu Hérend. 'Nírbo chomadas dam-sa
²⁴⁸⁰ sain,' bar Fergus, 'comrac ra gilla n-óc n-amulchach gan ulcha itir ⁊ ram dalta badessin,' Cid trá acht ¹a facessa¹ Medb Fergus co tromm, da fémmid gana comrac ⁊ gana comlund do gabáil do láim. Dessetar in n-aidchi sin and. Atraacht Fergus co moch arnabárach & tánic reme co áth in chomraic co airm i mbáe Cú Chulaind.
80ᵃ Atchonnairc Cú Chulaind dá ṡaigid é. 'Is fóenglinne | dotháet mo phopa Fergus dom ṡaigid-se. Ní fuil claideb i n-intiuch na lúe móre leis.'—Fír dó-som. Bliadain riasin sceol sa tarraid Ailill Fergus ic techt i n-óentaid Medba arsind lettir i Crúachain ⁊ a chlaideb arsind lettir 'na farrad, & tópacht Ailill in claideb assa
²⁴⁹⁰ intig ⁊ dobretha claideb craind dia inud, ⁊ dobert a bréthir ná tibred dó co tucad lá in chatha móir.—'Cumma limm itir, a daltáin,' bar Fergus, 'dáig giana beth claideb and so, nít ricfad-su ⁊ ní himmértha fort. Acht arinn airer ⁊ arinn altrom rabertus-sa fort ⁊ rabertatar Ulaid ⁊ Conchobor, teich romum-sa indiu i fiadnaisi
²⁴⁹⁵ fer nHérend.' 'Is lesc lim-sa innísin ám,' bar Cú Chulaind, 'teiched ria n-óenfer for Táin Bó Cúalnge.' 'Ní lesc ám duit-siu ón,' bar Fergus, 'dáig techfet-sa remut-sa inbaid bus chréchtach crólinnech tretholl tú bar cath na Tána, & á theichfet-sa m'óenur, teichfit fir Hérend uile.' Da baí dá mét rap áil do Choin Chulaind less
²⁵⁰⁰ Ulad do dénam co tucad a charpat d'indsaigid Con Culaind ⁊ co lluid 'na charpat & tánic i mmadmaim ⁊ i teiched ²ó feraib Hérend². Atchondcatar fir Hérend anísin. 'Ra theich romut! Ra theich remut, a Ferguis!' bar cách. 'A lenmain! A lenmain, a Ferguis!' bar Medb. 'Ná táet dít!' 'Aicce ón omm.' bar Fergus, 'nachas
²⁵⁰⁵ linub-sa secha so, dáig cid bec lib-si in cutrumma techid út rabertus-sa fair, ní thuc óenfer do feraib Hérend in neoch conarnecar ris ar Táin Bó Cúalnge. Is aire-sin ná co rísat fir Hérend timchell ar galaib óenfir, ní ricub-sa arís in fer cétna.'

Conid Comrac Fergusa and sin.

---

¹⁻¹ *read* asacessa *or* aracessa (*Windisch*)

²⁻² *better* ó Fhergus *as in* St 2496; uadha *Rec. III* (*cp.* Is and sin dolléci Cu Chulaind traigid for culu re Fergus LU 6695)

Cinnit Ferchon and so innossa 2510

FErchú Loṅgsech ésen do Chonnachtaib, baí-side bar gail ꝛ bar fogail Ailella ꝛ Medba. Án ló ra gabsatar ríge ní thánic fecht 'na ndúnud ná slúagad ná hairc ná écen ná écendál acht ac argain ꝛ ac indred a críchi ꝛ a feraind dia n-éis. Is and barrecaibsium i n-airthiur Aí in tan sain. Dá ḟer déc bá ssed a lín. Racúas 2515 dó-som óenḟer ac fostúd ꝛ ac immḟuirech cethri n-ollchóiced Hérend ó lúan taite ṡamna co taite n-imbuilc ac marbad ḟir ar áth cach laí díb ꝛ cét láech cach n-aidchi. Da mídair-sium a chomairle aice ra muntir. 'Cid bad ḟerr dún in chomairle dagénmais,' bar ésium, 'ná dul d'ḟópairt ind ḟir út fail ic fostúd ꝛ ac imḟuirech cethri n-oll- 2520 chóiced Hérend [ꝛ] a chend ꝛ a choscor do breith lind d'indsaigid[1] Ailella ꝛ Medba. Cid mór d'olcaib ꝛ d'écóraib dariṅgsem ri hAilill ꝛ ra Meidb, dagébam ar síd fair acht co táeth in fer sain lind.' Is hí-sein comairle ba nirt leó-som, & táncatar rempo go airm i mbaí Cú Chulaind, ꝛ and úair tháncatar, ní fír fer ná comlond óenḟir ra 2525 damsatar dó acht imsáiset na dá feraib déc fóe fa chétóir. Imsóe Cú Chulaind friu-som 'no ꝛ eiscis a dá cend déc díb fá chétóir & sádis dá lia déc leó i talmain. Acus atbert[2] cend cach fir díb bara líic & atbert[2] cend Ferchon Longsig 'no bar[a][3] líic. Conid Cinnit Ferchon Longsig áit i fargab Ferchú[4] Longsech a chend .i. Cennáit 2530 Ferchon.

IS and sin ra himráided ac feraib Hérend cia bad chóir do chomruc ꝛ do chomlund ra Coin Culaind ra húair na maitne muche arnabárach. Iss ed ra ráidsetar uile combad é Calatín Dána cona ṡecht maccaib fichet ꝛ a úa Glass mac Delga. Is amlaid ra 2535 bátar-saide neim ar cach ḟir díb ꝛ neim ar cach arm dá n-armaib, ꝛ ní theilged nech díb urchor n-imraill, ꝛ ní ḟuil bara fuliged nech díb, manbad marb a chétóir, rabad[5] marb ria cind nómaide.

80[b] Doragelta comada móra dóib arin comlund ꝛ arin comruc | do dénam, & ra gabsat do láim a dénam ꝛ bad ḟiadnaisi d'Ḟergus ra 2540 naidmthea sain, & ra fémmid tiachtain taris. Dáig iss ed ra ráidsetar corbo chomlund óenḟir leó Calatín Dána cona ṡecht maccaib fichet ꝛ a úa Glass mac Delga, dáig iss ed ra ráidset corbo ball dá ballaib a mac ꝛ corbo irrand dá irrandaib ꝛ combad[6] ra Calatín Dána sochraiti a chuirp fadessin. 2545

Tánic Fergus reme dochum a phupla ⁊ a muntiri ⁊ rabert a osnad
scísi bar aird. 'Is trúag lind in gním doníther imbárach and,' bar
Fergus. 'Garsa gním sain?' bar a munter. 'Cú Chulaind do marbad,'
bar ésium. 'Uch!' bar íat-som, 'cia marbas?' 'Calatín Dána,' bar
2550 ésium, 'cona secht maccaib fichet ⁊ a úa Glass mac Delga. Is amlaid
atát neim ar cach fir díb ⁊ neim ar cach arm dá n-armaib, ⁊ ní fuil
bara fuliged nech díb, munub marb a chétóir, nába marb ria cind
nómaide, & ní fuil digsed dá fiss dam-sa bad fiadnaisi don chomlund
⁊ don chomroc - daberad a fiss dam mar da mairbfithea Cú Chulaind,
2555 ná tibrind mo bennac[h]tain ⁊ m'eirred.'¹ 'Rachat-sa and,' bar
Fiachu mac Fir Aba. Dessetar and in n-aidchi sin. Atraacht
Calatín Dána co moch arnabárach cona secht maccaib fichet ⁊ a
úa Glass mac Delga, & táncatar rempo co hairm i mbáe Cú Chulaind,
& tánic 'no Fiacho mac Fir Aba. Ocus án úair ránic Calatín co
2560 airm i mbáe Cú Chulaind, tarlaicset a nnóe ngae fichet fair a chétóir
⁊ ní dechaid urchur n-imruill díb secha. Doringni Cú Chulaind
fáebarchless don scíath comdas ralatar uile coa mbolgánaib 'sin
scíath. Ac[h]t nírb urchur n-imruill dóib-sium sain, nír fulig ⁊
nír forderg gae díb fair-sium. Is and sin barróisc Cú Chulaind in
2565 claidiub assa intiuch Bodba d'imscothad na n-arm ⁊ d'immétrom-
mugud in scéith fair. I céin ra buí-seom aice-sain, ra ethsat-som
chuce ⁊ ra sáidsetar na nóe ndesndurnu fichet i nn-óenfecht ina
chend. Da chúrsatar-sun ⁊ ra chrommsatar leó é co tarla a gnúis
⁊ a aged - a einech ra grian ⁊ ra ganem inn átha. Rabert-sun a
2570 rucht míled bar aird - a iachtad n-écomlaind connach baí d'Ultaib
i mbethaid do neoch donárbo chotlud ná cúala. And-saic dariacht
Fiacha mac Fir Aba dá saigid ⁊ atconnaire aní sin, & tánic a ell
chondailbi fair, & barróisc in claideb asa intiuch Bodba ⁊ rabert
béim dóib coro scoth a nóe ndesndurnu fichet d'óenbulli díb co
2575 torchratar uile dara n-aiss ra dichracht ind fedma ⁊ in gremma i
rrabatar.
    Túargaib Cú Chulaind a chend ⁊ ra theilg a anáil ⁊ rabert a osnaid
scísi fair anechtair ⁊ rachonnaic intí ra fóir é. 'Is teóir i n-éim, a
derbchomalta,' bar Cú Chulaind. 'Cid teóir i n-éim duit-siu é,
2580 níba teóir a n-éim dúnni, dóig ra fuilemm trichait chét i n-as dech
clainne Rudraige i ndúnud ⁊ i llongphurt fer nHérend [⁊] rarbérthar
uile fa gin gae ⁊ chlaidib, cid bec lat-su in béim ra benas-sa, mad
dia festar forund é.' 'Tiur-sa bréthir,' bar Cú Chulaind, 'ó thúar-
gabusa mo chend ⁊ ára thelggius m'anál, acht mana derna badessin
2585 scél fort, nach nech díb-siút dagéna fadesta.' Is and sin imsóe
Cú Chulaind friu ⁊ ra gab bara slaide ⁊ bara slechtad coros cuir |

¹ add dó?

81ª úad 'na n-ágib minta ⁊ 'na cethramthanaib fodalta ar fut inn átha sair ⁊¹ siar. Ra étlá óenfer díb úad i mmunigin a retha i céin ra buí-sium ac díchennad cháich .i. Glas mac Delga. Ocus rabert Cú Chulaind sidi friss ⁊ tánic reme timchell pupla Ailella ⁊ Medba, ⁊ ní arnecair úad a ráda acht 'Fiach! fiach!' tráth rabert Cú Chulaind béim dó co tópacht a chend de. 'Is troit² ra báss risin fer út,' bar Medb. 'Gá fíach sút ra imráid, a Ferguis?' 'Nád fetar,' bar Fergus, 'acht meni dlessad fíachu do neoch 'sin dúnud ⁊ 'sin longphurt. Is íat ra boí ³ar[a] ari³. Acht atá ní chena,' bar Fergus, 'is fíach fola ⁊ feóla dó-som é. Atiur-sa bréthir chena,' bar Fergus, 'is innossa ra íctha a féich uile i n-óenfecht riss.'

Darochair Calatín Dána bán cóir sin ra Coin Culaind cona sécht maccaib fichet ⁊ a úa Glass mac Delga. Conid marthanach ar lár inn átha fóss in chloch 'ma ndernsat a sróengal ⁊ a n-imresón. Inad elta a claideb inti ⁊ a ṅglúni ⁊ a n-ullend ⁊ erlanna a sleg. Conid Fuil Iairn ra Áth Fir Diad aníar ainm inn átha. Is aire atberar ⁴Fuil [Iairn]⁴ ris dáig báe fuil dar fáebor and.

Conid Comrac Clainne Calatín connice sin.

2590
2595
2600
2605

Comrac Fir Dead in so

IS and sin ra imráided oc feraib Hérend cia bad chóir do chomlond ⁊ do chomrac ra Coin Culaind ra húair na maitni muchi arnabárach. Iss ed ra ráidsetar uile combad é Fer Diad mac Damáin meic Dáre, in mílid mórchalma d'feraib Domnand. Dáig ba cosmail ⁊ ba comadas a comlond ⁊ a comrac. Ac óenmummib darónsat⁵ gnímrada⁶ ⁷gaile ⁊ gascid⁷ do⁸ foglaim, ac Scáthaig ⁊ ac Úathaig ⁊ ac Aífe. Ocus ní baí immarcraid neich díb ac araile acht cless in gae bulga ac Coin Culaind. Cid ed ón ba coṅganchnessach Fer Diad ac comlund ⁊ ac comrac ra láech ar áth 'na agid-side.

Is and sin ra faíttea fessa ⁊ techtaireda ar cend Fir Diad. Ra érastar ⁊ ra éittchestar ⁊ ⁹ra repestar⁹ Fer Diad na techta sin & ní thánic leó dáig rafitir aní 'ma rabatar dó, do chomlond ⁊ do

2610
2615

---

¹ ⁊ *add. above line;* soir siar *St*   ² .i. is opund *in marg.*
³⁻³ ara airi-siomh *St*   ⁴⁻⁴ Fuil Iairn *St*   ⁵ daringsetar *interlined*
⁶ ceird *added in margin* (*cp.* Ag áonbhuime do-rónsatt a gcearda goile ⁊ gaisgidh d'foghlaim *St* 2609–10)   ⁷⁻⁷ gascid gaile *MS. with marks of transposition;* ⁊ *subscr.*   ⁸ dara *MS.*   ⁹⁻⁹ *read* ro opastar (*Windisch*)

²⁶²⁰ chomrac re charait, re chocle ⁊ re chomalta, ¹re Coin Culaind mac
Sualtaim¹, ⁊ ní thánic leó.   Is and sin faítte Medb na drúith ⁊ na
glámma ⁊ na crúadgressa ar cend Fir Diad ar co nderntaís teóra
áera sossaigthe dó ⁊ teóra glámma dícend go tócbaitís teóra bolga
bara agid, ail ⁊ anim ⁊ athis, ²murbud marb a chétóir combad marb
²⁶²⁵ re cind nómaide², munu thísed.   Tánic Fer Diad leó dar cend a
enig, dáig ba hussu les-sium a thuttim do gaib gaile ⁊ gascid ⁊
eṅgnama ná a thuttim de gaaib aíre ⁊ écnaig ⁊ imdergtha.   Ocus
á daríacht, ra fíadaiged ⁊ ra fritháled é & ra dáled lind soóla sochaín
somesc fair gurbo mesc medarchaín é, ⁊ ra gelta comada móra dó
²⁶³⁰ arin comlond ⁊ arin comrac do dénam .i. carpat cethri secht cumal
⁊ timthacht dá fer déc d'étgud cacha datha ⁊ comméit a feraind
de mín Maige hAí gan cháin ³[gan chobach, gan dúnad, gan
81ᵇ slúagad]³ | gan écendáil dá mac ⁊ dá úa ⁊ dá iarmúa go brunni
mbrátha ⁊ betha, ⁊ Findabair d'óenmnaí ⁊ in t-eó óir báe i mbrutt
²⁶³⁵ Medba fair anúas.

Is amlaid ra baí Medb 'gá ráda ⁊ rabert na briathra and ⁊ ra
[f]recair Fer Diad:

'Rat fia lúach mór mbuinne,
⁴rat chuit⁴ maige is chaille,
²⁶⁴⁰ ra saíre do chlainne
á 'ndiu co tí bráth.
A Fir Diad meic Damáin,
⁵eirggi guin is gabail,⁵
'tetha ás cech anáil,
²⁶⁴⁵ cid duit gana gabáil
aní gabas cách?'

F.D.   'Ní géb-sa gan árach
dáig ním láech gan lámach,
bud tromm form imbárach,
²⁶⁵⁰ bud fortrén in feidm.
Cú dán comainm Culand,
is⁶ amnas in urrand,
⁷ní furusa⁷ a fulang,
bud tairpech in teidm.'

¹⁻¹ re Fer ṅDiad mac ṅDamáin meic Dáre *MS.* (fri Coin Culainn *St*)
²⁻² *om. St., YBL*   ³⁻³ *end line cut off MS.;* cen cios na cain no coblach
⁊ cen dunad ⁊ cen sluaiged *St* 2629–30; can chain cen chobach *YBL, C*
⁴⁻⁴ *sic;* la cuid *St*   ⁵⁻⁵ *this line is superfluous; om. YBL;* atraig guin
is gabail *St;* atrac goin [is gabail] *C*   ⁶ budh *St, C*   ⁷⁻⁷ nibad **hurus** *St*

M. 'Rat fíat laích rat láma. ²⁶⁵⁵
noco raga ar dála,
sréin ⁊ eich ána
dabertatar[1] rit láim.
A Fir Diad inn ága,
dáig isat duni[2] dána. ²⁶⁶⁰
dam-sa bat fer gráda
sech cách gan nach cáin'.

F.Dead 'Ní rag-sa gan rátha
do chluchi na n-átha.
méraid co llá mbrátha ²⁶⁶⁵
go [m]bruth is co mbríg.
Noco géb ge ésti,
gera beth dom résci.
gan gréin ⁊ ésci
la muir ⁊ tír.' ²⁶⁷⁰

M. 'Gá chan duit a fuirech.
naisc-siu gorbat buidech
for deiss ríg is ruirech,
doragat rat láim.
Fuil sund nachat tuilfea. ²⁶⁷⁵
rat fia cach ní chungfea.
dáig rafess co mairbfea
in fer thic it dáil.'

F.D. 'Ní géb gan sé curu.
níba ní bas lugu. ²⁶⁸⁰
sul donéor mo mudu
i mbail i mbíat slúaig.
Dánam thorrsed m'ardarc.
cinnfet cuncup[3] comnart.
co ndernur in comrac ²⁶⁸⁵
ra Coin Culaind crúaid.'

---

[1] dobertar *St*, *C* (=do-bérthar), *sic leg.*
[2] cur *C*
[3] cincob *St*

M. 'Cid Domnall ná Charpre
ná Níamán án airgne,
gid íat lucht na bairddne,
2690   rot fíat-su ¹gid acht¹.
Fonasc latt ar Morand,
mad áill latt a chomall,
naisc Carpre mín Manand,
is naisc ²ar dá macc².'

2695 F.D. 'A Medb co mét mbúafaid,
nít chredb caíne núachair,
is derb is tú is búachail
ar Crúachain na clad.
Ard glór is art gargnert,
2700   domroiched sról santbrecc,
tuc dam th'ór is t'arget
dáig ro fairgged dam.'

M. 'Nach tussu in caur codnach
dá tiber delgg ndrolmach?
2705   Ó'ndiu co tí domnach
níba dál ba[s] sía.
A laích blatnig bladmair,
cach sét cáem ar talmain
dabérthar duit amlaid,
2710   is uili rot fía.   R.

'Finnabair na fergga,
rígan íarthair Elgga,
ar ndíth Chon na Cerdda,
a Fir Diad, rot fía'.   R.

2715 Is and sain rasiacht Medb máeth n-áraig bar Fer nDiad im
chomlond ┐ im chomrac ra sessiur curad arnabárach ná im chomlond
┐ im chomrac ra Coin Culaind a óenur dámbad assu leiss. Rasiacht
Fer Diad máeth n-áraig furri-si, nó andar leis, im chur in tsessir
chétna imna comadaib ra gellad dó do chomallud riss mad dá
2720 tóetsad Cú Chulaind leiss.
And sain ra gabait a eich d'Fergus ┐ ra hindled a charpat, & tánic
reme co airm ³[i mbaí Cú Chulaind dá innisin]³ dó sain. Firiss

───────────────

¹⁻¹ cen acht *YBL*   ²⁻² fora mac *C*   ³⁻³ *end line partially cut away;
tops of some letters left showing agreement with Stowe reading which is given*

82ª Cú Chulaind fálti riss. | 'Mo chen do thíchtu, a mo phopa Ḟerguis,' bar Cú Chulaind. 'Tarissi lim inní inn fálti, a daltáin,' bar Fergus, 'acht is dó radechad-sa dá innisin duit intí rotháet do chomlond ⁊ ²⁷²⁵ do chomruc rutt re húair na maitne muche imbárach.' 'Clunem-ni latt didiu,' bar Cú Chulaind. 'Do chara féin ⁊ do chocle ⁊ do chomalta, t'fer comchliss ⁊ comgascid ⁊ comgníma, Fer Diad mac Damáin meic Dáre, in mílid mórchalma d'feraib Domnand.' 'Attear¹ ar cobais,' bar Cú Chulaind, 'ní 'na dáil dúthracmar ar cara do thuidecht.' ²⁷³⁰ 'Is aire-sein iarum ale,' bar Fergus, 'ara n-airichlea ⁊ ara n-airelma, dáig ní mar chách conarnecar comlund ⁊ comrac riut for Táin Bó Cúalnge don chur sa Fer Diad mac Damáin meic Dáre.' 'Attú-sa sund ám,' bar Cú Chulaind, 'ac fostud ⁊ ac imfurech cethri n-oll-chóiced ṅHérend ó lúan taite ṡamna co tate imbuilg, & ní rucus ²⁷³⁵ traig techid re n-óenfer risin ré sin ⁊ is dóig lim ní mó bérat remi-sium.' Acus is samlaid ra baí Fergus² 'gá rád 'gá báeglugud ⁊ rabert na briathra ⁊ ra [ḟ]recair Cú Chulaind:

> 'A Chú Chulaind, comal ṅglé,
> atchíu is mithig duit éirge,   2740
> atá sund chucut ra feirg
> Fer Diad mac Damáin drechdeirg.'

C.C. 'Atú-sa sund, ní seól seṅg³,
> ac trénastud fer ṅHérend.
> Ní rucus for teched traig   2745
> ar apa chomlund óenfir.'

F. 'Amnas in fer dalae feirg
> as luss a chlaidib cródeirg,
> cnes coṅgna im Ḟer ṅDiad na ndroṅg,
> ris ní geib cath ná comlond.'   2750

C.C. 'Bí tost, ná tacair do scél,
> a Ḟerguis na n-arm n-imthrén,
> dar cach ferand, dar cach fond
> dam-sa nochon écomlond.'

F. 'Amnas in fer, fichtib gal,   2755
> nochon furusa a thróethad,
> nert cét 'na churp, calma in mod,
> nín geib rind, nín tesc fáebor.'

---

¹ *sic for* dar; dar ar mbreithir *YBL, C,* dar mo chubus *St*
² Fergus *follows* ga rád *with marks of transposition in MS.*
³ saeṅg *MS. corr. from* faṅg; seṅg *YBL, St, C*

C.C. 'Mad dia comairsem bar áth,
2760 missi is Fer Diad gascidgnáth.
níba é in scarad gan sceó,
bud ferggach ar fáebargleó.'

F. 'Rapad ferr lem andá lúag.
a Chú Chulaind chlaidebrúad.
2765 combad tú raberad sair
coscur Fir Diad díummasaig.'

C.C. 'Atiur-sa bréthir co mbáig
gen[1] com maith-se oc immarbáig.
is missi búadaigfes de
2770 bar mac ṅDamáin meic Dáre.'

F. 'Is mé targlaim na slúagu sair,
lúag mo sáraigthe d'Ultaib.
lim tháncatar á tírib
a curaid, a cathmílid.'

2775 C.C. 'Munbud Chonchobor 'na chess,
rápad chrúaid in comadchess,
ní thánic Medb Maige in Scáil
turus [ríam][2] bad mó congáir.'

F. 'Ra fail gním is mó bard láim,
2780 gleó ra Fer ṅDiad mac ṅDamáin.
Arm cruaid catut cardda raind
bíd acut, a Chú Chulaind.' A.

Tánic Fergus reme dochum ṅdúnaid ⁊ longphuirt. Luid Fer
Diad dochum a pupla ⁊ a muntiri & rachúaid dóib máeth n-áraig
2785 do tharrachtain do Meidb fair im chomlond ⁊ im chomrac ra sessiur
curad arnabárach ná im chomlund ⁊ im chomrac ra Coin Culaind
a óenur diambad assu leiss. Dachúaid dóib 'no máeth n-áraig
do tharrachtain dó-som for Meidb im chuir[3] in tsessir churad

---

[1] gon MS. cen St, C   [2] om. MS., supplied from YBL (St, C)
[3] sic: cp. supra l. 2718

chétna imna comadaib ra gellad dó do chomallad riss mad dá
táetsad Cú Chulaind leiss.   2790

Nírdar subaig sámaig ¹sobb[rónaig somenmnaig lucht puible
82ᵇ Fir Diad in n-aidhchi sin]¹ | acht rapsat dubaig dobbrónaig domen-
mnaig, dóig rafetatar airm condricfaitis na dá curaid - na dá
chliathbernaid chét co táetsad cechtar díb and nó co táetsaitís a
ndís, & dám² nechtar díb, dóig leó-som gombad é a tigerna féin. ²⁷⁹⁵
Dáig níba réid comlond *nó* comrac ra Coin Culaind for Táin Bó
Cúalnge.

Ra chotail Fer Diad tossach na haidchi co rothromm, & á thánic
deired na haidchi rachúaid a chotlud úad ᛅ raluid a mesci de.
Ocus da baí ceist in chomlaind ᛅ in chomraic fair, & ra gab láim ²⁸⁰⁰
ara araid ara ṅgabad a eocho ᛅ ara n-indled a charpat. Ra gab
in t-ara 'gá imthairmesc imme. 'Rapad ḟerr dúib ³[anad iná dul
and sin]³,' ar se in gilla. 'Bí tost dín, a gillai,' ar Fer Diad. Ocus
is samlaid ra boí 'gá rád ᛅ rabert na bríathra and ᛅ ra frecair in gilla:

'Tiagam issin dáil-sea   2805
    do chosnam ind ḟir-sea
    go rrísem in n-áth-sa,
    áth fors ṅgéra in Badb.
I comdáil Con Culaind,
    dá guin tre chreitt cumaiṅg   2810
    go rruca thrít urraind
    corop de bus marb.'

'Rapad ḟerr dúib anad,
    níba mín far magar,
    biaid nech diamba galar,   2815
    bar scarad bud snéid.
Techt i ndáil ailt Ulad,
    is dál dia mbia pudar,
    is fata bas chuman,
    mairg ragas in réim.⁴'   2820

'Ní cóir ana rádi,
ní hopair niad náre,
ní dlegar dín ále,
ní anfam fad dáig.
2825   Bí tost dín, a gillai.
Bid calma ár¹ síst sinni,
ferr teinni ná timmi,
²tiagam isin dáil.²'

Ra gabait a eich Fir Diad ⁊ ra indled a charpat & tánic reme co
2830 áth in chomraic, & ní thánic lá cona lánṡoilsi dó and itir. 'Maith a
gillai,' bar Fer Diad, 'scar dam fortcha ⁊ forgemen mo charpait
fóm and so coro tholiur mo thromthairthim súain ⁊ chotulta and
so, dáig níra chotlus deired na haidchi ra ceist in chomlaind ⁊ in
chomraic.' Ra scoir in gilla na eich. Ra díscuir in carpat fóe.
2835 Toilis a thromthairthim cotulta fair.

Imthúsa Con Culaind sunda innossa : ní erracht-side itir co tánic
láa cona lánṡoilse dó dáig ná hapraitis fir Hérend is ecla nó is úamun
dobérad ḟair mad dá n-éirged³. Ocus ó thánic láa cona lánṡolsi,
ra gab láim ar[a] araid ara ṅgabad a eocho ⁊ ara n-indled a charpat.
2840 'Maith a gillai,' bar Cú Chulaind, 'geib ar n-eich dún ⁊ innill ar
carpat dáig is mochérgech in láech ra dáil 'nar ṅdáil, Fer Diad
mac Damáin meic Dáre.' 'Is gabtha na eich, iss innilti in carpat.
Cind-siu and ⁊ ní tár dot gasciud.'

Is and sin cinnis in cur cetach clessamnach cathbúadach claideb-
2845 derg, Cú Chulaind mac Sualtaim, ina charpat. Gura gáirsetar imme
boccánaig ⁊ bánanaig ⁊ geniti glinne ⁊ demna aeóir, dáig dabertis
Túatha Dé Danand a ṅgáriud immi-sium combad móti a gráin ⁊
a ecla ⁊ a urúad ⁊ a urúamain in cach cath ⁊ in cach cathroí, in
cach comlund ⁊ in cach comruc i téiged.
2850   Nírbo chian d'araid Ḟir Diad co cúala inní ⁴[in fúaim ⁊ an fothram
83ᵃ ⁊ in fidrén ⁊ in toirm ⁊ in torann ⁊]⁴ | in sestanib ⁊ in sésilbi .i.
sceldgur na scíath cliss ⁊ slicrech na sleg ⁊ glondbéimnech na claideb
⁊ bressimnech ⁵in chathbarr⁵ ⁊ drongar na lúrigi ⁊ imchommilt
na n-arm, dechraidecht na cless, téteinmnech⁶ na tét ⁊ núallgrith
2855 na roth ⁊ culgaire in charpait ⁊ basschaire na n-ech ⁊ trommchoblach
in churad ⁊ in chathmíled dochum inn átha dá ṡaigid.

Tánic in gilla ⁊ forromair a láim fora thigerna. 'Maith a Ḟir

---

¹ = iar; ar YBL, iar C   ²⁻² supplied from l. 2820 above
³ go moch add. St   ⁴⁻⁴ end line cut off in MS., supplied from St
⁵⁻⁵ na ccathbharr St   ⁶ tedbemnech St

Diad,' bar in gilla, 'comérig ⁊ atáthar sund chucut dochum inn átha.'
Ocus rabert in gilla na briathra and:

'Atchlunim cul carpait   2860
ra cuiṅg n-álaind n-argait,
is fúath fir co forbairt
  ¹ás droich¹ carpait chrúaid.
Dar Bregross, dar Braine,
focheṅgat in slige,   2865
sech bun Baile in Bile
  is búadach a mbúaid.

'Is cú airgdech aiges,
is carptech glan geibes²,
is seboc sáer slaidess   2870
  a eocho fadess.
Is cródatta in cua,
is demin donrua,
rafess, níba tua,
  dobeir dún in tress.   2875

'Mairg bías isin tulaig
ar cind in chon cubaid,
barrarṅgert-sa ánuraid
  ticfad giped chuin.
Cú na hEmna Macha,   2880
cú co ndeilb cach datha,
cú crichi³, cú catha,
  dochlunim, rar cluin.'   At.

'Maith a gillai,' bar Fer Diad, 'gá fáth 'ma ra molais in fer sain
ó thánac ó[t] tig? Ocus is súail nach fatha conais dait a romét 2885
ros molais, ⁊ barairṅgert Ailill ⁊ Medb dam-sa go táetsad in fer
sain lemm, &⁴ dáig is dar cend lúage lochērthair lem-sa co llúath é
& is mithig in chobair.' Ocus rabert na briathra and ⁊ ra [f]recair
in gilla:

---

¹⁻¹ uas dreich YBL, os dreich C   ² gaibes C
³ creichi YBL, C   ⁴ omitt.

2890
'Is mithig in chabair,
bí tost dín, nach mbladaig,
nárbu gním ar codail
dáig ní bráth dar brúach.

2895
Má 'tchí churaid Cúalnge
co n-adabraib úalle,
dáig is dar cend lúage
lochērthair co llúath.'

'Má 'tchím curaid Cúalnge
co n-adabraib úalle,

2900
ní 'r teiched téit úanne
ac[h]t is cucaind tic.
Reithid is ní romall,
gid rogáeth ní rogand,
mar[1] usci d'forall

2905
ná mar thoraind tricc.'

'Súail nach fotha [conais][2]
a romét ras molaiss,
gá fáth 'ma ra thogais
ó thánac ó[t] tig?

2910
Iss innossa thócbait,
atát acá fúacairt,
ní thecat dá fúapairt
acht [mad][2] athig mith.'   M.

Nírbo chían d'araid Fir Diad dia mboí and co facca ní, in carpat
2915 caín cúicrind cethirrind[3] go llúth go llúais go lángliccus, go pupaill
úanide, go creit chráestana chráestírim chlessaird cholgfata churata,
ar dá n-echaib lúatha lémnecha ómair bulid bedgaig bolg[s]róin
uchtlethna beochridi blénarda basslethna cosscháela forttréna for-
ráncha fúa.   Ech líath leslethan lugléimnech[4] lebormongach
2920 fándara chuing don charput[5].   Ech dub dúalach dulbrass druim-
lethan fán chuing anaill.   Ba samalta ra sebacc dá chlaiss i lló
chrúadgaíthi ná ra sidi répgaíthi erraig i lló mártai dar muni
83[b] machairi ná ra tétag n-allaid arna chétglúasacht | do chonaib do
chétroí dá ech Con Culaind immon carpat, marbad ar licc áin
2925 tentidi co crothsat ⁊ co mbertsat in talmain ra tricci na dírma.

---

[1] leg. immar   [2] om. MS. supplied from YBL, C
[3] an leg. cethirriad?   [4] luthleimnech St   [5] charpait MS.

Acus daríacht Cú Chulaind dochum inn átha. Tarassair Fer
Diad barsan leith descertach ind átha. Dessid Cú Chulaind barsan
leith túascertach. Firis Fer Diad fáilte fri Coin Culaind. 'Mo
chen do thíc[h]tu, a Chú Chulaind,' bar Fer Diad. 'Tarissi lim
ní ind fálti mad costráthsa,' bar Cú Chulaind, '⁊ indiu ní dénaim ²⁹³⁰
tarissi de chena. Acus a Fir Diad.' bar Cú Chulaind, 'rapo chóru
dam-sa fálti d'ferthain frit-su ná dait-siu a ferthain¹ rum-sa, dáig
is tú daríacht in crích ⁊ in cóiced i tú-sa, & níra chóir duit-siu
tíchtain do chomlund ⁊ do chomrac rim-sa ⁊ rapa chóru dam-sa
dol do chomlond ⁊ do chomrac rut-sa, dáig is romut-sa atát mo ²⁹³⁵
mná-sa ⁊ mo meic ⁊ mo maccáemi, m'eich ⁊ m'echrada, m'albi ⁊
m'éiti ⁊ m'indili.' 'Maith a Chú Chulaind,' bar Fer Diad, 'cid rot
tuc-su do chomlund ⁊ do chomrac rim-sa itir? Dáig dá mbámmar
ac Scáthaig ⁊ ac Úathaig ⁊ ac Aífi, is tussu ba forbfer frithálma
dam-sa .i. ra armad mo slega ⁊ ra déirged mo lepaid.' 'Is fír ám ²⁹⁴⁰
sain ale,' bar Cú Chulaind. 'Ar óice ⁊ ar oítidchi donín-sea duit-siu, & ní
hí sin tuarascbáil bá tú-sa indiu itir, acht ní fil barsin bith láech nach
diṅgéb-sa indiu.'

Acus iss and sin ferais cechtar n-aí díb athc[h]ossán n-athgér
n-athcharatraid ráraile². Acus rabert Fer Diad na briathra and ²⁹⁴⁵
⁊ ra [f]recair Cú Chulaind:

'Cid ra[t] tuc, a Chúa,
do throit ra níaid núa?
Bud cróderg da chrúa
ás análaib th'ech. ²⁹⁵⁰
Mairg [tánic]³ do thurus,
bud atód ra haires,
ricfa a less do legess,
mad dá rís do thech.'

C.C. 'Dodechad ré n-ócaib ²⁹⁵⁵
im torc trethan trétaig,
re cathaib, re cétaib,
dot chur-su fán lind,
d'feirg rut is dot [f]romad
bar comrac cét conar ²⁹⁶⁰
corop dait bas fogal
do chosnom do chind.'

---

¹ fearthain *St, C*   ² *sic* = ri araile
³ *om. MS. supplied from YBL* (Bid mairg dit di turus *C*)

F.D. 'Fail sund nech rat méla,
    is missi rat géna
2965   dáig is dím facríth[1].
    C[r]onugud a curad
    i fiadnassi Ulad,
    gorop cian bas chuman,
      gorop dóib bus díth.'

2970   C.C. '[2]Car cinnas[2] condricfam?
    In ar collaib cneittfem?
    [3]Gid leind rarrficfam[3]
      do chomrac ar áth?
    Inn ar claidbib crúadaib,
2975   ná 'n ar rennaib rúadaib,
    dot ṡlaidi rit ṡlúagaib
      má thánic a thráth?'

F.D. 'Re funiud re n-aidche
    mádit éicen airrthe,
2980   comrac dait re Bairche[4]
      níba bán in gleó.
    Ulaid acot gairm-siu,
    [5]ra ṅgabastar aillsiu[5],
    bud olc dóib in taidbsiu,
2985   rachthair thairsiu is treó.'

C.C. 'Datrála i mbeirn ṁbáegail,
    tánic cend do ṡáegail,
    imbérthair ḟort fáebair,
    níba fóill in fáth.
2990   Bud mórglonnach bías,
    condricfa cach días,
    níba tóesech trías[6]
      tú á' ndiu go tí bráth.'

---

[1] glossed .i. tic   [2-2] Cair cindus YBL; cia indus C
[3-3] cia linn ara ficfam C   [4] glossed .i. sliab
[5-5] rotgabsad ar tfaillseo YBL   [6] triair YBL

F.   'Beir ass dín do robud,
      is tú is brassi for domon,   2995
      nít fía lúag ná logud,
      nídat doss ós duss.
      Is missi ratfitir,
      a chride ind eóin ittig,
      at gilla co ṅgicgil,   3000
      gan gasced, gan gus.'

C.C. 'Dá mbámmar ac Scáthaig
      a llus gascid gnáthaig,
      is aróen imréidmís,
      imthéigmís cach fích.   3005
      Tú mo chocne cride,
      tú m'aiccme, tú m'fine,
      ní fúar riam bad dile,
      ba dursan do díth.' |

84ᵃ   'Romór fácbai th'einech   3010
      connǎ dernam deibech,
      siul gairmes in cailech
      biaid do chend ar bir.
      A Chú Chulaind Cúalṅge,
      rot gab baile is búadre,   3015
      rot fía cach olc úanne,
      dáig is dait a chin.'   C.

'Maith a Fir Diad,' bar Cú Chulaind, 'nír chóir duit-siu tíachtain
do chomlund ⁊ do chomrac rum-sa trí indlach ⁊ etarchossaít Ailella
⁊ Medba. ⁊ cach óen tánic, ní ruc búaid ná bissech dóib ⁊ darochratar 3020
limm-sa. Acus ní mó béras búaid ná bissech duit-siu ⁊ rafáethaisiu
limm.'
      Is amlaid ra baí 'gá rád ⁊ rabert na briathra ⁊ ra gab Fer Diad
clostecht fris:

      'Ná tair chucum a laích láin,   3025
      a Fir Diad, a meic Damáin,
      is messu duit na mbía de
      contirfe[1] brón sochaide.

---

[1] *read* confirfe (*Windisch, M. O Daly*); *for* fodirfe ?

'Ná tair chucum dar fír cert,
3030 is lim-sa atá do thiglecht,
cid ná breth and dait nammá
mo gleó-sa ra míleda?

'[1]Nachat mucled ilar cless[1]
[2]girsat corcra conganchness[2],
3035 inn ingen asa taí oc báig,
níba lett, a meic Damáin.

'Findabair ingea[3] Medba,
gé beith d'fébas a delba,
in ingen gid cáem a cruth
3040 nochos tibrea re cétluth.

'Findabair ingen in ríg,
in dráth atberar a fír,
sochaide 'ma tart[4] bréic
⁊ [5]do loitt[5] do lethéit.

3045 'Ná briss form lugi gan fess,
ná bris chíg[6], ná briss cairdess,
[7]ná briss bréthir báig,[7]
ná tair chucum, a laích láin.   N.

'Ra dáled do choícait láech
3050 in ingen, ní dál dimbáeth;
is limm-sa [8]ra faíd[8] a llecht,
ní rucsat úaim acht crandchert.

'[9]Giara maess menmnach Fer Báeth[9]
acá mbaí teglach dagláech,
3055 gar úar gur furmius a bruth,
ra marbus din óenurchur.

'Srubdaire serb seirge a gal,
ba rúnbale na cét mban,
mór a bladalt ra baí than,
3060 nír anacht ór ná étgad.

[1-1] [N]ach[at mu]chleod ilar cless *C; read* imcloed?   [2-2] gersat cur co
congancnoss *C*   [3] ingen *C*   [4] *read* tarat (*Windisch*)   [5-5] ro loit *C*
[6] sith *C*   [7-7] na bris breith na bris baigh *C*   [8-8] ro faiti *C, read* ra faítte
[9-9] Ger amhus menmnach Fer Baeth *C*

'Dámbad dam ra naidmthea in bein[1]
[2]ris tib cend na cóiced cain[2],
nocho dergfaind-se do chlíab
tess ná túaid ná thíar ná tair.' N.

'Maith a Fir Diad', bar Cú Chulaind, 'is aire sin nára chóir duit-siu [3065]
tíachtain do chomlund ⁊ do chomruc rim-sa dáig dá mbámmar ac
Scáthaig ⁊ ac Úathaig ⁊ ac Aífe, is aróen imthéigmís cach cath ⁊
cach cathroí, cach comlund ⁊ cach comrac, cach fid ⁊ cach fásach,
cach dorcha ⁊ cach díamair.' Acus is amlaid ra baí 'gá ráda ⁊
rabert na briathra and: [3070]

r.[3] 'Ropar[4] cocle cridi,
ropar[4] cáemthe caille,
ropar[4] fir chomdéirgide,
contulmis tromchotlud
ar trommníthaib [3075]
i críchaib ilib echtrannaib;
aróen imréidmís,
imthéigmís cach fid,
forcetul fri Scáthaig.'

r.[3] 'A Chú Chulaind cháemchlessach'—bar Fer Diad, [3080]
'ra chindsem ceird comdána,
ra chlóiset cuir caratraid,
bocritha do chétguine,
ná cumnig in comaltus,
a Chúa nachat chobradar, [3085]
[5]a Chúa nachat chobrathar.[5]'

'Roíata atám amlaid seo badesta,' bar Fer Diad, '& gá gasced
ara ragam indiu, a Chú Chulaind?' 'Lat-su do roga gascid chaidchi
indiu,' bar Cú Chulaind, 'dáig is tú daríacht in n-áth ar tús.' 'Indat
mebair-siu itir,' bar Fer Diad, 'isna airigthib gascid daním mís ac [3090]
Scáthaig ⁊ ac Úathaig ⁊ ac Aífe?' 'Isamm mebair ám écin,' bar
Cú Chulaind. 'Masa mebair, tecam[6].'
Dachúatar bara n-airigthib gascid. Ra gabsatar dá scíath chliss
chomardathacha forro ⁊ a n-ocht ocharchliss ⁊ a n-ocht clettíni ⁊
a n-ocht cuilg ṅdét ⁊ a n-ocht ṅgothnatta néit. Imréitis úathu ⁊ [3095]

---

[1] ben C   [2-2] ris tibenn in coiced cain C   [3] in marg.
[4] ropdar MS. with d expunged; robatar St, robdar C
[5-5] sic, omitt.   [6] tiegam forra St

84ᵇ chuccu mar beocho | áinle[1]. Ní thelgtis nád amsitis. Ra gab cách díb ac díburgun araile dina clesradaib sin á dorblas na matne muche go mide medóin laí gora chlóesetar a n-ilchlessarda[2] ra tilib ⅂ chobradaib na scíath cliss. Giara baí d'febas in imdíburcthi ra

3100 boí d'febas na himdegla nára fulig ⅂ nára forderg cách díb bar araile risin ré sin. 'Scurem din gaisced sa fodesta, a Chú Chulaind,' bar Fer Diad, 'dáig ní de seo tic ar n-etergléod.' 'Scurem ám écin má thánic a thráth,' bar Cú Chulaind. Ra scoirsetar. Focherdsetar a clesrada úathaib i llámaib a n-arad.

3105 'Gá gasced i rragam ifesta, a Chú Chulaind?' bar Fer Diad. 'Let-su do roga gascid chaidche,' bar Cú Chulaind, 'dáig is tú doríacht in n-áth ar tús.' 'Tiagam iarum,' bar Fer Diad, 'barar slegaib sneitti snasta slemunchrúadi go súanemnaib lín lánchatut indi[3].' 'Tecam ám écin,' bar Cú Chulaind. Is and sin ra gabsatar

3110 dá chotutscíath chomdaiṅgni forro. Dachúatar bara slegaib snaitti snasta slemunchrúadi go súanemnaib lín lanchotut indi[3]. Ra gab cách díb ac díburgun araile dina slegaib á mide medóin laí go tráth funid nóna. Giara baí d'febas na himdegla ra buí d'febas ind imdíbairgthi goro fuilig ⅂ goro forderg ⅂ goro chréchtnaig cách

3115 díb bar araile risin ré sin. 'Scurem de sodain badesta, a Chú Chulaind,' bar Fer Diad. 'Scurem ám écin má thánic a thráth,' bar Cú Chulaind. Ra scoirsetar. Bacheirdset a n-airm úathu i llámaib a n-arad.

Tánic cách díb d'indsaigid araile assa aithle ⅂ rabert cách díb
3120 lám dar brágit araile ⅂ ra thairbir teóra póc[4]. Ra bátar a n-eich i n-óenscur in n-aidchi sin ⅂ a n-araid ic óentenid, & bognísetar a n-araid cossairleptha[5] úrlúachra dóib go frithadartaib fer ṅgona friu. Táncatar fiallach ícci ⅂ legis dá n-ícc ⅂ dá leiges, & focherdetar lubi ⅂ lossa ícci ⅂ slánsén ra cnedaib ⅂ ra créchtaib, rá n-áltaib
3125 ⅂ ra n-ilgonaib. Cach luib ⅂ cach lossa ícci ⅂ slánsén raberthea ra cnedaib ⅂ créchtaib, áltaib ⅂ ilgonaib Con Culaind, ra idnaicthea comraind úad díb dar áth siar d'Fir Diad, nár abbraitis fir Hérend, dá tuitted Fer Diad leisium, ba himmarcraid legis dabérad fair. Cach bíad ⅂ cach lind soóla socharchaín somesc daberthea ó feraib
3130 Hérend d'Fir Diad, da idnaicthea comraind úad díb dar áth fathúaith do Choin Chulaind, dáig raptar lia bíattaig Fir Diad andá bíattaig Con Culaind. Raptar bíattaig fir Hérend uli d'Fir Diad ar Choin Culaind do diṅgbáil díb. Raptar bíattaig Brega dano do Choin Culaind. Tictis dá acaldaim fri dé .i. cach n-aidche.

---

1 nó aille *written overhead;* *read* i lló áinle   2 = n-ilchlessrada   3 *sic*
4 *sic;* póga *St* (diaraile *add. St*)
5 cossair leptha *Dipl. Edn.* (da cosairleabaid *St*)

Dessetar and in n-aidchi sin. Atráchtatar go moch arnabárach ⁊ [3135]
táncatar rompu co áth in chomraic. 'Gá gasced ara ragam indiu, a
85ᵃ Fir Diad?' bar Cú Chulaind. | 'Lett-su do roga ṅgascid chaidchi.'
bar Fer Diad, 'dáig is missi barróega mo roga ṅgascid isind lathi luid.'
'Tiagam iarum,' bar Cú Chulaind, 'barar manaísib[1] móra murniucha
indiu, dáig is foicsiu lind don ág in t-imrubad indiu ²andá dond [3140]
imdíburgun inné². Gabtar ar n-eich dún ⁊ indliter ar carpait co
ndernam cathugud dar n-echaib ⁊ dar carptib indiu.' 'Tecam ám
écin,' bar Fer Diad. Is and sin ra gabsatar dá lethanscíath lán-
daṅgni forro in lá sin. Dachúatar bara manaísib móra murnecha
in lá sin. Ra gab cách díb bar tollad ⁊ bar tregdad, bar ruth ⁊ bar [3145]
regtad araile á dorblas na matne muchi go tráth funid nóna. Dámbad bés eóin ar lúamain do thecht tri chorpaib dóene, doragtaís
trina corpaib in lá sin, go mbértais na tóchta fola ⁊ feóla trina
cnedaib ⁊ trina créchtaib i nnélaib ⁊ i n-áeraib sechtair. Acus á
thánic tráth funid nóna, raptar scítha a n-eich ⁊ raptar mertnig [3150]
a n-araid ⁊ raptar scítha-som fadessin na curaid ⁊ na láith gaile.
'Scurem de ṡodain badesta, a Fir Diad,' bar Cú Chulaind, 'dáig isat
scítha ar n-eich ⁊ it mertnig ar n-araid, & in tráth ata scítha íat,
cid dúnni nábad scítha sind dano.' Acus is amlaid ra buí 'gá rád
⁊ rabert na briathra and: [3155]

'Ní dlegar dín cuclaigi',—bar ésiun—
'ra fomórchaib feidm;
curther fóthu a n-urchomail
áro scáich a ndeilm.'

'Scoirem ám écin má thánic a thráth,' bar Fer Diad. Ra scorsetar. [3160]
Facheirdset a n-airm úathu i llámaib a n-arad. Tánic cách díb
d'innaigid a chéile. Rabert cách lám dar brágit araile ⁊ ra thairbir
teora póc³. Ra bátar a n-eich i n-óenscur in aidchi sin ⁊ a n-araid
oc óentenid. Bógníset a n-araid cossairleptha⁴ úrlúachra dóib go
frithadartaib fer ṅgona friu. Táncatar fiallach ícci ⁊ leigis dá [3165]
fethium ⁊ dá fégad ⁊ dá forcomét in n-aidchi sin, dáig ní ní aile ra chumg-
etar dóib ra hacbéile a cned ⁊ a créchta, a n-álta ⁊ a n-ilgona acht iptha
⁊ éle ⁊ arthana do chur riu do thairmesc a fola ⁊ a fulliugu⁵ ⁊ a
ngae cró. Cach iptha ⁊ gach éle ⁊ gach orthana doberthea ra
cnedaib ⁊ ra créc[h]taib Con Culaind, ra idnaicthea comraind [3170]
úad díb dar áth síar d'Fir Diad. Cach bíad ⁊ cach lind soóla sochar-

---

¹ mánaisib *MS.*, *accent misplaced*   ²⁻² ina an t-imdiubhraccadh ané *St;*
read andá in t-imdíburgun inné   ³ *sic MS.*, poga *St*   ⁴ cossair leptha
Dipl. Edn. (da cosairleabaidh *St*)   ⁵ *sic MS., read* fuiligthe (a bfuilighthi *St*)

chaín somesc raberthea ó feraib Hérend do Fir Diad, ra hidnaicthea
comraind úad díb dar áth fothúaith do Choin Chulaind, dáig
raptar lia bíataig Fir Diad andá bíataig Con Culaind, dáig raptar
3175 bíattaig fir Hérend uile d'Fir Diad ar diṅgbáil Con Culaind díb.
Raptar bíataig Brega 'no do Choin Chulaind. Tictis dá acallaim
fri dé .i. cach n-aidche.
Dessetar in n-aidchi siu and. Atraachtatar co moch arnabárach,
⁊ táncatar rempo co áth in chomraic. Rachondaic Cú Chulaind
3180 mídelb ⁊ míthemel mór in lá sin bar Fer Diad. 'Is olc ataí-siu
indiu, a Fir Diad,' bar Cú Chulaind. 'Ra dorchaig th'folt indiu
85ᵇ ⁊ ra suamnig¹ do rosc ⁊ dachúaid do chruth | ⁊ do delb ⁊ do dénam
dít.' 'Ní 'r th'ecla-su ná ar th'úamain form-sa sain indiu ám,' bar
Fer Diad, 'dáig ní fuil i nHérind indiu láech ná diṅgéb-sa.' Acus
3185 ra buí Cú Chulaind ²ac écaíni ⁊ ac airchisecht [Fir Diad]² ⁊ rabert
na briathra and ⁊ ra [f]recair Fer Diad:

'A Fir Dīad, mása thú,
demin limm isat lomthrú,
tidacht ar comairli mná
3190 do chomlund rit chomalta.'

F.   'A Chú Chulaind, comall ṅgaíth,
a fíránraith, a fírlaích,
is éicen do neoch a thecht
cosin fót forsa mbí a thiglecht.'

3195 C.C.   'Findabair ingea³ Medba,
gia beith d'febas a delba,
a tabairt dait ní ar do seirc
ac[h]t do [f]romad do rígneirt.'

F.   'Fromtha mo nert á chíanaib,
3200 a Chú cosin cáemríagail,
nech bad chalmu noco closs,
cosindiu nocon fuaross.'

C.C.   'Tú fodera a fail de,
a meic Damáin meic Dáre,
3205 tiachtain ar comairle mná
d'imchlaidbed rit chomalta.'

¹ in stain MS., suanmig Facs. (and Windisch): luaimnig St   ² ² iga
eccaine samlaidh St, ica eccaine ⁊ ica aircisecht C   ³ sic; ingen St, C

F.   'Dá scaraind gan troit is tú.
      gidar comaltai. a cháemChú.
      bud olc mo briathar is mo blad
      ic Ailill is ac Meidb Chrúachan.'   3210

[C.C] 'Noco tard bíad dá bélaib
      ⁊ noco móo ro génair
      do ríg ná rígain can chess
      bara ndernaind-sea th'amles.'

F.   'A Chú Chulaind, tólaib gal.   3215
      ní tú acht Medb ¹rar marnestar¹.
      béra-su búaid ⁊ blaid.
      ní fort atát ar cinaid.'

C.C.   'Is cáep cró mo chride cain,
      ²bec nach rascloss² ram anmain.   3220
      ní comnairt limm. línib gal.
      comrac rit, a Fir Dïad.'   A.

³'Méid ataí-siu ac cesssacht form-sa indiu³. bar Fer Diad. 'gá
gasced fora ragam indiu?' 'Lett-su do roga gascid chaidchi indiu.'
bar Cú Chulaind, 'dáig is missi barróega in lathe luid.' 'Tiagam ³²²⁵
iaram,' bar Fer Diad, 'barar claidbib tromma tortbullecha indiu
dáig is facsiu lind dond ág inn imslaidi indiu andá ⁴dond imrubad⁴
indé.' 'Tecam ám écin,' bar Cú Chulaind. Is and sain ra gabsatar
dá leborscíath lánmóra forro in lá sain. Dochúatar bara claidbib
tromma tortbullecha. Ra gab cách díb bar slaide ⁊ bar slechtad. ⁵bar ³²³⁰
airlech ⁊ bar slechtad⁵, bar airlech ⁊ bar essorgain [araile]⁶ gomba
métithir ri cend meic mís cach thothocht ⁊ gach thinmi dobeired⁷
cách díb de gúallib ⁊ de slíastaib ⁊ de slinneócaib araile. Ra gab
cách díb ac slaide araile mán cóir sin á dorblass na matni muchi
co tráth funid nóna. 'Scurem do sodain badesta, a Chú Chulaind.' ³²³⁵
bar Fer Diad. 'Scorem ám écin má thánic a thráth.' bar Cú Chulaind.
Ra scorsetar. Facheirdsetar a n-airm úadaib i llámaib a n-arad.
Girbo chomraicthi dá subach sámach sobbrónach somenmnach.
⁸rapa scarthain⁸ dá ndubach ndobbrónach ndomenmnach a scarthain
in n-aidchi sin. Ní rabatar a n-eich i n-óenscur in n-aidchi sin. ³²⁴⁰
Ní rabatar a n-araid oc óentenid.

¹⁻¹ ron mairnestair C   ²⁻² beg nar shellus (sic) C,   beg nar sgaras St.
sic leg.   ³⁻³ om. St   ⁴⁻⁴ an t-imrubadh St, sic leg.
   ⁵⁻⁵ sic MS., omitt.   ⁶ a c[h]ele add. St   ⁷ do benadh St
   ⁸⁻⁸ rapa da scarthain MS.; rob sgart[h]ain St

Dessetar in n-aidchi sin and. Is and sin atraacht Fer Diad go moch arnabárach & tánic reme a óenur co áth in chomraic, dáig rafitir rap ésin lá etergleóid in chomlaind ⁊ in chomraic, ⁊ rafitir ³²⁴⁵ co táetsad nechtar de díb in lá sain and nó co táetsaitis a ndís. Is and sin ra gabastar-som a chatherriud catha ⁊ comlaind ⁊ comraic 86ª immi | re tíachtain do Choin Chulaind dá saigid. Acus ba don chatherriud chatha ⁊ chomlaind ⁊ comraic ra gabastar a fúathbróic srebnaide¹ sróil cona cimais d' ór bricc ²fria gelchness². Ra gabastar a ³²⁵⁰ fúathbróic ndondlethair ndegsúata tairrside³ immaich anechtair. Ra gabastar múadchloich móir méti clochi mulind tarrsi-side⁴ immuich anechtair. Ra gabastar a fúathbróic n-imdangin n-imdomain n-iarnaide do iurn athlegtha darin múadchloich móir méti clochi mulind ar ecla ⁊ ar úamun in gae bulga in lá sin. Ra gabastar a chírchath-³²⁵⁵ barr catha ⁊ comlaind ⁊ comraic imma chend barsa mbátar cethracha gemm carrmocail acá cháenchumtuch, arna ecur de chrúan ⁊ christaill ⁊ carrmocul ⁊ de lubib⁵ soillsi airthir bethad. Ra gabastar a sleig mbarnig mbairendbailc ina desláim. Ra gabastar a chlaideb camthúagach catha bara chlíu cona imdorn óir ⁊ cona muleltaib de ³²⁶⁰ dergór. Ra gabastar a scíath mór mbúabalchaín bara túagleirg a dromma barsa mbátar coíca cobrad bara taillfed torc taisselbtha bar cach comraid díb cenmothá in comraid móir medónaig do dergór. Bacheird Fer Diad clesrada ána ilerda ingantacha imda bar aird in lá sain nád róeglaind ac nech aile ríam, ac mumme ná ³²⁶⁵ ac aite ná ac Scáthaig nach ac Úathaig ná ac Aífe, acht a ndénum úad féin in lá sain i n-agid Con Culaind.

Daríacht Cú Chulaind dochum inn átha 'no & rachonnaic na clesrada ána ilerda ingantacha imda bacheird Fer Diad bar aird. 'Atchí-siu sút, a mo phopa Laíg, na clesrada ána ilerda ingantacha ³²⁷⁰ imda focheird Fer Diad bar aird, ⁊ bocotáidfer⁶ dam-sa ar n-úair innossa na clesrada út, ⁊ is aire sin mad forum-sa bus róen indiu, ara nderna-su mo grísad ⁊ mo glámad ⁊ olc do ráda rim gorop móite éir m'fír ⁊ m'fergg foromm. Mad romum bus róen 'no, ara nderna-su mo múnod ⁊ mo molod ⁊ maithius do rád frim gorop móti lim mo ³²⁷⁵ menma.' 'Dagéntar ám écin, a Chúcúc,' bar Láeg.

Is and sin ra gabastar Cú Chulaind dno a chatherriud chatha ⁊ chomlaind ⁊ comraic imbi & focheird clesrada ána ilerda ingantacha imda bar aird in lá sain nád róeglaind ac neoch aile ríam, ac Scáthaig ná ac Úathaig ná ac Aífe.

---

¹ srebraide *MS.* ²⁻² fa fri gelchness *MS.* fa *for* fria, *suprascript* i
forgotten; fria gelchneas *St;* *read* fria (=frisin fúathbróic) fri(a) gelchness
  ³ = tairrsi-side; tairsi sin *St* ⁴ ar-*compendium for* air ; tairsi sin *St*
  ⁵ *sic for* lecaib (do legaib soillsi *St*)
  ⁶ .i. fogebsa, *marg. gloss MS.;* do-gebha-sa *St*

Atchondairc Fer Diad na clesrada sain ⁊ rafitir go fuigbithea dó ³²⁸⁰
ar n-úair iat. 'Gá gasced ara ragam, a Fir Diad?' bar Cú Chulaind.
'Lett-su do roga gascid chaidchi,' bar Fer Diad. 'Tiagam far cluchi
inn átha iarum,' bar Cú Chulaind. 'Tecam ám,' bar Fer Diad.
Gi 'tubairt Fer Diad inní sein, is air is doilgiu leis daragad, dáig
rafitir iss ass ra forrged Cú Chulaind cach caur ⁊ cach cathmílid ³²⁸⁵
condriced friss bar cluchi inn átha. Ba mór in gním ám daringned
barsind áth in lá sain, na dá niad, na dá ánruith[1], dá eirrgi [2]íarthair
Eórpa, dá ánchaindil gascid Gáedel, dá láim thidnaicthi ratha ⁊
tairberta ⁊ túarastail íarthairthúascirt in domain, | [3]dá ánchaindil
gascid Gáedel[3] ⁊ dá eochair gascid Gáedel, a comraicthi do chéin ³²⁹⁰
máir tri indlach ⁊ etarchossaít Ailella ⁊ Medba. Da gab cách
díb ac díburgun araile dona clesraidib sin á dorbblass na matni
muchi go midi medóin laí and. Ó thánic medón laí, [4]ra feochraig-
setar[4] fergga na fer ⁊ [5]ra chomfaicsigestar[5] cách díb d'araile.

Is and sin cindis Cú Chulaind fecht n-óen and do ur inn átha go ³²⁹⁵
mbaí far cobraid scéith Fir Diad meic Damáin do thetractain[6] a
chind do búalad dar bil in scéith ar n-úachtur. Is and sin rabert
Fer Diad béim dá ullind clé 'sin scíath comdas rala *Coin Culaind*[7]
úad mar én bar ur inn átha. Cindis Cú Chulaind d'ur inn átha
arís co mbɛí far cobraid scéith Fir Diad meic Damáin do the- ³³⁰⁰
tarrachtain a chind do búalad dar bil in scéith ar n-úachtur. Rabert
Fer Diad béim dá glún chlé 'sin scíath gomdas rala *Coin Culaind*[8]
úad mar mac mbecc bar ur inn átha. Arigis Láeg inní sein. 'Amae
ale,' bar Láeg, 'rat chúr in cathmílid fail itt agid mar chúras
ben baíd a mac. Rot snigestar[9] mar snegair[10] cuip a lundu. Rat ³³⁰⁵
melestar mar miles mulend múadbraich. Ra[t] tregdastar mar
thregdas fodb omnaid. Rat nascestar mar nasces féith fidu. Ras
léic fort feib ras léic séig for mintu, connach fail do dluig ná do dúal
ná do díl ri gail ná ra gaisced go brunni mbrátha ⁊ betha badesta, a
siriti síabarthi bic,' bar Lóeg. ³³¹⁰

Is and sain atraacht Cú Chulaind i llúas na gaíthi ⁊ i n-athlaimi na
fandli ⁊ i ndremni in drecain [11]⁊ i nnirt inn aeóir[11] in tres fecht go
mbaí far comraid scéith Fir Diad meic Damáin do thetarrachtain a
chind da búalad dar bil a scéith ar n-úac[h]tur. Is and sin rabert
in cathmílid crothad barsin scíath [12]comdas rala *Coin Culaind* ³³¹⁵
úad[12] bar lár inn átha marbad é nach arlebad ríam itir.

---

[1] anradh *St, read* ánradh   [2] erridh *St*   [3-3] *repetition; omitt.* (da airsigh
engnamha *St*)   [4-4] ra feochraigesetar *MS.*   [5-5] ro comfoicsigetar *St*
[6] *sic, for* thetarrachtain   [7] *Cu Chulaind Dipl. Edn.*   [8] *Cu Chulaind Dipl. Edn.* (cur cuir dhe *St*)
[8] *Cu Chulaind Dipl. Edn.* (cur cuirestair C.C. de *St*)   [9] *sic, for* nigestar
[10] *sic, for* negair   [11-11] i nellaibh aeóir *St*   [12-12] gurro cuir C.C. uadha *St*

Is and sin ra chétríastrad im Choin Culaind goros lín att ⁊ infithsi
mar anáil i llés co nderna thúaig n-úathma[i]r¹ n-acbéil n-ildathaig
n-ingantaig de, gomba métithir ra fomóir ná ra fer mara in mílid
³³²⁰ mórchalma ós chind Fir Diad i certarddi.

Ba sé ²dlús n-imairic² darónsatar gora chomraicsetar a cind ar
n-úac[h]tur ˙ a cossa ar n-íc[h]tur ⁊ a lláma ar n-irmedón dar bilib
˙ chobradaib na scíath. Ba sé dlús n-imaric darónsatar goro dluigset
˙ goro dloingset a scéith á mbilib goa mbróntib. Ba sé dlús
³³²⁵ n-immaric darónsatar goro fillsetar ˙ goro lúpsatar ⁊ goro gúasaig-
setar a slega óa rennaib goa semannaib. Ba sé dlús n-imaric
darónsatar gora gársetar boccánaig ⁊ bánanaig ⁊ geniti glinni ⁊
demna aeóir do bilib a scíath ⁊ d'imdornaib a claideb ⁊ d'erlonnaib
a sleg. Ba sé dlús n-imaric darónsatar gora lásetar in n-abaind
87ᵃ assa curp ˙ assa cumac[h]ta gomba imda dérgthi | do ríg nó rígain
ar lár inn átha connach baí banna d'usci and acht muni siled ind
risin súathfadaig ˙ risin slóetradaig daringsetar na dá curaid ⁊ na
dá cathmílid bar lár inn átha. Ba sé dlús n-imaric darónsatar
goro memaid do graigib Gáedel [for]³ sceóin⁴ ⁊ sceinmnig, diallaib
³³³⁵ ˙ dásacht. goro maidset a n-idi ⁊ a n-erchomail, a llomna ⁊ a lleth-
renna. goro memaid de mnáib ˙ maccáemaib ⁊ mindóenib, midlaigib
˙ meraigib fer nHérend trisin dúnud síardess.

Bátar-sun ar fáebarchless claideb risin ré sin. Is and sin rasíacht
Fer Diad úair báeguil and fecht far Coin Culaind, ⁊ rabert béim
³³⁴⁰ din chulg dét dó gora folaig 'na chlíab go torchair a chrú 'na chriss
corb forrúammanda in t-áth do chrú a chuirp in chathmíled. Ní
faerlangair Cú Chulaind aní sein ára gab Fer Diad bara bráthbalc-
bémmennaib ˙ fótalbémmennaib ⁊ múadalbémmennaib móra fair,
˙ conattacht in ngae mbulga bar Láeg mac Riangabra.—Is amlaid
³³⁴⁵ ra baí-side ra sruth ra indiltea ˙ i lladair ra teilgthea. Álad óengac
leis ac techt i nduni ⁊ tríchu farrindi ri taithmech, ⁊ ní gatta a curp
duni go coscairthea immi.

Acus á'tchúala Fer Diad in ngae mbolga d'imrád, rabert béim
din scíath sís d'anacul íchtair a chuirp. Boruaraid Cú Chulaind
³³⁵⁰ in certgae delgthi do lár a dernainni dar bil in scéith ⁊ dar brollach
in chonganchnis gorbo róen in leth n-alltarach de ar tregtad a chride
'na chlíab. Rabert Fer Diad béim din scíath súas d'anacul úac[h]tair
a chuirp giarb í in chobair iar n-assu. Da indill in gilla in ngae
mbolga risin sruth ˙ ra [f]ritháil Cú Chulaind i lladair a chossi,

---

¹ mar (= mair) *add. later:* n-uathmair *St*
²⁻² *letter erased after* dlus (? i): *cp.* dlus an imairg *St* 3143, 3146, 3151.
³ *Cp. Stowe,* cur meabaidh do groidibh fer nErenn for dophar ⁊ for dasacht
(3151–2) ⁴ scréoin *MS.*

⁊ tarlaic róut n-urchoir de bar Fer ṅDiad co ndechaid trisin ³³⁵⁵
fúathbróic n-imdaṅgin n-imdomain n-iarnaide do iurn athlegtha
go rróebris in múadchloich máir méiti clochi mulind i trí, co ndechaid
dar timthirecht a chuirp and gorbo lán cach n-alt ⁊ cach n-áge de
dá ḟorrindib. 'Leór sain badesta ale,' bar Fer Diad, 'darochar-sa
de ṡein. Acht atá ní chena, is t[r]én¹ unnsi asdo deiss. Acus ³³⁶⁰
nírbo chóir dait mo thuttim-sea dot láim.' Is amlaid ra boí 'gá
rád ⁊ rabert na briathra:

'A Chú na cless cain,
   nír dess dait mo guin.
Lett in locht rom len,   ³³⁶⁵
   is ḟort ra ḟer m'ḟuil.

'Ní lossat na troich,
   recait bernaid ṁbraith.
Is galar mo guth,
   uch! doscarad scaith.   ³³⁷⁰

'Mebait m'asnae fuidb,
   mo chride-se is crú.
Ní mad² d'ḟerus báig,
   darochar, a Chú.'   A.

Rabert Cú Chulaind sidi dá ṡaigid assa aithle ⁊ ra iad a dá láim ³³⁷⁵
tharis, ⁊ túargaib leiss cona arm ⁊ cona erriud ⁊ cona étgud dar áth
fathúaid é gombad ra áth atúaid ra beth in coscur ⁊ nábad ra áth
87ᵇ aníar ac feraib Hérend. | Da léic Cú Chulaind ar lár Fer ṅDiad and
⁊ darochair nél ⁊ tám ⁊ tassi bar Coin Culaind ás chind Ḟir Diad and.
Atchonnaic Láeg anísin & atráigestar fir Hérend uile do thíchtain ³³⁸⁰
dá ṡaigid. 'Maith a Chúcúc,' bar Láeg, 'comérig badesta ⁊ daroisset
fir Hérend dar saigid ⁊ níba cumland óenḟir démait dúinn á darochair
Fer Diad mac Damáin meic Dáre lat-su.' 'Can dam-sa éirgi, a
gillai,' bar ésium, '⁊ intí darochair limm.' Is amlaid ra baí in gilla
'gá rád ⁊ rabert na briathra and ⁊ ra [ḟ]recair Cú Chulaind: ³³⁸⁵

'Érig, a árchu Emna,
   córu a chách duit mórmenma.
Ra láis dít Fer ṅDiad na ndroṅg,
   debrad! is crúaid do chomlond.'

---

¹ tren *St*   ² *letter after* mad *obscure in stain* (mad ro ferus baig *YBL, St*)

3390
'Gá chana dam menma mór?
Ram immart báeis ⁊ brón
[a]ithle¹ inn échta doringnius
iss in chuirp ra chrúadchlaidbius.'

3395
'Níra chóir dait a chaíniud,
córu dait a chommaídium.
Rat rácaib in rúad rinnech
caíntech créchtach crólindech.'

3400
'Dá mbenad mo lethchoiss sláin
dím is cor benad mo lethláim,
trúag nach Fer Diad boí ar echaib
tri bithu 'na bithbethaid.'

3405
'Ferr leó-som na ndernad de
ra ingenaib Cráebrúade,
sessium d'éc, tussu d'anad,
leó ní bec bar mbithscarad.'

'Án ló thánac a Cúalnge
i ndiaid Medba mórglúare.
is ár daíni lé co mblaid
ra marbais dá míledaib.

3410
'Níra chotlais i ssáma
i ndegaid da mórthána;
giarb úathed do dám malle
mór maitne ba moch th'éirge.'   E.

Ra gab Cú Chulaind ac écaíne ⁊ ac airchisecht Fir Diad and ⁊
3415 rabert na briathra:
'Maith a Fír Diad, bá dursan dait nach nech dind fiallaig rafitir
mo chertgnímrada-sa gaile ⁊ gascid ra acallais re coimríac[h]tain
dúin comrac n-immairic.
'Ba dirsan dait nach Láeg mac Riangabra rúamnastar comairle
3420 ar comaltais.
'Ba dirsan duit nách athesc fírglan Ferugsa forémais.
'Ba dirsan duit nach Conall cáem coscarach commaídmech
cathbúadach cobrastar ²comairle ar comaltais.²

---

¹ d'aithli St
²⁻² omitt., these three words erroneously repeated from ll. 3419-20

¹Dáig ní adiartaís² ind ḟir sein de ḟessaib ná dúlib ná dálib ná briathraib brécingill ban cendḟind Connacht¹. Dáig raḟetatar in[d] ³⁴²⁵ ḟir sin ná gigne géin gabas gnímrada cutrumma commóra Connachtaig³ rut-su⁴ go brunni ṁbrátha ┐ betha, eter imbeirt scell ┐ scíath, eter imbeirt gae ┐ chlaideb, eter imbeirt ṁbrandub ┐ ḟidchell, eter imbeirt ech ┐ charpat.

Níba lám laích lethas⁵ cárna caurad mar Ḟer ṅDiad nel[inn]⁶ ³⁴³⁰ ṅdatha. Níba búriud berna Baidbi béldergi do scoraib sciathcha scáthbricci. Níba Crúachain ⁷cossénas [nó] gébas⁷ curu cutrumma rut-sa go brunni ṁbrátha ┐ betha badesta, a meic drechdeirg Damáin,' bar Cú Chulaind.

Is and sin ra érig Cú Chulaind ás chind Ḟir Diad. 'Maith a Ḟir ³⁴³⁵ Diad,' bar Cú Chulaind, 'is mór in brath⁸ ┐ in trécun dabertatar 88ᵃ fir Hérend fort do thabairt do chomlund ┐ do chomruc | rum-sa dáig ní réid comlund ná comrac rum-sa bar Táin Bó Cúalnge.'

Is amlaid ra baí 'gá rád ┐ rabert na briathra:

'A Ḟir Diad, ardotchlóe brath, ³⁴⁴⁰
dursan do dál dédenach,
tussu d'éc, missi d'anad,
sírdursan⁹ ar sírscarad.

'Mad dá mmámar¹⁰ alla anall
ac Scáthaig búadaig búanand, ³⁴⁴⁵
¹¹[in]dar lind go bruthe bras¹¹
nocho bíad ar n-athchardes.

'Inmain lemm do ruidiud rán,
inmain do chruth cáem comlán,
inmain do rosc glass glanba¹², ³⁴⁵⁰
inmain t'álaig¹³ is¹⁴ t'irlabra.

'Nír chiṅg din tress tinbi chness,
nír gab feirg ra ferachas,
níra choṅgaib scíath ás leirg láin
th'aidgin-siu, a meic deirg Damáin. ³⁴⁵⁵

---

¹⁻¹ *in MS. this sentence follows* go brunni mbrátha ┐ betha: *order given above is that in St* ² aiderdais *St* ³ i Connachtaibh *St* ⁴ friomsa *St* ⁵ *sic, for* letras (*Windisch*); letra*st*air *St* ⁶ nél *MS., for* ndelinn (*Meyer*) ⁷⁻⁷ choisénas no gebas *St* ⁸ bráth *MS.* ⁹ sir-*in marg. for erased passage in text;* ba dursan *St* ¹⁰ *sic;* mbamar *St* ¹¹⁻¹¹ indar linn cusin mbrath mbras *St* ¹² b *made* ⁿ d; glanbdha *St; in marg. with ref. marks to* glanba, nó gregda ¹³ *in marg. with ref. marks* nó t'álle. ¹⁴ is *om. St*

'Ní tharla rumm sund co se
á bacear[1] Óenfer Aífe,
da mac samla—galaib gliad—
ní fuaras sund, a Fir Diad.   A.

3460   'Findabair ingea[2] Medba,
gé beith d'febas a delba,
is gat im ganem ná im grían
a taidbsiu duit-siu, a Fir Diad.'   A.

Ra gab Cú Chulaind ac fégad Fir Diad and. 'Maith a mo phopa
3465 Laíg,' bar Cú Chulaind, 'fadbaig Fer nDiad badesta ⁊ ben a erriud
⁊ a étgud de go faccur-sa in delg ara nderna in comlund ⁊ in comrac.'
Táinic Láeg ⁊ ra fadbaig Fer nDiad. Ra ben a erriud ⁊ a étgud de,
⁊ rachonnaic [Cú Chulaind] in delg. Acus ra gab 'gá écaíne ⁊ 'gá
airchisecht ⁊ rabert na briathra:

3470   '[3]Dursan a eó óir[3],
a Fir Diad na ndám,
a balcbémnig búain,
ba búadach do lám.

'Do barr bude brass
3475   ba cass, ba caín sét,
do chriss duillech máeth
immut táeb gut éc.

'Ar comaltus caín
radarc[4] súla saír,
3480   do scíath go mbil óir,
th' fidchell ba fíu maín.

'Do thuittim dom láim
[5]tucim nárb é chóir[5],
nírba [6]chomsund chaín[6],
3485   [3]dursan a eó óir[3].'   D.

'Maith a mo phopa Laíg, 'bar Cú Chulaind, 'coscair Fer nDiad
fadesta ⁊ ben in ngae mbolga ass, dáig ní fétaim-se beith i n-écmais

---

[1] do-cer St   [2] ingon St   [3]—[3] Dursan an t-eó óir St
[4] fadarc MS. radharc St   [5]—[5] tuicim narb í an chóir St
[6]—[6] sic, for chomlund caín (Windisch); cumtus cáin St

m'airm.' Tánic Láeg ⁊ ra choscair Fer ṅDiad & ra ben in ṅgae
ṁbolga ass. Acus rachonnaic-sium a arm fuilech forderg ra táeb
Fir Diad ⁊ rabert na briathra: 3490

'A Fir Diad, is trúag in dál
t'acsin dam go rúad robán,
missi gan m'arm do nigi,
tussu it chossair chróligi.

'Mád dá mmámar[1] allá anair   3495
ac Scáthaig is ac Úathaig,
nocho betis beóil bána
etraind is airm ilága.

'Atubairt Scáthach go scenb
a athesc rúanaid roderb: 3500
"Érgid uli don chath chass,
barficfa Germán Garbglass."

'Atubart-sa ra Fer[2] ṅDiad
⁊ ra Lugaid lánfíal
⁊ ra mac ṁBaetáin ṁbáin   3505
techt dún i n-agid Germáin.

'Lodmar go haille in chomraic
ás leirg Locha Lindformait,
tucsam chethri chét immach
a indsib na n-áthissech.   3510

'Dá mbá-sa is Fer Diad inn áig
i ndorus dúne Germáin, |
ro marbusa Rind mac Níuil,
ro marb-som Rúad mac Forníuil.

'Ra marb [3]Fer Báeth[3] arin leirg   3515
Bláth mac Colbai chlaidebdeirg,
ro marb Lugaid, fer duairc dían,
Mugairne Mara Torrían.

'Ro marbusa ar ṅdula innund
cethri choícait férn[4] ferglond,   3520
ro marb Fer Diad, duairc in drem,
Dam ṅDreimed is Dam ṅDílend.

88ᵇ

---

[1] mbámar *St*   [2] *sic*   [3-3] Fer Diad *St*   [4] r *made on* á, n *re-inked;* ferco *St*

'Ra airgsem dún ṅGermáin ṅglicc
ás ḟargi letha[i]n lindbricc,
3525   tucsam Germán i mbethaid
lind go Scáthaig scíathlethain.

'Da naisc ár mummi go mblad
ar cró cotaig is óentad,
conná betis ar ferga
3530   eter fini findElga.

'Trúag in maten maten máirt
ros bí mac Damáin díthráicht,
uchán! ¹do chara¹ in cara
dara² dálius dig ṅdergfala.

3535   'Dámbad and atcheind-sea th'éc
eter míledaib mórGréc,
ní beind-se i mbethaid dar th'éis,
gombad aróen atbailméis.

'Is trúag aní nar tá de,
3540   'nar ṅdaltánaib Scáth[ai]che,
missi créchtach ba chrú rúad,
tussu gan charptiu d'imlúad.

'Is trúag aní nar tá de,
'nar ṅdaltánaib Scáthaiche,
3545   missi créchtach ba chrú garb,
⁊ tussu ulimarb.

'Is trúag aní nar tá de,
'ṅar ndaltánaib Scáthaige,
tussu d'éc, missi beó brass,
3550   is gleó ferge in ferachas.'   A.

'Maith a Chúcúc,' bar Láeg, 'fácbam in n-áth sa fadesta.   Is roḟata
atám and.' 'Fáicfimmít ám écin, a mo phopa Laíg,' bar Cú Chulaind.
'Acht is cluchi ⁊ is gaíni lem-sa cach comlond ⁊ cach comrac darónus
i farrad chomlaind ⁊ comraic Ḟir Diad.'   Acus is amlaid ra baí
3555 'gá rád ⁊ rabert na briathra:

¹⁻¹ do-cer St, sic leg.   ² dar St

'Cluchi cách, gaíne cách
  co roich Fer ṅDiad issin n-áth.
Inund foglaim fríth dúinn,
  innund rograim ráth,
inund mummi máeth   3560
  ras slainni sech cách.

'Cluchi cách, gaíne cách
  go roich Fer Diad issin n-áth.
Inund aisti arúath dúinn,
  inund gasced gnáth.   3565
Scáthach tuc dá scíath
  dam-sa is Fer Diad tráth.

'Cluchi cách, gaíne cách
  go roich Fer Diad issin n-áth.
Inmain úatni óir   3570
  ra furmius ar áth.
A tarbga na túath,
  ba calma ná cách.

'Cluchi cách, gaíne cách
  go roich Fer Diad issin n-áth.   3575
In leóman lassamain lond,
  in tond báeth borr immar bráth.

'Cluchi cách, gaíne cách
  go roich Fer Diad issin n-áth.
Indar lim-sa Fer diḷ Diad   3580
  is am díaid ra bíad go bráth.
Indé ba métithir slíab,
  indiu ní fuil de acht a scáth.

'Trí díríme na Tána
  darochratar dom láma,   3585
formna bó, fer ⁊ ech
  roda slaidius ar cech leth.

'Girbat línmara na slúaig
  táncatar ón Chrúachain chrúaid,
mó trín is lugu lethi   3590
  ro marbus dom garbchluchi.

'Nocho tarla co cath cró,
níra alt Banba dá brú,
níra¹ chind de muir ná thír
3595   de maccaib ríg bud ferr clú.'   C.

Aided Fir Diad gonnici sin. |

89ᵃ ² A Nd sain daríachtatar óendóene d'Ultaib and so innossa
d'fortacht ⅂ d'fórithin Con Culaind .i. Senall Uathach ⅂ dá
mac Gégge .i. Muridach ⅂ Cotreb, & rucsatar leó é go glassib ⅂
3600 ³go aibnib [críchi Conaille] Murthemne³ do thúargain ⅂ do nige ⁴a
chneda ⅂ a chréchta, a álaid ⅂ a ilgona⁴ ⁵i n-agthib na srotha sain⁵ ⅂
na n-aband. Dáig dabertis Túatha Dé Danand lubi ⅂ lossa ícce ⅂
slánsén for glassib ⅂ aibnib críchi Conailli Murthemne do fortacht
⅂ do fórithin Con Culaind comtís brecca barrúani na srotha díb.
3605 Conid ed and so anmanda na n-aband legis sain Con Culaind:
⁶Sáis, Búain, Bithláin⁶, Findglais, Gleóir, Glenamain, Bedg, Tadg,
Telaméit, Rind, Bir, Brenide, Dichaem, Muach, Miliuc, Cumuṅg,
Cuilend⁷, Gáinemain, Droṅg, Delt, Dubglass.

I S and sain ra ráidset fir Hérend ri Mac Roth risin prímechlaig
3610 tec[h]ta d'foraire ⅂ do [f]reccomét dóib go Slíab Fúait arná
tíastais Ulaid gen robud gen rathugud dá saigid. Tánaic Mac Roth
reme 'no go Slíab Fúait. Nírbo chían do Mac Roth dia mbaí and co
n-facca ní in óencharptech i Slíab Fúait atúaid cach ṅdíriuch dá
ṡaigid. Fer díscir derglomnocht isin charput dá ṡaigid gan nach
3615 n-arm gan nach n-étgud itir acht bir iairn ina láim. Is cumma
congonad a araid ⅂ a eocho. Acus indar leis ní hé rafársed na slúago
'na mbethaid itir. Acus tánic Mac Roth co n-innisin in sceóil sin
go airm i mbáe Ailill ⅂ Medb ⅂ Fergus ⅂ mathe fer ṅHérend. Atfócht
Ailill scéla de ar rochtain. 'Maith a Meic Roth,' bar Ailill, 'in
3620 faccasa nech d'Ultaib ar slicht in tslúaig seo indiu?' 'Nád fetar-sa
ém,' ar Mac Roth, 'acht atchonnac ní óencharptech dar slíab Fúait
cach ṅdíriuch. Fer díscir derglomnocht isin charput gan nach
n-arm gan nach n-étgud itir acht bir iairn ina láim. Is cumma
congonand a araid ⅂ a eocho. Dar leis ní hé dafársed in slúag sa
3625 'na mbethaid itir.'

---

¹ nir | ra MS.   ² MS. much stained and rubbed, first fourteen lines
difficult to read   ³⁻³ co haiphnibh criche Conaille Murthemhne St
⁴⁻⁴ a cnedh ⅂ a crecht, a alta ⅂ a iolghon St   ⁵⁻⁵ a n-aghaidh na sroth sin St
⁶⁻⁶ sic Dipl. Edn., MS. stained; these are gen. forms; read with YBL Sás
Búan, Bithlán (Sas, Buas, Bithlán St)   ⁷ Coimleand St, C

'Cia bad dóig lat-su and sút, a Ferguis?' bar Ailill. 'Is dóig
lim-sa ém,' bar Fergus, 'combad é Cethern mac Fintain darossed
and.' Bá fír ám d'Fergus aní sin gombad é Cethern mac Fintain
darossed and. Acus doríacht Cethern mac Fintain dá saigid 'no.
Acus focress in dúnad ⁊ in longphort foraib ⁊ no ṅgonand cách **3630**
imme do cach aird ⁊ do cach airchind. Ra ṅgontar-som dano do
cech aird ⁊ do cech airchind. Ocus tánic úadib assa aithle, a fobach
⁊ a inathar fair anechtair, go hairm i mbaí Cú Chulaind dá ícc ⁊
dá leges, & conattacht liaig bar Coin Culaind, dá ícc ⁊ dá leges.
'Maith a phopa Laíg,' bar Cú Chulaind, 'dó dait-siu i ndúnad ⁊ **3635**
89ᵇ longphort fer ṅHérend & ráid | ri legib techt ass do legess Chethirn
meic Fintain. Natiur-sa bréthir ¹manu thísat¹ gid fó thalmain
beit nó i tig fo íadad, is missi conáirgeba bás ⁊ éc ⁊ aided forro sul
bus trásta imbárach manu thísat.' Tánic Láeg reme i ndúnad
⁊ i longphort fer ṅHérend ⁊ ra ráid ri legib fer ṅHérend **3640**
tíachtain ass do legess Chethirn meic Fintain. Nírbo réid
ám la legib fer ṅHérend aní sin, techta do leges a mbidbad
⁊ a námat ⁊ a n-echtrand. Acht atráigsetar *Coin Culaind* d'imbirt
báis ⁊ éca ⁊ aideda forro monu thíastaís. Dotháegat-som dano.
Cach fer díb mar dosroched barasfénad Cethern mac Fintain a **3645**
chneda ⁊ a chréchta, a álta ⁊ a fuli dó. Cach fer díb atdered: 'Níba
beo. Níba hindlega,' da benad Cethern mac Fintain béim dá
durn dess i tulchlár a étain dó go tabrad a inchind dar senistrib
a chlúas ⁊ dar comfúammannaib a chind dó. Cid trá acht
marbais Cethern mac Fintain go ráncatar cóic lega déc leis do legib **3650**
fer ṅHérend, ⁊ gid in cóiced líaig déc iss ind ṁbémmi ris ránic,
ac[h]t dorala-sain marb di múaid móir eter collaib na lega aile
ri ré cían ⁊ ri remes fata. Íthall líaig Ailella ⁊ Medba ba sed
a chomainm.
And sain conattacht Cethern mac Fintain liaig aile bar Coin **3655**
Culaind dá ícc ⁊ [d]á leges. 'Maith a phopa Laíg,' bar Cú Chulaind,
'dó dam-sa go Fíngin fáthlíaig, go Ferta Fíngin go Leccain Slébe
Fúait, co líaig Conchobuir. Ticed ass do leiges Chethirn meic
Fintain.' Tánic Láeg reme go Fíngin fáthlíaig go Ferta Fíngin
go Lecain Sléibi Fúait go líaig Conchobuir. Acus ra ráid ris taidecht **3660**
do leiges Chethirn meic Fintain. Tánic dano Fíngin fáthlíaig.
Acus and úair doríacht, barasfén Cethern mac Fintain a chneda
⁊ a chréchta, a álta ⁊ a fule dó.
'Fé²[ga latt dam in fuil se, a mmo phopa Fíngin].²' Fégais

---

¹⁻¹ *sic, omitt.*   ²⁻² *entry in marg., obliterated, save a few letters;*
Fec[h] dhamh an fhuil si a Fhinghin, ar Ceithiorn *St where the sentence follows*
Féchais Finghin na cnedha sin

3665 Fíngin in fuil sin. 'Fingal étrom indúthrachtach and so ale,' bar in
líaig, '⁊ nít bérad i mmucha.' 'Is fír ám ale,' bar Cethern. 'Domríacht-sa óenfer and. Tuidmaíle fair. Bratt gorm i filliud imme.
Delg n-argit isin brutt ása bruinne. Crommscíath go fáebur
chondúalach fair. Sleg c[h]úicrind inna láim. Faga faegablaige¹
3670 'na farrad. Dobert in fuil sain. Ruc-som fuil ṁbic úaim-se 'no.'
'Ratafetammar in fer sain ale,' bar Cú Chulaind. 'Illand Ilarchless
mac Fergusa sain, & níba dúthracht leis do thuttim-siu dá láim.
Ac[h]t rabert in ṅgúfargam sain fort arná hapraitis fír Hérend rapa
dá mbrath nó dá trécun muni thardad.'
3675   'Féga latt dam in fuil seo dano, a mmo phopa Fíngin,' bar Cethern.
Féchais Fíngin in fuil sin. 'Bangala² banúallach and so ale,' bar
in líaig. 'Is fír ám ale,' bar Cethern. 'Domríacht-sa óenben and.
Ben chaín bánainech leccanfata mór. Moṅg órbuide furri. Bratt

90ᵃ corcra ³gen daithi³ impi. | Eó óir isin brutt ósa brunni. Sleg díriuch
3680 drumnech ar derglassad 'na láim. Rabert in fuil sin form-sa.
Ruc-si fuil ṁbic úaim-se 'nó.' 'Ratafetammar in mnaí sin
ale,' bar Cú Chulaind. 'Medb ingen Echach Feidlig, ingen ardríg
Hérend, as í danríacht fán congraimmim sin. Bá búaid ⁊ choscor
⁊ commaídium lé gia dofaítheste-su dá lámaib.'
3685   'Fécha latt dam in fuil se 'no, a mo phopa Fíngen,' bar Cethern.
Féchais Fíngin in fuil sein. 'Galach dá fénned and so ale,' bar in
líaig. 'Is fír ám,' bar Cethern. 'Damríachtatar-sa días and. Dá
thodmaíle foraib. Dá bratt gorma i filliud impu. Delgi argait
isna brattaib ósa mbrunnib. Munchobrach argit óengil im brágit
3690 chechtar n-aí díb.' 'Rodafetammar in dís sein ale,' bar Cú Chulaind.
'Oll ⁊ Othine sain do sainmuntir Ailella ⁊ Medba. Ní thecat-sain
i nnóenden acht ra hirdalta gona duine do grés. Ba búaid ⁊ coscur
⁊ commaídium leó géa dofáethaisté-su dá lámaib.'
       'Fécha latt dam in fuil seo 'no, a mo phopa Fíngin,' for Cethern.
3695 Féchais Fíngin in fuil sain. 'Domríachtatar-sa días óac féinne and.
Coṅgraim n-án ferdaide forro. Cumaiṅg⁴ bir innium-sa cechtar
n-aí díb. Cumaṅg-sa⁵ in ṁbir sa trisin dara n-aí díb-sium.' Féchais
Fíngin in fuil sin. 'Dub ule in fuil seo ale,' ar in líaig. 'Trí[t]
chride dochúatar dait co nderna chrois díb trít chride, ⁊ ní
3700 furchanaim-sea ícc and so. Acht dogébaind-se dait-seo do lossaib
ícci ⁊ slánsén ní nachat bertais i mmucha.' 'Ratafetammar in dís
sain ale,' bar Cú Chulaind. 'Bun ⁊ Mecconn sain do sainmuntir
Ailella ⁊ Medba. Ba dúthracht leó géa dofáethaiste-su dá lámaib.'

---

¹ sic, foghablaighthe St   ² sic, bangal YBL, St, sic leg.
³⁻³ sic, read hi cendaithi (brat ... i cennfait impe YBL, ... i cendaith C)
⁴ = adcumaing   ⁵ = adcumangsa

'Fécha lat dam in fuil sea 'no, a mo phopa Fíngin,' ar Cethern. Féchais Fíngin in fuil sain. 'Dergrúathur dá [mac] ríg Caille and so [3705] ale,' ar in líaig. 'Is fír ám,' bar Cethern. 'Domríachtatar-sa dá óclách aigfinna abratgorma móra and go mindaib óir úasu. Dá bratt úane i forcipul impu. Dá chassán gelargit isna brattaib ása mbrunnib. Dá sleig cúicrinni inna lámaib.' 'It immaicsi na fuli dobertatar fort ale,' bar in líaig. 'It chráes dachúatar dait co [3710] comarnecgatar renna na ngae inniut, & ní hassu a ícc and so.' 'Ratafetammar in dís sain,' bar Cú Chulaind. 'Bróen ⁊ Brudni sain, meic theóra soillsi, dá mac ríg Caille. Bá búaid ⁊ choscur ⁊ chommaídib leó gia dofáethaiste-su leó.'

'Fécha latt dam in fuil sea 'no, a mo phopa Fíngin,' ar Cethern. [3715] Féchais Fíngin in fuil sain. 'Congas dá mbráthar and so ale,' ar in líaig. 'Is fír ám,' bar Cethern. 'Domriachtatar-sa días cétríglach and. Fuilt buide forro. Bruitt dubglassa fá loss i forcipul impu. Delgi duillecha do findruinu[1] isna brattaib ósa mbrunnib. Manaísi[2] lethanglassa 'na lámaib.' 'Ratafetammar in dís sain ale,' bar Cú [3720] Chulaind. 'Cormac Coloma ríg sain ⁊ Cormac mac Maele Foga do sainmuntir Ailella ⁊ Medba. Ba dúthracht leó géa dofáethaiste-su dá lámaib.'

'Fécha latt dam in fuil seo 'no, a mo phopa Fíngin,' ar Cethern. | Féchais Fíngin in fuil sain. 'Attach dá nderbráthar and so ale,' [3725] ar ín liaig. 'Is fír ám ale, ar Cethern. 'Domríachtatar-sa días máethóclách and it íat comchosmaile díb línaib. Folt cass[dond][3] bar indara n-aí díb. Folt cassbuide bar aile. Dá bratt úanide i forcipul impu. Dá chassán gelargit isna brattaib ása mbrunnib. Dá léni di slemainsíta buide fria cnessaib. Claidbi gelduirn ara [3730] cressaib. Dá gelscíath co túagmílaib argit findi[4] foraib. Dá sleig cúicrind go féthanaib argit óengil ina lámaib.' 'Ratafetammar in dís sain ale,' bar Cú Chulaind. 'Mane Máthremail sain ⁊ Mane Athremail, dá mac Ailella is Medba, ⁊ ba búaid ⁊ coscur ⁊ commaídium leó gé rofáethaiste-su dá lámaib.' [3735]

'Fécha lat dam in fuil sea, a mo phopa Fíngin,' bar Cethern. 'Domriachtatar días óac féinne and. Congraim n-écside forro[5], it é erarda ferdaide. Étaige allmarda ingantacha impo. Cumaing[6] bir innium-sa cechtar n-aí díb. Cumang-sa[7] trí chechtar n-aí díb-sium.' Féchais Fíngin in fuil sain. 'At amainsi na fuili [3740] rabertatar fort ale,' ar in líaig, 'go ndarubdatar féithe do chride inniut conda n-imbir do chride it chlíab immar ubull i fabull ná

90ᵇ

---

[1] sic; fiondruini St   [2] mánaisi MS.   [3] sic YBL   [4] sic
[5] forro misplaced in MS. and put after ferdaide (it é erarda ferdaide forro)
[6] = adcomaing YBL   [7] = adcomchusa YBL

mar chertli i fásbulg, connach fail féith itir icá immuluṅg, & ní
dergenaim-se ícc and so.' 'Ratafetamar in dís sain ale,' bar Cú
3745 Chulaind. 'Días sain de fénnedaib na hIrúade forróeglass d'óen-
toisc ó Ailill ┐ ó Meidb ar dáig do gona-su, dáig ní comtig beó dá
mbágaib do grés. Dáig ba dúthracht leó gé dofáethaiste-su dá
lámaib.'
'Fécha latt dam in fuil se 'no, a mo phopa Fíngin,' bar Cethern.
3750 Féchais Fíngin in fuil sain 'no. 'Imrubad meic ┐ athar and so ale,'
ar in líaig.' 'Is fír ám,' bar Cethern. 'Domríachtatar-sa dá fer
móra gaindelderca[1] and go mindaib óir órlasraig úasu. Erriud
rígdaidi impu. Claidbi órduirn intlassi bara cressaib go ferbolgaib
argit óengil go frithathartaib[2] óir bricc friu anechtair.' 'Ratafetamar
3755 in dís sain ale,' bar Cú Chulaind. 'Ailill ┐ a mac sain, Mane Condasgeib Ule. Ba búaid ┐ coscur ┐ commaídium leó géa rofáethaiste-su
dia lámaib.'
Fuli Tána connici sein.

' Aith a Fíngin, a fáthlíaig,' bar Cethern mac Fintain, 'gá
3760 **M** cumcaisi ┐ gá comairle doberi form-sa fadesta?' 'Iss ed
atderim-sea rit ám', bar Fíngin fáthlíaig, '[3]nír ármea[3] do
bú móra bar dartib issin bliadain se dáig [4]gia dosrine[4] ní tú ros
méla ┐ ní tharmnaigfet dait.' 'Is í sein cumcaisin ┐ comairli
dobertatar na lega aile form-sa acus is airchind ní ruc búaid ná
3765 bissech dóib ┐ darochratar lim-sa, ┐ ní mó béras búaid ná bissech
dait-siu ┐ dofaíthaisiu limm.' Acus dabretha trénlúa tarpech dá
choiss úad riss go tarla eter díb rothaib in charpait. 'Is dúaig
in lúa sengrintid sin ale,' bar Cú Chulaind. Gorop[5] de atá Úachtur
Lúa i Crích Roiss ó ṡein anall gosindiu.
91ª   Araí sein | barróega[6] Fíngin fáthlíaig a roga do Chetharn mac
Fintain: serglegi fada fair ┐ fortacht ┐ fórithin d'fagbáil assa athli
ná dergleges teóra lá ┐ teóra n-aidchi [7]go n-imre[7] féin a nert fora
námtib. Is ed ón barróega Cethern mac Fintain dergleges teóra
lá ┐ teóra n-aidchi go n-imred féin a nert fora námtib, dáig iss ed
3775 ra ráidestar-som ná faigbed[8] dá éis nech bud ferr leis dá athe nó
dá dígail andás badessin. Is and sin conattacht Fíngin fáthlíaig
smirammair for Coin Culaind do ícc ┐ do leigess Chethirn meic
Fintain. Tánic Cú Chulaind reme i ndúnud ┐ i llongphort fer
nHérend ┐ na fúair d'almaib ┐ d'étib ┐ d'indilib and tuc leis ass

---

[1] caindeldercai *YBL*   [2] féthanaibh *St*   [3-3] nirarmea *MS*. na rec *St*
[4-4] da recair *St*   [5] Conadh *St* (Is de atá . . *YBL*)   [6] do-rat *St, sic leg.*
[7-7] co n-imredh St   [8] *sic for* fáicbed; fuicfid *St*

íat. Acus dogní smirammair díb eter feóil ⁊ chnámaib ⁊ lethar. 3780
Acus tucad Cethern mac Fintain 'sin smirammair co cend teóra lá ⁊
teóra n-aidche & ra gab ac ól na smiramrach imme. Acus raluid in
smirammair and etera chnedaib ⁊ etera chréchtaib, dara áltaib ⁊ dara
ilgonaib. And sin atracht-som assin smirammair i cind teóra lá ⁊ teóra
n-aidche. Acus is samlaid attracht ⁊ clár a charpait re broind arná 3785
tuitted a fobach ⁊ a inathar ass.
    Is í sain amser luid a banchéile atúaid á Dún Da Bend ⁊ a chlaideb
lée dó .i. Finda ingen Echach. Tánic Cethern mac Fintain d'indsaigid fer nHérend. Acht atá ní chena, bertis robod reme-seom
d'Ítholl líaig Ailella ⁊ Medba. Dorala-saide marb de múaid móir 3790
eter chollaib na llega aile ra ré cían ⁊ ra remis fata. 'Maith a firu
Hérend', bar in líaig, 'daria Cethern mac Fintain dabar saigid
arna ícc ⁊ arna leges do Fíngin fáthlíaig, acus frithálter acaib é.'
    Is and sain faítsetar fir Hérend étgud Ailella ⁊ a imscing n-órda
immon corthe i Crích Ross combad fair no imbred Cethern mac 3795
Fintain a feirg ar tús ar torachtain. Atchondairc Cethern mac
Fintain aní sin, étgud Ailella ⁊ a imscing n-órda immun corthe i
Crích Ross, ⁊ andar leiss i n-écmaiss a fessa ⁊ a eólais bá sé Ailill
bóe and fodessin. Acus rabcrt side dá saigid ⁊ ra sáid in claideb
tresin corthe co ránic gonnice a irdorn. 'Bréc and so,' bar Cethern 3800
mac Fintain, '⁊ immum-sa rabertad in bréc sa. Acus atiur-sa
bréthir ná co fagaither acaib-si nech gabas in n-erriud rígdaide
út imme ⁊ in n-imscing n-órda, ná scér-sa láma riu 'cá slaide ⁊
icá n-essargain.' Rachúala sain Mane Andóe mac Ailella - Medba.
Acus ra gab in n-erriud rígdaide imme ⁊ in n-imscing n-órda & 3805
tánic reme tri lár fer nHérend. Lilis Cethern mac Fintain co mór
de. Acus tarlaic rót n-urchair dá scíath fair coro raind bil chondúail in
scéith i trí co talmain hé eter charpat ⁊ araid ⁊ eocho. And sain ra
théigsetar na slúaig imme da díb lethib co torchair accu issin
chalad i rrabe.   3810
    Conid Caladgleó Cethirn and sin ⁊ Fule Cethirn.

Fiacalgleó Fintain and so innossa

FIntan ésede mac Néill Niamglonnaig á Dún Da Bend. athair-
side Cethirn meic Fintain. Acus radeochaid-side do tharrachtain ainig Ulad ⁊ do dígail a meic barna slúagaib. Trí choícait 3815
bá sed a llín. Acus is samlaid táncatar-saide ⁊ dá gae for cach
n-óencrand leó, gae for renn ⁊ gae for erlond, gombad chumma ra
gontais do rennaib ⁊ d'erlonnaib na slúagu. Dobertatar teóra

catha dona slúagaib. Acus dorochratar a tri comlín leó, ⁊ torchratar
91ᵇ 'no munter | Fintain meic Néill acht Crimthann mac Fintain. Ro
hainced-saide fo amdabaig scíath la hAilill ⁊ la Meidb. Is and
sain ra ráidsetar fir Hérend nárbad athis d'Fintan mac Néill dúnad
⁊ longphort d'falmugud dó ⁊ a mac do lécud dó ass .i. Crimthann
mac Fintain, acus na slúaig do thigecht uidi laí for cúlu fathúaid
3825 doridisi ⁊ a gníma gascid do scur dona slúagaib ¹ar co tísed¹ chucu
do ló in mórchatha airm condricfaitis cethri ollchóicid Hérend for
Gárig ⁊ Ilgárig i cath Tána Bó Cúalnge feib ra tharṅgirset druídi
fer nHérend. Faímais Fintan mac Néill aní sin ⁊ ra léiced a mac
dó ass. Ra falmaiged dúnad ⁊ longphort dó. Acus lotar na slúaig
3830 ude lá² for cúlu fathúaid doridisi dá fastúd ⁊ dá n-immfuirech.
Is amlaid ra geibthe in fer de muntir Fintain meic Néill ⁊ in fer
d'feraib Hérend ⁊ beóil ⁊ sróna cáich díb i ndétaib ⁊ i fiaclaib a chéile.
Atchondcatar fir Hérend aní sein. 'Is é in fiacalgleó dún so,' bar
íat-som, 'fiacalgleó muntiri Fintain ⁊ Fintain badesin.'
3835 Conid Fiacalgleó Fintain and sain.

Ruadrucce Mind and so innossa

MEnd mac Sálcholgán éside ó Rénaib³ na Bóinne. Dá fer déc
ba sed a lín-saide. Is amlaid táncatar-saide ⁊ dá gae for cach
óencrand leó, gae for rend ⁊ gae for erlond combad chumma
3840 da gontais do rennaib ⁊ do erlonnaib na slúagu. Rabertsatar
teóra fúaparta dona slúagaib. Torchratar a trí comlín leó, & torc[h]-
ratar dá fer déc muntiri Mind. Ac[h]t ra gáet Mend féin calad gor⁴
rusti rúadderg fair. And sain ra ráidsetar fir Hérend: 'Is rúad
in rucce se,' bar íat-sum, 'do Mend mac Sálcholgán, a munter do
3845 marbad ⁊ do mudugud ⁊ a guin féin corop⁴ rusti rúadderg fair.'
Corop⁵ Ruadrucce Mind and so.
Is and sain ra ráidsetar fir Hérend nárbad athis do Mend mac
Sálcholgán dúnad ⁊ longphort d'falmugud dó ⁊ na slúaig do thecht
uide lá⁶ for cúlu fathúaid daridisi ⁊ a guin⁷ gascid do scor dona
3850 slúagaib go n-éirsed Conchobor assa chess nóenden co tucad cath
dóib for Gárig ⁊ Ilgárig feib ra tharngirsetar druídi ⁊ fádi ⁊ fissidi
fer ṅHérend.
Fáemais Mend mac Sálcholgán aní sein, dúnad ⁊ longphort
d'falmugud dó, & lotar na slúaig uide lá⁶ for cúlu doridi[si] ⁸dá
3855 fostud ⁊ dá immfuirech⁸.

¹⁻¹ no go dtísadh St, co tisad TBC²   ² laí St   ³ Reandaib St   ⁴ curbo St
⁵ conadh St   ⁶ laí St   ⁷ read gním? (ar gniom a laime do cosc diob
St 3799)   ⁸⁻⁸ read dá fostud ⁊ dá n-immfuirech? Cp. supra l. 3830

Airecur nArad and innossa

IS and sain daríachtatar chucu-som araid Ulad. Trí choícait
bá sed a llín. Rabertsatar teóra catha dona slúagaib.
Darochratar a trí comlín leó ⁊ torchratar na haraid barsin
róe i rrabatar.   3860
¹Corop Airecor ṅArad and sain¹. |

92ª   Bángleó Rochada and so innossa

R Eochaid mac Fathemain éside d'Ultaib. Trí choícait láech
ba sed a lín. Acus ra gab tilaig agid i n-agid dona slúagaib.
Atchondaic Findabair ingen Ailella ⁊ Medba anísein. Acus 3865
ra baí-si 'gá rád ria máthair ri Meidb: 'Ra charusa in láech út
úair chéin ám,' bar sí, '⁊ iss é mo lennán é ⁊ mo roga tochmairc.'
'Má ra charais, a ingen, fáe leis dádaig ⁊ guid fossad dún fair dona
slúagaib go tí chucaind do ló in mórchatha airm condricfat cethri
ollchóicid Hérend for Gárig ⁊ Ilgárig i cath Tána Bó Cúalnge.' 3870
Fáemais Reochaid mac Fathemain anísein ⁊ fáeiss ind ingen dádaig
leis.
    Rachúala sein airrí de Mumnechaib ra boí 'sin longphurt. Báe-
sium 'gá rád ria muntir: 'Banassa dam-sa ind ingen út uair chéin
ám,' bar ésium, '& is aire thánac-sa in slúaged sa don chur sa.' 3875
Cid trá acht airm i mbátar na secht n-airríg de Mumnechaib, iss
ed ra ráidsetar uile conid aire-sin táncatar. 'Cid dúnni 'no,' bar
íat-som, 'ná ragmais-ni do dígail ar mná ⁊ ar n-ainig arna Manib
fuil ac foraire dar éis in tslúaig ic Imlig in Glendamrach,'
    Is hí sin comairle ba nirt leó-som & atraac[h]tatar-som cona secht 3880
tríc[h]taib cét. Atracht Ailill dóib cona tríchait chét. Atraacht
Medb cona tríchait chét. Atraachtatar meic Mágach cona tríchtaib
chét. Atraacht in Galéoin ⁊ in Mumnig ⁊ popul na Temrach. Acus
fognithea etargaire eturru co ndessid cách díb i fail araile ⁊ i fail
a arm. Cid trá acht sul tarraid a n-etráin, torchratar ocht cét 3885
[láech]² lánchalma díb. Atchúala sain Findabair ingen Ailella
⁊ Medba in comlín sain d'feraib Hérend do thuttim trena ág ⁊
trena accais, ⁊ ro maid cnómaidm dá cride 'na clíab ar féile ⁊ náre.
Conid Findabair Slébe comainm ind inaid i torchair. Is and sain
ra ráidsetar fir Hérend: 'Is bán in gleó sa,' bar íat-som, 'do Reochaid 3890
mac Fathemain, ocht cét láech lánchalma do thuttim trina ág ⁊
trina accais ⁊ a dul féin cen fuligud gen fordergad fair.'
    Conid Bángleó Rochada and sain.

---

¹⁻¹ conid Airecor nArad innsin C   ² *sic St*

Mellgleó nÍliach and so innossa

3895 I Liach éside mac Caiss meic [Baicc][1] meic Rosa Rúaid meic
Rudraige. Racúas dó-saide cethri ollchóiceda Hérend oc argain
⁊ oc indred Ulad ⁊ Chruthni ó lúan taite samna co taite n-imbuilc,
& ra mídair-sium a chomairle aice rea muntir. 'Cid bad ferr dam-sa
in chomairle dogénaind ná techta d'fúapairt fer nHérend ⁊ mo
3900 choscur do chur díb remum ⁊ ainech Ulad do tharrachtain, ⁊ is
92ᵇ cumma géa rafóethus féin assa aithle.' | Acus is sí sin comairle ba
nirt leiseom. Acus ra gabait dó-som a dá sengabair chrína chrem-
manncha bátar for tráig do tháeib in dúnaid. Acus ra indled a
sencharpat forro cen fortga cen forgemne itir. Ra gabastar-som
3905 a garbscíath odor iarnaide fair co mbil chaladargit ina imthimchiull.
Ra gabastar a chlaideb ngarb nglasseltach nglondbémnech bara
chlíu. Ra gabastar a dá sleig chendchritháncha bernacha isin
charpat ina farrad. Ra ecratar a munter in carpat imme do
chlochaib ⁊ chorthib ⁊ táthleccaib móra. Tánic reime fán cóir sin
3910 d'indsaigid fer nHérend & is amlaid tánic ⁊ lebarthrintall a chlaip
triana charpat sís dó. 'Rapad maith lind ám,' ar fir Hérend, 'combad
hí sein túarascbáil fá tístais Ulaid uile dar saigid.'

Barrecgaib Dóche mac Mágach dó-som & firis fáilte friseom.
'Mo chen do thíchtu, a Ílíaich,' bar Dóche mac Mágach. 'Tarissi
3915 limm inní inn fálte,' bar Íliach[2], 'acht tair chucum mán úair innossa
in tráth scáigfit mo gala[3] ⁊ sergfait mo gala, corop tú benas mo chend
dím ⁊ nárop nech aile d'feraib Hérend. Acht maired mo chlaideb
acut chena do Laígaire.'

Ra gab-som dá armaib for feraib Hérend coro scáigsetar dó.
3920 Acus áro scáigsed a airm dó, ra gabastar de chlochaib ⁊ chorthib ⁊
táthleccaib móra bar feraib Hérend coro scáigsetar dó, ⁊ áro
scáigsetar dó, airm i mbered farin fer d'feraib Hérend, dabered
díanchommilt fair etera rigthib ⁊ a dernannaib co ndénad smirammair de eter feóil ⁊ chnámib ⁊ féthib ⁊ lethar. Corod[4] marthanach
3925 táeb ri táeb fós in dá smirammair, smirammair fogní Cú Chulaind
do chnámib chethra Ulad do leges Chethirn meic Fintain &
smirammair bogní Íliach do chnámib fer nHérend. Gorop[5] hí tres
dírím na Tána na torchair leis díb, gorop[6] Mellgleó nÍlíach and sain.

Is aire atberar Mellgleó nÍliach ris dáig de chlochaib ⁊ chorthib ⁊
3930 táthleccaib móra fogní-seom a gleó.

---

[1] om. MS., cp. infra 1. 3938 (Iliach mac Cais mic Factna mic Fecc mic Rosa
mic Rudraighi St)   [2] Ílíach MS.
[3] read airm (cp. St 3872–3 mar turnfas mo gal ⁊ mar racas sgeithlem ar
m'armaibh)   [4] sic; gurab St   [5] Conadh St   [6] = conid

Barrecaib Dóche mac Mágach dó-som. 'Nach é Iliach?' bar
Dóche mac Mágach. 'Is mé ám écin,' bar Íliach. 'Acht tair
chucum fodechtsa ⁊ ben mo chend dím & maired mo chlaideb acut
chena dot charait do Lóegaire.' Tánic Dóche dá saigid ⁊ tuc béim
claidib dó co tópacht a chend de.   3935
Conid Mellgleó Íliach gonici sein.

Oisligi Amargin i Taltin and so annossa

A Mairgin éside mac Caiss meic Baicc meic Rosa Rúaid meic
Rudraigi. Ruc-saide barna slúagaib ac techt dar Taltin síar ⁊
imsóe reme dar Taltin sathúaid íat, & tuc a ulli chlé fáe i ³⁹⁴⁰
Taltin, & ra ecratar a munter é de chlochaib ⁊ chorthib ⁊ táthleccaib
móra. Acus ra gab ac díburgun fer nHérend co cend teóra lá ⁊
teóra n-aidche.

Imthúsa Chon Ruí meic Dáire sund innossa

R Acúas dó-saide óenfer ac fostod ⁊ ac immfuirech chethri ³⁹⁴⁵
n-ollchúiced Hérend ó lúan taite samna co taite n-imboilg.
Acus ba dimbág laisium anísein ⁊ ba rochían leis bátar a
munter 'na écmais, ⁊ tánic reme do chomlund ⁊ do chomruc ra
Coin Culaind. Acus and úair ránic-sium go airm i mbaí Cú Chulaind,
rachonnaic-sium Coin Culaind and caíntech créchtach tretholl, ⁊ ³⁹⁵⁰
níro miad ⁊ níro maiss leisium comlund ná chomrac do dénam riss aithle
chomraic Fir Diad ar bíth arnábad mó bud marb Cú Chulaind dina
cnedaib | ⁊ dina créchtaib¹ dorat Fer Diad fair inn úair reime.
Acus gid ed tarcid Cú Chulaind dó-som comrac ⁊ comlund do
dénam ris-seom.   3955
Tánic Cú Ruí reime assa aithle d'indsaigid fer nHérend. Acus
and úair ránic-sium, atchondairc-sium Amairgin and ⁊ a uille chlé
fáe ri Taltin aníar. Tánic Cú Ruí ri feraib Hérend atúaid. Ra
ecratar a munter é de chlochaib ⁊ chorthib ⁊ táthleccaib móra &
ra gab ac díburgun ²i n-agid i n-agid² .i. i n-agid Amargin co'mbofrec- ³⁹⁶⁰
raitis na bairendlecca bodba i nnélaib ⁊ i n-aéraib úasu co ndénad
chét cloch di cach óenchloich díb. 'Ar fír do gascid fritt, a Chú

93ª

---

¹ insert do-bérad-som fair iná dina cnedaib ⁊ dina créchtaib as in St 3916 ;
omitted by homoeoteleuton
²⁻² sic; read agid i n-agid (aghaidh a n-aghaid fri hAimhirgin St)

Ruí,' ar Medb, 'scuir dún[1] díburgun dáig ní furtacht ná fórithin tic
dún de acht is mífurtacht tic dún de.' 'Tiur-sa bréthir,' bar Cú
3965 Ruí, 'ná scuriub-sa co brunni brátha ⁊ betha coro scuirea Amargin.'
'Scoirfet-sa,' bar Amargin, '⁊ geib-siu fort ná ticfa d'fortacht ná
d'fórithin fer nHérend ní bas mó.' Fáemais Cú Ruí anísein. Acus
tánic Cú Ruí reme d'indsaigid a chríche ⁊ a muntire.
Ráncatar dar Taltin síar risin ré sin. 'Ní hed ra nasced[2],' bar
3970 Amargin, 'itir gan na slúaig do díburgun arís.' Acus tánic riu
aníar ⁊ imsóe reme dar Taltin sairtúaid íat, & ra gab 'gá ndíburgun
ra ré cían ⁊ ra reimes fata.

Is and sain dano ra ráidsetar fir Hérend nábad athis d'Amargin
dúnad ⁊ longphort d'falmugud dó ⁊ na slúaig do thecht uide lá[3]
3975 for cúlu fathúaid doridisi dá fostud ⁊ dá n-imfuirech ⁊ a gním gascid
do scur dona slúagaib [4]ar co[4] tíssed chuccu do ló in mórchatha airm
condricfaitis cethri ollchóicid Hérend for Gárig ⁊ Ilgárig i cath
Tánad Bó Cúalnge. Fáemais Amargin anísin acus lotar na slúaig
uide lá[3] for cúlu fathúaid doridisi.
3980 Conid Ossligi Amargin i Taltin and sain.

Sírrobud Sualtaim and so innossa

SUaltaim éside mac Becaltaig meic Móraltaig, athair-side Con
Culaind meic Sualtaim. Rachúas dó-saide búadrugud a meic
ac comrac ra écomlund for Táin Bó Cúalnge .i. ri Calatín
3985 ṅDána cona secht maccaib fichet ⁊ rá húa ra Glass mac ṅDelga.
'Is do chéin gid so,' bar Sualtaim. 'In nem maides ná in muir
thráges ná in talam condascara ná inn é búadrugud mo meic-sea
so ac comrac ra écomlund for Táin Bó Cúalnge?' Bá fír ám do
Śualtaim anísein ⁊ raluid dá fis ár tain cenco dechaid a chétóir.
3990 Acus and úair ránic Sualtaim go airm i mbáe Cú Chulaind, ra gab
Sualtaim ac écgaíne ⁊ ac airchisecht de. Nírbo míad ⁊ nírbo maiss
ám ra Coin Culaind anísin, Sualtaim do écgaíne nó d'airchisecht
de, dáig rafitir Cú Chulaind, géara gonta ⁊ géara créchtnaigthe é,
nábad gress dá dígail Sualtaim. Ór is amlaid ra boí Sualtaim
3995 acht nírbo drochláech é ⁊ nírbo degláech acht múadóclách maith
ritacaemnacair[5]. 'Maith a mo phopa Śualtaim,' bar Cú Chulaind,
'dó duit-siu go hEmain go Ultu, & ráid ríu tec[h]t i ndíaid a tána
fadec[h]tsa, dáig nídam tualaiṅg-sea a n-imdecgail ní as mó for

---

¹ sic· MS., don St; read scuir don or scuir dún don ² formsa add. St
³ sic, laoi St ⁴⁻⁴ no co St ⁵ = atacaemnacair

bernadaib ┐ belgib críche Conaille Murthemne. Atú-sa m'óenur
i n-agid chethri n-ollchóiced ṅHérend ó lúan tate samna co taite ⁴⁰⁰⁰
n-imboilg ic marbad ḟir ar áth cach laí ┐ cét láech cach n-aidchi.

93ᵇ Ní damar fír fer dam ná comlond óenḟir ┐ ní | thic nech dom ḟortacht
ná dom ḟórithin. Is stúaga úrchuill congabat mo bratt torom.
Is suipp ṡesca fuilet im áltaib. Ní ḟuil finna fora tairised rind
snáthaite ádám[1] berrad gom bonnaib gan drúcht fola forrderge ⁴⁰⁰⁵
ar barrúachtur cach ḟindae acht in lám chlé ḟail ac coṅgbáil mo
scéith, & cid híside filet teóra coíca fuile fuirri. Acus munu díglat-
som a chétóir sein, ní dígélat co brunni ṁbrátha ┐ betha.'
Tánic Sualtaim reime forin Líath Macha d'óeneoch go robtaib
leis do Ultaib. Acus and úair ránic do tháeib na hEmna, rabert ⁴⁰¹⁰
na briathra sa and : ' Fir gontair, mná berdair, báe aegdair, a
Ultu', bar Sualtaim.

[2]Ní ḟúair [in frecra] ba leór leis[2] ó Ultaib ┐ dáig ná fúair, tánic
reme fa ḟordreich na hEmna. Acus rabert na briathra cétna and :
'Fir gontair, mná bertair, báe aegtair, a Ultu,' bar Sualtaim. ⁴⁰¹⁵
Ní ḟuair in frecra rabu leór leis ó Ultaib.—Is amlaid ra bátar Ulaid,
geiss d'Ultaib labrad rena ríg, geis don ríg labrad rena druídib.—
Tánic reme assa aithle for licc na ṅgíall i nEmain Macha. Rabert
na briathra cétna and : 'Fir gondair, mná berdair, báe aegtair.'
'Cia rodas gon ┐ cia rotas brat ┐ cia rodas beir ale ?' for Cathbath ⁴⁰²⁰
druí. ' Rabar n-airg Ailill ┐ Medb,' bar Sualtaim. ' Tuctha far
mná ┐ far meic ┐ far maccáemi, far n-eich ┐ far n-echrada, far n-albi
┐ far n-éiti ┐ far n-innili. Atá Cú Chulaind a óenur ac fostud ┐ ac
imfuirech cethri n-ollchóiced ṅHérend for bernaib ┐ belgib críche
Conaille Murthemne. Ní damar fír fer dó ná chomlund óenfir. ⁴⁰²⁵
Ní thic nech dá ḟortacht ná dá ḟórithin. [3]Ra gáet in mac. Raluid
a hāltaib [dó][3]. Is stúaga úrchuill coṅgabat a bratt taris. Ní
ḟuil finna ara tairissed rind snáthaite ódá berrad coa bonnaib cana
driúcht[4] fola forrderge co barrúachtur cach óenḟindae dó acht
in lám chlé ḟail ac coṅgbáil a scéith fair, & gid híside fuilet teóra ⁴⁰³⁰
coíca fuili fuirri. Acus manu dígailti-si a chétóir sein, ní dígéltai

94ᵃ go brun[n]i ṁbrátha ┐ betha.' | ' Is uissiu a bás ┐ a éc ┐ a aided
ind ḟir coṅgreiss in ríg samlaid,' for Cathbath druí. [5]'Is fír ám,'
uile annaide.[5] Tánic Sualtaim reme tria lunne ┐ anseirc dáig ná
fúair in [ḟ]recra ba leór leis ó Ultaib. And sein driuc[h]trais in ⁴⁰³⁵
Líath Macha ba Sualtaim & tánic reme fa urdreich na hEmna.

---

[1] adám MS.=óthá mo, Windisch (om berradh com bonn St)
[2]-[2] cp. infra l. 4016 (Ni fuair a fhreacra St)   [3]-[3] Cp. rogaet coṅdechaid
a altaib do YBL 2995-6   [4] sic   [5]-[5] read ' Is fír ám,' [ar Ulaid] uile
annsaide? (Cp. As fior emh, ar Ullta uile a n-ainfhecht St 3993-4)

Is and sain imsuí a scíath féin bar Sualtaim co tópacht bil a scéith
féin a chend de Sualtaim. Luid in t-ech féin bar cúlu arís i nEmain
⁊ in scíath barsinn eoch ⁊ in cend barsin scíath. Agus rabert cend
4040 Sualtaim na briathra cétna: 'Fir gondair, mná berdair, báe aegdair,
a Ultu,' bar cend Sualtaim. 'Romór bic in núall sa,' bar Conchobor,
'dáig nem úasaind ⁊ talam ísaind ⁊ muir immaind immácúaird,
acht munu tháeth in firmimint cona frossaib rétland bar dunadgnúis
in talman ná mono máe in talam assa thalamchumscugud ná
4045 mono thí inn fairge eithrech ochorgorm for tulmoing in bethad,
dobér-sa cach bó ⁊ cach ben díb cá lias ⁊ cá machad, co'aittc ⁊
co'adbai fadessin ar mbúaid chatha ⁊ chomlaind ⁊ chomraic.' Acus
iss and sain barrecgaim echlach dá muntir fadessin do Chonchobur,
Findchad Fer Bend Uma mac Fraechlethain, & ra ráid riss techt
4050 do thinól ⁊ do thóchostul Ulad. Acus is cumma barrurim bíu ⁊
marbu dó trí mesci a chotulta ⁊ a chessa nóenden, & rabert na
briathra :

‘ARdotraí, a Findchaid, ardotfáedim. Ní hadlicgi álsidi a fasnís
do ócaib Ulad. ¹[Ercc uaim co Derg]¹ co Dedaid gó inber; co
4055 Lemain; co Follach; co hIllaind go Gabair; co Dornaill Féic co
Imc[h]lár; co Derg Indirg; co Feidlimid Chilair² Chétaig go hEllond;
co Rígdond co Reochaid; co Lug[aid]³; co Lugdaig; co Cathbath
coa inber; co Carpre co hEllne; co Láeg coa thóchur; co Geimen
94ᵇ coa glend; co Senal[l] Úathach co Diabul n-Arda; | co Cethern
4060 mac Fintain go Carrlóig; co Tarothor; co Mulaig coa dún; cosin
rígfilid co Amargin; cosin nÚathaig mBodba; cosin Morrígain co
Dún Sobairche; co hEit; co Roth; co Fiachna có fert; co Dam
nDrend; co Andiaraid; co Mane Macbriathrach; co Dam nDerg;
co Mod; co Mothus; co Iarmothus; co Corp Cliath; co Gabarlaig
4065 i lLíne; co Eocho Semnech i Semne; co Celtchair mac Cuthechair
i lLethglais ; co hErrgi Echbél co Brí Errgi; co hUma mac Remarfessaig co Fedain Cúalnge; co Munremur mac Gerrcind co Moduirn;
co Senlabair co Canaind nGall; co Follomain; co Lugdaig; co ríg
mBuilg co Lugdaig Líne; co Búadgalach; co hAbach; co hÁne;
4070 co hÁniach; co Lóegair[e] Milbél coa breó; co trí maccaib Trosgail
co Bacc nDraigin; co Drend; co Drenda; co Drendus; co Cimm; co
Cimbil; co Cimmin co Fán na Coba; co Fachtna mac Sencha coa
ráith; co Sencha; co Sencháinte; co Briccni; co Briccirni⁴; co

1⁻1 sic St, om. LL (cp. Notfoidiu co Dedad YBL, Nod foidim co Dedad C)
2⁻2 sic = mac Ilair St; co Cilar YBL, co hIlar C   3 Lugaid St YBL
4 .i. filius Bricni, gloss added above the line

Brecc; co Búan; co Barach; co hÓengus m̃Bolg; co hÓengus mac
Lethi¹; ²co Mall ar fíach ar fénned²; co Brúachar co Slánge; co ⁴⁰⁷⁵
Conall Cernach mac Amargin co Midlúachair; co Coin Culaind
mac Sualtaim co Murthemne; co Mend mac Sálcholcán³coa rénaib³;
co tri maccaib Fiachnai, co Roṡs, co Dáre, co Imchaid co Cúaliñge;
co Connud mac Mornai co Callaind; co Condraid mac Amargin
coa ráith; co Amargin co Ess Rúaid; co lLáeg co lLéire; co Óengus ⁴⁰⁸⁰
Fer Bend Uma; co hOgma ñGrianainech co Brecc; co hEo mac
Forne; co Tollcend; co Súde, co Mag nÉola, co Mag ñDea; co Conla
Ṣáeb co h Ủarba; co Loegaire ⁴co hImmail⁴; co hAmargin Iarñgiun-
94ᶜ   naig co Taltin; | co Furbaide Fer Bend mac Conchobuir co Síl
co Mag nInis; co Causcraid Mend Macha mac Conchobuir co Macha; ⁴⁰⁸⁵
co Fíngin co Fíngabair; co Blae Fichet; co Blai m̃Briuga co Fésser;
co Eogan mac ñDurthacht co Fernmag; co hOrd co Serthig⁵; co
hOblán; co hObail co Culend⁶; co Curethar⁷; co Liana; co
hEthbenna; co Fer Néll; co Findchad Slébe Betha; co Talgobaind
co Bernos; co Mend mac Fir Chúaland co Maigi Dula; co hÍroll; ⁴⁰⁹⁰
co Blárine co hIalla ñIlgremma; co Ros mac nUlchrothaig co Mag
Nobla; co Ailill Find; co Fethen m̃Bec; co Fethen Mór; co Fergna
mac Findchona co Búrach; co hOlchar; co hEbadcha⁸; co Uathchar;
co hEtatchar; co Óengus mac Óenláme Gábe; co Ruadraig co
Mag Táil; ⁹co Beothag⁹; co Beothaig; co Briathraig coa ráith; ⁴⁰⁹⁵
co Nárithlaind co Lothar; co dá mac F̃eicge, co Muridach, co
Cótreib; co Fintan mac Néill Níamglonnaig co Dún Da Bend; co
95ᵃ   Feradach Find Fechtnach co Neimed Slébe Fúait; | co hAmargin
mac Ecelsalaig Goband co Búais; co Bunni mac Munremair; co
Fidach mac Doraire.'   ⁴¹⁰⁰

Níra dulig ám do F̃indchad in tinól ᚋ in tóchostul sain rabert
Conchobor riss do dénam. Dáig in neoch ra boí ó Emain sair ᚋ
ó Emain síar ᚋ ó Emain sathúaid rathóegat-saide ass a chétóir co
faítar i nEmain ra costud a ríg ᚋ ra bréthir a flatha ᚋ ra frithálim
comérgi Conchobuir. In neoch ra boí ó Emain sadess 'no ratháegat- ⁴¹⁰⁵
saide ass a chétóir ar slicht in tṡlúaig ᚋ i n-iñgenbóthur na tánad.

In cétna uide bachomluisetar Ulaid im Chonchobor co faítar i
nIraird Chullend in n-aidchi sin. 'Cid risa n-idnaidem-ni so
itir¹⁰, a f̃iro?' bar Conchobor. 'Anmaít-ni rit maccaib-siu,' bar
íat-som, ' ri Fiachaig ᚋ ra F̃iachna. Lotar úain ar cend Eirc ⁴¹¹⁰

---

¹ Leti *YBL, C*   ²⁻² co hAlamiach in fennich *YBL*, co Laimiach in
feindid *C*   ³⁻³ co Coirenda *YBL, C*   ⁴⁻⁴ co hImmiaille *MS.*, co hImpail
YBL, C   ⁵ *th add. over line* (co Seirid, co Serthe *YBL*, co Seirid *C*)
⁶ *a second* l *expunged*   ⁷ Cuirther *YBL*, Cuirechair *C*   ⁸ *sic, for*
Ebadchar (*Windisch*)   ⁹⁻⁹ *omitt.* (*Windisch*), *repetition*   ¹⁰ iiter *MS.*

meic Ḟeidilmthe Nóchruthaige, meic th'ingini-siu, mac-saide Carpri
Nia Fer, ar co tí co llín a ṡlúaig ⁊ a ṡochraite, a thinóil ⁊ a thóchostail
'nar sochraiti-ni din chur sa.' ' Tiur-sa bréthir,' bar Conchobor,
'nachas idnaidiub-sa and so ní bas mó ná co clórat fir Hérend mo
4115 chomérgi-sea assin león ⁊ assin chess i rraba, dáig ní ḟetatar fir
Hérend inad[1] beó-sa mad co se itir.'

And sain raluid Conchobor ⁊ Celtchair tríchu cét carpdech
n-imrindi co hÁth nIrmidi, & baralsat dóib and ocht fichti fer mór
do ṡainmuntir Ailella ⁊ Medba, ⁊ ocht fichti ban braiti accu. Bá
4120 sed a n-ernail do brait Ulad ben braiti i lláim cach ḟir díb. Éscis
Conchobor ⁊ Celtchair a n-ocht fichti cend díb ⁊ a n-ocht fichti ban
ṁbraiti. Áth nÍrmidi a ainm mad co sin. Áth Féinne a ainm ó
ṡain ille. Is aire atberar Áth Féinne riss dáig concomairnectar
inn óic ḟéinne anair ⁊ inn óic féinne aníar cathugud ⁊ imbúalad im
4125 urbrunni inn átha.

Tánic Conchobor ⁊ Celtchair for cúlu co faítar i nIrard Chullend
in n-aidchi sin i farrad Ulad. Búadris Celtchair and so innossa.

Is and sain rabert Celtchair na briathra sa innossa ac Ultaib i
nIraird Chullend in n-aidchi sin :

4130 .R[2]. ' Taible lethderg for ríg n-ágather. Án samlaide co
fodma féit. Deisme néomain im chét cráeb. Trícha chét n-arad.
Cét crúaid n-echdámach. Cét im chét drúad. Dar tús
imdesfíaid. Fer feraind im drumnib Conchobor. Faichlethar
cath claidid a ḟéinne. Gongáinethar cath for Gárich ⁊ Ilgárich
4135 issin matin sea monairther.'

Is í inn adaig cétna rabert Cormac Cond Longas mac Conchobuir
na briathra sa ac feraib Hérend ac Slemain Mide in n-aidchi sin:

'Amra maitne amra [mithisi][3]. Mescfaiter slúaig. Sáifiter slúaig
con máe re secht cléithe slúaig Ulad im Chonchobor. Cossénait
4140 a mná. Raseiset a n-éite for Gárig ⁊ Ilgárig isin matin sea monairther.'

Is hí inn adaig cétna rabert Dubthach Dáel Ulad na briathra sa
oc feraib Hérend i Slemain Mide in n-aidchi sin :

' Móra maitne maitne Mide. Móra ossud [ossud][3] Cullend.
Móra cundscliu cundscliu Chláthra. Móra echrad echrad Assail.
4145 Móra tedmand tedmand Tuath Bressi. Móra in chlóe clóe Ulad
im Chonchobor. Cossénait a mná. Raseisset a n-éiti for Gárig
⁊ Ilgárig isin matin se monairther.'

---

[1] sic for inda (Strachan)   [2] in marg.   [3] supplied from YBL, St

And sain confucht[r]aither¹ Dubthach trina chotlud coro mesc
ind Neamain barsin slóg co llotar i n-armgrith ba rennaib a sleg
95ᵇ ⁊ a fáebor co n-ébailt | cét laéch díb ar lár a ndúnaid ⁊ a llongphuirt ⁴¹⁵⁰
re úathgráin na gáre rabertatar ar aird. Cid trá acht ní hí sin
aidche ba sáime d'feraib Hérend fúaratar ríam ná híaram risin
tairchetul ⁊ risin tarngiri, risna fúathaib ⁊ risna haslingib facessa
dóib.

And sain atbert Ailill : 'Rasetarrad-sa ám,' bar Ailill, 'arggain ⁴¹⁵⁵
Ulad ⁊ Cruthni ó lúan tati samna co tate n-imbuilg. Tucsam a
mnáa ⁊ a meic ⁊ a maccáemi, a n-eich ⁊ a n-echrada, a n-albi ⁊ a
n-éiti ⁊ a n-indili. Barrallsam a tilcha dá n-éis co failet ina fántaib
comtís comarda síat. Is aire nachass idnaidib-sa and so ní bas
mó, acht tabrat chath dam-sa ar Maig Áe ²madi tecra leó². Ac[h]t ⁴¹⁶⁰
ci atberam-ni and so 'no, táeit³ nech d'farcsin maigi mór[f]arsing
Mide dá fiss in tecat Ulaid ind. Ocus má thecait Ulaid ind, ní
thechiub-sa da ráith itir, dáig ní robés ríg rotheched do grés itir.'
'Cia bad chóir do thecht and?' bar cách. 'Cia acht Mac Roth
in rímechlach and sút.'   ⁴¹⁶⁵

Tánic Mac Roth reime d'farcsi maigi mórfarsing Mide. Nírbo
chían do Mac Roth dá mbáe and co cúala inní, in fúaim ⁊ in fothrom,
in sestán ⁊ in sésilbi. Nír súail ní risbud samalta leiss acht marbad
hí in firmimint dothuitted bar dunegnúis in talman, ná marbad
hí ind fairrge eithrech ochargorm tísad for tulmoing in bethad, ⁴¹⁷⁰
ná marbad é in talam barrálad assa thalamchumscugud, ná marbad
hí ind fidbad ra thuitted cách díb i nglaccaib ⁊ gablaib ⁊ géscaib
araile. Cid trá acht barrafnit na fíadmíla barsin mag connárbo
réil tulmonga maige Mide fóthib.

Tánic Mac Roth co n-innisin [in] sceóil sein co airm i mbáe Ailill ⁴¹⁷⁵
⁊ Medb ⁊ Fergus ⁊ mathi fer nHérend. Dochúaid Mac Roth dóib
anísin. 'Cid and sút, a Ferguis?' bar Ailill. 'Ní handsa,' bar
Fergus. 'Is é fúaim ⁊ fothromm ⁊ fidréan atchúala-som,' bar
Fergus, ' toirm ⁊ torand, sestainib ⁊ sésilbi, at Ulaid barfópartatar
in fid, imdrong na curad ⁊ na cathmíled ac slaide ind feda cona ⁴¹⁸⁰
claidbib rena carpdib. Iss ed ón barraffind na fíadmíla barsin
mag connach réil tulmonga maige Mide fóthib.'

Fecht n-aill forréccaig Mac Roth in mag. Confacca ní, in
nglascheó mór ra ercc in comás eter nem ⁊ talmain. Andar leiss
batar⁴ indsi ás lochaib atchondaic ás fánglentaib na cíach. Andar ⁴¹⁸⁵
leis batar⁴ úama ursloicthi atchonnaic and i rremthús na cíach

---

¹ dofochtradar *YBL, C*
²⁻² *sic*=mad tacair leo; ma tacar doib *YBL*, mad ail leo *St* (*this passage
is given twice in C, taken from 2 versions; in first occurrence* (=*LL*) *mad tagra
leo; in second* (=*YBL*) *mad taccar doib*)   ³ i *subscr.*   ⁴ bátar *MS.*

cétna. Andar leis ba línanarta lín lángela ná bá snechta síthalta
ac snigi ratafarfáit and tri urdluich na cíach cétna, ¹ná andar¹ leis
ba éochain de ilénaib ilerda iṅgantacha imda, ná ba hilbrec[h]tnugud
96ª rétland roglan | i n-aidchi reóid roṡolais, nó ba haíble teined trichem-
rúaid. Atchúala ní, in fúaim ⁊ in fothrom ⁊ in fidréan, in toirm ⁊
in torand, in sestainib ⁊ in sésilbi. Tánic remi co n-innisin in
sceóil sin co hairm i mbaí Ailill ⁊ Medb ⁊ Fergus ⁊ mathi fer ṅHérend.
Dachúaid dóib anísein.

4195 'Cid and sút ale, a Ferguis?' bar Ailill. 'Ní handsa,' bar Fergus.
'Is é glascheó mór atchondaic-sium ra erc in comás eter nem ⁊
talmain imthinnsaitin anála na n-ech ⁊ na curad, smútgur in láir
⁊ lúathred na conar conas ecgaib ri seól ṅgaíthe úasu co nderna
tromchiaich treglaiss de i nnélaib ⁊ i n-áeraib.

4200 'Batar íat indsi ás lochaib atchonnaic-sium and, cind na cnocc
⁊ na tilach ás fánglentaib na cíach, cind na curad ⁊ na cathmíled
ósna carptib ⁊ na carpait archena. Batar íat úama urslocthi
atchondaic-sium and i rremthús na cíach cétna, beóil ⁊ sróna na
n-ech ⁊ na curad ac súgud gréne ⁊ gaíthe úathu ⁊ chuccu ra tricci
4205 na dírma.

'Batar² íat línanarta lín lángela atchondairc-sium and ná snechta
síthalta ac snigi, in t-úanbach ⁊ in chubrach curit glomraigi na sríana
bélbaigib na n-ech rúanaid rothend ri dremna ṅdírma. Ba hí éochain de
ilénaib ilerda iṅgantacha imda atchondaic-sium and, gand³ in láir ⁊
4210 adúachtur⁴ in talman curit na eich assa cossaib ⁊ assa crúib conas
ecgaib ra seól ṅgaíthi úasa.

'Is é in fúaim ⁊ fothromm ⁊ fidréan, toirm ⁊ torand, sestainib
⁊ sésilbi atchúala-som and, scellgur na scíath ⁊ slicgrech na sleg &
glondbéimnech na claideb ⁊ bressimnech na cathbarr, drongáir na
4215 lúrech ⁊ immchommilt na n-arm ⁊ dechairdecht na cless, tétimnech⁵
na tét ⁊ núallgrith na roth ⁊ baschaire na n-ech ⁊ culgaire na carpat
⁊ tromchoblach na curad ⁊ na cathmíled sund chucaind.

'Ba hé ilbrechtnugud rétland roglan i n-aidche roṡolais rotafárfáid-
sium and ná haíble tened trichemrúaid, súli cichurda adúathmara
4220 na curad ⁊ na cathmíled ásna⁶ cathbarraib caíni cummaidi
cumtachglana, lán din feirg ⁊ din baraind rabertaṭar leó risná
ragbad ríam ná híaram fír catha ná fornert⁷ comlaind ⁊ risná
gébthar co brunni brátha ⁊ betha.'

'Ní dénam robríg de,' bar Medb. 'Atethatar daglaích ⁊ degóic
4225 acainni dá n-acallaim.' 'Ní ármim-sea ón om, a Medb,' bar Fergus,

---

¹⁻¹ nandar *MS*. ² bátar *MS*. ³ deandgar *St*
⁴ adactur *MS*., uac[h]tar *St* ⁵ teidbemnech *St*
⁶ *sic, read* asna *or* assa (isna *St*) ⁷ fornirt *MS*

'dáig atiur-sa bréthir ¹nach raichnea¹ i nHérind nach i nAlpain slúag acallma Ulad á rasfecgat a fergga do grés.'

Is and sain ra gabsatar cethri ollchóiceda Hérend dúnad ⁊ longphort ac Cláthra in n-aidchi sin. Rà fácsatar fiallach foraire ⁊ freccométa úathu ra hagid Ulad ná tístais Ulaid gan robud gan rathugud dá ⁴²³⁰ saigid.

Is and sin raluid Conchobor ⁊ Celtchair trícha chét carptech
96ᵇ n-imrindi co ndessetar i Slemain Mide | dar éis na slúag. Acht ci atberam-ni and so, ní dessetar da ráith iter, acht ratháegat ass d'etarphurt do dúnud Ailella ⁊ Medba do thetarrac[h]tain a lláma ⁴²³⁵ d'furdergad re cách.

Nírbo chían do Mac Roth dá mbáe and co faccae ní, ²in n-echrad ṅdírecra ṅdermór² i Slemain Mide anairtúaid cach ndíriuch. Tánic reme go airm i mbáe Ailill ⁊ Medb ⁊ Fergus ⁊ mathi fer ṅHérend. Atfócht Ailill scéla de ar rochtain. 'Maith a Meic Roth,' bar ⁴²⁴⁰ Ailill, 'in faccasu nech d'Ultaib bar slicht in tslúaig seo indiu?' 'Nád fetar-sa ém,' ar Mac Roth, 'ac[h]t adchonnac-sa echraid³ ṅdírecra ṅdermóir i Slemain Mide anairtúaid cach ṅdíriuch.' 'Garsa lín na echraidi ale?' bar Ailill. 'Nád úatti lim trícho chét carpdech n-imrindi indi.i. deich cét ar fichit chét carpdech n-imrindi,' ⁴²⁴⁵ ar Mac Roth. 'Maith a Ferguis,' bar Ailill, 'cid latt-su ar ṁbúbthad-ni de smútgur ná do dendgur ná d'análfadaig mórslúaig mad gustráthsa acus ná fail latt lín catha dúnni acht sút?'

'Rolúath bic n-archessi forro,' bar Fergus '⁴dáig ro bífad co mbetis na slúaig níbad liriu ná mar rádit-sium.'⁴ 'Déntar comairle ⁴²⁵⁰ forbthe athgarit acainni de-side,' for Medb, 'dáig rofess rar fúabérad-ni in fer romór rogarg robruthmar út, Conchobor mac Fachtna Fáthaig meic Rossa Rúaid meic Rudraigi, ardrí Ulad ⁊ mac ardríg Hérend. Déntar dunibúali urslocthi do feraib Hérend ar cind Conchobuir ⁊ buiden trícho chét acá hiadad dá éis, ⁊ gabtar ⁴²⁵⁵ na fir ⁊ ná gondar itir, ⁵dáig ní mmó ná dán cimbeda rothóegat'.⁵—

Conid hí sin in tress bríathar is génnu ra ráded bar Táin Bó Cúalnge: Conchobor gana guin do gabáil ⁊ ⁶dán cimbeda do dénam⁶ dona deich cét ar fichit cét bátar 'na farrad de rígraid Ulad. Acus atchúala Cormac Cond Longas mac Conchobuir anísin ⁊ rafitir mani ⁴²⁶⁰ díglad a chétóir a mórbréthir bar Meidb ná dígélad go brunni ṁbrátha ⁊ betha.

---

¹⁻¹ nach bhfuil St (ni foigebthar YBL)   ²⁻² an eachraid ndifrecra
ndermair St.   ³ ecrait MS.   ⁴⁻⁴ sic; cp. uair bia [v.l. biaid] a fhios
agat-sa curap lia dhóibh na mar sutt St; rádit-sium = rádid-sium
⁵⁻⁵ cp. daigh ni mo ina ar ndiol do cimbedaibh tiegaid St
⁶⁻⁶ dán om. St (cimedha do genam)

Acus is and sin atraacht Cormac Cond Longas mac Conchobuir cona budin tríchat cét ¹d'ḟorddiglammad áig ⁊ urgaili¹ for Ailill ⁊ ⁴²⁶⁵ for Meidb. Atraacht Ailill cona tríchait chét dó-som. Atraacht Medb cona tríchait cét. Atraachtatar na Mani cona tríchtaib cét. Atraachtatar Meic Mágach cona tríchtaib cét. Atraacht in Galeoin ⁊ in Mumnig ⁊ popul na Temrach, ⁊ fogníthea etargaire eturru co ndessid cách díb ²i rail² araile ⁊ i fail a arm. Araí sein ragníad ra ⁴²⁷⁰ Meidb dunibúali ursloicthi ar cind Chonchobuir ⁊ buden trícho chét acá íadad dia éis. Daríacht Conchobor d'indsaigid na dunibúaled aursloicthi ⁊ ní rabi ic íarraid a dorais don tṡainruth itir, acht ra minaig beirn inaid míled ar urchomair a gnúsi ⁊ a agthi isin chath ⁊ bern chét dá leith deiss ⁊ bern cét dá leith chlí, ⁊ imsoí ⁴²⁷⁵ chuccu innond ⁊ ras mesc thall fora lár ⁊ torchratar ocht cét láech lánchalma lais díb. Acus tánic úadib assa athli gan ḟuligud gan ḟordergad fair co ndessid i Slemain Mide bar cind nUlad.

'Maith a ḟiru Hérend,' bar Ailill, 'táet nech úan d'ḟarcsi maige mórḟarsiṅg Mide dá ḟis cindas na hacgmi bá tecat Ulaid isin tulaig i ⁴²⁸⁰ Slemuin Mide, dá innisin dún túarascbáil a n-arm ⁊ a n-erriuda, a curad ⁊ a cathmíled ⁊ a clíathbernadach cét ⁊ a fíallach feraind.

97ᵃ Gardditi lind éistecht riss mad colléic.' | 'Cia doragad and?' bar cách. ' Cia acht Mac Roth in rímechlach,' bar Ailill.

Tánic Mac Roth reme co ndessid i Slemain Mide bar cind Ulad. ⁴²⁸⁵ Ra gabsat Ulaid ac tachim isin tulaig sin á dorbblais na matni muchi co tráth funid na nóna. Iss ed mod nárbo thornocht in talam fótho risin ré sin, cach droṅg díb imma ríg ⁊ cach buiden imma tóesech, cach rí ⁊ cach toísech ⁊ cach tigerna go lín a ṡlúag ⁊ a ṡochraite, a thinóil ⁊ a thóchostail fo leith. Cid trá acht doríacht- ⁴²⁹⁰ atar Ulaid uile re tráth funid nóna isin tulaig sin i Slemuin Mide.

Tánic Mac Roth reme go túarascbáil in chétna braini díb leis³ go airm i mboí Ailill ⁊ Medb ⁊ Fergus ⁊ mathi fer ṅHérend. Atfócht Ailill ⁊ Medb scéla de ar rochtain. 'Maith a Meic Roth,' bar Ailill, '⁴cindas na hecgmi [nó] na taicgme bá tecat Ulaid isin ⁴²⁹⁵ tulaig i Slemain Mide⁴?'

'Nád ḟetar-sa ám,' bar Mac Roth, 'ac[h]t⁵ tánic buden bruthmar brígach mórcháin isin tulaig sin i Slemuin Mide,' bar Mac Roth. 'Dóig ri farcsin ⁊ ri fégad trí⁶ tríchu cét indi. Barallsat⁷ a n-étaigi díb uile, concechlatar ḟirt fótbaig ba ṡuide a toísig. Óclach seta

---

¹⁻¹ do genamh aigh ⁊ iorgaili St   ²⁻² leg. i fail   ³ After leis lines 4280-81 supra inadvertently repeated: Tuarascbáil a n-airm ⁊ a n-erriuda, a curad ⁊ a cathmiled ⁊ a cliathbernada cét ⁊ a fiallach feraind
⁴⁻⁴ cp. caidi tuarascbail Uladh ic techt i Slemain Mide St   ⁵ act so amhain St   ⁶ om. YBL, St; omitt.   ⁷ rostellsad YBL, ro lasat C, ro cuirsit St

fata ¹n-airard n-ardmín¹ forúallach i n-airinuch na budni sin. ⁴³⁰⁰
Caíniu di flaithib in domuin ritacoemnacair² etera slúagaib, eter
urud ⁊ gráin ⁊ báig ⁊ chostud. Folt findbuide iss é cass dess drumnech
tóbach faride. Cuindsiu cháem chorcarglan leis. Rosc roglass
gossarda iss é cicharda adúathmar ina chind. Ulcha degablach
is sí buide úrchass³ bá smech. Fúan corcra corrtharach cáeicdiabuil ⁴³⁰⁵
imbi. Eó óir isin brutt ósa bruinne. Léine glégel chulpatach ba
dergintliud do dergór fria gelchness. Gelscíath go túagmílaib
dergóir fair. Claideb órduirn intlaissi isindara láim dó. Manaís⁴
lethanglass isin láim anaill. Dessid in láech sin i n-urard na tulcha
go toracht cách cuce, ⁊ dessetar a buden imbe. ⁴³¹⁰
  'Tánic buiden aile and dano isin tulaig cétna i Slemuin Mide,'
far Mac Roth. 'Tánaise dá tríchtaib cét atacaemnacair⁵ Fer
caín and dano i n-airinuch na budni sin caedessin. Folt findbuide
fair. Ulcha éicsi imchass imma smech. Bratt úanide i forcipul
imme. Cassán gelargit isin brut ósa brunni. Léni donderg míleta ⁴³¹⁵
ba dergindliud do dergór frí[a] gelchnes i caustul go glúnib dó.
Caindell rígthaige 'na láim go féthanaib argait ⁊ co fonascaib óir.
Is ingnad reba ⁊ ábarta dogní in tsleg fil 'na láim na óclaige.
Immireithet impe na féthana argit sechna fonascaib óir cachla céin
ó erlond gó indsma. ⁶I[n] céind aill⁶ dano it íat na fonasca óir ⁴³²⁰
immireithet sechna féthanaib argit ó indsma gó hirill. Scíath
bémmendach go fáebor chondúala fair. Claideb ço n-eltaib dét ⁊
co n-imdénam snáith óir bara chlíu. Dessid in láech sain for láim
in⁷ chlí ind óclaig thóesig tánic issin tulaig, ⁊ dessetar a buiden
imbe. Ac[h]t ci atberam-ni and so, ní destetar de ráith itir acht ⁴³²⁵
a nglúini fri lár dóib ⁊ imbel a scíath aca smechaib dóib ⁸a fat⁸
97ᵇ leó go léctar chucaind. Acht atá ní chena domfarfáit | formindi
mór issin óclach mór borrfadach is tóesech don budin sin.
  'Tánic buden aile and dano isin tulaig cétna i Slemuin Mide,'
for Mac Roth. 'Tánaise dá séitche eter lín ⁊ chostud ⁊ timthaige, ⁴³³⁰
Láech cáem cendlethan i n-airinuch na buidni sin. Folt dúalach
dondbuide fair. Rosc duillech⁹ dubgorm for folúamain ina chind.
Ulcha éicsi imchass is sí degablach imcháel imma smech. Bratt
dubglass ¹⁰ba loss¹⁰ i forcipul imme. Delg duillech de findruine
'sin brutt ósa bruinne. Léne gelchulpatach frí chness. Gelscíath ⁴³³⁵
co túagmílaib argait inti fair. Máeldorn findargait i n-intiuch

---

¹⁻¹ *sic. read* airard ardmín (*cp.* os é seada urard airdmhín *St*)
² = atacoemnacair   ³ = aurchass; imc[h]as *St*, erchas *YBL, C*
⁴ mánais *MS.*   ⁵⁻⁵ tanaisdi do trichait cet inti *St;* is tanasti dia
seitche *YBL*   ⁶⁻⁶ Icein daill *MS.;* In ccén n-aili *St*   ⁷ *omitt.;* for laimh
chlí *St*   ⁸⁻⁸ *better* ara fat *as in St*   ⁹ duilech *St*
¹⁰⁻¹⁰ *printed* baloss *Dipl. Edn.;* folus *YBL, C;* om *St, cp. supra l.* 3718

bodba fá choimm. Ture rígthige fria aiss. Dessid in láech sain
issind ḟirt ḟótbaig bad fiadnaisi dond óclach thóesech thánic isin
tulaig, ⁊ destetar a buiden imme. Acht ba binnithir lim ra fogor
4340 mendchrott i llámaib súad icá sírṡenmum bindḟogrugud a gotha ⁊
a irlabra inn óclaíg ac acallaim in óclaíg thóesig thánic issin tulaig
⁊ ac tabairt cacha comairle dó.'
'Cia sút ale?' bar Ailill ri Fergus. 'Ratafetammar ám ale.'
bar Fergus. ¹Is hé cétna láech cétrachlass¹ in fert fótbaig i n-urard
4345 na tulcha go toracht cách cuce Conchobor mac Ḟachtna Fáthaig
meic Rosa Rúaid meic Rudraigi, ardrí Ulad ⁊ mac ardríg Hérend.
Is é láech formend mór dessid fora láim chlí Conchobuir Causcraid
Mend Macha mac Conchobuir co maccaib ríg Ulad imme ⁊ co
maccaib ríg Hérend rafailet² ina farrad. Is hí in tṡleg atchondaic
4350 ina láim in Chaindel Chuscraid sein co féthanaib argit ⁊ go fonascaib
óir. Is bés don tṡleig sin nachis imrethet impe ríam ná híaram na
féthana argait sechna fonasca óir acht gar ré coscur écin, & is dóig
gombad gar ré coscur ros imreittís impe and so innossa.
'Is hé láech cáem cendlethan dessid issind ḟirt bad ḟiadnaissi
4355 don óclách thóesech thánic issin tulaig Sencha mac Ailella meic
Máilchló, so-irlabraid Ulad ⁊ fer sídaigthe ³slóig fer ṅHérend³. Acht
atiur-sa bréthir chena ní comairle mettachta ná midlaigechta rabeir
dá thigerna issin ló bága sa indiu, acht is comairle gaile ⁊ gascid ⁊
eṅgnama ⁊ mathiusa do dénam. Acht atiur-sa bréthir chena,'
4360 bar Fergus, 'is togaes⁴ dénma opre atraac[h]tatar im Chonchobor
i mmucha lá⁵ indiu and sain.' 'Ní dénam robríg díb,' bar Medb.
' Attethatar deglaích ⁊ dagóic acainni dá n-acallaim.' ' Ní ármim-
sea ón omm,' bar Fergus, 'acht atiur-sa bréthir nach ráichnea i
nHérind nach i nAlpain slúag acallma Ulad á rasfecat a fergga do
4365 grés.'
' Tánic buden aile dano isin tulaig cétna i Slemuin Mide,' bar
98ᵃ Mac Roth. 'Fer find fata mór | i n-airinuch na budni sin is ó
grísta gormainech. Folt dond temin fair is é slimthanaide bara
étun. Bratt forglass i filliud imme. Delg argit isin brutt ósa
4370 brunni. Léni gel manaísech fri chness. Cromscíath ⁶comḟaebur
chondúalach⁶ fair. Sleg cúicrinni⁷ 'na láim. Colg dét iarna
innud⁸.' 'Cia sút ale?' bar Ailill ri Fergus. 'Ratafetamar ám,'
bar Fergus. 'Is cur lám for debaid sin. Is cathmílid bar níth.

¹⁻¹ Is é in laoch fa ro clas *St; read*, . . . fá cetarachlass   ² filit *St*
³⁻³ sluaig Ulad *St*   ⁴ dagaes *YBL, C*, deghaos *St; read* dagaes   ⁵ laoi *St*
⁶⁻⁶ *read* co fáebur chondúalach (co faebar conduala *YBL, C*, co faobar
condualach *St*)   ⁷ coicrind *YBL, St*   ⁸ iomdha *St*

Is bráth bar bidbadu cách tánic and. Eogan mac Durthachta [1]a
fosta Fernmaige[1] atúaid and sin.'　　　　　　　　　　　　4375
'Tánic buden aile and isin tulaig cétna i Slemuin Mide,' for Mac
Roth. 'Ní gó ám is borrfadach forfópartatar in tulaig sin. Is
tromm in gráin, is mór [2]in t-urud[2] rabertatar leó. A n-étaige uile
dara n-aiss. Láech cendmár curata i n-airinuch na budni sin is
é cicharda úathmar. [3]Folt n-étrom ṅgrelliath fair[3]. Súle bude　4380
móra 'na chind. Bratt buide cáiclámach[4] imme. Delg óir buide
'sin brutt ósa bruinne. Léne bude chorrtharach frí chness. Gae
semnech slindlethan slegfota co mbráen fola dara fáebor ina
láim.' 'Cia sút ale?' bar Ailill ri Fergus. 'Ratafetamar ám ale
in láech sain,' bar Fergus. 'Ní imgab cath ná cathróe ná comlund　4385
ná comrac cách thánic and. Lóegaire Búadach mac Connaid Buide
meic Íliach [5]ó Immail[5] atúaid and sain.'
'Tánic buden aile and dano isin tulaig cétna i Slemuin Mide,'
for Mac Roth. 'Láech munremur collach i n-airinuch na buidni
sin. Folt dub tóbach fair. Gnúis chnedach chorcarda fúa. Rosc　4390
roglass lainnerda 'na chind. Gae súlech[6] go foscadaib úasu.
Dubscíath co caladbúalid findruini fair. Brat odorda bachuaslac[7]
imme. Bretnas bánóir isin brut ósa bruinne. Léine threbraid
síte fria chnes. Claideb co n-eltaib dét ⁊ co n-imdénam órṡnáith
[8]ara étaig[8] ïmmaig anechtair.' 'Cia sút ale?' bar Ailill ri Fergus.　4395
' Ratafetammar ám ale,' bar Fergus. 'Is cur lám for ugra sain.
Is tond romra bádes. Is fer trí ṅgreth. Is muir dar múru cách
thánic and. Munremur mac Gerrcind a Moduirn atúaid and sain.'
'Tánic buden aile and dano isin tulaig cétna i Slemuin Mide,'
for Mac Roth. 'Láech cetherlethan[9] comremar i n-airinuch na　4400
buidni sin is é anisc[10] odorda, is é derisc[11] tarbda. Crundrosc
[n-]odorda n-adardd ina chind. Folt bude rochass fair. Crundscíath
dérg co mbil chaladargait ina imthimchiull úasu. Gae slindlethan
slegfota 'na láim. Bratt ríabach imme. Eó uma isin brutt ása[12]
brunni. Léni chulpatach i caustul [13]gá forcnib[13] dó. Colg dét iarna　4405
chossliasait chlí.' 'Cia sút ale?' | ar Ailill ri Fergus. 'Rata-
fetammar ám ale,' bar Fergus. 'Is sond catha sain. Is búaid
cacha irgaile. Is fodb trescada cách thánic and. Connud mac
Morna ó Challaind atúaid and sain.'

98ᵇ

[1-1] ri Fernmaigi *YBL;* righ forusda Fernmaighi *St*　　　[2-2] in t-erfuath
YBL, in t-uruath *St*　　　[3-3] folt etrom greliath fair *YBL, sic leg.* (folt
gleliath ettrom fair *St*)　　　[4] caiclámach *MS.* coeclamach *St*
[5] o Impiul *corr. from* Impuil *YBL,* ó Imiol *St*　　　[6] suilech *YBL,* fuilech *St*
[7] fochlaidi *YBL,* fo casloi (v.l. fa chaslói) *St; cp. infra l.* 4519
[8-8] tara etach *YBL*　　　[9] cinnlethan *St*　　　[10] aindiosc *St*
[11] dirrisc *St*　　　[12] asa *MS.* = ósa　　　[13-13] cóa oircnibh *St*

⁴⁴¹⁰ 'Tánic buden aile and dano isin tulaig cétna i Slemuin Mide,' for Mac Roth. 'Ní gó ám is tailc ⁊ is tarbech forrópartatar in tulaig sin conro chrothsatar na slúaig conarnecar indi fora cind. Fer cáem gráta i n-airinuch na budni sin. Áldem de daínib in domuin eter chruth ⁊ deilb ⁊ dénam, eter arm ⁊ erriud, eter mét ⁊ ⁴⁴¹⁵ míad ⁊ masse, eter chreitt ⁊ gasced ⁊ chóra.' 'Ní gó ám ale,' bar Fergus, ' is hí [a] epert chomadas-som sain. Ní duí for lomma cách thánic and. Is bidba cáich. Is gus nád fulangar. Is tond anbthena bádes. Is luchair n-aga in fer álaind. ¹Feidilmid Chilair Chetail ó Elland atúaid and sain¹.'

⁴⁴²⁰ 'Tánic buden aile and dano 'sin tulaig cétna i Slemuin Mide,' for Mac Roth. 'Ní comtig láech is cháemiu ná in láech fail i n-airinuch na buidni sin. Folt tóbach dergbuide fair. Aiged fochaín² forlethan laiss. Rosc roglass gossarda is é caindelda gárechtach 'na chind. Fer cóir cutrumma is é fata fochóel ⁴⁴²⁵ folethan³. Beóil deirg thanaide leiss. Deóit niamda némanda. Corp gel cnesta. ⁴Cassán gelderg i fadi úasu⁴. Eó óir isin brutt ósa brunni. Léne de sról ríg ⁵ina dergfilliud⁵ de dergór fri gelchness. Gelscíath co túagmílaib dergóir fair. Claideb órduirn intlassi fora chlíu. Gae fata fáeborglass re faga féig fóbarta co súanemnaib ⁴⁴³⁰ loga co semmannaib findruine ina láim.' 'Cia sút ale?' bar Ailill ri Fergus. 'Ratafetammar ám ale,' bar Fergus. 'Is leth ṅglíad sain. Is galiud⁶ comlaind. Is londbruth n-archon cách tánic and. Reochaid mac Fathemain ó Rígdond atúaid and sain.'

'Tánic buden aile and dano isin tulaig cétna i Slemuin Mide,' ⁴⁴³⁵ for Mac Roth. 'Láech órainech⁷ remurslíastach i n-airinuch na budni sin. Bec nach remithir fer cach n-óenball de. Ní gó ám is fer co talmain,' all sé. 'Folt dond tóbach fair. Gnúis chorcra chrundainech fúa. Rosc mbrecht n-urard ina chind. Fer án athlam and samlaid co n-ócaib dubartacha dubsúlib⁸ co n-idna rúad ⁴⁴⁴⁰ lassamain co n-ábairt imtholta co saigit sech⁹ comlond do brissiud ar forlond, co tuidmech fóbarta fair, can chommairge Conchobuir aca itir.' 'Cia sút ale?' bar Ailill ri Fergus. 'Ratafetammar ám ale,' ar Fergus. '¹⁰Bá hitte di gail ⁊ di gasciud¹⁰ cách thánic and.

---

¹⁻¹ Fedhlimidh mac Ilair Chédaig Cualnge ó Callainn atuaidh inn sin *St;* Feidlimid Cilair Cetaig ann *YBL* ² fochael *YBL, St; sic leg.*

³ *sic, for* forlethan ⁴⁻⁴ *sic;* brat corcra hi forcibul imbi *YBL, St; sic leg.*

⁵⁻⁵ co ndergindled *YBL*, fo deirgindled *St* ⁶ clariud *YBL*, claired *St;* *read* cláriud ⁷ oirenech *YBL, St; sic leg.* ⁸ dobsúilb *MS.*, i *inserted subsequently and out of place* ⁹ secht *MS.*, tar *YBL, St*

¹⁰⁻¹⁰ *sic MS.;* baithi do gail ⁊ gaisced *YBL*, ba do gail ⁊ gaisced *St*

¹Báe itte di drúis ⁊ tarpige¹. ²Táthud do sluagaib ⁊ d'armaib².
Rind áig ⁊ imgona fer ṅHérend ³iar túaisciurt³, mo derbchomalta-sa ⁴⁴⁴⁵
fadessin, Fergus mac Leite ó Líne atúaid and sin.'
'Tánic buden aile and dano isin tulaig cétna i Slemuin Mide,' for
Mac Roth, 'is hí fossaid écsamail. Láech álaind éscaid i n-airinuch
na budni sin. Gormanart cáel corrtharach go stúagaib fíthi figthi⁴ féta
findruini, go cnappib dílsi deligthi derggóir for bernadaib ⁊ ⁴⁴⁵⁰
brollaig[ib]⁵ dó fri c[h]ness. Bratt bommannach co mbúaid cach
99ᵃ datha thariss. Caechruth⁶ óir | fair .i. a scíath fair. Claideb
crúaid catut colgdíriuch i n-ardgabáil churad bara chlíu. Sleg
díriuch drumnech ar derglassad 'na láim.' 'Cia sút ale?' bar
Ailill ri Fergus. 'Ratafetammar ám ale,' bar Fergus. 'Is roga ⁴⁴⁵⁵
rígfiled sain. Is rúathar rátha. Is rót do báre. Is tarbbech a
gal cách thánic and. Amargin mac Ecclsalaig Goband, in file
maith ó Búais atúaid.'
'Tánic buden aile and dano 'sin tulaig cétna i Slemuin Mide,'
for Mac Roth. 'Láech find buide i n-airinuch na budni sin. Find ⁴⁴⁶⁰
uile in fer sain eter folt ⁊ rosc ⁊ ulcha ⁊ abratchur ⁊ dechelt. Scíath
búaledach fair. Claideb órduirn intlassi bara chlíu. Sleg cúicrind
confaittnedar darin slúag uile ina láim.' 'Cia sút ale?' bar Ailill
ri Fergus. 'Ratafetammar ám ale,' bar Fergus. 'Inmain ám láech
side⁷ far⁸ túaith rarfánic and. Inmain bethir balcbéimnech. Inmain ⁴⁴⁶⁵
mathgamain mórglonnach fri hécratu ⁹cuncan ferglond fóparta⁹.
Feradach Find Fechtnach á Nemud Slébe Fúait atúaid and sain.'
'Tánic buden aile and dano isin tulaig cétna i Slemuin Mide,'
for Mac Roth. 'Días máethóclách i n-airinuch na budni sin. Dá
bratt úanide i forcipul impu. Dá chassán gelargait isna brattaib ⁴⁴⁷⁰
ása mbrunnib. Dá léne di slemunsítu buide fria cnessaib. Claidbi
gelduirn fora cressaib. Dá sleig cúicrind¹⁰ co féthanaib argait
óengil ina lámaib. Immáes bec eturru de sodain.' 'Cía sút ale?'
bar Ailill ri Fergus. 'Ratafetammar ám ale,' bar Fergus. 'Dá
ánrath¹¹ sain, dá óenmuntind, dá óenlosnaid¹², dá óenchaindill¹³, ⁴⁴⁷⁵
dá ching, dá churaid, dá chléthbriugaid, dá dreicg, dá thenid, dá
thuidmechtaid, dá deil, dá dána, dá dásachtach, dá threittell
Ulad imma ríg. ¹⁴Fiachaig ⁊ Fiachna¹⁴ and sain, dá mac Conchobuir
meic Fachtna meic Rossa Rúaid meic Rudraigi and sain.

---

¹⁻¹ baithi di druis ⁊ tairptigi *YBL*, ba do drús ⁊ tairptighi *St*
²⁻² tatha dí di sluagaib ⁊ airbrib *YBL*, ⁊ ba co n-armaib ⁊ aidmibh gona
bfer nErenn *St*   ³⁻³ ar túarsciurt *MS*.   ⁴ *om. St*   ⁵ brollachaib *St*
⁶ *sic, for* cúicroth; cuicroth *YBL*   ⁷ síde *MS*.   ⁸ = fo ar?
⁹⁻⁹ *sic;* co n-a[n]forlond fóbarta *St; sic leg.*   ¹⁰ coicrindi *St*
¹¹ óenrath *MS.,* anrath *YBL, St*   ¹² anloise *YBL,* anloisne *St*
¹³ ancoinnill *St*   ¹⁴⁻¹⁴ Fiachna ⁊ Fiacha *YBL, St*

4480 'Tánic buden aile and dano 'sin tulaig cétna,' for Mac Roth. 'Is bádud ar méit. 'Is tene [ar][1] rúadlossi. Is cath ar lín. Is ald ar nirt. Is bráth ar bláriud. Is torand ar tharpigi. Fer ferggach úathmar irggráin i n-airinuch na buidni sin. is é srónmar ómar ubullruisc. [2]Folt ṅgarb ngrelíath[2]. Bratt ríbáin imme. 4485 Cúalli iairn isin brutt ósa brunni congeib[3] ón gúalaind go araile dó. Léne garb threbraid[4] fri chness. Claideb [5]secht ṁbrattomon[5] do iurn athlegtha iarna tháebdruimm. Tilach dond fair .i. a scíath. Líathga mór co tríchait semmand trina[6] cró 'na láim. Cid trá acht ro lá dírna[7] dina cathaib ⁊ dina slúagaib ac déscin[8] in laích sin 4490 ⁊ a buden immi oc tíachtain 'sin tulaig i Slemuin Mide.' 'Cia sút ale?' bar Ailill ri Fergus. 'Ratafetammar ám ale,' bar Fergus. 'Is leth catha sain. Is cend n-imresna. Is cend ar gail. Is muir dar críchu cách thánic and. Celtchair Mór mac Uthechair á Lethglaiss atúaid and sain.'

4495 'Tánic buden aile and dano isin tulaig cétna i Slemuin Mide,' for Mac Roth. 'is hí baile bruthmar, is í éitig úathmar. Láech brúasach[9] bélmar i n-airinuch na budni sin. is hé lethgleóir leithinchind[10] lámfota. Folt dond rochass fair. Bratt dublúascach imme. Roth créda 'sin brutt ása brunni. Léni derscaigthi fri c[h]ness.

99b Claideb urfota fá choim. Manaís[11] murnech ina deiss. | Líathboccóit fair .i. a scíath.' Cía sút ale?' bar Ailill ri Fergus. 'Ratafetam[ar][12] ám,' bar Fergus. 'Is hé in leó lond lámderg sain. Is é in t-art amnas ágsidi forrges gail. Eirrge Echbél ó Brí Errgi atúaid and sain.'

4505 'Tánic buden aile and dano 'sin tulaig cétna i Slemuin Mide.' for Mac Roth. 'Fer mór bresta i n-airinuch na budni sin. Folt rúadderg fair. Súle rúadderga móra 'na chind. Sithithir ri cruimmthir meóir míled cechtar n-aí [13]na dá[13] rígrosc rúad romóra failet laiss. Bratt brecc imme. Scíath glass fair. Gae gorm 4510 tanaide úasa. Buiden fuilech fordergg imme [14]'s éssium[14] féin créchtach fuilech eturru ar medón fadessin.' 'Cía sút ale?' bar Ailill ri Fergus. 'Ratafetammar ám ale.' bar Fergus. 'Is é in dána díchondircil[15]. Is é [16]inn accil ómnach[16]. Is é [17]in lumne léitmenach[17].

---

1 om. MS.; cp. is tine ar aine YBL, St   2-2 sic MS.; folt garbh grendliath fair YBL, folt garb gleliath fair St   3 congabadh St   4 threbnaid MS. 5-5 secht ṁbrotha YBL, St   6 tria St   7 dirmae YBL, diorma mor St 8 déscid MS.   9 bruach YBL, bruachmhár St   10 lethanchend YBL 11 mánais MS.   12 ar-compendium om.   13-13 diná MS. (ceac[h]tar na da rigrosg romhora St)   14-14 Sessium Dipl. Edn. (cp. osse fuileach crechtach cadesin YBL, eisiomh fen crec[h]tach fuilech ar medon eatorra St) 15 dichonnarcell YBL, dic[h]oindirclech St   16-16 innacci lómnach MS.; an aicil luaimnech St   17-17 an luibne létmech St

Is é in robb rigthi. Is é ¹in Cholptha¹. Is é in búadgalach Balc.
²Is hé [in leó?] Luirg². Is é in búridach Berna. Is é in tarb ⁴⁵¹⁵
dásachtach. Mend mac Sálcholgán ³ó Rénaib na Bóinne³.'
'Tánic buden aile and dano 'sin tulaig cétna i Slemuin Mide,'
for Mac Roth. 'Laéch lecconfota odorda i n-airinuch na budni
sin. Foit dub fair. ⁴Sithballrad .i. cossa⁴. Bratt derg fa chaslaí
imme. Brettnas bánargait isin brutt ósa brunni. Léni línidi frí chness. ⁴⁵²⁰
Scíath chróderg co comraid (óir)⁵ fair. Claideb co n-irdurn argait
bara chlíu. Sleg uillech órchruí úasu.' 'Cía sút ale?' bar Ailill
ri Fergus. 'Ratafetamar ám ale,' bar Fergus. 'Fer trí ruitte sin, fer
trí raitti, fer trí rámata, fer trí mbristi, fer trí mbúada, fer trí
mbága. ⁶Fergna mac Findchonna rí Búraig Ulad atúaid and sain⁶. ⁴⁵²⁵
' Tánic buden aile and dano 'sin tulaig cétna i Slemuin Mide,'
for Mac Roth. 'Fer caín mór i n-airinuch na budni sin. Cosmail
ra Ailill n-ucut n-adrind n-inchoisc eter chruth ┐ ergnus ┐ gili, eter
arm ┐ erriud ┐ gail ┐ gasciud ┐ gart ┐ gnímrada. Scíath gorm co
cobraid óir. Claideb órduirnd bara chlíu. Sleg cóicrind co n-ór ⁴⁵³⁰
ina láim. Mind óir úasu.' 'Cía sút ale?' bar Ailill ri Fergus.
'Ratafetammar ám ale,' bar Fergus. 'Is forus ferdaide sain.
Fúaparta⁷ forlaind. Forbrisiud fer cách thánic and. Furbaidi
Fer Bend mac Conchobuir á Síl i mMaig Inis atúaid and sain.'
'Tánic buden aile and dano 'sin tilaig cétna i Slemuin Mide,' ⁴⁵³⁵
for Mac Roth, 'is hí fossud écsamail risna budnib aile. Aill bruitt
100ᵃ deirg. Aill bruitt | glaiss. Aill bruitt guirm. Aill bruitt úane.
Bláe bána buide it íat álle étroc[h]ta úasu. Undseo mac mbec
mbrecderg co mbrutt chorcra eturru bar medón badessin. Eó óir
isin brutt ósa brunni. Léne de sról ríg ba derggintliud de dergór fri ⁴⁵⁴⁰
gelchness. Gelscíat[h] go túagmílaib dergóir fair. Taul óir barsin scíath,
bil óir ina imthimchiull. Claideb órduirn bec bá choimm aice. Gae áith
étromm ⁸co foscathaib⁸ úasu.' 'Cía sút ale?' bar Ailill ri Fergus.
'Nád fetar-sa ám ale,' bar Fergus, 'innass na budni sin ná in mac
bec fil inti d'fácáil ri Ultaib dar m'éis, acht óen bad dóig lim-sa ⁴⁵⁴⁵
and comtis iat fir Themra im Ercc mac Fedilmi Nóchruthaigi,
mac-side Carpri Nia Fer, & más íat, ⁹nímo carat anairich and so⁹,

---

¹⁻¹ *something omitted before* Cholptha? *Cp.* in cathchuindich Colptai *YBL,*
in comlud Colptha *St; read* in cathchungid Cholptha? ²⁻² is hé Luirg
MS., whole sentence om. YBL and St ³⁻³ o Chorannaib *YBL,* ó Rendaibh
na Boindi *St* ⁴⁻⁴ *gloss interl. and repeated in marg.;* as e sithballradh *St*
⁵ *om. MS., sic St* ⁶⁻⁶ Fergnae mac Findchoime a Coronn sin *YBL,*
Ferccnadh mac Fiondchaimhe rigbrughaidh Uladh atuaidh ann sin *St*
⁷ *sic MS.; insert some such word as* fer *bef.* fúaparta, *or read with St* Is
fuapairt forlaind ⁸⁻⁸ co foscod *YBL, om. St*
⁹⁻⁹ *sic MS.,* nimmuscarat tairrid *YBL,* ni mus cara torae *St*

dóig 's ¹a dichmairc¹ a athar dodechaid in mac bec sain d'fórithin
a senathar din chur sa, ⁊ mad síat, bud muir conbáidfea dúib-si
⁴⁵⁵⁰ in buden sain dáig is ²tri [d]agin² na buidni sin ⁊ an meic bic rafail
inti conmáe foraib-si in cath sa don chur sa.' 'Cid de-side?' bar
Ailill. 'Ní handsa,' bar Fergus, 'dóig ní faccéga in mac bec sain
úath ná húamain 'gabar slaidi-si ⁊ 'gabar n-essarggain co tora lár
far catha chucaib. Concechlastar rucht claidib Conchobuir ³mar
⁴⁵⁵⁵ glimnaig n-archon³ i fathad ná mar leóman oc tech[t] fo math-
gamnaib. Concichre Cú Chulaind cethri múru móra de chollaib
dóene immon cath sechtair. ⁴Bát bágaig, bat condalbaig confúarcfet⁴
flaithe fer nUlad ar n-úair. Is ferda conbúrfet in damrad dermór
oc tessargain laíg a mbó issi[n] chath issin matin se imbárach.'
⁴⁵⁶⁰ 'Tánic buden aile and dano 'sin tulaig cétna i Slemuin Mide,' for Mac
Roth, 'nád úatti trícho chét indi. Fianna feochra forderga. ⁵Fir
gil glain guirm chorcarda⁵. Monga fata findbuidi. Gnúsi álle
étroc[h]tai. Ruisc réilli rígdaidi. Étaige lígda lendmassa. Deilge
órda airegda iar ndótib dendglana. Lénti síti srebnaide. Slega
⁴⁵⁶⁵ gorma glainidi. Scéith buide bémnecha. Claidbi órduirn intlassi
iarna sliastaib sudigthir⁶. Rotas triall brón búridach. Brónaig
uile eochraidi. Torrsig rurig rígdaide. Díllechta in slóg sorchaide
gana comsid costadaig imdíched a n-irúatha.' 'Cía sút ale?'
bar Ailill ri Fergus. 'Ratafetammar ám ale,' bar Fergus. 'At
⁴⁵⁷⁰ leómain londa sain. At glonna catha. Trícha cét Maige Murthemne |
100ᵇ and sain. Is ed dosgní cendchrom torsech n-anfálid cana ríg
n-aurraindi eturru badessin, can Choin Culaind costadaig coscaraig
claidebdeirg cathbúadaig.' 'Fail a mórabba ám dóib-sium sain,'
ar Medb, 'ciarsat cendchroimm torsig n-anfálid. Ní fuil olc ⁷nár
⁴⁵⁷⁵ dernsamar⁷ riu. Ratas airgsemar ⁊ ratas indrisem ó lúan tate
samna co taite n-imboilg. Tucsam a mná ⁊ a meicc ⁊ a maccáemi,
a n-eich ⁊ a n-echrada, a n-ailbi ⁊ a n-éiti ⁊ a n-indili. Barraeilseam⁸
a tailcha dá n-éis go failet ina fántaib comtís comartai síat.' '⁹Ní thá ní
nod maítte forro⁹, a Medb,' bar Fergus, 'dáig ní dernais d'olcaib
⁴⁵⁸⁰ ná d'écóraib friu ní ¹⁰nár urfuaith¹⁰ tóesech na degbuidne út fort,
¹¹dáig atbíth¹¹ cach fert ⁊ cach lecht, cach lia ⁊ cach ligi fuil adíu go
airther nHérend, is fert ⁊ is lecht, is lia ⁊ is ligi do degláech ⁊ do
degóc arna tuttim ra degthóesech na buidne út. Bochinmáir¹²

---

¹⁻¹ dichmairc *YBL*, a ndic[h]mairc *St*   ²⁻² tria nag *YBL*, tria daigin *St*
³⁻³ mar glimmaig n-archon *MS.*, amail gloim n-archon *YBL*, mar gloim
n-arc[h]on *St*   ⁴⁻⁴ Badh badach ⁊ badh condalbach dofuaircfet *St*
⁵⁻⁵ *om. St* ⁶ suidighthi *St*   ⁷⁻⁷ nach dernamair-ne St   ⁸ ro mursam *St*
⁹⁻⁹ nior maiti duit-si sin *St*   ¹⁰⁻¹⁰ .i. nar dígail (*a gloss interl. and repeated
in marg.*); naro diogail *St*   ¹¹⁻¹¹ *sic*; read dáig ar bíth   ¹² Cen mair *St*

rissa ṅgébat! Is mairg ara tochérat! Bud leór leth catha do
feraib Hérend síat ac cosnam a tigerna isin chath 'sin matin sea ⁴⁵⁸⁵
imbárach.'
'Atchúala-sa núall mór and,' for Mac Roth, 'risin cath aníar
¹nó risin cath anair¹.' 'Garsa núall sút ale?' bar Ailill ri Fergus.
'Ratafetammar ám ale,' bar Fergus. 'Cú Chulaind sain ac tríall
tíachtain docairt² dochum in chatha 'gá furmiáil ri fót fóenlaige i ⁴⁵⁹⁰
fFirt Sciach fa thúagaib ⁊ baccaib ⁊ refedaib³, ⁊ ní lécgat Ulaid ind
é ar bíth a chned ⁊ a chréchta dáig ní hinchomlaind ⁊ ní hinchomraic
aithle chomraic Fir Diad.'
    Ba fír ám do Fergus anísin. Ba sé Cú Chulaind sain 'gá furmiáil
ri fáet⁴ fóenlige i Firt Sciath ba thúagaib ⁊ baccaib ⁊ refedaib.³   ⁴⁵⁹⁵
And sain radechatar na dá banchánti a dúnud ⁊ longphurt fer
ṅHérend .i. Fethan ⁊ Collach, co mbátar oc fáschuí ⁊ oc fásguba
ás chind Con Culaind icá innisin dó madma bar Ultaib ⁊ marbtha
Conchobuir ⁊ tuittmi Fergusa i frithguin.

IS hí inn aidchi sin radechaid in Morrígu ingen Ernmaiss go ⁴⁶⁰⁰
mbaí oc indloch ⁊ oc etarchossaít eterna dá dúnad chechtarda.
Acus rabert-si na bríathra sa :
    R. 'Crennait brain braigte fer brunnid fer⁵ fuil. Mescthair
tuind. Fadbaib luind. Níthgalaib [iar⁶] luibnig. Lúth fiansa.
Fethal ferda. Fir Chrúachna. Scritha⁷ minardini. Cuirther ⁴⁶⁰⁵
cath ba chossaib araile. Eblait a rréim. Bo chin Ultu. Bó
mair Érno. Bo chin Ulto'.
    Iss ed dobert i clúais nÉrand, ní firfet a ṅglé fail fora cind.
    Is and sain rabert Cú Chulaind ra Láeg mac Riangabra: 'Ba
líag ám dait-siu, a mmo phopa Laíg,' bar Cú Chulaind, '⁸na derntá⁸ ⁴⁶¹⁰
eterna dá chath cechtarda indiu ní ná beth a fis acut dam-sa.'
'Cacha finnub-sa de ám, a Chúcúc⁹, bar Láeg, 'innisfithir duit-siu.
Acht undsea albín assin dúnud ⁊ assin longphurt aníar innossa
101ᵃ barsin mag. | Undsea chethirn ṅgilla 'na ndiaid dá fostud ⁊ dá
n-imfuirech. Undsea chethirn ṅgilla 'no assin dúnud ⁊ assin ⁴⁶¹⁵
loṅgphurt anair dá tetarrachtain.' 'Is fír ám sain ale,' bar Cú
Chulaind. 'Is mana mórglíad sain ⁊ is adbar ṅdegdebtha. Ragaid
in t-albín borsin mag ⁊ ¹condricfat in gillanraid. A condricfat ind

---

¹⁻¹ om. St; omitt. prob. scribal addition   ² sic MS., om. St; omitt?
(scribal error for foll. word, not expunged, Windisch)
³ réfedaib MS., refedhaibh St   ⁴ sic MS. = fót
⁵ omitt. (Windisch); om. YBL   ⁶ suppl. from YBL
⁷ cotascrith YBL   ⁸⁻ᵈ sic   ⁹ Chucucuc MS.

róe mór ba chétóir.¹' Ba fír ám do Choin Culaind anísin. Acus
4620 lotar in t-albín barsan mag & conráncatar in gillanrad. 'Cía
confirend in cath innossa, a mo phopa Laíg?' bar Cú Chulaind.
'Áes Ulath,' bar Láech, 'inund ⁊ áes ócbad.' 'Cindas confaegat²
ale?' bar Cú Chulaind. 'Is ferda confegat,' bar Láeg. 'Airm i
tát na láith gaile anair isin cath bérait toilg trisin cath síar. Airm
4625 i tát na láith aníar bérait toilg trisin cath sair.' 'Appraind nacham
fuil-sea do nirt beith eturru dom choiss de-side, dáig dá mbeind-sea
do nirt beith dom choiss, rapad réil ³mo thoilg-sea³ and sain indiu i
cumma cháich.' 'Cossan⁴ ar chena, a Chúcúc,' bar Láech. 'Ní tár dot
gasciud, ní haisc dot inchaib. Doringnis maith reme sút ⁊ dogéna 'na
4630 díaid.' 'Maith a mo phopa Laíg,' bar Cú Chulaind, '⁵todúsig do
Ultaib⁵ dochum in chatha fodesta, dáig is mithig dóib a thechta.'
Tánic Láeg ⁶⁊ todíuscis de Ultaib⁶ dochum in chatha, ⁊ rabert na
bríathra and:

R⁷. 'Coméirget ríg Macha mórglonnaig. Míannaigther Badb
4635 bó⁸ Immail, bar nertaib gal, bar cridib crú, bar tilaib téici. Turcbaid
in sním nítha, dáig ní fríth ra Coin Culaind comchosmail.
Cú gonben mían Macha mochtraide más ar búaib Cúalnge
coméirget.'

I S and sain atraachtatar Ulaid uile i n-óenfecht ra costud a rríg
4640 ⁊ ra bréthir a flatha ⊣ ra frithálim coméirgi bréithri Laíg meic
Riangabra. Acus is amlaid atrachtatar lomthornocht uile ac[h]t
a n-airm 'na llámaib. Cach óen dá mbíd dorus a phupla sair díb,
is triana phupaill síar théiged ara fat leis tíachtain timchell.
'Cinnas concoméirget Ulaid dochum [in chatha]⁹ innossa, a
4645 mo phopa Láeig?' bar Cú Chulaind. 'Is ferda concoméirget,'
bar Láeg. 'Lomthornocht uile,' bar Láeg. 'Cach óen dá tá
dorus a phupla sair díb, is triana phupaill síar téiged¹⁰ ara fat leis
tíchtain timchell.' 'Atiur-sa bréthir,' bar Cú Chulaind, 'is ¹¹degóir
éigmi¹¹ atrachtatar im Chonchobor i mmucha láe itráthsa and sain.'
4650 And sain atbert Conchobor ra Sencha mac Ailella: 'Maith a
mo phopa Sencha,' bar Conchobor, 'fostá Ulaid ⊣ ná léic [dochum]⁹

---

¹⁻¹ *LL seems corrupt here. No equivalent to passage in YBL. St (which
is translated) reads* Comraicfit an ghiollanrad anair ⊣ an gillanrad aniar
² e *with subscr.* a; *read* confegat   ³⁻³ mo tholc-sa *YBL,* mo tholg *St*
⁴ *sic MS., read* cosain *as supra l.* 2194   ⁵⁻⁵ toduisig Ulaidh *St*
(*cp.* asbert Cu Chulaind fria araid arna ndiuscad Ultu *YBL* 3466)
⁶⁻⁶ do greasacht Uladh *St*   ⁷ *in marg.*   ⁸ bu *YBL*
⁹ *om. MS. suppl. from St*   ¹⁰ tét *St*   ¹¹⁻¹¹ *read* degfóir éigmi ;
degc[h]obair éigme *YBL,* deccoir emge *St,* deagh- fhóir-eimhghe *P*

in chatha co tí nert don tṡeón ⁊ don tṡolud coro éirgea grían i
cléithib nimi, goro lína grían glenta ⁊ fánta ⁊ tulcha ⁊ tuaidibrecha
na Hérend.' Tarrasatar and co tánic nert don tṡeón ⁊ don tṡolud,
goro lín grían glenta ⁊ fánta ⁊ tulcha ⁊ tuadebrecha in chóicid. ⁴⁶⁵⁵
'Maith a mo phopa Ṡencha,' bar Conchobor, 'todíusig ¹de
Ultaib¹ dochum in chatha dáig is mithig dóib a thechta.' Todíuscais
Sencha ¹d'Ultaib¹ dochum in chatha. Rabert na bríathra and:

'Coméirget ríg Macha. Munter ḟial. Melat fáebair. Fégat
cath. Claidet búrach. Benat scíathu. Scítha labrai². Labra ⁴⁶⁶⁰
éiti. Éicni fastuda. Feochra costoda. Curther cath ba chos-
101ᵇ saib araile. Eblait | a réim. Bardanessat ⁊ bardalessat indiu.
Ibait deoga duirbbi fola. Línfaid cuma cridi rígan más ar
búaib Cúalṅgi coméirgit'.

Nírbo chían do Láeg dá mbáe and go facca inní, fir Hérend uile ⁴⁶⁶⁵
ac coméirge i n-óenḟecht ac gabáil a scíath ⁊ a ṅgae ⁊ a claideb ⁊ a
cathbarr ⁊ ac tulargain³ na mbuden rompu dochum in chatha.
⁴Da gabsat⁴ fir Hérend cách díb bar slaide ⁊ bar slechtad, for tóchtad
⁊ bar tinmi, for airlech ⁊ for essargain araile ri ré cían ⁊ ra reimes
fata. Is and sain ra iarfaig Cú Chulaind dá araid, do Láeg mac ⁴⁶⁷⁰
Riangabra, in tan ón⁵ baí nél solus barsin gréin: 'Cinnas confegar
in cath innossa, a mo phopa Láeig?' 'Is ferda confegar,' bar
Láeg. 'Cid condrualaind-sea mo charpat ⁊ Én ara Conaill a
charpat ⁊ giara thíasmaís i ndíb carptib ánd itte co araile iar
n-idnaib⁶ na n-arm, ní rossed crú ná roth ná fonnud ná fertas díb⁷ ⁴⁶⁷⁵
ara dlús ⁊ ara déinme⁸ ⁊ ara daiṅgne coṅgbaither a n-airm i llámaib
na mmíled itráthsa.' 'Appraind nacham ḟuil-sea do nirt beith
eturru de-side,' bar Cú Chulaind, 'dáig dá mbeind-se de nirt,
rapad réill ⁹mo thoilg-sea⁹ and sain indiu i cumma cháich,' bar
Cú Chulaind. 'Cossan¹⁰ ar chena, a Chúcúc,' bar Láeg. 'Ní tár ⁴⁶⁸⁰
dot gasciud. Ní haisc dot inchaib. Doriṅgnis maith reme sút.
Dagéna 'na díaid.'
    And sain ra gabsat fir Hérend fós bar slaide ⁊ bar slechtad, for
tochtad ⁊ for tinme, far airlech ⁊ for essargain araile [fri]¹¹ ré cían
⁊ fri reimes fata. And sain daríachtatar cuccu-som na nóecharptig ⁴⁶⁸⁵
de ḟénnedaib na hIrúade & in triar de choiss maróen riu, ⁊ níra
lúathiu na nóecharptig andá in triar de choiss.

---

¹⁻¹ Ulaidh *St*   ² lama *YBL*   ³ *sic MS., read* timmargain? (*for* tuargain
Windisch); imbain *St*   ⁴⁻⁴ ro gabsat *St*   ⁵ *om. St*
⁶ *sic, for* n-indaib (ar iondaibh *St*)   ⁷ lár *add. St*   ⁸ *sic, for* deimne
⁹⁻⁹ mo tholg *St*   ¹⁰ *sic, for* cosain   ¹¹ *om. MS., suppl. from St*

And sain daríachtatar chucu-som 'no ferchutredaig fer ṅHérend, & ba hed a ṅgním-sin[1] uile 'sin chath ar bíth gona Conchobuir 4690 diambad fair bad róen ⁊ ar bíth aṅcthe Ailella ⁊ Medba dámbad forro conmebsad.

Acus ba sed and so anmand na ferchutredach[2]: Trí Conaire Slébe Miss, trí Lussin Lúachra, trí Niad Choirbb Tilcha Loiscthe, trí Dóelḟir Deille, trí Dámaltaig Dercderce, trí Buidir Búase; trí Báeith Búagnige, trí Búageltaig Breg, trí Suibne Siúre, 4695 trí Échdaig Áne, trí Malléith Locha Érne, trí Abratrúaid Locha Rí, trí Meic Amra Essa Rúaid, trí Fiachaig[3] Feda Némain, trí Mane Murisce, trí Muridaig Mairge, trí Lóegaire Licci Derge, trí Broduind Berba, trí Brúchnig Cind Abrat, trí Descertaig Dromma Fornoc[h]ta, trí Find Findabrach; trí Conaill Collomrach, trí 4700 Carpri Cliach, trí Mane Mossud, trí Scáthglain Scáire, trí Échtaig 102ᵃ hErcce, trí Trenḟir Taíte, trí Fintain Femin, | [4]trí Rótanaig Raigne, trí Sárchoraig Suide Lagen, trí Etarsceóil Etarbáne, trí hAeda Aidne, trí Guare Gabla[4].

Is and sain atbert Medb ri Fergus: 'Ba bág ám dait-siu ga 4705 dobertha do greimm catha gan díchill lind indiu dáig r'indarbbad as da chrích ⁊ as t'orbba. Is acainne ḟúarais crích ⁊ ferand ⁊ forbba ⁊ mórmathius mór do dénam fort.' 'Dánam beth-sa mo chlaideb indiu ám,' bar Fergus, 'ra tescfaitis lim-sa bráigte fer for bráigte fer & dóte fer for dóte fer ⁊ forcléithe fer for forcléithe fer ⁊ cindu 4710 fer for óeib scíath combús[5] lir bommanna ega eter dá ráeib tírib imríadat echraide ríg cach ṁball sair ⁊ síar acum-sa de Ultaib indiu dánam beth-sa mo chlaideb.' Is and atbert Ailill rá araid badessin .i. ra Fer Loga: 'Domraiched craum[6] claideb choilles toind, a gillai,' bar Ailill. 'Nátiur-sa bréthir mad messu a bláth ná 4715 'lessugud lett indiu é andá in lá tucus barin lettir i Crúachnaib Aí, dá mbet fir Hérend ⁊ Alban acott anacul ḟorom indiu, nít ainset uile.' Tánic Fer Loga reime ⁊ tuc in claideb laiss ba búaid caíntaisceda ⁊ fo chaindil chaín lassamain & tucad in claideb i lláim Ailella & tuc Ailill i lláim Ḟergusa. Acus firis Fergus fálte 4720 risin claideb. 'Mo chen Caladbolg, claideb Leite,' bar ésium. '[7]Scíth á aí[7] óenḟir Bodba. Cía farsa n-immér-sa so?' bar Fergus. 'Arna slúagaib immut immácúaird,' bar Medb. 'Ná bered nech mathim ná hanacul inniu úait mani bera fírchara.' And sain gebis Fergus a gasced ⁊ tánic reime don chath. Gebis Ailill a 4725 gasced. Gebis Medb a gaisced & tánic don chath. Coro maid in cath fo thrí ríam rompo fathúaid condanimmart cúal gae ⁊ claideb

---

[1] sin omitt. (ba he a ngnimha [v.l. ngniomh] uile St)   [2] names are arranged in two columns in MS.   [3] Fiachaigaig MS.; Fiachaigh St   [4–4] sub. col.   [5] sic; ɔ symbol MS., comba St   [6] crib St   [7–7] scitha St, read scítha

for cúlu doridisi. Rachúala Conchobor anísin airm i mbáe 'na
inad chatha, in cath do maidm fo thrí ris atúaid. And sain atbert-
sum rá theglach badessin .i. ra crislach na Cráebrúade: 'Gebid-si
seo bic, a firo,' bar éseom, '.i. in t-inad a tó-sa go tiasur-sa dá fiss ⁴⁷³⁰
cía riasa maidend in cath fa thrí ruind atúaid bán cóir seo.' And
sain atbert a theglach-som: 'Gébmait-ni seo,' bar íat-sum, 'dáig
nem úasaind ┐ talam ísaind ┐ muir immuind immácúairt. Mono
tháeth ¹in firmimintni¹ cona frossaib rétland for dunignúis in
talman nó mani thí in farrgi eithrech ochargorm for tulmoiṅg in ⁴⁷³⁵
bethad nó ma[ni]² máe in talam, ní béram-ni mod n-ordlaig secha
so bar cúlu go brunni mbrátha ┐ betha go tísiu bar cúlu dorís
chucaind.'

Tánic Conchobor reme go airm i cúala in cath do maidm ³ba
102ᵇ thrí³ | ris atúaid, & gebid scíath ra scíath and .i. ra Fergus mac ⁴⁷⁴⁰
Róig, .i. in n-Óchaín Conchobuir cona cethri óeib óir ┐ ⁴cona cethri
sethrachaib derggóir⁴. And sain rabert Fergus trí balcbémmenda
Bodba issin nÓchaín Conchobuir goro géis a scíath for Conchobor.—
Ára géised scíath Conchobuir, ra géistis scéith Ulad uile.—Giaro
boí dá threisi ┐ dá tharpigi ra búail Fergus a scíath bar Conchobor, ⁴⁷⁴⁵
ra boí dá chalmacht ┐ dá churatacht ra choṅgaib Conchobor in
scíath connára chomraic ó in scéith ra hó Conchobuir cid itir.

'Amae, a firu!' bar Fergus. 'Cía concoṅgbathar scíath rum-sa
indiu 'si lló bága sa airm condrecgat cethri ollchóiceda Hérend bar
Gárig ┐ Ilgárig i cath Tána Bó Cúalnge?' 'Gilla iss ó ┐ iss imláne ⁴⁷⁵⁰
and so andáe ale, ┐ rap ferr máthair -┐ athair, fer rat indarb át chrích
┐ át ferand ┐ át forbba, fer rat chuir i n-adba oss┐fíadmíl┐sinnach, fer
nára léic leithet da gabail badéin dit chrích ná dit ferand dait, fer rat
chuir ar bantidnacul mná, fer rat sáraig im tríb maccaib Usnig do
marbad far th'einech fecht n-aill, fer rat dingéba indiu i fiadnaisi ⁴⁷⁵⁵
fer ṅHérend, Conchobor mac Fachtna Fáthaig meic Rossa Rúaid
meic Rudraigi, ardrí Ulad ┐ mac ardríg Hérend.'

'Immánic-sea ón omm,' bar Fergus. Acus tuc Fergus a dá
láim arin Caladbolg ┐ rabert béim de dar aiss síar goro chomraic
a fográin ri talmain, & da mídair a thrí bráthbémmenda Bodba da ⁴⁷⁶⁰
béim bar Ultaib comtís lir a mmairb andá a mbí. Dachonnaic
Cormac Cond Longas mac Conchobuir éside & rabert side d'indsaigid
Fergusa ┐ ra iad a dá láim thariss. ⁵'Aicclech nád aicclech sain⁵,
a mo phopa Ferguis. ⁶Náimdemail nád charddemail sain⁶, a mo

---

¹⁻¹ *sic*, an firmament *St*   ² muna *St*   ³⁻³ bad rí *MS.*, fo t[h]ri *St*
⁴⁻⁴ cona cethri sethrachaib do derggóir *MS.*, cona cet[h]ri sethnachaib
derccoir *St*   ⁵⁻⁵ foichleach nairfoichlech insin *YBL*, foichlech nad
nurfhoiclech *St*   ⁶⁻⁶ naimtidi in chairdine *YBL*, naimdighe nit
cairdine sin *St*

4765 phopa Ḟerguis. ¹Anchellach nád anchellach sain¹ a mo phopa Ḟerguis. Ná marbtar ⁊ ná mudaigter lett Ulaid trí bíthin do bráthbémmenda, acht imráid a n-einech 'si lló bága sa indiu.' 'Scuich bius ².i. úaim², a meic,' bar Fergus, 'dáig nída³ beó-sa meni benur mo thrí bráthbémmenda Bodba bar Ultaib indiu 4770 ⁴gorsat lir⁴ a mmairb andás a mbí.'

'Taí do lám⁵ go fáen ale,' bar Cormac Cond Longas, '⁊ tesc na tilcha dar cendaib na slúag, ⁊ bud dídnad dit ḟeirg.' '⁶Ráid ra Conchobor táet 'na inad catha didu.'⁶ Tánic Conchobor 'na inad catha.

4775 Is amlaid ra boí in claideb sain, ⁷claideb Fergusa⁷, claideb Leiti a sídib é. Inn úair ba háill búalad de 'no, ba métithir ra stúaig nimi i n-aeór é. Is and sain táeiss Fergus a láim go fáen dar cendaib na slúag goro thesc a trí cindu dina trí tulchaib, go failet 'sin ríasc ⁸bad ḟiadnaisi⁸. Gorop íat na trí Máela Mide and sain.

4780 Imthúsa Con Culaind and so innossa: rachúala-saide in nÓchaín Conchobuir ⁹'gá búalad d'Ḟergus⁹ mac Róig. 'Maith a mo phopa Laíg,' bar Cú Chulaind, '¹⁰cía conlinfadar¹⁰ in nÓchaín mo phopa Chonchobuir do thúarggain amlaid seo ⁊ messi ¹¹im bethaid¹¹?' 'Telggai boga fuile, formach n-áir ale,' bar Láeg. 'An fer Fergus 4785 mac Róig. Bacleth claideb carpait a ssídib. Rasíacht eochraide mo phopa Conchobuir cath.'

103ᵃ 'Oslaic go troit túaga, a gillai,' bar Cú Chulaind. | And sain focheird Cú Chulaind móroscur de co llotar a thúaga de go Mag Túaga i Connac[h]taib. Lotar a bacca de go Bacca i Corco M'rúad. 4790 Lotar na suipp ṡesca bátar 'na áltaib i cléthib aeóir ⊣ firmiminti feib is sía thiagait uiss¹² i lló áille nád bí gáeth. Ra gabsat a ḟuli ilgremma de gorbo lána tairchlassa ⊣ eittrigi in talman dá ḟulib ⊣ dá gáeib cró. Is é céternmas ṅgascid dariṅgni-sium ár n-érgi na dá banchánti bátar ac fáschuí ⊣ ac fásguba .i. Fethan ⊣ Cholla, 4795 barressairg cách díb da chind araile gorbo derg dá fuil ⊣ gorbo liath dá n-inchind¹³. Ní ḟargbad a arm 'na ḟarrad-som itir acht a charpat ammáin. Acus ra gab-som a charpat re aiss ⊣ tánic reme d'indsaigid fer ṅHérend ⊣ ra gab dá charpat ḟorro go rránic go airm i mbaí Fergus mac Róig. 'Táe ille, a mo phopa Ferguis,' bar ésium. Níra

---

¹⁻¹ ainbchellach ainbchellach *YBL*, anbceallach nat narbhceallach *St*
²⁻² *gloss written overhead* ³ níbam *St* ⁴⁻⁴ comba lia *St*
⁵ *sic, for* láim ⁶⁻⁶ Raid fri Concobar, ar Fergus, teacht ina ionadh catha *St*
⁷⁻⁷ claideb Fer*gusa bis MS.* ⁸⁻⁸ fo fiadnaisi ic fearaibh Erenn *St*, ba fiadnaise dferuibh Erenn *Ed, H* ⁹⁻⁹ da bualadh ic Fergus *St*
¹⁰⁻¹⁰ cia lamas *St* ¹¹⁻¹¹ i mbethaid *Dipl. Edn.: cp.* iom bethaid *St*
¹² uiseoga *St, sic leg.* (*cp.* scendit lasodain a suip sesca as a n-ardai eiret teiti uiseoc *YBL*) ¹³ in talamh *add. St*

[f]recair Fergus ór ní chúala. Atubairt-sium arís. 'Táe ille, a ⁴⁸⁰⁰
mo phopa Ferguis,' bar ésium, 'ná mani tháe ille, rat meliub mar
meles muilend múadbraich. Rat nigiub mar negair coipp a
lundu ¹.i. lind usci¹. Rat nasciub mar nasces féith fidu. Ras²
lécub fort feib ras léic séig far mintu.' 'Rommánic-sea ón omm,'
bar Fergus. ³'Cía conlinfadar³ na balcbríathra Bodba so do ⁴⁸⁰⁵
ráda frim-sa airm condrecgat cethri ollchóiceda Hérend for Gárig
⁊ Ilgárig i cath Tánad Bó Cúalnge?' 'Do dalta-su and so,' bar
ésium, '⁊ dalta Ulad ⁊ Choncobuir bar chena, Cú Chulaind mac
Sualtaim, & ro gellaisiu teiched remum-sa inbaid bad chréchtach
crólinnech tretholl mé for cath na Tána, dáig ra thechiusa romut-sa ⁴⁸¹⁰
ardo chomlond féin for Tánaid.'

Atchúala Fergus sain ⁊ ra impá ⁊ tucastar a thrí coscémmenda
láechda lánmóra, & óra impá-som, ra impátar fir Hérend uile.
Da maid d'feraib Hérend dar tilaig síar. Tarrassaid inn irgal im
chend Connacht. I mmedón laí⁴ tánic Cú Chulaind dochum in ⁴⁸¹⁵
chatha. Tráth funid nóna da maid din budin dédenaig de
Chonnac[h]taib dar tilaig síar. ⁵Nír diruais⁵ don⁶ charpat i lláim
Con Culaind risin ráe sin acht dorn dina bassaib⁷ immon roth ⁊
bass dina fertsib immon creitt, acht ic airlech ⁊ ic essargain cethri
n-ollchóiced nHérend risin ré sin.   ⁴⁸²⁰

And sain geibis Medb scíath díten dar éis fer nHérend. And
sain faítte Medb in Dond Cúalnge co coíca dá samascib imbe &
ochtor dá hechlachaib leiss timchell co Crúachain. Gipé rasossed,
gipé ná rossed, go rossed in Dond Cúalnge feib ra gell-si. Is and
drecgais⁸ a fúal fola for Meidb ⁹[⁊ itbert: 'Geib, a Ferguis,' bar ⁴⁸²⁵
Medb]⁹, 'scíath díten dar éis fer nHérend ¹⁰goro síblur-sa¹⁰ m'fúal
úaim.' 'Dar ar cubus,' ar Fergus, 'is olc in tráth ⁊ ní cóir a dénam.'
'Gid ed ní étaim-sea chena,' bar Medb, 'dáig nída¹¹ beó-sa ¹²meni
síblur-sa m'fúal úaim.'¹² Tánic Fergus ⁊ gebid scíath díten dar éis fer
nHérend. Siblais¹³ Medb a fúal¹⁴ úathi co nderna trí tulchlassa ⁴⁸³⁰
móra de ¹⁵co taille¹⁵ munter in cach thurchlaiss. Conid Fúal Medba
atberar friss.

Ruc Cú Chulaind furri ac dénam na huropra sain ⁊ níra gonastar-
sum; ní athgonad-sum 'na díaid hí. 'Ascaid dam-sa úait indiu,
a Chú Chulaind,' bar Medb. 'Gia ascaid connaige?' bar Cú ⁴⁸³⁵

¹⁻¹ *gloss added overhead*   ² *sic*   ³⁻³ cia lamhas *St*   ⁴ lá *MS.*
⁵⁻⁵ *sic, for* ní diruaraid, ní doruaraid? (Nir mair *St*)   ⁶ dar *MS.*, don *St*
⁷ *sic, read* hasnaib (*cp.* acht dorn dona hasnaib imon creit ⁊ dorn dona
fer[t]sib imon droch *YBL*, bas dona ferstib iman roth ⁊ dorn don cret imin
carpat *St*)   ⁸ do rala *St*   ⁹⁻⁹ om. *MS., suppl. from St*
¹⁰⁻¹⁰ coro sriblar-sa *St*   ¹¹ nibam *St*   ¹²⁻¹² muna dernar fúal *St*
¹³ sriubhlais *St*   ¹⁴ fúal *MS.*, fúal *St*   ¹⁵⁻¹⁵ co ttuillfedh *St*

Chulaind. 'In slúag sa bar th'einech ⁊ ardo chommairgi go rrosset
103ᵇ dar Áth Mór síar.' | '¹Gondnoim-sea ón omm¹,' bar Cú Chulaind.
Tánic Cú Chulaind i timchell ḟer ṅHérend ⁊ gebis scíath díten din
dara leith díb d'imdegail fer ṅHérend. Táncatar ferchutredaig fer
⁴⁸⁴⁰ ṅHérend din leith aile. Tánic Medb 'na hinad féin ⁊ gebis scíath
díten dar éis fer ṅHérend ⁊ rucsat leó bán cóir sin fir Hérend dar
Áth Mór síar.

And sain daríacht a chlaideb d'indsaigid Con Culaind ⁊ rabert
béim dona tríb Máelánaib Átha Lúain i n-agid na trí Máela Mide
⁴⁸⁴⁵ goro ben a trí cindu díb.

And sain ra gab Fergus ac tachim² in tṡlúaig ac dula a³ Áth
Mór síar. 'Rapa chomadas in lá sa indiu ám i ndíaid mná,'
⁴[ar Fergus]⁴ 'Condrecat lochta ra fulachta and so indiu'⁵ [bar
Medb] ra Fergus.⁵ 'Ra gattá ⁊ ra brattá in slúag sa indiu. Feib
⁴⁸⁵⁰ théit ⁶echrad láir⁶ rena serrgraig i crích n-aneóil gan chend cundraid
ná comairle rempo, is amlaid testa⁷ in slúag sa indiu.'

Imthúsa Medba sunna innossa, ra timsaigit ⁊ ra timmairgit fir
Hérend lé-si go Crúachain go factis gleicc na tarb.

IMthúsa in Duind Chúalnge sunda innossa, á 'tchonnaic-sén in
⁴⁸⁵⁵ tír n-álaind n-aneóil, rabert ⁸a thrí resse gémmend⁸ bar aird.
Atchúala in Findbennach Aí éside. Ní lamad míl firend géisecht
bud airdde ná gúasacht⁹ aci-side eter cethraib átha Aí uile, Áth
Moga ⁊ Áth Coltna, Áth Slissen ⁊ Áth ṁBercha, ⁊ túargab a chend
go díiṅg ⁊ tánic reme go Crúachain d'indsaigid in Duind Chúalnge.
⁴⁸⁶⁰ Is and sain ra ráidsetar fir Hérend cía bud ḟiadnaisi dona tarbaib.
Iss ed ra ráidset uile gombad é Bricriu mac Garbada.—Dáig
blíadain résin scél sa Tánad Bó Cúalnge tánic Bricri d'ḟaigde
Ḟergusa assin chóciud i n-araile, & ra ḟost Fergus ace é ic irnaide
ra sétaib ⁊ ra maínib. Acus darala eturru ic imbirt ḟidchilli ⁊
⁴⁸⁶⁵ Fergus & atrubairt-sium aithis móir ra Fergus. Dabert Fergus
béim dá durn dó-som ⁊ dind ḟir baí 'na láim goro thoilg in fer 'na
chind go róebriss cnáim ina chind. In fat ra bátar fir Hérend i
slúagiud na Tána, ésium 'gá leiges i Crúachain risin ré sin. Acus

---

¹⁻¹ Gabuim-si sin orm St   ² fēchain St, sic leg.   ³ tar St
⁴⁻⁴ suppl. from St   ⁵⁻⁵ ar Fergus St, but cp. is and asbert Meadb fri
Fergus: Correcad lochta ⁊ fulachta sund indiu, a Ferguis, ar si YBL
⁶⁻⁶ sic: echláir St, láir YBL   ⁷ atá St   ⁸⁻⁸ a tri gemeanna glordha St
⁹ sic, for gnusacht or gnúasacht? gnusachtach St

in lá tháncatar din tslúagud, is ésin lá ra érig-sium.—¹Dáig níra
choitchinníu Briccni dá charait andá dá námait.¹ Acus tucad far ⁴⁸⁷⁰
bernaid i fiadnaisi na ndam é.
Atchonnaic cách a chéile dina tarbaib ⅂ foċlassa búrach dóib and
⅂ fócerddetar in n-úir thairsiu. ²Ra chlaitar² in talmain dara
104ª formnaib ⅂ dara slinneócaib, ⅂ ra rúamnaigsetar a rruisc | ina
cendaib dóib immar cháera tenda tentide. Ra bulgsetar a n-óli ⁴⁸⁷⁵
⅂ a sróna mar bulgu goband i certchai, & rabert cách díb blascbéim
brátha d'indsaigid a chéile. Ra gab cách díb bar tollad ⅂ bar
tregdad ⅂ bar airlech ⅂ bar essorgain araile. And sain ra immir
in Findbennach meirbflech a astair ⅂ a imthechta ⅂ na sliged barin
Dond Cúalnge, ⅂ ra śáid adairc ina tháeb ⅂ brissis búrach fair. ⁴⁸⁸⁰
Iss ed rucsat a rrúathur go hairm i mbáe Bricni goro bertsatar
iṅgni na tarb ferchubat fir i talmain é arna bás.
    Conid Aided Bricni and sain.
    Atchonnaic Cormac Cond Longas mac Conchobuir anísin & ro
gabastar fogeist dárbo lán a glacc ⅂ rabert trí béimmenna don ⁴⁸⁸⁵
Dond Chúalnge ó ó go erboll. 'Nírap sét suthain suachnid dún
in sét sa,' bar Cormac, 'dáig ná tic de láeg a chomaísi badéin do
diṅgbáil.' Atchúala Dond Cúalnge annísein & báe ciall dunetta
aice ⅂ ra impá risin Findbennach & go 'mmarálaid dóib assa aithle
d'imbúalad ri ré cían ⅂ ri remis fota goro laig inn adaig bar feraib ⁴⁸⁹⁰
Hérend. Acus ára laig inn adaig, ní rabi ac feraib Hérend acht
éstecht ³re fúaim ⅂ re fothrom.³ Ra śirset na daim Hérind uili
in n-aidchi sin.

    Nírbo chían d'feraib Hérend dá mbátar and mochrád arnabárach
go faccatar in Dond Cúalnge dar Crúachain aníar ⅂ in Findbennach ⁴⁸⁹⁵
Aí ina ascarnaig ara bennaib ⅂ ara adarcaib. Ra éirgetar fir Hérend
⅂ ní fetatar cía dina tarbaib ra báe and. 'Maith a firu,' bar Fergus,
'léicid a óenur más é in Findbennach Aí fail and, & más é in Dond
Cúalnge, léicid a choscor leis. Natiur-sa bréthir is bec i ndernad
imna tarbaib i farrad na ndiṅgéntar innossa.' ⁴⁹⁰⁰
    Tánic in Dond Cúalnge. Tuc a dess ri Crúachain & ra fácaib
crúach dá óeib and, gorop⁴ de atá Crúachna Áe.
    Tánic reme go himmargain Átha Móir ⅂ ra fácaib a lón in
Findbennaig and, gorop⁵ de dá tá Áth Lúain.
    Tánic sair reme i crích Mide co Áth Troim goro fácaib a thromm ind ⁴⁹⁰⁵
Findbennaig and.

---

¹⁻¹ Agus as uime ro ordaighsiot Bricne samlaidh uair niorbo cotruime-
siomh dia c[h]araid inas dia namait *St*   ²⁻² ro claidheatar *St*
    ³⁻³ rea fuaim ⅂ rea ffotram *St*   ⁴ conad *St*   ⁵ conadh *St*

Túargaib a chend go díiṅg ⁊ ra chroth in Findbennach de fo
Hérind. Ra chuir a láraic de co Port Large. Ra chuir a chlíathaig
úad go Dublind rissa rátter Áth Cliath. Tuc a aged fathúaid assa
4910 aithle & tuc aichni far tír Cúalnge & tánic dá hindsaigid. Is and
ra bátar mnáa ⁊ meicc ⁊ mindóene ac coíniud in Duind Chúalnge.
Atchondcatar-som a thaul in Duind Chúalnge dá saigid. 'Taul
tairb chucaind!' bar íat-som. Conid de atá Taul Tairb ó ṡein
anall. |

104ᵇ   And sain imsóe in Dond Cúalnge fa mnáib ⁊ maccaib ⁊ mindóenib[1]
tíri Cúalnge & curis ár mór forro. Tuc a druim risin tilaig assa
aithle & ro maid cnómaidm dá chride 'na chlíab.

Gorop[2] a hús ⁊ a imthúsa ⁊ a deired na Tánad gonici sein.

4920 **B**Endacht ar cech óen mebraigfes go hindraic Táin amlaid seo
⁊ ná tuillfe cruth aile furri.

SEd ego qui scripsi hanc historiam aut uerius fabulam quibusdam
fidem in hac historia aut fabula non accommodo. Quaedam enim
ibi sunt praestrigia[3] demonum, quaedam autem figmenta poetica,
quaedam similia uero, quaedam non, quaedam ad delectationem
stultorum.

---

[1] mind | doenib *MS.*   [2] p *altered to* b *MS.*; conadh *St*
[3] *sic, for* praestigia

TRANSLATION

Here begins Táin Bó Cúalnge

ONCE upon a time it befell Ailill and Medb that, when their royal
bed had been prepared for them in Ráth Crúachain in Connacht,
they spoke together as they lay on their pillow. 'In truth, woman,'
said Ailill, 'she is a well-off woman who is the wife of a nobleman.'
'She is indeed,' said the woman. 'Why do you think so?' 'I think
so,' said Ailill, 'because you are better off today than when I married
you.' 'I was well-off before (marrying) you.' said Medb. 'It was
wealth that we had not heard of and did not know of.' said Ailill,
'but you were a woman of property and foes from lands next to
you were carrying off spoils and booty from you.' 'Not so was I,'
said Medb, 'but my father was in the high-kingship of Ireland,
namely Eochu Feidlech mac Find meic Findomain meic Findeoin
meic Findguill meic Rotha meic Rigéoin meic Blathachta meic
Beothechta meic Enna Agnig meic Óengusa Turbig. He had six
daughters: Derbriu, Ethne and Éle, Clothru, Mugain and Medb.
I was the noblest and worthiest of them. I was the most generous
of them in bounty and the bestowal of gifts. I was best of them
in battle and fight and combat. I had fifteen hundred royal
mercenaries of the sons of strangers exiled from their own land
and as many of the sons of native freemen within the province.
And there were ten men for each mercenary of these, ¹[and nine
men for every mercenary]¹, and eight men for every mercenary,
and seven for every mercenary, and six for every mercenary, and
five for every mercenary, ¹[and four for every mercenary]¹ and three
for every mercenary and two for every mercenary and one mercenary
for every mercenary. I had these as my standing household,'
said Medb, 'and for that reason my father gave me one of the

---

¹⁻¹ following St

137

provinces of Ireland, namely, the province of Crúachu. Whence
I am called Medb Chrúachna. Messengers came from Find mac Rosa
Rúaid, the King of Leinster, to sue for me, and from Cairbre Nia
Fer mac Rosa, the King of Tara, and they came from Conchobor
mac Fachtna, the King of Ulster, and they came from Eochu Bec.
But I consented not, for I demanded a strange bride-gift such as
no woman before me had asked of a man of the men of Ireland,
to wit, a husband without meanness, without jealousy, without
fear. If my husband should be mean, it would not be fitting for
us to be together, for I am generous in largesse and the bestowal
of gifts and it would be a reproach for my husband that I should
be better than he in generosity, but it would be no reproach if we
were equally generous provided that both of us were generous. If my
husband were timorous, neither would it be fitting for us to be
together, for single-handed I am victorious in battles and contests
and combats, and it would be a reproach to my husband that his
wife should be more courageous than he, but it is no reproach if
they are equally courageous provided that both are courageous.
If the man with whom I should be were jealous, neither would it
be fitting, for I was never ¹without one lover quickly succeeding
another¹. Now such a husband have I got, even you, Ailill mac
Rosa Rúaid of Leinster. You are not niggardly, you are not
jealous, you are not inactive. I gave you a contract and a
bride-price as befits a woman, namely, the raiment of twelve men,
a chariot worth thrice seven *cumala*, the breadth of your face in
red gold, the weight of your left arm in white bronze. Whoever
brings shame and annoyance and confusion on you, you have no
claim for compensation or for honour-price for it except what claim
I have,' said Medb, 'for you are a man dependent on a woman's
marriage-portion.' 'Not so was I,' said Ailill, 'but I had two
brothers, one of them reigning over Tara, the other over Leinster,
namely, Find over Leinster and Cairbre over Tara. I left the rule
to them because of their seniority but they were no better in
bounty and the bestowal of gifts than I. And I heard of no province
in Ireland dependent on a woman except this province alone, so
I came and assumed the kingship here in virtue of my mother's
rights², for Máta Muirisc the daughter of Mága was my mother.
And what better queen could I have than you, for you are the
daughter of the high-king of Ireland.' 'Nevertheless,' said Medb,
'my property is greater than yours.' 'I marvel at that,' said Ailill,

¹⁻¹ lit. without a man in the shadow of another
² following St

'for there is none who has greater possessions and riches and wealth than I, and I know that there is not.'

There were brought to them what was least valuable among their possessions that they might know which of them had more goods and riches and wealth. There were brought to them their wooden cups and their vats and their iron vessels, their cans, their washing-basins and their tubs. There were brought to them their rings and their bracelets and their thumb-rings, their treasures of gold and their garments, as well purple as blue and black and green, yellow and vari-coloured and grey, dun and chequered and striped. Their great flocks of sheep were brought from fields and lawns and open plains. They were counted and reckoned and it was recognised that they were equal, of the same size and of the same number. But among Medb's sheep there was a splendid ram which was the equivalent of a *cumal* in value, and among Ailill's sheep was a ram corresponding to him. From grazing lands and paddocks were brought their horses and steeds. In Medb's horse-herd there was a splendid horse which might be valued at a *cumal*. Ailill had a horse to match him. Then their great herds of swine were brought from woods and sloping glens and solitary places. They were counted and reckoned and recognised. Medb had a special boar and Ailill had another. Then their herds of cows, their cattle and their droves were brought to them from the woods and waste places of the province. They were counted and reckoned and recognised, and they were of equal size and equal number. But among Ailill's cows there was a special bull. He had been a calf of one of Medb's cows, and his name was Findbennach. But he deemed it unworthy of him to be counted as a woman's property, so he went and took his place among the king's cows. It was to Medb as if she owned not a penny of possessions since she had not a bull as great as that among her kine. Then Mac Roth the herald was summoned to Medb and she asked him to find out where in any province of the provinces of Ireland there might be a bull such as he. 'I know indeed,' said Mac Roth 'where there is a bull even better and more excellent than he, in the province of Ulster in the cantred of Cúailnge in the house of Dáire mac Fiachna. Donn Cúailnge is his name.' 'Go you there, Mac Roth, and ask of Dáire for me a year's loan of Donn Cúailnge. At the year's end he will get the fee for the bull's loan, namely, fifty heifers, and Donn Cúailnge himself (returned). And take another offer with you, Mac Roth: if the people of that land and country object to giving that precious possession, Donn Cúailnge, let Dáire himself come with his bull and he shall have the extent of his own lands in the level plain of Mag Aí and a chariot worth thrice seven *cumala*, and he shall have my own intimate friendship.'

Thereupon the messengers proceeded to the house of Dáire mac Fiachna. The number of Mac Roth's embassy was nine messengers. Then Mac Roth was welcomed in the house of Dáire. That was but right for Mac Roth was the chief herald of all. Dáire asked Mac Roth what was the cause of his journey and why he had come. The herald told why he had come and related the contention between Medb and Ailill. 'And it is to ask for a loan of the Donn Cúailnge to match the Findbennach that I have come,' said he, 'and you shall get the fee for his loan, namely, fifty heifers and the return of Donn Cúailnge himself. And there is somewhat besides: come yourself with your bull and you shall get an area equal to your own lands in the level plain of Mag Aí and a chariot worth thrice seven *cumala* and Medb's intimate friendship to boot.' Dáire was well pleased with that and (in his pleasure) he shook himself so that the seams of the flock-beds beneath him burst asunder, and he said: 'By the truth of my conscience, even if the Ulstermen object, this precious possession, Donn Cúailnge, will now be taken to Ailill and Medb in the land of Connacht.' Mac Roth was pleased to hear what [Mac] Fiachna said.

Then they were attended to and straw and fresh rushes were strewn underfoot for them. The choicest food was served to them and a drinking feast provided until they were merry. And a conversation took place between two of the messengers. 'In sooth,' said one messenger, 'generous is the man in whose house we are.' 'Generous indeed,' said the other. 'Is there among the Ulsterman any who is more generous than he?' said the first messenger. 'There is indeed,' said the second. 'More generous is Conchobor whose vassal Dáire is, for though all Ulstermen should rally round Conchobor, it were no shame for them.' 'A great act of generosity it is indeed for Dáire to have given to us nine messengers that which it would have been the work of the four great provinces of Ireland to carry off from the land of Ulster, namely, Donn Cúailnge.' Then a third messenger joined their conversation. 'And what are ye saying?' he asked. 'Yon messenger says that the man in whose house we are is a generous man. He is generous indeed, says another. Is there any among the Ulsterman who is more generous than he? asks the first messenger. There is indeed, says the second. Conchobor, whose vassal Dáire is, is more generous, and if all Ulstermen adhered to him it were indeed no shame for them. It was

generous of Dáire to give to us nine messengers what only the
four great provinces of Ireland could carry off from the land of
Ulster.' 'I should like to see a gush of blood and gore from the
mouth from which that (talk) comes, for if the bull were not given
willingly, he would be given perforce.'

Then Dáire mac Fiachna's butler came into the house with a
man carrying liquor and another carrying meat, and he heard what
the messengers said. He flew into a passion and laid down the
meat and drink for them, and he did not invite them to consume it,
neither did he tell them not to consume it. Thereafter he went
to the house where Dáire mac Fiachna was and said: 'Was it you
who gave that excellent treasure, the Donn Cúailnge, to the
messengers?' 'It was I indeed', said Dáire. 'Where he was given
may there be no (proper) rule, for what they say is true, that if
you do not give him of your own free will, you will give him by
force by reason of the armies of Ailill and Medb and the guidance
of Fergus mac Róig.' 'I swear by the gods whom I worship unless
they take him thus by force, they shall not take him by fair means.'
They spend the night thus until morning. Early on the morrow
the messengers arose and went into the house where Dáire was.
'Guide us, noble sir, to the spot where Donn Cúailnge is.' 'Not so
indeed,' said Dáire, 'but if it were my custom to deal treacherously
with messengers or travellers or voyagers not one of you should
escape alive.' 'What is this?' said Mac Roth. 'There is great cause
for it,' said Dáire. 'Ye said that if I did not give the bull willingly,
then I should give him under compulsion by reason of the army
of Ailill and Medb and the sure guidance of Fergus.' 'Nay,' said
Mac Roth, 'whatever messengers might say as a result of indulging
in your meat and drink, it should not be heeded or noticed nor
accounted as a reproach to Ailill and Medb.' 'Yet I shall not give
my bull, Mac Roth, on this occasion.'

Thus the messengers went on their way back and reached Ráth
Crúachan in Connacht. Medb asked tidings of them. Mac Roth
told her that they had not brought back his bull from Dáire. 'What
was the cause of that?' asked Medb. Mac Roth told her the reason
for it. 'There is no necessity to "smooth the knots", Mac Roth,
for [1]it was certain[1],' said Medb, 'that he would not be given freely
if he were not given by force, and he shall so be given.'

[1-1] lit. it was known.

Messengers went from Medb to the Maines to bid them come to Crúachu, the seven Maines with their seven divisions of three thousand, namely, Maine Máithremail, Maine Aithremail, Maine Condagaib Uile, Maine Míngor, Maine Mórgor and Maine Conda Mó Epert. Other messengers went to the sons of Mágu, namely Cet mac Mágach, Anlúan mac Mágach, Mac Corb mac Mágach, Baiscell mac Mágach, Én mac Mágach, Dóche mac Mágach and Scannal mac Mágach. These arrived, in number three thousand armed men. Other messengers went from them to Cormac Cond Longas mac Conchobuir and to Fergus mac Róig, and they too came, in number three thousand.

The first band of all had shorn heads of hair. Green cloaks about them with silver brooches in them. Next to their skin they wore shirts of gold thread with red insertions of red gold. They carried swords with white grips and handles of silver. 'Is that Cormac yonder?' they all asked. 'It is not indeed,' said Medb.

The second band had newly shorn heads of hair. They wore grey cloaks and pure white shirts next to their skins. They carried swords with round guards of gold and silver handles. 'Is that Cormac yonder?' they all asked. 'It is not he indeed,' said Medb.

The last band had flowing hair, fair-yellow, golden, streaming manes. They wore purple embroidered cloaks with golden inset brooches over their breasts. They had smooth, long, silken shirts reaching to their insteps. All together they would lift their feet and set them down again. 'Is that Cormac yonder?' they all asked. 'It is he indeed,' said Medb.

That night they pitched their camp and stronghold and there was a dense mass of smoke and fire (from their camp-fires) between the four fords of Aí, Áth Moga, Áth mBercna, Áth Slissen and Áth Coltna. And they stayed for a full fortnight in Ráth Crúachan of Connacht drinking and feasting and merrymaking so that (presently) their journey and hosting should be the lighter for them. And then Medb bade her charioteer harness her horses for her that she might go to speak with her druid to seek foreknowledge and prophecy from him.

When Medb came to where her druid was, she asked foreknowledge and prophecy of him. 'There are many who part here today from comrades and friends,' said Medb, 'from land and territory, from father and mother, and if not all return safe and sound, it is on me their grumbles and their curses will fall. Yet none goes forth and none stays here who is any dearer to us than we ourselves. And

find out for us whether we shall come back or not.' And the druid
said: 'Whoever comes or comes not back, you yourself will come.'

The driver turned the chariot and Medb came back. She saw
something that she deemed wonderful, namely, a woman coming
towards her by the shaft of the chariot. The girl was weaving a
fringe, holding a weaver's beam of white bronze in her right hand
with seven strips of red gold on its points (?). She wore a spotted,
green-speckled cloak, with a round, heavy-headed brooch in the
cloak above her breast. She had a crimson, rich-blooded[1]
countenance, a bright, laughing eye, thin, red lips. She had shining
pearly teeth; you would have thought they were showers of fair
pearls which were displayed in her head. Like new *partaing* were
her lips. The sweet sound of her voice and speech was as melodious
as the strings of harps plucked by the hands of masters. As white
as snow falling in one night was the lustre of her skin and body
(shining) through her garments. She had long and very white
feet with pink, even, round and sharp nails. She had long, fair-
yellow, golden hair; three tresses of her hair wound round her head,
another tress (falling behind) which touched the calves of her legs.

Medb gazed at her. 'And what are you doing here now, girl?'
said Medb. '(I am) promoting your interest and your prosperity,
gathering and mustering the four great provinces of Ireland with
you to go into Ulster for Táin Bó Cúailnge.' 'Why do you do
that for me?' said Medb. 'I have good reason to do so. I am a
bondmaid of your people.' 'Who of my people are you?' said
Medb. 'That is not hard to tell. I am Feidelm the prophetess
from Síd Chrúachna.' 'Well then, Feidelm Prophetess, how do
you see our army?' 'I see red on them. I see crimson.'

'Conchobor is suffering in his debility in Emain,' said Medb.
'My messengers have gone to him. There is nothing we fear from
the Ulstermen. But tell the truth, Feidelm. O Feidelm Prophetess,
how do you see our army?' 'I see red on them. I see crimson.'

'Cuscraid Mend Macha mac Conchobuir is in Inis Cuscraid in his
debility. My messengers have gone to him. There is nothing we
fear from the Ulstermen. But speak truth, Feidelm. O Feidelm
Prophetess, how do you see our army?' 'I see red upon them. I
see crimson.'

'Eogan mac Durthacht is at Ráth Airthir in his debility. My
messengers have gone to him. There is nothing we fear from the
Ulstermen. But speak truth to us, Feidelm. O Feidelm Prophetess,
how do you see our army?' 'I see red on them. I see crimson.'

---

[1] fair-faced St

'Celtchair mac Cuthechair is in his fortress in his debility. My messengers have reached him. There is nothing we fear from the Ulstermen. But speak truth, Feidelm. O Feidelm Prophetess, how do you see our army?' 'I see red on them. I see crimson.'

'I care not for your reasoning, for when the men of Ireland gather in one place, among them will be strife and battle and broils and affrays, in dispute as to who shall lead the van or bring up the rear or (first cross) ford or river or first kill swine or cow or stag or game. But speak truth to us, Feidelm. O Feidelm Prophetess, how do you see our army?' 'I see red on them, I see crimson.'

And Feidelm began to prophesy and foretell Cú Chulainn to the men of Ireland, and she chanted a lay:

> 'I see a fair man who will perform weapon-feats, with many a wound in his fair flesh. The hero's light is on his brow, his forehead is the meeting-place of many virtues.

> 'Seven gems of a hero are in his eyes. His spear-heads are unsheathed. He wears a red mantle with clasps.

> 'His face is the fairest. He amazes womenfolk, a young lad of handsome countenance; (yet) in battle he shows a dragon's form.

> 'Like is his prowess to that of Cú Chulainn of Muirtheimne. I know not who is the Cú Chulainn from Murtheimne, but this I know, that this army will be bloodstained from him.

> 'Four swordlets of wonderful feats he has in each hand. He will manage to ply them on the host. Each weapon has its own special use.

> 'When he carries his *ga bulga* as well as his sword and spear, this man wrapped in a red mantle sets his foot on every battle-field.

> 'His two spears across the wheel-rim of his battle chariot. High above valour (?) is the distorted one. So he has hitherto appeared to me, (but) I am sure that he would change his appearance.

'He has moved forward to the battle. If he is not warded off, there will be destruction. It is he who seeks you in combat. Cú Chulainn mac Sualtaim.

'He will lay low your entire army, ana he will slaughter you in dense crowds. Ye shall leave with him all your heads. The prophetess Feidelm conceals it not.

'Blood will flow from heroes' bodies. Long will it be remembered. Men's bodies will be hacked, women will lament, through the Hound of the Smith that I see.

Thus far the prophecy and augury, and the prelude to the tale, the basis of its invention and composition, and the pillow-talk held by Aílill and Medb in Crúachu.

This is the route of the Táin and the beginning of the hosting together with the names of the roads on which the men of the four great provinces of Ireland travelled into the land of Ulster: To Mag Cruinn, by way of Tuaim Móna, by Turloch Teóra Crích, by Cúl Sílinne, by Dubfid, by Badbna, by Coltan, across the river Shannon, by Glúine Gabur, by Mag Trega, by northern Tethba, by southern Tethba, by Cúil, by Ochain, by Uata northwards, by Tiarthechta eastwards, by Ord, by Slass, across the river Inneoin, by Carn, across Meath, by Ortrach, by Findglassa Asail, by Drong, by Delt, by Duelt, by Deland, by Selach, by Slabra, by Slechta which was cleared by swords for Medb and Ailill's passage, by Cúil Siblinne, by Dub, by Ochan, by Catha, by Cromma, by Tromma, by Fodromma, by Sláine by Gort Sláine, by Druimm Licci, by Áth nGabla, by Ardachad, by Feoraind, by Findabair, by Aisse, by Airne, by Aurthaile, by Druimm Salaind, by Druimm Caín, by Druimm Caimthechta, by Druimm mac nDega, by Eódond Bec, by Eódond Mór, by Méide in Togmaill, by Méide ind Eoin, by Baile, by Aile, by Dall Scena, by Ball Scena, by Ros Mór, by Scúap, by Timscúap, by Cend Ferna, by Ammag, by Fid Mór in Crannach Cúailnge, by Druimm Caín to Slige Midlúachra.

(After) the first day's march on which the hosts went, they spent that night in Cúil Silinne and Ailill mac Rosa's tent was pitched for him. The tent of Fergus mac Róich was on his right hand. Cormac Cond Longas mac Conchobuir was beside Fergus. Íth mac Étgaíth

was next, then Fiachu mac Fir Aba. then Goibnend mac Lurgnig.
Such was the placing of Ailill's tent on his right during that hosting,
and thus were the thirty hundred men of Ulster at his right hand so
that the confidential talk and discourse and the choicest portions
of food and drink might be nearer to them. Medb Chrúachan was
on Ailill's left with Findabair beside her. Then came Flidais
Fholtchaín, the wife of Ailill Find, who had slept with Fergus
on Táin Bó Cúailnge, and it was she who every seventh night on
that hosting quenched with milk the thirst of all the men of Ireland,
king and queen and prince, poet and learner. Medb was the last
of the hosts that day for she had been seeking foreknowledge and
prophecy and tidings, that she might learn who was loath and who
was eager to go on the expedition. Medb did not permit her chariot
to be let down or her horses to be unyoked until she had made a
circuit of the encampment.

Then Medb's horses were unyoked and her chariots were let down
and she sat beside Ailill mac Mágach. And Ailill asked Medb to find
out who was eager and who reluctant or loath to go on the hosting.
'It is useless for any to set out on it except for the one band ¹(namely,
the division of the Gailioin)¹' said Medb. 'What good service do
they do that they are praised above all others?' said Ailill. 'There
is reason to praise them' said Medb. 'When the others began
to pitch their camp, these had already finished making their bothies
and open tents. When the others had finished their bothies and
open tents, these had finished preparing food and drink. When
the others had finished preparing food and drink, these had finished
eating their meal. When the others had finished their meal,
these were asleep. Even as their slaves and servants surpassed
the slaves and servants of the men of Ireland, so their warriors
and champions will surpass those of the men of Ireland on this
occasion on the hosting.' 'All the better do we deem that,' said
Ailill, 'for it is with us they march and it is for us they fight.' 'It
is not with us they will go nor for us they will fight.' 'Let them
stay at home then,' said Ailill. 'They shall not stay,' said Medb.
'What shall they do then,' said Findabair, 'if they do not go forth
nor yet stay at home?' 'Death and destruction and slaughter I
desire for them,' said Medb. 'Woe betide him who speaks thus,'
said Ailill, 'because of their having pitched their tents and set up
their stronghold quickly and promptly.' 'By the truth of my
conscience,' said Fergus, 'only he who inflicts death on me shall

---

¹⁻¹ following St

inflict death on those men.' 'Not to me should you say that, Fergus,' said Medb, 'for my army is numerous enough to slay and kill you with the thirty hundred Leinstermen surrounding you. For I have the seven Maines with their seven divisions of thirty hundred and the sons of Mága with their division and Ailill with his division, and I myself have my household guard. Our numbers are sufficient to slay and kill you with the division of the Leinstermen around you.' 'It is not fitting to speak thus to me,' said Fergus, 'for I have here the seven underkings of the Munstermen with their seven divisions. Here too is a division of the best among the noble warriors of Ulster. Here are the finest of the noble warriors of the men of Ireland, the division of the Gailioin. I myself am bond and surety and guarantee for them since they came from their own lands, and me shall they uphold in this day of battle. Furthermore,' said Fergus, 'those men shall not be . . . I shall disperse yon division of the Gailioin amongst the men of Ireland so that not five of them shall be (together) in one place.' 'I care not,' said Medb, 'in what way they are, provided only that they are not in the close battle array in which they now are.' Then Fergus dispersed that division among the men of Ireland so that no five men of them were (together) in one spot.

Thereafter the hosts set out upon their march. It was difficult for them to attend to that mighty army, which set forth on that journey, with the many tribes and the many families and the many thousands whom they brought with them that they might see each other and know each other and that each might be with his familiars and his friends and his kin on the hosting. They said too in what manner it was fitting to go on that hosting. They said that they should go thus: with every troop around their king, with every band around their leader, every group around their chief, and every king and royal heir of the men of Ireland on his own mound apart. They discussed too who ought to guide them between the two provinces, and they said that it should be Fergus, because the hosting was a hostile hosting for him, for he had been seven years in the kingship of Ulster, and when the sons of Usnech had been slain in despite of his guarantee and surety, he had come from there, 'and he has been seventeen years in exile and in enmity away from Ulster.' Therefore it would be fitting that he should go before all to guide them. Then Fergus went before all to guide them, but a feeling of affection for the Ulstermen seized him and he led the troops astray to the north and to the south, and messengers went from him with warnings to the Ulstermen and he began to delay and hold back the army. Medb perceived this, and she reproached him and chanted the lay:

'O Fergus, what do we say of this? What manner of path is this which we go? Past every tribe we wander north and south.'

'O Medb, why are you perturbed? This is not anything which resembles treachery. O woman, the land I traverse belongs to the men of Ulster.'

'Ailill, the splendid, with his army, ²fears that you will betray him². Hitherto you have not given your mind to leading us on the right path.'

'Not to the disadvantage of the host did I go on each wandering road in turn, but to try and avoid thereafter Cú Chulainn mac Sualtaim.'

'It is wrong of you to betray our host, O Fergus mac Rosa Rúaid, for much wealth did you get here in your exile, O Fergus.'

'I shall not be in front of the army any longer,' said Fergus, 'but seek some one else to lead them.' Yet Fergus took his position in the van of the army.

The four great provinces of Ireland were on Cúil Silinne that night. A sharp premonition of the arrival of Cú Chulainn came to Fergus and he told the men of Ireland to be on their guard, for there would come upon them he who was the slashing lion and the doom of enemies and the foe of armies, the supporting leader and the slaughtering of a great host, the hand bestowing gifts and the flaming torch, to wit, Cú Chulainn the son of Sualtaim. And Fergus was thus prophesying the coming of Cú Chulainn, and he made the lay and Medb answered him:

'It is well for you to keep watch and ward with many weapons and many warriors. He whom we fear will come, the great and valiant one from Muirtheimne.'

2–2 following LU, St

'Kindly is that of you—a counsel of battle—O valiant
Mac Róig. Men and arms I have here on the spot to answer
Cú Chulainn.'

'Men and arms are expended in the fray, O Medb from Mag
Aí, against the rider of Liath Macha, every night and every
day.

'I have here in reserve warriors to fight and to plunder,
thirty hundred hostage chiefs, the warriors of the Gailioin.'

'Warriors from fair Crúachu, heroes from clear-robed
Lúachair, four provinces of fair Gaels—all these will defend
me from that one man.'

'He who has troops in Bairrche and Banna will draw blood
across the shafts of spears. Into the mire and sand he will
cast that division of the Gailioin.

'As swift as the swallow and as speedy as the harsh wind—
thus is my fair dear Cú in mutual slaughter above the breath
of his foes.'

'O Fergus, famed in song, let this message go from
you to Cú Chulainn, that it were prudent for him to be
silent for he shall be harshly checked in Crúachu.'

'Valiantly will men be despoiled in the land of Badb's
daughter. The Hound of the Smith—with shedding of
gore—will overthrow companies of goodly heroes(?).'

After that lay: the army of the four great provinces of Ireland
came eastwards over Móin Coltna that day and there met them
eight score deer. The army spread out and surrounded them
and killed them so that none escaped. Yet though the division
of the Gailioin were dispersed, only five deer fell to the men of
Ireland. The one division of the Gailioin carried off the (rest of
the) eight score deer.

It was on the same day that Cú Chulainn mac Sualtaim and
Sualtach Sídech, his father, arrived and their horses grazed around
the pillar-stone at Ard Cuillenn. Sualtaim's steeds cropped the
grass down to the soil north of the pillar-stone, Cú Chulainn's steeds

cropped the grass down to the soil and the bedrock to the south of the pillar-stone. 'Well, father Sualtaim,' said Cú Chulainn, 'I have a premonition that the army is at hand, so go for me with warnings to the Ulstermen that they stay not on the open plains but go to the woods and waste places and deep valleys of the province to evade the men of Ireland.' 'And you, my fosterling, what will you do?' 'I must go southwards to Tara to keep a tryst with the handmaiden of Feidilmid Noíchruthach with my own surety until morning.' 'Woe to him who goes thus', said Sualtaim 'and leaves the Ulstermen to be trampled underfoot by their enemies and by outlanders for the sake of going to a tryst with any woman.' 'I must go however, for unless I do, men's contracts will be falsified and women's words be verified.'

Sualtaim went with warnings to the Ulstermen. Cú Chulainn went into the wood and cut a prime oak sapling, whole and entire, with one stroke and, standing on one leg and using but one hand and one eye, he twisted it into a ring and put an ogam inscription on the peg of the ring and put the ring around the narrow part of the standing-stone at Ard Cuillenn. He forced the ring down until it reached the thick part of the stone. After that Cú Chulainn went to his tryst.

As for the men of Ireland, they came to the pillar-stone at Ard Cuillenn and began to survey the unknown province of Ulster. Now two men of Medb's household were always in the van at every encampment and hosting, at every ford and every river and every pass. And this they did so that no stain might come to the princes' garments in the crowd or crush of host or army. These were the two sons of Nera mac Nuatair meic Tacáin, the two sons of the steward of Crúachu. Err and Innell were their names, and Fráech and Fochnam the names of their charioteers.

The nobles of Ireland came to the pillar stone and began to survey the grazing which the horses had made around the stone and to gaze at the barbaric ring which the royal hero had left around the stone. And Ailill took the ring in his hand and gave it to Fergus and Fergus read out the ogam inscription that was in the peg of the ring and told the men of Ireland what the inscription meant. And as he began to tell them he made the lay:

'This is a ring. What is its meaning for us? What is its secret message? And how many put it here? Was it one man or many?

'If ye go past it tonight and do not stay in camp beside it, the Hound who mangles all flesh will come upon you. Shame to you if ye flout it.

'If ye go on your way from it, it brings ruin on the host. Find out, O druids, why the ring was made.

'It was the swift cutting (?) of a hero. A hero cast it. It is a snare for enemies. One man—the sustainer of lords, a man of battle (?)—cast it there with one hand.

'It gave a pledge (?) with the harsh rage of the Smith's Hound from the Cráebrúad. It is a champion's bond, not the bond of a madman. That is the inscription on the ring.

'Its object is to cause anxiety to the four provinces of Ireland—and many combats. That is all I know of the reason why the ring was made.'

After that lay: Fergus said: 'I swear to you that if ye flout that ring and the royal hero who made it and do not spend a night here in encampment until one of you make a similar ring, standing on one foot and using one eye and one hand as he did, even though (that hero) be hidden underground or in a locked house, he will slay and wound you before the hour of rising on the morrow, if ye flout it.' 'It is not that indeed that we would wish,' said Medb, 'that anyone should wound us or shed our blood after we have come to this unknown province, the province of Ulster. More pleasing to us that we should wound another and spill his blood.' 'We shall not set this ring at naught,' said Ailill, 'and we shall not flout the royal hero who wrought it, but we shall take shelter in this great wood in the south until morning. Let our encampment be made there.' Then the hosts advanced and with their swords they hewed down the wood to make a path for their chariots, so that Slechta is still the name of that spot where is Partraige Beca south-west of Cenannas na Ríg near Cúil Sibrilli.

Heavy snow fell on them that night. So deep it was that it reached to the shoulders of men, to the flanks of horses and to the shafts of chariots, so that the provinces of Ireland were all one level plain with the snow. But no tents or bothies or pavilions were set up that night. No preparation of food or drink was made. No meal or repast was consumed. None of the men of Ireland knew whether it was friend or foe who was next to him until the bright hour of sunrise on the morrow. It is certain that the men of Ireland had never experienced a night in encampment which held more discomfort and hardship for them than that night at Cúil Sibrilli. The four great provinces of Ireland came forth early on the morrow with the rising of the sun across the glistening snow, and they went forward from that district to another.

As for Cú Chulainn, however, he did not rise early until he ate a repast and meal and washed and bathed on that day. He told his charioteer to harness the horses and yoke his chariot. The charioteer harnessed the horses and yoked the chariot, and Cú Chulainn went into his chariot and they followed the track of the army. They found the trail of the men of Ireland going past them from one district to another. 'Alas, my friend Láeg,' said Cú Chulainn, 'would that we had not gone to our tryst with a woman last night. The least that one who is guarding a border can do is to give a warning cry or shout or alarm or tell who goes the road. We failed to announce it. The men of Ireland have gone past us into Ulster.' 'I foretold for you, Cú Chulainn,' said Láeg, 'that if you went to your tryst, such a disgrace would come upon you.' 'Go, Láeg, I pray you, on the track of the army and make an estimate of them, and find out for us in what number the men of Ireland went past us.' Láeg came to the track of the host and came in front of the track and to one side of it and went to the rear of it. 'You are confused in your reckoning, my friend Láeg,' said Cú Chulainn. 'I am indeed,' said Láeg. 'Come into the chariot and I shall make an estimate of them.' The charioteer came into the chariot. Cú Chulainn went on the track of the host and made an estimate of their numbers and came to one side and went to the rear. 'You are confused in your reckoning, little Cú,' said Láeg. 'I am not,' said Cú Chulainn, 'for I know in what number the hosts went past us, namely, eighteen divisions, but the eighteenth division was dispersed among the men of Ireland.'—Now Cú Chulainn possessed many and various gifts: the gift of beauty, the gift of form, the gift of build, the gift of swimming, the gift of horseman-ship, the gift of playing *fidchell*, the gift of playing *brandub*, the gift of battle, the gift of fighting, the gift of conflict, the gift of sight, the gift of speech, the gift of counsel, the gift of fowling(?), the gift of laying waste (?), the gift of plundering in a strange border.

'Good, my friend Láeg, harness the chariot for us and ply the goad for us on the horses. Drive on the chariot and turn your left-hand board to the hosts to see can we overtake them in the van or in the rear or in the middle. For I shall not live if a friend or foe among the men of Ireland fall not by my hand tonight.' Then the charioteer plied the goad on the horses. He turned his left board to the hosts and came to Taurloch Caille Móre north of Cnogba na Ríg which is called Áth nGabla. Then Cú Chulainn went into the wood and descended from his chariot and cut a forked pole of four prongs, whole and entire, with one stroke. He pointed it and charred it and put an ogam inscription on its side and cast it out of the back of his chariot from the tip of one hand so that two thirds of it went into the ground and but one third of it was above ground. Then it was that the two lads mentioned, the two sons of Nera mac Nuatair meic Tacáin, came upon him engaged in that task, and they vied with one another as to which of them would first wound him and first behead him. Cú Chulainn attacked them and cut off their four heads from them (and from their charioteers) and impaled a head of each man of them on a prong of the pole. And Cú Chulainn sent the horses of that band back by the same road to meet the men of Ireland, with their reins lying loose and the headless trunks red with gore and the bodies of the warriors dripping blood down on to the framework of the chariots. For he did not deem it honourable or seemly to take the horses or garments or arms from the bodies of those he killed. Then the hosts saw the horses of the band who had gone in advance of them and the headless bodies and the corpses of the warriors dripping blood down on the framework of the chariots. The van of the army waited for the rear, and all were thrown into panic.

Medb and Fergus and the Maines and the sons of Mágu came up. For this is how Medb was wont to travel; with nine chariots for herself alone, two chariots before her, two behind, two on each side and her chariot between them in the very middle. And the reason she used to do that was so that the clods of earth cast up by the horses' hooves or the foam dripping from the bridle-bits or the dust raised by the mighty army might not reach her and that no darkening might come to the golden diadem of the queen. 'What is this?' said Medb. 'Not hard to say,' they all answered. 'These are the horses of the band that went in advance of us and their headless bodies in their chariots.' They held counsel, and they decided that that was the track of a multitude and the approach of a great army and that it was the men of Ulster who came to them thus. And this is what they decided on: to send Cormac Conn Longes to find out who was at the ford, for if the Ulstermen were there, they would not kill the son of their own king. Then Cormac Conn

Longes mac Conchobuir came with thirty hundred armed men to
find out who was at the ford. And when he got there he saw only
the forked pole in the middle of the ford with four heads on it
dripping blood down the stem of the pole into the current of the
stream and the hoof-marks of the two horses, and the track of a
single charioteer and of a single warrior leading eastwards out of
the ford.

The nobles of Ireland came to the ford and they all fell to
examining the forked pole. They marvelled and wondered who
had wrought the slaughter. 'What name have ye for this ford
until now, Fergus?' said Ailill. 'Áth nGrena,' said Fergus, 'and
Áth nGabla shall be its name forever now from this forked pole.'
And he recited the lay:

> 'Áth nGrena will change its name because of the deed
> performed by the strong, fierce Hound. There is here a
> four-pronged forked branch to bring fear on the men of
> Ireland.

> 'On two of its prongs are the heads of Fraech and Fochnam—
> presage of battle! On its other two points are the heads
> of Err and Innell.

> 'What inscription is that on its side? Tell us, O druids fair.
> And who wrote that inscription on it? How many drove
> it into the ground?'

> 'Yon forked branch with fearful strength that you see there,
> O Fergus, one man cut—and hail to him!—with one perfect
> stroke of his sword.

> 'He pointed it and swung it back behind him—no easy
> exploit—and then flung it down that one of you might pluck
> it out of the ground.

> 'Áth nGrena was its name hitherto. All will remember it.
> Áth nGabla will be its name forever from that forked branch
> which you see in the ford.'

After the lay: Ailill said: 'I marvel and wonder, Fergus, who
would have cut the forked pole and slain so swiftly the four who

went before us.' 'Rather should you marvel and wonder at him who cut, whole and entire, the forked pole that you see with one stroke, who sharpened and pointed it and made a cast of it from the back of his chariot with the tip of one hand so that it went two thirds of its length into the ground and only one third is above it, and no hole was dug for it with his sword but it was driven in through the stony ground. It is tabu for the men of Ireland to go into the bed of this ford until one of you pluck out the pole with the tip of one hand even as he drove it in just now.' 'You are of our army, Fergus,' said Medb, 'so bring us the forked pole from the bed of the ford.' 'Let me have a chariot,' said Fergus. A chariot was brought to Fergus, and he gave a tug to the forked pole and made fragments and small pieces of the chariot. 'Let a chariot be brought to me,' said Fergus (again). A chariot is brought to Fergus and he gave a strong pull to the forked pole and made fragments and small pieces of the chariot. 'Bring me a chariot,' said Fergus. He tugged the pole with all his strength and shattered the chariot into pieces. As for the seventeen chariots of the Connachtmen, Fergus broke them all to fragments and small pieces and yet he could not draw the pole from the bed of the ford. 'Give over, Fergus,' said Medb, 'do not break any more of my people's chariots, for had you not been on this hosting now, we should already have reached the Ulstermen and had our share of booty and herds. We know why you are acting thus: it is to hold back and delay the host until such time as the Ulstermen recover from their debility and give us battle, the battle of the Táin.' 'Let a chariot be brought to me at once,' said Fergus. Then his own chariot was brought to Fergus, and Fergus gave a strong wrench to the forked pole and neither wheel nor pole nor shaft of the chariot creaked or groaned. As was the strength and bravery with which it was driven in by him who had driven it in, so was the might and valour with which the warrior drew it out—(Fergus), the gap-breaker of a hundred, the sledge hammer of smiting, the destructive stone of enemies, the leader of resistance, the enemy of multitudes, the destroyer of a mighty army, the blazing torch, the commander of a great battle. He drew it up with the tip of one hand until it reached the top of his shoulder and he put the forked pole in Ailill's hand. And Ailill looked at it. 'The fork seems all the more perfect to me,' said Ailill, 'in that it is a single cutting I see on it from top to bottom.' 'All the more perfect indeed,' said Fergus, and he began to praise (the forked pole) and made this lay about it:

'Here is the famous forked pole beside which harsh Cú Chulainn stood, and on which he left—to spite some one (of you)—the four heads of strangers.

'It is certain that he would not retreat from the forked pole at the approach of one man, strong and fierce. Though the bright Hound has left it, blood remains on its hard bark.

'Woe to him who will go eastward on the hosting to seek the cruel Donn Cúailnge. Heroes will be cut in pieces by the baneful sword of Cú Chulainn.

'No easy gain will be his strong bull for whom a fight will be fought with keen weapons. When every skull has been tormented, all the tribes of Ireland will weep.

'I have no more to say concerning the son of Deichtire, but men and women shall hear of this pole as it now stands.'

After that lay: Ailill said: 'Let us pitch our tents and pavilions, and let us prepare food and drink and let us make music and melody and let us eat and take food, for it is unlikely that the men of Ireland ever at any time experienced a night of encampment that held more hardship and distress for them than last night.' Their encampments were set up and their tents pitched. Food and drink was prepared by them, music and melody played, and they ate a meal. And Ailill asked Fergus a question: 'I marvel and wonder as to who would come to us on the marches and slay so swiftly the four who went in advance. Is it likely that Conchobor mac Fachtna Fáthaig the high king of Ulster would come to us?' 'It is not likely indeed,' said Fergus, 'for it is lamentable to revile him in his absence. There is nothing that he would not pledge for his honour's sake. For if it were he who had come, armies and hosts and the pick of the men of Ireland[1] who are with him would have come too, and even though the men of Ireland and the men of Scotland, the Britons and the Saxons were opposed to him in one place and one meeting and one muster, in one camp and on one hill, he would give them all battle, it is he who would win victory and it is not he who would be routed.' 'Tell me, then, who was likely to have come to us? Was it perhaps Cuscraid Mend Macha

---

[1] Ulster St

mac Conchobuir from Inis Cuscraid?' 'It was not likely,' said
Fergus, son of the high king. 'There is nothing he would not stake
for the sake of his honour, for if it were he who came, the sons of
kings and royal princes who are with him in mercenary service
would also come, and if there were before him in one spot and one
meeting and one muster, in one camp and on one hill, the men of
Ireland and the men of Scotland, the Britons and the Saxons, he
would give them all battle, it is he who would be victorious and it
is not he who would be routed.' 'Tell me, then, would Eogan mac
Durthacht the King of Fernmag come to us?' 'It was not likely
indeed for if it were he who came, the steady men of Fernmag would
come with him and he would give battle etc.' 'Tell me then who was
likely to come to us. Was it Celtchair mac Uthechair?' 'It was
not likely indeed. It is shameful to revile him in his absence. He
is the destructive stone of his enemies in the province, he is leader
of resistance to all, he is the Ulstermen's doorway of battle, and
if there were before him in one spot _ut ante_ together with all the men
of Ireland from west to east and from south to north, he would
give them battle, he would be victorious and not he would be
routed.'

'Tell me, then, who would be likely to have come to us?' 'Nay
who but the little lad, my fosterson and the fosterson of Conchobor.
Cú Chulainn na Cerdda (the Hound of Culann the Smith) he is
called.' 'Yes indeed,' said Ailill. 'I have heard you speak of that little
lad once upon a time in Crúachu. What is the age of that boy now?'
'It is not his age that is most troublesome indeed,' said Fergus, 'for
the deeds of that boy were those of a man when he was younger than
he is now.' 'How so?' said Medb. 'Is there among the Ulsterman
now his equal in age who is more redoubtable than he?' 'We do
not find there a wolf more bloodthirsty nor a hero more fierce nor
any of his contemporaries who could equal the third or the fourth
part of Cú Chulainn's warlike deeds. You do not find there,' said
Fergus, 'a hero his equal nor a sledge-hammer of smiting nor doom
of hosts nor a contest of valour who would be of more worth than
Cú Chulainn. You do not find there one that could equal his age
and his growth, his size and his splendour, his fearsomeness and his
eloquence, his harshness, his feats of arms and his valour, his bearing,
his attack and his assault, his destructiveness, his troublesomeness
and his tumultuousness, his quickness, his speed and his violence,
and his swift victory with the feat of nine men on each pointed
weapon[1] above him.' 'We make but little account of him,' said

1-1 following LU and St

Medb. 'He has but one body. He shuns wounding who evades capture. His age is reckoned as but that of a nubile girl nor will that youthful beardless sprite ye speak of hold out against resolute men.' 'We do not say so,' said Fergus, 'for the deeds of that little boy were those of a man when he was younger than he now is.'

Here begin the youthful deeds of Cú Chulainn

'For this boy was reared in the house of his father and mother at Airgdig in Mag Muirtheimne, and the stories of the youths of Emain were told to him. For this is how Conchobor spends his time of kingship since he assumed sovereignty: as soon as he arises, settling the cares and business of the province, thereafter dividing the day into three, the first third of the day spent watching the youths playing games and hurling, the second third spent in playing *brandub* and *fidchell* and the last third spent in consuming food and drink until sleep comes on them all, while minstrels and musicians are meanwhile lulling him to sleep. Though I am banished from him, I swear,' said Fergus, 'that there is not in Ireland or in Scotland a warrior the counterpart of Conchobor.

'The stories about the youths and boys in Emain were told to that lad, and the little lad asked his mother if he might go to play to the playing-field at Emain. "It is too soon for you, my son," said his mother, "until there go with you a champion of the champions of Ulster or some of the attendants of Conchobor to ensure your safety and protection from the youths." "I think it long (to wait) for that, mother," said the little boy, "and I shall not wait for it, but show me in what place lies Emain." "Far away from you is the spot where it lies," said his mother. "Slíab Fúait is between you and Emain." "I shall make a guess at it then," said he.

'The boy went forth and took his playthings. He took his hurley-stick of bronze and his silver ball; he took his little javelin for casting and his toy spear with its end sharpened by fire, and he began to shorten the journey (by playing) with them. He would strike his ball with the stick and drive it a long way from him. Then with a second stroke he would throw his stick so that he might drive it a distance no less than the first. He would throw his javelin and he would cast his spear and would make a playful rush after them. Then he would catch his hurley-stick and his ball

and his javelin, and before the end of his spear had reached the
ground he would catch its tip aloft in the air.

'He went on to the place of assembly in Emain where the youths
were.   There were thrice fifty youths led by Follomain mac
Conchobuir at their games on the green of Emain.   The little boy
went on to the playing-field into their midst and caught the ball
between his two legs when they cast it nor did he let it go higher
than the top of his knee nor go lower than his ankle, and he pressed
it and held it close between his two legs, and not one of the youths
managed to get a grasp or a stroke or a blow or a shot at it.   And
he carried the ball away from them over the goal.

'Then they all gazed at him.   They wondered and marvelled.
"Well, boys," said Follomain mac Conchobuir, "attack yon fellow,
all of you, and let him meet death at my hands, for it is tabu for
you that a youth should join your game without ensuring his
protection from you.   Attack him all together, for we know that he
is the son of an Ulster chieftain, and let them not make it a habit
to join your games without putting themselves under your protection and safeguard."

'Then they all attacked him together.   They cast their thrice
fifty hurley-sticks at the boy's head.   He lifted up his single play-
thing stick and warded off the thrice fifty sticks.   Then they cast
the thrice fifty balls at the little boy.   He raised his arms and his
wrists and his palms and warded off the thrice fifty balls.   They
threw at him the thrice fifty toy spears with sharpened butt.   The
boy lifted up his toy wooden shield and warded off the thrice fifty
spears.   Then he attacked them.   He threw fifty kings' sons of them
to the ground beneath him.   Five of them,' said Fergus, 'went
between me and Conchobor in the spot where we were playing
chess on the chess-board Cendcháem on the mound of Emain.
The little boy pursued them to cut them down.   Conchobor seized
the little lad by the arms.   "Nay, lad, I see that you do not deal
gently with the youths."   "I have good reason for that," said the
boy.   "Though I came from distant lands, I did not get the honour
due to a guest from the youths on my arrival."   "Why, who are
you?" asked Conchobor.   "I am little Sétanta mac Sualtaim,
the son of Deichtire your sister, and not through you did I expect
to be thus aggrieved."   "Why, my lad," said Conchobor, "do
you not know of the prohibition that the youths have, and that it
is tabu for them that a boy should come to them from outside and
not (first) claim their protection?"   "I did not know," said the
little boy, "and if I had known, I should have been on my guard

against them." "Well, lads," said Conchobor, "undertake the protection of the little boy." "We grant it indeed," say they.

'The little boy placed himself under the protection of the youths. Then they loosed hands from him but once more he attacked them. He threw fifty kings' sons to the ground beneath him. Their fathers thought that he had killed them but it was not so, he had merely terrified them with his many and violent blows. "Nay," said Conchobor, "why do you still attack them?" "I swear by my gods that until they in their turn all come under my protection and guarantee as I have done with them, I shall not lift my hands from them until I bring them all low." "Well, little lad, take on you the protection of the youths." "I grant it," said the little boy. Then the youths placed themselves under his protection and guarantee.

'A little boy who did that deed,' said Fergus, 'at the end of five years after his birth and overthrew the sons of champions and warriors in front of their own fort and encampment, there were no need of wonder or surprise that he should come to the marches and cut a four-pronged pole and kill one man or two men or three or four when his seventeen years are accomplished on Táin Bó Cúailnge.'

Then said Cormac Cond Longas, the son of Conchobor: 'The year after that that little boy did a second deed.' 'What deed was that?' asked Ailill. 'Culand the smith dwelt in Ulster. He prepared a feast for Conchobor and went to Emain to invite him. He told him to come with only a small number unless he could bring a few genuine guests, for neither land nor domain had he but only his sledge-hammers and his anvils, his fists and his tongs. Conchobor said he would bring with him to Culand only a small number.

'Culand came on to his fort to prepare food and drink. Conchobor remained in Emain until it was time to disperse when day drew to a close. The king put on his light travelling garb and went to bid farewell to the youths. Conchobor went to the playing-field and saw something that astonished him: thrice fifty boys at one end of the field and a single boy at the other end, and the single boy winning victory in taking the goal and in hurling from the thrice fifty youths. When they played the hole-game—a game which was played on the green of Emain—and when it was their turn to cast the ball and his to defend, he would catch the thrice fifty balls outside the hole and none would go past him into the hole. When it was their turn to keep goal and his to hurl, he would put the thrice fifty balls unerringly into the hole. When they played at pulling off each other's clothes, he would tear their thrice fifty mantles off them and all of them together were unable to take even the brooch out of his cloak. When they wrestled, he would throw the same thrice fifty to the ground beneath him and a sufficient number of them to hold him could not get to him. Conchobor began to examine the little boy: "Ah, my warriors," said Conchobor, "happy is the land from which came the little boy ye see, if his manly deeds were to be like his boyish exploits." "It is not fitting to speak thus," said Fergus, "for as the little boy grows, so also will his deeds of manhood increase with him." "Let the little boy be summoned to us that he may go with us to share the feast to which we are going." The little boy was summoned to Conchobor. "Well my lad." said Conchobor, "come with us to enjoy the feast to which we are going." "I shall not go indeed," said the little boy. "Why so?" asked Conchobor. "Because the youths have not yet had enough of play and games and I shall not go from them until they have had their fill of play." "It is too long for us to wait for you, little lad, and we shall not." "Go on ahead," said the little boy, "and I shall go after you." "You do not know the way at all, little boy," said Conchobor. "I shall follow the trail of the company and the horses and the chariots."

'Then Conchobor came to the house of Culand the smith. The king was served. and they were honoured according to rank and profession and rights and nobility and accomplishments. Reeds and fresh rushes were strewn beneath them. They began to drink and make merry. Culand asked Conchobor: "Good now, O King have you appointed anyone to follow you tonight to this stronghold?" "I have not," said Conchobor for he did not remember the little boy he had appointed to come after him. "Why so?" asked Conchobor. "I have a good bloodhound and when his dog-chain is taken off no traveller or wayfarer dares come into the same canton as he, and he recognises no one but myself.' His strength is such that he can do the work of a hundred." Then said Conchobor: "Let the bloodhound be loosed for us that he may guard the canton." His dog-chain was loosed from the bloodhound and he made a swift circuit of the canton and he came to the mound where he was wont to be while guarding the dwelling, and he lay there with his head on his paws. And wild, savage and fierce, rough, surly and battlesome was he who lay there.

'As for the youths, they remained in Emain until it was time for them to disperse. They went each of them to the house of his father and mother, or of his fostermother and fosterfather. But the little boy went on the track of the company until he reached the house of Culand the smith. He began to shorten the way as he went with his playthings. When he reached the green before the stronghold where Culand and Conchobor were, he threw away all his playthings in front of him except his ball alone. The bloodhound perceived the little boy and bayed at him, and the baying of the bloodhound was heard throughout all the countryside. And it was not a sharing out for a feast the hound was minded to make (of the boy) but rather to swallow him entire past the wall of his chest and the breadth of his throat and the midriff of his breast. The boy had no means of defence, but he made a cast of the ball and it went through the gaping mouth of the bloodhound and carried all his entrails out through the back way, and the boy then seized him by two legs and dashed him against the standing-stone so that he was scattered into pieces on the ground. Conchobor had heard the baying of the hound. "Alas, my warriors," said Conchobor, "would that we had not come to enjoy this feast." "Why so?" asked they all. "The little boy who arranged to come after me, my sister's son, Sétanta mac Sualtaim, has been killed by the hound." All the famous Ulstermen rose with one accord. Though the gateway of the dwelling was wide open, they all went to meet him out over the palisades of the stronghold. Though all reached him quickly, quickest was Fergus and he lifted the little boy from the ground on to his shoulder and brought him into the presence of Conchobor. And Culand came forth and saw his bloodhound lying in scattered pieces. His heart beat against his breast. He went across into the stronghold then. "I welcome your arrival, little boy," said Culand, "for the sake of your mother and your father, but I do not welcome your arrival for your own sake." "Why are you angry with the boy?" asked Conchobor. "Would that you had not come to consume my drink and eat my food, for my substance now is substance wasted, my livelihood a lost livelihood. Good was the servant you have taken from me. He used to guard my herds and flocks and cattle for me." "Be not angry at all, master Culand," said the little boy, "for I shall deliver a true judgment in this matter." "What judgment would you deliver on it, my lad?" said Conchobor. "If there is a whelp of that hound's breeding in Ireland, he will be reared by me until he be fit for action like his sire. I shall myself be the hound to protect Culand's flocks and cattle and land during that time." "A good judgment you have given, little boy," said Conchobor. "I would not have given a better myself," said Cathbad. "Why shall

you not be called Cú Chulainn (Culand's Hound) because of this ? "
" Nay," said the little boy, " I prefer my own name, Sétanta
mac Sualtaim." " Do not say that, lad," said Cathbad, " for the
men of Ireland and of Scotland shall hear of that name, and that
name shall be ever on the lips of the men of Ireland and of Scotland." " I am willing that it shall be my name," said the boy.
Hence the famous name of Cú Chulainn clung to him since he
killed the hound of Culand the smith.

'A little boy who performed that exploit,' said Cormac Cond
Longas, 'six years after his birth, who killed the bloodhound
with which hosts and armies dared not be in the same canton,
there were no need to wonder or marvel that he should come to
the marches and cut a four-pronged pole and kill one man or two
or three or four, now that his seventeen years are completed on
Táin Bó Cúailnge.'

'The little boy performed a third exploit in the following year
again,' said Fiachu mac Fir Aba. 'What exploit did he perform ? '
asked Ailill. 'Cathbad the druid was teaching his pupils to the
north-east of Emain, and eight pupils of the class of druidic learning
were with him. One of them asked his teacher what omen and
presage was for that day, whether it was good or whether it was ill.
Then said Cathbad that a boy who should take up arms (on that
day) would be splendid and famous but would be shortlived and
transient. Cú Chulainn heard that as he was playing south-west
of Emain, and he threw aside all his playthings and went to
Conchobor's sleeping chamber. " All good attend you, O king
of the warriors," said the little boy.—That is the speech of a person
making a request of someone.—" What do you ask for, little lad ? "
said Conchobor. " I wish to take arms," said the little boy. " Who
has advised you, lad ? " said Conchobor. " Cathbad the druid,"
said the little boy. " He would not deceive you, lad," said Conchobor. Conchobor gave him two spears and a sword and a shield.
The little boy shook and brandished the arms and shattered
them into small pieces. Conchobor gave him two other spears
and a shield and a sword. He shook and brandished, flourished
and waved them, and shattered them into small pieces. As for
the fourteen suits of arms which Conchobor had in Emain for the
youths and boys—for to whichever one of them should take arms
Conchobor would give equipment of battle and the youth would
have victory in his valour thereafter—that little boy made fragments and small pieces of them all.

' " Indeed these weapons are not good, father Conchobor,"
said the little boy, " none of them suits me." Conchobor gave
him his own two spears and his shield and his sword. He shook
and brandished and flourished and waved them so that the point
(of spears and sword) touched the butt, and yet he did not break
the weapons and they withstood him. " These weapons are good
indeed, " said the little boy, " they are suited to me. I salute
the king whose weapons and equipment these are. I salute the
land from which he came." Then Cathbad the druid came into
the tent and spoke. " Is yon boy taking arms ? " said Cathbad.
" He is indeed," said Conchobor. " Not by your mother's son
would I wish arms to be taken today," said Cathbad. " Why
is that ? Is it not you who advised him ? " said Conchobor.
" Not I indeed," said Cathbad. " What mean you, you distorted
sprite," said Conchobor, " have you deceived me ? " " Do not
be angry, father Conchobor," said the little boy, " for it is he who
advised me; for his pupil asked him what omen was for the day
and he said that a boy who took arms on this day would be
splendid and renowned but short-lived and transient." " I spoke
truth", said Cathbad. " You will be splendid and renowned but
short-lived and transient." " It is a wonderful thing if I am
but one day and one night in the world provided that my fame
and my deeds live after me." " Come, little lad, mount the
chariot now for it is the same (good omen) for you."

' He mounted the chariot, and the first chariot he mounted,
he shook and swayed around him and shattered it to pieces. He
mounted the second chariot and shattered it to pieces in the same
way. He made fragments of the third chariot also. As for the
seventeen chariots which Conchobor had in Emain to serve the
youths and boys, the little lad shattered them all to pieces and
they withstood him not. " These chariots are not good, father
Conchobor," said the little boy, " none of these suits me." " Where
is Ibar mac Riangabra ? " asked Conchobor. " Here," answered
Ibar. " Harness my own two horses for yon boy and yoke my
chariot." The charioteer harnessed the horses and yoked the
chariot. Then the little boy mounted the chariot. He rocked
the chariot around him and it withstood him and did not break.
" This chariot is good indeed," said the little boy, " and it is my
fitting chariot."

' " Well, little boy," said Ibar, " let the horses go to their
pasture now." " It is too soon yet, Ibar," said the little boy.
" Come on around Emain now for to-day is the first day I took

arms, that it may be a triumph of valour for me." They drove thrice around Emain. " Let the horses go to their pasture now, little boy," said Ibar. " It is too soon yet, Ibar," said the little boy. " Come on so that the boys may wish me well, for to-day is the first day I took arms." They went forward to the place where the boys were. " Is yon lad taking arms ? " they asked. " Yes indeed." " May it be for victory and first-wounding and triumph, but we deem it too soon that you took arms because you part from us in our games." " I shall not part from you at all, but it is with a good omen I took arms to-day." " Let the horses go to their pasture now, little boy," said Ibar. " It is still too soon, Ibar," said the little boy. " And this great road which goes past us, where does it lead ? " said the little boy. " Why do you bother about it ? " said Ibar. " You are an importunate fellow, I see, little lad," said Ibar. " I wish, fellow, to ask about the chief roads of the province. How far does it go ? " " It goes to Áth na Foraire on Slíab Fúait," said Ibar. " Do you know why it is called Áth na Foraire ? " " I do indeed," said Ibar. " A goodly warrior of the Ulstermen is always there, keeping watch and ward so that no warriors or strangers come to Ulster to challenge them to battle and so that he may be the champion to give battle on behalf of the whole province. And if poets leave Ulstermen and the province unsatisfied, that he may be the one to give them treasures and valuables for the honour of the province. If poets come into the land, that he may be the man who will be their surety until they reach Conchobor's couch and that their poems and songs may be the first to be recited in Emain on their arrival." " Do you know who is at that ford to-day ? " " I do indeed," said Ibar, " Conall Cernach mac Amargin, the heroic and triumphant, the finest of the warriors of Ireland," said Ibar. " Go on, fellow, that we may reach the ford." They drove forward in front of the ford where Conall was. " Is yon boy taking arms ? " asked Conall. " He is indeed," said Ibar. " May that be for victory and first-wounding and triumph," said Conall, " but we deem it too soon for you to take arms because you are not yet fit for action if he that should come hither needed protection, for you would be complete surety for all the Ulstermen, and the nobles of the province would rise up at your summons." " What are you doing here, master Conall ? " said the little boy. " I am keeping watch and ward for the province here, lad," said Conall. " Go home now, master Conall," said the boy, " and let me keep watch for the province here." " Nay,

little boy," said Conall. "You are not yet fit to meet a goodly
warrior." "Then I shall meanwhile go on southwards" said the
boy, "to Fertais Locha Echtrand to see if I might redden my
hands in the blood of a friend or an enemy to-day." "I shall
go with you to protect you, lad," said Conall. "that you may
not go alone to the marches." "Nay," said the boy. "I shall
indeed go with you," said Conall, "for the Ulstermen will censure
me if I let you go alone to the marches."

'His horses are harnessed for Conall and his chariot yoked, and
he went to protect the boy. When Conall came abreast of him,
the boy was certain that if (the chance of performing) a great
deed were to come his way, Conall would not let him do it. He
took from the ground a stone which filled his fist. He made a
cast at the yoke of Conall's chariot and broke it in two so that
Conall fell through the chariot on to the ground and his shoulder
was dislocated. "What is this, boy?" said Conall. "It was
I who cast a shot to see if my markmanship was straight and in
what way I shoot, and to see if I am the makings of a good fighter."
"A bane on your shot and a bane on yourself! Even if you leave
your head with your enemies now, I shall not go (with you) to
guard you any more." "That is exactly what I asked you,"
said he, "for it is tabu for you Ulstermen to proceed on your way
despite an insecure chariot." Conall came back again northwards
to Áth na Foraire.

'As for the little boy, he went south to Fertais Locha Echtrand.
He was there until the close of day. "If we might venture to
say so, little lad," said Ibar, "we would deem it time to go now to
Emain, for already for some time the serving of meat and drink
and the sharing out has been made in Emain. You have your
appointed place there between Conchobor's knees every day
you come there while my place is merely among the messengers
and jesters of Conchobor's household. I think it time for me
to go and scramble for a place with them." "Then harness the
horses for us." The charioteer harnesses the horses and the boy
mounted the chariot. "Well, Ibar, what mound is that mound
up there now?" "That is Slíab Moduirn," said Ibar. "And
what is that white cairn on the top of the mountain?" "That
is Findcharn Slébe Moduirn," said Ibar. "Yon cairn is pleasant,"
said the little boy. "It is pleasant indeed," said Ibar. "Come
on, fellow, to that cairn." "Well, you are an importunate boy,"
said Ibar, "but this is my first expedition with you. It will be
my last expedition for ever if once I reach Emain." However

they went to the summit of the hill. " Well now, Ibar," said the boy, " teach me (all the places of) Ulster on every side for I do not know my way at all about the territory of Conchobor." The driver pointed out to him all the places of Ulster all around him. He told him the names of the hills and plains and mounds of the province on every side. He pointed out the plains and strongholds and renowned places of the province. " Well now, Ibar," said the little boy, " what plain is that to the south of us which is full of retreats and corners and nooks and glens ? " " That is Mag mBreg," said Ibar. " Show me the buildings and renowned places of that plain." The driver showed him Temair and Tailtiu, Cleitech and Cnogba and Brug Meic in Óc and the fortress of the sons of Nechta Scéne. " Are not these the sons of Nechta who boast that the number of Ulstermen alive is not greater than the number of those Ulstermen who have fallen at their hands ? " " They are indeed," said the driver. " Come on to the stronghold of the sons of Nechta," said the little lad. " Woe to him who says that ! " said Ibar. " We know that it is a very foolish thing to say that. Whoever goes there," said Ibar, " it will not be I." " You shall go there alive or dead," said the boy. " Alive I shall go south," said Ibar, " but dead I know I shall be left at the stronghold of Nechta's sons."

' They went on to the stronghold and the boy leapt from the chariot on to the green. Thus was the green before the stronghold : there was a pillar-stone on it and around the stone an iron ring, a ring of heroic deeds, with an ogam inscription on its peg. And thus ran the inscription : if any man came on that green and if he were a warrior bearing arms, it was tabu for him to leave the green without challenging to single combat. The little boy read out the inscription and put his two arms around the stone, that is, the stone and its ring, and he pitched it into the pool and the water closed over it. " It seems to us," said Ibar, " that that is no better than that it should remain where it was, and we know that you will find on this green what you are looking for now, namely, symptoms of death and dissolution." " Well now, Ibar, settle the coverings and rugs of the chariot for me that I may sleep for a little while." " Woe to him who says that," said the driver, " for this is a land of enemies and not a green for pleasure." The driver arranged the rugs and skin-coverings of the chariot. The little boy fell asleep on the green.

' Then there came on to the green one of the sons of Nechta, Fóill mac Nechtain. " Do not unharness the horses, driver," said Fóill. " I do not attempt it at all," said Ibar, " their traces and reins are still in my hand." " Whose are these horses ? " said Fóill. " Conchobor's two horses," said the driver, " the two piebald-headed ones." " I recognise them as such, and what brought the horses here to the border of the marches ? " " A youthful lad of ours who took up arms," said the driver. " He came to the edge of the marches to display his form." " May that not be for victory or triumph," said Fóill. " Had I known that he was old enough to fight, his dead body would have returned north to Emain and he would not have returned alive." " He is not old enough to fight indeed," said Ibar, " and it is not meet even to say so to him. He is in (but) the seventh year from his birth." The little boy raised his head from the ground and passed his hand over his face, and he blushed crimson from head to foot. " I am indeed capable of action," said the little boy. " It pleases me better than that you should say that you are not." " It will please you (still) better if only we meet on the ford, but go and fetch your weapons for I see that you have come in cowardly fashion, unarmed, and I do not wound charioteers or messengers or those unarmed." The fellow hastened to fetch his weapon. " It behoves you to act warily with yon man, little lad," said Ibar. " Why is that ? " said the boy. " The man you see is Fóill mac Nechtain. No points nor weapons nor sharp edges harm him." " Not to me should you say that, Ibar," said the boy. " I shall take in hand for him my *deil cliss,* that is, the round ball of refined iron, and it will land on the flat of his shield and the flat of his forehead and carry out through the back of his head a portion of brain equal to the iron ball, and he will be holed like a sieve so that the light of the air will be visible through his head." Fóill mac Nechtain came forth. Cú Chulainn took in hand for him the *deil cliss,* and hurled it so that it landed on the flat of his shield and the flat of his forehead and took the ball's equivalent of his brains through the back of his head, and he was holed like a sieve so that the light of the air was visible through his head. And Cú Chulainn struck off his head from his neck.

' The second son, Túachall mac Nechtain, came forth on the green. " I see that you would boast of that deed," said Túachall. " Indeed I think it no cause for boasting to slay one warrior." " You will not boast of that now for you will fall by my hand." " Go and fetch your weapons for you have come in cowardly fashion, unarmed." The fellow hastened to fetch his weapons. " You should have a care for yon fellow, little lad," said Ibar.

" Why so ? " said the boy. " The man you see is Túachall mac
Nechtain. Unless you get him with the first blow or the first
cast or the first touch, you will never do so, so skilfully and craftily
does he move around the points of the weapons." "Not to me should
that be said, Ibar," said the boy. " I shall take in hand the great
spear of Conchobor, the venemous lance. It will land on the shield
over his breast, and having pierced his heart, it will crush through
a rib in the side that is farther from me. It will be the cast of
an outlaw not the blow of a freeman. From me he shall not get
until the day of doom any place where he may be cured or tended."
Túachall mac Nechtain came out on the green, and the boy threw
Conchobor's spear at him and it went through the shield over
his breast and crushed through a rib in the side farther from Cú
Chulainn after piercing his heart in his chest. Cú Chulainn struck
off his head before it reached the ground.

' Then came forth the youngest of the sons, Faindle mac
Nechtain, on to the green. " Foolish were they who fought with
you here." " Why is that ? " said the boy. " Come away down
to the pool where your foot will not touch bottom." Faindle
hastened on to the pool. " You should have a care for yon fellow,
little lad," said Ibar. " Why so ? " said the boy. " The man
you see is Faindle mac Nechtain, and he is so called because he
travels over water like a swallow or a squirrel. The swimmers of
the world cannot cope with him." " Not to me should that be
said, Ibar," said the boy. " You know our river Calland in Emain.
When the youths surround it to play their games on it and when
the pool is not safe, I carry a boy over it on each of my two palms
and a boy on each of my two shoulders and I myself do not wet
even my ankles as I carry them." They met upon the water and
the boy clasped his arms around Faindle (and held him) until the
water came up flush with him, and he dealt him a violent blow
with Conchobor's sword and struck his head from his trunk, letting
the body go with the current and taking with him the head.

' Then they went into the stronghold and pillaged the fort and
fired it so that its buildings were level with its outer walls. They
turned about on their way to Slíab Fúait and took with them
the three heads of the sons of Nechta.

' They saw in front of them a herd of wild deer. " What are
these numerous fierce cattle, Ibar ? " said the boy. " Are they
tame or are they deer ? " " They are deer indeed," said Ibar.
" That is a herd of wild deer which frequent the recesses of Slíab
Fúait." " Ply the goad on the horses for us, that we may catch

some of them." The charioteer plied the goad on the horses.
The king's fat horses could not keep up with the deer. The boy
dismounted and caught two swift, strong stags. He tied them
to the shafts and ropes and thongs of the chariot.

'They went forward to the mound of Emain. They saw a
flock of white swans fly past them. "What kind of birds are
those, Ibar ? " said the boy. "Are they tame or just birds ? "
"Just birds," said Ibar. "They are a flock of swans which come
in from the crags and rocks and islands of the ocean to feed on
the plains and level spots of Ireland." "Which would be the
more wonderful, to bring them alive to Emain or to bring them
dead, Ibar ? " said the boy. "More wonderful indeed to bring
them alive," said Ibar, "for not everyone can catch the living
birds." Then the boy cast a small stone at them. He brought
down eight of the birds. Then he cast a big stone and brought
down sixteen of the birds. "Bring hither the birds, Ibar," said
the boy. "I am in a predicament," said Ibar. "How is that ? "
said the boy. "I have good reason to say so. If I move from
where I am, the iron wheels of the chariot will cut me down, so
fierce and so powerful (?) and so strong is the pace of the horses.
If I stir at all, the stags' antlers will pierce and gore me." "Ah,
no true warrior are you, Ibar," said the boy, "for with the look
that I shall give the horses, they will not break their straight course,
and with the look that I shall give the deer, they will bow their
heads in awe and fear of me, and it will not matter to you even
if you stepped across their antlers." Then (Ibar) tied the birds
to the shafts and cords and thongs and strings and ropes of the
chariot.

'They went forward and came to Emain. Then Leborcham
perceived them. She was the daughter of Aí and Adarc. "A
single chariot-warrior is here," said Leborcham, "and terribly
he comes. He has in the chariot the bloody heads of his enemies.
There are beautiful, pure-white birds held (?) by him in the chariot.
He has wild, untamed deer bound and tied and fettered. If he
be not met tonight, the warriors of Ulster will fall at his hand."
"We know that chariot-warrior," said Conchobor. "It is the
little boy, my sister's son, who went to the marches and shed
blood there, but he has not had his fill of combat, and if he be
not met, all the warriors of Emain will fall by his hand." And
the plan they devised was this : to send the women-folk out to
meet the boy, thrice fifty women, that is, ten and seven score
women, all stark naked, led by their chieftainess, Scannlach, to

expose all their nakedness and shame to him. All the young women came forth and discovered all their nakedness and shame to him. The boy hid his face from them and laid his countenance against the chariot that he might not see the women's nakedness. Then the boy was lifted out of the chariot. He was placed in three vats of cold water to quench the ardour of his wrath. The first vat into which the boy was put burst its staves and hoops like the breaking of a nutshell about him. As for the second vat, the water would seethe several hand-breadths high in it. As for the third vat (the water grew hot in it so that) one man might endure it while another would not. Thereupon the boy's wrath abated, and his garments ¹were put¹ on him. His comely appearance was restored, and he blushed crimson from head to foot. He had seven toes on each of his feet and seven fingers on each of his hands. He had seven pupils in each of his royal eyes and seven gems sparkling in each pupil. Four dimples in each cheek, a blue dimple a purple, a green and a yellow. Fifty tresses of hair he had between one ear and the other, bright yellow like the top of a birch-tree or like brooches of pale gold shining in the sun. He had a high crest of hair, bright, fair, as if a cow had licked it. He wore a green mantle in which was a silver pin, and a tunic of thread of gold. The boy was placed between Conchobor's knees and the king began to stroke his hair.

' A little lad who did those deeds when he was seven years old, who overcame the champions and warriors by whom two thirds of the men of Ulster had fallen and had been unavenged until this boy arose, there were no need to wonder or marvel that he should come to the marches and kill one man or two or three or four when his seventeen years were completed at the time of the Cattle-raid of Cúailnge.'

Thus far then is some account of the youthful deeds of Cú Chulainn on the Cattle-raid of Cúailnge, together with the prologue of the tale and an account of the route and march of the host out of Crúachu.

The story proper is what follows now.

The four great provinces of Ireland came the next day eastwards over Cruinn, that is, (the) mountain (called Cruinn). Cú Chulainn went ahead of them. He met the charioteer of Órlám, the son of Ailill and Medb who was at Tamlachta Órláim to the north of

---

¹—¹ following St

Dísert Lochad, cutting chariot poles from a holly-tree in the wood.
' Well, Láeg,' said Cú Chulainn, ' boldly do the Ulstermen behave
if it is they who are thus cutting down the wood in front of the
men of Ireland. And do you stay here for a little while until
I find out who is cutting down the wood in this manner.' Then
Cú Chulainn went on and came upon the charioteer. ' What are
you doing here, lad ? ' asked Cú Chulainn. ' I am cutting the
chariot poles from a holly-tree here,' said the driver, ' for our
chariots broke yesterday hunting that famous deer, Cú Chulainn.
And by your valour, warrior, come to my help, lest that famous
Cú Chulainn come upon me.' ' Take your choice, lad,' said Cú
Chulainn, ' either to gather the poles or to strip them.' ' I shall
gather them for it is easier.' Cú Chulainn began to strip the
poles, and he would draw them between his toes and between
his fingers against their bends and knots until he made them
smooth and polished and slippery and trimmed. He would make
them so smooth that a fly could not stay on them by the time
he cast them from him. Then the charioteer looks at him.
' Indeed it seems to me that it was not a labour befitting you that
I imposed on you. Who are you ?' asked the driver. ' I am
the famous Cú Chulainn of whom you spoke just now.' ' Woe
is me ! ' cried the charioteer, ' for that am I done for.' ' I shall
not slay you, lad,' said Cú Chulainn, ' for I do not wound
charioteers or messengers or men unarmed. And where is your
master anyway ? ' ' Over yonder on the mound,' said the
charioteer. ' Go to him and warn him to be on his guard, for if
we meet, he will fall at my hands.' Then the charioteer went
to his master, and swiftly as the charioteer went, more swiftly
still went Cú Chulainn and struck off Órlám's head. And he
raised the head aloft and displayed it to the men of Ireland.

Then came the three Meic Árach on to the ford at Ard Ciannacht
to meet with Cú Chulainn. Lon and Ualu and Díliu were their
names ; Mes Lir and Mes Laig and Mes Lethair were the names
of their charioteers. They came to encounter Cú Chulainn because
they deemed excessive what he had done against them the previous
day, namely, killing the two sons of Nera mac Nuatair meic
Thacáin at Áth Gabla and killing Órlám, the son of Ailill and
Medb, as well and displaying his head to the men of Ireland.
(They came then) that they might kill Cú Chulainn in the same
way and bear away his head as a trophy. They went to the wood
and cut three rods of white hazel (to put) in the hands of their
charioteers so that all six of them together might fight with Cú

Chulainn.   Cú Chulainn attacked them and cut off their six heads.
Thus fell Meic Árach by the hand of Cú Chulainn.

There came also Lethan on to his ford on the Níth in the district
of Conaille Muirtheimne, to fight with Cú Chulainn.   He attacked
him on the ford.   Áth Carpait was the name of the ford where
they reached it, for their chariots had been broken in the fighting
at the ford.   Mulchi fell on the hill between the two fords, whence
it is still called Gúalu Mulchi.   Then Cú Chulainn and Lethan
met, and Lethan fell by the hand of Cú Chulainn who cut off his
head from his trunk on the ford, but he left it with it, that is, he
left his head with his body.   Whence the name of the ford ever
since is Áth Lethan in the district of Conaille Muirtheimne.

Then came the harpers of Caínbile from Ess Ruaid to entertain
them.   The men of Ireland thought that they had come from
the Ulstermen to spy on them, so the hosts hunted them vigorously
for a long distance until they escaped from them, transformed
into wild deer, at the standing-stones at Lia Mór.   For though
they were called the harpers of Caínbile, they were men of great
knowledge and prophecy and magic.

Then Cú Chulainn vowed that wherever he saw Medb, he would
cast a stone at her and it would not go far from the side of her head.
It happened as he said.   Where he saw Medb to the west of the
ford, he cast a stone from his sling at her and killed the pet bird
on her shoulder.   Medb went eastwards over the ford, and he
cast another stone from his sling at her east of the ford and killed
the pet marten which was on her shoulder.   Whence the names
of those places are still Méide in Togmaill and Méide ind Eóin,
and Áth Srethe is the name of the ford across which Cú Chulainn
cast the stone from his sling.

The four great provinces of Ireland came on the morrow and
began to ravage Mag mBreg and Mag Muirtheimne.   And there
came to Fergus, Cú Chulainn's fosterfather, a keen premonition
of the arrival of Cú Chulainn, and he told the men of Ireland to
be on their guard that night for Cú Chulainn would come upon
them.   And he praised him here again, as we have written above,
and chanted a lay :

‘ Cú Chulainn of Cúailnge will come upon you in advance
   of the heroes of Cráebrúad.   Men will be bloodily wounded
   because of the harrying of Mag Muirtheimne.

'For Cú Chulainn went a longer journey (than this), as far as the mountains of Armenia. He waged combat beyond his wont. He slaughtered the Amazons.

'More difficult was it for him to drive the sons of Nechta from their couches and to slay with one hand the hound of the smith—valourous deed !

'I have no more to say concerning Deichtere's son. I swear that, in truth, though you reach him not, he will come to you.'

After that lay : On the same day, the Donn Cúailnge came to Crích Mairgín and with him fifty heifers, and he pawed up the earth, that is, he cast the turf over him with his heels. On the same day the Mórrigu daughter of Ernmas came from the fairy-mounds and sat on the pillar-stone in Temair Cúailnge, warning the Donn Cúailnge against the men of Ireland. She began to speak to him and she said : 'Good now, O pitiful one, Donn Cúailnge, be on your guard, for the men of Ireland will come upon you and will carry you off to their encampment unless you take heed.' And she began to warn him thus and spoke these words aloud : *Nach fitir* etc.[1]

Then Donn Cúailnge came and advanced into Glenn na Samaisce in Slíab Culind with fifty of his heifers.

Here are some of the virtues of the Donn Cúailnge : He would bull fifty heifers every day. These would calve before the same hour on the following day, and those of them that did not calve would burst with the calves because they could not endure the begetting of the Donn Cúailnge. It was one of the virtues of the Donn Cúailnge that fifty youths used to play games every evening on his back. Another of his virtues was that he used to protect a hundred warriors from heat and cold in his shadow and shelter. It was one of his virtues that no spectre or sprite or spirit of the glen dared to come into one and the same canton as he. It was one of his virtues that each evening as he came to his byre and his shed and his haggard, he used to make a musical lowing which was enough melody and delight for a man in the north and in the south and in the middle of the district of Cúailnge. Those are some of the virtues of Donn Cúailnge.

---

[1] rhetoric not translated.

Then on the morrow the hosts came into the rocks and dunes (?) of Conaille Muirthemne. And Medb ordered that a shelter of shields should be placed over her lest Cú Chulainn should make a cast at her from hills or heights or mounds. However on that day Cú Chulainn did not succeed in wounding or attacking the men of Ireland in the rocks and dunes of Conaille Muirthemne.

The men of the four great provinces of Ireland spent that night in Réde Loche in Cúailnge and pitched their camps there. Medb told a handmaid of her household to go to the river and fetch her water for drinking and washing. Loche was the maid's name. Then Loche came, wearing the golden diadem of the queen on her head and accompanied by fifty women. And Cú Chulainn cast a stone at her from his sling and broke in three the golden diadem and killed the girl on the plain where she was. Whence is the name Réde Loche in Cúailnge. For Cú Chulainn had thought, for want of knowledge and information, that it was Medb who was there.

On the morrow the hosts went as far as the river Glais Cruind, and they tried to cross the Glaise but failed to do so. Clúain Carpat is the name of the first place where they reached it, and that spot is called Clúain Carpat because the Glaise carried a hundred of their chariots away to the sea. Medb asked of her people that a warrior from amongst them should go and test the depth of the river. A great and valiant warrior of Medb's household called Úalu, rose up and took on his back a huge rock, and he came to test the depth of the stream. And the river Glais swept him back, dead and lifeless, with his stone on his back. Medb ordered him to be brought up (out of the river) and his grave dug and his stone raised. Whence the name Lia Úaland in the district of Cúailnge.

Cú Chulainn kept very close to the hosts that day, inviting them to fight and do combat, and killed a hundred of their warriors, including Róen and Roí, the two historians of the Foray.

Medb ordered her people to go and fight and do combat with Cú Chulainn. ' It will not be I,' and ' It will not be I ', said one and all from the place where they were. ' No captive is due from my people. Even if he were, it is not I who would go to oppose Cú Chulainn, for it is no easy task to encounter him.'

The hosts proceeded along the side of the river Glàise since they were unable to cross it, and they reached the spot where the

Glaise rises in the mountain, If they wished, they could have gone between the Glaise and the mountain, but Medb did not permit it but (ordered them) to dig and hack a path for her through the mountain, so that it might be a reproach and disgrace to the Ulstermen. Since then Bernais Tána Bó Cúailnge is the name of that place, for afterwards the drove of cattle was taken through it.

The men of the four great provinces of Ireland encamped that night at Bélat Aileáin. Until then its name was Bélat Aileáin, but from that time its name was Glenn Táil, because of the great amount of milk which the herds and cattle yielded there to the men of Ireland. And Líasa Líac is another name for that place. It is so called because it was there that the men of Ireland built byres and enclosures for their herds and their cattle.

The men of the four great provinces of Ireland came on as far as Sechair. Sechair was the name of the river until then but Glas Gatlaig is its name ever since. It is so called because the men of Ireland brought their herds and cattle across it tied with withes and ropes, and when they had crossed, the hosts let their withes and ropes drift down the stream. Hence the name of Glas Gatlaig.

That night the men of the four great provinces of Ireland came and encamped in Druim Én in the district of Conaille Muirthemne, and Cú Chulainn took up his position close beside them at Ferta in Lerga. And that night Cú Chulainn waved and brandished and shook his weapons so that a hundred warriors among the host died of fright and fear and dread of Cú Chulainn. Medb told Fiachu mac Fir Aba of the Ulstermen to go and parley with Cú Chulainn and to offer him terms. ' What terms would be offered him ? ' asked Fiachu mac Fir Aba. ' Not hard to say,' answered Medb. ' He shall be compensated for the damage done to Ulstermen that he may be paid as the men of Ireland best adjudge. He shall have entertainment at all times in Crúachu and wine and mead shall be served to him, and he shall come into my service and into the service of Ailill for that is more advantageous for him than to be in the service of the petty lord with whom he now is.'—And that is the most scornful and insulting speech that was made on the Foray of Cúailnge, namely, to call Conchobor, the finest king of a province in Ireland, a petty lord.

Then came Fiachu mac Fir Aba to parley with Cú Chulainn. Cú Chulainn welcomed him. ' I trust that welcome.' ' You may well trust it.' ' To parley with you have I come from Medb.' ' What (terms) did you bring ? ' ' Compensation shall be

made to you for the damage done to the Ulstermen that you may
be paid as the men of Ireland best adjudge. You shall have
entertainment in Crúachu and be served with wine and mead.
And you shall enter the service of Ailill and of Medb, for that is
more advantageous for you than to be in the service of the petty lord
with whom you now are.' ' No, indeed,' said Cú Chulainn. ' I
would not exchange my mother's brother for another king.' ' Come
early tomorrow to Glenn Fochaíne to a meeting with Medb and
Fergus.'

Then early on the morrow Cú Chulainn came to Glenn
Fochaíne. Medb and Fergus came there too to meet him, and
Medb gazed at Cú Chulainn, and in her own mind she belittled
him for he seemed to her no more than a boy. ' Is that the famous
Cú Chulainn of whom you speak, Fergus ? ' asked Medb. And
Medb began to speak to Fergus and made the lay :

> ' If that is the fair Hound of whom ye Ulstermen speak, no
> man who faces hardship but can ward him off from the
> men of Ireland.'

> ' Though young the Hound you see there who rides over
> Mag Muirthemne, no man who places foot on earth but he
> will repel in single combat.'

> ' Let terms be taken from us to the warrior. He is mad if
> he violate them. He shall have half his cows and half
> his womenfolk, and let him change his way of fighting.'

> ' I wish that the Hound from great Muirthemne be not
> defeated by you. I know that if it be he, he fears no fierce
> or famous deed of arms.'

' Speak you to Cú Chulainn, Fergus,' said Medb. ' Nay,' said
Fergus, ' rather speak to him yourself, for ye are not far apart
in this glen, Glenn Fochaíne.' And Medb began to address Cú
Chulainn and chanted a lay :

> ' O Cú Chulainn renowned in song, ward off from us your
> sling. Your fierce famed fighting has overcome us and
> confused us.'

' O Medb from Múr mac Mágach, I am no inglorious coward. As long as I live I shall not yield to you the driving of the herd of Cúailnge.'

' If you would accept from us, O triumphant Hound of Cúailnge, half your cows and half your womenfolk, [1]you will get them from us through fear of you.[1]'

' Since I, by virtue of those I have slain, am the veteran who guards Ulster, I shall accept no terms until I am given every milch cow, every woman of the Gael.'

' Too greatly do you boast, after slaughtering our nobles, that we should keep guard on the best of our steeds, the best of our possessions, all because of one man.'

' O daughter of Eochu Find Fáil, I am no good in such a contention. Though I am a warrior—clear omen !—my counsels are few.'

' No reproach to you is what you say, many-retinued son of Deichtere. The terms are such as will bring fame to you, O triumphant Cú Chulainn.'

After that lay : Cú Chulainn accepted none of the terms that Medb asked of him. In that manner they parted in the glen and each side withdrew equally angry.

The men of the four great provinces of Ireland encamped for three days and three nights at Druim Én in Conaille Muirthemne. But neither huts nor tents were set up, nor was meal or repast eaten by them and no music or melody was played by them during those three nights. And every night until the bright hour of sunrise on the morrow, Cú Chulainn used to kill a hundred of their warriors. ' Not long will our hosts last in this manner,' said Medb, 'if Cú Chulainn kill a hundred of our men every night. Why do we not offer him terms and why do we not parley with him ? ' ' What terms are those ? ' asked Ailill. ' Let him be offered those of the cattle that have milk and those of the captives who are base-born, and let him cease to ply his sling on the men of Ireland and let him allow the hosts at least to sleep.' ' Who will

[1–1] translating from St C

go with those terms ? ' asked Ailill. ' Who else but Mac Roth, the messenger,' said Medb. ' I shall not go indeed,' said Mac Roth, ' for I do not know the way and I do not know where Cú Chulainn is.' ' Ask Fergus,' said Medb, ' it is likely that he knows.' ' I do not know,' said Fergus, ' but I should think that he might be between Fochaín and the sea, exposing himself to wind and sun after his sleeplessness last night when single-handed he slew and demolished the host.' It was as Fergus had said.

Heavy snow fell that night so that all the provinces of Ireland were one white expanse. And Cú Chulainn cast off the twenty-seven shirts, waxed and hard as boards, which used to be bound to his skin with ropes and cords so that his sense might not be deranged when his fit of fury came upon him. The snow melted for thirty feet around him on all sides, so great was the ardour of the warrior and so hot the body of Cú Chulainn, and the charioteer could not remain near him because of the greatness of the fury and ardour of the warrior and because of the heat of his body.

' A single warrior comes towards us, little Cú,' said Láeg. ' What kind of warrior ? ' asked Cú Chulainn. ' A dark-haired, handsome, broad-faced fellow. A fine brown cloak about him, a bronze pin in his cloak. A strong, plaited shirt next to his skin. Two shoes between his feet and the ground. He carries a staff of white hazel in one hand and in the other a one-edged sword with guards of ivory.' ' Well, driver,' said Cú Chulainn, ' those are the tokens of a messenger. That is one of the messengers of Ireland coming to speak and parley with me.'

Then Mac Roth arrived at the spot where Láeg was. ²' Whose vassal are you, fellow ? '² asked Mac Roth. ' I am vassal to the warrior up yonder,' said the driver. Mac Roth came to the spot where Cú Chulainn was. ²' Whose vassal are you, warrior ? '² asked Mac Roth. ' I am the vassal of Conchobor mac Fachtna Fáthaig.' ' Have you no information more exact than that ? ' ' That is enough for now,' said Cú Chulainn. ' Find out for me where I might find that famous Cú Chulainn whom the men of Ireland are hunting now on this hosting.' ' What would you say to him that you would not say to me ? ' asked Cú Chulainn. ' I have come from Ailill and Medb to parley with him and to offer him terms and peace.' ' What terms have you brought him ? ' ' All that are milch of the kine, all that are base-born among the

---

²⁻² reading with St

captives, on condition that he cease to ply his sling against the hosts, for not pleasant is the thunderfeat he performs against them every evening.' 'Even if he whom you seek were at hand, he would not accept the proposals you ask. For the Ulstermen, if they have no dry cows, will kill their milch cows for companies and satirists and guests, for the sake of their honour, and they will take their low-born women to bed and thus there will arise in the land of Ulster a progeny which is base on the side of the mothers.' Mac Roth went back. 'Did you not find him?' asked Medb. 'I found a surly, angry, fearsome, fierce fellow between Fochaín and the sea. I do not know if he is the famed Cú Chulainn.' 'Did he accept those terms?' ''He did not indeed.' And Mac Roth told them the reason why he did not accept. 'It was Cú Chulainn to whom you spoke,' said Fergus.

'Let other terms be taken to him,' said Medb. 'What terms?' asked Ailill. 'All the dry kine of the herds, all the noble among the captives, and let him cease to ply his sling on the hosts for not pleasant is the thunderfeat he performs against them every evening.' 'Who will go with those terms?' 'Who but Mac Roth.' 'I shall indeed go,' said Mac Roth, 'for now I know the way.' Mac Roth came to speak to Cú Chulainn. 'I have come now to speak with you for I know that you are the famous Cú Chulainn.' 'What (terms) did you bring with you then?' 'All the dry kine in the herd, all the nobly-born among the captives, and cease to ply your sling against the men of Ireland and let them sleep, for not pleasant is the thunderfeat you perform against them every evening.' 'I shall not accept those terms, for the Ulstermen will kill their dry kine for the sake of their honour, for Ulstermen are generous, and Ulstermen will be left without any dry cattle or any milch cattle. They will set their free-born women to work at querns and kneading troughs and bring them into slavery and servile work. I do not wish to leave after me in Ulster the reproach of having made slaves and bondwomen of the daughters of the kings and royal leaders of Ulster.' 'Are there any terms at all that you accept now?' 'There are indeed,' said Cú Chulainn. 'Do you tell me the terms then?' asked Mac Roth. 'I vow,' said Cú Chulainn, 'that it is not I who will tell them to you.' 'Who then?' asked Mac Roth. 'If you have within the camp,' said Cú Chulainn, 'some one who should know my terms, let him tell you, and if you have not, let no one come any more to me offering terms or peace, for whoever so comes, that will be the length of his life.' Mac Roth went back

and Medb asked him for news. ' Did you find him ? ' said Medb.
' I did indeed,' said Mac Roth. ' Did he accept ? ' asked Medb.
' He did not,' said Mac Roth. ' Are there any terms which he
accepts ? ' ' There are, he says.' ' Did he make known those
terms to you ? ' ' What he said,' answered Mac Roth,' was that
it will not be he who will tell you them.' 'Who then ? ' asked Medb.
' But if there is among us one who should know the terms he asks,
let him tell me, and if there is not, let no one ever again come near
him. But there is one thing I assert,' said Mac Roth, ' even if
you were to give me the kingship of Ireland I myself shall not go
to tell them to him.'

Then Medb gazed at Fergus. ' What terms does yonder man
demand, Fergus ? ' said Medb. ' I see no advantage at all for
you in the terms he asks,' said Fergus. ' What terms are those ? '
said Medb. ' That one man from the men of Ireland should fight
him every day. While that man is being killed, the army to be
permitted to continue their march. Then when he has killed that
man, another warrior to be sent to him at the ford or else the men
of Ireland to remain in camp there until the bright hour of sunrise
on the morrow. And further, Cú Chulainn to be fed and clothed
by you as long as the Foray lasts.'

' By my conscience,' said Ailill, ' those are grievous terms.'
' What he asks is good,' said Medb, ' and he shall get those terms,
for we deem it preferable to lose one warrior every day rather
than a hundred warriors every night.' ' Who will go and tell those
terms to Cú Chulainn ? ' ' Who but Fergus,' said Medb. ' No,'
said Fergus. ' Why not ? ' asked Ailill. ' Let pledges and
covenants, bonds and guarantees be given for abiding by those
terms and for fulfilling them to Cú Chulainn.' ' I agree to that,'
said Medb, and Fergus bound them to security in the same way.

Fergus's horses were harnessed and his chariot yoked, and his
two horses were harnessed for Etarcumul son of Fid and of Lethrinn,
a stripling of the household of Medb and Ailill. ' Where are you
going ? ' asked Fergus. ' We are going with you,' said Etarcumul,
' to see the form and appearance of Cú Chulainn and to gaze upon
him.' ' If you were to follow my counsel,' said Fergus, ' you
would not come at all.' ' Why so ? ' ' Because of your haughtiness
and your arrogance, and also because of the fierceness and the
valour and the savageness of the lad against whom you go, for
I think that there will be strife between you before ye part.'

' Will you not be able to make intervention between us ? ' said Etarcumul. ' I shall,' said Fergus, ' if only you yourself will not seek ¹(contention and strife)¹. ' I shall never seek that.'

Then they went forward to Cú Chulainn where he was between Fochaín and the sea, playing *búanbach* with his charioteer. And no one came into the plain unnoticed by Láeg and yet he used to win every second game of *búanbach* from Cú Chulainn. ' A single warrior comes towards us, little Cú,' said Láeg. ' What manner of warrior is he ? ' asked Cú Chulainn. ' It seems to me that the chariot of the warrior is as big as one of the greatest mountains on a vast plain. It seems to me that the curly, thick, fair-yellow, golden hair hanging loose around his head is as great as (the foliage of) one of the tall trees which stand on the green before a great fort. He wears a purple, fringed mantle wrapped around him with a golden, inlaid brooch in it. A broad, grey spear flashing in his hand. A bossed, scalloped shield over him with a boss of red gold. ²A long sword, as long as a ship's rudder,² firmly fixed and resting on the two thighs of the great, proud warrior who is within the chariot.' ' Welcome is the arrival to us of this guest,' said Cú Chulainn. ' We know that man. It is my master Fergus who comes.' ' I see another chariot-warrior coming towards us also. With much skill and beauty and splendour do his horses advance.' ' That is one of the youths of the men of Ireland, friend Láeg,' said Cú Chulainn. ' To see my form and appearance that man comes, for I am renowned among them within their encampment.' Fergus arrived and sprang from the chariot, and Cú Chulainn bade him welcome. ' I trust that welcome,' said Fergus. ' You may well trust it,' said Cú Chulainn, ' for if a flock of birds pass over the plain, you shall have one wild goose and the half of another. If fish swim into the estuaries, you shall have a salmon with the half of another. You shall have a handful of watercress and a handful of sea-weed and a handful of water-parsnip. If you must fight or do battle I shall go to the ford on your behalf and you shall be watched over and guarded while you sleep and rest.' ' Well indeed, we know what provisions for hospitality you have now on the Foray of Cúailnge. But the condition that you asked of the men of Ireland, namely, single combat, you shall have it. I came to bind you to that, so undertake (to fulfil) it.' ' I agree indeed, master Fergus,' said

---

¹—¹ translating St
²—² translating LU and St

Cú Chulainn. And he delayed no longer than that conversing lest the men of Ireland should say that Fergus was betraying them to his fosterling. His two horses were harnessed for Fergus and his chariot was yoked, and he went back.

Etarcumul remained behind him gazing at Cú Chulainn for a long while. 'What are you staring at, lad ?' said Cú Chulainn. 'I am staring at you,' said Etarcumul. 'You have not far to look indeed,' said Cú Chulainn. 'You redden your eye with that. But if only you knew it, the little creature you are looking at, namely, myself, is wrathful. And how do you find me as you look at me ?' 'I think you are fine indeed. You are a comely, splendid, handsome youth with brilliant, numerous, various feats of arms. But as for reckoning you among goodly heroes or warriors or champions or sledge-hammers of smiting, we do not do so nor count you at all.' 'You know that it is a guarantee for you that you came out of the camp under the protection of my master, Fergus. But I swear by the gods whom I worship that but for Fergus's protection, only your shattered bones and your cloven joints would return to the camp.' 'Nay, do not threaten me any longer thus, for as for the condition you asked of the men of Ireland, namely, single combat, none other of the men of Ireland than I shall come to attack you tomorrow.' 'Come on, then, and however early you come, you will find me here. I shall not flee from you.' Etarcumul went back and began to converse with his charioteer. 'I must needs fight with Cú Chulainn to-morrow, driver,' said Etarcumul. 'You have promised it indeed,' said the charioteer, 'but I know not if you will fulfil your promise.' 'Which is better, to do so tomorrow or at once tonight ?' 'It is my conviction,' said the driver, 'that though doing it tomorrow means no victory, yet still less is to be gained by doing it tonight, [1]for the fight is nearer[1]. 'Turn the chariot back again for me, driver, for I swear by the gods whom I worship never to retreat until I carry off as a trophy the head of yon little deer, Cú Chulainn.'

The charioteer turned the chariot again towards the ford. They turned the left board of the chariot towards the company as they made for the ford. Láeg noticed that. 'The last chariot-fighter who was here a while ago, little Cú,' said Láeg. 'What of him ?' said Cú Chulainn. 'He turned his left board towards us as he made for the ford.' 'That is Etarcumul, driver, seeking combat of me. And I did not welcome him because of the guarantee of

---

[1]—[1] for destruction is nearer tonight St

my fosterfather under which he came out of the camp, and not because I wish to protect him. Bring my weapon to the ford for me, driver. I do not deem it honourable that he should reach the ford before me.' Then Cú Chulainn went to the ford and unsheathed his sword over his fair shoulder and was ready to meet Etarcumul at the ford. Etarcumul arrived also. 'What are you seeking, lad ?' asked Cú Chulainn. 'I seek combat with you,' said Etarcumul. 'If you would take my advice, you would not come at all,' said Cú Chulainn. '(I say so) because of the guarantee of Fergus under which you came out of the encampment and not at all because I wish to protect you.' Then Cú Chulainn gave him a blow (*fotalbéim*) and cut away the sod from beneath the sole of his foot so that he was cast prostrate with the sod on his belly. If Cú Chulainn had so wished, he could have cut him in two. 'Begone now for I have given you warning.' 'I shall not go until we meet again,' said Etarcumul. Cú Chulainn gave him an edge-blow (*fáebarbéim*).' He sheared his hair from him, from poll to forehead and from ear to ear as if it had been shaved with a keen, light razor. He drew not a drop of blood. 'Begone now,' said Cú Chulainn, 'for I have drawn ridicule on you.' 'I shall not go until we meet again, until I carry off your head and spoils and triumph over you or until you carry off my head and spoils and triumph over me.' 'The last thing you say is what will happen, and I shall carry off your head and spoils and I shall triumph over you.' Cú Chulainn dealt him a blow (*múadalbéim*) on the crown of his head which split him to his navel. He gave him a second blow crosswise so that the three sections into which his body was cut fell at one and the same time to the ground. Thus perished Etarcumul, son of Fid and Leithrinn.

Fergus did not know that this fight had taken place. That was but natural, for sitting and rising, journeying or marching, in battle or fight or combat, Fergus never looked behind him lest anyone should say that it was out of fearfulness he looked back, but (he was wont to gaze) at what was before him and on a level with him. Etarcumul's charioteer came abreast of Fergus. 'Where is your master, driver ?' asked Fergus. 'He fell on the ford just now by the hand of Cú Chulainn,' said the driver. 'It was not right,' said Fergus, 'for that distorted sprite (Cú Chulainn) to outrage me concerning him who came there under my protection. Turn the chariot for us, driver,' said Fergus, 'that we may go and speak with Cú Chulainn.'

Then the charioteer turned the chariot. They went off towards the ford. ' Why did you violate my pledge, you distorted sprite,' said Fergus, ' concerning him who came under my safeguard and protection ? ' ' By the nurture and care you gave me, tell me which you would prefer, that he should triumph over me or that I should triumph over him. Moreover enquire of his driver which of us was at fault against each other.' ' I prefer what you have done. A blessing on the hand that struck him !'

Then two withes were tied round Etarcumul's ankles and he was dragged along behind his horses and his chariot. At every rough rock he met, his lungs and liver were left behind on the stones and rocks (?). Wherever it was smooth for him, his scattered joints came together around the horses. Thus he was dragged across the camp to the door of the tent of Ailill and Medb. ' Here is your youth for you,' said Fergus, ' for every restoration has its fitting restitution.' Medb came out to the door of her tent and raised her voice aloud. ' We thought indeed,' said Medb, ' that great was the ardour and wrath of this young hound when he went forth from the camp in the morning. We thought that the guarantee under which he went, the guarantee of Fergus, was not that of a coward.' ' What has crazed the peasant-woman ? ' said Fergus. ' Is it right for the common cur to seek out the bloodhound whom the warriors of the four great provinces of Ireland dare not approach or withstand ? Even I myself would be glad to escape whole from him.' Thus fell Etarcumul.

That is (the story of) the Encounter of Etarcumul and Cú Chulainn.

Then there rose up a great and valiant warrior of Medb's household, called Nath Crantail, and he came to attack Cú Chulainn. He scorned to bring with him any arms except thrice nine spits of holly which were sharpened, charred and pointed by fire. And Cú Chulainn was on the pond before him.—And as for the pond, it was not safe but there were nine spits fixed in it, and Cú Chulainn used not to miss a single spit of them.—Then Nath Crantail cast a spit at Cú Chulainn. Cú Chulainn stepped on to the upper point of the spit which Nath Crantail had cast. Nath Crantail cast a second spit. He cast a third spit and Cú Chulainn stepped from the tip of the second spit on to the tip of the last spit.

Then the flock of birds flew out of the plain. Cú Chulainn pursued them as (swift as) any bird, that they might not escape him but might leave him that evening's meal. For what sufficed and served Cú Chulainn on the Foray of Cúailnge was fish and fowl and venison. However Nath Crantail was sure that Cú Chulainn fled in defeat from him, so he went forward to the door of the tent of Medb and Ailill and lifted up his voice : ' This famous Cú Chulainn of whom ye speak,' said Nath Crantail, 'has fled in rout before me just now.' ' We knew,' said Medb, ' that that would happen, and that if only goodly heroes and warriors came to meet him, the young and beardless sprite would not withstand resolute men. For when a goodly warrior came to him, he did not hold out against him but was routed by him.' Fergus heard that and he was greatly grieved that any man should taunt Cú Chulainn with having fled. And Fergus told Fiachu mac Fir Aba to go and speak with Cú Chulainn. ' And tell him that it was seemly for him to attack the hosts as long as he performed deeds of valour upon them but that it were fitter for him to hide himself rather than to flee before a single warrior from among them.' Then Fiachu came to speak with Cú Chulainn. Cú Chulainn bade him welcome. ' I trust that welcome, but I have come to speak to you from your fosterfather Fergus. He said that it was seemly for you to attack the hosts as long as you did deeds of valour but that it were more fitting for you to hide yourself than to flee before a single man of their warriors.' ' Why, who among you boasts of that ? ' asked Cú Chulainn. ' Nath Crantail,' said Fiachu. ' Why, do you not know, you and Fergus and the nobles of Ulster, that I do not wound charioteers or messengers or folk unarmed ? No weapons had Nath Crantail, only a wooden spit, and I would not wound him until he had a weapon. Tell him to come to me here early in the morning tomorrow and I shall not flee from him.' It seemed long to Nath Crantail until it was bright day for him to attack Cú Chulainn. Early on the morrow he came to attack him. Cú Chulainn rose early on that day, and a fit of rage came on him, and he angrily cast a fold of his cloak around him so that it wrapped itself round the pillarstone, and he dragged the pillarstone out of the ground between himself and his cloak. And he knew nothing of this because of the greatness of his rage, and he became distorted. Then came Nath Crantail and said : ' Where is this Cú Chulainn ? ' ' Over yonder,' said Cormac Cond Longas mac Conchobuir. ' That is not how he appeared to me yesterday,' said Nath Crantail. ' Then repel yon warrior,' said Cormac, ' and it is the same as if you repelled Cú Chulainn.'

Then Nath Crantail came and cast his sword at Cú Chulainn, and it struck the pillarstone which was between Cú Chulainn and his cloak, and the sword broke on the pillarstone. Cú Chulainn jumped from the ground to the top of the boss of Nath Crantail's shield and dealt him a return blow past the top of the shield and cut off his head from his trunk. Quickly he raised his hand again and dealt him another blow on the top of the trunk and cut him into two severed parts down to the ground. Thus fell Nath Crantail by the hand of Cú Chulainn. Thereafter Cú Chulainn said :

' If Nath Crantail has fallen, there will be increase of strife. Alas that battle cannot now be given to Medb with a third of the host ! '

After that Medb with a third of the army of the men of Ireland proceeded as far north as Dún Sobairche and Cú Chulainn followed her closely that day. And Medb went to Cuib ahead of Cú Chulainn. And after he had gone northwards Cú Chulainn killed Fer Taidle, whence the place-name Taidle, and he killed the sons of Búachaill, whence the name Carn Mac mBúachalla, and he killed Lúasce in Leitre whence Leitre Lúasce. He killed Bó Bulge in his swamp, from which comes the name Grellach Bó Bulge. He killed Muir-themne on his hill whence the name Delga Muirthemne.

After that Cú Chulainn came southwards again to protect and guard his own land and territory, for it was dearer to him than the land and territory of any other.

Then there met him Fir Crandce, the two Artinnes and the two sons of Lecc and the two sons of Durcride, and the two sons of Gabal, and Drúcht and Delt and Dathen, Te and Tualang and Turscur, Torc Glaisse and Glas and Glaisne—these are the same as the twenty Fir Fochard. Cú Chulainn overtook them as they were pitching their camp ahead of the rest and they fell by him.

Then there met Cú Chulainn Buide mac Báin Blai from the land of Ailill and Medb, one of Medb's household. Twenty-four warriors (was the number of his company). Each man wore a mantle wrapped around him. Donn Cúailnge was driven hastily and forcibly in front of them after he had been brought from Glenn na Samaisce in Slíab Culind together with fifty of his heifers. ' Whence do ye bring the drove ? ' asked Cú Chulainn. ' From yonder mountain,' said Buide. ' What is your own name ? ' asked Cú Chulainn. ' (One who) loves you not, (who) fears you not,' said Buide. ' I am Buide mac Báin Blai from the land of

Ailill and Medb.' ' Here is this little spear for you,' said Cú
Chulainn. And he cast the spear at him. The spear landed in
the shield above his breast and crushed three ribs in the farther
side after piercing his heart, and Buide mac Báin Blai fell. Hence
the name Áth mBuide in Crích Rois ever since.

While they were thus engaged exchanging the two short spears—
for not at once did they finish—the Donn Cúailnge was carried
off hastily and forcibly from them to the encampment as any
cow might be taken. That was the greatest reproach and grief
and madness that was inflicted on Cú Chulainn in this hosting.

As for Medb, every ford at which she stopped is called Áth
Medbe. Every place where she erected her tent is called Pupall
Medba, and every spot where she planted her horse-whip is called
Bile Medba.

On this expedition Medb gave battle to Findmór the wife of
Celtchair in front of Dún Sobairche, and she slew Findmór and
ravaged Dún Sobairche.

After a fortnight the men of the four great provinces of Ireland
came to the encampment together with Medb and Ailill and the
men who were bringing the bull. But the bull's herdsman did
not allow them to carry off Donn Cúailnge, so despite him they
urged on (both bull and heifers) by beating their shields with
sticks, and drove them into a narrow pass, and the cattle trampled
the body of the herdsman thirty feet into the ground and made
small fragments of his body. Forgemen was his name.

Bás Forgaimin is the name of that tale in the Foray of Cúailnge.

When the men of Ireland reached one spot, together with Medb
and Ailill and the men who were bringing the bull to the camp,
they all said that Cú Chulainn would be no more valiant than
anyone else but for the strange feat he possessed, the javelin of
Cú Chulainn. Then the men of Ireland sent Redg, Medb's satirist,
to ask for the javelin. Redg asked for the javelin and Cú Chulainn
did not give it at once to him, that is, he was reluctant to give it.
Redg threatened to deprive Cú Chulainn of his honour. Then
Cú Chulainn cast the javelin after him and it lighted on the hollow
at the back of his head and passed through his mouth out on to
the ground, and he managed to speak only the words : ' Quickly
did we get this treasure,' when his soul parted from his body on

the ford. And since then that ford is called Áth Solomshét. And the bronze from the spear landed on the stream, whence is the name Umanshruth ever since.

The men of Ireland debated as to which of them should attack Cú Chulainn, and they all agreed that Cúr mac Da Lóth would be the right man to attack him. For such was Cúr that it was not pleasant to be his bedfellow or to be intimate with him, and they said that if it were Cúr who fell, it would mean a lightening of oppression for the hosts, and that if it were Cú Chulainn, it would be still better. Cúr was summoned to Medb's tent. ' What do they want of me ? ' asked Cúr. ' To attack Cú Chulainn,' said Medb. ' Ye think little of our valour, ye think it wonderful, when ye match me with a tender stripling such as he ! Had I myself known (why I was summoned), I should not have come for that. I should think it enough that a lad of his own age from among my household should go to oppose him on the ford.' ' Nay, it is foolish (?) to say that,' said Cormac Cond Longas mac Conchobuir. ' It would be a fine thing for you yourself were Cú Chulainn to fall by you.' ' Make ye ready a journey for me in the early morning tomorrow for I am glad to go. It is not the killing of yonder deer, Cú Chulainn, that will cause you any delay.' Early on the morrow, then, Cúr mac Da Lóth arose. A cartload of arms was brought by him to attack Cú Chulainn, and he began to try and kill him. Early on that day Cú Chulainn betook himself to his feats. These are all their names : [1]*uballchless, fóenchless, cless cletínech, tétchless, corpchless, cless cait, ích n-errid, cor ndelend, léim dar néim, filliud eirred náir, gai bulga, baí brassi, rothchless, cles for análaib, brúud gine, sían curad, béim co fommus, táthbéim, réim fri fogaist, dírgud cretti fora rind, fornaidm níad.*[1]

Cú Chulainn used to practice each of these feats early every morning, in one hand, as swiftly as a cat makes for cream (?), that he might not forget or disremember them. Mac Da Lóth remained for a third of the day behind the boss of his shield, endeavouring to wound Cú Chulainn. Then said Láeg to Cú Chulainn : ' Good now, little Cú, answer the warrior who seeks to kill you.' Then Cú Chulainn looked at him and raised up and cast aloft the eight balls, and he made a cast of the ninth

---

[1]—[1] It is impossible to translate most of these with any certainty as to the meaning.

ball at Cúr mac Da Lóth so that it landed on the flat of his shield
and the flat of his forehead and took a portion of brain the size
of the ball out through the back of his head. Thus Cúr mac Da
Lóth fell by the hand of Cú Chulainn.

'If your securities and guarantees now bind you,' said Fergus,
'send another warrior to meet yon man at the ford, or else remain
here in your camp until the bright hour of sunrise tomorrow, for
Cúr mac Da Lóth has fallen.' 'Considering why we have come,'
said Medb, 'it is all the same to us if we remain in the same tents.'
They remained in that encampment until there had fallen Cúr
mac Da Lóth and Lath mac Da Bro and Srub Daire mac Fedaig
and Mac Téora nAignech. Those men fell by Cú Chulainn
in single combat. But it is tedious to relate the prowess of each
man separately.

Then Cú Chulainn said to Láeg, his charioteer : 'Go, friend
Láeg, to the encampment of the men of Ireland and take a greeting
from me to my friends and my fosterbrothers and my coevals.
Take a greeting to Fer Diad mac Damáin and to Fer Dét mac
Damáin and to Bress mac Firb, to Lugaid mac Nóis and to Lugaid
mac Solamaig, to Fer Báeth mac Báetáin and to Fer Báeth mac
Fir Bend. And take a special greeting to my fosterbrother
Lugaid mac Nóis, for he is the only man who keeps faith and
friendship with me now on the hosting, and give him a blessing
that he may tell you who comes to attack me tomorrow.'

Then Láeg went forward to the encampment of the men of
Ireland and took a greeting to the friends and fosterbrothers of
Cú Chulainn, and he went too into the tent of Lugaid mac Nóis.
Lugaid bade him welcome. 'I trust that welcome,' said Láeg.
'You may do so,' said Lugaid. 'I have come from Cú Chulainn
to speak with you,' said Láeg, 'and he has sent you a true and
sincere greeting and wishes you to tell me who comes to attack
Cú Chulainn to-day.' 'The curse of his intimacy and familiarity
and friendship on him (who comes) ! It is his very own fosterbrother, Fer Báeth mac Fir Bend. He was taken just now into
Medb's tent. The girl Findabair was placed at his side. She it
is who pours goblets for him. She it is who kisses him at every
drink. She it is who serves him his meal. Not for all and sundry
does Medb intend the liquor which is served to Fer Báeth, for only
fifty wagon-loads of it were brought to the camp.'

Then Láeg went back to Cú Chulainn, crestfallen, sad, joyless and mournful. ' Crestfallen, sad, joyless and mournful my friend Láeg comes to me.' said Cú Chulainn. ' It means that one of my fosterbrothers comes to attack me.'—For Cú Chulainn disliked more that a warrior of the same training as himself should come to him rather than some other warrior.—' Good now, friend Láeg,' said Cú Chulainn, ' who comes to attack me to-day? ' ' The curse of his intimacy and brotherhood, of his familiarity and friendship be upon him ! It is your very own fosterbrother, Fer Báeth mac Fir Bend. He was taken just now into Medb's tent. The girl was placed at his side, and it is she who pours goblets for him. it is she who kisses him with every drink, it is she who serves his meal. Not for all and sundry does Medb intend the liquor which is served to Fer Báeth. Only fifty wagon-loads of it were brought to the camp.'

Fer Báeth waited not until morning but went at once to renounce his friendship with Cú Chulainn. Cú Chulainn adjured him by their friendship and intimacy and brotherhood, but Fer Báeth did not consent to relinquish the combat. Cú Chulainn left him in anger, and trampled a sharp shoot of holly into the sole of his foot so that it injured alike flesh and bone and skin. Cú Chulainn tore out the holly shoot by the roots and cast it over his shoulder after Fer Báeth, and he cared not whether it reached him or not. The holly shoot hit Fer Báeth in the depression at the nape of his neck and went out through his mouth on to the ground, and thus Fer Báeth died. ' That was indeed a good cast (*focherd*), little Cú,' said Fiacha mac Fir Aba. For he considered it a good cast to kill the warrior with the holly shoot. Whence is still the name Focherd Muirtheimne for the spot where they were.

' Go, friend Láeg,' said Cú Chulainn, ' and speak with Lugaid in the camp of the men of Ireland, and find out whether anything has happened to Fer Báeth or not [1][and ask him who will come against me tomorrow.' Láeg goes forward to Lugaid's tent. Lugaid welcomed him. ' I trust that welcome,' said Láeg. ' You may trust it,' said Lugaid. ' I have come to speak with you on behalf of your fosterbrother that you may tell me if Fer Báeth reached the camp.' ' He did,' said Lugaid, ' and a blessing on the hand that smote him for he fell dead in the glen a short time

[1] Stowe version

ago.' 'Tell me who will come tomorrow to fight against Cú
Chulainn.' 'They are asking a brother of mine to oppose him,
a foolish youth, proud and arrogant, but a strong smiter and a
victorious fighter. And the reason he is sent to fight him is that
he may fall by Cú Chulainn and that I might then go to avenge
his death on Cú Chulainn, but I shall never do that. Láiríne mac
Í Blaitmic is my brother's name. I shall go to speak with Cú
Chulainn about that,' said Lugaid. His two horses were harnessed
for Lugaid and his chariot was yoked to them. He came to meet
Cú Chulainn and a conversation took place between them. Then
said Lugaid : 'They are urging a brother of mine to come and
fight with you, a foolish youth, rough, uncouth, but strong and
stubborn, and he is sent to fight you so that when he falls by you,
I may go to avenge his death on you, but I shall never do so. And
by the friendship that is between us both, do not kill my brother.
Yet I swear, that even if you all but kill him, I grant you leave
to do so, for it is in despite of me that he goes against you.' Then
Cú Chulainn went back and Lugaid went to the camp.

Then Láiríne mac Nóis was summoned to the tent of Ailill and
Medb and Finnabair was placed beside him. It was she who used
to serve him goblets and she who used to kiss him at every drink
and she who used to hand him his food. 'Not to all and sundry
does Medb give the liquor that is served to Fer Báeth or to Láiríne,'
said Finnabair. 'She brought only fifty wagon-loads of it to
the camp.' 'Whom do you mean ? ' asked Ailill. 'I mean
that man yonder,' said she. 'Who is he ? ' asked Ailill. 'Often
you paid attention to something that was not certain. It were
more fitting for you to bestow attention on the couple who are
best in wealth and honour and dignity of all those in Ireland,
namely, Finnabair and Láiríne mac Nóis.' 'That is how I see
them,' said Ailill. Then (in his joy) Láiríne flung himself about
so that the seams of the flockbeds under him burst and the green
before the camp was strewn with their feathers.

Láiríne longed for the full light of day that he might attack
Cú Chulainn. He came in the early morning on the morrow and
brought with him a wagon-load of weapons, and he came on to
the ford to encounter Cú Chulainn. The mighty warriors in the
camp did not think it worth their while to go and watch Láiríne's
fight, but the women and boys and girls scoffed and jeered at his
fight. Cú Chulainn came to the ford to encounter Láiríne, but
he scorned to bring any weapons and came unarmed to meet him.
He struck all Láiríne's weapons out of his hand as one might deprive

a little boy of his playthings. Then Cú Chulainn ground and squeezed him between his hands, chastised him and clasped him, crushed him and shook him and forced all his excrement out of him until a mist arose on all sides in the place where he was. And after that he cast him from him, from the bed of the ford across the camp to the entrance of his brother's tent. However Láiríne never (after) rose without complaint and he never ate without pain, and from that time forth he was never without abdominal weakness and constriction of the chest and cramps and diarrhoea. He was indeed the only man who survived battle with Cú Chulainn on the Foray of Cúailnge. Yet the after-effects of those complaints affected him so that he died later.

That is the fight of Láiríne on the Foray of Cúailnge.

Then Lóch Mór mac Mo Febis was summoned to the tent of Ailill and Medb. ' What would ye with me ? ' asked Lóch. ' That you should fight with Cú Chulainn,' answered Medb. ' I shall not go on such an errand for I deem it no honour to attack a youthful, beardless stripling, and I do not intend that as an insult to him, but I have the man to attack him, namely, Long mac Emonis, and he will accept reward from you.' Long was summoned to the tent of Ailill and Medb, and Medb promised him great rewards, to wit, the clothing of twelve men in garments of every colour, a chariot worth four times seven *cumala*, Finnabair as his wedded wife, and entertainment at all times in Crúachu with wine served to him. Then Long came to meet Cú Chulainn and Cú Chulainn killed him.

Medb told her women-folk to go and speak to Cú Chulainn and tell him to put on a false beard of blackberry juice. The women came forward towards Cú Chulainn and told him to put on a false beard. ' For no great warrior in the camp thinks it worth his while to go and fight with you while you are beardless.' After that Cú Chulainn put on a beard of blackberry juice and came on to the hillock above the men of Ireland and displayed that beard to all of them in general.

Lóch mac Mo Febis saw this and said : ' That is a beard on Cú Chulainn.' ' That is what I see,' said Medb. She promised the same rewards to Lóch for checking Cú Chulainn. ' I shall go and attack him,' said Lóch.

Lóch came to attack Cú Chulainn and they met on the ford where Long had fallen. ' Come forward to the upper ford,' said

Lóch, ' for we shall not fight on this one.' For he held unclean the ford at which his brother had fallen. Then they met on the upper ford.

It was at that time that the Morrígan daughter of Ernmas from the fairy-mounds came to destroy Cú Chulainn, for she had vowed on the Foray of Regamain that she would come and destroy Cú Chulainn when he was fighting with a mighty warrior on the Foray of Cúailnge. So the Morrígan came there in the guise of a white, red-eared heifer accompanied by fifty heifers, each pair linked together with a chain of white bronze. The womenfolk put Cú Chulainn under tabus and prohibitions not to let the Morrígan go from him without checking and destroying her. Cú Chulainn made a cast at the Morrígan and shattered one of her eyes. Then the Morrígan appeared in the form of a slippery, black eel swimming downstream, and went into the pool and coiled herself around Cú Chulainn's legs. While Cú Chulainn was disentangling himself from her, Lóch dealt him a wound crosswise through his chest. Then the Morrígan came in the guise of a shaggy, russet-coloured she-wolf. While Cú Chulainn was warding her off, Lóch wounded him. Thereupon Cú Chulainn was filled with rage and] [1]wounded Lóch with the *ga bulga* and pierced his heart in his breast. ' Grant me a favour now, Cú Chulainn,' said Lóch. ' What favour do you ask ? ' ' No favour of quarter do I ask nor do I make a cowardly request.' said Lóch. ' Retreat a step from me so that I may fall facing the east and not to the west towards the men of Ireland, lest one of them say that I fled in rout before you, for I have fallen by the *ga bulga*.' ' I shall retreat,' said Cú Chulainn, ' for it is a warrior's request you make.' And Cú Chulainn retreated a step from him. Hence the ford has since then been known as Áth Traiged at the end of Tír Mór.

Cú Chulainn was seized by great depression that day for that he fought single-handed on the Foray of Cúailnge. And he ordered his charioteer Láeg to go to the men of Ulster and bid them come to defend their cattle. [2]And great dejection and weariness took possession of Cú Chulainn[2] and he uttered these verses :

---

[1] LL resumes
[2-2] translating St.

'Go forth from me, O Láeg. Let the hosts be roused. Tell them for me in strong Emain that each day in battle I am weary, and I am wounded and bloody.

'My right side and my left—hard to appraise either of them. It was no physician's hand which smote them . . .

'Tell noble Conchobor that I am weary, wounded sore in my side. Greatly has Dechtire's dear son, he of many retinues, changed in appearance.

'I am here all alone guarding the flocks, not only do I not let them not go, but neither can I hold them. In evil plight I am and not in good, as I stand alone at many fords.

'A drop of blood drips from my weapon. I am sorely wounded. No friend comes to me in alliance or to help, my only friend is my charioteer.

'If but few sing here for me, a single horn rejoices not. But if many horns make music, then the sound is sweeter.

'This is a proverb (known to) many generations : a single log does not flame. But if there were two or three, their firebrands would blaze.

'A single log is not easily burnt unless you get another to kindle it. One man alone is treacherously dealt with. A single millstone is ineffective.

'Have you not heard at every time that one man alone is treacherously dealt with ? I speak truth. But what cannot be endured is the harrying of a great army.

'However few the band, care is spent on them. The provision for an army is not cooked on a single fork—that is a similitude for it.

'I am alone before the host at the ford by the end of Tír Mór. I was outnumbered when attacked by Lóch together with Bodb, (according to) the prophecies of Táin Bó Regomna.

' Lóch has mangled my hips ; the shaggy, russet she-wolf has bitten me. Lóch has wounded my liver ; the eel has overthrown me.

' With my spearlet ⁷I warded off the she-wolf and destroyed her eye. I broke her legs at the beginning of this mortal combat.

' Láeg sent Aífe's spear downstream, a swift(?) cast. I threw the strong, sharp spear by which Lóch mac Emonis perished.

' Why do not the Ulstermen give battle to Ailill and the daughter of Eochu? While I am here in sorrow, wounded and bloody as I am.

' Tell the great Ulstermen to come and guard their drove. The sons of Mága have carried off their cows and divided them out amongst them.

' I pledge a pledge which holds, and has been fulfilled. I pledge by the honour of the Hound, that not one shall come to me as I stand alone.

' But vultures are joyful in the camp of Ailill and Medb. Sad are the cries . . . at their shout on Mag Muirthemne.

' Conchobor comes not forth until his numbers be sufficient. While thus he is not joyful, it is harder to reckon his anger'.

That is the Fight of Lóch Mór mac Mo Femis with Cú Chulainn on the Foray of Cúailnge.

Then Medb sent forth six together to attack Cú Chulainn, to wit, Traig and Dorn and Dernu, Col and Accuis and Eraíse, three druids and three druidesses. Cú Chulainn attacked them and they fell by him. Since the terms of fair play and single combat had been broken against Cú Chulainn, he took his sling and began to shoot at the host that day northwards from Delga. Though the men of Ireland were numerous that day, not one of them could turn southwards, neither hound nor horse nor man.

Then came the Mórrígu, daughter of Ernmas, from the elf-mounds in the guise of an old woman and in Cú Chulainn's presence she

milked a cow with three teats. The reason she came thus was
to be succoured by Cú Chulainn, for no one whom Cú Chulainn had
wounded ever recovered until he himself had aided in his cure.
Maddened by thirst, Cú Chulainn asked her for milk. She gave
him the milk of one teat. 'May this be swiftly wholeness for
me.' The one eye of the queen [which had been wounded] was
cured. Cú Chulainn asked her for the milk of (another) teat.
She gave it to him. 'Swiftly may she be cured who gave it.'
He asked for the third drink and she gave him the milk of the
(third) teat. 'The blessing of gods and non-gods be on you,
woman.'—The magicians were their gods and the husbandmen
were their non-gods.—And the queen was made whole.

Then Medb sent a hundred men together to assail Cú Chulainn.
Cú Chulainn attacked them all and they fell by his hand. 'It is
a hateful thing for us that our people should be slaughtered thus,'
said Medb. 'That was not the first hateful thing that came to
us from that man.' said Ailill. Hence Cuillend Cind Dúne is still
the name of the place where they were then and Áth Cró is the name
of the ford by which they were, and rightly so because of the great
amount of their blood and gore which flowed with the current
of the river.

Breslech Maige Muirthemne

The four provinces of Ireland pitched their camp at the place
called Breslech Mór in Mag Muirtheimne. They sent their share
of the cattle and booty on ahead southwards to Clithar Bó Ulad.
Cú Chulainn took his post at Ferta i lLergaib close beside them,
and his charioteer, Lóeg mac Riangabra, kindled a fire for him
on the evening of that night. Cú Chulainn saw far off, over the
heads of the four great provinces of Ireland, the fiery glitter of the
bright gold weapons at the setting of the sun in the clouds of
evening. Anger and rage filled him when he saw the host, because
of the multitude of his foes and the great number of his enemies.
He seized his two spears and his shield and his sword. He shook
his shield and brandished his spears and waved his sword, and
he uttered a hero's shout from his throat. And the goblins and
sprites and spectres of the glen and demons of the air gave answer
for terror of the shout that [1]he had uttered[1], and Nemain, the war

---

[1-1] following St.

goddess, brought confusion on the host. The four provinces of Ireland made a clangour of arms around the points of their own spears and weapons, and a hundred warriors of them fell dead that night of terror and fright in the middle of the encampment.

As Lóeg was there, he saw something : a single man coming straight towards him from the north-east across the camp of the four great provinces. 'A single man approaches now, little Cú,' said Lóeg. 'What manner of man is there ? ' said Cú Chulainn. ' An easy question : a man fair and tall, with his hair cut broad, curly, yellow hair. He has a green mantle wrapped about him with a brooch of white silver in the mantle above his breast. He wears a tunic of royal satin with red insertion of red gold next to his white skin and reaching to his knees. He carries a black shield with a hard boss of white bronze. In his hand a five-pointed spear and beside it a forked javelin. Wonderful is the play and sport and diversion he makes (with these weapons). But none accosts him and he accosts none, as if no one in the camp of the four great provinces of Ireland saw him.' 'That is true, my fosterling,' said he. 'That is one of my friends from the fairy mounds coming to commiserate with me for they know of my sore distress as I stand alone now against the four great provinces of Ireland on the Foray of Cúailnge.' It was indeed as Cú Chulainn said. When the warrior came to where Cú Chulainn was, he spoke to him and commiserated with him. ' Sleep now for a little while, Cú Chulainn,' said the warrior, ' your heavy slumber at the Ferta in Lerga till the end of three days and three nights, and for that space of time I shall fight against the hosts.'

Then Cú Chulainn slept his deep slumber at the Ferta in Lerga till the end of three days and three nights. It was right that the length of the sleep should correspond to the greatness of his weariness, for from the Monday before Samain exactly until the Wednesday after the festival of spring Cú Chulainn had not slept in that time, except when he dozed for a little while (leaning) against his spear after midday, with his head on his clenched fist and his clenched fist about his spear and his spear resting on his knee, but (he was) striking and cutting down and slaying and killing the four great provinces of Ireland during that time. Then the warrior put plants from the *síd* and healing herbs and a curing charm into the wounds and cuts and gashes and many injuries of Cú Chulainn so that Cú Chulainn recovered in his sleep without his perceiving it at all.

It was at this time that the youths came southwards from Emain Macha, thrice fifty of the kings' sons of Ulster together with Follomain mac Conchobuir, and they gave battle thrice to the hosts and three times their own number fell by them, but the youths fell too, all except Follomain mac Conchobuir. Follomain vowed that he would never go back to Emain until he should take with him Ailill's head and the golden diadem that was on it. That was no easy thing for him for the two sons of Beithe mac Báin, the two sons of Ailill's fostermother and fosterfather, came up with him and wounded him so that he fell by them. That is the Death of the Youths from Ulster and of Follomain mac Conchobuir.

Cú Chulainn however was in his deep sleep at Ferta in Lerga till the end of three days and three nights. He arose then from his sleep and passed his hand over his face and he blushed crimson from head to foot, and his spirit was strengthened as if he were going to an assembly or a march or a tryst or a feast or to one of the chief assemblies of Ireland. ' How long have I been now in this sleep, warrior ? ' said Cú Chulainn. ' Three days and three nights,' said the warrior. ' Woe is me ! ' said Cú Chulainn. ' Why is that ? ' said the warrior. ' Because the hosts have been left without attack for that space of time,' said Cú Chulainn. ' They have not (so been left) indeed,' said the warrior. ' Tell me, who has attacked them ? ' said Cú Chulainn. ' The youths came from the north, from Emain Macha, thrice fifty of the kings' sons of Ulster led by Follomain mac Conchobuir, and thrice they gave battle to the hosts in the space of the three days and three nights when you were asleep, and three times their own number fell by them and all the youths fell too except for Follomain mac Conchobuir. Follomain vowed etc.' ' Alas that I was not in my full strength, for had I been, the youths would not have fallen as they did nor would Follomain have fallen.'

' Strive on, little Hound, it is no reproach to your honour and no disgrace to your valour.' ' Stay here for us tonight, O warrior,' said Cú Chulainn, ' that we may together avenge the youths on the hosts.' ' I shall not stay indeed,' said the warrior, ' for though a man do many valourous and heroic deeds in your company, not he but you will have the fame or the reputation of them. There-fore I shall not stay, but exert your valour, yourself alone, on the hosts for not with them lies any power over your life at this time.'

' The scythed chariot, my friend Lóeg,' said Cú Chulainn, ' can you yoke it ? If you can yoke it and have its equipment, then yoke it, but if you have not its equipment, do not yoke it.'

Then the charioteer arose and put on his hero's outfit for chariot-driving. Of the outfit for chariot-driving which he put on was his smooth tunic of skins, which was light and airy, supple and of fine texture, stitched and of deerskin, which did not hinder the movement of his arms outside. Over that he put on his outer mantle black as raven's feathers.—Simon Magus had made it for the King of the Romans, and Darius gave it to Conchobor and Conchobor gave it to Cú Chulainn who gave it to his charioteer. The same charioteer now put on his helmet, crested, flat-surfaced, four-cornered, with variety of every colour and form, and reaching past the middle of his shoulders. This was an adornment to him and was not an encumbrance. His hand brought to his brow the circlet of red-yellow like a red-gold plate of refined gold smelted over the edge of an anvil, as a sign of his charioteering, to distinguish him from his master. In his right hand he took the long spancel of his horses and his ornamented goad. In his left he grasped the thongs to check his horses, that is, the reins of his horses, to control his driving.

Then he put on his horses the iron inlaid breastplates which covered them from forehead to forehand, (set) with little spears and sharp points and lances and hard points, so that every wheel of the chariot was closely studded with points and every corner and edge, every end and front of that chariot lacerated in its passage. Then he cast a spell of protection over his horses and over his companion so that they were not visible to anyone in the camp, yet everyone in the camp was visible to them. It was right that he should cast this spell, for on that day the charioteer had three great gifts of charioteering, to wit, *léim dar boilg, foscul ndíriuch* and *immorchor ndelind*.

Then the champion and warrior, the marshalled fence of battle of all the men of earth who was Cú Chulainn, put on his battle-array of fighting and contest and strife. Of that battle-array of fighting and contest and strife which he put on were the twenty-seven tunics worn next to his skin, waxed, board-like, compact, which were bound with strings and ropes and thongs close to his fair skin, that his mind and understanding might not be deranged when his rage should come upon him. Over that outside he put his hero's battle-girdle of hard leather, tough and tanned, made from the best part of seven ox-hides of yearlings, which covered him from the thin part of his side to the thick part of his arm-pit ; he used to wear it to repel spears and points and darts and lances and arrows, for they glanced from it as if they had struck against stone or rock or horn. Then he put on his apron of filmy silk with its border of variegated white gold, against the soft lower part of his body. Outside his apron of filmy silk he put on his dark apron of pliable brown leather made from the choicest part of four yearling ox-hides with his battle-girdle of cows' skins about it. Then the royal hero took up his weapons of battle and contest and strife. Of these weapons of battle were these : he took his ivory-hilted, bright-faced sword with his eight little swords ; he took his five-pronged spear with his eight little spears; he took his javelin with his eight little javelins ; he took his *deil chliss* with his eight little darts. He took his eight shields with his curved, dark-red shield into the boss of which a show-boar could fit, with its very sharp, razor-like, keen rim all around it which would cut a hair against the stream, so sharp and razor-like and keen it was. When the warrior did the ' edge-feat ' with it, he would cut alike with his shield or his spear or his sword. Then he put on his head his crested war-helmet of battle and strife and conflict, from which was uttered the shout of a hundred warriors with a long-drawn-out cry from every corner and angle of it. For there used to cry from it alike goblins and sprites, spirits of the glen and demons of the air, before him and above him and around him, wherever he went, prophesying the shedding of the blood of warriors and champions. There was cast over him his protective dress of raiment from Tír Tairngire brought to him from Manannán mac Lir, from the King of Tír na Sorcha.

Then his first distortion came upon Cú Chulainn so that he became horrible, many-shaped, strange and unrecognisable. His haunches shook about him like a tree in a current or a bulrush against a stream, every limb and every joint, every end and every member of him from head to foot. He performed a wild feat of contortion with his body inside his skin. His feet and his shins and his knees came to the back ; his heels and his calves and his hams came to the front. The sinews of his calves came on the front of his shins and each huge, round knot of them was as big as a warrior's fist. The sinews of his head were stretched to the nape of his neck and every huge, immeasurable, vast, incalculable round ball of them was as big as the head of a month-old child.

Then his face became a red hollow (?). He sucked one of his eyes into his head so that a wild crane could hardly have reached it to pluck it out from the back of his skull on to the middle of his cheek. The other eye sprang out on to his cheek. His mouth was twisted back fearsomely. He drew the cheek back from the jawbone until his inner gullet was seen. His lungs and his liver fluttered in his mouth and his throat. He struck a lion's blow with the upper palate [1]on its fellow[1] so that every stream of fiery flakes which came into his mouth from his throat was as large as the skin of a three-year-old sheep. The loud beating of his heart against his ribs was heard like the baying of a bloodhound . . . or like a lion attacking bears. The torches of the war-goddess, the virulent rain-clouds, the sparks of blazing fire were seen in the clouds and in the air above his head with the seething of fierce rage that rose above him. His hair curled about his head like branches of red hawthorn used to re-fence the gap in a hedge. Though a noble apple-tree weighed down with fruit had been shaken about his hair, scarcely one apple would have reached the ground through it but an apple would have stayed impaled on each single hair because of the fierce bristling of his hair above him. The hero's light rose from his forehead so that it was as long and as thick as a hero's whetstone[1]. As high, as thick, as strong, as powerful and as long as the mast of a great ship was the straight stream of dark blood which rose up from the very top of his head and became a dark magical mist like the smoke of a palace when a king comes to be attended to in the evening of a wintry day.

After Cú Chulainn had been thus distorted, the hero sprang into his scythed chariot with its iron points, its thin sharp edges, its hooks, its steel points, with its sharp spikes of a hero, its arrangement for opening, with its nails that were on the shafts and thongs and loops and fastenings in that chariot.

Then he performs the thunder-feat of a hundred and the thunder-feat of two hundred and the thunder-feat of three hundred and the thunder-feat of four hundred, and he stopped at the thunder-feat of five hundred for he thought that at least that number should fall by him in his first attack and in his first contest of battle against the four provinces of Ireland. And he came forth in this manner to attack his enemies, and took his chariot in a wide circuit outside the four great provinces of Ireland. And he drove the chariot heavily. The iron wheels of the chariot sank deep into the ground so that the manner in which they sank into the the ground (left furrows) sufficient to provide fort and fortress, for there arose on the outside as high as the iron wheels dikes and boulders and rocks and flagstones and gravel from the ground.

[1-1] translating St.
[1] fist St.

The reason why he made this warlike encircling of the four great provinces of Ireland was that they might not flee from him and that they might not disperse around him until he took revenge on them by thus pressing them (?) for the wrong done to the youths (of Ulster). And he came across into the middle of the ranks and threw up great ramparts of his enemies' corpses outside around the host. And he made the attack of a foe upon foes among them so that they fell, sole of foot to sole of foot, and headless neck to headless neck, such was the density of their corpses. Thrice again he went around them in this way so that he left a layer of six around them, that is the soles of three men to the necks of three men, all around the encampment. So that the name of this t?le in the Táin is Sesrech Breslige, and it is one of the three (slaughters) which cannot be numbered in the Foray, (the three being) Sesrech Breslige and Imslige Glennamnach and the battle at Gáirech and Irgáirech, except that on this occasion hound and horse and man suffered alike. Others say that Lug mac Eithlend fought along with Cú Chulainn at Sesrech Breslige.

Their number is not known nor is it possible to count how many fell there of the common soldiery, but their chiefs alone have been counted. Here follow their names : Two men called Cruaid, two called Calad, two called Cír, two called Cíar, two called Éicell, three called Cromm, three called Cur, three called Combirge, four called Feochar, four called Furachar, four called Cas ; four called Fota, five called Caur, five called Cerman, five called Cobthach, six called Saxan, six called Dauith, six called Dáire, seven called Rochaid, seven called Rónán, seven called Rurthech, eight called Rochlad, eight called Rochtad, eight called Rinnach, eight called Mulach, nine called Daigith, nine called Dáire, nine called Damach, ten called Fiac, ten called Fiacha, ten called Feidlimid. Ten and six score kings did Cú Chulainn slay in the Breslech Mór in Mag Muirtheimne, and a countless number besides of hounds and horses and women and boys and children and the common folk. For not one man in three of the men of Ireland escaped without his thigh-bone or the side of his head or one eye being broken or without being marked for life.

Cú Chulainn came on the morrow to survey the host and to display his gentle, beautiful appearance to women and girls and maidens, to poets and men of art, for he held not as honour or dignity the dark form of wizardy in which he had appeared to them the previous night. Therefore he came on that day to display his gentle, beautiful appearance.

Beautiful indeed was the youth who came thus to display his form to the hosts, Cú Chulainn mac Sualtaim. Three kinds of hair he had, dark next to the skin, blood-red in the middle and (hair like) a crown of red-gold covering them. Fair was the arrangement of that hair with three coils in the hollow at the back of his head, and like gold thread was every fine hair, loose-flowing, golden and excellent, long-tressed, distinguished and of beautiful colour, as it fell back over his shoulders. A hundred bright crimson twists of red-gold red-flaming about his neck. A hundred strings with mixed carbuncles around his head. Four dimples in each of his two cheeks, a yellow dimple and a green, a blue dimple and a purple. Seven gems of brilliance of an eye in each of his royal eyes. Seven toes on each of his feet, seven fingers on each of his hands, with the grasp of a hawk's claws and the grip of a hedgehog's claws in every separate one of them.

Then he puts on his dress for assembly that day. Of that raiment was a fair mantle, well-fitting, purple, fringed, five-folded. A white brooch of white silver inset with inlaid gold over his white breast, as it were a bright lantern that men's eyes could not look at for its brilliance and splendour. A tunic of silk next to his skin, bordered with edges and braidings and fringes of gold and of silver and of white bronze, reaching to the top of his dark apron, dark-red, soldierly, of royal satin. A splendid dark-purple shield he bore with a rim of pure white silver around it. He wore a golden-hilted ornamented sword at his left side. In the chariot beside him was a long grey-edged spear together with a sharp attacking dagger, with splendid thongs and rivets of white bronze. He held nine heads in one hand and ten in the other, and these he brandished at the hosts in token of his valour and prowess. Medb hid her face beneath a shelter of shields lest Cú Chulainn should cast at her on that day.

Then the women begged the men of Ireland to lift them up on platforms of shields above the warriors' shoulders that they might see Cú Chulainn's appearance. For they wondered at the beautiful, gentle appearance they beheld on him that day compared with the dark buffoon-like shape of magic that had been seen on him the night before.

Then Dubthach Dáel Ulad was seized with envy and spite and great jealousy concerning his wife, and he advised the hosts to

betray and abandon Cú Chulainn, that is, to lay an ambush around
him on every side that he might be killed by them.   And he spoke
these words :

> ' If this is the distorted one, there will be corpses of men
> because of him, there will be cries around courts.   Men's
> feet will be . . . ravens shall eat ravens' food.

> ' Stones shall be erected over graves because of him.   There
> will be increase of kingly slaughter.   Unlucky are ye that
> battle with the wild one reached you on the slope.

> ' I see the wild one's form.   Nine heads he carries [1]among
> his cushions[1].   I see the shattered spoils he brings, and
> ten heads as treasured triumph.

> ' I see how your womenfolk raise their heads above the battle.
> I see your great queen who comes not to the fight.

> ' If I were your counsellor, warriors would be (in ambush)
> on all sides that they might shorten his life, if this is the
> distorted one.'

Fergus mac Róig heard this, and it grieved him that Dubthach
should advise the hosts to betray Cú Chulainn.   And he gave
Dubthach a strong and violent kick so that he fell on his face
outside the group.   And Fergus brought up against him all the
wrongs and injustice and treachery and evil deeds that he had
ever at any time done to the men of Ulster.   And he spoke these
words then :

> ' If it is Dubthach Dóeltenga, he draws back in the rear
> of the host.   He has done nothing good since he slaughtered
> the womenfolk.

> ' He performed an infamous and terrible deed of violence—
> the slaying of Fiacha mac Conchobuir.   Nor was fairer
> another deed that was heard of him—the slaying of Cairbre
> mac Fedlimtne.

---

[1]-[1] LU

' It is not for the lordship of Ulster that the son of Lugaid
mac Casruba contends.   This is how he treats men : those
he cannot kill he sets at loggerheads.

' Ulster's exiles do not wish that their beardless boy should
be killed.   If the men of Ulster come to you, they will turn
back your herds.

' All your cattle will be driven afar before the Ulstermen
if they rise (from their sickness).   There will be deeds of
violence—mighty tales— and queens will be tearful.

' ¹(Men's corpses will be trampled underfoot)¹.   Men's feet
will be in ravens' abode (?).   Shields will lie flat on the
slopes.   Furious deeds will increase.

' I see that your womenfolk have raised their heads above
the battle.   I see your great queen—she comes not to the
combat.

' The unvalorous son of Lugaid will not do any brave or
generous deed.   No king will see lances redden if this is
Dubthach Dóeltenga.'

Thus far the Scythed Chariot.

Then a bold warrior of the Ulstermen called Óengus mac Óenláime
Gábe came up with the hosts, and he drove them before him from
Moda Loga, which is now called Lugmud, to Áth Da Fhert on
Slíab Fúait.   Learned men say that if they had come to Óengus
mac Óenláime Gábe in single combat, they would have fallen by
his hand.   However that it is not what they did, but an ambush
was made around him on every side and he fell by them at Áth
Da Fhert on Slíab Fúait.

Here now is the tale Imroll Belaig Eóin

Then came to them Fiacha Fíaldána of the Ulstermen to have
speech with the son of his mother's sister, namely, Mane Andóe of
the Connachtmen, and he came accompanied by Dubthach Dóel

---

¹⁻¹ LU etc.

Ulad. Mane Andóe moreover came accompanied by Dóche mac
Mágach. When Dóche mac Mágach saw Fiacha Fíaldána, he
cast a spear at him straightaway and it went through his own
friend Dubthach Dóel Ulad. Fiacha cast a spear at Dócha mac
Mágach and it went through his own kinsman Maine Andóe of the
Connachtmen. Then said the men of Ireland : ' A badly aimed
cast,' said they, ' was what befell the men, each of them wounding
his own friend and relation.' So that is the miscast at Belach
Eóin. And another name for it is Another Miscast at Belach Eóin.

Here now is the tale Tuige im Thamon

Then the men of Ireland told Tamon the jester to put on Ailill's
garments and his golden crown and to go on the ford in front of
them. So he put on Ailill's garments and his golden crown and
came on the ford in front of them. The men of Ireland began
to scoff and shout and jeer at him. ' It is the covering of a stump
(tamon) for you, Tamon the jester,' said they, ' to put on you
Ailill's garments and his golden crown.' So (that story) is (called)
Tuige im Thamon, the Covering of a Stump. Cú Chulainn saw
Tamon, and it seemed to him, in his ignorance and want of information, that it was Ailill himself who was there, and he cast a stone
at him from his sling and killed him on the ford where he was.
    So that (the place) is Áth Tamuin and (the story is called) Tuige
im Thamon.

The four great provinces of Ireland encamped at the pillar-stone
in Crích Roiss that night. Then Medb asked the men of Ireland
for one of them to fight and do battle with Cú Chulainn on the
morrow. Every man of them kept saying : ' It will not be I who go.'
' It will not be I who leave my place. No captive is owing from
my people.'
    Then Medb asked Fergus to go to fight with and encounter Cú
Chulainn, since she was unable to get the men of Ireland to do so.
' It would not be fitting for me,' said Fergus. ' to encounter a
young and beardless lad, my own fosterling.' However when
Medb begged Fergus so urgently, he was unable not to undertake
the fight. They remained there that night. Fergus rose early
on the morrow and came forward to the ford of combat where
Cú Chulainn was. Cú Chulainn saw him coming towards him.

'With weak security does my master Fergus come to me. He has no sword in the sheath of the great scabbard.' Cú Chulainn spoke truly.—A year before these events Ailill had come upon Fergus together with Medb on the hillside in Crúachu with his sword on the hill beside him, and Ailill had snatched the sword from its sheath and put a wooden sword in its place, and he swore that he would not give him back the sword until he gave it on the day of the great battle.—'I care not at all, my fosterling,' said Fergus, 'for even if there were a sword in it, it would not reach you and would not be wielded against you. But for the sake of the honour and nurture I and the Ulstermen and Conchobor gave you, flee before me to-day in the presence of the men of Ireland.' 'I am loath to do that,' said Cú Chulainn, 'to flee before one man on the Foray of Cúailnge.' 'You need not shrink from doing so,' said Fergus, 'for I shall flee before you when you shall be covered with wounds and blood and pierced with stabs in the battle of the Táin, and when I alone shall flee, then all the men of Ireland will flee.' So eager was Cú Chulainn to do whatever was for Ulster's weal that his chariot was brought to him and he mounted it and fled in rout [1]from the men of Ireland.[1] The men of Ireland saw that. 'He has fled from you ! He has fled from you, Fergus ! ' said all. 'Pursue him, pursue him, Fergus,' said Medb, 'let him not escape from you.' 'Not so indeed,' said Fergus, 'I shall not pursue him any farther, for though ye may belittle that flight I put him to, yet of all who encountered him on the Foray of Cúailnge not one man of the men of Ireland did as much. So I shall not meet that man again until the men of Ireland meet him in turn in single combat.'

That is (called) the Encounter of Fergus.

Here now is (the story) Cinnit Ferchon.

Ferchú Loingsech was of the Connachtmen. He was engaged in fighting and harassing Ailill and Medb. From the day these assumed rule, he came not to their encampment on expedition or hosting, in straits or need or hardship, but spent his time plundering and pillaging their borders and lands behind their backs. At that time he happened to be in the eastern part of Mag nAí. Twelve men was the number of his band. He was told that one man had been holding back and checking the four great provinces of Ireland

---

[1]-[1] from Fergus St; before Fergus LU, *sic leq.*

from the Monday at the beginning of Samain until the beginning
of spring, slaying one man of their number at a ford every day
and a hundred warriors every night.   Ferchú took counsel with
his men.   ' What better plan could we carry out,' said he, ' than
to go and attack yonder man who is checking and holding back
the four great provinces of Ireland and to bring back with us his
head in triumph to Ailill and Medb.   Though we have done many
wrongs and injuries to Ailill and to Medb, we shall obtain peace
thereby if that man fall by us.'   That is the plan they decided on.
And they came forward to the place where Cú Chulainn was, and
when they came, they did not grant him fair play or single combat
but all twelve of them attacked him straightaway.   However
Cú Chulainn fell upon them and forthwith struck off their twelve
heads.   And he planted twelve stones for them in the ground
and put a head of each one of them on its stone and also put Ferchú
Loingsech's head on its stone.   So that the spot where Ferchú
Loingsech left his head is called Cinnit Ferchon that is, Cennáit
Ferchon (the Headplace of Ferchú).

Then the men of Ireland debated as to whom they should send
to fight and do combat with Cú Chulainn at the hour of early
morning on the morrow.   They all agreed that it should be Calatín
Dána with his twenty-seven sons and his grandson Glas mac Delga.
Now there was poison on each man of them and poison on each
weapon that they carried ; none of them ever missed a throw,
and anyone whom one of them wounded, if he died not at once,
would die before the end of nine days.   Great rewards were
promised them for this fight and they undertook to engage in it.
This agreement was made in the presence of Fergus but he was
unable to dispute it ; for they said that they counted it as single
combat (that) Calatín Dána and his twenty-seven sons and his
grandson Glas mac Delga (should all engage in the fight), for they
asserted that his son was (but) one of his limbs and one of his
parts and that the issue of his own body belonged to Calatín Dána.
   Fergus came forward to his tent and followers and heaved a
sigh of weariness.   ' We are sad for the deed to be done to-morrow,'
said Fergus.   ' What deed is that ? ' asked his followers.   ' The
killing of Cú Chulainn,' said he.   ' Alas ! ' said they, ' who kills
him ? '   ' Calatín Dána,' said he, ' with his twenty seven sons and
his grandson Glas mac Delga.   There is poison on every man of
them and poison on each of their weapons, and there is none

whom one of them wounds but dies before the end of nine days if he do not die at once. And there is no man who should go to witness the encounter for me and bring me news if Cú Chulainn should be killed, to whom I would not give my blessing and my gear.' ' I shall go there,' said Fiachu mac Fir Aba. They remained there that night. Early on the morrow Calatín Dána arose with his twenty-seven sons and his grandson Glas mac Delga, and they advanced to where Cú Chulainn was, and Fiachu mac Fir Aba came too. And when Calatín reached the spot where Cú Chulainn was, they cast at him at once their twenty-nine spears nor did a single spear miss its aim and go past Cú Chulainn. Cú Chulainn performed the ' edge-feat ' with his shield and all the spears sank half their length into the shield. Not only was that not a misthrow for them but yet not a spear wounded him or drew blood. Then Cú Chulainn drew his sword from its warlike scabbard to lop off the weapons and so to lessen the weight of his shield. While he was so doing, they went towards him and all together they smote his head with their twenty-nine clenched right fists. They belaboured him and forced his head down so that his face and countenance met the gravel and sand of the ford. Cú Chulainn uttered his hero's cry and the shout of one outnumbered and no Ulsterman alive of those who were awake but heard him. Then Fiachu mac Fir Aba came towards him and saw how matters were, and he was filled with emotion on seeing a man of his own folk in danger. He drew his sword from its warlike scabbard and dealt a blow which lopped off their twenty-nine fists at one stroke and they all fell backwards; so intense was their effort, so tight their grip.

Cú Chulainn raised his head and drew his breath and gave a sigh of weariness, and then he saw the man who had come to his help. ' It is timely aid, my fosterbrother,' said Cú Chulainn. ' Though it be timely aid for you, it will not be so for us, for though you think little of the blow I struck, yet if it be discovered, the three thousand men of the finest of Clann Rudraige that we number in the camp of the men of Ireland will be put to the sword.' ' I swear,' said Cú Chulainn, ' now that I have raised my head and drawn my breath, that unless you yourself make it known, not one of those yonder shall tell of it henceforth.' Then Cú Chulainn fell upon them and began to strike them and to cut them down, and he scattered them around him in small pieces and divided quarters, east and west throughout the ford. One of them, Glas mac Delga, escaped by taking to his heels while Cú Chulainn was beheading the rest, and Cú Chulainn rushed after him, and Glas came round the tent of Ailill and Medb and only managed to say ' fiach, fiach '[1] when Cú Chulainn struck him a blow and cut off his head.

' They made quick work of yon man,' said Medb. ' What debt did he speak of, Fergus ? ' ' I do not know,' said Fergus, ' unless perhaps some one in the camp owed him debts and they were on his mind. However,' said Fergus, ' it is a debt of flesh and blood for him. I swear indeed,' said Fergus, ' that now all his debts have been paid in full to him.'

Thus fell at Cú Chulainn's hands Calatín Dána and his twenty-seven sons and his grandson Glas mac Derga. And there still remains in the bed of the ford the stone around which they fought and struggled and on it the mark of their sword hilts and of their knees and elbows and of the hafts of their spears. And the name of the ford is Fuil Iairn to the west of Áth Fhir Diad. It is called Fuil Iairn because swords were bloodstained there.

Thus far the Encounter with the Sons of Calatín.

The Encounter with Fer Diad

Then the men of Ireland considered what man should be sent to fight with Cú Chulainn in the hour of early morning on the morrow. They all said that it should be Fer Diad mac Damáin meic Dáire, the brave warrior from Fir Domnand. For similar and equal was their (power of) fighting and combat. With the same fostermothers, Scáthach and Úathach and Aífe, had they learnt the arts of valour and arms, and neither of them had any advantage over the other save that Cú Chulainn possessed the feat of the *ga bulga*. However, to counterbalance this Fer Diad had a horn-skin when fighting with a warrior on the ford.

Then messengers and envoys were sent for Fer Diad. Fer Diad refused and denied and again refused those messengers and he did not come with them, for he knew what they wanted of him, which was, to fight with his friend and companion and foster-brother, Cú Chulainn mac Sualtaim, and so he came not with them. Then Medb sent the druids and satirists and harsh bands for Fer Diad that they might make against him three satires to stay him and three lampoons, and that they might raise on his face three

---

[1] i.e. debt, debt, but intended for Fiachu, Fiachu, the name of Cú Chulainn's rescuer.

blisters, shame, blemish and disgrace, so that he might die before
the end of nine days if he did not succumb at once, unless he came
(with the messengers).   For the sake of his honour Fer Diad came
with them, for he deemed it better to fall by shafts of valour and
prowess and bravery than by the shafts of satire and reviling and
reproach.   And when he arrived, he was greeted with honour
and served, and pleasant-tasting, intoxicating liquor was poured
out for him until he was intoxicated and merry.   And great
rewards were promised him for engaging in that fight, namely, a
chariot worth four times seven *cumala*, the equipment of twelve
men in garments of every colour, the equal of his own domains
in the arable land of Mag nAí, freedom from tax and tribute, from
encampment and expedition and exaction for his son and his
grandson and his great-grandson to the end of time, Findabair
as his wedded wife, and in addition the golden brooch in Medb's
mantle.

As Medb made these promises, she spoke the (following) words
and Fer Diad answered her :

' You shall have a reward of many bracelets and your share
of plain and forest together with freedom for your posterity
from to-day for ever, O Fer Diad mac Damáin.   You shall
have beyond all expectation (?). Why should you not
accept what others accept ? '

' I shall not accept it without surety, for no warrior without
skill in casting am I.   It will be an oppressive task for me
to-morrow, great will be the exertion.   A Hound called
also (of) Culann, hard is the task, it is not easy to resist
him.   Great will be the disaster.'

' You shall have warriors as guarantee.   You shall not go
to assemblies.   Into your hand shall be given fine steeds
and their bridles.   O valourous Fer Diad, since you are a
fearless man, you shall be my confidant before all others
and free of all tribute.'

' I shall not go without sureties to engage in the battle of
the ford.   Its memory will live on till doomsday in full
vigour and strength.   I shall not accept (guarantees other
than) sun and moon, sea and land . . . '

' What avails you to delay it ? Bind it, as may please you,
by the right hand of kings and princes who will go surety
for you. . . You shall have all that you ask, for it is certain
that you will kill the man who comes to encounter you.'

' Without six sureties—let it not be less— I shall not accept
(these conditions) before performing my exploits there
where there are hosts. Were I to have my wish, I shall decide,
though I am not equal, to fight with brave Cú Chulainn.'

' Domnall or Cairbre or bright Niamán of plundering, even
the bardic folk, you will have as sureties however. Take
Morand as a security, if you wish for its fulfilment, take
gentle Cairbre Manand and take our two sons.'

' O Medb, great in boastfulness ! the beauty of a bridegroom
does not touch you. You are assuredly the master in
Crúachu of the mounds. Loud your voice, great your
fierce strength. Bring me satin richly variegated, give me
your gold and your silver, for you have offered them to me.'

' Are you not the chief hero to whom I shall give my circular
brooch ? From to-day until Sunday, no longer shall the
respite be. O strong and famous warrior, all the finest
treasures on earth shall thus be given to you, you shall
have them all.'

' Finnabair of the champions, the queen of the west of Inis
Elga, when the hound of the Smith has been slain, you shall
have, O Fer Diad.'

Then Medb took sureties from Fer Diad that he should fight
with six heroes on the morrow, or if he deemed it preferable, fight
with Cú Chulainn alone. And Fer Diad took sureties from her,
as he believed, that she should send those six heroes to fulfil the
conditions that had been promised to him if Cú Chulainn were
to fall at his hands.

Then his horses were harnessed for Fergus and his chariot yoked
and he came forward to where Cú Chulainn was that he might
tell him how matters were. Cú Chulainn made him welcome.
' Welcome is your coming, my master Fergus,' said Cú Chulainn.
' I deem that welcome trustworthy, my fosterling,' said Fergus.

' But the reason I have come is to tell you who comes to meet you and fight with you at the hour of early morning tomorrow.' ' Let us hear it from you then,' said Cú Chulainn. ' Your own friend and companion and fosterbrother, the man who is your equal in feats of arms and prowess and great deeds, Fer Diad mac Damáin meic Dáire, the brave warrior of Fir Domnand.' ' By my conscience,' said Cú Chulainn, ' it is not to encounter him we wish any friend of ours to come.' ' That is why,' said Fergus, ' you should be on your guard against him and prepare for him, for not like the rest who encountered you and fought with you on the Foray of Cúailnge at this time is Fer Diad mac Damáin meic Dáire,' ' I have been here, however,' said Cú Chulainn, ' checking and holding back the four great provinces of Ireland from the Monday at the beginning of Samain until the beginning of spring, and in all that time I have not gone a step in retreat before a single man. Still less shall I retreat, I think, before this man.' And as Fergus spoke thus putting him on his guard, he said these words and Cú Chulainn answered him :

' O Cú Chulainn—clear covenant— I see that it is time for you to rise. Fer Diad mac Damáin of the ruddy countenance comes here to meet you in his wrath.'

' I am here—no easy task—strongly holding back the men of Ireland. I never retreated a step to avoid encounter with a single opponent.'

' Fierce is the man who wreaks his anger with his blood-red sword. Fer Diad of the many followers has a horn-skin against which no fight or combat can prevail.'

' Be silent, argue not this matter, O Fergus of the mighty weapons. Over every land and territory, there is no fight against odds for me.'

' Fierce is the man—scores of deeds of valour—it is not easy to overcome him. There is the strength of a hundred in his body, brave is the hero. The points of weapons pierce him not, the edge of weapons cuts him not.'

' If I and Fer Diad of well-known valour were to meet at a ford, it would not be a fight without fierceness (?) ; our sword-fight would be wrathful.'

' I should prefer above reward, O Cú Chulainn of the red sword, that you should be the one to take the spoils of proud Fer Diad eastwards.'

· I vow and promise, though I am not good in vaunting, that I shall be the one to triumph over the son of Damán mac Dáre.'

' It was I who, in requital for the wrong done me by Ulstermen, collected the forces from the east. With me their heroes and warriors came from their own lands.'

' Were Conchobor not in his debility, the meeting would be hard. Medb of Mag in Scáil has never come on a more uproarious march.'

· A greater deed now awaits your hand—the fight with Fer Diad mac Damáin. Have with you O Cú Chulainn, a weapon harsh and hard and famed in song.'

Fergus came forward to the encampment. Fer Diad went to his tent and his followers and told them how Medb had obtained from him a covenant whereby he would fight and encounter six heroes on the morrow or else fight and encounter Cú Chulainn alone if he should prefer. He told them too that he had obtained from Medb a covenant whereby she should send the same six heroes to fulfil the promises that had been made to him if Cú Chulainn should fall by him.

That night the men in Fer Diad's tent were not cheerful, tranquil, joyful or merry, but they were sad, sorrowful and downhearted. For they knew that when the two heroes, the two battle-breaches of a hundred, encountered each other, one of them would fall or both would fall, and if it were one of them, they believed that it would be their own lord, for no easy matter was it to fight and encounter Cú Chulainn on the Foray of Cúailnge.

Fer Diad slept heavily at the beginning of the night and when the end of the night was come, his sleep departed from him and his drunkenness left him, and anxiety concerning the fight preyed upon him. He ordered his charioteer to harness his horses and to yoke his chariot. The charioteer began to dissuade him. ' It were better for you to stay here than to go there,' said the driver. ' Hold your peace, lad,' said Fer Diad. And as he spoke, he said these words and the servant answered him :

' Let us go to this encounter to contend with this man, until we reach the ford above which the war-goddess will shriek. Let us go to meet Cú Chulainn, to wound him through his slender body, that a spear-point may pierce him so that he may die thereof.'

' It were better for you to stay (here). No smooth speech will ye exchange. There will be one to whom sorrow will come. Your fight will be short. An encounter with a noble of the Ulstermen is one from which harm will come. Long will it be remembered. Woe to him who goes on that course ! '

' Not right is what you say, for diffidence is not the business of a warrior and we must not show timidity. We shall not stay here for you. Be silent, lad. We shall presently be brave. Better is stoutness than cowardice. Let us go to the encounter.'

Fer Diad's horses were harnessed and his chariot was yoked, and he came forward to the ford of combat though as yet day with its full brightness had not come. ' Well, lad,' said Fer Diad, ' spread the coverings and rugs of my chariot beneath me that I may sleep a heavy fit of slumber here, for I did not sleep during the last part of the night with anxiety about the fight.' The servant unharnessed the horses and unyoked the chariot, and Fer Diad slept his heavy fit of slumber on it.

As for Cú Chulainn now, he rose not until day had dawned on him with its full brightness lest the men of Ireland should say that it was fear or cowardice that caused him to do so if he rose (early). But when day came with its full brightness, he bade his charioteer harness his horses and yoke his chariot. ' Good my lad,' said Cú Chulainn, ' harness our horses for us and yoke our chariot, for an early riser is the warrior appointed to meet us, namely, Fer Diad mac Damáin meic Dáire.' ' The horses are harnessed, the chariot is yoked. Mount the chariot then. There is no reproach to your valour.'

Then Cú Chulainn mac Sualtaim mounted his chariot, the blow-dealing, feat-performing, battle-winning, red-sworded hero, and around him shrieked goblins and sprites and fiends of the glen and demons of the air, for the Túatha Dé Danand used to raise a cry about him so that the fear and terror and horror and fright that he inspired might be all the greater in every battle and field of conflict and in every encounter to which he went.

Not long was Fer Diad's charioteer there when he heard something : a noise and a clamour and an uproar, a tumult and thunder, a din and a great sound, namely, the clash of shields, the rattle of spears, the mighty blows of swords, the loud noise of helmet, the clang of breastplate, the friction of weapons, the violence of feats of arms, the straining of ropes, the rumble of wheels and the creaking of the chariot, the hoof-beats of the horses and the deep voice of the hero and warrior as he came to the ford to meet him.

The servant came and laid his hand upon his master. ' Well, Fer Diad,' said the servant, ' arise for they are coming to you at the ford.'　And the servant spoke these words :

' I hear the sound of a chariot with fair yoke of silver ; (I perceive) the form of a man of great size rising above the front of the strong chariot.　Past Bregros and past Braine they advance along the road, past the tree-stump at Baile in Bile, victorious is their triumph.

' A clever Hound drives, a bright chariot-fighter harnesses, a noble hawk lashes his steeds towards the south.　Bloodstained is the Hound.　It is sure that he will come to us. We know—let there not be silence about it—that he comes to give us battle.

' Woe to him who is on the hill awaiting the worthy Hound. Last year I foretold that he would come at some time, the Hound of Emain Macha, the Hound with shape of every colour, the Hound of spoils, the Hound of battle.　I hear him and he hears us.'

' Well, lad,' said Fer Diad, ' why have you praised that man ever since you left your house ?　It is almost a cause of strife that you should have praised him so highly.　But Ailill and Medb have prophesied to me that that man would fall by me, and since it is for reward, he shall be destroyed shortly by me.　And now it is time for help.'　And he spoke these words and the servant answered him :

' It is time now for help. Be silent, do not praise him. It was no deed of friendship, for he is not doom over the brink(?). If you see the hero of Cúailnge with his proud feats, since it is for reward, he shall soon be destroyed.

' If I see the hero of Cúailnge with his proud feats, he does not flee from us but towards us he comes. Though skilful, he is not grudging. He runs and not slowly, like water from a high cliff or like a swift thunderbolt.'

' So much have you praised him that it is almost a cause of a quarrel. Why have you chosen him since you came forth from your house? Now they appear, now they are challenging him. None come to attack him save cowardly churls.'

Not long was Fer Diad's charioteer there when he saw something : a beautiful, five-edged, four-wheeled chariot (approaching) with strength and swiftness and skill, with a green awning, with a framework of narrow compact opening, in which feats were exhibited, a framework tall as a sword-blade, fit for heroic deeds, behind two horses, swift, high-springing, big-eared, beautiful, bounding, with flaring nostrils, with broad chests, with lively heart, high-groined, wide-hoofed, slender-legged, mighty and violent. In one shaft of the chariot was a grey horse, broad-thighed, small stepping, long-maned. In the other shaft a black horse, flowing maned, swift-coursing, broad-backed. Like a hawk to its prey (?) on a day of harsh wind, or like a gust of the stormy spring wind on a March day across a plain, or like a furious stag newly roused by hounds in the first chase—so were the two horses of Cú Chulainn in the chariot, as if they were on a bright, fiery flagstone, so that they shook the earth and made it tremble with the speed of their course.

Cú Chulainn arrived at the ford. Fer Diad remained on the southern side of the ford, Cú Chulainn stayed on the northern side. Fer Diad made Cú Chulainn welcome. ' Welcome is your coming, Cú Chulainn,' said Fer Diad. ' Until now I trusted that welcome,' said Cú Chulainn, ' but today I trust it no more. And Fer Diad,' said Cú Chulainn, ' it were fitter that I should welcome you rather than that you should welcome me, for it is you who have come to the country and province in which I dwell, and it was not right for you to come and fight with me, rather should I have gone to fight with you, for (driven) before you are my womenfolk and youths

and boys, my horses and steeds, my droves and flocks and herds.'
' O Cú Chulainn,' said Fer Diad, ' what caused you to come and
fight with me ? For when we were with Scáthach and Úathach
and Aífe, you were to me a serving-man who used to prepare my
spears and dress my couch.' ' That is true indeed,' said Cú
Chulainn, ' because of my youth and lack of age I used to act thus
for you. But that is not how I am today indeed for there is not
in the world a warrior whom I shall not drive off.'

And then each of them reproached the other bitterly as they
renounced their friendship, and Fer Diad spoke these words and
Cú Chulainn answered him :

> ' What has led you, little Hound, to fight with a strong
> champion ? Your flesh (?) will be blood-red above the
> steam of your horses. Woe to him who comes as you do !
> It will be (as vain as) the kindling of a fire from a single
> stick of firewood. If you reach your home, you will be in
> need of healing.'

> ' I have come, a wild boar of the herd, before warriors, before
> troops, before hundreds, to thrust you beneath the waters
> of the pool. In anger against you and to prove you in a
> many-sided encounter so that harm may come to you as
> you defend your life.'

> ' There is here one who will crush you. It is I who will slay
> you, for it is I who can. The defeat[1] of their hero in
> the presence of the Ulstermen, may it long be remembered,
> may it be to them loss.'

> ' How shall we meet ? Shall we groan oyer corpses ? On
> what pool shall we fight as we meet on the ford ? Shall it
> be with hard swords or with strong spear-points that you
> will be slain before your hosts if the time has come ? '

> ' Before sunset, before night, if you are in straits, you attack.
> When you fight at Bairche, the battle will not be bloodless.
> The Ulstermen are calling you. A cancer (?) has attacked
> them. Evil will be the sight for them. They will be
> utterly defeated.'

---

[1] reading *cronugud*

' You have come to the gap of danger. The end of your life
is at hand. Sharp weapons will be wielded on you, it will
be no gentle purpose. It will be a great champion who
will slay you. We two shall meet. You shall not be the
leader of three men from now until Doomsday.'

' Leave off your warning. You are the most boastful man on
earth. You shall have neither reward nor remission for
you are no hero overtopping others. I it is who know you,
you with the heart of a bird. You are but a nervous lad
without valour or force.'

' When we were with Scáthach, by dint of our usual valour
we would fare forth together and traverse every land.
You were my loved comrade, my kin and kindred. Never
found I one dearer. Sad will be your death.'

' Too much you neglect your honour that we may not do
battle, but before the cock crows, your head will be impaled
on a spit. O Cú Chulainn of Cúailnge, frenzy and madness
have seized you. All evil shall come to you from us for
yours is the guilt.'

'Well, Fer Diad,' said Cú Chulainn, ' it was not right for you to
come and fight with me by reason of the strife and dissension
stirred up by Ailill and Medb, and all who came thus got neither
success or profit but they fell by me, and neither shall you have
success or profit from it and you will fall at my hands.' As he
spoke, he said these words and Fer Diad hearkened to him :

' Do not draw near me, O valiant warrior, Fer Diad son of
Damán. You will fare the worse for it. It will bring sorrow
to many.

' By just truth, come not near me, for I am the one destined
to bring you to your grave. Why was not my prowess[1]
directed solely against you ?

' Let not many feats overcome (?) you, though you the
hornskinned are bloodstained. The maid of whom you
boast will not be yours, O son of Damán.

---

[1] lit. my contest with warriors

' Findabair, the daughter of Medb, though great her beauty. that maid though fair, you shall not wed.

' Findabair, the king's daughter, when the truth of the matter is told, she played many men false, she destroyed such as you.

' Break not unknowing your oath to me.　Break not compact, break not friendship.　Break not word and promise.　Come not towards me, O valiant warrior.

' To fifty warriors the maid was pledged—a wise pledge indeed.　Their death came through me, from me they got only justice dealt by a spear.

' [1]Though fierce and proud was Fer Báeth[1] with his household of goodly warriors, yet I soon quelled his pride and slew him with one cast.

' Bitter was the lessening of Srubdaire's valiant deeds, Srubdaire who was the darling of a hundred women.　Once his renown was great but neither gold nor fine raiment saved him.

' If it were to me that she had been affianced, the woman [2]in whom all the fair province delights[2], I would not wound your breast, in the south or in the north, in the west or in the east.'

' O Fer Diad,' said Cú Chulainn, ' that is why it was not right for you to come and fight with me.　For when we were with Scáthach and Úathach and Aífe, we used to go together into every battle and field of contest, into every fight and combat, into every wood and wasteland, every secret place and hidden spot.'　And as he spoke he said these words :

' We were loving friends.　We were comrades in the wood. We were men who shared a bed.　We would sleep a deep sleep after our weary fights in many strange lands.　To- gether we would ride and range through every wood (when we were) taught by Scáthach.

---

[1]–[1] reading *Ger amnas menmnach Fer Báeth*
[2]–[2] reading with C

'O accomplished Cú Chulainn,' said Fer Diad, 'we have learnt the same art. They have overcome the bonds of friendship. Your wounds have been paid for. Remember not our fosterage together. O Hound, it is of no avail to you.'

'Too long have we been like this now,' said Fer Diad, 'and what weapons shall we use today, Cú Chulainn?' 'Yours is the choice of weapons until night today,' said Cú Chulainn, 'for you were the first to reach the ford.' 'Do you remember at all,' said Fer Diad, 'the choice feats of arms which we practised with Scáthach and Úathach and Aífe?' 'I remember them indeed,' said Cú Chulainn. 'If you do, let us have recourse to them.'

They had recourse to their choicest feats of arms. They put on two shields marked with emblems and took their eight *ocharcles* and their eight javelins and their eight ivory-hilted blades and their eight battle-darts. These would fly from them and to them like bees on a fine day. They cast no weapon which found not its aim. Each of them began to cast these weapons at the other from the twilight of early morning until the middle of the day, and they blunted their many weapons against the curved surfaces and bosses of the shields. Despite the excellence of the casting, the defence was so good that neither of them wounded or drew blood from the other during that time. 'Let us lay aside these weapons now, Cú Chulainn,' said Fer Diad, 'since not by them comes the decision between us.' 'Let us do so indeed if the time has come,' said Cú Chulainn. They ceased then and gave their weapons into the hands of their charioteers.

'What weapons shall we use now, Cú Chulainn?' said Fer Diad. 'Yours is the choice of weapons until night,' said Cú Chulainn, 'since you were the first to reach the ford.' 'Let us take then,' said Fer Diad, 'to our polished, sharpened, hard, smooth spears with their thongs of hard flax.' 'Let us do so indeed,' said Cú Chulainn. Then they took on them two hard, equally strong shields and they had recourse to the polished, sharpened, hard, smooth spears with their thongs of hard flax. Each of them fell to casting the spears at the other from the middle of the day till the evening. Despite the excellence of the defence, so good was their mutual casting that during that time each of them bled and reddened and wounded the other. 'Let us cease from this now, Cú Chulainn,' said Fer Diad. 'Let us do so indeed if the time has come,' said Cú Chulainn. They ceased then and gave their weapons into the hands of their charioteers.

Then each of them went towards the other and put an arm around the other's neck and kissed him thrice. That night their horses were in one paddock and their charioteers at one fire, and their charioteers made litter-beds of fresh rushes for them and on them pillows for wounded men. Then came folk of healing and curing to heal and cure them, and they put herbs and healing plants and a curing charm into their wounds and cuts, their gashes and many stabs. Of every herb and healing plant and curing charm which was applied to the wounds and cuts, the gashes and many stabs of Cú Chulainn, an equal amount was sent westwards by him across the ford to Fer Diad lest the men of Ireland should say, if Fer Diad fell by him, that it was because of the advantage Cú Chulainn had over him in healing. Of every food and every palatable, pleasant, strong drink which was brought from the men of Ireland to Fer Diad, an equal portion was sent northwards from him across the ford to Cú Chulainn, for the purveyors of food to Fer Diad were more numerous than those of Cú Chulainn. All the men of Ireland were purveyors of food to Fer Diad that he might ward off Cú Chulainn from them. The men of Bregia were purveyors to Cú Chulainn. They used to come to him daily, that is, every night.

They remained there that night. They arose early on the morrow and came forward to the ford of combat. 'What weapons shall we use today, Fer Diad ? ' said Cú Chulainn. ' Yours is the choice of weapons until night,' said Fer Diad, ' since I had choice of weapons on the day that is past.' ' Let us then,' said Cú Chulainn, ' take to our great long spears today, for we think that thrusting with the spears today will bring us nearer to a decisive victory than the casting of missiles did yesterday. Let our horses be harnessed for us and our chariots yoked that we may fight from our horses and chariots today.' ' Let us do so indeed,' said Fer Diad. Then they put on two broad, strong shields that day. They had recourse to the great long spears that day. Each of them began to pierce and wound, to overthrow (?) and cast each other down (?) from the twilight of early morning until sunset. If it were usual for birds in flight to pass through men's bodies, they would have gone through their bodies that day and carried lumps of flesh and blood through their wounds and cuts into the clouds and the air outside. And when evening came their horses were weary and their charioteers tired, and the heroes and champions themselves were weary too. ' Let us cease from this now, Fer Diad,' said Cú Chulainn, ' for our horses are weary and our charioteers are tired, and when they are weary, why should we also not be weary ? ' And as he spoke he said these words :

> ' We are not bound to endure the swaying (of the chariots),' said he, ' straining against giants. Let their spancels be put on the horses, for the noise of battle is over.'

' Let us cease indeed if the time for it has come,' said Fer Diad. They ceased. They gave over their weapons into the hands of their charioteers. Each of them came towards the other. Each put an arm around the other's neck and kissed him thrice. That night their horses were in one paddock, their charioteers at one fire. Their charioteers made for them litter-beds of fresh rushes with the pillows of wounded men on them. Physicians and doctors came to examine and watch them and to attend on them that night, for, because of the dreadfulness of their wounds and gashes, of their cuts and many stabs, all they could do for them was to apply spells and incantations and charms to them to staunch the bleeding and haemorrhage and to keep the dressings in place. Of all the spells and incantations and charms which were applied to the wounds and gashes of Cú Chulainn, an equal portion was sent by him westwards across the ford to Fer Diad. Of all the food and palatable, pleasant, strong drink which was brought from the men of Ireland to Fer Diad, an equal amount was sent by him northwards across the ford to Cú Chulainn. For Fer Diad's purveyors of food were more numerous than those of Cú Chulainn as all the men of Ireland were purveyors of food to Fer Diad for warding off Cú Chulainn from them, but only the men of Bregia were purveyors of food to Cú Chulainn. They used to come and converse with him daily, that is, every night.

They remained there that night. They rose early on the morrow and came forward to the ford of combat. Cú Chulainn saw that Fer Diad had an ill and gloomy appearance on that day. ' Your appearance is not good today, Fer Diad,' said Cú Chulainn. ' Your hair has grown dark today and your eye dull, and you are changed from your usual form and figure.' ' Not because I fear or dread you am I thus today however.' said Fer Diad, ' for there is not in Ireland today a warrior I shall not repel.' And Cú Chulainn was lamenting and pitying (him), and he spoke these words and Fer Diad answered :

'O Fer Diad, if this is you, sure I am that you are one utterly doomed, that you should come at a woman's behest to fight with your fosterbrother.'

'O Cú Chulainn—wise fulfilment—O great hero, great warrior ! A man must make this journey to the sod whereon is his grave.'

'Findabair the daughter of Medb, however beautiful her form, was given to you not for love of you but to prove your noble might.'

'My might is long since proven, O Hound of the gentle rule. None braver has been heard of or found until today.'

'You are the cause of all that happens, O son of Damán mac Dáire, that you should come at woman's behest to cross swords with your fosterbrother.'

'Should I part from you without a fight, O gentle Hound, though we are fosterbrothers, my word and my name would be held in ill esteem by Ailill and Medb of Crúachu.'

'He has not yet put food to his lips nor has he yet been born of king or bright queen for whom I would (consent to) do you harm.'

'O Cú Chulainn—many deeds of valour—not you but Medb betrayed us. You will have victory and fame. Not on you is our guilt.'

'My brave heart is a clot of blood. My life has almost left me. No equal fight do I deem it to encounter you, Fer Diad.'

'However much you belittle me today,' said Fer Diad, 'what weapons shall we use ? ' 'You have the choice of weapons until night today,' said Cú Chulainn, 'for it was I who chose them yesterday.' 'Let us then,' said Fer Diad, 'take our heavy, hard-smiting swords today, for we think that the mutual striking with swords today will bring us nearer to a decisive victory than did the thrusting with spears yesterday.' 'Let us do so indeed,' said

Cú Chulainn. Then they took up two great, long shields that day. They wielded their heavy, hard-smiting swords. Each of them began to smite and hew, to slaughter and slay each other, and every portion and piece that each hacked from the shoulders and thighs and shoulder-blades of the other was as big as the head of a month-old child. Each of them kept on smiting the other in this way from the twilight of early morning until evening. ' Let us cease from this now, Cú Chulainn,' said Fer Diad. ' Let us cease indeed if the time for it has come,' said Cú Chulainn. So they ceased and gave over their weapons into the hands of their charioteers. Though two cheerful, tranquil, happy and joyful men had met there, their parting that night was the parting of two sad, unhappy, dispirited ones. That night their horses were not in the same paddock nor their charioteers at the same fire.

They remained there that night. Then Fer Diad rose early on the morrow and came alone to the ford of combat, for he knew that this was the decisive day of the fight, and he knew too that one of them would fall in the fight that day or that both would fall. Then before Cú Chulainn came to meet him, he put on his battle equipment. Of that battle equipment was his filmy satin apron with its border of variegated gold which he wore next to his fair skin. Outside that he put on his apron of supple brown leather, and outside that a great stone as big as a millstone, and outside that stone, through fear and dread of the *ga bulga* that day, he put his strong, deep, iron apron made of smelted iron. On his head he put his crested helmet of battle which was adorned with forty carbuncle-gems, studded with red enamel and crystal and carbuncle and brilliant stones from the eastern world. In his right hand he took his fierce, strong spear. He set at his left side his curved battle-sword with its golden hilt and guards of red gold. On the arching slope of his back he put his huge, enormous fair shield with its fifty bosses into each boss of which a show boar could fit, not to speak of the great central boss of red gold. That day Fer Diad exhibited many and wonderful and brilliant feats of arms which he had not learned from anyone before that, neither from fostermother nor fosterfather, not from Scáthach nor Úathach nor Aífe, but he invented them himself on that day to oppose Cú Chulainn.

Cú Chulainn too came to the ford and he saw the many brilliant, wonderful feats of arms performed by Fer Diad. ' You see yonder, my friend Láeg, the many brilliant, wonderful feats performed by Fer Diad, and in due course now all those feats will be directed

against me. Therefore if it be I who am defeated this day, you
must incite me and revile me and speak evil of me so that my ire
and anger shall rise the higher thereby. But if it be I who inflict
defeat, you must exhort me and praise me and speak well of me
that thereby my courage rise higher.' ' It shall so be done indeed,
little Cú,' said Láeg.

Then Cú Chulainn too put on his battle-equipment and performed
that day many brilliant, wonderful feats which he had not learned
from any other, not from Scáthach nor from Úathach nor from Aífe.

Fer Diad saw these feats and knew that they would in due course
be directed against him. ' What feat of arms shall we perform
today, Fer Diad ? ' said Cú Chulainn. ' Yours is the choice until
nightfall,' said Fer Diad. ' Let us perform the " feat of the ford "
then,' said Cú Chulainn. ' Let us do so indeed,' said Fer Diad.
But though he said that, it was the feat he deemed it hardest to
encounter for he knew that it was at the ' feat of the ford ' that
Cú Chulainn overthrew every champion and every warrior he
encountered. Great was the deed that was done on the ford
that day, the two heroes, the two champions and the two chariot-
fighters of western Europe, the two bright torches of valour of
the Irish, the two bestowers of gifts and rewards and wages
in the northwestern world, the two mainstays of the valour
of the Irish coming from afar to encounter each other through
the sowing of dissension and the stirring up of strife by Ailill and
Medb. Each of them began to cast these weapons at each other
from the twilight of early morning until midday, and when midday
came, the rage of the combatants grew fiercer and they drew
closer to each other.

Then for the first time Cú Chulainn sprang from the brink of
the ford on to the boss of Fer Diad's shield, trying to strike his
head from above the rim of the shield. Then Fer Diad gave the
shield a blow with his left elbow and cast Cú Chulainn off like a
bird on to the brink of the ford. Again Cú Chulainn sprang from
the brink of the ford on to the boss of Fer Diad's shield, seeking
to strike his head from above the rim of the shield. Fer Diad
gave the shield a blow with his left knee and cast Cú Chulainn
off like a child on to the brink of the ford. Láeg noticed what
was happening. ' Alas ! ' said Láeg, ' your opponent has chastised
you as a fond mother chastises her child. He has belaboured you
as flax (?) is beaten in a pond. He has ground you as a mill grinds
malt. He has pierced you as a tool pierces an oak. He has bound
you as a twining plant binds trees. He has attacked you as a

hawk attacks little birds, so that never again will you have a claim or right or title to valour or feats of arms, you distorted little sprite,' said Láeg.

Then for the third time Cú Chulainn rose up as swift as the wind, as speedy as the swallow, as fierce as the dragon, as strong as the air, and landed on the boss of Fer Diad's shield, seeking to strike his head from above the rim of the shield. Then the warrior shook the shield and cast off Cú Chulainn into the bed of the ford as if he had never leapt at all (?).

Then occurred Cú Chulainn's first distortion. He swelled and grew big as a bladder does when inflated and became a fearsome, terrible, many-coloured, strange arch, and the valiant hero towered high above Fer Diad, as big as a *fomóir* or a pirate.

Such was the closeness of their encounter that their heads met above, their feet below and their hands in the middle over the rims and bosses of the shields. Such was the closeness of their encounter that they clove and split their shields from rims to centres. Such was the closeness of their encounter that they caused their spears to bend and turn and yield to pressure from points to rivets. Such was the closeness of their encounter that sprites and goblins and spirits of the glen and demons of the air screamed from the rims of their shields and from the hilts of their swords and from the butt-ends of their spears. Such was the closeness of their encounter that they forced the river from its usual course and extent, and a couch might have been prepared for king or queen on the floor of the ford for not a drop of water remained there except what might drip there with the wrestling and trampling of the two heroes and champions on the floor of the ford. Such was the closeness of their encounter that the horses of the Irish went mad and· frenzied and broke their spancels and shackles, their ropes and traces, and women and boys and children and those unfit to fight and the mad among the men of Ireland broke out through the camp south-westwards.

By this time the two combatants were at the edge-feat of swords. Then Fer Diad caught Cú Chulainn unguarded and dealt him a blow with his ivory-hilted blade which he plunged into Cú Chulainn's breast. And Cú Chulainn's blood dripped into his belt and the ford was red with the blood from the warrior's body. Cú Chulainn brooked not this wounding for Fer Diad attacked him with a succession of deadly stout blows, and he asked Láeg for the *ga bulga*.—Such was the nature of the *ga bulga* : it used to be set downstream and cast from between the toes : it made one wound as it entered a man's body but it had thirty barbs when one tried to remove it and it was not taken from a man's body until the flesh was cut away about it.

And when Fer Diad heard the mention of the *ga bulga*, he thrust down the shield to shelter the lower part of his body. Cú Chulainn cast the fine spear from off the palm of his hand over the rim of the shield and over the breast-piece of the horn-skin so that its farther half was visible after it had pierced Fer Diad's heart in his breast. Fer Diad thrust up the shield to protect the upper part of his body but that was help that came too late. The charioteer sent the *ga bulga* downstream. Cú Chulainn caught it between his toes and made a cast of it at Fer Diad. And the *ga bulga* went through the strong, thick apron of smelted iron and broke in three the great stone as big as a millstone and entered Fer Diad's body through the anus and filled every joint and limb of him with its barbs. 'That suffices now,' said Fer Diad. 'I have fallen by that (cast). But indeed strongly do you cast from your right foot. And it was not fitting that I should fall by you.' As he spoke, he uttered these words :

'O Hound of the fair feats, it was not fitting that you should slay me. Yours is the guilt which clung to me. On you my blood was shed.

'Doomed men who reach the gap of betrayal do not flourish. Sad is my voice. Alas ! heroes (?) have been destroyed.

'My ribs like spoils are broken. My heart is gore. Would that I had not fought ! I have fallen, O Hound.'

Then Cú Chulainn hastened towards him and clasped him in his arms and lifted him up with his weapons and armour and equipment and took him northwards across the ford so that his spoils might be to the north of the ford and not to the west with the men of Ireland. Cú Chulainn laid Fer Diad on the ground there and as he stood over Fer Diad a swoon and faintness and weakness came upon him. Láeg saw that and he feared that all the men of Ireland would come and attack Cú Chulainn. 'Come, little Hound,' said Láeg, ' arise now for the men of Ireland will come to attack us and it will not be single combat that they will grant us since Fer Diad mac Damáin meic Dáire has fallen at your hands.' ' What avails it me to arise now, fellow,' said Cú Chulainn,

' considering the man who has fallen by me.' As the servant spoke, he said these words and Cú Chulainn answered him :

> ' Arise, O war-hound of Emain.   High courage befits you more
> ' than ever.   You have cast off Fer Diad of the hosts,   God's
> doom !   Your fight was hard.'

> ' What avails me high courage ?   Madness and grief have hemmed me in, after the deed I have done and the body that I have wounded harshly with my sword.'

> ' It was not fitting for you to mourn him.   Fitter for you to boast in triumph.   The strong man armed with spears has left you mournful, wounded, bleeding.'

> ' Even had he cut off a leg from me or a hand, I still grieve that Fer Diad who rode on steeds is not living for ever.'

> ' The maidens of the Red Branch are better pleased at what has been done, that he should die and you should live, though they do not deem it a small thing that you two should be parted for ever.

> ' Since the day you left Cúailnge in pursuit of the brilliant Medb. all that you have killed of her fighters she deems indeed a famous carnage.

> ' You have not slept peacefully in pursuit of your great herd. Though your company was few, yet many a morning you rose early.'

Cú Chulainn began to lament for and commiserate with Fer Diad then and he spoke these words :
' Alas, Fer Diad, sad for you that you spoke not with one of the company who knew of my great deeds of valour and arms before we met together in conflict !
' Sad for you that Láeg mac Riangbra did not put you to shame with counsel about our comradeship !
' Sad for you that you did not agree to the clear advice of Fergus !
' Sad for you that Conall the fair. triumphant. exultant, victorious Conall, did not help you !

'For those men do not follow the messages or desires or sayings of the false promises of the fairhaired women of Connacht. For those men know that there will not be born among the Connachtmen a being to perform deeds equal to yours, in the wielding of shields and bucklers, of spears and swords, in the playing of chess and draughts, in the driving of horses and chariots.

'There will not be a hero's hand to hack warrior's flesh like that of Fer Diad, the shapely scion. The breach made by the red-mouthed war-goddess will not be dug up (?) for encampments full of shimmering shields. It will not be Crúachain that will contend for or obtain covenants equal to yours till the very end of life now, O red-cheeked son of Damán ! ' said Cú Chulainn.

Then Cú Chulainn rose and stood over Fer Diad. ' Ah Fer Diad,' said Cú Chulainn, ' greatly did the men of Ireland betray and abandon you when they brought you to fight and do combat with me, for to contend and do battle with me on the Foray of Cúailnge was no easy task.'

As he spoke, he said these words :

' O Fer Diad, you have been betrayed. Alas for your last meeting where you have died while I remain ! Alas for ever for our long parting !

' When we were yonder with Scáthach the victorious, we thought that till great doomsday our friendship would not end.

' Dear to me was your splendid blush. dear your perfect and fair form, dear your bright clear eye. dear your bearing and your speech.

' There never strode to flesh-rending fight, there never grew wrathful in his manliness, there never held shield upon the wide slope, one like unto you. warlike son of Damán.

' I have never met such as you until now, since the only son of Aífe fell ; your peer in deeds of battle I found not here. O Fer Diad.

' Findabair, the daughter of Medb, though great her beauty. it is as vain to show her now to you, O Fer Diad, as to bind a withe around sand or gravel.'

Then Cú Chulainn began to gaze at Fer Diad. ' Well now, my friend Láeg,' said Cú Chulainn, ' strip Fer Diad and take off his armour and his clothes that I may see the brooch for the sake of which he did battle.' Láeg came and stripped Fer Diad. He took his armour and clothing from him and (Cú Chulainn) saw the brooch and began to mourn for Fer Diad and to commiserate him, and he spoke these words :

> ' [1]Alas for the golden brooch[1], O Fer Diad of the hosts ! O strong and valiant smiter, victorious was your arm.

> ' Your thick yellow hair was curly—a fair jewel. Your girdle, supple and ornamented, was around you until your death.

> ' Our true comradeship was a delight for the eye of a nobleman. Your shield with its golden rim, your chess-board worth much treasure.

> ' That you should fall by my hand I acknowledge was not just. Our fight was not gentle. [1]Alas for the golden brooch !![1] '

' Well, my friend Láeg,' said Cú Chulainn, ' cut open Fer Diad now and remove the *ga bulga* for I cannot be without my weapon.' Láeg came and cut open Fer Diad and removed the *ga bulga*. And Cú Chulainn saw his bloodstained, crimson weapon lying beside Fer Diad and spoke these words :

> ' O Fer Diad, it is sad that I should see you thus, bloodstained yet drained of blood, while I have not as yet cleansed my weapon of its stains and you lie there in a bed of gore.

> ' When we were yonder in the east with Scáthach and with Úathach, there would not be pale lips between us and weapons of battle.

> ' Sharply Scáthach spoke her strong firm command : " Go ye all to the swift battle. Germán Garbglas will come."

---

[1]–[1] translating St

' I said to Fer Diad and to generous Lugaid and to [Fer Báeth] the son of fair Báetán that we should go to meet Germán.

' We went to the rocks of battle above the sloping shore of Loch Lindfhormait. Four hundred we brought out from the Islands of the Victorious.

' When I and valiant Fer Diad stood before the fort of Germán, I killed Rind mac Níuil and he slew Rúad mac Forníuil.

' On the battle-field Fer Báeth killed Bláth son of Colba of the red sword, and Lugaid, the stern and swift, slew Mugairne from the Tyrrhene Sea.

' After going in I slew four hundred wrathful men. Fer Diad slew Dam Dreimed and Dam Dílend—a stern company.

' We laid waste the fort of wise Germán above the wide. many-coloured sea. We brought Germán alive to Scáthach of the broad shield.

' Our fostermother imposed on us a pact of friendship and agreement that we should not grow angry with the tribe of fair Elg.

' Sad was the battle, that slaughtering battle in which the son of Damán was struck down in weakness. Alas ! the friend to whom I served a drink of red blood has fallen.

' Had I seen you die amidst the warriors of great Greece, I should not have survived you, we should have died together.

' Sad what befalls us, the fosterlings of Scáthach. I am wounded and covered with red gore while you no longer drive chariots.

' Sad what befalls us, the fosterlings of Scáthach. I am wounded and covered with red gore while you lie dead.

' Sad what befalls us, the fosterlings of Scáthach, you dead. I alive and strong. Valour is an angry combat.'

' Well, O little Cú,' said Láeg, ' let us leave this ford now. Too long have we been here.' ' We shall leave it indeed, friend Láeg,' said Cú Chulainn. ' But to me every battle and contest I have fought seems but play and sport compared with my fight against Fer Diad.' And as he spoke, he said these words :

' Game was all and sport was all until it came to my meeting with Fer Diad on the ford. The same instruction we had, the same power of guarantee (?). The same tender foster-mother we had whose name is beyond all others.

' All was play and sport compared with my meeting with Fer Diad on the ford. The same nature we had, the same fearsomeness, the same weapons we used to wield. Scáthach once gave two shields to me and to Fer Diad.

' All was play and sport compared with my meeting with Fer Diad on the ford. Beloved was he, the golden pillar, whom I laid low on the ford. O strong one of the tribes, you were more valiant than all others.

' All was play and sport compared with my meeting with Fer Diad on the ford, the furious, fiery lion, the wave, wild and swelling, like the day of doom.

' All was play and sport compared with my meeting with Fer Diad at the ford. I thought that beloved Fer Diad would live after me for ever. Yesterday he was huge as a mountain, today only his shadow remains.

' Three uncountable bands there fell by my hand on the Foray. The finest men, the finest cattle and horses I slaughtered on every side.

' Though numerous the army which came from stout Crúachu, yet I slew more than a third of them and less than half with the rough plying of my weapons.

' There has not come into the centre of battle, nor has Banba ever nurtured, nor has there travelled over land or sea any king's son more famous than Fer Diad.'

Thus far the Tragic Death of Fer Diad.

There came now to help and succour Cú Chulainn a few of the Ulstermen, namely, Senall Uathach and the two Maic Fécce, Muiredach and Cotreb. They took him to the streams and rivers of Conaille Muirthemne to wash and cleanse his wounds and his stabs, his cuts and many sores, against the current of those streams and rivers. For the Túatha Dé Danann used to put herbs and healing plants and charms on the streams and rivers in Conaille Muirthemne to help and succour Cú Chulainn, so that the streams used to be speckled and green-surfaced from them.

These are the names of the rivers which healed Cú Chulainn : Sás, Búan, Bithlán, Findglais, Gleóir, Glenamain, Bedg, Tadg, Telaméit, Rind, Bir, Brenide, Dichaem, Muach, Miliuc, Cumuṅg, Cuilenn, Gainemain, Drong, Delt, Dubglass.

The men of Ireland told Mac Roth, the chief herald, to go to keep watch and ward for them on Slíab Fúait lest the Ulstermen should come upon them unawares. So Mac Roth came to Slíab Fúait. Not long was he there when he saw a single chariot-warrior on Slíab Fúait coming straight towards him from the north. In the chariot was a man, stark-naked, with neither weapon nor garment save only an iron spit in his hand with which he pricked alike his charioteer and his horses, and it seemed to him as if he would never reach the hosts while they were still alive. Mac Roth brought these tidings to the place where were Ailill and Medb and Fergus with the nobles of the men of Ireland. Ailill asked news of him on his arrival. ' Well, Mac Roth,' said Ailill, ' have you seen any one of the Ulstermen on the track of this host today ? ' ' I know not indeed,' said Mac Roth, ' but I saw a solitary chariot-fighter coming straight across Slíab Fúait. In the chariot there is a man, stark-naked, with no garment or weapon at all except for an iron spit in his hand with which he pricks alike both his charioteer and his horses, for it seemed to him that he would not reach this host in time to find them alive.'

' Who would you think was yonder, Fergus ? ' said Ailill. ' I think,' said Fergus, ' that it would be Cethern mac Fintain coming there.' It was true for Fergus that it was Cethern mac Fintain arriving there. Then Cethern mac Fintain reached them, and the fort and encampment was overthrown (?) on them and he wounds them all around him in every direction and on all sides. He too is wounded from all sides and points. Then he came from them, with his entrails and intestines hanging out, to the place

where Cú Chulainn was being cured and healed, and he asked
Cú Chulainn for a physician to cure and heal him. ' Well, my
friend Láeg,' said Cú Chulainn, ' go to the encampment of the
men of Ireland and tell their physicians to come forth and cure
Cethern mac Fintain. I swear that though they be hidden underground or in a locked house I shall inflict death on them before
this hour tomorrow if they do not come.' Láeg came forward to
the encampment of the men of Ireland and bade their physicians
come forth and cure Cethern mac Fintain. The physicians of
the men of Ireland thought it no pleasant task to come and cure
one who was to them a foe and an enemy and an outlander, but
they feared that Cú Chulainn would inflict death on them if they
did not come. So they came. As each man reached him, Cethern
mac Fintain would show him his wounds and his gashes, his sores and
his bleeding cuts. To each man who would say : ' He will not
live. He cannot be cured.' Cethern mac Fintain would deal a
blow with his right fist in the middle of his forehead and drive
his brains out through the orifices of his ears and the joinings of
his skull. However, Cethern mac Fintain slew up to fifteen of the
physicians of the men of Ireland. As for the fifteenth man, only
a glancing blow reached him, but he lay unconscious in a heavy
swoon among the corpses of the other physicians for a long time.
His name was Íthall, the physician of Ailill and Medb.

Then Cethern mac Fintain asked Cú Chulainn for another
physician to heal and cure him. ' Well now, friend Láeg,' said
Cú Chulainn, ' go for me to Fíngin the seer-physician, the physician
of Conchobor at Ferta Fíngin in Lecca Slébe Fúait, and let him come
hither to heal Cethern mac Fintain.' Láeg came on to the seer-
physician Fíngin at Ferta Fíngin in Lecca Slébe Fuait and told
him to come and cure Cethern mac Fintain. So Fíngin the
seer-physician came, and when he had come, Cethern mac Fintain
showed him his wounds and his stabs, his gashes and his bleeding
cuts.

' Examine this wound for me, master Fíngin,' said Cethern.
Fíngin examined the wound. ' This is a slight wound given un-
willingly by one of your own blood,' said the physician, ' and it
would not carry you off prematurely.' ' That is true indeed,'
said Cethern. ' One man came to me there. He had a crest of hair.
He wore a blue cloak wrapped around him. A silver brooch in
the cloak over his breast. He carried a curved shield with scalloped edge ; in his hand a five-pointed spear and beside it a small
pronged spear. He dealt this wound and he got a slight wound

from me too.' 'We know that man.' said Cú Chulainn. 'That was Illand Ilarchless the son of Fergus, and he had no desire that you should fall by his hand but gave that mock-thrust at you lest the men of Ireland should say that he was betraying or abandoning them if he did not give it.'

'Examine this wound also for me, master Fíngin', said Cethern. Fíngin examined the wound. 'This is the deed of a proud woman,' said the physician. 'That is true indeed,' said Cethern. 'There came to me there a woman, tall, beautiful, pale and long-faced. She had flowing, golden-yellow hair. She wore a crimson, hooded cloak with a golden brooch over her breast. A straight, ridged spear blazing in her hand. She gave me that wound and she too got a slight wound from me.' 'We know that woman.' said Cú Chulainn. 'It was Medb the daughter of Eochu Feidlech, the high-king of Ireland, who came in that wise. She would have deemed it victory and triumph and cause for boasting had you fallen at her hands.'

'Examine then this wound for me, master Fíngin,' said Cethern. Fíngin examined the wound. 'This is the attack of two champions,' said the physician. 'It is true indeed,' said Cethern. 'Two men came to me there. They had crests of hair. Two blue cloaks wrapped about them. Silver brooches in the cloaks above their breasts. A necklace of pure white silver round the neck of each of them.' 'We know those two men,' said Cú Chulainn. 'They were Oll and Othine, members of the household of Ailill and Medb. They never go into battle that they are not assured of wounding a man. They would deem it victory and triumph and cause for boasting that you should fall at their hands.'

'Examine this wound for me now, master Fíngin,' said Cethern. Fíngin examined that wound. 'Two warriors came to me there of splendid, manly appearance. Each of them thrust a spear in me and I thrust this spear through one of them.' Fíngin examined that wound. 'This wound is all black,' said the physician. 'The spears went through your heart and crossed each other within it. and I prophesy no cure here, but I would procure for you some herbs of healing and curing so that the wounds should not carry you off prematurely.' 'We know these two,' said Cú Chulainn. 'They were Bun and Mecconn of the household of Ailill and Medb. They desired that you should fall at their hands.'

'Examine this wound for me now, master Fíngin,' said Cethern. Fíngin examined the wound. 'This was the bloody onset of the two sons of the king of Caill.' 'That is true,' said Cethern. 'There came to me two warriors, fair-faced, dark-browed, tall, with golden crowns on their heads. Two green mantles wrapped about them. Two brooches of white silver in the mantles over their breasts. Two five-pronged spears in their hands.' 'Very numerous are the wounds that have inflicted on you,' said the physician. 'Into your gullet the spears went and their points met within you, nor is it easy to work a cure here.' 'We know these two,' said Cú Chulainn. 'They are Bróen and Brudne the sons of three lights, the two sons of the king of Caill. They would think it victory and triumph and cause for boasting if you should fall by them.'

'Examine this wound for me, master Fíngin,' said Cethern. Fíngin examined that wound. 'This was the attack (?) of two brothers,' said the physician. 'That is true indeed,' said Cethern. 'There came to me two choice warriors. They had yellow hair. Dark-grey, fringed cloaks wrapped about them. Leaf-shaped brooches of white bronze in the mantles over their breasts. Broad, shining spears in their hands.' 'We know those two,' said Cú Chulainn. 'They are Cormac Coloma Ríg, and Cormac mac Maele Foga of the household of Ailill and Medb. They would have wished you to fall at their hands.'

'Examine for me this wound, master Fíngin,' said Cethern. Fíngin examined that wound. 'This was the attempt of two brothers,' said the physician. 'It is true indeed,' said Cethern. 'There came to me two youthful warriors, both alike. One had curling brown hair, the other curling yellow hair. Two green mantles were wrapped around them and two brooches of bright silver were in the mantles over their breasts, Two shirts of smooth, yellow silk next to their skin. Bright-hilted swords at their girdles. Two bright shields they carried, ornamented with animal designs in silver. Two five-pronged spears with rings of pure white silver they bore in their hands.' 'We know those two,' said Cú Chulainn. 'They were Maine Máithremail and Maine Aithremail, two sons of Ailill and Medb. They would deem it victory and triumph and cause for boasting if you should fall at their hands.'

'Examine this wound for me, master Fíngin,' said Cethern. 'Two warriors came to me there. A brilliant appearance they had and they were tall and manly. They wore strange, foreign clothes. Each of them thrust a spear into me and I thrust a spear into each of them.' Fingin examined the wound. 'Severe are the wounds they have inflicted on you,' said the physician. 'They have severed the sinews of your heart within you so that your heart rolls about in your breast like an apple in movement (?) or like a ball of thread in an empty bag, and there is not a sinew supporting it at all, and I cannot effect a cure here.' ' We know those two,' said Cú Chulainn. ' They are two of the warriors of Irúath who were chosen expressly by Ailill and Medb that they might kill you, since not often does anyone survive their attack. For they desired that you should fall at their hands.'

' Examine this wound for me, master Fíngin,' said Cethern. Fíngin examined that wound. ' This was the thrust of a father and son,' said the physician. ' It is true indeed,' said Cethern. ' There came to me two tall men, with shining eyes, with golden diadems flashing on their heads. They wore kingly raiment. Gold-hilted, ornamented swords at their girdles with scabbards of pure white silver and rings[1] of variegated gold outside them.' ' We know those two,' said Cú Chulainn. ' They were Ailill and his son Maine Condasgeib Uile. They would deem it victory and triumph and cause of congratulation if you had fallen at their hands.'

Thus far the Wounds of (Cethern on) the Táin.

' Well then, Fíngin, seer-physician,' said Cethern mac Fintain, ' what remedy and advice do you give me now ? ' ' What I say to you,' said Fíngin the seer-physician,' ' is that you should not exchange[1] your great cows for yearlings this year, for if you do, it is not you who will enjoy them and they will not profit you. ' That is the remedy and advice the other physicians gave me, and it is certain that it brought them no advantage or profit but they fell by me, and neither shall it bring advantage or profit to you for you will fall by me.' And Cethern gave him a strong, violent kick so that he landed between the two wheels of the chariot. ' Wicked is that old man's (?) kick,' said Cú Chulainn. Whence the name of Úachtar Lúa in Crích Rois from that day until today.

Nevertheless Fíngin Fáithlíaig gave his choice to Cethern mac Fintain : either a long illness and afterwards help and succour, or else a temporary healing during three days and three nights that he might then exert all his strength against his enemies. Cethern chose a temporary healing of three days and three nights that he might himself exert all his strength against his enemies,

---

[1] translating St

for, as he said, he would leave behind him no one he would better like to take vengeance for him than himself. So then Fíngin Fáithlíaig asked Cú Chulainn for a marrow-mash to cure and heal Cethern mac Fintain. Cú Chulainn proceeded to the encampment of the men of Ireland and brought from there all he found of their herds and flocks and droves, and made of them a mash, flesh and bones and hides all together. And Cethern was placed in the marrow-mash for the space of three days and three nights, and he began to soak up the marrow-mash which was about him. And the marrow entered into his wounds and gashes, his sores and many stabs. Then after three days and three nights he arose from the marrow-mash, and thus it was that he arose : with the board of his chariot pressed to his belly to prevent his entrails from falling out.

That was the time when his wife Finda daughter of Eochu came from the north, from Dún Da Benn, bringing him his sword. Cethern mac Fintain came towards the men of Ireland. However he gave a warning of his coming to Íthall, the physician of Ailill and Medb. Íthall had lain unconscious in a heavy swoon among the corpses of the other physicians for a long space of time. ' O men of Ireland,' said the physician, ' Cethern son of Fintan will come to attack you now that he has been cured and healed by Fíngin Fáithlíaig, so make ready to answer him.' Then the men of Ireland put Ailill's garments and his golden crown on the pillar-stone in Crích Rois that Cethern mac Fintain might first wreak his rage on it when he arrived. Cethern saw Ailill's garments and his golden crown on the pillar-stone, and for want of information he thought that it was Ailill himself who was there. He made a rush at it and drove the sword through the pillar-stone up to its hilt. ' This is a trick,' said Cethern, ' and against me it has been played, and I swear that until there be found among you some one to put on that royal dress and golden crown I see yonder, I shall not cease to smite and slaughter them.' Maine Andóe, the son of Ailill and Medb, heard this, and he put on the royal dress and golden crown and advanced through the midst of the men of Ireland. Cethern pursued him closely and made a cast of his shield at him, and the scalloped edge of the shield cut him in three to the ground together with chariot and charioteer and horses. Then the armies attacked Cethern on both sides and he fell at their hands in the spot where he was.

Those are the tales of Caladgleó Cethirn and Fuile Cethirn.

Here follows Fiacalgleó Fintain.

Fintan was the son of Niall Niamglonnach from Dún Da Benn, and the father of Cethern. And he came to avenge the honour of the Ulstermen and to take revenge for his son's death on the hosts. Thrice fifty was the number of their band, and they came with two spear-heads on every shaft, a spear-head on the point and a spear-head on the butt, so that they wounded the hosts alike with points and butts. They gave battle three times to the hosts and three times their own number fell by them, and there fell also all the people of Fintan mac Néill except Crimthann the son of Fintan who was saved by Ailill and Medb under a shelter of shields. Then the men of Ireland said that it would be no disgrace to Fintan mac Néill to evacuate the encampment for him and that his son Crimthann should be allowed to go free with him, while the hosts should withdraw a day's march to the north and he should cease to attack the hosts until such time as he should come to them on the day of the great battle when the four great provinces of Ireland should meet at Gáirech and Ilgáirech in the Foray of Cúailnge, as had been prophesied by the druids of the men of Ireland. Fintan mac Néill agreed to this and his son was set free to him. The encampment was evacuated for him and the hosts retreated a day's journey northwards again, checking and holding themselves back. And each man of Fintan's people and each man of the men of Ireland were found with the lips and nose of each of them in the teeth of the other. The men of Ireland noticed this and said : ' This is the tooth-fight for us, the tooth-fight of Fintan's people and of Fintan himself.'

So that is Fiacalgleó Fintain.

Here follows Ruadrucce Mind

Mend mac Sálcholgán was from Réna na Bóinne. His force numbered twelve men. They had two spear-heads on each shaft, a spear-head on the point and a spear-head on the butt, so that they wounded the hosts alike with points and butts. They attacked the hosts three times and three times their own number fell by them, and twelve of Mend's people fell. But Mend himself was wounded grievously so that he was reddened and bloodstained. Then said the men of Ireland : ' Red is this shame for Mend mac Sálcholgán, that his people should be killed and destroyed and he himself be wounded until he is reddened and bloodstained.'

This is Ruadrucce Mind.

Then the men of Ireland said that it were no disgrace for Mend mac Sálcholgán if the encampment were cleared for him and if the hosts went back a day's journey to the north again, provided that he should cease to attack the hosts until Conchobor recovered from his debility and gave them battle at Gáirech and Ilgáirech, as the druids and prophets and seers of the men of Ireland had foretold.

Mend mac Sálcholgán agreed that the encampment should be vacated. The hosts withdrew a day's journey to the north again, checking and staying themselves.

Here follows Airecur nArad.

Then the charioteers of the Ulstermen came to them, three fifties in number. They gave battle three times to the host and three times their own number fell by them, and the charioteers fell on the level spot on which they stood.

That is Airecor nArad.

Here follows Bángleó Rochada.

Reochaid mac Faithemain was of the Ulstermen. His force numbered one hundred and fifty, and he took up his position on a hillock opposite the host. Findabair, the daughter of Ailill and Medb, noticed that, and she said to her mother Medb: ' I loved yonder warrior long ago and he is my beloved and my chosen wooer.' ' If you loved him, my daughter, spend tonight with him and ask him for a truce for us with the host until he come to us on the day of the great battle where the four great provinces of Ireland will meet at Gáirech and Ilgáirech at the battle of the Foray of Cúailnge.' Reochaid mac Faithemain agreed to that and the girl spent that night with him.

One of the underkings of Munster who was in the camp heard of this and said to his people : 'That girl was betrothed to me long ago and that is why I have come now upon this hosting.' However, as for the seven underkings of Munster, they all said that that was why they had come. ' Why then,' said they, ' should we not go to take vengeance for the woman and for our honour on the Maines who are keeping guard in the rear of the host at Imlech in Glendamrach ? '

That was the plan they decided upon and they arose with their seven divisions of three thousand. Then Ailill rose to oppose them with his three thousand. Medb rose with her three thousand, and the sons of Mágu with their divisions. The Gaileóin and the Munstermen and the people of Tara rose. Intervention was made between them so that each man sat next to the other and beside his weapons. Yet before the intervention was accomplished, eight hundred valiant men from among them had fallen. Findabair, the daughter of Ailill and Medb, heard that this number of the men of Ireland had fallen because of her and on account of her, and her heart cracked like a nut in her breast through shame and modesty. Findabair Slébe is the name of the spot where she died. Then said the men of Ireland : ' Bloodless is this fight for Reochaid mac Faithemain, since eight hundred valiant soldiers have fallen because of him but he himself has escaped without a wound and without shedding his blood.'

That is Bángleó Rochada.

Mellgleó Iliach.

Íliach was the son of Cas mac Baicc meic Rosa Rúaid meic Rudraige. He was told how the four great provinces of Ireland had been plundering and laying waste Ulster and Pictland from the Monday at the beginning of Samain until the beginning of spring, and he took counsel with his people. ' What better plan could I devise than to go and attack the men of Ireland and win victory over them and avenge the honour of Ulster ? It matters not if I myself fall thereafter.' And that was the plan he decided on. His two old, decrepit, mangy horses which were on the strand beside the fort were harnessed for him, and his old chariot without any rugs or covering was yoked to the horses. He took up his rough, dark-coloured, iron shield with the rim of hard silver around it. On his left side he put his rough, heavy-smiting sword with grey guard. He took his two gapped, shaky-headed spears in the chariot beside him. His people filled his chariot around him with stones and rocks and great flagstones. In this wise he came forward towards the men of Ireland with his private parts hanging through the chariot. ' We should like indeed,' said the men of Ireland, ' if it were thus that all the Ulstermen came to us.'

Dóche mac Mágach met him and welcomed him. ' Welcome is your arrival, Íliach,' said Dóche mac Mágach. ' I trust that welcome,' said Íliach, ' but come to me presently when my weapons are exhausted and when my valour has diminished so that you may be the one to behead me and not any other man of the men of Ireland. But keep my sword for Láegaire.'

Íliach plied his weapons on the men of Ireland until he had exhausted them, and when his weapons were exhausted, he attacked the men of Ireland with stones and rocks and great flagstones until they too were exhausted, and when they were finished, wherever he could seize one of the men of Ireland, he would crush him swiftly between his arms and his hands and make a marrow-mash of him, flesh and bones, sinews and skin all together. And the two marrow-mashes still remain side by side, the one which Cú Chulainn made from the bones of the Ulstermen's cattle to cure Cethern mac Fintain and the one which Íliach made from the bones of the men of Ireland. So that all those who fell at the hands of Íliach are (called) one of the three uncountable (slayings) of the Táin, and that tale is called Mellgleo nÍliach.

It was called Mellgleó nÍliach because he fought his fight with stones and rocks and great flagstones.

Dóche mac Mágach met him. ' Is not this Íliach ? ' said Dóche. ' It is I indeed,' said Íliach, ' but come to me now and cut off my head and keep my sword for your friend Láegaire.' Dóche came to him and with a stroke of the sword cut off his head.

Thus far Mellgleó Íliach.

Oislige Amargin in Tailtiu.

Amairgin was the son of Cas mac Baicc meic Rosa Rúaid meic Rudraige. He overtook the hosts going westwards over Tailtiu and he turned them and drove them northwards over Tailtiu. He lay on his left elbow in Tailtiu and his people furnished him with stones and rocks and great flagstones and he fell to pelting the men of Ireland for three days and three nights.

Concerning Cú Ruí mac Dáire.

Cú Ruí was told that a single man had been holding the four great provinces of Ireland in check from the Monday at the beginning of Samain until the beginning of spring. He was grieved by this and he thought that his people had been without him for too long, so he came forward to do battle and combat with Cú Chulainn.

When he reached the place where Cú Chulainn was, he saw him lying there groaning, wounded and stabbed, and he scorned to do battle or combat with him after Cú Chulainn's fight with Fer Diad lest Cú Chulainn should die not so much of the wounds and gashes which he would inflict on him as of those which Fer Diad had already inflicted on him. Nevertheless Cú Chulainn offered to engage in battle with Cú Ruí.

Cú Ruí went forward then to the men of Ireland and when he got there, he saw Amairgin lying on his left elbow to the west of Tailtiu. Cú Ruí came to the north of the men of Ireland. His people furnished him with stones and rocks and great flagstones and he began to hurl them directly against Amargin so that the warlike battle-stones collided in the clouds and in the air over their heads and each stone was shattered into a hundred pieces. ' By the truth of your valour, Cú Ruí,' said Medb, ' cease from this stone-throwing, for it is no help to us but a hindrance.' ' I swear,' said Cú Ruí, ' that I shall not cease till the day of doom until Amargin cease too.' ' I shall do so', said Amargin, ' and do you undertake not to come again to help and succour the men of Ireland.' Cú Ruí agreed to that and went away to his own land and his own people.

By this time they had gone westwards past Tailtiu. ' It was not the agreement I made,' said Amargin, ' not to cast stones at the host again.' So he came to the west of them and turned them before him to the north-east past Tailtiu and began to pelt them for a long time.

Then said the men of Ireland that it would be no dishonour for Amargin if they vacated the encampment and the hosts went back a day's journey northwards holding themselves in check, and that Amargin should cease to attack the hosts until he came to them on the day of the great battle where the four great provinces of Ireland would meet at Gáirech and Ilgáirech in the battle of the Foray of Cúailnge. Amargin agreed to that and the hosts withdrew a day's journey northwards once more.

That is Oislige Amargin in Tailtiu.

The Long Warning of Sultaim.

Sualtaim was the son of Becaltach mac Móraltaig and the father of Cú Chulainn. He was told of the distress of his son fighting against odds with Calatín Dána and his twenty-seven sons and

his grandson Glas mac Delga. ' This is from afar,' said Sualtaim.
' Is it the sky that cracks or the sea that ebbs or the earth that
splits or is it the distress of my son against odds on the Foray of
Cúailnge ? ' Sualtaim spoke truly indeed, and he went to Cú
Chulainn presently though he did not go at once. When Sualtaim
came to where Cú Chulainn was, he began to lament and commiserate with him. Cú Chulainn liked not that Sualtaim should
lament and pity him, for he knew that though he was wounded
and injured Sualtaim would be no protection to avenge him. For
the truth was that Sualtaim was not a coward but neither was he
a valiant fighter but only a middling one. ' Well now, father
Sualtaim,' said Cú Chulainn, ' go to the Ulstermen in Emain and
tell them to go now after their cattle, for I am unable to protect
them any longer in the gaps and passes of the land of Conaille
Muirthmene. I have stood alone against the four great provinces
of Ireland from Monday at the beginning of Samain until the
beginning of spring, killing one man at the ford every day and a
hundred warriors every night. Fair play is not granted to me
nor single combat, and no one comes to help or succour me. Bent
hoops of fresh hazel keep my mantle from touching me. Dry
wisps of tow are stuffed in my wounds. From the crown of my head
to the soles of my feet there is not a hair whereon the point of a
needle could rest but has a drop of crimson blood on its very tip,
except alone my left hand which is holding my shield, and even
that hand has thrice fifty wounds on it. And unless they take
vengeance for that at once, they will never do so until the brink
of doom.'

Sualtaim set forth on the Líath Macha as his only horse, to take
these warnings to the Ulstermen. And when he reached the side
of Emain, he spoke these words : ' Men are slain, women carried
off, cattle driven away, O Ulstermen!'

He got not the answer that sufficed him from the Ulstermen,
and so he came forward opposite Emain and spoke the same words
there : ' Men are slain, women carried off, cattle driven away,
O Ulstermen!' He got not the answer that sufficed him from the
Ulstermen.—This is how it was with the Ulstermen : it was tabu
for them to speak before their king and it was tabu for the king
to speak before his druids.—Sualtaim came forward then to the
stone of the hostages in Emain Macha. He spoke the same words
there : ' Men are killed, women carried off, cattle driven away!'
' Who kills them and who carries them off and who drives them
away ? ' said Cathbath the druid. ' Ailill and Medb have ravaged

you,' said Sualtaim. ' Your women-folk and your sons and your
youths have been carried off, your horses and your steeds, your
herds and your flocks and your cattle. Cú Chulainn alone is
checking and holding back the four great provinces of Ireland
in the gaps and passes of Conaille Muirthemne. Fair play is not
granted to him nor single combat, and no one comes to aid or
succour him. The youth has been wounded, blood has drained
from his wounds. Bent hoops of fresh hazel hold his mantle
over him. There is not a hair from his head to his feet on which
the point of a needle could stand but has a drop of bright red
blood on its tip, save only the left hand which holds his shield
and even that hand bears thrice fifty wounds. And unless ye
avenge this at once, ye will never avenge it until the end of doom
and life.' ' More fitting is death and destruction for the man
who so incites the king,' said Cathbath the druid. ' That is true
indeed,' said all the Ulstermen. Sualtaim went his way in anger
and wrath since he got not the answer which sufficed him from
the Ulstermen. Then the Líath Macha reared under Sualtaim
and came forward opposite Emain, and his own shield turned on
Sualtaim and its rim cut off his head. The horse itself turned
back again into Emain, with the shield on the horse and the head
on the shield. And Sualtaim's head spoke the same words. ' Men
are slain, women carried off, cattle driven away, O Ulstermen!'
said the head of Sualtaim. ' A little too loud is that cry,' said
Conchobor, ' for the sky is above us, the earth beneath us and the
sea all around us, but unless the sky with its showers of stars fall
upon the surface of the earth or unless the ground burst open in
an earthquake, or unless the fish-abounding, blue-bordered sea
come over the surface of the earth, I shall bring back every cow
to its byre and enclosure, every woman to her own abode and
dwelling, after victory in battle and combat and contest.' Then
a messenger of his own household met Conchobor, to wit, Findchad
Fer Bend Uma mac Fráechlethain, and Conchobor bade him go
and assemble and muster the men of Ulster. And even as he
enumerated the quick and the dead for him in the intoxication
of his trance and his sickness, he said these words :

' Arise, O Findchad, I send you forth. It is not desirable to
neglect to tell it to the warriors of Ulster. (Go from me to Derg)[1] '
to Dedaid in his inlet ; to Lemain ; to Follach ; to Illaind at
Gabar ; to Dornaill Féic at Imchlár ; to Derg Indirg ; to Feidilmid

---

[1] foll. St

Chilair Chétaig at Ellonn ; to Rigdonn, to Reochaid ; to Lugaid ;
to Lugdaig ; to Cathbath in his inlet ; to Cairbre at Ellne ; to
Láeg at his causeway ; to Geimen in his valley ; to Senall Úathach
at Diabul Arda ; to Cethern mac Fintain at Carrlóg ; to Tarothor ;
to Mulach in his fort ; to the royal poet Amairgin ; to Úathach
Bodba ; to the Morrígan at Dún Sobairche ; to Eit ; to Roth ;
to Fiachna at his mound ; to Dam Drend ; to Andiaraid ; to
Maine Macbriathrach ; to Dam Derg ; to Mod ; to Mothus ;
to Iarmothus ; to Corp Cliath ; to Gabarlach in Líne ; to Eochu
Semnech in Semne ; to Celtchair mac Cuthechair in Lethglais ;
to Errge Echbél in Brí Errgi ; to Uma mac Remarfessaig in Fedan
Cúailnge ; to Munremur mac Gerrcind in Moduirn ; to Senlabair
in Canann Gall ; to Follomain ; to Lugaid ; to Lugaid Líne king
of Bolg ; to Búadgalach ; to Abach ; to Áne ; to Ániach ; to
Lóegaire Milbél at his fire (?) ; to the three sons of Trosgal at
Bacc Draigin ; to Drend ; to Drenda ; to Drendus ; to Cimm ;
to Cimbil ; to Cimmin at Fán na Coba ; to Fachtna mac Sencha
in his rath ; to Sencha ; to Sencháinte ; to Briccne ; to Briccirne ;
to Brecc ; to Búan ; to Barach ; to Óengus Bolg ; to Óengus
mac Lethi ; [1]to Alamiach the warrior[1] ; to Bruachar in Slánge ;
to Conall Cernach mac Amargin at Midlúachair ; to Cú Chulainn
mac Sualtaim in Muirthemne ; to Mend mac Sálcholcán at Réna ;
to the three sons of Fiacl.na, Ross, Dáire and Imchaid, in Cúailnge ;
to Connud mac Morna in Callann ; to Condraid mac Amargin in
his rath ; to Amargin in Ess Rúaid ; to Láeg at Léire ; to Óengus
Fer Bend Uma ; to Ogma Grianainech at Brecc ; to Eo mac
Forne ; to Tollcend ; to Súde at Mag nÉola and Mag nDea ; to
Conla Sáeb at Úarba ; to Lóegaire at Ráith Imbil ; to Amargin
Iarngiunnaig in Tailtiu ; to Furbaide Fer Bend mac Conchobuir
at Síl in Mag nInis ; to Causcraid Mend Macha mac Conchobuir
in Macha ; to Fíngin at Fíngabar ; to Blae Fichet ; to Blai Briuga
at Fesser ; to Eogan mac Durthacht at Fernmag ; to Ord at
Serthe ; to Oblán ; to Obail at Culend ; to Curethar ; to Liana ;
to Ethbenna ; to Fer Néll ; to Findchad of Sliab Betha ; to
Talgoba at Bernas ; to Mend mac Fir Chúaland of Mag Dula ; to
Íroll ; to Blárine at Ialla Ilgremma ; to Ros mac Ulchrothaig in
Mag Nobla ; to Ailill Find ; to Fethen Bec ; to Fethen Mór ; to
Fergna mac Findchona in Búrach ; to Olchar ; to Ebadcha[r] ;
to Uathchar ; to Etatchar ; to Óengus mac Óenláme Gábe ;

---

[1] [1] foll. YBL

to Ruadri at Mag Táil ; to Beothach ; to Briathrach in his rath ; to Nárithlaind to Lothor ; to Muridach mac Feicge and Cotreib mac Feicge ; to Fintan mac Néill Níamglonnaig in Dún Da Bend ; to Feradach Finn Fechtnach in Neimed of Slíab Fúait ; to Amargin mac Ecelsalaig Goband by Búas ; to Buinne mac Munremair ; to Fidach mac Doraire.'

It was not difficult, however, for Findchad to make that muster and assembly which Conchobor had ordered. For those who were east of Emain and west of Emain and north of Emain came forth at once and spent the night at Emain at the behest of their king and the command of their prince, awaiting the recovery of Conchobor. But those who were south of Emain set forth at once on the track of the host along the road beaten out by the hooves of the cattle.

On the first stage of the journey on which the Ulstermen set forth with Conchobor, they spent the night at Irard Cuillend. ' What do we wait for here, O men ? ' said Conchobor. ' We await your sons,' said they, ' Fiacha and Fiachna. They have gone from us to fetch Erc, the son of your daughter Fedlimid Nóchruthach and of Cairbre Nia Fer, that he may come to our army at this juncture with his full muster and assembly, his full gathering and levy.' ' I vow,' said Conchobor, ' that I shall not await them here any longer until the men of Ireland hear that I have recovered from the weakness and debility in which I was, for the men of Ireland do not know yet if I am still alive.'

Then Conchobor and Celtchair went to Áth nIrmide with thirty hundred chariot-fighters armed with spears, and there they met eight score big men of the household of Ailill and Medb with eight score captive women. One captive woman held prisoner by each man of them, that was their share of the plunder of Ulster. Conchobor and Celtchair struck off their eight score heads and freed their eight score captives. Áth nIrmide was the name of that place until then, but it is called Áth Féinne ever since. The reason it is called Áth Féinne is because the warriors of the war-band (fian) from the east and the warriors of the war-band from the west met there in battle and contest on the brink of the ford.

Conchobor and Celtchair came back and spent that night in Irard Cuillend beside the men of Ulster. The trance of Celtchair follows here.

Then Celtchair uttered these words among the Ulstermen in Irard Cuillend that night : Taible lethderg etc.

In the same night Cormac Cond Longas, the son of Conchobor, spoke these words among the men of Ireland in Slemain Mide : *Amra maitne* etc.

In the same night Dubthach Dáel Ulad spoke these words among the men of Ireland in Slemain Mide : *Móra maitne* etc.

Then Dubthach awoke from his sleep and the Nemain brought confusion on the host so that they made a clangour of arms with the points of their spears and their swords, and a hundred warriors of them died on the floor of their encampment through the fearsomeness of the shout they had raised. However that was not the most peaceful night ever experienced by the men of Ireland at any time, because of the prophecies and the predictions and because of the spectres and visions which appeared to them.

Then said Ailill : ' I have succeeded in laying waste Ulster and the land of the Picts from the Monday at the beginning of Samain until the beginning of spring. We have carried off their women-folk, their sons and their youths, their horses and steeds, their flocks and herds and cattle. We have levelled their hills behind them into lowlands, so that they might be of equal height. Wherefore I shall not wait here for them any longer, but let them give me battle on Mag Aí if it so please them. And yet though we say this, let some one go forth to reconnoitre the broad plain of Meath to see whether the Ulstermen come thither, and if they do, I shall in no wise retreat, for it is not the good custom of a king ever to retreat.' ' Who should go there ? ' said they all. ' Who but Mac Roth, the chief messenger yonder.'

Mac Roth came forward to reconnoitre the great plain of Meath. Not long was he there when he heard a noise and a tumult and a clamour. It seemed to him almost as if the sky had fallen on to the surface of the earth, or as if the fish-abounding, blue-bordered sea had swept across the face of the world, or as if the earth had split in an earthquake, or as if the trees of the forest had all fallen into each other's forks and bifurcations and branches. However the wild beasts were hunted across the plain (in such numbers) that the surface of the plain of Meath was not visible beneath them.

Mac Roth came to report that to where Ailill was with Medb and Fergus and the nobles of the men of Ireland. He related those tidings to them. ' What was that, Fergus ? ' asked Ailill. ' Not difficult to tell,' said Fergus. ' The noise and clamour and tumult that he heard, the din and the thunder and the uproar, were the Ulstermen attacking the wood, the throng of champions and warriors cutting down the trees with their swords in front of their chariots. It was that which hunted the wild beasts across the plain so that the surface of the plain of Meath is not visible beneath them.'

Once more Mac Roth scanned the plain. He saw a great grey mist which filled the void between heaven and earth. He seemed to see islands in lakes above the slopes of the mist. He seemed to see yawning caverns in the forefront of the mist itself. It seemed to him that pure-white linen cloths or sifted snow dropping down appeared to him through a rift in the same mist. He seemed to see a flock of varied, wonderful, numerous birds, or the shimmering of shining stars on a bright, frosty night, or the sparks of a blazing fire. He heard a noise and a tumult, a din and thunder, a clamour and uproar. He came forward to tell those tidings to where were Ailill and Medb and Fergus and the nobles of the men of Ireland. He told them these things.

' What was that, Fergus, ? ' asked Ailill. ' Not difficult to tell,' said Fergus. ' The grey mist he saw which filled the void between earth and sky was the expiration of the breath of horses and heroes, and the cloud of dust from the ground and from the roads which rises above them driven by the wind so that it becomes a heavy, deep-grey mist in the clouds and in the air.

' The islands in lakes which he saw there, and the tops of hills and mounds rising above the valleys of the mist, were the heads of the heroes and warriors above their chariots and the chariots themselves. The yawning caverns he saw there in the forefront of the same mist were the mouths and nostrils of horses and heroes, exhaling and inhaling the sun and the wind with the swiftness of the host.

' The pure-white linen cloths he saw there or the sifted snow dropping down were the foam and froth that the bits of the reins cast from the mouths of the strong, stout steeds with the fierce rush of the host. The flock of varied, wonderful, numerous birds which he saw there was the dust of the ground and the surface of the earth which the horses flung up from their feet and their hooves and which rose above them with the driving of the wind.

' The noise and the tumult, the din and the thunder, the clamour and the outcry which he heard there was the shock of shields and the smiting of spears and the loud striking of swords, the clashing of helmets, the clangour of breastplates, the friction of the weapons and the vehemence of the feats of arms, the straining of ropes, the rattle of wheels, the trampling of the horses' hoofs and the creaking of chariots, and the loud voices of heroes and warriors coming towards us here.

'The shimmering of shining stars on a bright night that he saw there, or the sparks of a blazing fire, were the fierce, fearsome eyes of the warriors and heroes from the beautiful, shapely, ornamented helmets, eyes full of the fury and anger with which they came, against which neither equal combat nor overwhelming number prevailed at any time and against which none will ever prevail until the day of doom.'

'We make little account of it,' said Medb. 'Goodly warriors and goodly soldiers will be found among us to oppose them.' 'I do not count on that, Medb,' said Fergus, 'for I pledge my word that you will not find in Ireland or in Alba a host which could oppose the Ulstermen when once their fits of wrath come upon them.'

Then the four great provinces of Ireland made their encampment at Clártha that night. They left a band to keep watch and guard against the Ulstermen lest they should come upon them unawares.

Then Conchobor and Celtchair set forth with thirty hundred chariot-fighters armed with spears and halted in Slemain Mide in the rear of the host. But though we say 'halted,' they did not halt completely, but came forward presently to the encampment of Medb and Ailill in an attempt to be the first to shed blood.

Not long was Mac Roth there when he saw something : a great and numerous troop of horsemen coming straight from the north-east to Slemain Mide. He went to where Ailill and Medb and Fergus and the nobles of the men of Ireland. Ailill asked tidings of him when he arrived. 'Well now, Mac Roth,' said Ailill, 'did you see anyone of the Ulstermen on the track of the host today ?' 'I know not indeed,' said Mac Roth, 'but I did see a great and numerous troop of horsemen coming directly from the north-east to Slemain Mide.' 'How many in number are the horsemen ?' said Ailill. 'Not fewer, it seemed to me, than thirty hundred chariot-fighters armed with spears,' said Mac Roth. 'Well, Fergus,' said Ailill, 'why did you try to frighten us just now with the dust and the smoke and the panting of a great army while that is all the battle force you have for us ?'

'A little too soon do you disparage them,' said Fergus, 'for perhaps (?) the army is more numerous than Mac Roth says.' 'Let us make a good plan swiftly concerning this,' said Medb, 'for

it was known that yonder huge, fierce, vehement man would attack
us, Conchobor son of Fachtna Fáthach mac Rosa Rúaid meic
Rudraige, the high-king of Ulster and the son of the high-king of
Ireland.   Let the men of Ireland be drawn up in open array to
face Conchobor with a force of thirty hundred closing it in from
the rear, and let the men be taken prisoner but not wounded for
[1]those who come number no more than the prisoners we need[1].—
That is one of the three most satirical sayings of Táin Bó
Cúailnge, (to suggest) that Conchobor should be captured unwounded
and that the thirty hundred princes of Ulster who accompanied
him [1]should be taken prisoner.[1]—Cormac Cond Longas, the son
of Conchobor, heard that and he knew that, if he did not take
vengeance at once on Medb for her boastful speech, he would never
avenge it until the very end of doom and life.

Then Cormac Cond Longas rose up with his force of thirty
hundred to wage war and battle on Ailill and Medb.   To meet
him rose Ailill with his thirty hundred, and Medb rose with her
thirty hundred.   The Maines arose with their thirty hundreds
and meic Mágach with their thirty hundreds.   The Leinstermen
and the Munstermen and the people of Tara rose up.   The combatants were separated and each man of them sat down beside
the other and near by his weapons.   Nevertheless Medb drew up
a hollow array to face Conchobor with a force of thirty hundred
men closing in the rear.   Conchobor came to this array of men
and in no wise sought a way of entry, but cut a breach broad enough
for a soldier opposite his face and his countenance, and cut a breach
broad enough for a hundred on his right hand and another breach
for a hundred on his left, and he turned in on them and wrought
confusion in their midst and eight hundred valiant warriors of
them fell at his hands.   Then he came from them, unwounded
and unhurt, and took up his station in Slemain Mide, waiting
for the Ulstermen.

'Come now, ye men of Ireland,' said Ailill, 'let some one of us
go to reconnoitre the broad plain of Meath to find out in what
fashion the Ulstermen come to the hill in Slemain Mide and to
give us an account of their arms and equipment, their heroes and
soldiers and their battle-champions and the people of their land.
To listen to him will be all the more pleasant for us now.'   'Who
should go there?' asked they all.   'Who but Mac Roth, the
chief messenger,' said Ailill.

---

[1]-[1] translating St

Mac Roth came forward and took up his station in Slemain Mide to await the Ulstermen. The Ulstermen began to muster on that hill (and continued doing so) from the twilight of early morning until sunset. In all that time the ground was hardly bare of them (as they came) with every division round its king, every band round its leader, and every king and every leader and every lord with the full number of his own particular forces and his army, his muster and his gathering. However before the hour of evening sunset all the Ulstermen had reached that hill in Slemain Mide.

Mac Roth came forward to the place where were Ailill and Medb and Fergus and the nobles of the men of Ireland, bringing an account of· the first band. Ailill and Medb asked tidings of him on his arrival. ' Well now, O Mac Roth,' said Ailill, ' in what guise and fashion do the men of Ulster come to the hill in Slemain Mide ? '

' I know only this indeed,' said Mac Roth. ' There came a fierce, powerful, well-favoured band on to that hill in Slemain Mide. It seems, if one looks at it, as if it numbered thirty hundred. They all cast off their garments and dug up a mound of turf as a seat for their leader. A warrior, slender, very tall, of great stature and of proud mien, at the head of that band. Finest of the princes of the world was he among his troops, in fearsomeness and horror, in battle and in contention. Fair yellow hair he had, curled, well-arranged, ringletted, cut short. His countenance was comely and clear crimson. An eager grey eye in his head, fierce and awe-inspiring. A forked beard, yellow and curly, on his chin. A purple mantle fringed, five-folded, about him and a golden brooch in the mantle over his breast. A pure-white, hooded shirt with insertion of red gold he wore next to his white skin. He carried a white shield ornamented with animal designs in red gold. In one hand he had a gold-hilted, ornamented sword, in the other a broad, grey spear. That warrior took up position at the top of the hill and everyone came to him and his company took their places around him.

' There came also another band to the same hill in Slemain Mide,' said Mac Roth. ' ¹It numbered almost thirty hundred.¹ A handsome man in the forefront of that same band. Fair yellow hair he had. A bright and very curly beard on his chin. A green mantle wrapt around him. A pure silver brooch in the mantle over his breast. A dark-red, soldierly tunic with insertion of

---

¹⁻¹ translating St

red gold next to his fair skin and reaching to his knees. (A spear like) the torch of a royal palace in his hand, with bands of silver and rings of gold. Wonderful are the feats and games performed by that spear in the warrior's hand. The silvern bands revolve round the golden rings alternately from butt to socket, and alternately the golden rings revolve round the silvern bands from socket to thong. He bore a smiting shield with scalloped rim. On his left side a sword with guards of ivory and ornament of gold thread. That warrior sat on the left hand of the warrior who had first come to the hill, and his company sat around him. But though we say that they sat, yet they did not really do so, but knelt on the ground with the rim of their shields at their chins, in their eagerness to be let at us. And yet it seemed to me that the tall, fierce warrior who led that company stammered greatly.

'There came still another company to the same hill in Slemain Mide,' said Mac Roth. 'Almost the same were they as the preceding one in number and appearance and apparel. A handsome, broad-headed warrior in the van of that company. Thick, dark-yellow hair he had. An eager, dark-blue, restless eye in his head. A bright and very curly beard, forked and tapering, on his chin. A dark-grey, fringed cloak wrapt about him. A leaf-shaped brooch of white bronze in the cloak over his breast. A white-hooded shirt next to his skin. A white shield with animal ornaments of silver he carried. A sword with rounded hilt of bright silver in a warlike scabbard at his waist. (A spear like) the pillar of a palace on his back. This warrior sat on the turfy mound in front of the warrior who had come first to the hill and his company took up their positions around him. But sweeter I thought than the sound of lutes in the hands of expert players was the melodious tone of the voice and speech of that warrior as he addressed the warrior who had come first to the hill and gave him counsel.'

'Who are those?' asked Ailill of Fergus. 'We know them indeed,' said Fergus. 'The first warrior for whom the sodded mound was cast up on the top of the hill until they all came to him was Conchobor mac Fachtna Fáthaig meic Rosa Rúaid meic Rudraige, the high-king of Ulster and the son of the high-king of Ireland. The great stammering hero who took up his position on the left of Conchobor was Causcraid Mend Macha, the son of Conchobar, with the sons of the Ulster princes around him and the sons of the kings of Ireland who are with him. The spear with silver bands and rings of gold that Mac Roth saw in his hand is called the Torch of Causcraid. It is usual with that spear that the silver bands do not revolve around the golden rings except shortly before some victory, and not at any other time, and it is likely that it was just before victory that they revolved just now.

· The handsome, broad-headed warrior who sat on the mound in front of the warrior who had first come to the hill was Sencha mac Ailella meic Máilchló, the eloquent speaker of Ulster, the man who appeases the armies of the men of Ireland. But I pledge my word that it is not counsel of cowardice or fear that he gives his lord today in this day of battle, but counsel to act with valour and bravery, courage and might. And I pledge my word too,' said Fergus, 'that those who rose up around Conchobor in the early morn today are goodly men who can carry out (such) deeds.' 'We reck little of them,' said Medb. 'There will be found among us goodly heroes and goodly warriors to answer them.' 'I count not on that,' said Fergus, 'but I swear that you will not find in Ireland or in Alba an army which can answer the Ulstermen when once their fits of wrath have come upon them.'

'There came still another band to the same hill in Slemain Mide,' said Mac Roth. 'A big, tall, fair man in the forefront of that band, fiery and swarthy-faced. Brown, dark hair he had smooth and fine over his forehead. A grey cloak wrapt around him and a silver pin in the cloak over his breast. A white . . . shirt next to his skin. A curved shield with a scalloped rim he bore. A five-pronged spear in his hand. An ivory-hilted sword over his shoulder.' 'Who was that ?' said Ailill to Fergus. 'We know him indeed,' said Fergus. 'The man who came there is the starting of strife, a warrior for conflict, doom of enemies. That was Eogan mac Durthachta from the north, [1]the steadfast ruler of Farney[1].'

'There came another band to the same hill in Slemain Mide,' said Mac Roth. 'In truth boldly they made for that hill. Great is the horror and vast the fear which they brought with them. Their garments were all cast back. A big-headed, valiant warrior in the van of that company, and he was fierce and fearsome. Fine grizzled hair he had. Great yellow eyes in his head. A yellow mantle of the breadth of five hands around him. A pin of yellow gold in the mantle over his breast. A yellow, bordered shirt next to his skin. In his hand a rivetted spear, broad-bladed and long-shafted, with a drop of blood on its edge.' 'Who was that ?' asked Ailill of Fergus. 'We know that hero indeed,' said Fergus.

[1-1] translating St

' He who came there shuns not battle nor battlefield nor conflict.
That was Lóegaire Búadach mac Connaid Buide meic Iliach from
Ráith Immil in the North.

' There came still another company to the same hill in Slemain
Mide,' said Mac Roth. ' A thick-necked, corpulent warrior in
the van of that company. He had black, cropped hair and a scarred,
crimson countenance. A grey, bright eye in his head. A bloodstained[1] spear shimmering above him. A black shield with hard
rim of white bronze he bore. A dun-coloured mantle [1]of curly
wool[1] around him. A brooch of white gold in the mantle over
his breast. A plaited shirt of silk next to his skin. A sword with
guards of ivory and ornament of thread of gold over his garments
on the outside.' ' Who was that ? ' asked Ailill of Fergus. ' We
know him indeed,' said Fergus. ' He that came there is the
starting of strife ; he is the stormy wave which drowns ; he is a
man of three shouts ; he is the sea pouring over ramparts. That
was Munremur mac Gerrcind from Modorn in the north.'

' There came still another company to the same hill in Slemain
Mide,' said mac Roth. ' A broad[2], bulky warrior in the van of
that company, . . . and dusky-coloured, fierce and bull-like. A
round eye, dull and haughty, in his head. Yellow, very curly
hair he had. A round, red shield he bore aloft, with a rim of hard
silver around it. In his hand a broad-bladed, long-shafted spear.
A striped cloak he wore with a brooch of bronze in the cloak over
his breast. A hooded shirt reaching to his calves. An ivory-hilted
sword on his left thigh.' ' Who was that ? ' Ailill asked Fergus.
' We know him indeed,' said Fergus. ' He is a prop of battle. He
is victory in every conflict. The man who came there is an instru-
ment which pierces. That was Connud mac Morna from Callann
in the north.'

' There came yet another company to the same hill in Slemain
Mide,' said Mac Roth. ' Vigorously and violently, in truth, did
they make for that hill and shook the forces that had arrived there
before them. A handsome and noble man in the van of that
company. Most beautiful of the men of the world was he, in
shape and form and make, in arms and equipment, in size and
dignity and honour, in figure and valour and proportion.' ' That
is indeed no lie,' said Fergus. ' That is his fitting description. He
who came there is no foolish one in bareness. He is the enemy of

---

[1]-[1] translating St
[2] broad-headed St

all. He is the force which cannot be endured. He is a stormy wave which engulfs. The glitter of ice is that handsome man. That was Feidilmid Chilair Chetail from Ellann in the north.'

'There came still another band on to the same hill on Slemain Mide,' said Mac Roth. 'Not many heroes are more beautiful than the hero in the forefront of that band. Cropped, red-yellow hair he had. His face was narrow below and broad above. An eager, grey eye, glittering and gay, in his head. A shapely, well-proportioned man, tall, slender-hipped, broad-shouldered. Thin red lips he had and shining, pearl-like teeth. A white, seemly body. [1]A purple cloak wrapt about him[1]. A golden brooch in the cloak over his breast. A shirt of royal silk with a hem of red gold next to his white skin. A white shield with emblems of animals in red gold on it he bore. At his left side an ornamented sword with golden hilt. In his hand a long spear with shining edge and a sharp aggressive javelin with splendid thongs, with rivets of white bronze.' 'Who was that?' asked Ailill of Fergus. 'We know him indeed,' said Fergus. 'He who came there is in himself the half of a battle ; he is the dividing of a combat ; he is the wild fury of a watch-dog. That was Reochaid mac Faithemain from Rígdond in the north.'

'There came still another band to the same hill in Slemain Mide,' said Mac Roth. 'A hero brawny-legged, thick-thighed, in the forefront of that band. Every one of his limbs is almost as thick as a man. In truth, he is every inch a man,' said he. 'Brown, cropped hair he had, and a ruddy, round countenance. An eye of many colours high in his head. A splendid swift man was he thus, accompanied by contentious, black-eyed warriors, with red, flaming banner, with self-willed behaviour, avoiding equal combat to vanquish overwhelming numbers, with the releasing (?) of an attack upon him and without any protection from Conchobor.' 'Who was that?' asked Ailill of Fergus. 'We know him indeed,' said Fergus. He who came there [was full ?] of valour and prowess, of hot bloodedness and violence. He is a consolidator of hosts and weapons. He is the point of perfection in battle and combat of the men of Ireland in the north, my own foster-brother, Fergus mac Leite from Líne in the north.'

'There came still another band on to the same hill in Slemain Mide,' said Mac Roth. ' one which was steady and outstanding. A

---

[1]–[1] translating YBL, St

handsome, lively hero in the forefront of that band. Next to his skin
a fine, fringed garment of blue cloth with plaited, intertwined fine
loops of white bronze and strong, splendid buttons of red gold on
its slashes and its breast. A mantle of many pieces with the
choicest of colours wrapt about him. Five concentric circles of
gold, to wit, his shield, he bore. At his left side a sword, hard,
tough and straight, held in a high heroic grasp. A straight, ridged
spear blazing in his hand.' ' Who was that ? ' asked Ailill of
Fergus. ' We know him indeed,' said Fergus. ' He is the choicest
among royal poets. He is an attack on a fort. He is the way
to the goal. Violent is the valour of him who came there, Amairgin
mac Ecelsalaig Goband, the noble poet from Búas in the north.'

' There came still another company to the same hill in Slemain
Mide,' said Mac Roth. ' A fair, yellow-haired warrior in the van
of that company. Fair in all points was that man, hair and eye
and beard and eyebrows and garments. A rimmed shield he bore.
At his left side a gold-hilted, ornamented sword. In his hand a
five-pronged spear which flashed above the whole host.' ' Who
was that ? ' asked Ailill of Fergus. ' We know him indeed, said
Fergus. ' Beloved is that warrior who came into our territory to
us. Beloved is that strong-smiting hero, beloved that bear which
performs great deeds against enemies with the overwhelming
violence of his attack. That was Feradach Find Fechtnach from
Nemed in Sliab Fúait in the north.'

' There came still another company to the same hill in Slemain
Mide,' said Mac Roth. ' Two youthful warriors at the head of
that company. Two green cloaks wrapt about them and two
brooches of white silver in the cloaks over their breasts. Two
shirts of smooth, yellow silk next to their skin. Swords with
white hilts at their girdles. Two five pronged spears with bands
of pure white silver in their hands. A slight difference of age
between them.' ' Who are those ? ' asked Ailill of Fergus.
' We know them indeed,' said Fergus. ' They are two
men of valour, two equally strong-necked ones, two equally
bright flames, two equally bright torches, two champions, two
heroes, two chief hospitallers, two dragons, two fires, two scatterers,
two brave scions, two doughty ones, two fierce ones, the two
beloved by the Ulstermen around their king. Those are Fiacha
and Fiachna, two sons of Conchobor mac Fachtna meic Rosa
Rúaid meic Rudraigi.'

' There came still another company to the same hill,' said Mac
Roth, ' in size like the overwhelming sea, in red blazing like fire,

in numbers a battalion, in strength a rock, in combativeness like doom, in violence like the thunder. A wrathful, terrible, fearsome man at the head of that company. He was big-nosed, big-eared and with prominent eyes. Rough, grizzled hair he had. A striped cloak he wore and in that cloak over his breast an iron stake which reaches from shoulder to shoulder. A rough, plaited shirt next to his skin. Along the side of his back a sword of refined iron, tempered seven times in the heat. A brown mound, to wit, his shield, he carried. A great, grey spear with thirty rivets through its socket in his hand. But the battalions and hosts were thrown into confusion on seeing that warrior surrounded by his company advancing to the hill in Slemain Mide.' 'Who was that ?' said Ailill to Fergus. 'We know him indeed,' said Fergus. 'He is half a battle (in himself), he is a leader of strife, he is a chief in valour. The man who came is the sea pouring across boundaries. That was Celtchair Mór mac Uthechair from Lethglais in the north.'

'There came still another band to the same hill in Slemain Mide,' said Mac Roth, 'and they were strong and fierce, hateful and fearsome. A big-bellied, big-mouthed hero at the head of that band, with bright cheeks (?), with broad head, with long arms. Brown, very curly hair he had. A black swinging mantle he wore with a round brooch of bronze in the mantle over his breast. A splendid shirt next to his skin. A very long sword at his waist. A large spear in his right hand. A grey buckler, to wit, his shield, he bore.' 'Who was that ?' asked Ailill of Fergus. 'We know him indeed,' said Fergus. 'He is the lion, fierce, with bloodstained paws. He is the bear, violent and terrible, that overcomes the valiant. That was Eirrge Echbél from Brí Eirrgi in the north.'

'There came still another company to the same hill in Slemain Mide,' said Mac Roth. 'A huge and splendid man in the van of that company. Red hair he had and great red eyes in his head, and each of his great royal eyes was as long as a warrior's finger. A variegated mantle he wore. A grey shield he carried. A slender blue spear he held aloft. Around him was a company, bloodstained and wounded, while he himself was wounded and bloody in their very midst.' 'Who was that ?' asked Ailill of Fergus. 'We know him indeed,' said Fergus. 'He is the bold and ruthless one. He is the awe-inspiring eagle. He is the strong spear. He is the goring beast (?) He is [the fighter] of Colptha. He is the victorious one of Baile. He is [the lion ?] of Lorg. He is the loud-

voiced hero from Berna. He is the mad bull. That was Mend
mac Sálcholgán from Réna na Bóinne.'

'There came still another company to the same hill in Slemain
Mide,' said Mac Roth. 'A long-cheeked, sallow-faced man at the
head of that company. Black hair he had and long legs. He
wore a red cloak of curly wool with a brooch of pale silver in the
cloak over his breast. A linen shirt next to his skin. A blood-red
shield with a boss of gold he carried. At his left side a sword
with hilt of silver, and aloft he carried an angular spear with socket
of gold.' 'Who was that ?' asked Ailill of Fergus. 'We know
him indeed,' said Fergus. 'He is the man of three paths, the
man of three roads, the man of three highways, the man of three
routs, the man of three triumphs, the man of three combats. That
was Fergna mac Findchonna the chief of Búrach Ulad in the north.'

'There came still another band to the same hill in Slemain
Mide,' said Mac Roth. 'A great, comely man at the head of that
band. He was like to Ailill yonder, the keen one who can restrain,
in appearance and dignity and brightness, in arms and equipment,
in valour and prowess, in generosity and great deeds. A blue
shield with golden boss he carried. At his left side a gold-hilted
sword. In his hand a five-pronged spear with gold. A golden
diadem on his head.' 'Who was that ?' asked Ailill of Fergus.
'We know him indeed,' said Fergus. He is manly steadfastness.
He is an assault on overwhelming forces. He who came there is
the vanquishing of men. That was Furbaide Fer Bend, the son of
Conchobor, from Síl in Mag nInis in the north.'

'There came still another company to the same hill in Slemain
Mide,' said Mac Roth, 'and they were steadfast and unlike the
other companies. Some wore red mantles and some grey. Some
wore blue mantles and others green. Overmantles of white and
yellow, beautiful and brilliant, above them. There is in their very
midst a little freckled lad in a crimson cloak with a golden brooch
in the cloak over his breast. A shirt of royal satin with insertion
of red gold next to his white skin. A white shield with animal
designs in red gold he bore and on the shield was a boss of gold
and around it a rim of gold. A small sword with golden hilt he
had at his waist. Aloft he held a light sharp spear which shim-
mered.' 'Who was that ?' said Ailill to Fergus. 'I know not
indeed,' said Fergus, 'that I left behind me with the Ulstermen
such a company as that or the little lad who is with them, and yet
I should think it likely that they were the men of Tara with Erc
the son of Fedilmid Nóchruthach, who is also the son of Cairbre

Nia Fer, and if it is they, . . . for this little lad has come on this occasion to succour his grandfather without asking permission of his
father, and if it is they, this company will overwhelm you like the
sea, for it is by reason of this company and the little lad among
them that ye will be defeated on this occasion.' ' How is that ? '
asked Ailill. ' Not difficult to say,' answered Fergus, ' for this
little lad will experience neither fear nor dread when slaying and
slaughtering you until he comes to you into the middle of your
army. The noise of Conchobor's sword shall be heard like the
baying of a watchdog . . . or like a lion attacking bears. Outside
the line of battle Cú Chulainn will cast up four great ramparts of
men's corpses. Filled with affection for their own kin, the chiefs
of the men of Ulster will in due course smite (you). Bravely will
those powerful bulls roar as they rescue the calf of their own cow
in the battle on the morrow's morn.'

' There came still another company to the same hill in Slemain
Mide,' said Mac Roth, ' which numbered no less than thirty hundred.
Fierce, bloodstained warrior bands. Fair, clear, blue and crimson
men. They had long, fair-yellow hair, beautiful, brilliant counten-
ances, clear kingly eyes. Shining, beautiful garments they wore.
Wonderful, golden brooches on their bright-hued arms. Silken,
fine-textured shirts. Shining, blue spears they carried. Yellow,
smiting shields. Gold-hilted ornamented swords are set on their
thighs. Loud-voiced care has come to them. Sad are all the
horsemen (?). Sorrowful are the royal leaders. Orphaned the
bright company without their protecting lord who used to defend
their borders.' ' Who are these ? ' asked Ailill of Fergus. ' We
know them indeed,' said Fergus. ' They are fierce lions. They
are champions of battle. They are the thirty hundred from Mag
Muirtheimne. The reason they are downcast, sorrowful and
joyless, is because their territorial king is not among them, namely,
Cú Chulainn, the restraining, victorious, red-sworded, triumphant
one.' ' They have good cause,' said Medb, ' to be downcast,
sorrowful and joyless, for there is no evil we have not wrought on
them. We have plundered them and we have ravaged them
from the Monday at the beginning of Samain until the beginning
of spring. We have carried off their women and their sons and
their youths, their horses and their steeds, their herds and their
flocks and their cattle. We have cast down their hills behind
them on to their slopes until they were of equal height.' ' You have
no reason to boast over them, Medb,' said Fergus, ' for you did
no harm or wrong to them that the leader of that goodly band

yonder has not avenged on you, since every mound and every grave, every tombstone and every tomb from here to the eastern part of Ireland is a mound and a grave, a tombstone and a tomb for some goodly hero or for some brave warrior who fell by the valiant leader of yonder band. Fortunate is he whom they will uphold ! Woe to him whom they will oppose ! They will be as much as half a battle force against the men of Ireland when they defend their lord in the battle tomorrow morning.'

'I heard a great outcry there,' said Mac Roth, 'to the west of the battle [1]or to the east of the battle[1].' 'What outcry was that ? ' asked Ailill of Fergus. 'We know it indeed,' said Fergus. 'That was Cú Chulainn trying to come to the battle when he was being laid prostrate on his sick-bed in Fert Sciach, with wooden hoops and restraining bands and ropes holding him down, for the Ulstermen allow him not to come there because of his wounds and gashes, for he is unfit for battle and combat after his fight with Fer Diad.'

It was as Fergus said. That was Cú Chulainn being laid prostrate on his sick-bed in Fert Sciach, held down with hoops and restraining bands and ropes.

Then there came out of the encampment of the men of Ireland two female satirists called Fethan and Collach, and they pretended to weep and lament over Cú Chulainn, telling him that the Ulstermen had been routed and that Conchobor had been killed and that Fergus had fallen in the fight against them.

It was on that night that the Morrígu daughter of Ernmas came and sowed strife and dissension between the two encampments on either side, and she spoke these words : *Crennait brain* etc.

She whispered to the Érainn that they will not fight the battle which lies ahead.

Then said Cú Chulainn to Láeg mac Riangabra : 'Alas for you, my friend Láeg, if between the two battle-forces today anything should be done that you would not find out for me.' 'Whatsoever I shall find out concerning it, little Cú,' said Láeg, 'shall be told to you. But see a little flock coming from the west out of the encampment now on to the plain. There is a band of youths after them to check and hold them. See too a band of youths

---

[1]–[1] omitt. ; om. St

coming from the east out of the encampment to seize them.' 'That is true indeed,' said Cú Chulainn. 'It is the omen of a mighty combat and a cause of great strife. The little flock will go across the plain [1]and the youths from the east will encounter those from the west[1].' It was as Cú Chulainn said : The little flock went across the plain and the youths met. 'Who gives battle now, my friend Láeg ? ' asked Cú Chulainn. 'The people of Ulster,' said Láeg, 'that is, the youths.' 'How do they fight ? ' asked Cú Chulainn. 'Bravely do they fight,' said Láeg. 'As for the champions who come from the east to the battle, they will make a breach through the battle-line to the west. As for the champions from the west, they will make a breach through the battle-line to the east.' 'Alas that I am not strong enough to go afoot among them ! For if I were, my breach too would be clearly seen there today like that of the rest.' 'Nay then, little Cú,' said Láeg, 'it is no disgrace to your valour and no reproach to your honour. You have done bravely hitherto and you will do bravely hereafter.' 'Well now, friend Láeg,' said Cú Chulainn, 'rouse the Ulstermen to the battle now for it is time for them to go there.'

Láeg came and roused the Ulstermen to the battle, and he spoke these words : '*Coméirget ríg Macha mórglonnaig*' etc.

Then all the Ulstermen rose together at the call of their king and at the behest of their lord and to answer the summons of Láeg mac Riangabra. And they all arose stark naked except for their weapons which they bore in their hands. Each man whose tent door faced east would go westwards through his tent, deeming it too long to go around.

'How do the men of Ulster rise for battle now, friend Láeg ? ' asked Cú Chulainn. 'Bravely do they rise,' answered Láeg. 'All are stark naked. Each man whose tent-door faces east rushed westwards through his tent, deeming it too long to go around.' 'I pledge my word,' said Cú Chulainn, 'that their rising around Conchobor now in the early morn is speedy help in answer to a call of alarm.'

Then said Conchobor to Sencha mac Ailella : 'Good my master Sencha, hold back the men of Ulster, and do not let them come to the battle until omens and auguries are strongly in their favour and until the sun rises into the vaults of heaven and fills the glens and slopes, the hills and mounds of Ireland.' There they remained

1-1 trans. St

until a good omen was strengthened and sunshine filled the glens and slopes and hills and mounds of the province.

'Good my master Sencha,' said Conchobor, 'rouse the men of Ulster for battle for it is time for them to go.' Sencha roused the men of Ulster for the fight, and he spoke the words : ' *Coméirget ríg Macha* ' etc.

Not long was Láeg there when he saw all the men of Ireland rising together and taking up their shields and their spears and their swords and their helmets, and driving the troops before them to the battle.　The men of Ireland began each of them to strike and smite, to hew and cut, to slay and slaughter the others for a long space of time.　Then Cú Chulainn asked Láeg, his charioteer, when a bright cloud covered the sun : ' How are they fighting the battle now, my friend Láeg ? ' ' Bravely they fight,' said Láeg. ' If we were to mount, I into my chariot and Én, the charioteer of Conall, into his chariot, and if we were to go in two chariots from one wing of the army to the other along the tips of their weapons, not a hoof nor a wheel nor an axle nor a shaft of those chariots would touch the ground, so densely, so firmly and so strongly are their weapons held in the hands of the soldiers now.' ' Alas that I have not the strength to be among them ! ' said Cú Chulainn, ' for if I had, my attack would be clearly seen there today like that of the rest.' ' Nay then, little Cú,' said Láeg, ' it is no disgrace to your valour and no reproach to your honour. You have done bravely hitherto and you will do bravely hereafter.'

Then the men of Ireland began again to strike and smite, to hew and cut, to slay and slaughter the others for a long space and time.　There came to them then the nine chariot-fighters of the warriors of Irúad and the three men on foot together with them, and the nine chariot-riders were no swifter than the three on foot.

Then there came to them also the *ferchuitredaig*, the triads of the men of Ireland, and their sole function in the battle was to slay Conchobor if he should be defeated and to rescue Ailill and Medb if it were they who were overcome.　And these are the names of the triads : the three Conaires from Slíab Mis, the three Lussins of Lúachair, the three Niad Choirbb from Tilach Loiscthe, the three Dóelfers from Dell, the three Dámaltachs from Loch Derg-derge, the three Bodars from the river Búas, the three Báeths from the river Buaidnech, the three Búageltachs of Mag mBreg, the three Suibnes from the river Suir, the three Échtachs from

Áine Cliach, the three Mailléths from Loch Éirne, the three Abratrúads from Loch Ríb, the three Mac Amras from Es Rúaid, the three Fiachas from Fid Némain, the three Maines from Muiresc, the three Muiredachs from Mairge (?), the three Lóegaires from Lec Derg, the three Brodonns from the river Barrow, the three Brúchnechs from Cenn Abrat, the three Descertachs of Dromm Fornochta, the three Finns from Findabair, the three Conalls from Collamair, the three Cairbres from Cliu, the three Maines of Mossud (?). the three Scáthglans of Scár (?). the three Échtachs of Eirc, the three Trénfers of Taite (?), the three Fintans from (Magh) Femen, the three Rótanachs from (Mag) Raigne, the three Sárchorachs of Suide Laigen, the three Etarscéls of Étarbán, the three Aeds from (Mag) nAidne, the three Guaires from Gabail.

Then said Medb to Fergus : ' It were indeed fitting for you to give us your aid unstintingly in fighting today, for you were banished from your territory and your land and with us you got territory and land and estate and much kindness was shown to you.' ' If I had my sword today,' said Fergus, ' I would cut them down so that the trunks of men would be piled high on the trunks of men and arms of men piled high on arms of men and the crowns of men's heads piled on the crowns of men's heads and men's heads piled on the edges of shields, and all the limbs of the Ulstermen scattered by me to the east and to the west would be as numerous as hailstones between two dry fields (?) along which a king's horses drive, if only I had my sword.' Then said Ailill to his own chariot-eer, Fer Loga : ' Bring me quickly the sword that wounds men's flesh, O fellow. I pledge my word that if its condition and preserva-tion be worse with you today than on the day when I gave it to you on the hillside at Crúachna Aí, even if the men of Ireland and of Alba are protecting you against me today, not all of them will save you.' Fer Loga came forward and brought the sword in all the beauty of its fair preservation, shining bright as a torch, and the sword was given into Ailill's hand. And Ailill gave the sword to Fergus and Fergus welcomed the sword : ' Welcome to you, O Caladbolg, the sword of Leite,' said he. ' Weary are the champions of the war-goddess. On whom shall I ply this sword ? ' asked Fergus. ' On the hosts that surround you on all sides,' said Medb. ' Let none receive mercy or quarter from you today except a true friend.' Then Fergus seized his arms and went forward to the battle. Ailill seized his arms. Medb seized her arms and came to the battle and three times they were victorious in the battle northwards until a phalanx of spears and swords forced them to retreat again. Conchobor heard from his place in the battle-line that the battle had three times gone against him in the north. Then he said to his people, the intimate household of the Cráebrúad : ' Take up for a short time, my men, the position in which I am so that I may go and see who is thus victorious three times to the north of us.' Then said his household : ' We shall do so, for heaven is above us and earth beneath us and the sea all around us, and unless the firmament with its showers of stars fall upon the surface of the earth, or unless the blue-bordered fish-abounding sea come over the face of the world, or unless the earth quake, we shall never retreat one inch from this spot until such time as you come back to us again.'

Conchobor came forward to where he had heard the rout of battle against him three times in the north, and against the shield of Fergus mac Róig he raised his shield, the Óchaín Conchobuir, with its four golden corners and its four coverings of red gold Then Fergus gave three strong, warlike blows on the Óchaín Conchobuir and Conchobor's shield groaned.—Whenever Conchobor's shield groaned, the shields of all the Ulstermen groaned.— Strongly and violently as Fergus struck Conchobor's shield, even as stoutly and as bravely did Conchobor hold the shield, so that the corner of the shield did not even touch Conchobor's ear.

' Alas, my men ! ' said Fergus, ' who holds his shield against me today in this day of conflict where the four great provinces of Ireland meet at Gáirech and Ilgáirech in the battle of the Foray of Cúailnge ? ' ' There is a man here younger and mightier than you, and whose father and mother were nobler, one who banished you from your land and territory and estate, one who drove you to dwell with deer and hare and fox, one who did not permit you to hold even the length of your own stride in your land and territory, one who made you dependent on a woman of property, one who outraged you on one occasion by slaying the three sons of Usnech despite your safeguard, one who today will ward you off in the presence of the men of Ireland, namely, Conchobor mac Fachtna Fáthaig meic Rossa Rúaid meic Rudraigi, the high king of Ulster and the son of the high king of Ireland.'

' That has befallen me indeed,' said Fergus. And Fergus grasped the Caladbolg in both hands and swung it back behind him so that its point touched the ground, and his intent was to strike three terrible and warlike blows on the Ulstermen so that their dead might outnumber their living. Cormac Cond Longas, the son of Conchobor, saw him and he rushed towards Fergus and

clasped his two arms about him. ' Ready; yet not ready (?),
my master Fergus. Hostile and not friendly is that, my master
Fergus. Ungentle but not heedful (?) is that, my master Fergus.
Do not slay and destroy the Ulsterman with your mighty blows,
but take thought for their honour on this day of battle today.'
' Begone from me, lad,' said Fergus, ' for I shall not live if I strike
not my three mighty, warlike blows upon the Ulstermen today
so that their living outnumber their dead.'

' Turn your hand level,' said Cormac Cond Longas, ' and strike
off the tops of the hills over the heads of the hosts and that will
appease your anger.' ' Tell Conchobor to come then into his battle-
position.' Conchobor came to his place in the battle.

Now that sword, the sword of Fergus, was the sword of Leite
from the elf-mounds. When one wished to strike with it, it was
as big as a rainbow in the air.—Then Fergus turned his hand level
above the heads of the hosts and cut off the tops of the three hills
which are still there in the marshy plain as evidence. Those are
the three Máela of Meath.

Now as for Cú Chulainn, when he heard the Óchaín Conchobuir
being struck by Fergus mac Róig, he said : ' Come now, my friend
Láeg, who will dare thus to smite the Óchaín of Conchobor my
master while I am alive ? ' ' This huge sword, as big as a rainbow,
sheds blood, increase of slaughter,' said Láeg. ' It is the hero
Fergus mac Róig. The chariot sword was hidden in the fairy
mounds. The horsemen (?) of my master Conchobor have reached
the battlefield.'

' Loosen quickly the wooden hoops over my wounds, fellow,'
said Cú Chulainn. Then Cú Chulainn gave a mighty spring and
the wooden hoops flew from him to Mag Túaga in Connacht. The
bindings of his wounds went from him to Bacca in Corco M'ruad.
The dry wisps of tow which plugged his wounds soared into the
uppermost air and firmament as high as larks soar on a day of
fair weather when there is no wind. His wounds broke out afresh
and the trenches and furrows in the earth were filled with his
blood and the tents from his wounds. The first exploit which
Cú Chulainn performed after rising from his sickbed was against
the two female satirists, Fethan and Colla, who had been feigning
to weep and lament (over him). He dashed their two heads
together so that he[1] was red with their blood and grey with their
brains. None of his weapons had been left beside him save only

---

[1] the ground St

his chariot. And he took his chariot on his back and came towards the men of Ireland, and with his chariot he smote them until he reached the spot where Fergus mac Róig stood. ' Turn hither. my master Fergus,' said Cú Chulainn. Fergus did not answer for he did not hear him. Cú Chulainn said again : ' Turn hither, my master Fergus, or if you do not, I shall grind you as a mill grinds goodly grain, I shall belabour you as flax-heads (?) are belaboured in a pool, I shall entwine you as a woodbine (?) entwines trees, I shall swoop on you as a hawk swoops on little birds.' ' That has befallen me indeed,' said Fergus. ' Who will dare to speak those proud, warlike words to me here where the four great provinces of Ireland meet at Gáirech and Ilgáirech in the battle of the Foray of Cúailnge ? ' ' Your own fosterson,' said Cú Chulainn, ' and the fosterson of Conchobor and of the rest of the men of Ulster, Cú Chulainn mac Sualtaim, and you promised that you would flee before me when I should be wounded, bloody and pierced with stabs in the battle of the Táin, for I fled before you in your own battle on the Táin.'

Fergus heard that, and he turned and took three mighty, heroic strides, and when he turned, all the men of Ireland turned and were routed westwards over the hill. The conflict was centred against the men of Connacht. At midday Cú Chulainn had come to the battle. It was sunset in the evening when the last band of the men of Connacht fled westwards over the hill. By that time there remained in Cú Chulainn's hand only a fistful of the spokes around the wheel and a handful of shafts around the body of the chariot, but he kept on slaying and slaughtering the four great provinces of Ireland during all that time.

Then Medb covered the retreat of the men of Ireland and she sent the Donn Cúailnge around to Crúachu together with fifty of his heifers and eight of Medb's messengers, so that whoever might reach Crúachu or whoever might not, at least the Donn Cúailnge would arrive there as she had promised. Then her issue of blood came upon Medb (and she said : ' O Fergus, cover) the retreat of the men of Ireland that I may pass my water.' ' By my conscience,' said Fergus, ' It is ill-timed and it is not right to do so.' ' Yet I cannot but do so,' said Medb, ' for I shall not live unless I do.' Fergus came then and covered the retreat of the men of Ireland. Medb passed her water and it made three great trenches in each of which a household can fit. Hence the place is called Fúal Medba.

Cú Chulainn came upon her thus enagaged but he did not wound her for he used not to strike her from behind. ' Grant me a favour today, Cú Chulainn,' said Medb. ' What favour do you ask ? ' said Cú Chulainn. ' That this army may be under your protection and safeguard till they have gone westwards past Áth Mór.' ' I grant it,' said Cú Chulainn. Cú Chulainn came around the men of Ireland and covered the retreat on one side of them to protect them. The triads of the men of Ireland came on the other side, and Medb came into her own position and covered their retreat in the rear. In that fashion they took the men of Ireland westwards past Áth Mór.

Then Cú Chulainn's sword was given to him and he smote a blow on the three blunt-topped hills at Áth Luain, as a counter-blast to the three Máela Mide, and cut off their three tops.

Then Fergus began to survey[1] the host as they went westwards from Áth Mór. ' This day was indeed a fitting one (for those who were) led by a woman,' said Fergus . . . said Medb to Fergus. ' This host has been plundered and despoiled today. As when a mare goes before her band of foals into unknown territory, with none to lead or counsel them, so this host has perished today.'

As for Medb, she gathered and assembled the men of Ireland to Crúachu that they might see the combat of the bulls.

As for the Donn Cúailnge, when he saw the beautiful strange land, he bellowed loudly three times. The Findbennach of Aí heard him. Because of the Findbennach no male animal between the four fords of all Mag Aí, namely, Áth Moga and Áth Coltna, Áth Slissen and Áth mBercha, dared utter a sound louder than the lowing of a cow. The Findbennach tossed his head violently and came forward to Crúachu to meet the Donn Cúailnge.

Then the men of Ireland asked who should be an eye-witness for the bulls, and they all decided that it should be Bricriu mac Garbada.—A year before these events in the Foray of Cúailnge, Bricriu had come from one province to another begging from Fergus, and Fergus had retained him in his service waiting for his chattels and wealth. And a quarrel arose between him and Fergus as they were playing chess, and Bricriu spoke very insultingly to Fergus. Fergus struck him with his fist and with the chessman that he held in his hand and drove the chessman into

---

[1] reading St

his head and broke a bone in his skull. While the men of Ireland were on the hosting of the Táin, Bricriu was all that time being cured in Crúachu, and the day they returned from the hosting was the day Bricriu rose from his sickness.—(And the reason they chose Bricriu in this manner was) because he was no fairer to his friend than to his enemy. So Bricriu was brought to a gap in front of the bulls.

Each of the bulls caught sight of the other and they pawed the ground and cast the earth over them. They dug up the ground (and threw it) over their shoulders and their withers, and their eyes blazed in their heads like distended balls of fire. Their cheeks and nostrils swelled like smith's bellows in a forge. And each collided with the other with a crashing noise. Each of them began to gore and to pierce and to slay and slaughter the other. Then the Findbennach Aí took advantage of the confusion of the Donn Cúailnge's journeying and wandering and travelling, and thrust his horn into his side and visited his rage on him. Their violent rush took them to where Bricriu stood and the bulls' hooves trampled him a man's length into the ground after they had killed him.

Hence that is called the Tragical Death of Bricriu.

Cormac Cond Longas, the son of Conchobor, saw this happening and he took a spear which filled his grasp and struck three blows on the Donn Cúailnge from his ear to his tail. ' No wonderful, lasting possession may this chattel be for us,' said Cormac, ' since he cannot repel a calf of his own age.' Donn Cúailnge heard this for he had human understanding, and he attacked the Findbennach, and for a long time and space they fought together until night fell on the men of Ireland. And when night fell, all the men of Ireland could do was to listen to their noise and their uproar. That night the bulls traversed the whole of Ireland.

Not long were the men of Ireland there early on the morrow when they saw the Donn Cúailnge coming past Crúachu from the west with the Findbennach Aí a mangled mass on his antlers and horns. The men of Ireland arose and they knew not which of the bulls was there. ' Well now, men,' said Fergus, ' leave him alone if it is the Findbennach Aí, and if it is Donn Cúailnge, leave him his triumph. I swear that what has been done concerning the bulls is but little in comparison with what will be done now.'

The Donn Cúailnge arrived. He turned his right side to Crúachu and left there a heap of the liver of the Findbennach. Whence the name Crúachna Áe.

He came forward to the brink of Áth Mór and there he left the loin of the Findbennach. Whence the name Áth Luain.

He came eastwards into the land of Meath to Áth Troim and there he left the liver of the Findbennach.

He tossed his head fiercely and shook off the Findbennach over Ireland. He threw his thigh as far as Port Lárge. He threw his rib-cage as far as Dublind which is called Áth Clíath. After that he faced towards the north and recognised the land of Cúailnge and came towards it. There there were women and boys and children lamenting the Donn Cúailnge. They saw the forehead of the Donn Cúailnge coming towards them. ' A bull's forehead comes to us ! ' they cried. Hence the name Taul Tairb ever since.

Then the Donn Cúailnge attacked the women and boys and children of the territory of Cúailnge and inflicted great slaughter on them. After that he turned his back to the hill and his heart broke like a nut in his breast.

So far the account and the story and the end of the Táin.

A blessing on every one who shall faithfully memorise the Táin as it is written here and shall not add any other form to it.

But I who have written this story, or rather this fable, give no credence to the various incidents related in it. For some things in it are the deceptions of demons, others poetic figments ; some are probable, others improbable ; while still others are intended for the delectation of foolish men.

NOTES TO TEXT

1-146 No part of this introductory section is found in Rec. I TBC which opens abruptly with the mustering of the Connacht forces and the description of the Dubloinges under Cormac. No reason for the expedition is explicitly given in Rec. I, and the first reference to the bull occurs after the list of places through which the Connacht army marched: *A Findabair Chúalngi is ass fodailte in tslóig Hérend fón cóiced do cuingid in tairb* (LU 4609-10). We cannot tell whether Rec. III contained the Introduction, as the beginning of the tale up to l. 1177 of the present text is lacking. Much of the material of this Introduction is found in other tales e.g. the names of the six daughters of Eocho Feidlech and of the two brothers of Ailill; the statement that Medb insisted on a husband *cen neóit, cen ét, cen omun* and that such a one was Ailill; the two bulls, the Finnbennach of Connacht and the Donn Cúailnge of Ulster which were destined to meet in battle. All these points are found in other texts and would have formed part of the traditional lore of any compiler. Thurneysen ascribes the composition of the Introduction to the compiler of the LL version, Bearbeiter C. This, however, must remain conjecture. All we can assert is that someone—Thurneysen's Bearbeiter C or another— wove all these strands of tradition together and invented a contention between Medb and her husband to explain why she coveted the Donn Cúailnge and mustered the forces of the four provinces of Ireland to invade Ulster and gain possession of the famous bull. As regards this *immarbáig*, between Ailill and Medb, Thurneysen in his analysis of Táin Bó Fraích (Held. p. 290 n.2) suggested that it was inspired by the passage in TBFr where Ailill calls for his jewels: *Tucaid mo shéotu damsa huili! ol se* (TBFr² 288). But no reason is given for Ailill's counting of his treasures at this point except to explain how it was that he missed the ring he had given to Findabair. Far closer to the LL *immarbáig* is the long passage in Táin Bó Flidais II, the version contained in the Glenmasan MS. Bricne returns to Medb from Oilill Finn and sings at great length the praises of Oilill's household (Celt. Rev. II, pp. 28, 30, 32, 100, 102, 104). *Is fír duit-si gurab maith tech Oilella Finn, ar Medb, agus gidedh as ferr mo tech-sa go mór ana sé. Is ferr gaisged mo laoch agus mo lath ngaili. Is lia mh'urradha agus mo deóraid. Is lia mo macaim agus mo bandtracht. Is lia mo sheóid agus mo maeine. Is lia mo chruid agus mo chetra* etc. etc. (loc. cit. pp. 32, 100). We may note the words *urrada, deóraid; seóit, maeine; cruid, cethra*. In TBFlid II this long passage is likewise called *immarbáig*; *Ba fergach Medb de sin .i. fa tech sa doman do chur tar a tech féin. Do neimdligis, a Bricni, bar Medb, imarbaid do cur am cenn* (loc. cit. p. 28); and at the end of the contention: *Do leigedar secha iar sin an imarbaid* (loc. cit. p. 104). In TBFlid II there are also other passages which are in agreement with the LL-Táin in points where LL differs from Rec. I, and in Rec. III TBC there are some points which agree with the tradition of TBFlid II against that of the other TBC recensions. We must assume either (*a*) that TBFlid II was largely influenced by the LL-Táin and by Rec. III TBC, or (*b*) that TBFlid II, though found only in late MSS., drew like LL- TBC and Rec. III TBC on a variant version of TBC (x). I am inclined to the second theory. The importance given to Medb and to Connacht by thus introducing her and Ailill at the opening of TBC suggests that this x-version of TBC may have been one which was written from the Connacht point of view. Though largely concerned with the activities of Fergus and the Dubloinges in Connacht, the late version of TBFlidais seems clearly to have been compiled by a Connacht man.

2 *arrecaim* An example of the confusion of preverbs in this text. Here *ar-* for *do-*.

comrád ..⁝.:d cherchailli Again infra 277. Here the genitive followed by another genitive is lenited, as in the current language. Pedersen notes this simply as a Mid. Ir. example of aspiration after a nom. sing. m. noun (KZ 35.436). In MU we have an instance in *Sencha Mór . . . sobérlaid fher in talman* (MU² 757), but the aspiration seems to be optional, not obligatory. Stowe has *comradh cintt cearcailli* (2) and *comradh cind chearcailli* (280); and in Togail Troí we find *cend gascid clainne hAdaim* (LL 32856).

3 *is maith ben ben dagfir* Similarly infra 113, *is maith fer fer in taige i táam*, beside *is maith fer in taige i tám* 106, the first sentence meaning 'The owner of the house in which we are is a generous man', the second 'The owner of the house is generous'. Windisch needlessly suggested that in *is maith ben ben dagfir* the first *ben* was superfluous. In Stowe the reading is *Maith an ben bean deigfir* (4), identical in meaning with the LL sentence. For this use of the definite article cp. *Is maith in láech Mac Cecht* LU 7228 (BDD). Stokes translated this as 'Good is the hero M.Ċ.', but, as in Mod. Ir., it means 'A goodly hero is M.C.' Later in the same text, when Conaire is beheaded, his head speaks: *Maith fer Mac Cecht | Fó fer Mac Cecht* (LU 7950). The omission of the article in *is maith ben ben dagfir* is more or less the same usage as the omission of the article with substantives defined by a relative clause. The sentence was perhaps felt to be *is maith ben as ben dagfir*. Cp. *Huar gaeth gaeth domanicc uait, a ben* RC xiv 243, 245.

6 *.in lá thucus-sa thú* Example of Mid. Ir. aspiration of verb in relative position. Other examples infra: *in tú thuc in sét?* 125; *cia thic* 535; *gid moch this* 1622. Note also aspiration of pronoun object. Contrast *fégaim-se tussu* 1609 (= *fecaim thusa* St 1651).

8 *ar bantincur mnáa* As Thurneysen pointed out (Studies in Early Ir. Law p. 20 n. 2) legal exactness is not to be expected of a storyteller. Here the legal phrase is loosely used and *bantinchur* is applied to the property of a woman before marriage. Similarly infra 48.

14 *sé ingena* In Cath Bóinde the six daughters of Eochu Feidlech are also named (Ériu II 174-6).

21 *Bátar ocum sain* In MS. *Bátar | ocum sain*. Stowe reads *Batar-sin agam*. Windisch missed *ocum* at the beginning of the line in the LL text. He also omitted *agam* in the Stowe reading.

24ff. This passage gives us the background for TBC, *aimser na cóicedach*, when, according to the pseudo-historians, there was no king of Ireland, only provincial kings (*cóicedaig*). Here four of the provinces are mentioned, *Lagin*, *Temair*, *Ulaid* and *Mumu*. In the text Medb and her allies are called *ceithre cóicid Hérend*. The opening lines of Rec. I give a similar picture: *Tarcomlad sloiged mór la Connachtu .i. la hAilill ⁊ la Meidb, ⁊ hetha húaidib cossna tri choíced aili* (LU 4480-2).

26 *ó Eochaid Bic* that is, Eochaid Bec mac Cairpre, king of Clíu. Stowe here has *ó Eochaidh mac Luchta* (28), which seems to be the better reading. In Cath Ruis na Ríg Eochu mac Luchta is king of all Munster. The official view of the historians was that there were two provinces of Munster, 'the fifth of Cú Roí' and 'the fifth of Eochu mac Luchta'. Cú Roí appears later in the Táin but he is not called king of a *cóiced*. LL's introduction of Eochaid Bec here seems to be a slip, perhaps a reminiscence of Táin Bó Dartada. See Ir.T. II² 189. Eochu Bec mac Cairpre is mentioned also in the Dindshenchus of Áth Cliath Medraige (Rennes Dinnschenchus §61, Bodleian Dinnshenchus §26). In LL-TBC Cairbre Nia Fer is reckoned as king of Tara which is treated as a province distinct from Lagin. Similarly Cairpre is called *rí Temrach* in CRR (LL 23192). See infra note 46.

27ff. As Tomás Ó Máille pointed out in an article entitled Medb Chruachna in ZCP xvii 129ff., the material in this passage is found in fuller detail in the tale Cath Bóinde, Ériu II 174ff., otherwise called Ferchuitred Medba (Anecd. V. 18). Medb's husbands are there enumerated as Conchobor, Fidech mac Féicc, Tinne mac Connrach, Eochaid Dála of the Fir Chraíbe in Connacht, and finally Ailill mac Máta. It is to be noted that of those who sought Medb in marriage in the present tale, only Conchobor agrees in the list of husbands. Ó Máille suggests that the compiler of LL- TBC drew on the source of the tale Cath Bóinde. But such traditions concerning Medb must have been commonly known.

27 *is mé ra chunnig . . . nára chunnig* Ra for *ro* is very common throughout the text: *Ra rimit ⁊ ra hármit ⁊ ra achnit* 60, *ra iarfacht* 90–91, *anra chansat* 122, *ra ráidsebair* 135 etc.

27 *in coibchi n-ingnaid* Here again the term *coibche* is misused. *Coibche* was the bride price paid to a woman's father. Again ll. 39–40 Medb speaks of giving Ailill *cor ⁊ coibche*, as if it were to the man that the *coibche* should be paid.

28 *fer cen neóit...* In Cath Bóinde it is stated that it was *geis* for Medb to marry a man unless he was *cen étt, cen omun, cen neoitt* (Ériu II.182. 3–4). Similarly in TBFlidais II, Ailill, discovering Medb and Fergus together, is tempted to kill Fergus, but, it is explained, spares him because he is *gan neóid, gan éd, gan omun* (Celt. Rev. I. 228. 11–13).

37 *dáig ní raba-sa . . .* See Ó Máille, ZCP xvii 129ff. where Medb is taken to represent the sovreignty of Ireland, and this reference to her numerous husbands to imply that there were always rivals for the sovreignty.

39 *nírsat neóit . . .* Here *nírsat* is syntactically a present. Other examples of this artificial literary formation occur infra: *nírsa eólach* 1475, *girsat corcra conganchness* 3034, *ciarsat cendchroimm* 4574. See my note in Celtica vii p. 38-42.

43 *ní fuil diri* Mid. Ir. aspiration after *ní*. Similarly *ní chúala* 48, *ní thibér-sa* 139 etc.

44 *fer ar tincur mná* The man who married an heiress, *banchomarba*, was *fer for bamtinchur*. In general the wife's honour-price was one half of her husband's, but in a *bantinchur* marriage the husband's honour-price was reckoned according to that of his wife. Such a husband had half the honour-price of his wife. Here again the storyteller is inexact. (See Thurneysen ZCP xx. 208-9). Note that, though Medb reproaches Ailill with being

fer ar tinchur mná, in the event it is proved that Ailill's property is as great as that of Medb and that in addition he now possesses the Donn Cúailnge.

46 *Find for Lagnib ⁊ Carpre for Temraig* There was some uncertainty about Cairbre Nia Fer. He is included as king of Lagin in the list of *cóicedaig* (see EIHM 178, and ib. n. 2) but here as in CRR he is represented as a brother of Finn and of Ailill and is king of Tara. According to T. F. O'Rahilly this was a Laginian invention, and taken up by the redactor of the LL-Táin and CRR (EIHM 179). Here the king of Tara is reckoned as a provincial king together with Find in Lagin, Eochu in Munster and Ailill in Connacht. The Leinstermen and Munstermen and *popul na Temrach* are named later in the narrative but Cairbre does not appear. Cairbre's son Erc is represented as bringing the men of Tara to help the Ulaid 'without his father's leave'.

47 *Léicsius* Similarly *gabsus* 49. M. O Daly takes these as forms with a double ending and compares *scérdait* 2422 (Ériu xiv. 61. 1102–3).

47 *nip[tar] ferra* A plural form *ferra* occurs three times in Airec Menman Uraird Maic Coise (15–16 cc. MSS.): *dus in mbatar seoit ba ferra isin uaim andas a tucadh este* Anecd. II 57. 19–20; 53. 14; 62. 8. Cp infra 2100 where equative *lir* is treated as a positive with plural *liri*: *Giambtar liri fir Hérend.*

48 *ni chúala chúiced . . .* Mid. Ir. aspiration of object. Cp. *fer find firfes chless* 234; *dobérad chloich* 1273, *focheird chloich* 1274; *dobeir phóic* 1878, etc.

49 *i tunachus mo máthar* Tunachus does not seem to be attested elsewhere, but Stowe's *a dualgus mo mhathar* (53), which is translated, gives the required meaning. (Windisch has *tanachus* of the Facsimile, and the word is given with the meaning 'succession' in RIA Contribb.). Possibly *tunachus* is no more than a scribal misreading of a partly illegible *tochus*. Cp. *Gabais iarom Conchobur righi n-Ulad iar sin ar thochus a máthar ⁊ a athar* RC vi. 178. 131. *Tochus* means 'property, possessions'.

50 *dáig ar bith . . .* This double conjunction is common in this text. Cp. infra 648, 1226, 1558 etc. As Rudolf Hertz pointed out (ZCP xix. 257), this form of the conjunction seems to be confined to recensions of TBC. There are, however, only two instances of it in Rec. I TBC (LU 6301, 6406–7), and both occur in the section Breslech Maige Murthemne, which as Thurneysen has noted, is later in style and language than the rest of Rec. I, and which is identical except for small details in all recensions of TBC. In Stowe we get no instance of *dáig ar bith*; it has been replaced by *uair* in practically every occurrence of LL. Neither does *dáig ar bith* occur in Rec. III. In the Breslech Maige Murthemne passage the first occurrence infra 2149 is *dóig* in Rec. III, and the second 2227 is *ar* in Rec. III; *uair* in Stowe in both instances. In the newly-discovered O'Curry MS (C), the first occurrence in the Breslech passage is *dáig fo bith* (1196) and the second *dóigh ar bith* (1297). The conjunction *ar bith* (= *fo bith*) occurs once in the LL– text: *ar bith cét carpat ruc in Glassi dib co muir* 1352. The preposition *ar bith* is common in the LL text e.g. 1638, 1645, 4592 etc. It does not occur in the Stowe version; the construction is altered or another preposition substituted, e.g. for *ar bith ainig Fergusa* 1645 Stowe has *ar dáig oinigh Fergusa* (1685), and for *ar bith gona Conchobuir* 4689 Stowe substitutes *do guin Concobair* (4796).

58 *fornasca* for *ordnasca* (*ordu* + *nasc*), 'thumb-rings'.

60 *a murthréta* (= a *móirthreda* St 64) Again 66 (= *a mallthréda* St 67). The meaning of *mur-* is doubtful, but it seems to be an intensive prefix. Cp. *muirech*, 'leader, lord'.

72 *Findbennach* 80 *Dond Cúalnge* In these two bulls were re-incarnated two swineherds of the *síd*. Before becoming bulls they had passed through many other re-incarnations. An account of these is given in the tale De chophur in Dá Muccida (Ir. T. III¹ 235–247). The only references to the magical nature of the bulls are to be found in LL-TBC, not in Rec. I; they are an enumeration of Dond Cúalnge's attributes (infra 1320–33) and the sentence *Atchúala Dond Cúalnge anni sein & báe ciall dunetta aice* (4888). Dond Cúalnge is not mentioned by name in Rec. I (where he is called merely *in tarb*) until after the Fer Diad episode when the titles of the episodes which follow are enumerated, among them *A us in Duib Cúalngi for Tain* (TBC² 2752). Again towards the end of the text: *Focairt chos in Duind Chúailgni ar adairc a chele* (TBC² 3656–7). Findbennach is mentioned by name in the same passage (3650) and earlier in the text in a poem spoken by Dubthach: *sliagud n-imdub arubtha | fri Findbend mná Ailellá* (LU 4665–6).

76 *conscomarc* for *im(m)comairc* (*fiafraighes de* St 81). Similarly the corrupt form *conscodarc* 142 where Stowe reads *ro fhiarfaigh*.

80 *Tó duit-siu . . . Tó = dó.* Cp. infra 1242, 1649, 1653 etc. *Tó, dó* is the preposition *do* with 3 s. m. and n. pronouns, 'to it', in the sense of 'hither'. It is used as an equivalent of an imperative, 'go, be off'. Cp. *Tó duit-siu ar cend t'arm* 1115 = *Eirg-siu ar cend t'arm* 1096. *Dó* is used adverbially in *Lotar .. . na echlacha dó co tech Dáre* 88, *Lotar na echlacha ar cúl dó samlaid* 141, 'went off, went away'.

86–7 *cardes mo [s]liasta-sa fessin* In TBFlidais II Medb offers to Fer Diad and others as an inducement to fight against Ailill Find *righe na Gamanraidhi . . . agus feis a Cruachain do gres agus cairdes mo shliasda-sa fos* (Celtic Review III. 124). Where in the present text Medb offers Fer Diad her golden brooch (2634–5), in TBC² she promises *comaid dom sliasaidsea* (2232).

91 *cid dobretha* Throughout the text *dobreth* and *dobretha* are used as pret. 3 s. active.

99 *cid an ni ra Ultaib* (*gidh olc gidh maith la hUlltaibh é* St 108–9). The phrase *is ni limm* (indef. pron. n.), 'I think it important, advisable', is usual but the use of the article here together with the reading of Stowe suggests that *ni* is a substantive, 'thing', and that an adjective has dropped out.

101 *ra ráde* for *anra ráde Ba maith dano la Mac Roth in fregra sin d'faghail ó Dhairi* St. 110–111.

104 *co mbátar búadirmesca* for *combtar búadirmesca* or *co mbátar for búadirmesca.* Cp. St. which here reads *go mbatar for buaidhir measgtha ⁊ meraighthi* (113).

106 *is maith fer in taige* Windisch translates 'ein guter Mann' and the translation 'wealthy' is suggested in the Contributions s.v. *maith* (e), but *maith* is here used in the sense of 'generous'. Cp. *Bá maith iarom inti Guairi* LU 9724.

109 *cid immi gabtais . . .* The omission of *no* is noteworthy. Contrast St. which here reads *cidh ime-siomh no iaddaois Ulaidh uile* (119). Again infra 192 *Gilidir snechta sniged fri óenaidchi.*

112 *And sain dano conarraid in tres echlach comrád forru* In Mid. Ir., though not in O. Ir., the construction is relative when part of the sentence other than subject or object is brought forward for emphasis by an *is-* clause. Thus infra 2198 *ní fair bias a nós.* Cp. *Is and sin bias* . . . LU 2464, 2495, *Is·and sin icfas* . . . ib. 2286. Here (*is*) *and sain* prefixes relative *con-* (for *do-*) to the verb. Cp. infra *And sain confucht[r]aither Dubthach* (4148). As I suggested in my note on the preverb *con-* in LL- TBC (Ériu xx. 108 f.), other preverbs besides *con-* seem to have been used in relative position. This I believe, may be in part the explanation of the interchange of preverbs which is so marked a feature of the LL- Táin, and, to a lesser extent of LL- MU and of CRR. It is noteworthy, for instance, how often such interchange occurs in (Mid. Ir.) relative position. Thus in the present text *bar-* for *do-*: *Is and barrecaib-sium i n-airthiur Ai an tan sin* (2514–5). (Such forms have spread to non-relative position, as e.g. 3913, 3931). So too we get *fo-* for *do-* in relative position: *Cia maith fogniat* 318; *is aire fognid Medb sin* 583; *cluichi puill fognithi for faithchi na Emna* 834. Again we find *ro-*, *ra-* for *do-*: *Is and sin radechaid chucu-som Fiacha Fialdána* 2448; *And sain radechatar* . . . 4596. (Cp. *Is aire condeochatar...* 1249, with *con-* for *do-*). With these instances from the present text compare also: *Is and barecmaing dóib sin bith ar múr Temra Lúachra* MU² 362; *And barrecaim Ailill flathbriugaid dosum and* LL 22977 (CRR). There are instances of this interchange of prefix in relative position in Rec. III TBC: *in torannchleas boghní Cu Chulainn* (ZCP viii. 544. 20, 25); *Is annsin tainic Eadurcomhal* . . . ⁊ *baregaim comradh do* ⁊ *da ara charbaid* (ib. 551. 23), where *is annsin* is understood before *baregaim* also.

119 *Nírb uráil limm* . . . The same phrase occurs in MU: *Nírb uráil lim lom cró* ⁊ *fola issin mbél tacras sin* (MU² 369).

127 *aní rádit* Possibly something has been omitted after these words in LL. Compare St which reads: *Is fíor a raidit na hechlacha. Cret adeirit? ar Daire. Aderit, ar in t-oclach, muna ttuctha ar áis* etc. (132–3).

128–9 *ra móreólas Fergusa meic Róig* Róig written here with long vowel. The older form is *mac Roig* (*Roaig, Roïch*), disyllabic, but in Mid. Ir. the form is usually *Mac Róig.* In LL 28125–6 we get *Fergus mac Róig: Assail mcic Umóir.*

129 *Dothuṅg mo deo* Cp. infra *Dothongusa mo dee* 808: a mid. Ir. corruption of *tongu. Tongu-sa na dé* 1630.

130 *Fessit* A denominative verb from *fess*, vn. of *foaid*, 'spends the night.'

138 *dot lind-su* ⁊ *dot bíud* The force of the prep. *de* here is clearly brought out in Stowe by the use of a prep. phrase: *do druim do leanna-sa* ⁊ *do bidh* (146).

144 *féth dar fudba* This seems to be a proverbial expression meaning that Mac Roth's bad news need not be glossed over.

145–6 One would expect this line to read *dobértha* or *co tibertha ar écin cen co ttuctha ar áis.* Cp. supra 135–6.

148 *na secht Mani* The Maines are mentioned not here in Rec. I, but later (LU 4640–43) when Medb enumerates her forces (= 335ff. in this text). Maine Andóe is included among them in Rec. I.

150 *Mane Conda Mó Epert* Obviously a mistake for *Mane Andóe* ⁊ *Mane Mó Epert.*

151 *co maccaib Mágach* In Rec. I Ailill mac Mágach instead of Scandal.

155-6 ⁊ co Fergus mac Róig His name is not mentioned here in Rec. I nor in Stowe (which drew on Rec. I). LU reads: co Cormac Cond Longas mac Conchobair cona thrib cétaib boi for condmiud la Connachta (4484-6); similarly Stowe: co Corbmac Con Loinges mac Concobair ⁊ gusin dubloinges (162-3).

157 forthi The original meaning of forthi seems to be 'cape' or 'over-mantle'. Cp. BDD² 673. Here used for long hair. O'Clery with this passage in mind gives foirtchi bearrtha .i. monga dubha no gruaga dubha—an attempt to explain the unknown forthi, or, more likely, taken by O'Clery from Stowe TBC which here reads fuilt dhubha forra (166).

158 Lénti . . . ba tórniud See Zimmer in ZCP iii 285f. where he suggests that ba tórniud of LL is a Mid Ir. attempt at the older fo thairinniuth of LU here. He suggests that the LU form is for fodorintliud (= odor-), which he translates 'mit braunem Einschlag'. The suggestion put forward by Windisch is much more likely. For the second band LU has lénti co ndercintliud co horcnib sis (4493-4). Cp. infra Léne... fo derggindliud do derggór 2143, 4306-7, 4316.

168 I nn-óenfecht dostor(g)baitis a cossa Here the infixed pronoun -s- functions as a relative.

177 Sochaide scaras . . . In Rec. I Medb addresses this speech not to her druid but to her charioteer who thereupon turns the chariot to the right as a good omen.

179 forom-sa combenfat Co n- prefixed to a verb in relative position. Similarly is tria glaslec clochi conindsmad 634, Dámbad forro conmebsad 4690-1.

180 ni théit . . . as diliu The antecedent of as is omitted. Such omission is of common occurrence in the text, and it is noteworthy that the younger version Stowe invariably inserts an antecedent in such sentences. Here Stowe reads: ní théid amach . . . neach as annsa linn (193).

182 Cipé [tic] nó ná tic St reads Cibé na ticc (195). One would expect the subjunctive. Cp. infra Gibé dig 1064, Cipé ti 1538 etc.

184 for fertais in charpait I take this to be the chariot of Feidelm, not that of Medb. In Rec. I, after the description of Feidelm comes the sentence: Gaisced lasin n-ingin ⁊ dá ech duba foa cárput (LU 4522), and in a later passage in YBL (but not in LU), referring to Feidelm we find co n-accadar in carpat ⁊ ingen alaind and (TBC² 277).

184-5 'na farrad ina dochum In LU, co n-accatar in n-ingen macdacht remib (4511); in St only ina farradh (198). In a note to St 198 I suggested that ina farrad here meant '(coming) towards her', and that LL should read 'na farrad .i. ina dochum, and I quoted examples of i farrad meaning 'towards'. But perhaps it is unnecessary to take ina dochum as a gloss; there are other examples of a double use of a prepositional phrase. Compare Bui Conchobur mac Neusa aidqi n-ann ina chotlud confacco ni ind oiccbein chuici ina dochumb (Tochm. Ferbe, Ir. T III² 548. 1-2). and Co n-acca a cheile chuice iarsind tsossad ina arrud cheana (Tochm. Étaíne, Ériu 8. 176. 2-3). (I take ina arrud in the latter sentence to mean 'towards him'). An instance of i farrad = 'towards' occurs in CRR: ⁊ na gabad mórslúag na farrad LL 23091.

185 ic figi chorrthairi i.e. weaving threads in a magical manner, to enable her to prophesy the coming battles. See Stowe TBC n. 199 and Addendum.

186   *ina déssaib*   Dpl. of *días*, 'point, end'?   The words are omitted in St.   Possibly *ina dessaib* is no more than a scribal error for *ina desra*, a repetition of *ina láim deiss*.   (Cp. the scribal error *dessa* for *desra* infra 2217). In LU 6059 we get the spelling *fora desraid*, a form which a scribe might carelessly copy as *dessaib*.

190   *Cosmail do núapartaing . . . Partaing*, some bright red substance. Windisch suggests 'Parthian leather dyed scarlet' (p. 28 n. 3).   Generally used in this stereotyped phrase referring to red lips (*dergithir p. a beoil* LU 9273), but also used of some red ornament (TBFr[2] 101).

191   *a llámaib sirsúad*   (Contrast *i llámaib súad* infra 4340).   In Mid. Ir. there are occasional instances of aspiration after the dative plural.   Again infra *fria chlessaib chluchi* 928, *im chailaib choss Etarcomla* 1680.   See Pedersen KZ 35. 439 and V. Hull, Language 26. 274–5.   The instance *i n-ilgonaib Chon Culaind* 2165 exemplifies the aspiration of a proper name in the genitive. The text, however, abounds in examples of the non-aspiration of gen. proper name after the dative plural e.g. *i ndiamraib Sléibi Fúait* 1152, *ria curadaib Cráebrúade* 1286 etc.

195–6   *Trilis aile combenad foscad fria colptha   Benaid foscad fri . . . ,* lit. 'casts a shadow against', that is, 'lightly touches'.   *Co n-* used as a relative particle.   Cp. supra 179.

197   *Forrécacha   For-* for *do-*.   A reduplicated preterite of *do-éccai*.   Cp. infra 4183 *forrécaig*, and in YBL *doreccacha* (TBC[2] 3107).

204   *Cia facci*   As can be seen from the readings of Rec. I and Stowe *cia* here is for *co*, geminating conjunct. interrog. particle, meaning 'how?' *Co acci* LU 4530, 4540, 4544, *co fhaice* St 226, 230.

207   *'na chess noinden*   (Again *ina chess* 212, 216, 221).   The *cess noinden* was a mysterious disability which incapacitated all the Ultonians except women and boys and Cú Chulainn (*Treide forna bíd noenden di Ultaib .i. meic ⁊ mnaa ⁊ Cú Chulaind* LL 14580).   Here then Medb implies that her foray into Ulster will be successful since the Ulster heroes are suffering from the *cess noinden*.   In one account of the origin of the *cess noinden* which is given in the tale Noenden Ulad ⁊ Emuin Macha (LL 14547–14585), it was said to be the result of the curse of Macha, forced to compete, when pregnant, in a race with the king's horses: *Iac[h]taidsi dano lia hasait.   Nach duine rodachúala son foscerded i cess cóic laa ⁊ cethri aidche . . . Nert mna siúil ba hed no bíd la cach fer di Ultaib fri saegul nonbair isind noe[n]den* (LL 14575–79). On the basis of this tale the *cess noinden* has been taken by scholars as a survival of the practice known as couvade.   (See Thurneysen, Heldensage p. 359, M-L. Sjoestedt, Dieux et Héros des Celtes p. 39).   In a recent article (Éigse x. 286–99) Tomás Ó Broin disputes this theory and suggests that the *cess noinden* is a tradition of the vegetation ritual and represents the death of winter decay and the spring rebirth.   His theory would better explain the implication in the Táin that the *cess* lasted from Samain until the beginning of spring.

207   *connice*, 'to him'.   More usually *connice sin*, but again 1241 *Dó duit-siu connice.*

234–275   *Atchiu fer find . . .*   This poem is interpolated (in rasura) in hand H in LU.   See Thurneysen, Heldensage p. 236 and ZCP ix. 430–32. YBL is lacking here.   Thurneysen (Heldensage 121) suggests that the interpolated poem takes the place of a rhetorical passage and one which was obviously shorter than the poem which has been squeezed into the space afforded by erasure. This is the only H interpolation in LU which is also found in Rec. II, if we except the brief insertion of the name Flidais (LU 4623).

234 *fer find* O'Clery, quoting *fear fionn*, glosses .i. *fear beag*. But *find* in the sense of 'blessed' or 'handsome' seems more suitable here.

242 *Ro fail gnúis* . . . An absolute use of *ro fail* = *fail*, as used in preceding stanza. *Fail* at the beginning of a line is very common in Irish verse. LU has *Dofil gnúis as grato dó* (4554), which is quoted in RIA Contribb. s.v. *do-fil* as '*do-fil do* (of possession), "falls to" ', but here LU *dofil* = LL *ro fail*. In Eg. 1782 the line reads *Fil gnúis is gratai do atchiu*, whence Stowe's *Do fail gnuis as grata dhó at-cíu* (246). Again infra 2779 we get *Ra fail gnim is mó bard láim*, where Stowe has *Do fuil gniom is mó frit laimh* (2803) and YBL has *Ita feidm is mó ad laim* (TBC² 2411). Examples in prose occur. In CRR: *Ro fail ám a morabba damsa sain ale* (LL 22647). It is used absolutely three times in the prose of Rec. III TBC: *Acht rofil derbrathir agumsa* RC xiv. 262 §48, 263 §52, 266 §62. See infra note 4349. *Do-fil* with the meaning 'is at hand, approaches', does not occur in the LL- text. It occurs several times in Rec. I: *Tofil slóg mór tarsa mag cucund* (TBC² 1440–1) etc.

261 Printed *ard ás gail* Diplomatic Edition. In LU the two lines read *Ardaslig tar fonnad clé | cotagoin in riastarthe* (4574–5), 'He smites them over the left wheel of the chariot. The distorted one wounds them'. A verb seems called for here. I would suggest *ardasgail* (*ardascail*), '(which) he looses, casts'. For this use of *ar-* infixing a pronoun in a rel. sentence see Ériu xx. 110.

266 *is é farsaig As e dob soigh* St 268. In LU the line reads *Dóich lim iss e dobobsáig* (sic) (4580) and Eg. 1782 has . . . *dobursáigh*.

282 *for Cúl Silinni* This is the place in Co. Roscommon where the army spent the first night of their expedition (297–8) and which is mentioned again in a passage found only in Rec. II (infra 393ff.). The name is *Cúil Sibrinne* in LU (4594). It is to be distinguished from *Cúl Siblinni* 288. See Index, Stowe TBC.

284 *for Cuil [Siblinni], for Ochain, for Uatu fothúaid.* These names are misplaced here and are partly repeated 288. After *Slechta* LU reads *for Cuil Sibrinni, for Ochuind fadess, for hUatu fathúaid* (4599).

287 *for Slechta conselgatar claidib* . . . 'which swords cut down'. *Con-* here used as relative. Similarly Stowe: *for Sleachta .i. ainm an ionaid consealgatar cloidhme re nOilill 7 re Medhb* (293), but LU has merely the words *for Sléchta selgatar* (4598). See infra 512–5.

288 *for Ochun* Read *for Ochain* as supra 284 *For Ochuind fadess* LU 4599. Stowe has *for Ógan*, a misreading.

288 *for Catha* seems to be a misreading for *for hUatu fathúaid* (LU 4599); cp. supra 284. *For Cromma* is a misreading of *for Commur fadess* LU 4600. These two names are omitted here in Stowe.

295–6 *for Fid Mór i Crannaig Cúalñgi* Here LU inserts the names of the two rivers which play a part in the story later: *for Fid Mór, for Colbtha, for Crond i Cúalngi* (4608). Similarly Stowe 302. The LL reading *i Crannaig Cúalñgi* looks like a distortion of *for Crond i Cúalngi*. Stowe, as often elsewhere, combines the readings of LU and LL: *for Crand i cCuailnge .i. i cCrandaig Chualann* (sie) (302).

297-8 *for Cúil Siblinni* The *b* of *Siblinni* is faint or effaced here in MS.; we should read *Silinni* as supra 282 and again infra 393. (The name is *Cúil Sibrinne* here in LU 4615, and there is a gloss on the name in hand H *.i. Loch Carrcin ⁊ o Silind ingin Madchair ro ainmniged* at its earlier occurrence 4594). See Gwynn's note on Turloch Silinde, MD III p. 547. In the prose dindshenchus of Loch mBlonac it is called *Cúil Silinde a Muigh Ai* (ib. p. 546).

299ff. In LU 4619ff. Fergus, Cormac and Fiacha mac Fir Febe are mentioned also in this list, and in addition Conall Cernach. Then come Medb and Findabair, and Flidais has been added to these two in hand H. The two others named here in LL, Íth mac Étgaíth and Gobnend mac Lurgnig, are mentioned as being with Fergus in exile in TBFlidais II of the Glenmasan MS. (Celt. Rev. I. 208, II 202, 306). Neither Íth nor Gobnend plays any part in later events in the LL- Táin. It seems probable that the names were taken by the LL- compiler from the source of TBFlidais II. The reference to Flidais which occurs here agrees closely with a passage in TBFlidais II.

306-7 *Flidais . . . arna feis la Fergus ar Táin Bó* Cúalnge For *ar Táin Bó* Cúalnge, which is omitted in Stowe, we should read *ar Táin Bó Flidais* (cp. TBFlid. LU 1629f.). Corresponding to this passage 306-7 in LL, LU has merely the H interpolation: *fora laim sidi Flidais* (4623). The interpolation is not in Eg. 1782 nor in YBL.

307-8 *is si no bered . . . ingalad* For this I read *a ngalad,* 'their burning thirst'. For *galad* with this meaning see note in Celtica vii p. 45. In the LU TBFlidais there is a passage describing Flidais's help: *Is de sin no geibed Flidais cach sechtmad láa di feraib Herend do bóthorud dia thoscid* [read *toscid*] *ocon Táin* (LU 1631-2).

312 *Ní arlacair* (*Níor leic* St 320). An analogical deponent formation. It occurs again infra 1368.

318 *Cia maith fogníat* (*do-gniett* St 327) See note 112.

328 *dáig is lind imthiagat ⁊ is erund conbágat* I take *im-* and *con-* to be relative prefixes here. LU reads *Is airiund arbáget dano* (4632), with *ar-* prefixed in relative position. Stowe has *uair is linn tiagaid* (338) but was confused by *conbágat.* See note 338-9 Stowe TBC.

322 *ar abba dúnad . . . do gabáil* One would expect *dúnad* and *longphort* to be in the genitive. This seems to be an early example of the verbal noun phrase taken as a whole and the nouns left uninflected. Cp. infra 1852 *ar apa a fáth táncammar,* and 2746 *ar apa chomlund óenfir* (where Stowe has *ar aba comluinn aoinfhir* 2770 and TBC² reads *ar apa comruicc fri henfer* 2438, and the editors note that the *i* of *comruicc* has been added below).

344 ⁊ *lim congébat Gaibid la,* 'stands by, sides with'. Another instance of *con-* prefixed to a simple verb in relative position. Stowe changes to ⁊ *is lim cuingenait* (359).

345-6 *Níbat ecra . . .* In LU the passage reads *Acht ni thacersa ani sin, ol Fergus, arlifimni na hócu chena conná gébat forsin tslóg* (4648-9), *-tacér* fut. of *do-accair.* Windisch, no doubt influenced by the LU reading, translates the LL passage as 'Diese Männer sollen nicht (Gegenstand einer) Argumentation sein'. (He prints *ni batecra,* but presumably read *ni bat tecra*). But *tacra,* 'pleading', does not seem to be used in this sense. The whole sentence is omitted in Stowe, obviously because the compiler did not understand it. Windisch leaves the gloss untranslated as unintelligible. Perhaps we could read, *ecna(i)*, visible, manifest', for *ecra, níbat ecna(i) ind fir sin dam* (cp. *ba airecnai (.i. ba follus) dond aes batar oc forcsin ona longaib in tsoillsi* LU 7004), and then take *ní deceltar* as a gloss on *ecna*.

348 *case chruth Case* for *ciaso*. Stowe has *cibé cruth* (363) which is what one would expect.

350 *nád boi* ... Cp. Stowe which here reads *cona boi coicer i n-enionadh dib* (365). For *nád* = *conná* see infra note 2360.

351ff. This passage, which is not in Rec. I, is carelessly written or copied in LL. In Stowe it is condensed and better expressed.

354 *combad chách* Better perhaps if ⁊ were inserted before *combad*. I take *combad* to be the past subj. of the copula and the words from *chách* to *tslúagud* to be the predicate. See Pedersen, KZ 35. 322 §8 where he discusses the gloss *act bad chách dar ési aréli .i. nabad immalle labritir* Wb. 13ᵃ5. Stokes translated as 'but let every one be after the other' but Pedersen takes *chách dar ési aréli* to be the predicate of the copula and would translate 'es soll der eine nach dem anderen sein'. So here 'that (their order) might be each man with his companions and friends . . .' Alternatively we might emend *combad* to *co mbed*, subst. verb. For aspiration of *cách* subject after subst. verb cp. *is samlid inso nobiad chách* Wb 9ᵃ25.

371–390 There are six stanzas here in LU and they are in different order, those spoken by Medb coming together at the beginning and three stanzas spoken by Fergus following.

379 *Ardattágadar Ar-* for *ad-*. Both LU and Stowe have *Atotágathar*. The LU reading *Atotágathar dia mbrath | Ailill Aie lia slúagad* (4707–8) gives a better meaning, 'A. and his army fear that you will betray them'. In Mid. Ir. *ar-* is sometimes used to infix a pronoun. Cp. infra *ardotchlóe brath* 3440, *ardotfaedim* 4053. The scribe was influenced by such forms when he wrote *ardattágadar*.

384 *biurt-sa* So Diplomatic Edition. M. O Daly suggested reading present tense, *biur-sa*, which would agree with LU (*tiagu cech fordul a húair* 4721) and with Stowe (*beirim cach fordal ar úair* 397).

392 *ríam remán rempo* The phrase *ríam remán (remáin)* is used throughout the text to strengthen a form of the preposition *re n-*. It is invariably omitted in the Stowe version.

393–435 None of this passage is found in Rec. I but it is in Stowe. That it is an interpolation can be seen from the opening words *Bátar . . . bar Cúil Sílinni in n-aidchi sin*, i.e. in Cúil Sílinni in Roscommon where they had spent the first night of their expedition (supra 297–8). Yet by this time Fergus had led them north and south out of their way (367). One might possibly take it that Fergus had brought them back to their starting point, but that is hardly likely. In LU 4686 we are told that the army had spent the night *i nGránairud Tethba túascirt (.i. Gránard indiu* M), and then Fergus led them astray to the south. The LL interpolation would have been better inserted immediately after l.350.

394 *gérmenma* The same phrase is used again 446, 1281. See note 407 Stowe TBC, and note on *meanma*, 'presentiment', 'premonition' by Tomás de Bhaldraithe, Éigse vii. p. 155; note on Sc.-G. *meanmhainn* with same meaning by Aonghus Mac Mhathain, ib. p. 216.

421 *consrengfa, con-* for *do-* 422 *concicher, con-* for *fo-*. I take both verbs to be relative, the antecedent being *buidnech* 'he who possesses bands',

428 Stowe reads *A Fherguis dá roicit rainn* (441).

437 ⁊ *da 'marallsatar dóib ocht fichtiu oss* Stowe has *Tarla doibh ocht ficit oss allaidh* (451). Windisch takes the verb to be for *do-immarallsatar*, perf. 3 pl. of *do-imm-chuiredar*, but the verb does not seem to occur elsewhere. It would seem to be a scribal confusion of *do-cuirethar* and *imm-cuirethar*. One would expect *co 'marallsatar* . . . (without ⁊), (but note infra ⁊ *go 'mmarálaid dóib* 4889); or ⁊ *doralsatar dóib* . . . (cp. infra & *baralsat dóib* and *ocht fichti fer mór* 4118.) The form *fichtiu, fichti*, apl., may be a reminiscence of the older use of *do-cuirethar*, trans., in impersonal construction.

439 *garbat discailtig Garbat* seems to be intended for 3 p. pret. with *gé*. Cp. Stowe's reading: *gérbo sgailtech tricha cet na nGailian* (473).

441 It is explained here in Rec. I that all the Ulstermen were suffering from the *cess noínden* except Cú Chulainn and his father: *Bátirside hi cess calléic acht Cu Chulaind* ⁊ *a athair .i. Sualtaim* (LU 4690).

441 *Sualtach* Elsewhere in the text, when written in full, the form is *Sualtaim*, gen. id. See infra 3981ff. *Subaltach* in Stowe.

443–4 *ic Ard Chuillend* In LU 4692 *Iraird Cuillend* glossed *Crossa Cail*, that is, Crossa Kiel west of Kells, Co. Meath.

449–50 *cid amgéna* = *imgéna* (cp. spelling *amthiagat* 1591, *amsoi* 804). Confusion of preverbs, *im-* for *do-*. Cp. in Rec. I, *Imgéna fir limsa* ⁊ *daglaechdacht, or Cu Chulaind* (LU 5532) = *Dogéna* . . . The sentence is printed *cid am gena* in Diplomatic Edition.

460 *Scibis*, here used transitively. *Sáidhis in idh* St 474.

468–70 *dá mac Nera* . . . In LU 4728ff. this passage reads: *Tecait tra co mbátár i nIraird Chuillend. Eirr* ⁊ *Inell, Foich* ⁊ *Fochlam a nda ara. Cethri meic Iraird meic Ánchinne.* (The words from *a nda ara* to the end of the sentence are written by M in rasura and spread out to fill the line). In the margin is a note, also in hand M: *nó cethri meic Nera meic Núado meic Taccain ut in alis libris inuenitur.* The marginal note is lacking in both YBL and Eg. 1782 which give merely the four names without mentioning their father's name. Nera is mentioned twice in the Glenmasan MS. (Celt. Rev. I 228. 14, 298. 5) where he is called *Nera mac Niaduil.* (Compare Stowe 484 *Nera mac Nietair mic Dathain*). According to the Glenmasan MS. it was to Nera that Ailill gave Fergus's sword after he had taken it from its scabbard in Crúachu. (In LU 5400f. it was to Cuillius, and infra 4713 it was to Fer Loga).

471–2 *oc fégad [na] ingelta* Printed *oc fégad in gelta* by oversight in Diplomatic Edition.

472–3 *oc fégad ind idi barbarda* The adj. *barbarda* is here used referring to the ogam inscription which was unintelligible to all except Fergus and the druids. The etymological meaning of E. barbarian, Ir. *barbarda* is 'speaking a foreign language'. Note in Bedell's translation of the Bible: *An tan do chúaidh Israel amach ón Négipt, tigh Iácob ó na daoinibh barbartha* Ps. 114.1, 'When Israel went out of Egypt, the house of Jacob from a people of strange language'. Cp. also *ar ni tucci a mbéelre asbiur, is he[d] asbéra iarum barbár inso* Wb. 12ª6. In a later passage infra 870 *barbarda* is used in the sense of 'fierce, wild'.

502 *Atbiur-sa mo bréthir frib Doberim-si* St 518. *Atbiur* for *dobiur.*

506 *condirgife, Con-* for *do-, do-áirgife,* fut. 3 s. of *do-áirci.*

513 *baslechtat (ro selgatár* LU 4769, *ro gerrsat* St 530). This seems to be 3 pl. pres. ind. of a denominative verb from pass. pret. of *slig-,* 'cut down'. Better perhaps read as *baslechtsat,* with *ba-* intended for *ra-, ro-.*

529 *in n-echrad* for *in n-echraid.*

533-4 *Iss ed is lugu condric ó neoch* The phrase is *ro-icc ó neoch,* 'one can (do something)'. Here in relative position *con-* and the infix which is commonly found with this verb.

535 *ni tharnic úan do rád.* There is obviously an anacoluthon here. The phrase *égim nó iachtad nó urfócra nó a rád cia thic 'sin sligid* goes both with *Iss ed is lugu condric* and with *ni tharnic úan do rád.* Otherwise we should have something like *ni tharnic úan sin* (or *an mét sin*) *do rád.* Cp. Stowe which changes to *Ni ria[cht] éncuitt dibh-sin uaindi* (552), 'We did not manage (to utter) any of them'.

536 *Forairngert-sa* for *dorairngert-sa.* Cp. *farairṅgert* infra 2886.

540 *baluid* for *doluid.*

541 *Mesc fri árim fort* again 545. *Fri = fria.* Printed *Mesc for árim* in both instances in Diplomatic Edition, following the MS. A similar scribal error, *f* with an overhead stroke for *fi* occurs in LL-Tcgail Troí: *Uair roptar mó a n-ancridi ⁊ a n-écora, a fuachtana ⁊ a frithberta Troiánna fri* (*fi* MS.) *Greco co ssin anda Gréc for* (*f* with overhead stroke MS.) *Troianno* (LL 31037). The Stowe version reads *As mesc let lea airemh sin* (560), *As mesg let lea airemh an t-airdmes sin* (564-5), while LU has *conid mesc fria rím* (4794).

546 *forfetar-sa* According to Bergin (Ériu xii. 222-3) this is a Mid. Ir. transformation of *ro-fetar, -fetar,* where the verb is treated as a compound and *ro* infixed. We might however explain the form as an instance of interchange of prefix. Cp. *Cia forindfad dún cá crích ina fuilem?* MU² 287. Many of the instances quoted by Bergin are relative or follow conjunctions which took relative *-n-* in the older language. Here *dáig forfetar-sa,* and again infra *dáig ar bíth foretatar-som* 2149; *úair foretammarni* LU 8553.

548-9 *ilbúada ilarda imda* In Rec. I only three gifts are named: *buaid roisc ⁊ intliuchta ⁊ airdmessa* (LU 4791). One would have expected *búaid airdmessa* to be included in the LL list.

555 *Cuindscle* The noun *cuindscle, cuindscleo,* 'attack', occurs commonly in the phrase *ina chétchuindscle.* Cp. infra 2302-3 *ina chétchumscli.* No other instance of a verb *cuindscle* seems to be attested.

568 *Impádar* A corrupt Mid. Ir. deponent form. It occurs again infra 1256.

577 *focuirethar i n-artbe Artbe* vn. of *air-di-ben,* 'destroys'. *Focuirethar* used impersonally, and the phrase taken to be synonymous with *focuirethar i n-armgrith. Armgrith,* lit. 'a clatter of arms', hence great commotion, a panic, among armed men. Cp. infra 2134-5. The whole sentence is omitted by the Stowe compiler—a sure proof that the idiom was unfamiliar to him.

582   ⁊ *carpat eturru ar medón cadesin*   Windisch supplies *a* before *carpat*. *Cadesin* then refers back to *a carpat*, 'and her own chariot in the middle of them.' This is what is given by the Stowe compiler who has ⁊ *a carpat fen ar medhon etorra* (604–5).   Possibly, however, *cadesin* goes with *eturru*; the phrase *eturru ar medón fadessin* occurs twice infra 4511, 4539.   Or there may have been a phrase *ar medón fadesin*, 'in the very middle', with *fadesin* qualifying an adverbial phrase.   Cp. *innossae fen* CCath. 4080, *in lá sin fen* ib. 4463, 4530.   For a different explanation of the phrase see R.I.A. Dict. s.v. *fadéin* III. 1.

582   *fognid* for *dognid, fo-* for *do-* in relative position.

588–9   *gondat Ulaid ardastánic* (*comadh ict Ulaidh táinic ann* St 611). One would expect *do-da-ánic* or *do-das-ánic*, 'who came to them', but 3 pl. rel. infix may here be used for 3 s. neut. which is often used with this verb, *do-n-ánic* or *do-d-ánic*.   For *ar-* prefixed to a relative verb with infixed pronoun cp. *Ní hé sút cruth ardomfarfaid-se indé* 1740.

592   Cormac is not named here in Rec. I.

594   *á ránic* (MS. *a ránic*) *Agus ō rāinic* St 617.

601   *Áth nGrena*   In LU 4795–6 the name is *Áth Grena* but it is *Áth Grencha* ib 6101.   It is *Áth nGrencha* whenever it is referred to in Stowe. Hogan gives Áth Grena as a ford of the Boyne, but as Gwynn has pointed out (MD iv. 394) it is said in TBC to be north of Knowth on the north bank of the Boyne, and seems to have been a little east of Slane.

604–627   This poem is not in Rec. I.   It is found with four additional verses in the Metrical Dindshenchus (iv. 78–80) where it was edited by Gwynn from LL, Stowe and five other MSS.   According to the Dindshenchus poem Fráech, Fochnam, Err and Innell were the four charioteers of Órlám, son of Ailill and Medb.   They are called *maic Auraird maic Ainchinne*.   Aurard's death is also told and his two horses, Cruan and Cnámrad, are named.   There is also a prose dindshenchus on Áth Gabla ocus Urard (RC xvi. 155–6).   The extra verses in the metrical dindshenchus are obviously based on this prose account which also mentions Aurard's death, gives the names of his two horses and calls the four slain by Cú Chulainn the four charioteers of Órlám.   Both prose and poem agree with LL in the names of the four men, but agree with LU in the name of their father.   It is possible that the points in which they differ from the known TBC recensions were derived from a variant version of TBC.   But, as Gwynn suggested (loc. cit. p. 395), it is more likely that the account of Aurard's death was merely an attempt to explain the place-name Irard Cuilenn.   Gwynn also suggested (loc. cit. p. 145) that the passage in LU 4728 written in rasura by hand M, which calls the men the four sons of Irard mac Ánchinne (see supra note 468–70) was derived from the Dindshenchus.   The reference to Órlám in the Dindshenchus may have been no more than a reminiscence of LU 5824–8 where Cú Chulainn is said to have killed Órlám's charioteer.   In AS 4912 two horses are named Err and Indell—a curious confused reminiscence of the dindshenchus with the names of two of the charioteers substituted for those of the horses mentioned.

629   *i traiti se*   Again infra 693.   *Traite* (iā. f.), 'quickness'.   One would expect *a thraite se*, with possessive.   The corresponding passage in LU is *Is machtad, ol Ailill, a thraite ro bíth in cethror* (4810).   Cp. also *a luas ro*

gonadh an cethrar St 652. The adverbial phrase is *i traiti*, 'quickly'. Here we have confusion of the adverbial phrase and the noun used with possessive adverbially. Perhaps *i traiti* was felt to be one word, an adverb, and *i traiti se* was based on a phrase like *amlaid se*.

634 *is tria glaslec clochi conindsmad Con-* prefixed to relative verb. Cp. Stowe: *as tria glaislecaibh cloch ro hiondsmadh* (658).

639 *frosse (feidm fiorlaidir* St 663) Pl. of *fras, (frais)* used in the sense of 'attack, effort'. See RIA. Dict. s.v. *fras* (d).

641 *feirtche (feidm fiorlaidir* St 665) The word does not seem to occur elsewhere. Cp. 2 *firt* RIA Dict.

644 *Airm i mbátar na secht carpait déc* Lit. 'Where the seventeen chariots were'. Generally such a phrase means no more than 'As for the seventeen chariots'. (Cp. the use of the copula in such a phrase as *Cid mi fadéin* infra 1692-3 = *Cid mē fēn ann* St 1732-3). The phrase occurs again infra 938, 967, 3876; *airm i tát* 4623-4. Cp. *Coí hi fil in tríchait cét isin tig, ni ermadair nech dib a suidi nach a ligi lasin triar* BDD² 1375; *Baili i rrabatar cúicc mili .c. 7 .x. c. in cach mili, ni thérna dib acht oenchóicer namá* ib. 1493.

687 *Dáig is comtig ara fagbaitis . . .* In Stowe *uair as deimhin nach bfuaratar . . .* (712). Cp. supra *is comtig conná fuaratar . . .* (522). The sense here requires a negative in the LL reading. Windisch suggested reading *arná* for *ara*, but better perhaps would be *ni comtig ara fagbaitis*, 'it is uncertain, unlikely, that they would get (experience) . . .', *-fagbaitis* being 3 pl. past subj. With *ni comtig* compare *écomtig*, 'unusual, unwonted' (gl. inusitam Sg. 138ᵃ13). Compare also a passage in Cath Ccatharda 5080-82: *As ing mas eol doib a n-airm d'imurcur. Madh a n-imluadh immorro i n-agaid miled, ni combtig* (v.l. *comtig*) *co mbad eol daib.* The second sentence is translated by Stokes: 'But as to their movement against soldiers, it is not often that they have knowledge'. Better: 'But as for wielding them against soldiers, it is doubtful that they would know (how to do so).' With this meaning of *ni comtig* cp. infra *ni comtig láech is cháemiu* 4421 (*Ni coimdig laech bad chaime* TBC² 3242), lit.'Not usual is a handsomer hero',well paraphrased by the Stowe compiler as *Ni hiomda laoch as ailli ⁊ is caime ina in laoch fil i tossach na buidne sin* (4195-6). With *ara n-* for *co n-* cp. a similar use with the negative of *rofitir*, that is, a verb expressing doubt: *Ni fitir ara mbeth and* Trip.² 1428; *ár ni fitirsium ara tisad slúag secha i tír* LL 31364-5 (TTr.²). See infra note 2105-6.

678 *riam nó iarum* lit. 'before or after', hence 'never at any time.' A literal translation of the phrase would sometimes make nonsense in English e.g. infra 2405.

702 *Nír dóig Nár dóig* 709, 712. Stowe has *Ni doich* in all instances, which is what one would expect, the question *dóig innar tisad?* standing for *In dóig inar tised?* as in l. 693. Cp. *Nád dóig* 694.

724 *Ni airgem . . . Ni airge* 726, 728. The corresponding forms in LU are *Ni fairgébasu, Ni fuircebasu* (4843, 4838); in Stowe *Ni fuichi-si, Ni fuigi* (749, 751, 755). The LU forms are fut. 2 s. of *fo-ricc*, 'finds', while Stowe changes to the verb *fo-gaib*. The LL forms are pres. ind. of *air-icc*.

728 *basad bas* subj. rel. + suff. pron. 3 s.

728-34 The same eulogy of Cú Chulainn is pronounced by Emer in Fled Bricend: *Ni faigbistar fer and conmestar a aes ┐ a ás ┐ a anius* etc. (LU 8445-54).

750 *rádis . . . ar co ndigsed* The meaning seems to be 'asked if he might go'. The corresponding passage in LU has *Gudid Cú Culaind dia máthair dī a lécud dochum na macraide* (4862). The Stowe compiler changed to *adubairt co rachadh fen da cluic[h]i eatorra* (777).

756 *airm indas fil* (*áit i bhfuil* St 783). After the preposition *i dofil* is used a few times in this text: *gia airm inda fil* 1475-6, (= *i bfuil* St), *in t-imned mór anam uil-sea* (= *indam fuil-sea*) 2150 (= *i bfuilim* St). After the preposition + relative *tá* is the usual form in LL: *assin magin i tú* 1170, *ní hi sin tuarascbáil bá tú-sa* 2942, *dia tá* 1763, *airm i tát* 4623-4. Exceptionally in the Introduction *airm i fail tarb* (78).

763 *cona berad = conda berad.* One would expect *band* before *niba lugu* (*co mberedh bann narbo lugha* St 790). The form *berad* may be for imperfect, but is probably subjunctive here.

766 *tráth congebed . . . Con-* used relatively after the conjunction *tráth.* Here Stowe reads *an trath do gabadh a barr etarbuas* (794).

766 *etarla* The meaning of the word is obscure. See RIA Dict. s.v. Characteristically the Stowe compiler omits *etarla.*

770 *ecrais cid in liathróit* One would expect *ecrais cach óenliathróit* here, and then the infixed pronouns in the passage might be taken as plural. But *furri* 773 is against this reading. The verb *ecraid* means 'arranges, sets in order', and a plural object would suit the sense better. Compare LU 4876-7: *Focherdat dano a liathróite uli fairseom ┐ nos gaibseom cech oenliathróit ina ucht* = ll. 784-5 infra.

771 *ris = ros* (influence of *nis arlaic*).

788 *Scarais Scaraid* used for *doscara.* Stowe has *trasgrais* (820). Again 804 = *trascrais* St 836.

793 *in maccrad* Nom. for acc.

800 *combeind = no beind Con-* sometimes replaces *no* in the conditional and imperfect in Mid. Ir. See note 835. Cp. the Stowe reading: *Da bfesaind, do beind ina bfaitches ┐ ina n-oirchill* (832).

806 *bretha* for *dobretha* as in l. 805, pass. form for active. *Ba doich laa n-aithrib is bas tuc doibh ┐ niba hedh on acht uathbas ┐ cridenbhas do-rat iompa* St 837-8.

808 *Dothongu-sa = tongu* Cp. *dotongu* TBFr² 183, and supra 129 *dothung.* See RIA Contribb. s.v. 1 *do-toing.*

815 *coro scart* Here perhaps for *conroscart,* con- for do- in relative position. Cp. infra 1207-8 *Mac bec . . . barroscart na curaid,* with bar- for do- in relative. But *scaraid* is used for *doscara* supra 788, 804.

822 *Ciaso gnim* Printed *Cia so gnim* Diplomatic Edition.

829 *a fiallgud* (*a edgadh* St 861) The word is not attested elsewhere. Windisch suggests that it is from *fíal-étgud.*

832 *Dobered . . .* This is for *No bered.* The equivalent passage in Stowe is *┐ an t-aonmhac ic breth an baire forna tri coicait mac* (864-5).

835 *congeibed = no geibed.* Here *con-* replaces *no* in the imperfect. Another instance 841 *concured.* Contrast *no chuired* 837 and *no benad-som* 839. In a poem in Suibne Geilt we get the following: *Gach áenúair ro linginnse | co mbinn ar in lár | co faicinn* (v.l. *confaicinn*) *in cremthannán | this ac creim na cnám* (Early Ir. Lyrics 46. 42). Gerard Murphy (Glossary s.v. 3 *co n-* notes *co faicinn* as an 'unexpected' use of the conjunction. We should read *confaicinn*; it is Mid. Ir. *con-* before imperfect. In Togail na Tebe we get other instances: ⁊ *concarad gach duine atchid e* (1543–4); *Concinged-sum ara echaib* (1553).

842 *Arrópart = forópart* The change of preverb misled the Stowe compiler here for he writes *Adubairt Concobar ag faigsin in mic bicc: A oga* etc. (875).

847 *Conágart* Pret. active 3 s. for passive *conaccrad.* Again infra 1821.

853 *rag-sa* (*Ragatsa in far ndíaid* LU 4997, *rachad-sa inbur ndiaigh* St 884). Again infra 1019. The form *rag-sa* occurs SR 1588, 1663, and in a passage of Rec. I TBC not found in LL: *Ragsa conda tuc* (LU 4946).

854 *far ndiaid* For *infar ndiaid,* the reading of LU and Stowe. Cp. *Nibá mo chuitse immurro dorega far cend* TTr 338 = *Niba hé mo chuitsea daraga infar cend* TTr² 862; *far ndiaid* Trip.² 1651, 1653; *biaid sibh ar mhaidin bhar* (v.l. *ionnbhar*) *ttenál* Ériu I. 18. 40, The omission of the preposition *i* is very common in such phrases in late texts, e.g. in Bedell's Bible, Gallagher's Sermons, and also in the modern dialects.

863–4 *ni laimthanoch tasciud . . . do fír chúardda* Windisch and editors of Diplomatic Edition read *ni laimthe nech tasciud . . . do fírchuardda,* and Windisch suggests supplying *chur* before *fírchuardda. Laimthenach,* 'to be ventured or dared'. Usually with the preposition *la* (*uair nogo laimthenac la nech beith go nuradus* BCrólige 34 n. 7, quoted in Contribb. s.v.), but here with *do,* for, as D1. Dillon has pointed out to me, *do fír chúardda nó imthechta la nech beith go nuradus* is dative of *fer cúardda nó imthechta.* Cp. *des imthechta* supra 134.

866 *Oslaicther dún dond árchoin* Windisch takes *dún* here as = 'stronghold, court'. The Stowe compiler understood it otherwise, and I have translated as in Stowe: *Sgaoilter duinn don choin sin* (897). Cp. infra *Cóir duit arechus dúin fris sút* 1098–99, 1116, 1133, where *dúin* has the force of 'for my sake, I pray you'.

867 *fochuir lúathchúaird* Probably *fochuir* was intended for *ro chuir,* as *fogab* supra 760 for *ro gab.* Stowe has *ro chuir luathcuairt* (899).

883 *di fobaig inathair* (*d'abach* ⁊ *d'ionathar* St 914) *Fobach,* a variant of *ab(b)ach,* 'entrails'.

891 *cid ellom condránic cách, luaithium conarnic F. Con-* prefixed to verb *ránic* in relative position, with the neuter infix common in this verb. *Conarnic* is similar in meaning with confusion of *ro-icc* and *ar-icc.* Stowe here reads *Cid ullamh ráinic . . . luaithi* ⁊ *lanaibeile rainic F.* (923–4). The same sentence occurs again infra 1243–4 (= 1281–2 St).

900 *fer muntiri* For the meaning 'famulus, servant' cp. the gloss on Moses Wb. 33ᵃ5, *rubu fersom muintere maith dano du Ísu, nirbu choimdi imurgu,* 'he was a good servant to Jesus, but he was not a master'. This corrected reading of the gloss is given by Bergin ZCP xvii 224.

900 *Concométad . . . Con-* prefixed to imperfect.

910 *dáig concechlabat* Reduplication of the verb kept with an *f*-future ending. In the older language *dáig* was followed by relative *-n-*. Here *con-* shows relativity. Cp. infra 1260, 4123.

925 *ciaso sén* Printed *cia so sén* in Diplomatic Edition.

926 *mac bec congébad gasced* Another instance of relative *con-*. (Contrast infra 958 *mac bec no gébad gasced and*). Here LU has *óclách no gébad gaisced* (5039) and Stowe *Cidhbe mac og do gebadh culaidh gaisgidh* (959–60).

930–1 *Aithesc dano cungeda . . .* This is obviously a parenthetical remark. The greeting *Cech maith duit* is the equivalent of *co raib maith agat*, that is, thanks rendered, here thanks given in expectation of a favour. The Stowe compiler did not take the sentence as a parenthesis but put it in the mouth of Conchobar: *Aithesc cuingeadha neth o nech sin, ar Concobar, ⁊ cid iarra, a mic bic?* (964–5). Cp. in Mesca Ulad the sentence *Fálti fir connaig ascid sin*, in answer to *Mochen bithchen do thichtu* (MU² 61, 93).

947 *conarnic* Cp. St: *co rainic a foghrainne fria n-iorlainn* (979). *Conarnic* here is for *co comarnic*.

948 *ros fulgetar dó* The infixed pronoun has no apparent meaning here. (*ro fuilngeatar dó* St 980).

951 *Airm conagab sút?* for *In airm-congab sút?* (*Conagab* seems no more than a scribal error for *congab*. Cp. *Congab* 985, 1007). *Con-* in relative position. Stowe reads *An airm sut gabas in mac bec?* (984).

961 *Amra brig* Lit. 'wonderful power', i.e. 'a marvellous thing'. *Amra brigi son* LU 5059 . The phrase is often used ironically, as, for instance, in Imram Máeldúin (RC ix. 488), and is so taken here by the Stowe compiler who renders it as *Bec a brig liom-sa sin . . . gen go rabar* etc. (966).

963 *na cétna* Read *a cétna?*

973 *Geib lat mo dá ech* *Gabtar lat mo dha each fen* St 1007–8 Similarly infra 1042 *Geib lat dún echrad* where again Stowe has *Gaptar lat duin* (1075). Cp. *fairc-siu lett renna aéoir* MU² 218.

987 *ba romoch lind congabais armu* *Con-* for *ro* to show relativity. *As romoch lindi do gapais arma* St 1022. Contrast infra 1009 *Bad romoch lind ra gabais armu.*

987 *fo bith do deligthi ruind* The only instance of the preposition *fo bith* in the text. It has been taken over in this passage by the Stowe compiler: *fó bhith do deligthi frind* (1022–3). Elsewhere in the LL text the preposition is *ar bíth* (1645, 1646, 4592 etc).

988 *do seón congabsa* For *congabus-sa?* Another instance of relative *con-*. Stowe reads *Do shen do gabus arma* (1024).

997–8 *corop é in láech [sin] conairr comrac* *Con-* for *do-* in relative position. *Conairr*, 3 s. pres. subj. of *do-áirci* confused with that of *do-airicc*. Cp. 999 *corop é conairr séta ⁊ máine.*

1012 *rit báig* Cp. *ar báig aenlaithi*, 'upon one day's summons' Lebor na Cert² 16.211, *teacht le báig ríg Caisil*, 'to go at the summons of the King of Cashel' ib 213.

1015 *no léicfe dam-sa* For *na lléicfe*. Better perhaps *na léic-se* (*f* confused with *s* by scribe?) for other versions have imperative here: *Rom leicsea* LU 5085, *leicc dam-sa* St 1047.

1030 *dia fis dús* . . . Read perhaps *dia fis scél dús* . . . , a tautological phrase which occurs LU 6043. Or omit *dús*.

1036 *dochuaid-se* This would seem to be a scribal error for *dochuaidside* But Bergin seems to have taken this to be *dochuaid sé* (see DF III cix). The only instances of the analytic form of the verb in the LL–Táin are *nách cúala tú* 2048 and *clóechlád sé* 1424, both occurring in interpolated poems.

1078 *coro thurthaind* past subj. 1 sg. of *do-airret*, 'gets, attains to' Again 1080 *taurthais*, an s-preterite. Changed in Stowe to *co ccodlar began* . . . *tuitis an mac beac 'na chodladh* (1111, 1114).

1087 *máethmaccáem congab armu* Relative con-. *Maothmacaom og do gab arm* St 1122.

1093–4 *doringni rothmol* . . . (*rothnuall* St). The exact meaning of the phrase is obscure. It is used in this text to describe Cú Chulainn arising from sleep (here and infra 2178–9) and recovering from a fit of anger (infra 1198–9).

1096 *acht condrisem* (*acht co risam* St 1131) *Acht co n-* with infixed *-d-*, as often with compounds of *icc*. Cp. *in lucht condránic frit* infra 1131, and *Condránic dóib* 1142.

1097 *midlachda* 'cowardly', but with the overtone 'unarmed, not bearing arms as a warrior should'.

1104 *triana chúladaib Cúlad* is 'one of the two strong muscles on which the head is supported' according to Dr. M. A. O'Brien (Celtica III. 168–9). It is an old compound of *cúl* and *féth*, an older form of *féith*, 'muscle, sinew'. A form *cúilfhéithe* occurs in Cath Catharda 3685, translated 'back-sinews(?)' by Stokes. Another instance occurs in Smaointe Beatha Chríost 4777: *Gurab aire sin do gherr Dia in c[h]irt cuilfeite na pecach*, where *cúilfhéithe* = cervix in the Vulgate. Cp. also *baccaige cuilfeithe*, tr. 'lameness of a back-tendon', in Mediaeval Treatises on Horses, ed. Brian Ó Cuív, Celtica II, 34, 38, 40.

1120 *non imrend* The verb *imm-beir* (*imrid*) used reflexively with the meaning 'move about'. Again infra 3742.

1124 *Bad* (= *ba*) *aurchor deóraid sin ⁊ níba hicht urraid Icht* = 'a blow'. See Contribb. s.v. 2 *icht* (b). LU reads *Bid lám deóraid dó* (5137). See ZCP xvii. 70–71 for a note on the LU reading by Vernam Hull who takes *lám deóraid*, 'the hand of a hostile stranger' to be a proverbial expression meaning 'death, destruction.'

1157 *co 'mafaccatar* Here *imm-acci* has merely the sense of *ad-cí*. *Confacatar* St 1195.

1163–4 *ní cách conairg na eóin beóa do gabáil* We might take *conairg* to be from *dofairget, targaid*. Cp. *do-airg* Alex. 610. But the readings of other recensions suggest that it may be intended for 3 s. pres. ind. of *do-airicc, -tairic* (cp. form *-tairc* infra 1552), or it may be for 3 s. pres. ind. of *do-áirci* (cp. Mid. Ir. *do thairg* ZCP viii. 539). Here again con- for do- in relative position. Cp. the LU equivalent passage, *ní cach óen condric* (*darric* YBL), *samlaid* (5165), 'not everyone can do so', *ro-icc*, 'reaches, attains to' with prefixed *con-* in relative position and the common infixed *-d-*. *Ro-icc ó x*, 'x can do it', is also used. Cp. the reading of Eg. 1782 *Ni cach oen ondarricc samlaid*, and in the following sentence in this passage of LU we get *a mmarb immorro ni fil úadibseom ó nach ric* (*ri* YBL) 'who cannot do it'. Stowe paraphrases as *ni faghthar neach beres a mbeo les* 1201.

1166–7 *Cenglais* . . . This sentence referring to the birds should come at the end of the paragraph. Ibar collected them only after he had been reassured by Cú Chulainn. Thus after l. 1176 LU reads *Dognith són iarom, conrig Cu Chulaind inna esse ⁊ tecmala in t-ara ina heónu*. *Conreraig Cu Chulaind iar sin, inna heónu di thétaib ⁊ refedaib in carpait* (5185–7) and in Stowe 1212–3 *Tet ier[amh] gurro thinoil na heona ⁊ cenglais iet ier sin d'fertsib ⁊ d'iallaibh ⁊ do theudaibh in carpait*. The confusion seems to have arisen because Ibar collected the birds and Cu Chulainn tied them to the chariot.

1178–8 *Leborcham* Similarly in Rec. III it is Leborcham who sees them arrive (ZCP viii. 538 9). In Rec. I it is the watchman who catches sight of them (*in dercaid i nEmain Macha* LU 5192).

1180 *ic imuarad aice* (*ac iomfhuaradh aige* St 1217; *oc folúamain uassa* LU 5189).

1182 *Rodafetammar* . . . Again Rec. III agrees with LL and it is Conchobor who identifies Cú Chulainn (ZCP viii, 538. 15). In Rec. I Conchobor merely orders that naked women be sent to meet him (LU 5197).

1189 *Scandlach* In Rec. III nine of the women are named and they include Scandlach (ZCP viii. 538). In LU Mugain is named; 'or Férach' is added in hand M. Neither Mugain nor Férach is included in the nine named in Rec. III.

1189 *do thócbáil* . . . *Do-fócaib* used in the sense of 'shows, exhibits'. Similarly *túargbatar* 1190.

1195 *configfed* Another example of *con-* for *no*, here before conditional. The use of the conditional is peculiar to LL. Rec. I, Stowe and Rec. III use past tense: *fichis dornaib de* (i.e. from C.C.) LU 5202; *ro fhiuch dorn uaisdi* (i.e. above the vat) St 1231; *fiuchais dorn uaisdi* ZCP viii. 538. Possibly *configfed* is for *configed*, imperfect, here. The imperfect is commonly used in descriptions. Cp. following sentence.

1196 *fer fos foilnged* . . . *Fo* of verb repeated. Imperfect as often in description. Cp. *eo óir ina brut rosaiged a gúalaind for cach leth* LU 10663 10663 (Tochm. Étaíne), and in the present text (with *no* omitted) *Gilidir snechta sniged fri óenaidchi taidlech a cniss* (192–3).

1199ff. There is no description of the boy here in Rec. I, but it occurs in Rec. III. The passage seems to have been introduced here in LL (and Rec. III) from a later passage describing Cú Chulainn, 2344ff, which is common to all recensions. It is noteworthy that Rec. III follows the later passage even more closely than LL, introducing the description with the words *Rob alainn amh in mac tucadh anis annsin da fegad* (= 2344 infra) and describing his clothes in words identical with those of the later passage (ZCP viii. 538–9).

1203–4 *amal chír mbethi* Compare a passage in MU² 586–7: *cunid samalta ra cír mbethi ra dered fagamair no ra bretnasaib bánóir glantaitnem a fhuilt,* where the words *ra dered fagamair* make clear that a russet-yellow colour was implied here.

1204 *mar bó ataslilad* The condit. 3 s. of *ligid,* where one would expect a past subjunctive. The preverb *ad-* is used to infix the pronoun which refers to *máel* (f.). Cp. a passage in Siaburcharpat Con Culaind LU 9265 where of Cú Chulainn's hair is said *Atá lim is bó roda lig.* This might refer to the smoothness of Cú Chulainn's hair, but I take it as meaning that he had 'a cow's lick', that is, an upstanding tuft of hair in front. See infra note 3667.

1204 *delg n-argait indi* Windisch translates 'eine Nadel von Silber darin', but we should have *and* not *indi* if that were the meaning. Stowe alters to *Delcc airgit isin brat osa bruinne* (1241-2). In a later passage of the present text we get *dá gelsciath co túagmilaib argit findi foraib* (3731), translated by Windisch as 'von weissen Silber', i.e. as if it read *argit find.* We might perhaps take *findi,* (*f*)*indi* in these two instances to be g. of *finne,* 'brightness', but such a construction seems unusual; *delg finnargit* or *co túagmilaib finnargit* would better express such a meaning. Cp. two other instances in the text: *Tiagam . . . barar slegaib sneitti . . . go súanemnaib lín lánchatut indi* 3109 and repeated 3111. Here the Stowe compiler omits the phrase *go súanemnaib . . .* in both occurrences, a sure indication that he found some difficulty in it. Windisch again took *indi* in these two instances as = 'daran'.

1208 *na curaid ⁊ na cathmílid* This refers to the three sons of Nechta Scéne.

1218 *dar Cruind* Presumably the gloss was added by the LL compiler to distinguish the name from that of the river Glais Cruind (infra 1349). Strangely all versions give a different place-name here: *tar Iraird Culend* LU 5247 (*tar Airdd* YBL), *dar Duib atuaid* Rec. III, ZCP viii. 539.

1219 *conarnaic* (again 1225) seems to stand for *co comarnaic* Cp. supra 947. Stowe paraphrases to *co dtarla . . . dó* in both instances.

1224 *fón samlaid se* (*fon samail sin* St 1261) Again infra 1225; *fón samlaid sin* 1254, 1694, *fón samlaid cétna* 1564. This use of *samlaid* as a noun is common in the present text. It occurs once in another LL text, Aided Guill ⁊ Gairb (12817).

1227 *ro mebtatar . . .* The incident here referred to is omitted in the LL recension (and in Rec. III), but is told in Rec. I (LU 5212 ff.). The Connacht forces reach Mag Mucceda. Cú Chulainn cuts down and places a tree in their path with an ogam inscription ordering them not to pass until a warrior should leap across it. Thirty chariots are broken in the attempt. Still another short passage, written in rasura by hand H, occurs in LU 5243-5, telling how they broke the shafts of the chariots. The wording of this interpolation is like that of the LL passage: *Mór in cuitbind dúib, ol Medb, can tophund na erri angceóil ucut fil co for nguin. Doberatsom iarom topund fair iar sin* [*sini* MS.] *coro brisiset fertsi a carpat oca.*

1229-30 *in Cú Chulaind* Here, and again 1514, 1523, 1712, printed *in Chu Chulaind* in the Diplomatic Edition, but inconsistently printed *in Cú Chulaind* 1237, 1410, 1738. In all these instances the MS. has *in .c̄c̄.* I have read this as *in Cú Chulaind* throughout; in the MS. we find written in full *in Cú* 248, 1413, 1417.

1236 *ni h'opair* . . . That *h'* here stands for *th'* seems to be shown by readings of the other version: *Nip si th'opar comadas* . . . LU 5261; *Nip si hopuir chomadais* . . . Eg. 1782; *Ní hí h'obair comadas* . . . St 1274; *Ní hí th'urobair fein* . . . Rec. III (ZCP viii. 540 9). It is printed *ni hopair* Dipl. Ed. Note also *trah* 1324.

1238 *Ro mairc-se ón ém* We should probably read *Rom mairc-se*, 'alas for me!' *Romairgge són ém* LU 5264 but *Rommaircge* YBL. Cp. infra *Ron marg-sa* 2183 (*Ron mairgsea* LU 6361) where C has *Rom mairc-si deside* (1255).

1239 *Nád bia* 1 sg. fut. of *benaid*.

1242 *connice* One would expect *connice sain*, but cp. supra 207, 217, 222·

1242 *urtha* < *air-fo-eth*. Cp. LU 4694 *orthá .i. eirg co rrobad do Ultaib* Co has been omitted; read *urtha co robud dó*, 'go with a warning to him, take him a warning'. (*tabair robadh dó romam-sa* St 1279-80). Possibly *urtha* here may mean 'take, bring'. Cp. use of *ticc = do-beir*, and see note on W. *dyfod, mynd* = 'bring' by P. Mac Cana, *Ériu* xx. 215-7.

1243 *ar dogné* Possibly the scribe began to write *ara ndéna*, changed to *dogné*, 2 s. pres. subj. used as imperative, and then forgot to expunge *ar*. Elsewhere we get pres. subj. with *ro* as jussive *ara nderna-su* 3272, 3273. Or read *ardogné = ardagné*, with *ar-* used to infix pronoun?

1244-5 *Cid lúath condránic* . . . See supra n. 891. Cp. Stowe which reads *gid luath rainic in giolla, is luaithi rainic Cu Chulainn* (1281-2).

1247 *tri meic Árach* In Rec. I *tri meic Gárach* (LU 5277), and similarly Stowe 1285. In the Dindshenchus of Dún Ruissarach they are mentioned by name and their father is called Gárach mac Fomhuir do Domhnannchaibh: *Iss iat a thri mic sin robitha for Tana la Coinchulaind .i. Lon ˥ Diliu ˥ Uala a n-anmanna* (MD iv. 290).

1248 *Úal[u]* The same scribal error, *Ual* for *Ualu*, in Rec. III (ZCP viii. 540. 27). *Ualu* Rec. I and Stowe.

1249 *Is aire condeochtar* . . . Relative *con-* for *do-*. *Is aire tangatar* . . . St 1287.

1253 *coro gontais-[s]ium* . . . ˥ *go ructais* These verbs are dependent on *Is aire condeochatar*.

1255 *condristais* . . . *gliaid* (*go ccuirdis gliaid* St 1294) Co + 3 p. past subj. of *ro-icc* with infixed *-d-*, as is usual after *co n-* in this verb. 'That they might attain to fighting, that they might fight'.

1256 *Impádar* See supra note 568. *Iompoidis* St 1294.

1259 *Barrópart* We should expect a plural verb here. Read *barropartatar*? Cp. Stowe which reads *do-chuadar forin ath* (1299).

1260 *áit mal connairnechtatar*, and again infra 1351 *áit mal connarnectar*, '(in) the place they reached'. I take this to be for relative *con-* + pret. 3 p. of *air-icc = ro-icc*. There seems to have been confusion between *ro-icc* and *air-icc*, as, for example, *conarnic* 1245 = *condránic* 1244. The word *mal* does not seem to occur elsewhere. *Mail* is found in *Marait na clocha is tir thair mail ro bámmar 'nar cathaib* LL 29385 (Festschr. W. Stokes 9. 24), with a variant *bhail a mbamar*. Here we might take the reading to be *ait mail (bhail) connarnectar*. Or possibly *áit mal* is for *áit amal*. *Feib* and *ama(i)l* are synonymous, but both are used in e.g. *na bia Emain feib amal conairnecus*

LL 12830. Similarly *do réir amal ba bés dona geintib* TTr² 484. In the later language these become *fé mar, do réir mar*. So *áit amal* might give the modern *áit mar*. Note, however, that *amail mar* occurs in the later language. The Stowe compiler misunderstood here and took the verb to be *conricc*: *ainm in atha arar chomraicsetar* (1300). The second occurrence (infra 1351) was omitted in Stowe, for obviously there *conricc* was not the verb needed.

1260 *dáig conmebdatar* See supra note 910.

1261–2 *Is and sin focera Mulchi . . .* The Diplomatic Edition prints this sentence *Is and sin focera Cu Chulaind Muilchi . . .*, with a footnote to *Cu Chulaind*: 'supply *gualaind*, om. by homoeoteleuton'. But I have followed Windisch in deleting .c̄c̄. The Stowe reading supports this: *is ann sin ro tuit Mulcha forsan tulaigh* (1303). *Focera* is for *do-cer*, 3 s. pret. of *do-tuit*. Cp. infra 3457 *á bacear Óenfer Aife*. The passage *Is and sin focera . . . béus* (1261–2) should come at the end of the paragraph, as it does in the other recensions. The two fords referred to are Áth Carpait and Áth Lethan.

Mulchi is called the charioteer of Lethan in Rec. I and Rec. III. In Rec. III it is not Cú Chulainn who fights with Mulchi but Láeg mac Riangabra, and the death of Mulchi is called Láeg's one exploit in the course of the Táin (*Aeinēcht Laeigh ar Tanaidh*, ZCP viii. 540z).

1270 *giarsa = ciaso, cia* + 3 s. pres. indic. of copula. The same form is found in Rec. III here: *gersa Cruiti Cainbhili aderthea riu* (ZCP viii. 453. 24).

1276 *in petta toqmalláin* In LU called both *togmall* (5289, 5292) and *togán* (5296). *Togán* in Stowe (1315, 1316), *togmall* in Rec. III (ZCP viii. 540. 4). See DF III Glossary s.v. *togán*. Cp. also ZCP xix. 126–7 where Thurneysen has a note expressing disagreement with the translation of *togmall* as 'weasel'. The place-names Méide in Togmaill and Méide in Eóin are explained in LU: *is for gúalaind Medba bátar immalle eter togán ⁊ én, ⁊ is a cind bentatár na urchora dib* (5295–6).

1280–1300 This passage is not in Rec. I, but it occurs in Rec. III where it comes later, after the death of Medb's handmaid (ZCP viii, 542. 18–36). It is a doublet of an earlier passage (supra 393–435), which again is not in Rec. I. (We cannot tell if the earlier passage was also in Rec. III as that part of the tale is lacking). That the scribe himself recognised the repetition seems to be shown by the words *amal ra scríbsam remaind*. Note how the same phrase is used in both passages: *Tánic gérmenma géribrach Con Culainn do Fergus*.

1283 *fatchius* Probably the words *do dénam* should be added. Cp. supra 395.

1285 *Damb ró Dob ro* St 1324, *Ba rua* Rec III (ZCP viii. 542). There is an extra quatrain in this poem in Rec. III.

1289–92 Here Cú Chulainn's exploits are said to be like those of Hercules. For the spelling *cichloiste*, for *cichloiscthe*, see Pedersen VG I. 420 (*sk + th > st*). Similarly *cichloiste* TTr. 1686, *cichloistib* ib. 1734–5; *cigloisti* ZCP iii. 38 §6. A late example of *-st-* for *-scth-* occurs once in Bedell's translation of the Bible: *gur ofráladur iodhbuirt loiste* Judges 21.4 (but elsewhere in Bedell spelt *loiscthe* or *loisge*).

1302 *crich Margín* The geographical directions are different in Rec. I. The bull was in Slíab Culind with his herdsman Forgemen, and cast off the

150 boys from his back, killeᵤ two thirds of them, then went and pawed the ground in Tír Marrcéni (LU 5329-33). In LL the opposite is the case. The bull is in Crích Margín and goes thence to Slíab Culind. Rec. III agrees with LL.

1304   *in Mórrigu*   In a later passage (infra 4600) spelt *Morrigu*. In LU 5320 *in Mórrigan*; so too Stowe. Both forms are in Rec. III. The first component of the name is said to be cognate with O.H.G. *mara* and AS. *maere* (Eng. nightmare, Fr. cauchemar), but scribes probably took it as the adj. *mór*, for it is frequently accented.

1306-9   *Maith a thrúaig* . . .   Not in Rec. I, but in Rec. III. In Stowe these sentences come after the rhetoric.

1307   *dáig ar [bíth] dotroset* . . . = *daig ticfaidh Cú Chulainn dabur n-iondsaigi anocht* St 1321.   Printed *dáig ar dotroset* in Diplomatic Edition. Alternatively we could read *dáig ardotroset*, with *ar-* before infixed pronoun.

1311-17   In ZCP xxiii pp. 142-8 Wolfgang Krause has given a translation of this rhetoric as it is in Rec. I.

1320-33   The attributes of the bull are not given in Rec. I, but are enumerated in Rec. III. They are also given in C, but with different wording.

1334   *im ailib ⁊ im airtraigib crichi Conaille Murthemne*   Here Stowe reads *im shailib ⁊ im dortaib* (1370) but in the second occurrence *im shailibh iomdortaib* (1374).   The reading of Rec. I suggests that in LL we have a misunderstanding of a place-name: *Ni rubai Cu Chulaind nech eter na Sailiu Imdorthi hi crich Conaille co rancatár Cuailngi* (LU 5334). With the LL phrase cp. infra 1682.

1341-48   This incident is very briefly told in Rec. I (LU 5340-3). Rec. III agrees here with LL in the details given: Loche goes to fetch water for drinking and washing; she is wearing Medb's diadem; Cú Chulainn casts a stone from his sling which breaks the diadem in three (ZCP viii. 543. 10-17). None of these details is given in Rec. I.

1345   *co rróebriss*   This form is due to analogy with *co rróemid (maidid)* and other reduplicated perfects where *ro-* with reduplication vowel becomes *roi-, róe-*. It occurs again infra 1745.

1349-71   In this passage only one river is named in LL, Glaiss Chruind, but in Rec. I two are mentioned: Glaiss Chruind where Úalu was drowned and at the source of which Medb made the men of Ireland dig the pass, Bernas Bó Cúalngi (LU 5365-72), and the river Colptha in which their chariots were lost, giving the name Clúain Carpat (LU 5380-2). Rec. III here agrees with Rec. I (ZCP viii. 543. 4. 17).

1350   *forfémdetar a techt*   In Rec. I and Rec. III we are told that Glaiss Chruind had overflowed its banks to impede the progress of Medb's army (LU 5362-3, ZCP viii. 543. 4-6). Similarly the river Colptha rose against them (LU 5381, ZCP viii. 543. 13). In an earlier passage in Rec. I, Dubthach chants a lay, prophesying that the river Glaiss Chruind will rise to prevent the army from entering Murthemne (LU 4663 ff.). Later in Rec. I Cú Chulainn himself calls upon the river to help him and repeats a stanza of Dubthach's prophecy (LU 5512 ff.). The LL compiler suppresses here, as occasionally elsewhere, the reference to magic help given to Cú Chulainn.

1352 *in Glassi* This was the river Colptha, according to Rec. I (LU 5380–82).

1355 *in Glaiss* This refers to Glaiss Chruind (LU and Rec. III). Here again the details in LL are found also in Rec. III (ZCP viii. 543. 6–12), while Rec. I merely has *Atá a lecht ⁊ a lia forsin tsligi ocon glais .i. Lia Ualand á ainm* (LU 5367–8).

1362–5 This passage occurs only in LL (and Stowe). A similar passage occurs later in all three recensions, prefacing the account of Cú Chulainn's meeting with Fergus (infra 2474–7, LU 6682–5, RC xv. 205 §§202–4). It is obviously out of place here in LL; the single combats have not yet begun.

1366 *fri táeb na Glassi* This was Glais Chruind, according to Rec. I.

1368 *bacóistis* Ba- for do- (*docóestis* LU 5370). Here again the detail (*in sliab do chlaidi ⁊ do letrad rempi*) is not in Rec. I, but in Rec. III we find *gomadh iad a dair ⁊ a modhaid féin do claidhed in sliab rempe* (ZCP viii. 543. 31).

1372–7 Here LL differs from Rec. I. In Rec. I Glend Dáil Imda (= Glend Táil LL) was reached after the army had crossed through Bernas Bó Cúailnge and Botha was the name of that place after they had set up their tents there (LU 5378–80). Then they went on around the river Colptha to its source and to Bélut Alióin between Cúailnge and Conaille which was called Liasa Liac after they had made sheds there for their calves (LU 5380–85). In Rec. III there is confusion; Bélid Ailién, Liasa Liag, Glenn Dáil and Botha are all given as names for the same place. If the compiler of LL had before him both Rec. I and the forerunner of Rec. III, it is easy to see how his confusion of Glais Colptha and Glais Chruinn, Bélat Aileáin and Glend Táil and Liasa Liac arose.

1378 *Glass Gatlaig* In Rec. I (but not in Rec. III) it is said that Glaiss Gatlaig rose in flood to impede them (LU 5386). This explains why they were unable to ford the river.

1383ff. These incidents—the mission of Fiachu and the meeting of Medb and Cú Chulainn at Glenn Fochaíne—do not occur in Rec. I where only the missions of Mac Roth (three, not two as in LL) are given (LU 5592–5622). In LL Fiachu's terms are those of Mac Roth's first embassage in Rec. I.

The LL passage is clearly an interpolation here; the opening sentence (1283–4) is repeated after the interpolation (1464–5). In the interpolated passage a hundred men are said to have died of fright when Cú Chulainn brandished his weapons, but in the later passage, in agreement with Rec. I, Cú Chulainn is said to have killed a hundred men a night for three nights. Part of the interpolation seems to have borrowed from the section Breslech Maige Murthemne; note the sentence *gabais C.C. acond Ferta i lLerga 'na firfocus* 1384–5 = 2124–5 infra, and the account of a hundred men dying of fright = 2135–6 infra.

This interpolation is also found in Rec. III but there Fiachu's mission comes at the end of the negotiations between Mac Roth and Cú Chulainn, where it fits more naturally into the narrative. In the poems the terms to which reference is made are those offered to Cú Chulainn later in the narrative by Mac Roth. Only one poem is given in Rec. III (*A Chú Chulaind cardda raind*, but 5 stanzas instead of 7), the other being summarised briefly in the prose.

In C the meeting of Medb and Cú Chulainn has been inserted in what is otherwise the Rec. I narrative, but there is no mention of Fiachu in C. Thus where LU reads *Gabais [Cú Chulaind] tabaill dóib a hOchainiu ina farrad. Bid dímbúan ar slóg la Coin Culaind in cruth sa,* ol Ailill (5584–6), C reads *Gaboid a tabhoilt a Fochaine ina farrad. Tic Medb la Fergus do dexin Con Culaind* (401–3), leading into the meeting at Glenn Fochaíne and the two poems. Then, after the sentence *lotar ass tre coimferg leth ar leth* (= LL 1462–3), continuing the LU narrative: *Bid dimbuan ar sluacch la Coin Culaind an crut sai,* or Ailill (456–7). In C, as in Rec. I, three missions of Mac Roth are given.

1389 *Ciarso choma* = *Ciaso choma.* Again infra 1471, 1517, 1548. *Carsa coma* St. 1425. A form *carsad* occurs frequently in this passage in Rec. III: *carsad comhadha* (ZCP viii. 544. 22), *carsad coma* (ib. 545. 13; 546. 10, 18, 37; 547. 13, 22).

1394–7 *Conid si briathar* . . . This sentence is not in Rec. I, but it is in Rec. III and in C. In Rec. III it occurs on the second mission of Mac Roth (ZCP viii. 546. 15–17). In C it comes after Ailill has laid down terms for the first mission.

1405–6 *Acus ar co tis* . . . At first sight it seems as if something had been omitted before *ar co tis.* But in Tochmarc Étaíne we get a similar use of a subjunctive, giving the import of a message and corresponding to what would be an imperative in the message itself: *Luid an Mac Oc co Dian Cecht:* '*Co ndeachaidis liumsa*', ar sé, '*do tesarcain mo aidi*' (Ériu xii. 148 §10, where the editors suggest in a footnote that something has been omitted). In the equivalent passages in Stowe TBC and Rec. III a verb of saying has been introduced with verbal noun construction: *Adubrad rit, ar Fiacha, toidecht co moch go Glend Fochaoine i n-aires Medba* (St. 1442–3); *Nacha d'aedaigecht tanag-sa acht da radha rit-sa techt a ngleann* (Rec. III, ZCP viii. 548. 9–11). Cp. a similar use of the subjunctive in *ar co festa-su dam-sa* . . . 1500, and ⁊ *ar co n-epertha-su frim-sa cia dotháet* 1873 (the last sentence being changed in Stowe 1908 to ⁊ *adubairt frit go n-eberta friom-sa* . . .). We may take LL *Acus ar co tis* . . . (and in Tochmarc Étaíne, YBL *co ndeachaidis* . . .) as elliptical, some such phrase as *Is ed as áil do Meidb* or *Is aire thánac-sa chucut* being understood before the subjunctive. Cp. in Rec. I TBC, *Dodeochadsa o Findabair ar do chendso co ndechais dia hacallaim* (LU 5939–40). The *acus* presents no difficulty; in dialogue *acus* frequently begins a sentence, as e.g. supra 990 and infra 2202. Cp. MU² 171, 409, and BDD (*Ocus iar sin cia acca and?* 786 etc.).

1409 *cessis a menma fair* For this idiom see Stowe TBC, note 1447–8.

1424 ⁊ *clóechlád sé* is *claochlaid sé* St 1462, ⁊ *coemcloidedh se* C 418.

1425 *gana chlothar.* Stowe and C have a better reading: *Fó liom gin co clotar uaibh* St 1463; *Fó lium cenco claiter uaib* C 419.

1444 *rot biad dáig is ecengal* Windisch takes *ecengal* to be a misspelling of *écendál,* but I have translated the reading of Stowe and C.

1449 *Is romór a nad-maide* I take the verb to be a *nnod-maide,* 'what you boast of'. Cp. *a nnod-ail,* 'that which she rears', Anecd. III. 28. 9. For this use of *-d-* as a neuter relative pronoun see Thurn. Gram. §425 and

Strachan, Ériu I 172. In the present text we find two other examples of this pronoun with *maídid* used relatively: *Cid ón, cia nod maíd acaib-si sin?* (1726) (= *Cia ro maidi sin?* LU 5743, *Cid ón, cia maidhess agaibh-si sin?* St 176'); *Ni tha ni nod maitte forro, a Medb* (4578–9). In Stowe we have *As romor a natmaidhi* (1488) which at first sight might seem to contain the 2 s. reflexive pronoun. But in O. Ir. the reflexive verb was followed by the preposition *i*, e.g. *nitta ni indot moíde* Wb. 24ᵃ30. We should expect *Romór ani indot maíde*, unless it were possible to take *a* as preposition 'in', with omission of antecedent. The reflexive verb is not found outside the Glosses, however, and it would be very strange to find an example in such a late text as Stowe TBC. In LU, for instance, we get non-reflexive trans. verb: *Ced ed no maidedsom* (5743), 3 s. past subj. Windisch took *a nad-maide* to be from O. Ir. *admidiur*, but this is wrong. For the spelling *nad-* for *nod-* cp. supra 1264–5 *nad n-ácaib leis*.

1473 *léiced écin cotlud dona slúagaib* Perhaps we should read *cotlud écin*, but *écin* is often used = 'at any rate, however'. Cp. *Tairr isin linde iarum ecin conacamar do snam*, YBL version of TBFraích, ed. Byrne and Dillon p. 19 l. 17, (where *écin* is omitted in the LL version of TBFr.).

1475 *nirsa eólach etir* Here *nirsa* is syntactically a present = *Nim eolach itir* St 1516, *Ni heolach mhisi mara bfuil se* Rec. III (ZCP viii. 544. 28). These forms occur also elsewhere in Rec. III. For 1438 supra *nidam drochláech*, Rec. III has *nirsum drochlaech* (ZCP viii 548. 18); for 2019 infra *condam créchtach crólinnech* Rec. III has *orsam crechtach crolindtech* (RC xv. 69 §94).

1478 *eter Ḟochain ⁊ muir* Similarly Rec. III but in Rec. I *Is and bad dóig la Fergus bith Con Culaind i nDelga* (LU 5594).

1495–6 *Ciarsat comainm céli-siu* Some corruption here and 1498 in LL. *Cia dian celi-si* St 1537, 1540. Indirect speech in LU: *Imcomairc Mac Roth do Láeg cia diambo chéli* (5597). See note Celtica vii. 38–42.

1501–2 *immna n-egat fir Hérend* Read *imma n-agat* as in Scéla Conchobuir (LL 12532). Cp. *in gilla imma n-ágar sund* (referring to Cú Chulainn) LU 6117. M. O Daly (Ériu xiv. 110) takes the verb to be *agid imm* 'makes a commotion about'. Perhaps we may take it to mean, 'drives around, pursues,' like *imm-aig*. Cp. the phrase used by Órlám's charioteer referring to the fight against Cú Chulaind: *ic taffund na hailiti urdairce út .i. Con Culaind* (1227). Windisch took the verb to be *éigid imm*, 'raises a cry (of alarm) about'. Stowe misunderstood and wrote *in Cu Chulainn oirderc sin imomnaigit fir Erenn* (1543).

1509 *grissaib* This word must mean something like 'companies'. Stowe has *do grésaibh ⁊ do glamaibh ⁊ damaibh ⁊ aoidedhaib* (1551–2), which is translated. Perhaps the original reading was *grinnib*, here contaminated by the preceding *gressaib*?

1534 *ni mé adféta dúib* Again 1542 *ná ba é dosféta dúib*. *Adféta* and *dosféta* are intended as 3 s. fut. of *ad-fét*. Windisch takes them to be subjunctive, but cp. M. O Daly, Ériu xiv. 112. 2801. In simple periphrasis the relative clause is not subjunctive after a negative. Cp. *Ni bu fua réir fesin boi-som* Ml. 14ᵇ13. But after a negative denying the existence of the antecedent, e.g. after the negative of the substantive verb, a subjunctive is required: *Ni fil d'feraib Hérend ti imbárach dit fópairt acht missi* (infra 1621–2).

1535–6 *Má tá ocaib . . . rofessad* The antecedent of *rofessad* is omitted. Similarly 1543. The modern version, as usual, inserts *nech*: *Ma ta linne ar medon nech atberad* St 1589.

1555 *a bíathad ┐ a étiud Con Culaind* Not in the terms of Rec. I; but in Rec. III: *a bhiathadh ┐ a eidedh in fad bheas ar tánaidh* (ZCP viii. 547. 31). But later in Rec. III Medb asks Fergus if Cú Chulainn will remit any of the terms, and Fergus answers: *Ní chuinneocha se a biathadh no a eidedh oraib-si* (ZCP viii. 549. 2–3). The LL- compiler omitted this, though elsewhere in LL we are told that Cú Chulainn depended for his food on what fish or birds or deer he could kill, and that only the men of Bregia supplied Cú Chulainn with food while the men of Ireland supplied Fer Diad.

1555 *forin tánaid se* Similarly gen. *tánad* 4106, 4807, 4862, for regular *tána*. An analogical change of declension.

1567 *Cid imluid-siu* In Mid. Ir. the pret. st--- *l--i!* is u..... to form a verb. Here *im-* seems to be used as a relative particle. Stowe reads *Cia leith teigi-si*? . . . *Teigim let-sa* (1611–12).

1570 *Rafétad* (*Rofétfainn* St 1617) *Rafétad* seems to be for *rafétfat*, 1 s. fut. of Mid. Ir. *fétaid*. *Ro-* perhaps due to the influence of *ro-fetar*. Cp. *cach olc as mó rofétat . . . iss ed dogniat* LU 2379.

1575 *A mboi* Insert *airm* before *a mboi*.

1576 *eter Fochain ┐ muir* All recensions vary on this point; Rec. I has *Delga* (LU 5632). Rec. III has *a crich Rois* (ZCP viii. 549).

1577 *ni théiged* . . . Pedersen twice compares this sentence with the common instances of omission of antecedent (VG II. 183 n. 2; KZ 35. 395). But there is no relative verb here and the sentence is not on a par with such a sentence as *Má tá ocaib . . . rofessad* 1535–6. It is likely that a subject has dropped out after the verb. Cp. the readings of Stowe and Rec. III.

1589f. The second chariot-rider is not described or identified here in Rec. I, for immediately after the sentence *Claideb sithidir loi churaig fora dib sliastaib* (LU 5639), Cú Chulainn interrupts Láeg to say that the great scabbard is empty, and to tell how Fergus had been found with Medb and a wooden sword substituted for Fergus's own sword (LU 5639–44). This episode is told later in LL (infra 2487–91). Like LL, Rec. III omits the flash-back here, and describes the second chariot-rider and gives the reason for his coming with Fergus (ZCP viii. 550).

1590–1 *Is lór n-árgigi . . . amthiagat Am-* = *imm*, here used as relative particle. Cp. *daig is chucaind imthigit* TBC² 2554.

1597 *Dia tuinne* . . . In LU 5646, *dia tí* and in Stowe 1638 *dia snaidhi*. This *-tuinne* must be for 3 s. pres. subj. of **do-sná* (with influence of *do-sní*?). YBL and Egerton 1782 here read respectively *Dia tonda iasc . . .* (TBC² 1167) and *Dia tonná iasc* (ZCP ix. 148z).

1599 *Dámsat éicen* (*éicni* MS.) This sentence is taken in the Diplomatic Edition (and in Windischs's edition of LL TBC) as *Damsat comrac* etc., with *éicni* to be omitted. Presumably the editors took it that the scribe copied *éicni* from l. 1597. I take it that *éicen* was the original reading and that the scribe carelessly spelt *éicni* as in the preceding line. Possibly a better reading would be *Dámsat éicen comrac nó chomlond [do dénam]*. Cp. *Isim écensa techt*

LU 4694, *Am écensa tocht* ib. 8015; *Isam écen-sa . . . comrac fri C.C.* infra 1624–5. *Dámsat* is 2 s. pres. subj.

1600 *co táthais do súan* ⁊ *do chotlud* Windisch takes the verb to be *s*-subj. 2 s. of *do-autat, do-etat,* but Pedersen takes it as 2 s. pres. subj. of *do-tuit.* Stowe alters to *in gcen ber it c[h]odladh* (1642).

1605–6 Here LL agrees with Rec. III against Rec. I. *Is annsin tainic Ferghus uadha nach abradais fir Erenn gomad aga mbrath no aga tregan da mbeth ni bad sia ag imagallaib re Coin Culainn* (ZCP viii. 550). The same sentence occurs again in Rec. III when Lugaid mac Nóis goes to speak with Cú Chulainn (RC xiv. 264. §56), but is not in the equivalent Stowe passage.

1611 *immonderca súil i sodain duit* In R.I.A. Contribb. the verb is taken as perfective form of *imm-décci,* 'looks around'. Windisch translates tentatively as 'Das Auge soll dir genau hinsehen'. But I am inclined to take the verb as *imm-derga* (with confusion of reflexive pron. *imma n-?*). The equivalent passage in Stowe does not help for the Stowe compiler, probably not understanding the LL sentence, changed to ⁊ *tabair suil torum* ⁊ *lean Fergus* (1653). But in Rec. III we have *Nirbo dergta súil fri sodhain duidsi sin* (ZCP viii. 551). The transitive form of *imm-derga* presents a difficulty. Perhaps we should read *Ní 'monderca súil* (acc.) *i sodain duit,* 'you do not redden your eye by that' (i.e. looking at C.C. who is quite near is no eye-strain). Or take *immon derca* as for *immus-derca?* Rec. I reads *Mós tairchella ém súil tar sodain* (LU 5654). 'An eye soon takes in that' (alluding to Cú Chulainn's small size).

1626–8 *acht ni búaid . . . is mó is dimbúaid* This is the idiom *acht ní . . . ní . . . ,* 'not only not . . . but not . . .', though here the second negative is expressed by *dimbúaid.* Cp. infra 3995.

1639–43 Not in LU, but to this passage corresponds one sentence in YBL (TBC² 1193) and in Eg. 1782: *Ara chind dun sis dond ath co fiasmar, or Cu Chulaind.* The passage is in Rec. III (ZCP viii. 552. 3–5).

1647 *fotalbéim (fótbém* St 1687); 1651 *fáebarbéim co commus (bém co ccomus* St 1692); 1658 *múadalbéim (bém* St 1697). There must have been some technical meaning for these compounds, but except for *béim co commus* it is impossible to get the exact significance. Similarly supra 806–7 we get *tulbéim, múadbéim* and *fotalbéim,* and again infra 3342–3. *Béim co commus* is enumerated among Cú Chulainn's feats (*béim co fommus* infra 1837 = *béim co commus no co fomus* LU 5972–3). For the meaning cp. *Nosberr in gilla mail fair cosin claideb .i. bem co fomus* Ériu I. 118 §9 (Death of Conla). None of the blows are specifically named here in Rec. I.

1650 *condrísam* here and 1654 seems to be for *co comairsem.* Stowe has *co ccomraicem* (1691).

1662–5 Not in Rec. I, but in Rec. III: *Ón ló do gab Ferghus airm laeich ina laimh nir féghusdair ara ais riam ina 'na dheghaidh acht mana tegmadh neach aird i n-aird ris* (ZCP viii. 552. 31–33).

1670–1 *ar condrisam = ar co risam (co ndec[h]am d'acallaimh Con Culainn* St 1710).

1677 *iarfaig-siu a gilla-som* Here *iarfaig* is transitive, meaning 'question, enquire of'. Other recensions have *iarfaig de.* In a passage in Rec. I TBC which has no equivalent in Rec. II we get *Ro iarfacht Medb in buachaill dóig leiss cáit i mbai in tarb* (LU 5359), but in the C version this reads *Ro ierfacht Medb din boachaill* . . . (185). In Esnada Tige Buchet (LL): *Ro iarfaig fecht and inti Cormac in n-ingin*: *Cia tai, a ingen?* or *Cormac* (Fingal Rónáin 516-8).

1680-1   So Achilles bound the corpse of Hector to his chariot, Iliad xxii. 395ff.

1691   *ciarso dúal = ciaso dúal*. *Carsa dúal*, Rec. III (ZCP viii. 553).

1692   *Cid mi fadéin*   For *mi* cp. supra 1445, and *in tan ba gilla bec mi* TTr. 1262.

1696ff.   The Nath Crantail episode is the one which varies most in the three recensions. (See Thurneysen, ZCP ix, 434). The Rec. I version is complicated, and contains features reminiscent of the Fer Báeth and Láiríne Mac Nóis episodes as well as a passage which seems to have been taken from Cú Chulainn's fight with Lóch Mór. Thus Lugaid mac Nóis goes to warn Cú Chulainn of Nath Crantail's coming attack (LU 5721-4). When Nath Crantail sees Cú Chulainn, he refuses to fight with a beardless boy, so Cú Chulainn gets Láeg to make him a false beard, *uilche smerthain* (ib. 5755-9). A detail which does not occur in Rec. II and Rec. III: Nath Crantail pleads with Cú Chulainn to let him return to the camp to tell his twenty-four sons where his treasures are hidden (ib. 5768-75). In Rec. I two explanations of Nath Crantail's inability to recognise Cú Chulainn are possible: it may have been the bulkiness of the *corthe cloichi* beneath his cloak, or more probably, it may have been that Nath Crantail had not previously realised Cú Chulainn's youthfulness.

In LL the whole episode is simplified and tidied up. When Nath Crantail's sword is shattered against the pillar-stone beneath Cú Chulainn's cloak in Rec. I, Cú Chulainn is distorted. But in LL the distortion of Cú Chulainn comes before Nath Crantail's failure to recognise him. So that in LL we have also a choice of two explanations for the non-recognition of Cú Chulainn, either his unwonted bulkiness or his distorted appearance.

Stowe follows LL, but omits both the pillar-stone episode and Cú Chulainn's distortion. In the opening paragraph Stowe shows affinities with Rec. III.

In Rec. III part of the episode is wanting. Judging from the account of the fight which has survived (RC xiv. 256), the *corthe cloichi* episode was omitted in Rec. III also. There is no reference to Cú Chulainn's fowling, no mention of a pool, and in this recension the final meeting with Nath Crantail is quite different.

1697-8   *Nír fiú leis* . . .   Only in LL (and Stowe) is it explained that Nath Crantail scorned to use weapons against Cú Chulainn.

1698   *trí noí [m]bera culind*   Similarly St 1741. In Rec. III we also get *beris trí naí mbera cruaidhcuilinn lais* (ZCP viii. 554. 12). But in Rec. I *berid .ix. mbera culind . . . laiss* (LU 5726-7), and subsequently in that recension the nine holly-spits are accounted for.

1699   *follscaide* and *forloiscthi* are synonymous. In Rec. I only *follscaidi* (LU 5727); in Rec. III only *faillsgidhi*. Perhaps *forloiscthi* (part. of *for-loisci*) was originally a gloss on *follscaide* (part. of *fo-loisci*).

1699-1701   *Acus is and* . . . *dib*   This passage is not in Rec. I or Rec. III. (The first sentence alone has been taken over from LL by Stowe: *Is ann sin ro boi Cu Chulainn forsan lind fora c[h]ionn* 1742–3). Windisch suggests that the passage was originally a marginal gloss. It seems as if the LL compiler misunderstood his original here and assumed that even before Nath Crantail's arrival there were nine spits in the pond, presumably to make it safer, that is, so that Cú Chulainn could step from one to the other when crossing the water. Something must, however, be omitted here in LL. The *iall én* is abruptly introduced. Cp. LU which, instead of *Is and boi Cú Chulaind forsin lind* etc., has *Is and boi Cú i sudiu oc foroim en ┐ a carpat inna farrad* (5727–8).

1704 *for ind úachtarach in bera contarlaic* 'on the tip of the spit which he (N.C.) had thrown.' Windisch and the editors of the Diplomatic Edition of LL omit *contarlaic* here, taking it to be dittography. But I take it to be another instance of *con-*, relative prefix.

1708 *iss ed arfurad ┐ arfognad Coin Culaind* . . . Here the preverb *ar-* is used in relative position, like *con-* elsewhere in the text. I take the verbs to be *feraid* and *fo-gni*. The abbreviation .c̄c̄. for *Coin Culaind* occurs elsewhere in the text e.g. 1501, 3950 etc. Windisch and the editors of the Diplomatic Edition read *Cú Chulaind*, and Windisch translates 'Denn das ist es, was Cuchulinn zu beschaffen und zu bezorgen pflegte, Fisch und Geflügel und Wildpret' (p. 258. 4–5). Stowe paraphrased to *uair is edh ba betha do C[h]oin Culainn for Tanaidh iascach ┐ énac[h] ┐ esfheoil* (1750).

1720 *bith forsna slúagaib* Windisch translates 'bei den Schaaren zu sein', but the Stowe reading here suggests that the meaning is 'to be engaged in attacking them': *beith ag selg forna sluagaibh* (St 1761–2). Cp. in Brislech Mór Maige Murthemni, *Ataat fir Herend form sund ┐ atú forro dano* (LL 13967), said by Cú Chulainn when he refused to give up his spear.

1720 *cian gar* . . . 'while, during the time that . . .' For this meaning see note 1822 Stowe TBC. Here Stowe has *in gcen do-rinde* (1762), and for *cian gar doringnis* 1724, Stowe has *in fat do-righnis* (1765).

1735 *focheird fathi ferge* Perhaps the word *ferga* l. 1734 was recopied here as *ferge*. Cp. the equivalent passage in Rec. I: *Focheird fáthi imbi iar cathais na haidchi* LU 5448.

1740 *Ní hé sút cruth ardomfarfaid-se indé* *Ar-* used to infix pronoun as elsewhere in this text e.g. *ardotchlóe brath* 3440, *ardotrai, a Findchaid* 4053. (Printed *ar domfarfaidse* Dipl. Edn.).

1758 *co luid Medb i nGuiph* Compare the readings of LU and Stowe. The name Guiph does not seem to occur elsewhere, but Rec. III has a form of the place-name corrupt like that of LL: *Doluidh Medb i lLaith Ghuifi re Coin Culaind* (RC xiv. 256 §6), and in a later passage (§11): ┐ *ag taidhecht atuaid do Choin Culainn adorchair Guifi les conadh de ata Laith Guifi.* Cuib is apparently Mag Coba, now the barony of Iveagh, Co. Down. In the Rennes Dindshenchus of Áth Lúain, we find in a passage referring to the same foray of Medb: *Is iarum doluidh [Medb] co ceithrib coicedhaib Erenn hi crich nUlad ┐ Fergus d'eolus rempu co riacht Magh Coba* (RC xv. 465 §66).

1759 *comdar techt fathúaith* The suggested reading, *conid ar techt fathúaith*, is supported by Rec. III which here has four times the phrase *is ag taidhecht atuaid do Choin Culainn do mharbhasdair se* . . . followed by the names of those he killed. Note, however, that in Rec. III Cú Chulainn killed Meic Búachalla etc. when he was returning from the north while here in LL we have 'when he had gone north'.

1759 *Fer Taidle* . . . Of the names given here only one, Meic Búachalla, is common to the three recensions. In Rec. I are also named Nath Coirpthe, Marc, Meille, Bodb and Bogaine (5813–6), and all these, together with Redg and Buide, were killed in Cuib. In Rec. III Cú Chulainn is said to have killed Meic Búachalla, Guifi, Fir Cranchai and Buide on his way back to his own land from the north. The Redg episode in Rec. III comes much later in the narrative, between Cú Chulainn's meeting with Fergus (infra 2478–2508) and the fight with Ferchú Loingsech.

1767 *Is and sin tra forecmangaid Firu Crandce* . . . Dr. Knott (RIA Dict. s.v. *for-ecmaing*) suggests that this is a spurious 3 s. pres. form. Cp. *Is ann sin tarla do Fir Cruinice* St 1805, with the variant *tarlatar do Firu Cruince*, which suggests that the original reading may have been *Is and sin forecmangait* (*for-* for *do-* after *Is and sin*) *dó Firu Crandce*, with acc. *firu* for nom. In a later passage in the C version we have *Agus is iatt sin fioru Cronice ar is a Croinic hi Fochartt ron bithe* (997–8) = ⁊ *is iat fir Chrónige* etc. LU 6167. In Rec. I we are told that *Fir Focherda* were killed in Mag Murthemne (LU 5817ff.).

1767–71 The corresponding names in Rec. III are garbled: *Is iad seo anmanda fer Cranncha gan dighbhail .i. da Artuir, da mac Leghi, ᵈa Bheoch-raidhi, da Dhurchraidhi .i. Drucht ⁊ Dealt ⁊ Taidhen, Tedhi ⁊ Tula[ng], Trasgur ⁊ Tulghlaisi, gorubfi[che Fer Fo]cherda annsin .i. dech ndeogmairi ⁊ [.x.]fhindchada, is edh rosbi C.C. ar gabail dunaidh ⁊ longphairt dferaib Erind* (RC xiv. 257 §12). In Rec. I no names are given in the corresponding passage which reads *Iar tiachtain iarom geogain [Cú Chulaind] firu Crochine (nó Croiniche, hand M) .i. Focherda. .xx. fer focherd de. Dosnetarraid oc gabáil dúnaid dóib .x. ndeogbaire ⁊ .x. fénnide.*

Thurneysen has dealt at some length with this passage of LL (ZCP ix. 426–9, Die Überlieferung der TBC). It is one of the passages on which he based his belief (1) that the compiler of the LL- version knew and made use of the expanded LU-version, i.e. LU with H-interpolations, and (2) that Rec. III was a free re-telling of the LL-version with some echoes of Rec. I included here and there. It will be noted that in LL only seventeen names are given for the twenty men, while in Rec. III eight men are given and seven names for these eight. The names in Rec. III seem to derive from an original like that of the LL-compiler while the last sentence of the passage is reminiscent of the LU-reading. In a later episode in Rec. I, which is an interpolation in hand H in LU, the Fir Focherda or Fir Chrónige are again referred to, and are again said to have been slain by Cú Chulainn. There, however, they number only fourteen, and the fourteen are named (LU 6164–8). In this interpolated passage they are the fourteen men who, at Medb's instigation, lie in wait for Cú Chulainn when Medb has invited him to a peaceful meeting 'in Ard Aignech, today called Fochaird' (LU 6158–9). This interpolated passage is of course a doublet of the earlier one. It occurs also in Rec. III where it agrees practically word for word with the LU-passage (RC xv. 65 §73). According to Thurneysen (loc. cit p. 427 n. 2) it was taken into *eh*, the forerunner of Rec. III, from a MS. of the expanded LU-version.

There are, Thurneysen says, two possible explanations for the fact that the names of the Fir Focherda are given here in LL though not in Rec. I.

(1) The H-interpolater knew the LL-version as we now have it, took from this passage the names of the Fir Focherda and inserted them into the later episode of Medb's meeting with Cú Chulainn, reducing the number to fourteen. Thurneysen rightly rejects this explanation. It would stand, he says, only if elsewhere also the H-interpolator drew from the LL-version what it has in excess of LU. (He believes, for instance, that if the H-interpolator had used the LL-version he would surely not have omitted the 'Pillow-talk' which forms such a pleasing introduction to the tale). (2) Much simpler, Thurneysen says, is the second explanation. The compiler of the LL-version knew the expanded LU-version (i.e. LU with H-interpolation). He unified the narrative, suppressing the second occurrence of the Fir Focherda and transferring the names to the first occurrence. Against this theory, however, is the fact that, except for the poem *Atchiu fer find firfes chless* supra 234ff., there is nothing in the LL-version corresponding to the long interpolations in hand H in LU. Neither have the names in the H-passage been taken over exactly (*Dá Glas Sinna da mac Buccridi. Da Ardáin da mac Licce. Da Glas Ogma da mac Cruind. Drucht ⁊ Delt ⁊ Dathen. Téa ⁊ Tascur ⁊ Tualang. Taur ⁊ Glese* LU 6161–3). A third explanation may be offered. The H-interpolator, the compiler of the LL-version and the compiler of Rec. III all drew on another version of TBC (*x*). In the passage in LU containing the first reference to the Fir Focherda there is no interpolation by hand H. But for that passage both LL and Rec. III compilers drew on the variant version which here gave the names of the men. The words *deich ndeogbaire ⁊ deich fénnide* were omitted by the LL-compiler, included by the compiler of Rec. III. The second reference to the Fir Focherda occurs in a long H-interpolation (LU 6131–6206), and of this passage lines 6133–6192, 6200–6206 are found practically unchanged in Rec. III. Instead of concluding with Thurneysen that these lines were taken by the compiler of *eh* from a MS. of the expanded LU-version, I suggest that the H-interpolator and the compiler of *eh* here drew on the same variant version(s) which I call *x*. So too I would explain the poem *Atchiu fer find firfes chless*. In LU it has been written on an erased passage in hand H. It was taken by the H-interpolator and by the LL-compiler from a variant version of TBC.

In the later passage interpolated by H in LU (6159–63), and in the same passage in Rec. III, only fourteen men are named and they are called *cethri fir deac Fócherda* (LU 6167). In the recently discovered O'Curry MS. (C) the scribe, no doubt remembering the earlier incident, changes to *fiche fer Focherdai* (997–8), and *cethri goi deac* of LU 6165 becomes in C *fiche gai*. Only fourteen names are given in this passage in LU, *Dá Glas Sinna da mac Buccridi* (6161), for instance, being only two men. But the C scribe takes them to be four men, *dá Glas Sinna* and *dá mac Buccridi*, and similarly for the next two pairs, thus arriving at a total of twenty and justifying his change from fourteen to twenty.

1770 *Basnetarraid* Ba- for do-.

1778 *Nit charadar* Here -*caradar* for -*cara*, 'for the sake of the jingle', according to Strachan (Deponent Verb p. 59 n. 1).

1784 *ic clóechchlód na dá chertgae* This phrase inadvertently inserted by the LL compiler is not in LU, nor has it been taken over by Stowe. It is an obvious proof that the LL-compiler had access to the variant version used in Rec. III. *Maith a Bhuidhi, bhar Cu Chulainn, tairr romaind bharsin áth sa sis go cláechlomais da urchur dha chele. Tangadar forsin ath ⁊ da chlaechlodar dha urchur ann ⁊ darochair Buidhi bharsan ath gorub atha Buidhi a ainm dha es. Cidh tra acht gérbha ghairid do bhadar in dana churaidh sin ⁊ in dana chathmhilidh ag cláechlodh in dana urchur sin etarru, rugadar in t-ochtur fear mor in Donn Cúailghni* (RC xiv. 258. §§16, 17). There is no exchange of casts in the description of the fight in LU or LL; Buide is slain by a spear (*certgae*) cast by Cú Chulainn.

1785   *ni fo chétóir conarnic úadib*   Here *con-*, relative particle, replaces *do, t-*. Cp. *ni tharnaic úad acht a rád* infra 1812. Or possibly the verb *ar-icc* is the one used here as in *ni arnecair úad a rádha acht . . .* 2591. This phrase is omitted here in Stowe.

1789–91   According to LU these places were in Cuib (LU 5824–5).

1792–4   The three recensions differ here. In Rec. I Medb spent a fortnight ravaging the province, then fought with Findmór, wife of Celtchair mac Uthidir, plundered Dún Sobairche and took fifty women prisoners (LU 5820–3). In Rec. III Medb devastated Cruithne, Cúalnge and the land of Conall Cernach, spent a night before Dún Sobairche, took Findmór prisoner with fifty women, and tortured and hanged them (RC xiv 258 §18). Only in LL are we told that Medb killed Findmór (Stowe omits the words ⁊ *ro marb Findmóir*), but Rec. III gives the most savage account of this incident. The name of Celtchair's wife is more usually given as Findabair. According to the Banshenchus (LL 16749–54) Celtchair was married three times, his wives being Doruama, Findabair and Bríg Brethach. (Bríg Brethach is the one referred to in Aided Celtchair LL 13725). Findabair's connection with Dún Sobairche is mentioned in a poem on that place LL 2109–2168: *Áit i mbai ind rigan rúanaid | Cheltchair chuanaig cauir chéilig | Findabair find a glégen* (2117–9). Celtchair's home was Dún Da Lethglais or Dún Lethglaise (cp. infra 4065–6, 4494 and *Dún Da Lethghlass dún ciúin Celtchair* LL 16753).
    In the prose dindshenchus of Áth Lúain (RC xv. 465 §66) Medb is said to have captured in Dún Sobairche not Celtchair's wife but the wife of Conall Cernach. (Her name is not given here, but she is called *Lendabair ingen Eógain meic Derthacht* LU 8234 and LL 16730). In the metrical dindshenchus of Áth Luain (MD III 366–74 = LL 20868ff.) there is no mention of Dún Sobairche; Medb ravages Cúalnge and takes prisoner Conall Cernach's wife (loc. cit. 368. 41–44).

1797   *ni arlaic . . .*   Stowe makes this clearer by adding *do breth leo* after *Dond Cúalnge*.

1798   *co ndasrimmartatar . . .*   Windisch takes the infixed pronoun to be intended for masc. sing. and translates 'Da Drängten sie ihm (den Stier) mit Speerschaft auf Schild auf ihn (den Hirten).' But the Stowe reading (*gur cursit na sloig i mbernaidh n-iomcumaing fair in Donn Cuailnge cusna halmaib batar uime la crand for sciath* 1832–3) suggests that we can take the infixed pronoun as plural. referring to the bull and his heifers, *na halma* referred to later in the same passage. I take *fair* to be the dativus incommodi; it might be better if it were put, as in Stowe, after *i mbernaid cumaing*. In LU the passage reads: *Acht gabais a mbúachaill a tarb dib conid timachtatár taris i mbernai cumaing la crand for sciathu, conid bertatar cossa na slabrai triasin talmain* (5827–9).

1808 *nád tard dait[h] Cú Chulaind in clettin dó* (*nad tarddait Cu Chulaind* . . . Windisch and Diplomatic Edition). The word *tard-* is at the end of a line with a faint hyphen (faint because of an attempt at obliteration ?). The word *dait* at the beginning of the next line is clear in the MS. I take it to be for *daith*, 'swift, prompt', used adverbially = *co daith*. Very probably *co* has dropped out here. Cp. for the omission at the line-end *mar beocho | áinle* infra 3096. Both Rec. III and Stowe have an adverb in this sentence : *Ni thard C.C. fochedóir dho é* (RC xv. 207 §218); *Ni tuc C.C. in cleitin dó ar tús* (St 1842). The sentence which follows in LL I take to be a gloss. No doubt the compiler felt that (*co*) *daith* required an explanation. (*Daith* is twice glossed *ésgaid* by O'Clery). *Ni sain* ┐ . . . is the equivalent of *Is inann* ┐ . . . Otherwise explained in RIA Contribb. s.v. 1. *sain* (c), under the influence of Windisch's mistranslation: 'Er hielt es für nichts Besonderes . . . ihn hinzugeben'. Windisch was further misled by the Facsimilist who read ┐ *nadescaind*, taking the flourish of *d* for *n*-stroke.

For *nád* in ┐ *nád tard* and ┐ *nád éscaid*, see Pedersen (VG II 254) who notes that there is a tendency to use the dependent negative after *ocus* and quotes *Ní ind fessin eirbthi* ┐ *nach do du-aisilbi na nni do-gni acht is do Dia* Ml. 51ᵇ12.

1813 *tráth conroscar* . . . Printed *tráth coro scar* . . . in the Diplomatic Edition. Windisch takes the sentence to be *Is solom dún in sét sa tráth*, and *tráth* to be for *trá*. Again infra 1822–3 he reads as *Is amra lib tráth. Is ro máethmaccáem* etc. So too in Contribb., s.v. *trá*, these two instances are given as = *trá*. I take *tráth* to be the conjunction in both cases. With the sentence *ni tharnaic úad acht a rád* . . . compare infra *ni arnecair úad a ráda acht Fiach! Fiach! tráth rabert Cú Chulainn béim dó* (2591–2). For *tráth* followed by relative *con-* as in *tráth conroscar* . . . cp. supra *ni roiched bun a bunsaige lár tráth congebed a barr etarla etarbúas* (765–6).

1816ff. Before the Cúr episode a long section has been inserted by hand H in LU. (This section is not in YBL, nor in Rec. III). First Lugaid goes to parley with Cú Chulainn, then Fergus. Then comes the meeting with Findabair, followed by three episodes which are doublets of later passages. Then once more Lugaid goes to ask Cú Chulainn for a truce. These passages are also in Eg. 1782, but for part there is the loss of a page of the MS. As well as these passages, the first half of Comrac Cúir, up to the enumeration of the feats, is in hand H in LU (5834–5967).

1821–30 All this conversation is lacking in Rec. III, as well as the enumeration of the feats.

1822 *Is cert ar [ṁ]búaid lib* To this corresponds in LU (5959) *Is cert in brig doberid dún*, 'Ye make little (?) account of me', taking *cert* to be O'Davern's *cert .i. beg* (70). Cp. *certga*, 'a small spear'. The sentence is omitted in Stowe which has merely *Is boeth a macasamla sin do s[h]amail frim-sa* (1857). Here *is amra lib* seems to be used ironically. Possibly we should read *Is cert ar ṁbúaid lib .i. is amra lib*.

1825 *is acca a rád* The corresponding passage in LU reads *Ecca sin, for Cormac Cond Longas* (5961). *Acc, acca*, 'nay, not so'. Here the negative particle seems to be used as an adjective.

1829 *Ní hed nobar furgfe sib* Here we have both infixed and independent pronoun. *Niba hedh fhuireochas sib* St 1863.

1840 Not in LU. The words *ar lus . . . cróich* are omitted in the Stowe rendering. I have translated them according to a note by Vendryes, Ét. Celt. iv. 317–8 but the force of the simile is obscure to me. *ar lus = a llos.* Vendryes suggests that *lethláim* is to be corrected to *lethláime*. Alternatively we might read *ar lu[a]s 'na lethláim*?

1843 *Láeg* In Rec. I it is Fiacha mac Fir Febe who speaks to Cú Chulainn (LU 5978), but in Rec. III, as here, it is Láeg (RC xiv. 259). In Rec. I Fiacha's intervention is explained: Cú Chulainn is so absorbed in practising his feats that he does not notice Cúr's attack. This passage has been taken over from Rec. I by the Stowe compiler.

1858ff. In Rec. I Láeg is sent to greet Lugaid mac Nóis and to find out who will come to encounter Cú Chulainn next day (LU 5991–3). No others in the Connacht camp are named. This passage comes later in Rec. III, after the death of Fer Báeth. Láeg is then told to greet Cú Chulainn's friends and foster-brothers but Lugaid mac Nóis alone is named.

1895-6 *& conattecht Cú Chulaind in caratrad . . . friss* Here we have confusion of two verbs *condieig* and *adteich*. The equivalent passage in LU reads *Attaich C.C. friss a chomaltus* (6018–9), 'C.C. appealed to him by his fosterbrotherhood'. The same confusion infra 2474–6. This arose when in Mid. Ir. a relative construction occurred after such a phrase as *(Is) and sin.* See note in Celtica vii. 42-4.

In Stowe this sentence reads *Ro cuimnigh Cu Chulainn in comann ⁊ in caratrad boi etarra* (1929–30). In the Glossary to Stowe TBC I give the verb as *cuimnigid*, 'remembers, recalls', which fits the passage well: 'C.C. reminded him of their friendship'. But it is noteworthy that in Mid. Ir. we sometimes find the spelling *cuimnig* for *cuinnig*. There are, for instance, three occurrences of *ro cuimnigh* for *ro cuinnig* in Lorgaireacht an tSoidhigh Naomhtha, 1941, 2131, 3712, and two instances in Scél Saltrach na Rann (Celtica iv p. 34) 480–482. Assuming a continuing confusion of *condieig* and *adteich*, we might take the Stowe reading to be for *ro cuinnig (conattecht = (con)ataig)* : *Ro cuinnig C.C. [friss] in comann ⁊ in caratrad boi etarra*, 'He invoked the friendship between them . Again infra 3084 *ná cumnig in comaltus* might well be for *ná cuinnig in comaltus*, 'do not invoke our friendship', which fits the context well.

1898 *fosnessa Fo-* for *do-*, *do-nessa*, 'tramples', with infix. pron. 3 s.f. anticipating *sleig*. LU reads *Fornessa sleig culind isin glind hi coiss Con Culaind* (6021–2), (to be read *Fornessa [Fer Baeth] sleig culind . . .*?). In Stowe the sentence is altered to *go ttarla slegh cuilinn i mbontt a c[h]oisi* (1931), while in Rec. III, in part illegible here, it seems as if it were Fer Báeth who threw the stake at Cú Chulainn (. . . *easairg F.B. sleigh cuilind a [m]bond . . . do goro chumaisg thrithi eter fheoil ⁊ [leath]ar, feithib eter cnamhaib [leg. eter fheithib ⁊ cnamhaib]* RC xiv. 260 §§37-8). The sentence which follows here in LL, *Tarngid . . . assa frémaib* suggests that to the LL compiler the holly-stake was growing in the ground and Cú Chulainn trod on it. In Rec. III Cú Chulainn tears the holly-stake out of his sinews and bones, his flesh and skin (§38), not out by its roots from the ground. This is to be expected if Fer Báeth had cast the stake. But the LL reading seems to be the better one; it is hard to see how his opponent could have driven a stake into the sole of Cú Chulainn's foot.

1901 *Dotarl*aic . . . For *dosfarlaic*? Windisch read as *Dotarla in sleg*, no doubt influenced by the Stowe reading *go ttarla in t-urchar i cclasaib a c[h]uil*.

1911 *in ráinic Fer Baoth* . . . To be read *in ráinic [ní] Fer Baoth* as supra 1907, omitting *an longphort* which is the reading of H[1] and P.

1940 *tuc L. bogadh ⁊ bertnugadh fair* Cp. supra 98–9.

1954 *gurbo ceó aéerda an ceat[h]araird i mboi* Here LU reads *con sephaind a channebor ass combo buadartha in t-ath día chacc ⁊ combo thruallnethe aér na cethararda dia dendgur* (6075–7) and Rec. III *condasesairg a chainnebhur as gorbha cheó trúailnithi tormosach i mbuí do mhothor ina mhedhor* (RC xiv. 265). In his attempt to moderate the description, the Stowe compiler failed to make much sense. Read *gurbo ceó truaillnithe aér na cethararda* omitting *i mboí*?

1967 Long mac Emónis was brother to Lóch Mór mac Febis. The difference in names is explained in LU in a footnote to l. 6103 (in a later hand, not H): *Lóch Mór mac Mo Femis (nó Emonis* above line) *nó mac Mo Febis (.i. Febes ainm a mathar* above line).

1989ff. Cú Chulainn's fight with Lóch and the Morrígan is essentially the same in all versions but differs in detail. In Rec. I (LU 6196–6214) the Morrígan appears (1) as an eel and Cú Chulainn breaks her ribs; (2) in the form of a she-wolf and he wounds one of her eyes; (3) in the form of a hornless, red heifer and he breaks her legs. In Rec. III the order of appearance is (2), (1), (3), and in (1) Cú Chulainn breaks the head of the eel and not her ribs (§83). It will be noted that only one wound inflicted by Cú Chulainn on the Morrígan is mentioned in Stowe, and it is not the she-wolf which loses an eye but the white, red-eared heifer.

1993 *co coicait samasc uimpi* This detail is not in Rec. I, but is in Rec. III (§84).

1994–6 *Dobertsat in banntracht gesa* . . . Not in Rec. I but in Rec. III (§85).

2012 *gebis athrechus mór Coin Culaind* (σ̄c MS.) This is the usual idiom. Similarly *Ro gab toirsi mor na hapstalu* PH 5118. Contrast 2014–5 *ro gab-sum merten ⁊ athscis forru*. (Should *forru* be omitted here, or is the meaning that his weariness and depression was caused by them i.e. the Ulstermen?) Cp. *ro gab ét ingen Roduib* MD iii 94. 13 and *Do gob nairi an ingen an* ib. iv. 10. 13. An impersonal use of *gaibid* with two accusatives occurs in Seirgligi Con Culaind: *ro gab etere mór ⁊ drochmenmain in n-ingin* (SC[2] 764 and n.)

2016–2091 This poem is obviously an interpolation, taken by the compiler of LL-TBC from that version of TBC which was the forerunner of Rec. III. It has been taken by Stowe from LL, but we also find it in the O'Curry MS. (C) and of course in Rec. III. In LU we have a short poem of four stanzas *M'óenurán dam ar étib* (LU 6216–31), which comes before Lóch is killed by the *ga bulga*, not, as this poem does in the other versions, after his death. Some of the lines of the LU poem agree with those in LL (6216, 6222–3, 6225–7), but in LU the poem is not addressed to Láeg. In C the poem shows influence of the LU-poem, and one stanza has been taken over

(LU 6220–3). It will be noted that the poem as it stands in LL gives the version of the fight with Lóch and the Morrígan which occurs in the prose account of it given in Rec. III.

2022　*Ní lám Fíngin roda slaid*　Read *rodas*. This refers to Fíngin Fáithlïaig (infra 3659ff.), the prototype of a great physician. The meaning is that the hand which smote his sides was the hand of a fighting warrior not that of a healing physician. A similar expression occurs in Tochmarc Ferbe describing the fighting of Maine Mórgor, son of Ailill and Medb. *Nír bo lám lega la Mani inn uair sin ocus dorochratar nonbur Fomorach dia chétscundscli* (sic) *a oenur* (318–9).

2036　*Mad úathad dochanat form*　Rec. III reads *Mádh aenfher concana corm* (leg. *form*), with rel. use of *con-*.

2038　*mad ilar ceól* . . .　MS. reads *mad ilar corn* with *corn* cancelled by overstroke and *ceól* in margin. Both Stowe and C have *mad iolar corn*, but Rec. III *maith* (leg. *madh*) *ilar ciúil* (RC xv. 70).

2047　*noco modmar cach n-óenbró*　This seems to be a proverbial saying. *Ní nodmar dina nach aenbro* (Tochmarc Ailbe, ZCP xiii. 281. 20).

2052　*lín in chaire*　Here Rec. III has *Madh úathadh luchd a chaire*, 'the contents of the cauldron' (?).

2057　*'gund áth i cind Tíri Móir*　In the prose account in Rec. III we are told that Cú Chulainn and Lóch met *bharsan áth n-uachtarach eter Mhédhi ⁊ Ghédi i cind Tíri Moir* (RC xv. 66 §78).

2058　*co lleith Bodba*　I take *co lleith* as = *i lleith*, 'with, together with'. (*niterpi i lled nach áili* Wb. 1ᵈ10, 'thou dost not trust thyself with anyone else'), but I have not found any other instance of *co lleith* with this meaning.

2060　*Ra lettair Lóch mo dá lón*　This follows not the account given in the LL-text but that of Rec. III: *In comad do bhi-siun ag ursglaigi na saidhi, geoghain Lóch tremhid a dhá lúan é* (§80).

2061　*rom tesc in tsód garb glasród*　This again refers to the prose account in Rec. III where the she-wolf bites Cú Chulainn's arm: *tesgais a dhóid Chon Culainn* (§80). This incident in Rec. III agrees with the prophesied account in TBRegamna (Ir.T. II. 247).

2062–3　To these lines correspond in the account of the fight in Rec. III. *Bhocerd [in esconga] curu ⁊ snadhmanna eter dhibh cosaibh . . . do Choim Culainn condrochair faen fotharsna . . . ⁊ suil ráinig les érghi, geoghain Lóch tremhid a thromaibh é* (§81). In Stowe (here representing LL) we have *ro ghon Lóch urt[h]arsna é tre c[h]ompar a c[h]léb* (supra 2000).

2065　*a rosc*　In the Stowe account this wound—the only one mentioned —was inflicted on the Morrígan in the form of a heifer.

2066　*a gerr gara*　No reference to this wounding in the Stowe account. It is mentioned in Rec. I (LU 6214) and in Rec. III (§86).

2069　*ba seol faethe*　*Bha seól saithi* Rec. III (§94.13) gives perhaps a better reading. We might take *saithe* as from *sáeth*, 'a distressful course'. Windisch suggests *saithe*, 'a swarm of bees', taking it to refer to the many points of the *ga bulga*, but *saithe* will not fit the verse (*saithe: Aífe*).

2071　*dar thoeth Lóch* (*la ttaoth* St, *le taoth* C, *gor bhath* Rec. III). Windisch took the verb to be subjunctive, but it seems to be an instance of the fut. stem *toeth-* of *do-tuit* used as general stem of the verb.

2095   *And sain faitti Medb in sessiur úadi   Is and sin faoidis Medb sesior uaithe* St 2133. The form *faitte* used as 3 s. pres. ind. occurs again infra 2114 (= *do cuir St*), 2621 (= *faoidhis* St) and 4822 (= *cuires* St). As M. O Daly notes (Ériu xiv. 34.75), *faitti* is from *faidith-i*, 3 s. pres. ind. with suffixed pronoun 3 s.m. The form has become petrified and the force of the suffixed pronoun completely lost. In *And sain faitte Medb in Dond Cúalnge* (4822) we might take *faitte* as including a suffixed pronoun were it not that there are no other instances of the suffixed pronoun in the text. In LU-TBC the suffixed pronoun occurs, and the LL *faitti* might arise from a misunderstanding of such a sentence as *Foítisi techta úaidi* (LU 6147). (*Techta* here is singular; mentioned again as *in techtairi .i. Traigtrén* l. 6150. Cp. *Is and sain ra faidestar Média techta co Iasón. Flichta ingen Rófaid a hainm in techtai* LL 31123-4). A pret. pass. pl. occurs infra 2617 : *Is and sin ra faittea fessa ⁊ techtaireda.* In CRR *faitti* is used as in LL-TBC: *Is and confaitti Conall fessa ⁊ techta uad* (LL 22773), with relative *con-* prefixed. In CRR too a petrified form *faitti* is used as 2s. imperative three times: *Acus faitti fessa ⁊ techta uaitsiu chena cot chairdib écmaissi* (LL 22737, 22742, 22846).

2100   *Giambtar liri*   In Mid. Ir. *cia* sometimes nasalizes the copula: *ciambad trom leis* TTr. 388 *ciambad áil do* ib. 2050. Early Mod. Ir. *gémadh.* Here *lir* is not equative but is treated as a positive and inflected for the plural.

2103-2113   The healing episode is better told in Ṛec. I. In both Rec. I and Rec. III the sentence *Bendacht dee ⁊ andee fort* comes after the Morrígan has given Cú Chulainn the first drink. In Rec. I she is healed (1) of the wound in her head, (2) of her eye, (3) of her broken legs, while in Rec. III the eye and head only are mentioned, the third healing being a vague *bhá slán si bha chetoir* (§110). It is curious that Rec. I should here agree with Rec. III in mentioning the healing of her head. In the account of her wounding given in Rec. I, it was her ribs which were crushed. (See Introd. to Stowe Táin p. xxi n. 2). Noteworthy is the account of the healing in Version C: first her eye is healed (1145), then her legs (1146) and finally her ribs (*combo slan a hasnai* 1147). In Cóir Anmann §149 an account of the healing is given, agreeing almost word for word with LL. There too, as in LL, only the healing of the Morrígan's eye is referred to.

2105-6   *dáig ni gonad Cú Chulaind nech ara térnad co mbeth cuit dó féin 'na legius,* 'for none whom C.C. wounded ever recovered until he himself had aided in his cure' (lit. 'C.C. wounded no one who recovered . . .', mistranslated by Windisch, and also in Contribb. s.v. *do-érni,* 'from whom he escaped'). Here *ara n-* is used like *co n-* before the relative verb. This use of *ara n-* seems to occur in Mid. Ir. when the main verb is negative. (We might compare the use of the conjunction *ara n-* for *co n-* in similar cases. See note 687 supra). The construction must have seemed strange to the Stowe compiler for he altered the sentence to *úair ni thernód nech dá ngonadh C.C. no co mbeth cuid dó féin ina leighes* (2144-5), which is exactly the meaning of the LL-sentence. (In the Introduction to Stowe TBC p. xlv (12) I mistakenly suggested that Stowe here offered a better reading than the LL-text).

A similar construction is found in LL-Mesca Ulad: *Ní rabi d'esbaid nach do chumaid ar duni d'Ultaib riam ara tucad da ari acht co facced Róimid*

Rigónmit (MU² 699–701). The editor prints *a ra tucad*, and quotes the sentence in the Glossary s.v. *a n-*, 'that which, what', and translates *d'esbaid . . . a ra tucad da ari* 'as much of want . . . as he would heed' which comes close to the meaning. I should translate: 'No Ulsterman ever suffered loss or grief that he took heed of if only he saw R.R.' i.e. 'No Ulsterman took heed of any loss or grief that he suffered etc.' Again in CRR: *Dáig ba demin leo ni fil inad i faicfithé gnúis Chonaill ara teichfithe and* (LL 23108), '. . . that nowhere where C. was seen would men flee'. Hogan took *ara* to be preposition + relative and translated 'on which men would flee', but if a preposition were called for one would expect *i teichfithe*. Compare also in a marginal quatrain in LB: *Nis fil uaib d'oc na do shean | . . . | ar a* (read *ara*) *ndercfa ar gnúis Meic Dé | mine dernai in iar[mér]ge* (Celtica II 29), 'None of you . . . will gaze on the face of the Son of God'. In the last two instances Mid Ir. indicative replaces older subjunctive. In the present text we get the older usage in e.g. *Ní fil d feraib Hérend ti imbárach dit fópairt acht missi* (1621–2) beside the later *Ní fil ni nád gellfad dar cend a enig* (695–6). The verb *-térnád* is most likely imperfect indicative here.

2121 *Breslech Maige Murthemne* See Stowe TBC note 2159. This long descriptive passage is practically identical in all three recensions. (There is a long omission here in YBL and Eg. 1782 does not go beyond LU 6053, the opening of Comrac Láiríne meic Nóis). According to Thurneysen, the style and language of this section is later in date of composition than the rest of the tale.

2122–3 *isin Breslig Móir* According to Rec. III (§113), they encamped at *Áth Aladh Fhind i Muig Murthemne*. See Introd. Stowe TBC p. xix.

2146 *ni saig nech fair . . .* Translated in RIA Contribb. as 'He attacks no one and no one attacks him'. But the phrase seems to be the equivalent of *ni saig nech comrád fair*. Cp. *Saigis C. coir comraid fair* CCath 2882–3, *niro saigh immorro comradh forru* ib 4205. This is the meaning which the Stowe compiler gave the phrase when he paraphrased as *Ni dhēnann nech urān fair ๅ ni dēnand-som urān for nech* (2184).

2149 *in t-imned mór* Nom. where one would expect acc. Similarly LU 6302 and Stowe 2187. Rec. III reads *or doigh foreadarson ind nith mor inarfuilimsea* (RC xv. 73 §120).

2153 After the words *airchisis de* there is a passage in LU and in Rec. III in which the warrior identifies himself as Lug mac Ethlend, Cú Chulainn's father from the fairy mounds (LU 6305–9). He sings a rhetoric over Cú Chulainn as he sleeps (ib. 6315–31).

2155 *firbat-sa . . .* In LU *fifatsa* (6312) and in Rec. III *fifadsa*, but in C *firfatsai* (1208). This is 1 s. fut. of *feraid*, with Mid. Ir. stem *fir-*. More usually with object *cath, comlann* etc., but here used without object.

2173 *úair dofárthetar dá mac Beithe meic Báin* Slightly different in Rec. III: *air dharadadh minn righ Ailella eter dhá mac Bhethi mic Bháin, dá mac mhuimi ๅ aidi dho Ailill* (§127); and in the repetition (= 2190 infra), *úair thugadh mind oir Ailella ar cairthi eter dha mac Bhethi mic Bháin ๅ rogheoghnadar Fallamain* (§129).

2202 Before the words *Ocus in carpat serda . . .* Rec. III mentions the departure of Lug (*Is annsin da imtkigh in t-óglach sidhech úaithibh ๅ nir fhedadar cá conair imarlodh* §131). In Rec. III there is no later reference to Lug such as we find in the text of LU (6616–7) and in a marginal note in LL (infra 2321–2). Then in Rec. III Cú Chulainn addresses Láeg: *Maith a m*[o] *phoba a Laigh, bhar C.C., tegum imall*[e] *do dhighail na macraidhi arna slúaghaibh. Rachaidsea let on, bhar Láegh* (§132). Possibly some such sentences have been omitted in the other recensions. The *carpat serda* is introduced very abruptly.

2221   *congebethar dóib* (*co ngebethar dóib*, Diplomatic Edition). A relative use of the preverb *con-*. The verb is not deponent in Stowe which reads *congabadh doib* (2262). Compare infra 2237 *congabad dó* (so printed in Diplomatic Edition) and 2363 *condriced*.

2251   *ina thaul tárla* The meaning of *tárla* is obscure. Similarly *ina tul tárla* LU 6431, St 2289, but in Rec. III: *ina thul tarla in sgeth mhóir mileta sin* (§138). See Glossary BDD² s.v. *tarlethar*. Compare in a passage of the Fer Diad episode in YBL which has no equivalent in LL: *Cromsciath . . . fair . . . co taillfed osairchosair cethora ndroṅg ndeichenbair fa thairrlethar in sceith fil foro thairr sceo thaullethan inn oclaich* (TBC² 2338–41).

2252   *contescfad finna* Relative use of *con-*.

2254   *is cumma imthescad . . .* Similarly LU 6434. This may be the verb *imm-tesc*, but more probably it is an instance of the preverb *im-* used, like *con-*, in relative position. Stowe reads *is cuma no teascadh . . .* (2292) while Rec. III has *cuma* (*cuman* MS.) *congonadh-son . . .* (§138). Cp. infra 3615–6 *Is cumma congonad a araid ⁊ a eocho*. So too *is cumma congáiritis* 2257 is the simple verb *gáirid* with *con-* prefixed in relative position. Cp. infra 2845–6 *Gura gáirsetar imme boccánaig ⁊ bánanaig*.

2259–60   *Ro chres . . .* I have translated this verb as a passive, but, it was probably intended for 3 s. active like the preceding verbs *ro gabastar, ro gab*. See infra note 3630. Stowe, however, has *Ro cuired a cealtar comga tairis* (2297).

2260–1   *don tlachtdillat Tíre Tairṅgire dobretha dó ó Manannán mac Lir ó ríg Thíre na Sorcha* Here LU reads *don tlachtdillat Tíre Tair*[n]*gire dobretha o aiti druidechta* (6440–1), while Stowe combines the readings of both LU and LL.

2261.   *ó ríg Thíre na Sorcha* This line is quoted in the Contribb. s.v. 2 *sorchae*, 'light, brightness', as if *Tír na Sorcha* meant 'Land of Brightness', a synonym of *Tír Tairngire*. But *Sorcha, Tír na Sorcha*, was one name for Syria. See Gadelica I 274–5. It is used as a synonym for Arabia in a passage of Cathcharpat Serda LL 189ª, where, in a description of Cú Chulainn's equipment we get the following: *Dofuil cathbarr cirach clárach comecartha do gemmaib solusglana do chumtuch ingantach tiri Arabiae .i. crichi na Sorcha. A ferand min Manannain doberthe dosom do ascid flatha* (LL 24889–91). It seems likely that it was from this passage that the LL-compiler took the name for the Táin. He probably misunderstood the last sentence and took *a ferand min Manannáin* to be synonymous with *crich na Sorcha* in the first. (Cathcharpat Serda is obviously a mnemonic exercise and would be known to the compiler). Note the reading of Rec. III here: *don tlacht idalta Thiri Tairngiri dobrethea i n-asgaid dho-son o Mhanannann mac Lir o righ na Sorcha* (§140). Where jewels or golden ornaments are referred to in the tales, they are often said to have come from the East. Cp. infra 3256–7. Of Hector's sword is said in the LL-Togail Troí: *Conrottacht in t-imdorn fessin d'ór órlasrach Arabia* (32359).

2273 *cúach cera* The meaning of this phrase is obscure.

2279 *béim n-ulgaib* The word *ulgab, ulgam* is obscure. It occurs only in this phrase.

2282 *immar glimnaig árchon i fotha* Stowe reads *mar glaim n-arcon i fothoch* (2321) and Rec. III *amail gloimnigh n-árchon i fathoch* (§149). The word *fotha(d), fothach* is obscure. Cp. *samalta ri gláim con allmaraig i fathod* MU² 611 and Glossary s.v. The phrase occurs again infra 4555.

2298 *cona tharngib gaithe* Similarly Rec. I, Rec. III and Stowe. The meaning of *gaithe* is obscure. Windisch tentatively translates 'mit seinen stechenden Nägeln', and connects the word with *gaoth*, 'a dart' (i.e. *gae*).

2299 Here follows in LU (6483–91) a description of the chariot and horses. C also contained this passage, but is defective at this point, only a word here and there being legible. The passage is not in Rec. III.

2318 *impu fá mórthimchell* (MS. *impu fa*) Similarly LU 6512, but Stowe has *iompa moirthimchiol*. C has *impaib foa mortimcell* (1341). I take *fo mórthimchell* as = *im mórthimchell*, and *fá* as = *foa*. Cp. *dorat mur . . . | di ór imma imthimchell* SR 1032. More usually *i timchell* and *i n-imthimchell*. So infra 2365 *ina imthimchiull*.

2346 *mind órbuide ardatuigethar* The preverb *ar-* here used to infix a pronoun in a relative verb. Cp. supra 1740. In LU 8599–8600 (FB) we get *Mind n-óir budi in folt fordatuigithar*. Cp. also *imdatuigethar* ib 7731 (TBDD). Rec. I here agrees with LL, but Rec. III has *mind órda nothuigithir* (§174) while Stowe alters to *mionn orbuidhe ica thuighi don toep amuich* (2385).

2347 *concuirend Con-* as relative prefix.

2360 *nád chumgaitis súli dóeni déscin* In a relative clause one would expect . . . *do déscin*. LU reads the same with *déicsin* for *déscin*. Stowe (2396–7) has *marbadh lochrann lansolusta nach bfedais daoine a déchsain* (with v.l. *dfechsain* for *a déchsain*). Rec. III omits *immar bad lócharnn lánaolusta* and reads *Delg finnargaid isin brut osa bhanbhruindi conach cumgaidis súile daíni a desgain ar ghleórdhacht ┐ ar ghloinidhecht* (§175), while the O'Curry MS. has *imar lochrand lansolusda nad cumccdais suiliu doeine a dexin* (1386–7). The last two readings suggest that *nád* in LL is for *conná*, consecutive, and that we should read *a déscin, a déicsin* in LL and LU respectively. For *nád* = *conná* see supra 350. In TBC² we find *Benaid Fergus tri bemind fair nad comairnic cid bic a sceith dosom fora* [? read *fria*] *cend* (3581–2) = *connára chomraic* infra 4747. In LL-Togail Troí, in the account of the death of Castor and Pollux: *co n-erracht anfud anfóill dind fairind cora tuargit ri tirib nad fess aided dóib* (31907–8), 'so that their destruction was unknown' (Stokes). Cp. also in Scél Saltrach na Rann: *Tuc in muriall n-én a haeor nad betis cen feoil* (Celtica iv. 34. 497), for *conná betis*.

2361·4 LU differs somewhat: *Cliabinar sróil sirecda ré chness co ngebethar dó co barrúachtar a dondfúathroci donddergi mileta do sról ríg* (6557–9). Rec. III reads: *Is amhlaidh do bi sé ┐ cliabinar sirig fria chness arna imthacmaisi dho chresaibh ┐ do chimsaibh ┐ do chorrtharaib óir ┐ airgid ina imthimcheall*

(§175). The LL reading combines Rec. I and Rec. III. *Arna imthacmaṅg massi*: *massi* is intended as a correction of *imthacmaṅg* to *imthacmassi.*

2363 *condriced* 'which reached'; *co roiched . . . dó*, 'so that it reached' St 2399. Rel. *con-* in LL and LU has been replaced with the conjunction *co* in Stowe.

2369. *Laigis Medb a hainech* Here *laigid* for *fo-lugi*, or *foilgid*. Cp. supra 1191. Rec. I reads *Follaig immorro Medb a hainech* (LU 6567). Stowe has *Cuiris . . .*

2376 *Is and sin ra gab ét . . .* There is no reference to Dubthach's jealousy in Rec. I, but these sentences introducing the verse have been taken into C (1398–1401). Dubthach's jealousy is explained in Rec. III and in Stowe. Thus: ⁊ *ro ataigh bean Dubthaigh a tógbháil do thaidhbriudh chrotha Chon Culainn. Ó'dchúala Dubthach inni sin, gabais éd ⁊ ealcmhairecht é* (§§183, 184); *Is ann sin ro gab ed . . . Dubthach Daol Ulad ima mnaire Coin Culainn uair doconnairc a bean ic drem risna fiora do dechain Con Culainn* (St 2411–3).

2380–2400 This poem is omitted in Stowe, possibly because the LL readings show corruption here and there (e.g. extra line in first stanza, repetition of *fóendelach*).

2384 *ri brainessu Ess*, 'food'. The word occurs in Cormac's Glossary s.v. *iasc* where it is glossed by *biad*. Acc. pl. *essu* here. See Stokes, KZ 37. 250.

2387 *Ni maith fararlith in cath* I have followed the translation of Windisch who takes *fararlith* as *far-* infixed pron. and *arlith* as pret. *ar-luid*. The LU reading, 'Not well do ye fight the battle', gives perhaps a better meaning here.

2394 *a n-aidche* = *a n-aigthe* See Middle-Irish Pronunciation, Hermathena 1926, p. 189.

2398 *da betis* for *no betis*. LU reads *biad slóg imme di cach leith* (6587). In Rec. III the order of lines is *Biaid sluaigh uime do gach leth | diamadh mé badh chomarledh* (§193).

2401–5 In Rec. I this passage comes after the second poem and is followed by a series of short rhetorics spoken by Ailill, Medb and others (6621–47). C, however, inserts this passage here (1422–4), partly in the wording of LU, partly in that of LL, and gives the rhetorics immediately after the second poem (1461–87).

2406–2437 The Stowe compiler omitted all but the first two stanzas of this poem. In the LU fragment of Mesca Ulad the first three stanzas are quoted (LU 1458–65) but the eighth line there reads *guin Mani meic Fédelmtheó.*

2409 *ó geguin in n-ingenraid* This is a reference to Dubthach's having killed the maidens of Ulster in revenge for the betrayal of Meic Uisnig (Ir.T. I 76 §16). There is a detailed account of this slaughter in the Glenmasan MS. which gives the names of the women slain (Celtic Review I 216, II 118).

2411–3 *guin Fiachach meic Conchobuir . . . guin Charpri meic Fedilmtheo* In Rec. III *guin Fiachaig mic Conchobair . . . guin Dairi mic Fhedhlimtheo* (§ 198), a reading which agrees with the version of Oided Mac nUisnig in the Glenmasan MS. : *Agus tarladar an dias uasal ard-macam-sa chuca .i.*

Fiacha mac Conchobair agus Daire mac Fedlimthi agus do marb Dubthach iat 'na ndis (Celtic Review I 212; see also II 118).

2439ff. This and the following two short sections are not in Rec. III. According to the Glenmasan MS. (Celtic Review I 208) Óengus mac Óenláime Gábe was with the Dubloinges in Connacht. Rec. III TBC sometimes shows agreement with Glenmasan MS. tradition, and that may account for the omission of this incident in Rec. III. In the present text, however, Óengus is included in the list of Ulstermen roused by Findchad Fer Bend Uma (infra 4094).

2441-2 *co Áth Da Fert i Sléib Fúait* The words *i Sléib Fúait* are not in Rec. I. As Thurneysen noted they must be wrong here. Cp. at the end of TBC[2]: *dobert a etan frisin tealaig oc Ath Da Ferta. Is de ita Etan Tairb i Muig Muirrthemni* (3674-5). Perhaps a confusion with *Áth na Foraire*? But cp. *Áth dhá Fherta i Sliabh Fuaid* L. Cl. A.Buidhe p. 21.

2442 *Iss ed marimat eólaig Asberat ind eólaig* LU 6678, *Is edh atberit eolaigh* St 2437. In the Contribb. *marimat* is taken as 3 p. pres. ind. of *imm-rimi* with this passage the only example quoted. We may perhaps take *ma-* (*ba-*) to be another instance of the confusion of preverbs (which, as noted, is common in relative verbs), standing here for *do-*, the verb being the common *do-rími*. Cp. *Is ed dorime in senchus* Hib. Min. 164; *amal dorrimet na scribenna* PH 6281.

2442-4 *dámmad ar galaib óenfir dosfístá Óengus,* 'if they had come to O. in single combat . *Dosfístá* is imperf. subj. pass. of *do-icc* with direct object *Óengus.* The infixed pronoun has a relative function. (In Mid. Ir. a relative construction would be usual here). A similar use of the passive of *conricc* is found in the corresponding passage of LU: ⁊ *asberat ind eólaig imneblaid* [read *immusneblaid* as in YBL] *ríam remáin co tíastain* [*tiachtain* C, *tiastis* YBL] *fo chlaideb oc Emain Macha acht bid ar galaib óenfer conristá friss* LU 6678-80, 'if only they met him in single combat'. The words *ríam remáin reime ar galaib óenfer* in LL are superfluous. They seem to have come from a misreading of a version like Recension I.

2451-2 *Á danaccaig* (*A dan accaig* Diplomatic Edition). A late form of the preterite of *ad-cí* (prototonic), with *do* used to infix a pronoun, 'when he saw him'. Possibly influenced by *danécci*, 'he looks at him'. For the form *-accaig* cp. in LL-Togail Troí: *co faccaig féin ó súlib in fer n-aurdairc atchualaig o chluasaib* (LL 31838).

2458-9 *Imroll aile . . .* In Rec. I a second incident also entitled Imroll Belaig Eóin follows here (6655-68). It has nothing in common with the first incident beyond the name of Mane, but the title is explained by the opening words *Tiagait na sloig do Beluch Eúin.* If the LL sentence refers to the second incident, we should read ⁊ *Imroll aile Belaig Eóin ainm do scél aile dano.*

2460 *Tuige im Thamon* This incident is very briefly told in Rec. I. The same motif is found in an earlier incident in LL (but not in LU), namely, in the killing of Medb's handmaid (supra 1341-46).

2462-3, 2464 *bad fiadnaissi dóib* The phrase *bad fiadnaise do x* occurs frequently in this text, and has been taken as the copula with *fiadnaise* as predicate. See M. O Daly, Ériu xiv. 98 and RIA Dict. s.v. *fiadnaise.* In this passage Stowe reads *forin áth ina fiadnaise* (2554-5), 'on the ford in

front of them'. Elsewhere in the LL-text we get e.g. *i fiadnaisi Conchobuir*
892–3, *i fiadnaisi fer ṅHérend* 2494–5. I suggest that *bad* in *bad fiadnaissi
dóib* is not the copula but the preposition *fa*. The phrase would be *fa fiadnaisi
do x*, 'in x's presence', as contrasted with the synonymous *i fiadnaisi x*. (Note
that *i fiadnaisi do x* does not occur). There are two instances of the phrase
in the Battle of Mag Rath: *gur marbustar G.G. . . . . a dalta ba fiadnaisi dó*
(200. 21-22); *coro innis a aithesc ocus gur thagair a thechtairecht ba fiadnaisi
dóib* (178. 15–16). The phrase *i bfiadnaisi x* is common elsewhere in MR.
The spelling *bad* for *ba = fa* in LL is peculiar, but it may be no more than
an idiosyncracy of the LL scribe.

In such a sentence as *goro thesc a tri cindu dina tri tulchaib go failet 'sin
riasc bad fiadnaisi* (infra 4778–9), *bad* cannot be the copula. If it were we
should have *as fiadnaise*. In the Stowe version (MSS. H¹ and Ed) this
sentence reads . . . *go bfuilit sin ba fiadnaise dferuibh Erenn* (4833), with v.l.
co bfhuilitt sin i fhiadnuisi fher nErenn. The phrase *ba fiadnaisi do x* occurs
a few times in Rec. III: ꝛ *ba fiadhnaisi do Medb do raidh-siun sin* §67 (again
§§4, 65). Cp. in the Macgnímrada section of LU *Hi fiadnaise Bricriu ucut
dorónad* (4928). In one instance in the present text we may have the conditional of the copula: *Is and sain ra ráidsetar fir Hérend cia bad fiadnaisi
dona tarbaib* (4800), '. . . who would be a witness for the bulls, i.e. the bulls'
fight'. But we might emend even that instance to *cia [bud cóir] bad [= fa]
fiadnaisi dona tarbaib*. Cp. Stowe which here reads *cia badh cóir i bfiadnuise
gleaca na dtarb* (5015).

2469 *bosréothi cloich . . . fair* Windisch printed as *bosréthi* with a
footnote indicating that what seems like an *o* has been added under *é*. The
word is printed *bosréothi* in the Diplomatic Edition.

In this passage LU reads *Srédis C.C. cloich fair* (6671), while Stowe
changes to *Telcis cloich . . . fair* (2460). *Srédis* of LU (*Sreidis* YBL, *Srethis* C)
is 3 s. pret. of *sréid (sreid)*, used in the sense of 'casts', hurls'. In his paper on
'Old-Irish *sernaid* and related forms' in Ériu xviii (pp. 85–101), Prof. C.
Watkins takes the verb *sreid* to be a back formation from such forms of
sernaid as pret. pass. *sreth*. He notes that the meaning 'hurls, casts' for
sreid is found preponderantly in TBC (loc. cit. p. 96). The *-d-* of *srédis* and
the *-th-* in such forms as *sraithi* (LU 5799), *sraithius* (LU 5342) i.e. pret. *srui*
+ suff. pron., serve to break the hiatus. *Sreid* occurs only once in LL-TBC:
conid Áth Srethe comainm ind átha dara sredestar C.C. in cloich assa thabaill
(1278–9). The form *bosréothi* is peculiar, but similar forms occur five times
in Rec. III. Thus for the last quoted passage in LL Rec. III has: ꝛ *Loch
Sreoidh ainm in locha, doigh is uimi aderar Loch Sreoidh ris doigh basredh
C.C. cloch inn* (ZCP viii. 542. 8). One would expect *basreoidh* rather than
basredh here, and in fact that form occurs earlier in the same passage:
Basreoidh C.C. cloc[h] eile bar Meidhbh (loc. cit. 542. 4) = *Dobretha cloich
assa thabaill furri béus* (supra 1276). Again *mara faca-sun Medhbh, basreo
cloch . . . fuirthi* (loc. cit. 542. 3) = *port indas facca Meidb, focheird chloich
furri* (supra 1273–4). Finally where LL reads *Focheird C.C. de na secht
cneslénti fichet* (supra 1481–2), Rec. III has *basredha a hshecht cneisléinti
fichet . . . dhe* (loc. cit. 545. 30). In this last occurrence the MS. reads *basreodha*
with punc. del. under the *o*. If we ignore the deletion, we get a form near to
LL *bosréothi*. Perhaps a by-form **sréoid* existed beside *sréid*. (For an
association of *sreöd*, 'sneeze', with *sreth* etc., see Prof. Watkin's article p. 94).

In the Franciscan MS. of AS we get the phrase *sreó n-urchair* (acc.) as v.l. to *ró n-urchair, rót n-urchair*. The editor of RIA Contribb. S suggests that *sreó* is vn. of *sréid*. The prefix *ba-* in the Rec. III forms might stand for *ro-* or *do-* (we get, for instance, *ba-gni, bho-luidh* in that text), or possibly there may have been influence of *fo-ceird*, 'throws, casts' (which is spelt *bho-cherd* in Rec. III §81). The occurrence of *bosréothi* in the LL-text suggests that the compiler of LL-TBC had access to the original of Rec. III, which, we may assume, contained such verbal forms as *basreódha* etc.

2467   *Corop Tuigi im Thamon and sain*   See note in Celtica vii p. 33·7. on *gurab*, present indicative of the copula.

2474-6   *And sin conattec[h]t* . . .   Another instance of the confusion of the preterite of *condieig* and *adteich*. In Stowe here we get the later form of *adteich* used as a simple verb: *Is ann sin ro ataigh Medhph fir Erenn . . .* (2465). Again infra *And sin conattecht Medb Fergus* (2478).

Here the recensions differ somewhat. In Rec. I it is not Medb but Cú Chulainn who challenges some one of the men of Ireland to fight: *Táet nech uaib ar mo chendsa, ar Cú Chulaind oc Ath Da Ferta. Nipa messe* etc. (LU 6682-5). In Rec. III: *And sin ra himraidhedh ag feraibh Erenn cia bhadh chóir do chomlonn ⁊ do chomrag re Coin C.* etc. (§§202-204).

2479   *ar ros fémmid firu Hérend*   The construction is strange. Windisch and Dr. Knott (in RIA Dict.) take it to be an impersonal use of *fo-émid* with accusative. Dr. Knott translates 'the men of Ireland had refused (had failed?) to do so'. This is the only example of such a usage with *fo-émid* given in the RIA Dictionary. Cp. the Stowe reading: *ō ro ēmgeatar fir Erenn dul do chomrac friss* (2470), which seems to have suggested the translation given. The usual construction is *fo-émid* as a personal active verb with vn. or its equivalent, e.g. *dáig fosrémdetar a techt* 1366, 'since they were unable to go across it'. We might possibly take the sentence to be *ar ros fémmid [Medb] firu Hérend*, 'for M. was unable to get the men of Ireland to do so'. An example of a personal object in an elliptical sentence of this kind is quoted in the RIA Dict.: *Tuc M. a máthair a gentlidecht; forémdid im[morro] a athair* (patre in malis perseverante) Lat. Lives 91. Again in the Metrical Dindshenchus: *Is and ro erig Find féin | in trath rofemmid in Féin* (MD iv. 358. 95-6). Gwynn translated 'Uprose then Find himself when the Fianna shirked the fight', and suggested reading *ón Féin* 'to avoid treating *féin* as put for *fian* for sake of rhyme'. Perhaps we might translate 'when he was unable to get the Fian to undertake the fight'.

2481-2   *Cid trá acht a facessa* . . .   Windisch suggested reading *asacessa* or *aracessa* for *a facessa*, that is, either a form of *asa-gúisi*, 'desires' (see Glossary SC²), or of *ar-ceissi*, 'pities' ('complains'). He translates the sentence as 'Indessen Medb beklagte sich schwer über Fergus, dass er es verweigere, und ihren Kampf und ihren Streit nicht übernehme'. Neither *asacessa* nor *aracessa* suits here. *Ar-ceissi* with direct object means 'pities, commiserates'; *asa-gúisi* gives no good sense with the remainder of the sentence. In the RIA Dictionary *da fémmid* in this passage has been taken as the vn. of *fo-émid, for-émid*, and the passage is translated 'for refusing to undertake the combat'. (Similarly V. Hull, ZCP xxix. 377 n. 1, who takes *da fémmid* to be for *dia fémmid*). But this construction with *cen* does not occur elsewhere, and if it were the verbal noun, one would expect it to come immediately after Fergus and to get something like *Fergus da fémmid* [? *féimded*] *in chatha*. It seems as if this passage is corrupt and must be emended. It is possible that *a facessa* may be intended for a 3s. active, See note 2259–60 supra. Or we might omit *Medb* and read *á* (= *ó*) *fogessa* (or *ro gessa*) *Fergus co tromm*, 'since Fergus was earnestly entreated'. In the corresponding passage in Rec. I the verb *guidid* occurs twice: *Is and gessa do Fergus mac Roich techt ara c[h]end-som. Opaid-side dano dul ar cend a daltai .i. Con Culaind. Dobreth fin do ┐ ro mesca[d] co trén ┐ ro guded (ro ges* C) *im dula isin comrac. Téit ass iarom ó ro bás oca etarguide co tromda* (LU 6686–9). The last sentence corresponds to our passage 2481–2. Cp. also *Gessa Lóch* LU 6104 = *Do geassa Loch* YBL. *Da fémmid* I take to be another instance of confusion of preverb. Cp. in a later passage infra *& ra fémmid tiachtain taris* (2541). It is perhaps worth noting the Stowe reading (2472–4) for this passage: *Ciodh thra acht baoi da tromdhacht ro aslaigh Meadhbh Fearghus nar fhéad Fearghus gan an comhlann do ghabháil do láimh* (written by a relieving scribe with more modern spelling). In other passages of this text where LL has the verb *fo-émid, for-émid*, Stowe has the negative of *fétaid*, e.g. *forfémdetar a techt* 1350> *nior fedsat teacht tairsi* St 1385, *dáig fosremdetar a techt* 1366 > *uair nior fedatar tairsi* St 1403. So here the Stowe compiler undoubtedly took *da fémmid* as = *nir fhét*.

2494   *teich romum-sa*   Both LU and Rec. III have here *Teilg traigid dam* (LU 6693, Rec. III §210). This is one of the points of resemblance in wording between Rec. I and Rec. III which led Thurneysen to postulate a connection between the two recensions (ZCP viii. 528–9). In fact, however, Rec. III elsewhere in this incident is much closer to LL. Thus in Rec. III: *Is lesg limsa sin*, bar *C.C.*, *.i. techedh ré n-aenfher a slúaighedh mhór T.B.C.* (§212), which is not in LU; and *And sin bholuidh C.C. ina charbad uadha* (§213) where there is no mention of a chariot in LU. So too the sentence *Dáig cid bec lib-si . . .* has no equivalent in LU but in Rec. III we have *gidh beg techidh tugusa air, ni thugsabhair-si cethri coigidh Erenn a urdail air ar Táin Bó Cualgne* (§214).

2506   *in neoch conarnecar ris*   Conarnecar < *conrice*, an analogical deponent form. Cp. *Ní mar chách conarnecar comlund ┐ comrac riut . . . Fer Diad* 2732–3. Contrast in LU in a different passage (at the end of Láríne's fight) *di neoch cotranic friss ar Tana* (6080).

2507   *timchell* = *ar timchell*, 'in turn'. A similar use is found in a interpolated passage in Stowe TBC: *Is amlaidh beide co maidin anosa ┐ cach cethracha fer diobh timchiol i ngliaidh fria aroile* (4680).

2510ff.   *Cinnit Ferchon*   A marginal note here in LL: *in aliis hoc ante Comrac Fergusa*. But this does not apply to any known recension. This incident in LL illustrates well the expansion in Rec. II of the earlier version. In Rec. I it is told very briefly (LU 6701–4). Rec. III is also an expanded version, the first part agreeing with LL, but the latter part incorporating portion of the Calatín Dána episode (which is not contained in Rec. III) and including a quatrain spoken by Cú Chulainn (§227) which is not in any other version.

2534–5   *Calatín Dána, Glass mac Delga*   In YBL *Gaile Dana* and *Glas mac Delgna* (TBC² 2184, 2185), *Gala Danu* and *Glass mac Delccu* (C 1566–8). In a later reference to this encounter in the section Sírrobud Sualtaim, where LL has the correct number: *ri ʋalatín nDána cona secht maccaib fichet ┐ rá húa, ra Glass mac ṅDelga* (3984–5), Rec. I refers to *da mac dec Gaile Dana* (TBC² 2975). Possibly this is due to confusion with the fight against Ferchú Loingsech. We note that in the Rec. III account of the episode Cinnit Ferchon a detail of the fight with Calatín Dána and his sons has been incorporated, namely, the driving of their spears half their length into Cú Chulainn's shield (RC xv. 207 §§225–6). Rec. III ends with Cinnit Ferchon, but the fight with Calatín is mentioned as the next episode (*Comrag Cailitin gona cloind ad neasu sund* RC xv. 208 §230). The Calatín Dána episode is much expanded and embellished in the LL-version. Neither Fergus's plaint (2546–55) nor the flight of Glas mac Delga (2585–92) is in Rec. I. C has taken the second of these passages, the flight of Glas mac Delga, and inserted it clumsily into the Rec. I version.

2537–8 *ni fuil bara fuliged nech díb . . . rabad marb* Rabad for *nábad*. The scribe inadvertently took *ni fuil bara fuliged nech díb* as if it were *cach óen bara fuliged . . .* Stowe, as usual, inserts an antecedent: *Ni biodh nech ara bfuiligedh fer diobh nacha* (v.l. *nach ba*) *marp i ccetoir no re cionn nomaidhi é* (2534–6).

2540–1 ⁊ *bad fiadnaisi d'Fergus ra naidmthea sain* See supra note 2462. This is translated in RIA Dict. as if the verbs were subjunctive, 'that F. should be witness to the compact'; similarly Windisch. But *bad* is for *ba (fa)* and *ra naidmthea* is pret. pass. Cp. the Stowe reading ⁊ *ba i fiadnaise Fergusa ro snadmadh é* (2538).

2546ff. This passage does not occur here in Rec. I, but in the Fer Diad episode (TBC² 2312–2317) a passage occurs with wording almost identical. It is found also in Thurneysen's Version IV of the Fer Diad episode (ZCP x. 278 §12).

2547. *in gnim doníther imbárach* Present tense also in the Fer Diad passage of YBL referred to in the last note: *Tainic Fergus coa pupall. Truag limsa in gnim dognither isin maidin sea imbarach* (TBC² 2313), and here in Stowe: *an gniomh do-gnithar amarach* (2544). Cp. a passage in Oided Chloinne Uisnig, Glenmasan MS., where Lebarcham refers to *an gnim doniter anocht a nEmain*, meaning the treachery that will take place (Celtic Review I 122). With this use of the present of *do-gní*, compare *tic* used with future meaning.

2564 *Is and sin barróisc . . .* Again 2573. This verb seems to be pret. of *do-icsa (tiscaid)*. Stowe has *ro bean, do bean* (2562, 2571). *Bar-* for *do-* in relative position. This pret. form does not occur elsewhere. In an earlier passage in Rec. I (which has no equivalent in LL) we get pres. ind.: *Tánisca Cuillius asa thruaill* (LU 5405).

2580 *Dóig ra fuilemm trichait chét . . .* Stowe has here *doich atamait tricha cet do maithiph Cloinne Rudriaghe i llongphort bfer nErenn* (2576–7). For *ra* with *fuil* see infra note 4349. In the older language the conj. *dáig* (*dég*) was followed by a nasalizing relative clause. In the present text there are examples of relative *con-* used after *dáig* e.g. *dáig conmebdatar a carpait* (1260–1). A similar use of *ro* is found in CRR: *Noco glinnigthi duitsiu sain ale, bar Ailill, acht is foenglinni daig ron fuilet in Galian* ⁊ *Lúaigne na Temrach and sain bardo chind* (LL 22981–2).

2548 *Garsa gnim* Garsa for *ciarsa* = *ciaso*.

2554 *mar da mairbfithea* = *mar no mairbfithea*. Stowe has *an muirbfidhi Cu Chulainn isin comlonn sin* (2551).

2591 *ní arnecair úad a rádha acht* . . . *Arnecair*, analogical deponent form of *ar-icc*. In this idiom *do-airicc* is common. Cp. supra 9348–9, *ní tharnaic úad acht a rád* . . . The Stowe equivalent passage here is *ní tairnic leis do rada acht Fíacha*, Fíacha (2587), which exemplifies the construction with *acht* found in the later language. Cp. *Ní cóir do bheith ionnta acht seacht n-orthana* Eochairsgiath 41. See note in Desiderius p. 265.

2006ff. *Comrac Fir Dead* See Introduction to Stowe TBC §6 pp. xxiv-xxix.

2611 *Ac óenmummib* . . . The Diplomatic Edition prints *daronsat ceird gnimrada gaile ⁊ gaiscid, ceird* being added in the margin with caret marks. I take *ceird* as intended by the scribe as an alternative reading to *gnimrada*, just as he wrote *daringsetar* above *darónsat*. Cp. the Stowe reading. In Tochmarc Emere Fer Diad is mentioned as one of Cú Chulainn's fellow-students with Scáthach (LU 10433). In TBC² there is no reference here to Scáthach, Uathach and Aífe.

2614–5 *conganchnessach* In Togail Troí² *conganchnes* is used in the sense of 'invulnerable' where Hector says *Biam congancnes ic comrac fri cech fer úaidib. Ní chomraicfet a n-airm frimsa ar feabas na hersclaide* (768–770).

2617–8 *Ra érastar* As Strachan noted (Deponent Vb. p. 63) deponent endings in the s- Preterite are very common in this section, Comrac Fir Diad.

2643 *eirggi guin is gabáil* This line, which is superfluous, is obscure in meaning and has not been translated.

2644 *ás cech anáil* Windisch suggests that the meaning is 'über jede Erwartung', and compares *Rot biat . . . feib dothaiset latt anáil* LU 9124–5 (FB).

2655–70 These two stanzas are not in YBL. In C (which here = LL) ll. 2667–8 read *Nocha geb ge eisce | ge rabtha dom freisci* (1674–5). Stowe reads like LL.

2675 *Fuil sund nachat tuilfea Fil sund nachad fuirfe* TBC² 2271; *Fuil sunn noch[at t]uilbthe* C 1682; *Fuil sonn nech rot tuillfea* St 2672. I have tentatively translated as if the verb were *do-aile*, 'blames, disparages'. Cp. vn. *toil* in *Ní toil dot gasciud* LL 12579. Compare also infra 2963 *Fail sund nech rat méla*, and the Stowe reading here, which suggests that we might emend the line to *Fuil sund nech not tuilfe*, 'Here is one who will bring shame on you'.

2687 *Cid Domnall ná Charpre* 'Whether (it be) D. or C. , *ná* = *nó*.

2690 *rot fiat-su gid acht* The reading of TBC², *cen acht*, gives good sense. With the LL reading cp. *ani ná roich lam cid acht | is écen dam a dút[h]racht* LU 3935–6 (SC). The meaning of *cid acht* is not immediately apparent. It can hardly be the *cadecht, cedacht* of the Glosses (gl. nondum Ml. 19ᵇ4, 30ᵇ15) and of Críth Gablach (87–8). Perhaps 'though it be difficult', a development of *acht*, 'but'? Or simply 'however'?

2715 *rasiacht Medb máeth n-áraig bar Fer nDiad* Again 2784, 2787. The exact meaning of *máeth n-áraig* is doubtful. See note on the word Celtica iii. 323, and compare *Ro érig Molling cuce ⁊ ro naidm in Trínoit ⁊ in cethar soscela comdeta fair. Ocus táirthis báeth n-araig fair* RC xiii. 106 §138 (Bóroma). The Stowe compiler avoided *máeth n-áraig*, substituting for it *ratha ⁊ urradha* in the first occurrence (2737) and *cuir ⁊ ratha* (2810, 2812) for the other instances.

2729 *Attear ar cobais* *Attear* has been taken as for *atbiur* by Windisch and M. O Daly. But in that case we should expect *mo chobais*. Cp. *Dobēr mo c[h]obais cen chāin* TBC² 2399. I take it to be for *Dar ar cobais*. Cp. *dar lim, atar lim*.

2730 *ní 'na dáil* . . . Here YBL is corrupt: *nochon an andail is dech lend noragmais* (TBC² 2563), but C gives the correct reading *nocha 'na dail* . . . (1763-4). Perhaps we should read *ní 'nar ndáil* . . . here in LL which would give a better meaning. Note nom. *cara* in the vn. clause dependent on *dúthracmar*.

2731 *ara n-airichlea ⁊ ara n-airelma* The first verb is 2 s. pres. subj. of *ar-foichlea (airichlid)*. The second seems to be a compound of *ellam*.

2732 *ní mar chách conarnecar comlund ⁊ comrac riut* . . . *Fer Diad* Here *conarnecar* is a 3 s. deponent form = *condránic*. Cp. supra 2506 *in neoch conarnecar ris*. Here *comlund ⁊ comrac* are cognate accusatives as in such a sentence as *In comrac sa condrancatar fir Herenn ⁊ Alban* RC xiii. 458 §55 (Mag Mucrime).

2746 *ar apa chomlund óenfir* In TBC² and Stowe *ar aba comlainn einfir*. See supra note 332.

2751 *Bi tost = Bi it thost* Again 2803, 2825. *Bi tast* TBC² 2391.

2768 *gen com maith-se* *Gen gob maith-si* St 2796.

2784 *rachúaid = adchúaid*. Similarly *dachúaid* 2787.

2793 *airm condricfaitis* . . . Here YBL has *baile i comrecdais* (TBC² 2432) and C has *ait hi comricfitis* (1831), while Stowe reads *áit i ccomraicfidis* (2817). There are several instances in the LL text of the omission of the prep. *i* before *con-ricc* in this phrase: *airm condricfaitis* 3826, 3976-7 where Stowe has in both cases *airm i ccomraicfidis*. Again *airm condricfat* 3869 (no equivalent in Stowe).

2800 *da bai* In this section note also *do loitt* 3044, *da gab* 3291, *da indill* 3353, *da léic* 3378.

2814 *níba mín far magar* See note on the word *magar* by Gwynn, Ériu iii. 190. Originally *magar* meant 'spawn, small fish' (*magur .i. min-iasc* Corm.), figuratively 'bait, allurement', finally 'smooth talk'. The same phrase occurs in a poem in the Glenmasan MS.: *Nochar mín an madar* (v.l. *magar*), Celt, Rev. III 304. Is it possible that we have here in LL and in the Glenmasan MS. a different word from *magar*, 'spawn'?

2834 *na eich* Nom. pl. for acc. pl. Again 2840.

2837 *dáig ná* . . . A final conjunction. More commonly *ar dáig co, ar dáig ná*. Stowe here reads *arnach abradais* (2837).

2851 *in sestanib* Stowe here has *in troimsestán*. *Sestanib* occurs again 4179, 4192.

2872 *in cua* *Cua* is disyllabic. Meyer (Contribb.) takes this as *Cua*, a shortened form of *Cú Chulainn*. Cp. infra the poem beginning *Cid ra[t] tuc, a Chúa* (2947) which in TBC² is *Can ticiseo, a Chua* (2617). In TBC² there is an introductory passage to the poem: *⁊ adubairt Per Diad ri Coin Culaind: Can ticiseo, a Chu [read Chua]? ar se. Daig cua ainm na claine isin* *tsengaidilc* ⁊ *secht imleasan batar i rigrosc Con Culainn. Da mac imleasan dibsidi* ⁊ *siat claena* (2610-12).

2913 *acht* [*mad*] *athig mith* In TBC² *acht mad aigith meith* (2607). Windisch takes *mith* (*meith*) as gs. of *meth*, 'decay, failure', and translates as 'derbe Männer'.

2940 *ra armad . . . ra déirged   Ra* for *no* before imperfect. Cp. *ra forrged* 3285, *ra gontais* 3817-8, *Ára géised . . . ra géistis* 4744.

2951 *mairg* [*tánic*] *do thurus* Possibly a better reading would be that of C. here: *Bid mairg duit do thurus*, the future tense agreeing with the rest of the stanza.

2952 *bud atód ra haires* O'Clery has *airis .i. aithinne* but see Ériu xix. 112-3 where Prof. David Greene suggests *aires* = 'a bundle of firewood'. This is a proverbial expression denoting a vain undertaking. Thus *Tigedhus do bheith gan mnaoi . . . as adúd re hénoires* Buile Shuibhne² 835. Cp. also *ba hatod ri hoentenid* LL 19621, translated by Gwynn (MD iv. 53) as 'It was kindling fire from a single spark'.

2963-9 This stanza is not in TBC². It is, however, in C, and there too, as in LL, a line is wanting. C has also an extra (partly illegible) stanza (2006-13).

2965 *dáig is dim facrith* Dr. Knott suggests (RIA Dict. s.v. *fo-cren*) that *facrith* may be pass. pret. of *fo-cren*. Cp. *bocritha do chétguine* infra 3083.

2966 *conugud* Similarly C. I suggest that this is a scribal error for *cronugud*, 'rebuking, chiding'. In LB *cronugud* is twice misspelt *conugud* (PH 2647, 3847).

2970 *Car* = *cair*, interrogative.

2972 *gid leind rarrficfam* The line is omitted in YBL. C gives an intelligible reading here.

2083 Here YBL has *rotgabsad ar tfaillseo* (TBC² 2645). We might emend to *ro gabsatar t'faill-siu*, 'they have come upon you unguarded'. But the sense does not suit.

2990 *bad mórglonnach bias* This line is glossed by O'Clery: *bhias .i. ghonfas, bud mórglonnach bhias .i. budh moirghniomhach ghonfas tú.*

3000 *at gilla co ṅgicgil* The meaning of *gigil* is uncertain. For examples see Ériu v. 70.

3021 *rafáethaisiu* Printed *ra fáethaisiu* in Diplomatic Edition. But *ra* here is for *da-, do-*. The editors of the Diplomatic Edition elsewhere print *rabert* = *dobert, radechaid* = *dodechaid, racuas* = *atcuas*, but *geara foethus féin* 3901, *gero faethaiste-su* 3735, 3756. Cp. *gia dofaethaiste-su* 3684, 3693 etc.

3034 *girsat Girsat* here is syntatically a present. Cp. supra note 1475. C has a better reading for this line: 'though you are a warrior with a hornskin'.

3046 *ná bris chig* See Ét. Celt. iii 373 where Dr. M. A. O'Brien takes *cíg* here to be synonymous with *cotach*, 'treaty of friendship', and to derive from the expression 'sugere mammellas' in the Confessio Patricii. See also Ériu xi. 94-6. In a poem on the end of the world from Laud 615, occurs the sentence *Báithfither cích ocus cotach* (ZCP viii. 195).

3047 For this line read *ná briss bréthir, ná bris báig.*

3053 This line is obviously corrupt. C gives a better reading. Read *amnas* for *amus* in C?

3062 *ris tib* . . . Vendryes (Ét. Celt. iii 42) suggests reading *ris tib gen na cóiced cain*. But in that case *gen* would be nom. Better *ris tib gen in cóiced cain*, 'in whom the fair province delights'.

3067 *is aróen imthéigmis* This may be the verb *imm-téit* as the Contribb. take it to be, but it may be imperfect of *téit* with *im-* used as relative prefix. Cp. *aróen contiagmais ar chae* LL 6393 (printed . . . *co tiagmais* Diplomatic Edition). Similarly l.3077 *aróen imréidmis*.

3082 C has *ro cloisim cur caratraid* (2143), which suggests *ra chlóisem* as a better reading here.

3108–9, 3111 *go súanemnaib lín lánchotut indi* For *indi* see supra n. 1205. One would expect *lánchotait*.

3122 *cossairleptha* Acc. pl. of *cossairlebaid*. For this plural form cp. Early Ir. Lyrics 45.9.

3122 *go frithadartaib fer ṅgona* *Gona* gen. of vn. *guin*. Cp. *Ro choitéltais fir gona*, 'wounded men would sleep', Early Ir. Lyrics 48.14. Stowe has the participle: *go bfrithadartaibh bfer ngonta friu* (2951).

325 *cach lossa icci* So too *cach iptha* . . . ⁊ *gach orthana* 3169. For examples of this occasional use of *cach* with a plural noun see DF III Glossary s.v. *gach*.

3156 *cuclaige* vn. of *con-clich-*, 'dash, toss', means 'a shaking, tossing'. It is often applied to the shaking of chariots: *co nderṅsat na heich cuclaigi moir fon carput roim in caeirig cun rala in ri asin carput* Lismore Lives 2325.

3157 *ra fomórchaib feidm* The *fomóraig* must here denote the horses of Fer Diad and Cú Chulainn; it is the only noun to which *fóthu* in the next line can refer. In the independent version of the Fer Diad fight (denoted by Thurneysen as Version IV or F), the passage equivalent to ll. 3152–9 is as follows: *Is and sin adbert Fer Diad: Ad sgithe ar n-eich* ⁊ *ar n-aroidh,* ⁊ *an ni marus d'iarsma na bFomorach agoinn ar n-eich, ciodh duinne an trath budh eadh nach ba headh inde?* ⁊ *adbert an ran[n] sa: Ni dleaghor dhinn cuiglaighi | re Fomhorchoibh feidhm | cuirther futha a n-urcomail | mar as glainn do deilb* (ZCP x. 290–1). Here it is obvious that the redactor has taken *fomóraig* to mean the horses, 'our horses are all that remain to us of the Fomóraig'. Underlying the various forms *fomoir, fomóir, fomórach*, there seems to have been the idea of monstrosity or physical disfigurement. Cp. LL 5ᵃ21 and LU 124. In Mid. Ir. *fomóir* commonly means giant (cp. infra 3319) but it is also used of the centaurs (St. Ercuil 571, 2570). That *fomoir (fomóir)* had sometimes reference to mythical beings like the centaurs can be deduced from the name of their king *Eochu Echenn (Echcend) rí na Fomaire* (Rawl. B 156ᵃ35, Corp. Gen. Hib. p. 269), alias *Eochaid Cenn Mairc* (EIHM p. 291). Compare the *goborchind* of LU 124. From the meaning Centaur or Horsehead to gigantic horse or magic steed is not too hard a step. Reading *re Fomórach feidm* would give us a better meaning. Unfortunately it is against all manuscript tradition.

3168–9 *do thairmesc* . . . *a ngae cró* Windisch, following Stokes's translation of *ga cró* as 'a shooting pain of death, a deadly pang', translates 'um . . . ihre tödlichen Schmerzen zu stillen'. For my suggested translation see Éigse ix. 183.

3182   *ra suamnig* Possibly we should read *ra ruamnig*; the MS. is stained here. Cp. infra ⁊ *ra rúamnaigsetar a rruisc* (4874).

3220   *bec nach rascloss* M. O Daly suggests that *rascloss* is by dissimilation for -*rascras* (Ériu xiv. 62). Stowe's reading has been translated.

3238   *Girbo chomraicthi . . .* Again 3290. Cp. *is bud crúaid in comraicthe* LL 6893. *Comraicthe* = 'meeting'.

3264   *nád róeglaind* Again 3278. (*Naro foglaim* St 3083, 3097). The 3s. perf. of *fo-gleinn*.

3267   *rachonnaic* Printed *ra chonnaic* Dipl. Edn., but probably a printer's error for the verb is printed *rachondaic* ll. 3179 and 3950.

3270   *bocotáidfer dam-sa* M. O Daly (Ériu xiv. 52) takes this to be a deponent future form 1sg. of *ad-cota*. But the gloss *fogeb-sa* is probably intended for the whole phrase. Cp. Stowe which reads ⁊ *do gebha-sa* (v.l. *do ghebhsa*) *na clesa ut ar n-uair* (3089). Further on in this same passage we get *Atchondairc F.D. na clesrada sain* ⁊ *rafitir go fuigbithea dó ar n-úair iat* (3280), pass. secondary future, = *go bfuighthi dó-somh fen ar uair iet* St 3097. I take *bocotáidfer* to be intended for a passive, future. Read *bocotfaider* for *adcotfaider*. For Middle Irish *f*-future forms of this verb see note by Vernam Hull, ZCP xxix. 175–6.

3284   *is air is doilgiu leis daragad* Daragad for *no ragad*. As *air fa doilge les dul* St. 3102.

3298   *comdas rala Coin Culaind úad* Again 3302, 3315. (In the first two instances the editors of the Diplomatic Edition print Cu Chulaind, in the third Coin Chulaind (sic)). I take the verb to be *ro-lá*, -*rala*; the prepositional pronoun *úad* seems to support this. Cp. also the Stowe reading. *Comdas* is a Mid. Ir. distortion of *conas* (*condas*). It occurs supra 2562 *comdas ralatar* (= *go dtarladar* St 2560) and also in Rec. III: *Tarlaig ro n-urchair uadha . . . gomdas easairg in sleigh culind ina culadhaibh do* (RC xiv. 261 §38). We might take *comdas* (*condas*) here as a Mid Ir. usage of -*dos*-, -*das*- for 3s. m. pronoun and the sentence to be for *condid rala . . .* with anticipatory pronoun. Cp. *Is é Hercoil tuc a hinis Creit in tarb ṅdásachtach . . . conas tuc for commus rig Gréc* TTr. 425. But often in Mid. Ir. *condos*, *condas* the pronoun has lost its force and is meaningless. Strachan (Ériu I. 175) quotes examples of *conus* used in this way (= *co*) from PH, and it is very common in later texts. Thus in Rec. I where LU has *co torchair tara aiss* (6026), the O'Curry MS., giving the same version, reads *conus torchair tara ais* (859) = *co torchair Fer Báeth amlaid* supra 1902–3. So too when in the present text we have *co tarla darráib ra budin anechtair* (2403), Rec. III has *conas tarla ré buidhin anechtair é* (RC xv. 204 §195). In Mid. Ir. too *coros* sometimes = *coro*. Thus *coros ort in petta ṅ-eóin* 1274 (if the fem. infixed pron. is not intended for masc.). Cp. *goro ort in petta togmalláin* 1276.

With the passage 3295 ff. compare a similar description in Loinges Mac nDuíl Dermait Ir.T. II¹ 184. 254–61, where Cú Chulainn leaps three times on to the rim of his opponent's shield and is three times cast off before he finally throws the *ga bulga*.

3307   *Ras léic fort feib ras léic séig* The fem. reflexive pronoun suits *séig* which is m. or f., but the first instance is for masc. pronoun, *ra lléic*.

The force of the reflexive pronoun was not obvious to the compiler. Compare another instance infra 4803-4, with *ras lécub* for *nom lécub*.

3416　*marbad é nach arlebad ríam itir* Windisch suggested that the verb was 3s. sec. fut. of *air-ling*. M. O Daly disagrees and would divide *nachar lebad*. Perhaps we should read *nach arleblad*?

3319　*ra fomóir ná ra fer mara* The mythical Fomoire play an important part in the legendary history of Ireland. The sing. *fomoir* has been derived from *fo* + *muir*, but this derivation has been rejected by some scholars. The forms *fomóir* and *fomórach* have been influenced by the adj. *mór*. Here the word *fomóir* seems to combine the two meanings of 'giant' and 'pirate' or 'invader from the sea.'

3321　*dlús n-imairic* Possibly we should read *dlús in imairic* as in Stowe's *dlus an iomaircc* 3142, 3146, 3151. In Version IV of the Fer Diad episode we have *ba he dlus na hiomghona . . . baoi do dlus an iomaircc* (edited to *a n-iomaircc*), ZCP x. 293. 27, 29, 34.

3323-4　*goro dluigset ⁊ goro dloingset* See Strachan, Verbal System of SR p. 63. Similarly *co ndluigtis ⁊ co ndlongtis . . . na catha . . . risin ndírim ndlúith* TTr. 1574. The original paradigm with *-n-* in the present stem later split into two stems *dlong-* and *dluig-*. Cp. *ro sernad ⁊ ro srethad an tsealg ar gach leith lais* Lorg. an tSoidhigh Naomhtha 2989-90.

3334　*diallaib* This word does not occur elsewhere. Perhaps dpl. of *diall*, vn. of *do-ella*, 'turns aside, deviates'?

3349　*Boruaraid Cú Chulaind in certgae . . .* Stowe reads *Telgis Cú Chulainn an certgae . . .* (3262) which is what is translated. *Boruaraid* seems to be 3s. pret. of a compound of *reth-*. See *at-rethim* Contribb. (Meyer) and note on *atnuarat* by Bergin, Ériu xii. 229. Bergin suggests that in the phrase *atnuarat Ferchess di gai mór*, 'F. struck him with a great spear' (RC xiii. 436) the verb has some such meaning as 'runs into him'. We should expect the LL-reading here to be *Boruaraid C.C. din chertgae*, but it is to be noted that where the Laud copy of the Expulsion of the Déssi reads *atnuarith side din tsleig* Ériu iii. 136 the Rawlinson copy has *atroeraid Oengus in tsleig triit*, which is the construction we have here. The equivalent passage in TBC² reads *Atnuara* (read *Atnuarat*) *Cu Chulaind cusann gai osin sciath curro bris a cleith nasna conla triana cride Fir Diad* (2692-3). Windisch took *delgthi* to be for *teilgid-i* (suffixed pronoun), beginning a new sentence. I take it as an adjective, *deiligthe*, or perhaps for *deilgthide*, 'thorny, sharp', with *-de* omitted because of confusion with following *do*. Stowe reads *Telgis C.C. an certage boi 'na laimh do lar a boisi* [v.l. *dhearnoinne*] *tar bile in sgeith* etc. (3262-3), and Version iv of the Fer Diad episode has *And sin adbert C.C. an certgha do lar a boisi dF.D.* (ZCP x. 296).

3380　*atráigestar* This is 3s. pret. of *ad-águr*. The sentence was misunderstood by the Stowe compiler who carelessly read as *atrachtatar* and rendered the LL sentence as *Ro ergeter fir Erend do techt dar n-iondsaighi, a Chu Chulainn, ar sé, ⁊ érig as do néall* etc. (3294-6). Windisch, no doubt influenced by the Stowe reading, also misunderstood: 'Laeg sah dies und die Männer von Irland erhoben sich alle um zu ihn zu kommen'. One would expect *firu Hérend* but cp. supra 2730.

3416–34   For this passage see Bergin, Ériu xii. 200–201. It is, he notes, a poor imitation of the archaic rhetoric in Aided Con Culainn LL 14257–14273, where the older construction is preserved.

3430   *níba lám laich* . . . Meyer discusses this passage Misc. Hib. II 13 and Sitzungsberichte xlv pp. 1045–6. See note 3341–2 Stowe TBC.

3440   *ardotchlóe brath* (Printed *ardot chlóe* Diplomatic Edition). Here the preverb *ar-* is used to infix a pronoun. Cp. infra *Ardotrai, a Findchaid, ardotfáedim* 4053. See Ériu xx. p. 110.

3444   *Mad dá mmámar* for *mad dia mbámar* In Mid. Ir. *Mad dá (dia)* often takes the place of *má* + subjunctive. Cp. supra *mad dia ti taris* (1422). Heɪe exceptionally *mad* is prefixed to *dá (dia)* = 'when', with indicative. Again infra 3495. One would expect some such reading as *Mad tú dá mbámar anair.* Cp. *Dia mbāmar mad tū leis oc foglaim bindiussa* (ZCP iii. 249. 64, Tochm. Emire).

3446   *go bruthe bras* A better reading in Stowe which is translated.

3452   *Nír ching din tress tinbi chness*   *Nior cinc do[n] treas tinmidh cnes* (St 3365). Windisch and M. O Daly takes *tinbi* to be for 3s. pres. ind. of *do-indben*, 'cuts, hacks'. In RIA Contribb. it is suggested that *tinbi* may be vn. with *cness* gpl. Compare *Laimthech a des tindben cét* SC² 425; *Mo claideb derg tinbi cēt* Ir.T. II¹ 185. 289.

3470   *Dursan a eó óir* In RIA Dict. Fasc. E, s,v, 1 *eó*, this is taken as a figurative use of *eó*, 'a salmon', for a hero or leader, here vocative. But Stowe gives a better reading. Cp. supra 2034–5 where Fer Diad is promised *in t-eó óir báe i mbrutt Medba.* We may take it, as the Stowe compiler obviously did, that *eó* here is O'Clery's *eo .i. dealg.*

3502   *Germán Garbglass* Earlier in the Fer Diad episode in TBC² the fight with Germán Garbglas is recalled by Fer Diad's charioteer (TBC² 2491–5). The same passage is in Thurneysen's Version IV of the Fer Diad episode (ZCP x. 283).

3520   *cethri choicait férn ferglond* Stowe's *fercc* is perhaps better. Cp. 2 *ferg* RIA Dict. (a figurative use of 1 *ferg* ?). Cp. also 2 *fern*, 'a man', poetic or cryptic term.

3599   *dá mac Gégge* . . . (*da mac Gedhe* St 3510, *da mac Fice* TBC² 2736). For *dá mac Gégge* we should read *dá mac Féige*. They are mentioned again in the list of Ulster hereos when Findchad is sent *co dá mac Feicge, co Muridach, co Cotreib* (4096–7). In TBC² they are not in this list of heroes, but they appear earlier, in the fight against Gaile Dána and his sons where they are said to have helped Cú Chulainn (2194–5).

3602   *Dáig dabertis* . . . This detail is peculiar to the LL- version; TBC² merely says that they took Cú Chulainn *do icc ⁊ bualad a crecht do uscib Conailli* (2737–8). For herbs strewn on the surface of the water in which the wounded were bathed see First Battle of Moytura, Ériu viii. 34 §32.

3615–6   *Is cumma congonad* . . . Again 3623–4 *Is cumma congonand* . . . Example of relative *con-*. Also in TBC² which has *Is cuma congoin* . . . (2756), but Stowe reads *as cuma do gonadh* . . . (3527) and *as cuma gonas* . . . (3532).

3616   *ni hé rafársed* . . . , 3624 *ni hé dafársed* . . . Conditional 3s. of *do-airret*.

3630 *focress in dúnad . . . foraib* (The sentence is omitted in Stowe). This seems to mean that the encampment was overturned, cast into disorder. Cp. *co tarat a tech forthu* LL 14522; *roláiset . . . a tech tara cend* MU² 1003. The verbs *fo-ceird, fo-cuirethar* and *do-cuirethar* are sometimes confused (*Tochur tar cenn* BDD² 503). In the context here *focress* pass. pret. may be a spurious formation, intended for 3s. act. Cp. *Foscress Find fo thalmain traig* MD III. 130, 'F. embedded them a foot under earth'.

3638 *is missi conáirgeba bás . . . forro* 3s. fut. of *do-áirci* (with influence of *gaibid*). Relative *con-* for *do-*.

3653 *Íthall* The name in LL comes from a misreading of a later passage in Rec. I where he is referred to as *in lliaigh hi thall* (TBC² 2881), 'yonder physician'. He is not named in TBC², nor is he said to be the physician of Ailill and Medb. From *Íthall* in LL, the Stowe compiler devised the name *Idhal* (3577, 3705).

3665 *Fingal* This is glossed by Stowe as *cnedh tuc duine dot coirpfhine fort* (3588). Both Fergus and Fintan were of Clann Rudraige.

3667 *Tuidmaile fair* The meaning of *tuidmaile* is obscure. I am reluctant to follow the Contributions in their suggestion of 'a bald patch'. Baldness in a son of Fergus seems strange. Stowe here reads *urmaoile fair* (3590) while TBC² has *tri tuith fair* (2798). Again in the present text 3687–8 *Dá thodmaile foraib* where TBC² has *Tuid maile foraib* (2804) and the phrase is omitted in Stowe. In RIA Contribb. Fasc. T s,v. *tuid*, it is suggested that *tuid* means 'a patch', *tuid maile*, 'a bald patch', and in Fasc. A s.v. *airmaile*, Stowe's *urmaoile fair* is translated 'he was bald in front (?)'. O'Clery made many wild guesses at the meaning: *Tuadmhaoile .i. monga don táobh budh thuaidh orra; tuadhmhaoile fair .i. folt slim nó tiugh nó cas*. It is noteworthy, however, that O'Clery does not take *-maoile* to mean 'baldness'. In the translation of this passage of TBC in Manners and Customs III. 98, *tuidmaile fair* is left untranslated, but for *dá thodmaile foraib* O'Clery's gloss has obviously been taken. It is translated 'with two glossy curled heads of hair'. See Windisch's note p. 612.

The noun *máel* f. is generally taken to mean 'a close-cropped head of hair'. But *máel* in some contexts seems to mean 'an upstanding tuft of hair, a crest of hair', Anglo-Irish 'quiff' (coiffe). This sense, for example, would better fit the passage 1204–5 supra, *Máel glé find fair mar bó ataslilad*. Cú Chulainn's hair has already been described in the preceding sentence, and is certainly not 'close-cropped'. We find also *súasmáel* for which the meaning 'crest of hair' would fit. Cp. *súasmháel .i. folt . . . súas* H. 3.18 (quoted in Contribb. s.v. *súas*). *Laech and . . . suas maeldub* [read *suasmael dub*] *demis fair. Atá lim is bó roda líg* LU 9264–5 (with v.l. Egerton 88 *Laoch and . . . Suasmael dub deimsidhe fair suidhe. Ata lim is bo rodlelaig* Anecd. III 50. 6–7). Again MU² 688 *Óenfher eturru, súasmáel dubrintach fair*. So too *putrall* seems to mean 'a tuft of hair above the forehead.' *At-condarc-sa and triar allmarda co pudrallaib imgerra urarda* MU² 773; *pudralla gerra garbha* Lism. Lives 4568. In the passage of TBC² dealing with Cethern's wounds we get *Putrall maile duibe for cach n-ae* (2835) which in C reads *Puttrall dubh for cech n-ae* (2368). A diminutive of *putrall, puirtleóg*, 'a tuft', is given by Peter O Connell, O'Reilly has *puirleóg*, 'crest, tuft'. The development of meaning is from 'tonsure, baldness' to that part of the hair left uncut. Compare *berrad*, 'tonsure, shaving', glossed *mullach a chinn* (M. and C. 107) and used in the present text with that meaning (infra 4005, 4028) and also with the meaning 'a head of hair' (*berrtha núa leó* 161, *berrtha lethna leó* 165). Cp. also LU 7584, 7503, 7139 (BDD). In one instance *berrad* seems, like *máel*, to mean 'a tuft of hair'; *Berruth cirdub for gach fir dib, indar lat is bo ro leluig cech ae* (Ir. T. III 239. 138). Possibly this too is the meaning of *berrad* in the passage supra 2141, *Berrad lethan lais. Folt casbuide fair.* (Recension III omits the second phrase). I suggest that *tuidmaile* (? *tuid maile*) in TBC may have the meaning 'crest of hair'.

3671 *Illand Ilarchless* See Stowe TBC note 3595.

3679 *Bratt corcra gen daithi impi* Windisch translates 'Ein purpurner Mantel ohne Farben um sie' which he explains as 'ohne jede andere Farbe', but the reading is obviously corrupt. TBC[2] has *Brat tlachtgorm corcarrda hi cennfait impe* (2791) while Stowe changes to *Lene do srol rig fria gelchnes* (3603), omitting the brooch.

3686 *Galach dá fénned* This is glossed by Stowe as *dies feindedh tuc in dá cneidh si i n-áoncneidh fort* (3610-11).

3691-2 *Ní thecat-sain* . . . Lit. 'They come into battle only with the assurance of always killing a man', i.e. 'whenever they come into battle they are assured of killing a man'. *Ni teaccait-sin i ccath na i comlonn nach ba demin leó neach do ghuin do ghrés* St 3617-8; *Nico[n] tiagait a noindin itir nach erdalta gona duine bis leo-so[m] it e nodgonad* TBC[2] 2806-7. In the last instance too *irdalta* (*airdalta*) is used as a substantive, 'certainty'. Cp. *áirithe, áiride*. Gordon Quin translates the TBC[2] sentence: 'They do not undergo *noinden* at all; whatever man-slaying has been decided on by them (i.e. Ailill and Medb), it is they (i.e. Oll and Oichni) who carry it out.' This implies that Oll and Oichni are of the Ulaid. The meaning 'warlike gathering, fray' is given in Contributions s.v. 2 *noinden*, but the word is doubtful. I have translated the Stowe reading.

3696 *Cumaing* . . . 3697 *Cumang-sa* . . . Intended for *ad-cumaing, ad-cumang* (hist. pres.). Again 3738, 3739. *Ad-cumaing*, means 'strikes'; in the present context it seems to mean 'thrusts'. Cp. the Stowe reading: *coro saidhsiot bior cach fir iondam-sa 7 do shaidhesa bior in cach fer diob-sium* (3621).

3702 *Bun 7 Mecconn* The same names are given in BDD to two of the king's guards: *Bun 7 Meccun* . . . *da mac Maffir Thuill* (LU 7672-3).

3709 *It immaicsi na fuli* The adj. is spelt *imfoicsi* where it occurs in Stowe (3665), *imoicsi* in TBC[2] 2842 with *f* inserted later between *m* and *o*. See Stowe TBC Glossary s.v. *im(m)-*.

3716 *Congas dá mbráthar Cungas da derbraithre* TBC[2] 2844. The word *congas* is obscure. Meyer suggested *con + gus*, 'equal valour'. See J. Carney, Poems of Blathmac, ITS xlvii, p. 118, n. 97-8, who suggests *con+gnás*, literally 'the being together of two brothers'. One would have expected *congal* rather than *congas* here. Cp. in the same section in TBC[2] *Congal tri niath anisiu* (2818), and note *fingal, bangal, galach* supra. Some word meaning 'attack' or 'onset' is required.

3718 *fá loss Los*, 'tail, end'. The translation is conjectural. The expression occurs again infra 4334. The Stowe compiler omitted it in both cases.

3731 *argit findi* See supra note 1204.

3741 *go ndarubdatar féithe do chride inniut* TBC² reads *imruidbiset féthi do chridi indit* (2828) < *im(m)-díben*. The LL reading I take to be for *go ndaruidbetar* < *do-fuiben*.

3745 *forróeglass* This is pass. perf. of *do-eclainn*, with *for-* for *do-* in relative position.

3746 *ni comtig beó* . . . In Stowe *ni gnath do nech terno beo oa ngonaibh* (3670-1).

3752 *gaindelderca* for *caindelderca* I have taken this to mean 'bright-eyed'; cp. *mnáib dergdercaib* LU 10377, 'women whose eyes are red (from weeping)'. *Caindelda* is used of shining eyes infra 4423. On the other hand Windisch takes the second element to be the adj. *derg* and translates 'lichtrote Männer'. In LU 10006 *A Condlai Rúaid muinbrec caindelderg barrbude* is translated by Pokorny as 'O tapferer Conle . . . rot schimmernd wie eine kerze' (ZCP xvii. 196).

3754 *go frithathartaib Adart* is generally used meaning a rest for a weapon: *a scéith fora ndelgnaib, a claidbe fora n-adartaib* LL 12965. Stowe gives a better reading.

3760 *gá cumcaisi* The vn. of *com-ad-ci*, used of a leech's inspection or examination of his patients. Here used with the meaning 'decision after examination, diagnosis', or perhaps 'remedy' which seems to be the meaning in Aisl. M. Conglinne 125. 30.

3761-2 *nír armea* . . . *gia dosrine* The meaning of this passage is clearer in TBC² and in Stowe: *Nitad tabartha bui duit ar dartaib indosa* (TBC² 2856), 'You should not exchange your cows now for yearlings'; *Na rec do ba mora da tabairt ar dairtibh isin mbliadain si, uair da recair, ni tú chaithfes iet* (St 3682-4), 'Do not sell your grown cows this year in exchange for yearlings, for if you do, you will not live to enjoy them.' *Nír armea* would seem to be 2s. pres. subj. of *áirmid* (*ad-rimi*). For the second verb M. O Daly suggests reading *dosrime* (*do-rimi*). Windisch translates 'Du sollst nicht deine grossen Kuhe zahlen für jahrige Kühe in diesem Jahre'. But *áirmid* followed by the preposition *ar* is not used in the sense required here; it means 'counts as', e.g. *áirmid ar nemni* . . . , 'he counts as nothing . . .' The Stowe compiler seems to have taken the LL verb as *ren-*, which suggests the emendation *ni riae* and *cia dosriae* (*nosriae*) (omitting the *n*-stroke of *dosrine* we should get a form near to *dosriae*). C here combines the readings of YBL and LL: *Nido[t] tabarthai ba doitt ar dairtib andossai ar cia dobertha ni tú féin rodus meala* (2390-91).

3770 *barróega Fingin* . . . Pret. and perf. of *do-goa* with *bar-* for *do-*. Here the scribe probably began to write the later sentence *Is ed ón barróega Cethern*. Cp. YBL which has *Is and sin iarum erpais in leig togu do* (TBC² 2865).

3776-80 *Is and sin conattacht Fingin fáthliaig smirammair for Coin Culaind* . . . *Tánic Cú Chulaind reme* . . . A contradictory passage in LL. Cú Chulaind was lying wounded (cp. infra 3950, 4003-7) and could not have killed the cattle. In TBC² the passage reads: *Is iarum condiacht Cú Chulaind smiur don leig dia frebaid. Dorigne smirchomairt di chnamaib ina cethra frisi comairnig* (2868-9), 'Then Cú Chulainn asked for marrow for the leech that he might cure him. He made a marrow-mash of the bones of the cattle he encountered'. This is ambiguous. If we read *forin liaig* for *don leig* and insert *Fingin* after *do-rigne*, it would make good sense; the physician provides the remedy. On the other hand the C compiler writes *Dorigne Cu Chulaind smir [d]on liaig do cnamhaib na cetra frisa comarnaic* (2404–5). But the C compiler probably had a copy of Recension II before him.

3777 *smirammair* Usually translated as 'marrow-bath'. Windisch takes it to be from *smir + ammar, ommar*, 'trough, tub'. Cp. the spelling *smirombair* TBC² 2870. But the gender, and the gen. form *smiramrach* (3782 and Stowe 3701), cast doubt on the derivation, and the expressions *ac ól na smiramrach imme* (3782) and *raluid in smirammair and* (3782–3) suggest that *smirammair* is merely a synonym of *smirchomairt*. Stowe has *do-gni smioramair ⁊ smiorcomairt dioph* (3699). Cp. in the fight between Hercules and Antaeus in CCath.: ⁊ *faiscid iter a rightea ⁊ a taeba co nderni smirammair da cnamaib i mmedhon a croicind* (2983–4). Where YBL has *Co rochotail la co naidche iar n-ol in smero* (TBC² 2871–2), C has . . . *ier n-ol an smerai ⁊ ierna fotracadh and* (2406). We may compare the bath prepared for the wounded Fráech: *Déntar fothrocud lib don [fh]ir sa .i. enbruithe n-úrshaille ⁊ cárna samaisci do indarggain fo thál ⁊ béuil ⁊ a thabairt issin fothrucud* (TBFr.² 248–51).

3788 *Finda ingen Echach Inda ingen Eachach Salbuidhi* St 3725. She is called *Find Becc* in TBC², gen. *Finde Becce* 2876–7 (but gen. *Findbéice* C 2411). In Talland Étair she is called *Findbec* (LL 13428), similarly in Fled Bricend (LU 8408).

3789–10 *Bertis robod reme-seom d'Ítholl* (Printed *Bertis robod remeseom. Ditholl* etc. in Diplomatic Edition, following MS. which has a point after *remeseom* and a capital *D*). Here the LL compiler or scribe seems to have misunderstood his original. We might emend by omitting the preposition before *Ítholl*. In TBC² it is the physician who gives warning of Cethern's arrival to the men of Ireland: *Berid robud remi isin dunad in liaig hi thall adrolla uadhsom remi. Robai marb* [in margin] *eter collaib na legi naile* (2079–82) = '. . . He had been lying unconscious among the corpses of the other physicians'. In C *marb* (incorrectly) comes earlier; *In liaig marb hi tall.* I take *bertis* in the LL reading to be a late formation for 3s. pret. of *beirid*. The form *bert(a)is* is very common in Rec. III e.g. ZCP viii. 542, 543 (2).

3801 *rabertad* Here used as pret. pass. of *do-beir*. Stowe reads *Brey do-ronadh cucam-sa so* (3715).

3808–9 *And sain ra théigsetar Ra théigsetar* is taken in the Contribb. s.v. *téit* as a preterite formed from *tiag-*. Exactly the same form occurs in a similar context in CRR: *Is and sin dolluid-sium . . . ⁊ ra theigsetar na slúaig immisium do dib leithib 'na ⁊ bognitha guin galann de co torchair accu* (LL 23026–7). *Do-tét* rather than *téit* would seem to be the verb intended here, *ra théigsetar* for *do-théigsetar*, based on the imperfect of *do-tét*. Stowe has *tángatar* and the rest of the sentence is like that in CRR: *Tangatar uime as cech aird ⁊ do ronsat guin galann de co ttorcair leó forin ccaladh sin* (3722–3). One would have expected the verb *iadaid* here as in TBC² (*Ro iadh in slog imbi-som iarum* 2890). Possibly some confusion arose with a spelling like *doíagat* which occurs twice in LU: *Doíagat side fona hegme* 4964, *Doíagat arna bárach do Cholptu* 5380. This is for *dothíagat* which is the form in the corresponding passages in TBC² (477, 904). *Ro iadsat* might have been taken as *Doíagat*. Cp. in TBFr.² 266 *Dothíagat na mná immi*, and in particular we may note in the Glenmasan MS. *Agus ro iadhutar-som uime-sium agus do ronsat guin galann dei* (Celt. Rev. III. 116. 15–16).

3817 *Gombad chumma ra gontais . . . Ra* for *no.* Cp. infra 3839–40 *da* for *no, combad chumma da gontais . . .*

3825 *ar co tísed Ar co* used for *nó co* or *co.* Again infra 3976.

3832 *i fiaclaib a chéile* Here Stowe adds *ar dtaircsin a n-arm* to explain why Fintan's men had to fight with their teeth.

3837 *ó Rénaib na Bóinne* TBC² has *for Boaind hi Correndaib.* The name is *Coirenna, Coranna* but the LL form is due to a misreading of *Correndaib* as *co rrendaib* (Thurneysen, Heldensage 194n.). In Tochustul Ulad where TBC² has *co mac Salcolca co Coirenda,* LL again reads *co Mend mac Sálcholcán coa rénaib* (4077); and *o Chorannaib* TBC² 3339 is in LL *ó Rénaib na Bóinne* 4516).

3842 *ra gáet Mend féin calad* Similarly TBC². *Calad* used adverbially. Cp. *Eirig, a bean, fég calad,* Ériu v 38, translated 'look closely'. Stowe changes the sentence to *Ro gonadh é fén co mór* (3795–6).

3842 *gor = corbo (curbo ruisti ruadhdercc é uile* St 3796). With this use of the preposition *for* compare *curba thaí tastadach for Ultaib* MU² 874.

3856 *Airecor nArad* The name is explained in Stowe. In the C manuscript: *Conid Aircor nArad innsin on airegar dobertsat forna trensluagaib* (2522–3). See Stowe TBC note 3894–3900.

3874 *Banassa* for *fonassa,* pret. pass. (pl. for s.) of *fo-naisc. Do naiscedh* St 3818.

3879 *ic Imlig in Glendamrach* Here YBL has *i nGlenn Domain* (2923). The incident, however, is entitled *Imslige Gleanndomnach* (2926), *imslige,* vn. of *imm-slig (Docertar and secht cét i n-imsligi Glindi Domain).* From the title *Imslige Glenndomnach* in TBC² the LL compiler made the place-name *Imlech in Glendamrach.* Cp. supra 2321.

3913 *Barrecgaib Dóche . . .* In TBC² we are told that Dóche checked the rabble who were jeering at the naked man *(Is and ro choisc Doche mac Maghach in daescorslog ocon chuitmead* 2940) and that it was for this reason that Íliach singled him out. Lóegaire was the grandson of Íliach.

3922 *airm i mbered . . .* Printed *airm imbered . . .* in Diplomatic Edition (to be translated 'when any weapons that he might ply . . . were exhausted'?). Windisch suggested that *airm* should be deleted; he took *imbered* as imperfect of *immbeir* and translated 'liess er (seinem Zorn) an dem Manne . . .' Cp. the Stowe passage here: *Agus ó thairnic dó sin, do beredh arin bfer ba neasa dhó d'feraib Erenn ⁊ do-bered diencomailt* etc. (3878–9), where *do bered = no bered.*

3929–30 *Is aire atberar Mellgleó nÍliach ris . . .* In Contribb. it is suggested that *mellgleó* is a compound of *mell,* 'a large ball, a round mass' (here referring to the stones).

3937 *Oisligi Amargin Aislingi nAmirgin* is the title in Rec. I and the passage opens with the words *Dosbiudc asa aislingi* (TBC² 2956–7). *Oislige* in Rec. II seems to be merely a misreading of *Aislinge,* the meaning of which was not apparent when no reference was made to Amargin's dream. One would have expected a rhetoric sung by Amargin in his sleep to occur here. Later in TBC² we get the words *Aislingi Dubthaich so* before a rhetoric (3085f.).

C has as title *Oisligi Amairgin ind[s]o nó Aislinccthi Amergain,* and the rest of the passage agrees with TBC².

3951–2 *aithle chomraic Fir Diad* . . . No reference to Fer Diad here in Rec. I.

3983–5 *Rachúas dó-saide* . . . *ra Glass mac ṅDelga* Stowe adds ⁊ *re Fer Diad ier sin* (3947–8). Note that Cú Chulainn was not wounded seriously in the fight against Cailitín Dána and his sons, so that the Stowe addition is called for here. Cp. supra 3950–2. In TBC² we get *rocluinethar* . . . *buad-rugud a meic Con Culaind fri da mac dec Gaile Dana* (2973–5). Similarly in C (2545). This seems to come from a confusion with the fight against Ferchú Loingsech and his twelve followers. It is noteworthy that Rec. III shows a somewhat similar confusion of the two episodes, for a passage from the fight with Cailitín Dána and his sons in Rec. II occurs in Rec III in the fight with Ferchú Loingsech (RC xv. 207 §226).

3989 *raluid* for *luid* (*Is ann sin tet Subaltach d'fios Con Culainn* St 3950). Similarly supra 2799 and infra 4027.

3993 *géara gonta* ⁊ *géara créchtnaigthe é* M. O Daly (Ériu xiv. 46) takes these verbs as past subj. passive, with *ra* (*ro*) for *no*. Cp. *Cia nongonta* TBC² 2979. But we might take *géara* as = *gérbo* (*ciarbo*) + past participles. Cp. supra *niro* 3951 = *nirbo*. In Togail Troí we get *ar ciaro maccaom som arai n-aisi, robo cathmilid arai n-engnama* (1588–9), Stokes edits to *cia ro(p)*, but obviously *ciaro* is for *ciarbo* here.

3994 *gress* For *gress* meaning 'protection' see Ériu xi. 95–6.

3995 *acht nirbo* . . . ⁊ *nirbo* For this use of *acht* see RIA Contribb. s.v. *acht* (j).

4003 *stúaga úrchuill* The sentence is expanded in Stowe to explain the use of these hoops: ⁊ *atá do met mo chnedh* ⁊ *mo chrecht conach fuilngim mo earradh no mo ededh do buain re mo chneas conadh tuagh* (v.l. *tuagha*) *urchuill congbas mo brat os mo chionn* (3965–7).

4016–7 Cp. *Óen do gessib Ulad labrad ríana ríg* ⁊ *óen do gessib in ríg labrad ríana druídib* MU² 234–5. In TBC² *re n-, ria n-* has been confused with *fri* (*ri*): *Ba airmert di Ultaib, ni labrad nedh dib acht fri Conchobar, ni labrad Conchobar acht ressna tríb druídib* (2988–9).

4020 *Cia rodas gon* ⁊ *cia rotas brat* ⁊ *cia rodas beir ale?* I take the verbs to be 3s. pres. ind. with *ro-* for *no-*. In the corresponding passage in TBC² they are present indicative: *Cistabrata, cisdagata, cisdabeir?* (2989–90) and in an earlier passage in LU *ciche brata, ciche áig, ciche goin?* (5563–4) while Stowe has present relative: *Cia gonas, cia ṅeres?* (3981).

4027 *raluid a háltaib* [*dó*] *Raluid* for *luid* as supra 3989. For the translation of this sentence see Stowe TBC note 3986.

4029 *driúcht* See note on uncertainty of the quality of initial *tr-, cr-,* by Dr. M. A. O'Brien, Celtica iii. 184.

4033 *coṅgreiss* Relative *con-.* Cp. *Ba hisa* (*hussa* C), *ol in drai, a bas ind fir rogresi* (*ro greis* C) *ind ríg* TBC² 2996; *As coir bas* . . . *do tabairt don ti gresis in righ mar sin* St 3991–2.

4039 *rabert cend Sualtaim na briathra cétna* The motif of a severed head speaking is common. Cp. Cormac's Glossary s.v. *orcc* (*cenn Lomnai drúith*), and Reicne Fothaid Canainne, Fianaigecht 8. It forms the chief incident in

Cath Almaine (RC xxiv 60). In Togail Bruidne Da Derga Conaire's severed head speaks to Mac Cécht. See also Anne Ross: Severed Heads in Wells, Scottish Studies vi, pts. 1, 2.

4049 In TBC² he is the son of Conchobor (3008-9). In Stowe he is called *Fionnchaidh Fer Bend mac Traiglethain* (4007). See Stowe TBC note 4007.

4050 *is cumma barrurim* Interchange of preverb, *bar-* (*for-*) for *do-* in relative position. This is 3s. perf. of *do-rími*, 'enumerates'. 4051 *tri mesci a chotulta ⁊ a chessa nóenden.* See Stowe TBC n. 4009.

4053 *Ardotrai . . . ardotfáedim* Here the preverb *ar-* is used to infix a pronoun. Cp. supra *ardotchlóe brath* (3440). TBC² here has *Atroi, a Findchaid, not foidiu* (3012) while Stowe reads *Attraicc, a Fhiondchaidh, rod faidim* (4012).

4053-4 *ní hadlicgi álsidi a faisnis do ócaib Ulad* In the MS, there is a point after *álsidi*; evidently the scribe interpreted this as the end of a sentence. This sentence is not in YBL. In Stowe it reads *Ni hadlaic ailseda. Aisnes d'ocaibh Uladh* (4012-3), which, as I have suggested in the Glossary to Stowe TBC, might be read *Ni hadlaic ailsed a aisnes*, 'it is not desirable to neglect to tell it, we must not neglect . . . '. C, influenced no doubt by a version like Stowe, has here *Atrai, a [Findchaid] oir nid adluicc ailsidh a aisneis di occaib Ulad* (2579-80). See Contrib. s.v. *adlaic* where it is suggested that *adlicgi = adilccne*, 'necessity'.

4054-4100 This catalogue contains many names not found elsewhere in the text. But of the nineteen leaders described later in the section Toichim na mBuiden (so-called in TBC² 3097), fifteen are mentioned here. Omitted naturally is Conchobor himself, but so also are Fergus mac Leite, Fiacha and Fiachna, sons of Conchobor, and Erc mac Carpri, Some of the names seem to have been invented, perhaps in an attempt to extend the list, e.g. triplets such as *co Mod, co Mothus, co Iarmothus* and *co Ebadchar, co Uathchar, co hEtatchar*. It is consequently not always easy to distinguish personal from place names. The order of names is not the same in YBL. In Stowe many of the names are corrupt and the list breaks off less than halfway through with the words *et reliqui*.

4055 *co Follach; co hIllaind* Here YBL has *co Fallach, co hIllann mac Fergosa* (TBC² 3013), but C has *co Fallach mac Fergusai, co Iollann* (2583). Illann mac Fergusa was in Connacht with his father. See supra 3671-2. Stowe puts the name later in the list and has merely *co hAillinn* (4016).

4058-9 *co Geimen coa glend* Cp. *Ocus Gemen ón gorm-glind* Metr. Dind. iv. 202. Glend Gaimen is near Dungiven, Co. Derry.

4059 *co Senal(l) Uathach* Mentioned among the Ulstermen supra 3598.

4059-60 *co Cethern mac Fintain* Also in TBC² (3018). Cethern's death has been told supra 3809-10. But note that in Rec. II Conchobor enumerates both living and dead (supra 4050-51).

4061 *Cosin rígfilid co Amargin* This presumably is the same as Amargin Iarnguinnaig 4083-4 and Amargin mac Ecelsalaig 4098-9. Note also *co Amargin co Ess Rúaid* 4080.

4068-9 *co Lugdaig co ríg mBuilg* In TBC² *co Lugaid ri Fer mBolc* (3027); in Stowe *co Lugaidh righ bFer mBolca* (4021).

4073 *co Briccni* This cannot refer to Bricne Nemthenga mac Carbad for he was in Connacht with Fergus. See infra 4861ff. TBC² has *co Briccir, co Bricirne* (3033).

4074 *co hÓengus ṁBolg = co hÓengus Fer m[B]olg* TBC² 3035. In the Battle of Airtech Aengus Fer mBolcc or *rí Fer mBolg* is mentioned but there he is on the side of Connacht (Ériu viii. 187).

4074-5 *co hÓengus mac Lethi = co hOengus mac Leti* TBC² 3034. In TBC² (and in Stowe) he is followed by *Fergus mac Leiti* (TBC² 3034, St 4024). Fergus mac Leiti appears later in Toichim na mBuiden; Windisch suggested that his omission in LL is due to a slip.

4076 *Conall Cernach* See Introduction Stowe TBC pp. xii–xiii.

4078 *co tri maccaib Fiachnai* . . . They are not named in TBC² which reads merely *co tri macu Fiachna* (3036) but they appear later in Toichim na mBuiden in TBC², and from TBC² they were taken into the Stowe version (4382–92). They do not appear later in LL.

4096 *co dá mac F̃eicge* . . . Mentioned earlier in the text as *dá mac Gégge* (3598-9). The names are not in this catalogue in TBC².

4101 *Níra dulig Nírbo* Stowe and TBC². Cp. supra 3951.

4103 *rathóegat-saide ass* 4105-6 *ratháegat-saide ass Ra–* for *do–*. Here printed *ra thoegat-saide, ra thaegat-saide* Dipl. Edn. but infra 4234 it is printed as *rathaegat ass.* This form is for 3 pl. pres. indic. of *do-tét,* here a historic present. Cp. *dotháegat* LU 4660 (TBC), 7924 (BDD).

4107 *In cétna uide bachomluisetar Ba-* for *do-,* 3 pl. pret. of *do-cumlai.* Cp. *a laa documlásat* TBC² 22.

4114 *ná co clórat* A new analogical form from the 1 sg. pres. subj. *clóor.* Changed in Stowe to *no co ccluinit* (4037).

4117 *And sain raluid* . . . *Raluid* for *luid, ra* being added below the line. Both Stowe and TBC² have *luid* here.

4118 *baralsat dóib Ba-* for *do-.*

4123 *dáig concomairnectar* . . . In the older language *dáig* was followed by relative *n-.* Here *con-,* relative, is used. Cp. supra 910, 1260. A similar use of *con-* for *do-* with *feb* occurs in BDD: *feb condrarngertsom feissin* (BDD² 1406), v.l. *feib dorair[n]gertsom. Dáig concomairnectar . . . cathugud ⁊ imbúalad.* Compare supra *ní mar chách conarnecar comlund ⁊ comrac ríut . . . Fer Diad* (10173), and *Bid garb an comhracsa condricfad fir Asia ⁊ Éorpa* TTr. 905, *Condráncatar iarum comrac deisi* ib. 1563.

4148 *And sain confucht[r]aither Con-* for *do-* in relative position. *Do-fochtra* has deponent inflexion here.

4151 *rabertatar* Here Stowe has singular *dobert* (4061) which refers back to *ind Nemain.* This perhaps makes better sense.

4155 *Rasetarrad-sa Ra-* for *do-,* 1s. pret. of *do-etarrat.*

4158 *Barallsam a tilcha dá n-éis* The same sentence again infra 4577-8, *Barraeilseam a tailcha dá n-éis.* Quiggin (Die Lautliche Geltung p. 48) takes *barrallsam* to be from *do-alla,* and is followed by Windisch who translates 'Wir haben hinter ihnen ihre Hügel weggenommen.' Both forms may be for *doralsam (do-cuirethar* in the sense 'casts down , cp. *docuirethar dar cend).* Quiggin was influenced by a later passage where LL read *Barallsat a n-étaigi díb uile* (4298 for which the equivalent in YBL is *Rostellsad a n-etaigi diib* (TBC² 3143). Note however that in that passage C has *rolasat* and Stowe *ro cuirsit.*

4163 *ní robés ríg rotheched* The *ro-* prefix here has little meaning but is used for alliteration.

4165 *in rimechlach* Again 4283. Confusion of lenited initial *p-* and *f-*. *Echlach* is feminine. Cp. *mo rímchuit in bargen út* LL 6689. In the late versions, Rec. III and Stowe, this lenited form gave rise to *rígheachlach.*

4173 *barrafnit* Stowe reads *ro taifnid* (4081). *Barrafnit* is Mid. Ir. pret. pass. of *do-seinn*, with preverb *bar-* (*for-*) for *do-*. Cp. *dosroiffnetar* LL 18322. Again in the sentence *Iss ed on barraffind* . . . 4181 we have 3s. perf. = *Is edh ro thaffainn* St 4087.

4179 *at Ulaid barfópartatar* The 3 pl. pret. of *do-fúapair*, with *bar-* for *do-*.

4183 *forrécaig* This appears to be an *s-* pret. of *do-éccai* formed from reduplicated preterite. Cp. *forrécacha Medb furri* supra 197.

4188 *ratafarfáit* Again *rotafárfáid-sium* 4218. In Diplomatic Edition printed *rata farfáit, rota fárfáidsium*. Both verbs are in relative position and contain an infixed pronoun, 'which appeared to him'. Cp. supra *Ní hé sút cruth ardomfarfaid-se indé*. *Ratafarfáit* may be for *do-da-farfáit*. Cp. *ni siabrae rodatánic* (= *ro-dat-tánic*) LU 9302.

4188 *tri urdluich na cíach* (= *tresan ciaigh* St 4093). It is suggested in RIA Contribb. s.v. *airdluige* that *urdluich* here is for *urdluige* (*ar-dloing*). I have translated it accordingly. Possibly *urdluich* might be for *urdluith* (*urdlúith*) from *dluith* (*dlúith*), 'compact, dense', here used as substantive = 'dense mass'.

4189 *ba éochain de ilénaib* Again 4208–9 *Éochain* is for *féochaine*, coll. from *fiach*, and means 'ravens, flocks of birds.' Stowe has *fechine* in the first occurrence (4094) and *na fechini* (pl.) in the second (4113). Cp. in a similar description of the sods cast up by horses' hooves: *Indar lat is feochuine Herend fil uasa* LL 14122, with a marginal gloss *nó fiaich.*

4198 *conas ecgaib* Again 4210–11. I take this to be for *condas ecgaib*, the pronoun being reflexive. Perhaps a by-form of *do-fócaib*. Cp. RIA Contribb. s.v. *do-ecaib*. Windisch suggested that the correct spelling was *conas-uccaib*, referring to his Wörterbuch s.v. *con-ucbaim*. The verb is omitted in both instances in Stowe.

4226 *nach raichnea* . . . See Contribb. s.v. 2 *aithnid*, where a note by Plummer is quoted suggesting that this form is 2s. pres. ind. of *aithnid* (older *ad-gnin*) with *ro-*, 'that you cannot find = that you will not find'. The same sentence occurs again 4363–5.

4227 *á rasfecgat* For *ó do-s-ecat*. Again infra 4364 *á rasfecat.*

4234 *ratháegat ass* *Ra-* for *do-*.

4237–8 *in n-echrad ṅdirecra ṅdermór* Here *echrad* seems to be treated as a neuter, but cp. 4242–3.

4249 *Rolúath bic n-archessi forro* *Roluath bic arcesi forra* St 4150. The editors of the Diplomatic Edition suggest that *n-archessi* is for *no archessi*. The verb intended is *ceisid* (*for*), and *ar-* may be taken as a relative prefix as in *Is airiund arbáget* TBC² 170 = *is erund conbágat* supra 328. *Rolúath archessi* is treated as a nasalising relative clause. The *n-* is superfluous.

4249–50   *dáig ro bífad co mbetis* . . . Changed in Stowe, no doubt because unintelligible to the compiler. *Ro bífad* seems at first sight to be due to a confusion between the spellings of *benaid* and the substantive verb. M. O Daly (Ériu 14. 86) takes *ro bífad* as 3s. sec. fut. of substantive verb but notes that 'the meaning calls for a past subj. or perf. past subj.: "it might be" or "it could be" '. But unless we take *dáig* here as = *dóig* such a construction as *dáig no beth co mbetis* would not be possible. (*Dóig* = *dáig* occurs supra 2580). Perhaps LL's original read *dóig* (conj.) *robad dóig co mbetis* . . . with *robad* 3s. sec. fut. of copula? The sentence seems to call for some form of the copula + an adjective denoting possibility or probability. On the other hand we could emend more drastically by omitting *ro bífad* (intended to be expunged but overlooked?), and by taking *dáig* to be for *dóig*, 'likely, possible'. The original might have read *Dóig co mbetis* . . .

4250   *nibad liriu*   Here, as above *giambtar liri* 2100, *lir* is treated as a positive, with the comparative *liriu*. See Bergin, Ériu xiv. 140–1.

4251–2   *rar fúabérad-ni* = *co n–ionnsochadh* . . . *sinne* St 4152.

4256   *dán cimbeda*   Again 4258. The phrase is obscure. Stowe has been translated.

4264   *d'forddiglammad áig* ⁊ *urgaili*   The meaning is obviously that of the Stowe reading. *Forddiglammad* based on the verb *fordiuclammaid*, 'devours, swallows', here 'used loosely in the meaning "wage (war)" ' (RIA Dict. s.v.).

4279   *cindas na hacgmi bá tecat Ulaid*   Again 4294 *cindas na hecgmi* [*nó*] *na taicgme bá tecat Ulaid*, where Stowe omits the doubtful words and has *ciondus tiegait* (4178). Both *acgmi* (*ecgmi*) and *taicgme* seem to be the gen. of a fem. word. A form *eccim* masc. occurs in Togail Troí (*heccim* ⁊ *écosc* TTr. 280, *indas ind eccmi* ib. 1041 = *indas in toichme* TTr.² 498). Windisch (TBC p. 730 n. 1) suggests that the forms *acgmi, taicgme* are intended as genitives of *ecmong, tecmong*. But apart from the gender, the meaning does not suit here. We might read *cindas* . . . *na taichmi* 4294. Originally neuter, *tochim* is masc. as in *indas in toichime* (sic) TBC² 2939 (Mellgleó nÍliach) but fem. in the same passage in C: *indus na toichmi* (2507–8). *Eccim, accim* (?) I take to be another compound of *céimm*.

4318   *'na láim na óclaige*   *Óclach* here treated as a fem. *ā*-stem. Cp. *echlach*. Earlier we get *fónd óclaig* 1580–1 and *dond óclaig móir* 1586.

4319   *Immireithet* for *imreithit*. (*Imrethid* St 4264, where this passage comes later). Probably influenced by the following sentence where the verb is relative.

4321   *ó indsma gó hirill*   *Irell*, the thong binding spearhead to shaft. Stowe has *o indsma co irlainn* (4267).

4339–42   *Acht ba binnithir* . . . Not in TBC².

4349   *co maccaib ríg Hérend rafailet ina farrad* (. . . *filit ina fharradh* Stowe 4262). The form *rafailet* is relative here. Cp. supra n. 2580. Similarly in Cath Ruis na Ríg: *Ráid dam risna tri coicait senorach* [⁊] *senlaech ro failet ina ligi aisi* (LL 22849). Again in the present text we get sing. *rafail: dáig is tri* [d]*agin na buidni sin* ⁊ *in meic bic rafail inti* . . . (4550) (= . . . ⁊ *in mic bic fil eatorra* Stowe 4581); and in CRR: *Cid in dirram slóig móir ro fail it [d]egaid?* (LL 22979). There is one instance of *rofuil* used relatively in Rec. III, referring to Fergus's scabbard with its wooden sword: *indtech na laidhedh moiri rofuil leis* (ZCP viii. 550. 5). In TBC² *dafil* occurs once in relative position: *cech conair imateit in milig morglonnach dafil isin c[h]arput* (2576, Fer Diad episode). (Elsewhere in TBC² *dofil, dafil* is used absolutely with the meaning 'is at hand, approaches', e.g. 1022, 1192, 1440, 1716). Possibly the absolute use of *fail*, 'there is', *failet*, 'there are', which occurs in the present text (238, 240, 241, 340, 341) suggested the need of a particle to distinguish it when used relatively. Or the use of *dafil*, relative as in TBC² 2576, might very well give rise to *rofail, rafail* in the present text, where *ra-*, for *do-* is frequently found in relative construction. Relative *dofail, rofail* beside *fil, fail* might give rise in turn to the absolute use of *dofail, rofail* beside the usual *fil, fail*. See supra note 242. But an absolute use of *dofil* may have preceded the relative use. In Cathcharpat Serda LL 189ª we get *Dofuil 'sin charput sin laéch foltfind* (24873) and *Dofuilet tri fuilt arin laech sin* (24875).

4351–3   *nachis imrethet . . . ros imreittis*   Here the simple verb *imrethid*, used with reflexive pronoun.

4355–6   *Sencha . . . fer sidaigthe slóig fer ṅHérend*   The Stowe reading is perhaps better. In Mesca Ulad Sencha is called *sobérlaid fher in talman ⁊ fer sidaigthe slúaig Ulad* (MU² 757–8).

4373   *Is cur lám for debaid*   The same expression 4396, *Is cur lám for ugra.* Compare in the Glosses: *is cor lame ar dodced buith oc airbiathad sainte* Wb. 29ᵇ18, 'It is inviting misfortune to be feeding covetousness'.

4374–5   *Eogan mac Durthachta a fosto. Fernmaige*   Cp. supra 709–10, *ticfaitis fosta fer Fernmaige leiss*, which I have translated 'the steady men of F.' But in Aided Guill ⁊ Gairb: *co tech Eogain meic Durthacht co fosud fer Fernmaige* (LL 12999), which seems to mean 'to the abode of the men of F.' Here we might have expected *a fosud fer Fernmaige*. I have, however, translated the reading of Stowe.

4391   *Gae súlech go foscadaib*   *Gae súlech* may mean a spear with *súil* or *súili*, but the meaning of *súil* in this context is doubtful. I have translated the simpler Stowe reading. *Foscad* is used of gleaming or reflected light. Again infra 4543 *gae . . . go foscathaib*. Cp. *ri foscthib lassamna . . . na lecg logmar* TTr. 1664.

4397   *fer tri ṅgreth*   Cp. in the poem *Diambad messe bad rí reil* LL 147ᵇ; *Trí gáre buada do ríg | . . . | gair choscuir chruaid | gair molta muaid, gair im fleid* (18699–18702). See Measgra Dánta 36. 1–4, and note p. 86.

4401   *anisc*   The meaning is doubtful.

4416   *Ní dui for lomma*   I have taken *lomma* as = 'bareness, nakedness', but the precise meaning of the sentence is obscure. In an additional note TBC p. 1117 Windisch writes: 'Kein Narr bei nackten (Frauen)? Die Schönheit des Fedilmid wird besonders betont.' In RIA Contribb. s.v. *loman*, 'cord, rope', the variant of the late MS. H is quoted, *ni dui for lomain*, and *dui for lomma* is translated 'a spancelled fool'. But the exact meaning of that I find doubtful also.

4426   *Cassán gelderg i fadi úasu*   LL seems to be corrupt here. *Cassán* means 'a pin, brooch', and here we should expect a description of the *brat* which is referred to in the next sentence, where the brooch is specified as *eó óir*. *I fadi* seems to mean 'wrapped' and would apply to a cloak (cp. *brat . . . i faithi imbi* TBC² 3200, in description of Láegaire Búadach). But *úasu* cannot refer to either cloak or brooch.

4432 *Is galiud comlaind* I have taken the reading of Stowe and TBC². In another passage in TBC², the praise of Cú Chulainn just before the Macgnímrada, the term *cláriud comlainn* is applied to Cú Chulainn (LU 4841–2). O'Clery has *cláired .i. roinn.*

4443–4 *Bá hitte di gail . . . Báe itte di drúis . . .* The text is obviously corrupt here. Stowe alters to *Ba do gail ⁊ do gaisced . . . in cách tainic ann* (4221–2), but this is a doubtful construction. YBL's *baithi* is probably intended for *bai* + suffixed pronoun, but the use of a past tense is strange here, and the sense requires some such word as *lór* or *mór* after the verb. Emending the LL text we might get *Baithi [lór] di gail . . . Baithi [lór] di drúis . . .*, or, emending more drastically, *Táthai lór di gail* etc.

4444 *Táthud do slúagaib . . .* I have given the translation suggested in RIA Contribb. s.v. *táthud*, but for the meaning given one would expect *táthud slúag ⁊ arm.* The corrupt reading of TBC² here suggests an emendation to *Táthai [lór] do slúagaib ⁊ d'armaib.*

4462–3 *Sleg cúicrind confaittnedar darin slúag uile* M. O Daly (Ériu xiv 37) takes *confaittnedar* to be a deponent 3s. pres. ind. (I note that in the Contribb., s.v. *do-aitni, confaittnedar* is given as 3 pl. pret. which is impossible in the context). Strachan does not include this form in his list of deponent verbs in LL–TBC. There is no equivalent for the sentence in YBL, but Stowe reads *Slegh coicrinn ina laimh co fuitnenn* (v.l. *go faithneann*) *tarin sluag uile friaa ais* (4378). Hence O'Clery's gloss: *faitnenn .i. an taithneamh .i. go bfaithneann .i. go ttaitneann.* I believe that the LL reading is due to dittography and that we should take it as *confaittne darin slúag,* 'which shines over the host', *con-* for *do-* in relative position and inorganic *f* inserted. The Stowe form is based on the LL reading.

4474–5 *Dá ánrath* (*óenrath* MS.) The prefix *óen-* here is for *án-*. Probably the same holds good for the next three nouns. Stowe has *an-* in all three. *Ánchaindel,* 'bright torch', occurs commonly as an epithet for a warrior. Cp. supra 3288, 3289. Alternatively we might read *dá anchaindill* etc., taking *an-* as an intensive prefix.

4483 *irggráin* An adjective here. We might read as *úathmar a irggráin,* except that such a phrase does not occur in any of the other descriptions The usual form of the adjective is *airgráinne, airgránda.* TBC² and Stowe read *Loech garbaineiuch huathmar i n-airinach na buidne sin.*

4486 *Claideb secht mbrattomon Bruth,* a mass of glowing metal in a furnace, *bruth-damna* > *brattamna.* We should expect *secht mbrattamna* here. Cp. *claideb deich mbrattamna* TTr. 1546.

4489 *ro lá dírna dina cathaib* An impersonal use of *fo-ceird dírna* (*dírma*) *di,* meaning 'casts into confusion'. For the idiom see note Hermathena 1931 p. 4.

4497 *is hé lethgleóir* Hessen gives the meaning of *lethgleóir* as 'bright in one eye', basing the translation on O'Clery's gloss *.i. leathshúil ghlóire* which was perhaps no more than a guess on O'Clery's part). In Stowe

TBC there is a description here among the Ulstermen of Genonn Gruadsolus mac Cathbad. He is said to have *drech lethan lethgabar* (4487, 4500). Here the meaning of *lethgabar*, *leth* + *gabor*, 'white, bright', seems to be 'bright-cheeked' = *gruadsholus*. *Lethgleóir* is apparently a synonym of *lethgabar*. For *lethgabra* BDD 1150 the Egerton reading is *lethgleora*: *it é lethgabra amal Conall Cernach*. In the same tale Conall Cernach is described: *Gilithir snechta indala grúad dó, breicdeirgithir sian a ngruád n-aile* (960–1), and in Fled Bricrend is said of him *Drech lethderg lethgabur laiss* (LU 8636).

4508 Possibly to be read as *cechtar n-ai dina rigrosc[aib] ruad[a] romóra*.

4525 *rí Búrig Ulad* Cp. supra 4092–3 *co Fergna mac Findchona co Bárach*. So in Mesca Ulad Fergna is called *ri Búrig Ulad* (MU[2] 326). But Stowe here in TBC calls him *Fergna mac Findchaime rigbrughaidh Uladh* (4436). Thurneysen (Heldensage p. 206 n. 3) and Watson (MU[2] n. 325) take *rí Búrig* to be a corruption of *rígbriuga*, 'royal hospitaller'.

4547 Combining the readings of LL and TBC[2] we might emend to *ní 'amuscarat a n-airig and so,* 'their leaders do not separate here', which unfortunately does not make much sense in the context.

4549 *bud muir conbáidfea* Here *bad* = *ba*, future, and *con-* is relative. Again in the next sentence *conmáe foraib-si*. Stowe keeps the relative *con-* in both sentences, *conbaidfidh, conmaidfe* (4580, 4582).

4552 *ní faccéga* Mid. Ir. 3s. fut. of *ad-ci*.

4557–8 *Bát bágaig . . . confúarcfet* Relative *con-* for *do-*; the verb is *do-fúairc*, with Mid. Ir. *f-* future. Here Stowe has the usual construction, 3s. copula + sing neut. adj.: *Badh* (= *ba*) *badach* ⁊ *badh cóndalbach do-fuaircfet . . .* (4588). In the next sentence here we get this construction, *Is ferda conbúrfet . . .* (again with relative *con-*). There are occasional instances of this use of plural copula and plural adjective: *Batar doichi ém fir maithi do saigid a chéle don bruidin sin innocht* BDD 633–4; *Do-thoet Conall fa chettoir co Temraigh* ⁊ *robtar luaithi rangator techta Ulad* ZCP xiv. 223. 4; and in an interpolated passage of Stowe TBC: *At annsa dia n-eccmaltibh baramhail forra . . .* ⁊ *it use dia ngnaithchibh* (4662–3).

4559 *oc tessargain laíg a mbó* The reference is to Erc, the son of Conchobor's daughter, Feidlimid Noíchruthach. Stowe does not continue the metaphor of *damrad*, but has *ic teasarrcain a n-úa* (4590). In TBC[2] the passage equivalent to *Bát bágaig . . . imbárach* comes before the sentence *Concechlastar rucht claidib Conchobuir*.

4560–86 This band is not described in TBC[2]. Corresponding, to this description is a long passage in TBC[2] (3383–3404): Mac Roth says that it would take too long to describe all who came to the hill. Conall Cernach did not come, he says, nor did the three sons of Conchobor. (See Stowe TBC Introduction xii). Neither did Cú Chulainn who had been wounded fighting against odds (that is, in the fight with Gaile Dána and his sons). There seems to be some corruption here in TBC[2] for, immediately after the sentence *Ni thanic tra Cu Chulaind and iarna chrechtnugud i nn-ecomlund* (3387–8), Mac Roth goes on to describe Cú Chulainn's horses and chariot, Cú Chulainn himself and his charioteer. They are then identified by Fergus: *Ni anse, or Fergus. Cú Chulaind mac Soaltaim a sidaib* ⁊ *Loeg mac Riangabra a arae Con Culaind insain, or Fergus* (3402–4).

4561–8 This passage is in rhythmical form, seven-syllabled lines ending in a trisyllable.

4561–2 *Fir gil glain guirm chorcarda* Omitted here in Stowe. Perhaps better omitted, for the sentence makes little sense.

4568 *gana comsid costadaig* Again 4572 *can Choin Culaind costadaig coscaraig claidebdeirg cathbúadaig.* For this use of a fem. form of the adjective in the acc. sing. masc. see note by Dr. M. A. O'Brien, Celtica II 346–8.

4568 *a n-irúatha* It is suggested in Contribb. s.v. *irúatha* that this should read *a n-iruda*. *Irud*, 'edge, margin'. Cp. SC² 268.

4571 *n–anfálid* Again 4574. This seems to be a petrified sing. acc. form. See Myles Dillon, Nominal Predicates in Irish, ZCP xvi. 353.

4578–9 *Ní thá ní nod maitte forro* It is suggested in Contribb. that this should read *not maitte* (reflex. pron. 2s.), but such a construction with *for* does not seem possible. See supra n. 1449. Windisch takes the verb to be 2s. imperfect, but it is past subjunctive, here used after a negative (Strachan, Subj. Mood §73). A past subjunctive following a pres. indic. (or future) in the main clause is common. Cp. supra *Ní fuil digsed da fiss dam-sa* 2553, *Má tá ocaib 'sin dún . . . rofessad na coma* 1535–6.

4583 *Bochinmáir* In Stowe *cén mair* (4709). Possibly a confusion of *mochen, fochen* and *céin mair*. In a rhetoric infra 4606 *Bo chin Ultu = cen mair hUlltaib* TBC² 3424. But *bo-* is probably the preterite of the copula used modally here. Cp. a similar use of the copula before *mad-génair* in Early Irish Lyrics 30. 6: *Ba ma-ngénar do mac Dimma.*

4590 *'gá furmiáil* Again 4594. In Stowe *ga uirmeail* (4715, 4719). For this word see Glossary Stowe TBC s.v. *uirmeail.*

4591 *i bFirt Sciach* Again 4595. In both instances Stowe reads *a Firt Sciach ingine Deagadh* (4715, 4719).

4597 *Fethan ⁊ Collach* Here Stowe has *Fethan ⁊ Cuillech*, v.l. *Collach* (4724). At the second occurrence infra 4794 the names are given as *Fethan ⁊ Cholla*, and Stowe reads *Fethan ⁊ Culladh*, v.l. *Colla* (4916–7).

4609–10 *Ba líag . . . na dernta* To be read as *dá ndernta*. Cp. *na dernta* supra 1644–5. Or read *ná dernta*, and take as an anacoluthon. The scribe overlooked *Ba líag ám dait-siu*, 'It were a pity for you', and carried on the sentence as if the opening words had been something like *Ba chóir dait-siu.* The sentence was altered by the Stowe compiler: *A Laig, ar se, na dentar isin cath aniú ní na bia a fios leat dam-sa* (4730–1).

4620–1 *Cia confirend in cath* In Stowe *Cia fichus an cath* (4731). Relative *con-*. Similarly 4622, 4623, 4671, 4672.

4630 *todúsig do Ultaib* Again 4632, 4656, 4657–8. This use of *do, de* with *do-diuschi* seems to be confined to the LL- text.

4642 *Cach óen dá mbid . . .* Lit. 'to whom there was . . . ' *Cach óen dá tá . . .* 4646. Similarly in TBC², *Inti dia da dorus a pupaill sair* (3475) A different preposition in Stowe: *Cech aen ica mbid . . .* (4758), *Cech aon ica dtá . . .* (4761).

4648–9 *degóir éigmi* Windisch read as *degúair éimgi*, and translated 'eine gute Stunde geeigneter Zeit'. Marstrander suggested the reading *degfhóir éigme*, taking *éigme* as gen. of *éigem*, 'a cry of distress, a summons'.

4685 *na nóecharptig de fénnedaib na hIrúade* Stowe reads *naoi ccairptigh d'feindedaib na hIoruaidhi* (4793), while TBC² has *ind noi carpait di feindidib inna Iruaithi* (3536).

4686–7 *nira lúathiu* *Niorbo luaithi* St 4794.

4688 *ferchutredaig fer ṅHérend* For these see Stowe TBC notes 4795, 4798–4811. It is to be noted that in TBC² it is not the *ferchuitredaig* who act as bodyguard for Ailill and Medb, but the warriors from Irúath (3538–40).

4690–1 *dámbad forro conmebsad* Relative *con-* for *no,* and sec. fut. form for imperfect subjunctive. Stowe reads *dámad forra do maidfidi* (v.l. *do maoidfedh*) *an cath* (4797).

4704 *Ba bág ám duit-siu . . .* In Stowe *Ba coir duit-si . . . do grem catha do tabairt co dutrachtach linde* (4812–3). *Bág* is used similarly in an earlier passage in TBC²: *Ba bag doibsom an toitim oc tesorcain a neiti* (3435). Windisch translates the LL passage as 'Es wäre wahrhaftig ein Gegenstand des Rühmens für dich . . .' I have translated the Stowe reading.

4708 *bráigte* The length-mark is omitted in the MS. In the later language there was a tendency to shorten *á* before *gt,* but it would be unsafe to conclude that this tendency existed ca. A.D. 1100. See note by T. F. O'Rahilly, Ériu xiii. 129.

4710 *combús* The reading must be corrupt here. We should expect *combad* or *comtis,* sec. fut. Perhaps *combús* is a scribal misreading of *combtis* ? Cp. the TBC² equivalent passage: *Ma nobith em mo claideb acomsa, ol Fergus, beitis lir leamsa cendae fer . . . andate bommann ega* (3543–4); and Stowe: *ro teasfaidhi liom braigde fer for braigdibh fer . . . comba* (= *combad*) *lir bomanda egha cech mball soir* ⁊ *siar agam-sa d'Ultaibh* (4815–8).

4748 *Cia concoṅgbathar sciath rum-sa . . . Cia haccaibh congbas a sciath friom-sa* St 4859. Relative *con-,* here prefixed to a compound verb, though generally *con-* is found only with simple verbs. Cp. *Maith a ṅgein concoimpred* and RC xiii. 452 §41.

4761 *comtis lir . . . Lir* here used as comparative. Again *gorsat lir* 4770.

4763 *Aicclech nád aicclech sain* For this idiom cp. *Mét mo boithe— bec nád bec* Early Irish Lyrics 8.10a, translated by Gerard Murphy as 'The size of my hut—small yet not small'; and *comraic nát chomraic a mbarr,* SC² 503.

4768 *nída beó-sa* Pres. ind. used with future meaning. Again infra 4828.

4770 *gorsat lir . . . Gorsat* here seems to be pres. subjunctive.

4778–9 *go failet 'sin riasc bad fiadnaisi* The sentence is probably unfinished. Elsewhere in the text *bad fiadnaisi* is always followed by *do* + noun. The Stowe readings suggest that the words *d'feraib Hérend* have been omitted here.

4782 *Cia conlindfadar . . .* Again infra 4805. Relative *con-.* The form *-linfadar* is for *-lilmadar,* 3s. fut. of *ro-laimethar.* In an earlier passage in TBC² (Macgnímrada), where LU 5185 has *noco lemaithir a glúasacht,* YBL reads *nocho linfaithir.*

4784–6 The translation of this passage is tentative. The Stowe compiler omitted it and inserted in its place a few sentences which have no equivalent here or in TBC².

4793 *dá gáeib cró* Not in Stowe nor in TBC². For the suggested translation, 'tents (of wounds)', see Éigse ix. 181–6.

4795 *barressairg* Preverb *bar-* for *do-*. The verb is *do-essuirg*, 'smites', **di-ess-org* Pedersen VG II 588.

4795–6 *gorbo derg . . .* Stowe's *in talam* must be added. If we leave the sentence unamended, it means that Cú Chulainn was grey with their brains and red with their blood. TBC² has *combo liath ceachtar de de inchind a seitche* (3616–7), 'so that each of them was grey with the other's brains'.

4817 *nír dirúais* This verb seems to be intended for the preterite of *do-fuarat*. Perhaps the *dar* which follows comes from some such form in LL's exemplar as *nír diruarastar, don* before *charpat* being omitted by an oversight. Cp. *do neoch rothiruarthestar din churp doenna* RC xxv. 242. 8. The meaning is clear. Stowe has *Nir mair don charpad*.

4824–32 This incident is not in TBC² which has merely *Tarraid C.C. iarum Meidb oc teacht isin cath* (3639). But a similar incident is related in the Glenmasan MS. Táin Bó Flidais, in the account of the pursuit of Fergus and Medb and their troops by the Gamanrad (Celtic Review iv 208. 23–30).

4826 *goro siblur-sa* Again 4829 *meni siblur-sa*. This has been taken as pres. subj. of *silid*. Stowe has a denominative verb *sriblaid* (*sriball*), *coro sriblar-sa* (4956). In a passage of the Fer Diad episode which occurs only in TBC² we get *Is and dorala Medb ic sriblad a fuail* (2477). In Aided Echach Meic Maireda a pret. 3s. *siblais* (LU 2940) occurs beside *silis* (ib. 2954), and an imperfect subjunctive *arnár siblad* (ib. 2949). It seems likely that *silid* and *sriblaid* were confused or that suprascript *i* was read as *i* in LL here.

4834 *ni athgonad-sum . . .* Ath- seems to have no force here. The use of the imperfect is strange. It is perhaps due to the passage in TBC²: *Rosnanacht iarum huair nad gonad mna* (3641–2). Stowe omits the sentence.

4848 *Condrecat lochta ra fulachta and so indiu* The meaning of this sentence is obscure. It is perhaps a proverbial saying with a play on words. TBC² reads *Correcad lochta ┐ fulachta* (3646), and Stowe *Condregat re fulachta ann so aniu* (4984).

4850 *echrad láir* Omit *echrad*. Or read, perhaps, as *echrad .i. láir rena serrgraig*.

4856 *Ní lamad . . .* Cp. the closing words of De Chophur in dá Muccida, referring to the two bulls: *Nicon rabai la Connachto agh ro lamad geim lasan adh thiar. A cumna cetno dana nicon raba la hUlltu agh ra llamad geim laissin agh thair* (Ir. T. III. 243).

4857 *gúasacht Gnúsachtach,* the form here in Stowe, is given in Contribb. as meaning 'bellowing, lowing'. A form *gnuasacht* which is not in Contribb. also occurs: *Do righnios gnuasacht go nuaill | re sprocht croidhe is re hanmhuain =* 'rugiebam a gemitu cordis mei', Penitential Psalms in Irish Verse, Éigse viii. 52.

4864 *darala eturru* For this quarrel see Echtra Nerai RC x. 212.

4869–70 *Dáig nira choitchinniu* . . . *Nira* for *nírba*. Cp. *écoitchenn*, 'biassed'.

4880 *brissis búrach fair* I take *búrach* here to mean 'rage, anger', and the phrase to be synonymous with e.g. *ro immir a búrach fair*. Cp. *cor imbir a búrach forru amal dam ndamgaire* TTr.² 1159.

4882 *ferchubat fir* 'A man's length'. A confusion of *cubat* and *comfat*.

4885 *fogeist* *Fogast* was some kind of spear. Stowe has *fidhslat* here (5043), 'a wooden rod'.

4886 *Nírap* . . . Taken by Windisch as preterite of the copula. It is pres. subjunctive. *Narab* St 5045.

4889 & *go 'mmarálaid dóib* . . .*d'imbúalad* Stowe reads ⁊ *co 'mborala dhoiph* . . . *ic iombualadh* (5050). In the Contribb. the verb *imm-cuirethar* (*do*) is taken to mean 'befalls, happens (to)'. But if that were the meaning, one would expect *imbúalad*, nom., not *d'imbúalad*. I take *imma-* here to be the reciprocal preverb and the meaning to be 'they met in conflict, they engaged in battle', that is, more or less the same meaning as *dorala eturru*. Cp. *imma-tarlae dóib* SCMD² §8 and Thurneysen's note in the Glossary.

4918 *a hús* ⁊ *a imthúsa* ⁊ *a deired* In translating *a hús* as 'account' I follow Windisch: 'So war seine Geschichte und seine Schicksale und das Ende der Táin'. Cp. in the list of titles for the various sections in TBC² 2747-52: *Aus* [? read *a us*] *in Duib Cualngni*. O'Cléry gives *ús .i. iomthús nó sgéla*, *ús na* [read *in*] *Duinn Cuailgne dia thir .i. sgéla nó imtheachta*. But possibly *a hús* here = *a thús*, contrasting with *a deired*. For *h* = *th* see supra n. 1236. So too *tabhair haigid* (= *th'aigid*) *form* in a gloss in hand H LU 2981.

INDEX OF PERSONS

345

Éle (daughter of Eocho Feidlech) 14.

Én (ara Conaill) 4673.

Én mac Mágach 152.

Eo mac Forne 4081–2.

Eocho Bec ó Eochaid Bic 26.

Eocho Feidlech 11; gen. Echach 1453, 2073, Echach Feidlig 3682.

Eocho Semnech 4065.

Eogan mac Durthacht 216, 708, 4087, Eogan mac Durthachta 4366–4374.

Eraíse 2096. A druidess.

Ercc mac Feidilmthi 4546; gen. Eirc meic Feidilmthe Nóchruthaige 4111.

Err mac Nera 469; gen. Eirre 611.

Errge Echbél 4066, 4496–4503.

Ethne (daughter of Eocho Feidlech) 14.

Etarcumul mac Feda ⁊ Lethrinne 1566–1695; gen. Etarcomla 1643, 1666, 1680, 1695.

trí Etarsceóil 4702.

Etatchar 4094.

Fachtna mac Sencha 4072.

Faindle mac Nechtain 1130, 1131, 1134–5.

deich Féic 2332.

Feidelm Banfáid 202ff., in banfáid Feidelm 271.

Feidelm (Feidilmid) Noíchruthach (daughter of Conchobor) gen. Feidilmthi Noíchruthaige 450–1, Feidilmthe Nóchruthaige 411, Fedilmi Nóchruthaige 4546.

Feidlimid Cilair Chétaig 4056, F. Chilair Chetail 4410–4419.

deich Feidlimid 2332.

cethri Feochair 2327.

Fer Báeth mac Baetáin 1863, 3515, mac Baetáin 3505.

Fer Báeth mac Fir Bend 1863–4, 1876–1905, 3053.

Fer Dét mac Damáin 1862.

Fer Diad mac Damáin 1861, 2609ff.

Feradach Find Fechtnach 4098, 4459–4467.

Ferchú Longsech 2511–2531, gen. Ferchon Longsig 2529, –30, –31.

Fergna mac Findchonna 4092–3, 4517–4525.

Fergus mac Lete 4446.

Fergus mac Róig 333 et passim, gen. Fergusa meic Róig 128, 137, 299 etc., voc. a Ferguis 371 etc., a Ferguis (: guiss) 616, a Fergus (: i fus) 390.

Fer Loga (ara Conchobuir) 4713, 4717.

Fer Néll 4089.

Fer Taidle 1759.

Fethan (bancháinte) 4597, 4794.

Fethan Bec 4092 Cp. Fethen Bec mac Amairgin LL23154 (CRR).

Fethen Mór 4092 Cp. Feithen Mór mac Amairgin LL23149 (CRR).

Fiacha Fialdána 2448, 2452, 2454.

Fiacha mac Conchobuir 4110, Fiachaig 4478, gen. Fiachach meic Conchobuir 2411.

deich Fiachaig 2332.

trí Fiachaig 4696.

Fiachna 4062.

Fiachna mac Conchobuir 4110, 4478.

Fiachu mac Fir Aba 301, 922, 1388, 1390, 1398, 1718, 1903–4, 2556, 2559, 2572.

Fidach mac Doraire 4100.

Find mac Rosa Rúaid 24, 46. Rí Lagen.

trí Find 4699.

Finda ingen Echach (Sálbuidi) 3788.

Findabair ingen Ailella ⁊ Medba 306, 330, 1877, 1932, 1939, 2711, 3037, 3195, 3460, 3886.

in Findbennach Aí 72, 4856, 4879, 4898, gen. in Findbennaig 4904, 4906.

Findchad Fer Bend Uma mac Fráechlethain 4049, 4101, voc. a Findchaid 4053.

Findchad Slébe Betha 4089.

Findmór (ben Celtchair) acc. Findmóir 1792, 1793.

Fíngin Fáthlíaig 3657ff., g. Fíngin 2022, co Fíngin co Fíngabair 4086.

trí Fintain 4701.

Fintan mac Néill Niamglonnaig 3813–3835, 4097.

Flidais Foltchaín (ben Ailella Find) 306.

Fochnam (ara) 470, gen. Fochnáim 609.

Fóill mac Nechtain 1082–3, 1100, 1106.

Scandlach 1189.

Scáthach 3499, ac Scáthaig 2612, 2939, 3091, 3496, fri Scáthaig 3079, gen. Scáthaiche 3540, 3544, Scáthaige 3548.

trí Scáthglain 4700.

Senall Úathach 3598, 4059.

Sencha mac Ailella meic Máilchló 4355, 4650, 4656, 4658. co Sencha 4073.

co Sencháinte 4073.

Sétanta mac Sualtaim (= Cú Chulaind) 796, 888, 909.

Simón Druí (Simon Magus) 2209–10.

Srub Daire mac Fedaig 1855, Srubdaire 3057.

Sualtach Sídech (= Sualtaim) 442–3.

Sualtaim athair Con Culaind 452, 456, S. mac Becaltaig meic Móraltaig 3982–4041, gen. Sualtaim 444, 3981, 4040.

trí Suibne 4694.

co Talgobaind 4089.

Tamun drúth 2461–2472, gen. Tamuin 2472, voc. a Thamuin drúith 2466.

Tarathor 4060.

Te 1769.

Tollcend 4082.

Torc Glaisse 1769.

Traig 2096. A druid.

trí Trénfir 4701.

Túachall mac Nechtain 1112, 1113, 1126, Túachail mac Nechtain 1117.

Tualang 1769.

Turscur 1769.

Úalu 1354, gen. Úaland 1358.

Úalu mac Árach 1248.

Úathach ac Úathaig 2613, 2939, 3091, 3496.

Úathach Bodba 4061.

Uathchar 4093.

Uma mac Remarfessaig 4066–7 (Cp. Uma mac Remanfisig MU² 99, 620).

INDEX OF PEOPLES, PLACES AND RIVERS

Glend na Samaisce 1318, dat. a Glind na S. 1775.

Glend Táil 1373.

Gleóir 3606. *A river in Conaille Muirthemne.*

Glúine Gabur 283.

Gort Sláne 289.

Gréic gen. mórGréc 3536. *The Greeks, Greece.*

Grellach Bó Bulge 1762.

Gúalu Mulchi 1262.

Guiph 1758 (i nGuiph — i Cuib LU, St).

co hIalla nIlgremma 4091.

Ilgáirech See Gáirech.

Imchlár 4056.

Imlech in Glendamrach 3879. (See note).

Immail 4083, 4387. Cp. hi Raith Impail TBC² 2932.

Indeóin 285. *Dungolman River* (IER 1912. 605).

Inis Cuscraid 211, 702.

Iraird Cuillend i nIraird Chullend 4108, 4129. *Glossed Crossa Cail in LU = Crossa Kiel six miles due west of Kells, Co. Meath.*

Irgáirech = Ilgáirech 2321.

Irúad gen. Irúade 3745, 4686. *Norway. Here prob. = some mythical northern land.* (See Windisch, p. 625 n.8 and Contribb. s.v. Irúait).

Laigin gen. Lagen 24, d. Lagnib 39, 46. *Leinstermen, Leinster.*

Lecc Derg gen. Licci Derge 4697.

Lecca Slébe Fúait co Leccain S.F. 3657, co Lecain S.F. 3660.

Léire 4080.

Leittre Lúasce 1761.

Lethglais 4066, 4494. *Now Downpatrick, Co. Down.*

Lia Mór ic Líic Móir 1270.

Lia Úaland 1358.

Liana 4088.

Líasa Líac 1375.

Líne 4065, 4446. Cp. Mag Líne, *Moylinny.*

Loch Éirne gen. Locha Érne 4695. *L. Erne, Co. Fermanagh.*

Loch Lindfhormait gen. Locha Lindformait 3508.

Loch Rí gen. Locha Rí 4695-6. *L. Ree on the Shannon.*

Lorg gen. Luirg 4515.

Lúachair 417, gen. Lúachra 4692.

Lugmad Lugmud 2441. *Louth.*

Macha gen. ríg Macha 4635, 4659, co Macha 4085.

Máela Mide na trí Máela Mide 4779, 4844.

Máeláin Átha Lúain dat. Máelánaib Á.L. 4844.

Mag nAí gen. Maige Aí 86, 96, 2632, ar Maig Áe 4160. *Plain in Roscommon.*

Mag mBreg 1057, gen. Maigi Breg 1281. *Plain between Liffey and Boyne.*

Mag Cruinn 281.

Mag nDea 4082.

Mag nÉola 4082.

Mag nDula co Maigi Dula (co *omitt.*) 4090.

Mag nInis 4085, 4534.

Mag in Scáil gen. Maige in Scáil 2777.

Mag Nobla (? nDobla) 4091-2.

Mag Murthemne 1418, gen. Maigi M. 1281, 1288, 4570. *Plain S. of Dundalk. Cú Chulainn's district.*

Mag Táil 4095.

Mag Trega 283. *Moytra in Longford.*

Mag Túaga 4788-9. *In Connacht.*

Mairge (gen.) 4697.

Méide ind Eóin 293, 1277-8.

Méide in Togmaill 293, 1277.

Mide 285, gen. 4143, 4162, 4166 etc. *Meath.*

Midlúachair 4076.

Míliuc 3607. *A river in Conaille Muirthemne.*

Moda Loga dat. a Modaib Loga 2441 (= Lugmad).

Modorn co Moduirn 4067, a Moduirn 4398.

Móin Coltna 437.

Mossud (gen.) 4700.

Muach 3607. *A river in Conaille Muirthemne.*

Tiarthechta 284.

Tilach Loiscthe gen. Tilcha L. 4692–3.

Timscúap for Timscúaib 295.

Tír Mór gen. Tíri Móir 2011, 2057.

Tír na Sorcha gen. Tíre na Sorcha 2261. (See note).

Tír Tairngire gen. Tíre Tairṅgire 2260. *Land of Promise.*

Tromma 289.

Tuaim Móna Tóm Móna 281.

Tuath Bressi 4145.

Tuatha Dé Danand 2847, 3602.

Turloch Teóra Crích 281.

Úachtur Lúa 3768–9. *In Crích Roiss.*

Úarba 4083.

for Uatu 284. (See note).

Ulaid 109, 116, 889, gen. Ulad 26, 111, 199, acc. Ultu 208, 213, 217, dat. Ultaib, 99, 107, 114. *Ulstermen, Ulster.*

Umanshruth 1815.

ADDENDA

p. xxix   *Add as footnote to l. 18:* Studies in Irish Literature and History, chap. II.

p. 290   *Add after n. 951:* The LL-reading implies that Cathbad is addressing the boy, 'your mother's son'. The reading of LU and Stowe gives a better meaning 'his mother's son'; Cathbad is still speaking to Conchobor.

CORRIGENDA

p. 14 l. 494 *For* Furópair *read with MS.* Furopair (= airopair).

p. 20 l. 702 *For* Fergus, mac ind ardríg *read* Fergus. 'Mac ind ardríg. (*The reference is to Cuscraid mac Conchobuir*).

p. 151 l. 9 *For* It gave a pledge etc. *read* Truly an exceptional feat [done] with the harsh rage of the Smith's hound.

l. 30 *For* where is Partraige Beca etc. *read* where (now) live Partraige Beca on Cúil Sibrilli south-west of Cenannus na Ríg.

p. 157 l. 2 *For* Fergus, son of the high king. *read* Fergus. '(He is) the son of the high king.

p. 165 ll. 7, 34 *For* first-wounding *read* slaughter of hundreds.

p. 294 l. 5 *For* 1324 *read* 1322.